Dairy Science and Technology

Second Edition

FOOD SCIENCE AND TECHNOLOGY

A Series of Monographs, Textbooks, and Reference Books

76. Food Chemistry: Third Edition, *edited by Owen R. Fennema*
77. Handbook of Food Analysis: Volumes 1 and 2, *edited by Leo M. L. Nollet*
78. Computerized Control Systems in the Food Industry, *edited by Gauri S. Mittal*
79. Techniques for Analyzing Food Aroma, *edited by Ray Marsili*
80. Food Proteins and Their Applications, *edited by Srinivasan Damodaran and Alain Paraf*
81. Food Emulsions: Third Edition, Revised and Expanded, *edited by Stig E. Friberg and Kåre Larsson*
82. Nonthermal Preservation of Foods, *Gustavo V. Barbosa-Cánovas, Usha R. Pothakamury, Enrique Palou, and Barry G. Swanson*
83. Milk and Dairy Product Technology, *Edgar Spreer*
84. Applied Dairy Microbiology, *edited by Elmer H. Marth and James L. Steele*
85. Lactic Acid Bacteria: Microbiology and Functional Aspects, Second Edition, Revised and Expanded, *edited by Seppo Salminen and Atte von Wright*

86. Handbook of Vegetable Science and Technology: Production, Composition, Storage, and Processing, *edited by D. K. Salunkhe and S. S. Kadam*
87. Polysaccharide Association Structures in Food, *edited by Reginald H. Walter*
88. Food Lipids: Chemistry, Nutrition, and Biotechnology, *edited by Casimir C. Akoh and David B. Min*
89. Spice Science and Technology, *Kenji Hirasa and Mitsuo Takemasa*
90. Dairy Technology: Principles of Milk Properties and Processes, *P. Walstra, T. J. Geurts, A. Noomen, A. Jellema, and M. A. J. S. van Boekel*
91. Coloring of Food, Drugs, and Cosmetics, *Gisbert Otterstätter*
92. *Listeria*, Listeriosis, and Food Safety: Second Edition, Revised and Expanded, *edited by Elliot T. Ryser and Elmer H. Marth*
93. Complex Carbohydrates in Foods, *edited by Susan Sungsoo Cho, Leon Prosky, and Mark Dreher*
94. Handbook of Food Preservation, *edited by M. Shafiur Rahman*
95. International Food Safety Handbook: Science, International Regulation, and Control, *edited by Kees van der Heijden, Maged Younes, Lawrence Fishbein, and Sanford Miller*
96. Fatty Acids in Foods and Their Health Implications: Second Edition, Revised and Expanded, *edited by Ching Kuang Chow*
97. Seafood Enzymes: Utilization and Influence on Postharvest Seafood Quality, *edited by Norman F. Haard and Benjamin K. Simpson*
98. Safe Handling of Foods, *edited by Jeffrey M. Farber and Ewen C. D. Todd*
99. Handbook of Cereal Science and Technology: Second Edition, Revised and Expanded, *edited by Karel Kulp and Joseph G. Ponte, Jr.*
100. Food Analysis by HPLC: Second Edition, Revised and Expanded, *edited by Leo M. L. Nollet*
101. Surimi and Surimi Seafood, *edited by Jae W. Park*
102. Drug Residues in Foods: Pharmacology, Food Safety, and Analysis, *Nickos A. Botsoglou and Dimitrios J. Fletouris*
103. Seafood and Freshwater Toxins: Pharmacology, Physiology, and Detection, *edited by Luis M. Botana*
104. Handbook of Nutrition and Diet, *Babasaheb B. Desai*
105. Nondestructive Food Evaluation: Techniques to Analyze Properties and Quality, *edited by Sundaram Gunasekaran*
106. Green Tea: Health Benefits and Applications, *Yukihiko Hara*
107. Food Processing Operations Modeling: Design and Analysis, *edited by Joseph Irudayaraj*
108. Wine Microbiology: Science and Technology, *Claudio Delfini and Joseph V. Formica*
109. Handbook of Microwave Technology for Food Applications, *edited by Ashim K. Datta and Ramaswamy C. Anantheswaran*

110. Applied Dairy Microbiology: Second Edition, Revised and Expanded, *edited by Elmer H. Marth and James L. Steele*
111. Transport Properties of Foods, *George D. Saravacos and Zacharias B. Maroulis*
112. Alternative Sweeteners: Third Edition, Revised and Expanded, *edited by Lyn O'Brien Nabors*
113. Handbook of Dietary Fiber, *edited by Susan Sungsoo Cho and Mark L. Dreher*
114. Control of Foodborne Microorganisms, *edited by Vijay K. Juneja and John N. Sofos*
115. Flavor, Fragrance, and Odor Analysis, *edited by Ray Marsili*
116. Food Additives: Second Edition, Revised and Expanded, *edited by A. Larry Branen, P. Michael Davidson, Seppo Salminen, and John H. Thorngate, III*
117. Food Lipids: Chemistry, Nutrition, and Biotechnology: Second Edition, Revised and Expanded, *edited by Casimir C. Akoh and David B. Min*
118. Food Protein Analysis: Quantitative Effects on Processing, *R. K. Owusu- Apenten*
119. Handbook of Food Toxicology, *S. S. Deshpande*
120. Food Plant Sanitation, *edited by Y. H. Hui, Bernard L. Bruinsma, J. Richard Gorham, Wai-Kit Nip, Phillip S. Tong, and Phil Ventresca*
121. Physical Chemistry of Foods, *Pieter Walstra*
122. Handbook of Food Enzymology, *edited by John R. Whitaker, Alphons G. J. Voragen, and Dominic W. S. Wong*
123. Postharvest Physiology and Pathology of Vegetables: Second Edition, Revised and Expanded, *edited by Jerry A. Bartz and Jeffrey K. Brecht*
124. Characterization of Cereals and Flours: Properties, Analysis, and Applications, *edited by Gönül Kaletunç and Kenneth J. Breslauer*
125. International Handbook of Foodborne Pathogens, *edited by Marianne D. Miliotis and Jeffrey W. Bier*
126. Food Process Design, *Zacharias B. Maroulis and George D. Saravacos*
127. Handbook of Dough Fermentations, *edited by Karel Kulp and Klaus Lorenz*
128. Extraction Optimization in Food Engineering, *edited by Constantina Tzia and George Liadakis*
129. Physical Properties of Food Preservation: Second Edition, Revised and Expanded, *Marcus Karel and Daryl B. Lund*
130. Handbook of Vegetable Preservation and Processing, *edited by Y. H. Hui, Sue Ghazala, Dee M. Graham, K. D. Murrell, and Wai-Kit Nip*
131. Handbook of Flavor Characterization: Sensory Analysis, Chemistry, and Physiology, *edited by Kathryn Deibler and Jeannine Delwiche*

132. Food Emulsions: Fourth Edition, Revised and Expanded, *edited by Stig E. Friberg, Kare Larsson, and Johan Sjoblom*
133. Handbook of Frozen Foods, *edited by Y. H. Hui, Paul Cornillon, Isabel Guerrero Legarret, Miang H. Lim, K. D. Murrell, and Wai-Kit Nip*
134. Handbook of Food and Beverage Fermentation Technology, *edited by Y. H. Hui, Lisbeth Meunier-Goddik, Ase Solvejg Hansen, Jytte Josephsen, Wai-Kit Nip, Peggy S. Stanfield, and Fidel Toldrá*
135. Genetic Variation in Taste Sensitivity, *edited by John Prescott and Beverly J. Tepper*
136. Industrialization of Indigenous Fermented Foods: Second Edition, Revised and Expanded, *edited by Keith H. Steinkraus*
137. Vitamin E: Food Chemistry, Composition, and Analysis, *Ronald Eitenmiller and Junsoo Lee*
138. Handbook of Food Analysis: Second Edition, Revised and Expanded, Volumes 1, 2, and 3, *edited by Leo M. L. Nollet*
139. Lactic Acid Bacteria: Microbiological and Functional Aspects: Third Edition, Revised and Expanded, *edited by Seppo Salminen, Atte von Wright, and Arthur Ouwehand*
140. Fat Crystal Networks, *Alejandro G. Marangoni*
141. Novel Food Processing Technologies, *edited by Gustavo V. Barbosa-Cánovas, M. Soledad Tapia, and M. Pilar Cano*
142. Surimi and Surimi Seafood: Second Edition, *edited by Jae W. Park*
143. Food Plant Design, *edited by Antonio Lopez-Gomez; Gustavo V. Barbosa-Cánovas*
144. Engineering Properties of Foods: Third Edition, *edited by M. A. Rao, Syed S.H. Rizvi, and Ashim K. Datta*
145. Antimicrobials in Food: Third Edition, *edited by P. Michael Davidson, John N. Sofos, and A. L. Branen*
146. Encapsulated and Powdered Foods, *edited by Charles Onwulata*
147. Dairy Science and Technology: Second Edition, *Pieter Walstra, Jan T. M. Wouters and Tom J. Geurts*

Dairy Science and Technology

Second Edition

Pieter Walstra
Jan T. M. Wouters
Tom J. Geurts

A CRC title, part of the Taylor & Francis imprint, a member of the Taylor & Francis Group, the academic division of T&F Informa plc.

Published in 2006 by
CRC Press
Taylor & Francis Group
6000 Broken Sound Parkway NW, Suite 300
Boca Raton, FL 33487-2742

Printed in the United States of America on acid-free paper
10 9 8 7 6 5 4 3 2 1

International Standard Book Number-10: 0-8247-2763-0 (Hardcover)
International Standard Book Number-13: 978-0-8247-2763-5 (Hardcover)
Library of Congress Card Number 2005041830

Library of Congress Cataloging-in-Publication Data

Walstra, Pieter.
Dairy science and technology / Pieter Walstra, Jan T.M. Wouters, T.J. Geurts.--2nd ed.
p. cm. -- (Food science and technology ; 146)
Rev. ed. of: Dairy technology / P. Walstra ... [et al.]. c1999.
Includes bibliographical references.
ISBN 0-8247-2763-0 (alk. paper)
1. Dairy processing. 2. Milk. 3. Dairy products. I. Wouters, Jan T. M. II. Geurts, T. J. (Tom J.) III. Dairy technology. IV. Title. V. Food science and technology (Taylor & Francis) ; 146.

SF250.5.D385 2005
637'.1--dc22 2005041830

Taylor & Francis Group
is the Academic Division of T&F Informa plc.

Visit the Taylor & Francis Web site at
http://www.taylorandfrancis.com

and the CRC Press Web site at
http://www.crcpress.com

Preface

The primary theme of this book is the efficient transformation of milk into high-quality products. This needs a thorough understanding of the composition and properties of milk, and of the changes occurring in milk and its products during processing and storage. Moreover, knowledge of the factors that determine product quality, including health aspects and shelf life, is needed. Our emphasis is on the *principles* of physical, chemical, enzymatic, and microbial transformations. Detailed manufacturing prescriptions and product specifications are not given, as they are widely variable.

Aimed at university food science and technology majors, the book is written as a text, though it will also be useful as a work of reference. It is assumed that the reader is familiar with the rudiments of food chemistry, microbiology, and engineering. Nevertheless, several basic aspects are discussed for the benefit of readers who may be insufficiently acquainted with these aspects. The book contains no references to the literature, but suggestions for further reading are given.

The book is made up of four main parts. Part I, "*Milk*," discusses the chemistry, physics, and microbiology of milk. Besides providing knowledge of the properties of milk itself, it forms the basis for understanding what happens during processing, handling and storage. Part II, "*Processes*," treats the main unit operations applied in the manufacture of milk products. These are discussed in some detail, especially the influence of product and process variables on the (intermediate) product resulting. A few highly specific processes, such as churning, are discussed in product chapters. In Part III, "*Products*," integration of knowledge of the raw material and of processing is covered for the manufacture of several products. The number of dairy products made is huge; hence, some product groups have been selected because of their general importance or to illustrate relevant aspects. Procedures needed to ensure consumer safety, product quality, and processing efficiency are also treated. Part IV, "*Cheese*," describes the processes and transformations (physical, biochemical, and microbial) in the manufacture and ripening of cheese. Here, the processes are so specific and the interactions so intricate that a separate and integrated treatment is needed. It starts with generic aspects and then discusses specific groups of cheeses.

Several important changes have been introduced in this second edition. The reasons were, first, to improve the didactic quality of the book and, second, to make it more useful as a reference source. More basic and general aspects are now treated, especially physicochemical and microbiological ones. Part I has been substantially enlarged, one reason why the title of the book has been broadened. The nutritional aspects of milk components are now included, and those of some products are enlarged. A section on milk formation has been added.

Naturally, the text has been updated. Moreover, several parts have been reorganized or rewritten. Factual information has been increased and partly moved to an Appendix.

Pieter Walstra
Jan Wouters
Tom Geurts
Wageningen, The Netherlands

Acknowledgments

First, we want to stress that much of the present book derives from the substantial contributions that our then-coauthors, Ad Noomen, Arend Jellema, and Tiny van Boekel, made to the first edition. We are grateful that we could benefit from their extensive expertise.

Several people have provided information and advice. Professors Norman Olson (University of Wisconsin, Madison), Marie Paulsson (Lund University, Sweden), and Zdenko Puhan (Technical University, Zürich, Switzerland) scrutinized (parts of) the first edition and gave useful advice. We consulted several colleagues from our department, from NIZO Food Research (Ede, the Netherlands), and from the Milk Control Station (Zutphen, the Netherlands). We also received information from the following Dutch companies: Campina (Zaltbommel and Wageningen), Carlisle Process Systems (formerly Stork, Gorredijk), Friesland Foods (Deventer), and Numico (Wageningen). We thank all of the people involved for their cooperation and for the important information given.

Contents

Part I
Milk

Part II
Processes

Part IV
Cheese

Part V
Appendix and Index

Part I

Milk

1 Milk: Main Characteristics

Milk is defined as the secretion of the mammary glands of mammals, its primary natural function being nutrition of the young. Milk of some animals, especially cows, buffaloes, goats and sheep, is also used for human consumption, either as such or in the form of a range of dairy products. In this book, the word milk will be used for the 'normal' milk of healthy cows, unless stated otherwise. Occasionally, a com-parison will be made with human milk.

This chapter is meant as a general introduction. Nearly all that is mentioned — with the exception of parts of Section 1.2 — is discussed in greater detail in other chapters. However, for readers new to the field it is useful to have some idea of the formation, composition, structure, and properties of milk, as well as the variation — including natural variation and changes due to processing — that can occur in these characteristics, before starting on the main text.

1.1 COMPOSITION AND STRUCTURE

1.1.1 Principal Components

A classification of the principal constituents of milk is given in Table 1.1. The principal chemical components or groups of chemical components are those present in the largest quantities. Of course, the quantity (in grams) is not paramount in all respects. For example, vitamins are important with respect to nutritive value; enzymes are catalysts of reactions; and some minor components contribute markedly to the taste of milk. More information on milk composition is given in Table 1.3.

Lactose or milk sugar is the distinctive carbohydrate of milk. It is a disaccharide composed of glucose and galactose. Lactose is a reducing sugar.

The *fat* is largely made up of triglycerides, constituting a very complicated mixture. The component fatty acids vary widely in chain length (2 to 20 carbon atoms) and in saturation (0 to 4 double bonds). Other lipids that are present include phospholipids, cholesterol, free fatty acids, monoglycerides, and diglycerides.

About four fifths of the *protein* consists of casein, actually a mixture of four proteins: α_{S1}-, α_{S2}-, β-, and κ-casein. The caseins are typical for milk and have some rather specific properties: They are to some extent phosphorylated and have little or no secondary structure. The remainder consists, for the most part, of the milk serum proteins, the main one being β-lactoglobulin. Moreover, milk contains numerous minor proteins, including a wide range of *enzymes.*

The *mineral substances* — primarily K, Na, Ca, Mg, Cl, and phosphate — are not equi-valent to the salts. Milk contains numerous other elements in trace

TABLE 1.1
Approximate Composition of Milk

Component	Average Content in Milk (% w/w)	Range[a] (% w/w)	Average Content in Dry Matter (% w/w)
Water	87.1	85.3–88.7	—
Solids-not-fat	8.9	7.9–10.0	—
Fat in dry matter	31	22–38	—
Lactose	4.6	3.8–5.3	36
Fat	4.0	2.5–5.5	31
Protein[b]	3.3	2.3–4.4	25
casein	2.6	1.7–3.5	20
Mineral substances	0.7	0.57–0.83	5.4
Organic acids	0.17	0.12–0.21	1.3
Miscellaneous	0.15	—	1.2

Note: Typical for milks of lowland breeds.

[a] These values will rarely be exceeded, e.g., in 1 to 2% of samples of separate milkings of healthy individual cows, excluding colostrum and milk drawn shortly before parturition.

[b] Nonprotein nitrogen compounds not included.

quantities. The salts are only partly ionized. The *organic acids* occur largely as ions or as salts; citrate is the principle one. Furthermore, milk has many *miscellaneous components*, often in trace amounts.

The total content of all substances except water is called the content of dry matter. Furthermore, one distinguishes solids-not-fat and the content of fat in the dry matter.

The chemical composition of milk largely determines its nutritional value; the extent to which microorganisms can grow in it; its flavor; and the chemical reactions that can occur in milk. The latter include reactions that cause off-flavours.

1.1.2 Structural Elements

Structure can be defined as the geometrical distribution of the (chemical) components in a system. It may imply, as it does in milk, that the liquid contains particles. This can have important consequences for the properties of the system. For instance, (1) chemical components are present in separate compartments, which can greatly affect their reactivity; (2) the presence of particles greatly affects some physical properties, like viscosity and optical appearance; (3) interaction forces between particles generally determine the physical stability of the system; and (4) the separation of some components (fat and casein) is relatively easy.

Figure 1.1 shows the main structural elements of milk. Of course, the figure is schematic and incomplete. Some properties of the structural elements are given

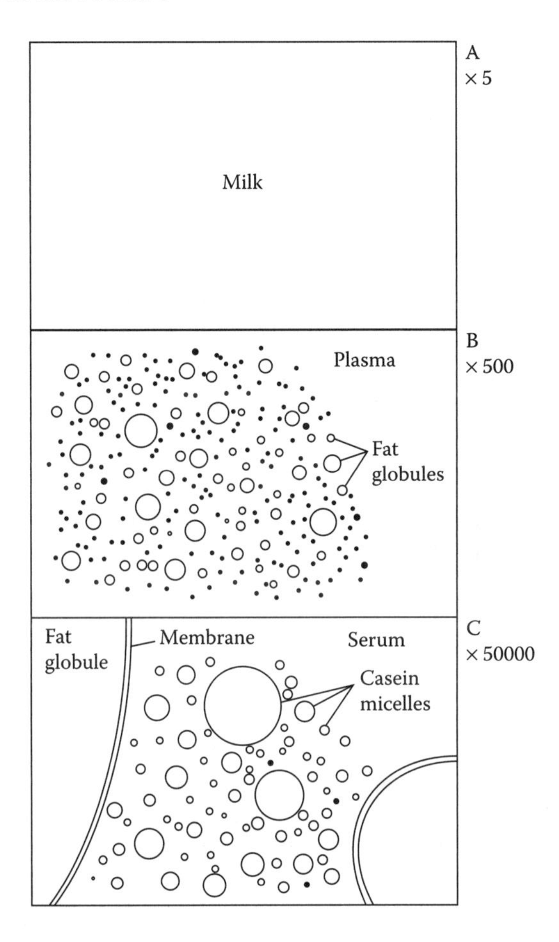

FIGURE 1.1 Milk viewed at different magnifications, showing the relative size of structural elements (A) Uniform liquid. However, the liquid is turbid and thus cannot be homogeneous. (B) Spherical droplets, consisting of fat. These globules float in a liquid (plasma), which is still turbid. (C) The plasma contains proteinaceous particles, which are casein micelles. The remaining liquid (serum) is still opalescent, so it must contain other particles. The fat globules have a thin outer layer (membrane) of different constitution. (From H. Mulder and P. Walstra, *The Milk Fat Globule*, Pudoc, Wageningen, 1974.)

in Table 1.2, again in a simplified form; the numerical data mentioned are meant only to define orders of magnitude. The table clearly shows that aspects of colloid chemistry are essential for understanding the properties of milk and the many changes that can occur in it. All particles exhibit Brownian motion; they have an electrostatic charge, which is negative at the pH of milk. Their total surface area is large.

Fat globules. To a certain extent, milk is an oil-in-water emulsion. But the fat globules are more complicated than emulsion droplets. In particular, the surface layer or *membrane* of the fat globule is not an adsorption layer of one

TABLE 1.2
Properties of the Main Structural Elements of Milk

	Milk			
		Plasma		
			Serum	
	Fat Globules	Casein Micelles	Globular Proteins	Lipoprotein Particles
Main components	Fat	Casein, water, salts	Serum protein	Lipids, proteins
To be considered as	Emulsion	Fine dispersion	Colloidal solution	Colloidal dispersion
Content (% dry matter)	4	2.8	0.6	0.01
Volume fraction	0.05	0.1	0.006	10^{-4}
Particle diameter[a]	0.1–10 μm	20–400 nm	3–6 nm	10 nm
Number per ml	10^{10}	10^{14}	10^{17}	10^{14}
Surface area (cm^2/ml milk)	700	40000	50000	100
Density (20°C; $kg \cdot m^{-3}$)	920	1100	1300	1100
Visible with	Microscope	Ultramicroscope		Electron microscope
Separable with	Milk separator	High-speed centrifuge	Ultrafiltration	Ultrafiltration
Diffusion rate (mm in 1h)[a]	0.0	0.1–0.3	0.6	0.4
Isoelectric pH	~3.8	~4.6	4–5	~4

Note: Numerical values are approximate averages.

[a] For comparison, most molecules in solution are 0.4 to 1 nm diameter, and diffuse, say, 5 mm in 1 h. 1 mm = 10^3 μm = 10^6 nm = 10^7 Å.

single substance but consists of many components; its structure is complicated. The dry mass of the membrane is about 2.5% of that of the fat. A small part of the lipids of milk is found outside the fat globules. At temperatures below 35°C, part of the fat in the globules can crystallize. Milk minus fat globules is called *milk plasma*, i.e., the liquid in which the fat globules float.

Casein micelles consist of water, protein, and salts. The protein is casein. Casein is present as a caseinate, which means that it binds cations, primarily calcium and magnesium. The other salts in the micelles occur as a calcium phosphate, varying somewhat in composition and also containing a small amount of citrate. This is often called colloidal phosphate. The whole may be called calcium-caseinate/calcium-phosphate complex. The casein micelles are not micelles in the colloid-chemical sense but just 'small particles.' The micelles have an open structure and, accordingly, contain much water, a few grams per gram of casein. *Milk serum*, i.e., the liquid in which the micelles are dispersed, is milk minus fat globules and casein micelles.

Serum proteins are largely present in milk in molecular form or as very small aggregates.

Lipoprotein particles, sometimes called milk microsomes, vary in quantity and shape. Presumably, they consist of remnants of mammary secretory cell membranes. Few definitive data on lipoprotein particles have been published.

Cells, i.e., leukocytes, are always present in milk. They account for about 0.01% of the volume of milk of healthy cows. Of course, the cells contain all cytoplasmic components such as enzymes. They are rich in catalase.

Table 1.3 gives a survey of the average composition and structure of milk.

1.2 MILK FORMATION

Milk components are for the most part formed in the mammary gland (the udder) of a cow, from precursors that are the results of digestion.

Digestion. Mammals digest their food by the use of enzymes to obtain simple, soluble, low-molar-mass components, especially monosaccharides; small peptides and amino acids; and fatty acids and monoglycerides. These are taken up in the blood, together with other nutrients, such as various salts, glycerol, organic acids, etc. The substances are transported to all the organs in the body, including the mammary gland, to provide energy and building blocks (precursors) for metabolism, including the synthesis of proteins, lipids, etc.

In ruminants like the cow, considerable predigestion occurs by means of microbial fermentation, which occurs for the most part in the first stomach or *rumen*. The latter may be considered as a large and very complex bio-fermenter. It contains numerous bacteria that can digest cellulose, thereby breaking down plant cell walls, providing energy and liberating the cell contents. From cellulose and other carbohydrates, acetic, propionic, butyric and lactic acid are formed, which are taken up in the blood. The composition of the organic acid mixture depends on the composition of the feed. Proteins are broken down into amino acids. The rumen flora uses these to make proteins but can also synthesize amino acids from low-molar-mass nitrogenous components. Further on in the digestive tract the microbes are digested, liberating amino acids. Also, food lipids are hydrolyzed in the rumen and partly metabolized by the microorganisms. All these precursors can reach the mammary gland.

Milk Synthesis. The synthesis of milk components occurs for the greater part in the *secretory cells* of the mammary gland. Figure 1.2 illustrates such a cell. At the basal end precursors of milk components are taken up from the blood, and at the apical end milk components are secreted into the lumen. Proteins are formed in the endoplasmic reticulum and transported to the Golgi vesicles, in which most of the soluble milk components are collected. The vesicles grow in size while being transported through the cell and then open up to release their contents in the lumen. Triglycerides are synthesized in the cytoplasm, forming

TABLE 1.3
Composition and Structure of Milk[a]

FAT GLOBULE

Glycerides	
triglycerides	40 g
diglycerides	0.1 g
monoglycerides	10 mg
Fatty acids	60 mg
Sterols	100 mg
Carotenoids	0.3 mg
Vitamins A, D, E, K	
Water	60 mg
Others	

MEMBRANE

water	+
protein	700 mg
phospholipids	250 mg
cerebrosides	30 mg
glycerides	+
fatty acids	15 mg
streols	15 mg
other lipids	
enzymes	
alkaline phosphatase	
xanthine oxidase	
many others	
Cu	4 μg
Fe	100 μg

CASEIN MICELLE

Protein	
casein	26 g
proteose peptone	+
Salts	2 g
Ca	850 mg
phosphate	1000 mg
citrate	150 mg
K, Mg, Na	
Water	~80 g
Enzymes	
lipase	
plasmin	

LEUKOCYTE

Many enzymes
e.g., catalase
Nucleic acids
Water

LIPOPROTEIN PARTICLE

lipids
protein
enzymes
water

SERUM

Water	790 g	***Organic acids***		***Proteins***	
		citrate	1600 mg	casein	+
Carbohydrates		formate	40 mg	β-lactoglobulin	3.2 g
lactose	46 g	acetate	30 mg	α-lactalbumin	1.2 g
glucose	70 mg	lactate	20 mg	serum albumin	0.4 g
others		oxalate	20 mg	immunoglobulins	0.8 g
		others	10 mg	proteose peptone	+
Minerals				others	
Ca, bound	300 mg	***Gases***		***Nonprotein nitrogenous***	
Ca, ions	90 mg	oxygen	6 mg	***compounds***	
Mg	70 mg	nitrogen	16 mg	peptides	+
K	1500 mg	***Lipids***		amino acids	50 mg
Na	450 mg	glycerides	+	urea	250 mg
Cl	1100 mg	fatty acids	20 mg	ammonia	10 mg
phosphate	1100 mg	phospholipids	100 mg	others	300 mg
sulfate	100 mg	cerebrosides	10 mg	***Enzymes***	
bicarbonate	100 mg	sterols	15 mg	acid phosphatase	
		others		peroxidase	
Trace elements				many others	
Zn	3 mg	***Vitamins, e.g.***		***Phosphoric esters***	~300 mg
Fe	120 μg	riboflavin	2 mg	***Others***	
Cu	20 μg	ascorbic acid	20 mg		
many others					

[a] Approximate average quantities in 1 kg milk. *Note*: The water in the casein micelles contains some small-molecule solutes.

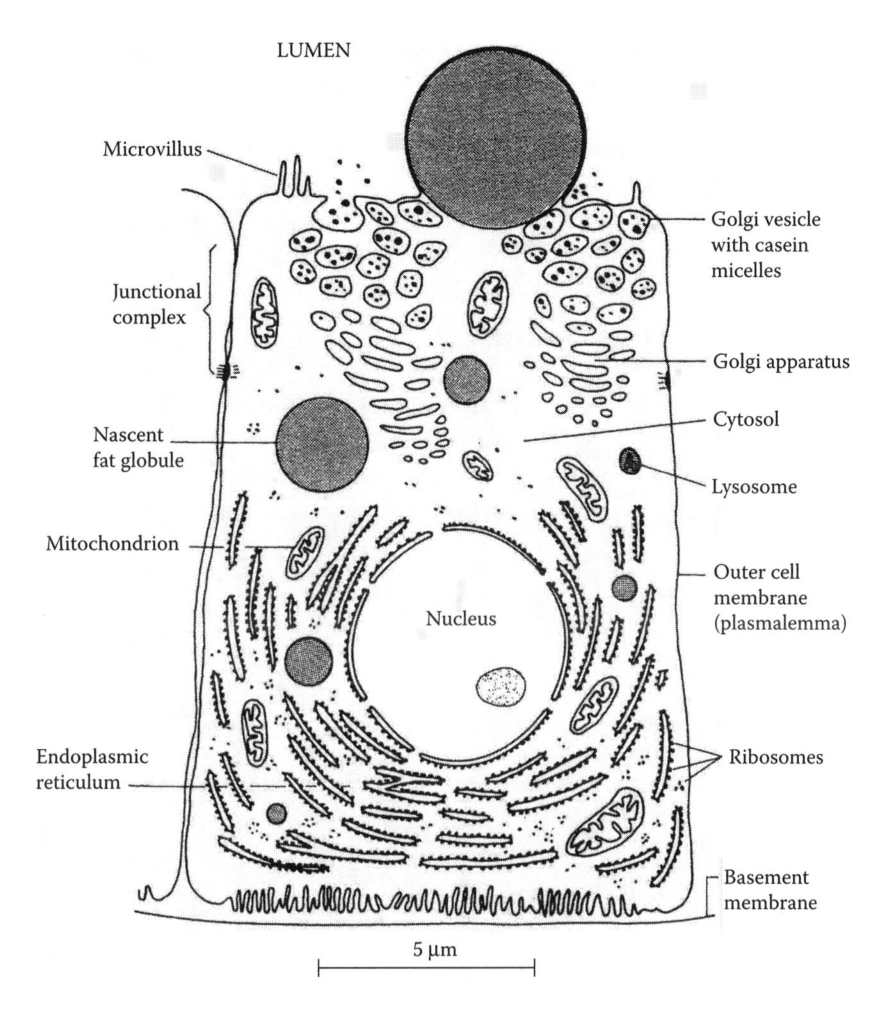

FIGURE 1.2 Stylized diagram of a mammary secretory cell. Below is the basal part, on top the apical part of the cell. The cell is bounded by other secretory cells to form the glandular epithelium. See text for further details. (From P. Walstra and R. Jenness, *Dairy Chemistry and Physics*, Wiley, New York, 1984. With permission.)

small globules, which grow while they are transported to the apical end of the cell. They become enrobed by the outer cell membrane (or plasmalemma) while being pinched off into the lumen. This type of secretion is called *merocrine*, which means that the cell remains intact.

Table 1.4 gives some information about the synthesis of specific components. Most are synthesized in the cell. Others are taken up from the blood but, generally, not in the same proportion as in the blood; see, especially, the salts. This means that the cell membranes have mechanisms to reject, or allow passage of, specific components. Some substances, notably water and small lipophilic molecules, can

TABLE 1.4
Synthesis of Important Milk Components

Milk Component		Precursor in Blood Plasma		Synthesis of Component		
Name	Concentration (% w/w)	Name	Concentration (% w/w)	In the Secretory Cell?	Specific for Milk?	Specific for the Species?
Water	86	Identical	91	No	No	No
Lactose	4.7	Glucose[a]	0.05	Yes	Yes	No
Protein						
Caseins	2.6			Yes	Yes	Yes[b]
β-lactoglobulin	0.32	Amino acids	0.04	Yes	Yes	Yes
α-lactalbumin	0.12			Yes	Yes	Yes
Lactoferrin	0.01			Yes	No	Yes
Serum albumin	0.04	Identical	3.2	No	No	Yes
Immunoglobulins	0.07	Most are identical	1.5	No	No	Yes
Enzymes	Trace	Various	—	Yes[c]	No[c]	Yes
Lipids						
Triglycerides	4	Acetic acid 0.01; β-Hydroxy butyric acid 0.006; Acylglycerols		Partly	Partly	
Phospholipids	0.03	Some lipids	0.3			
Citric Acid	0.17	Glucose[a]	0.05	Yes	No	No
Minerals		Identical		No	No	No
Ca	0.13		0.01			
P[d]	0.09		0.01			
Na	0.04		0.34			
K	0.15		0.03			
Cl	0.11		0.35			

[a] Glucose can also be formed in the secretory cells from some amino acids.
[b] All proteins are species specific, but comparable proteins occur in the milk of all ruminants.
[c] Is not true for all enzymes.
[d] In various phosphates.

pass the cell more or less unhindered. Some other components, such as serum albumin and chlorides, can 'leak' from the blood into the milk by passing through the spaces between secretory cells. Also, some leukocytes somehow reach the lumen. Finally, cell remnants, such as part of the microvilli depicted in Figure 1.2 and tiny fragments of cytoplasm that occasionally adhere to a fat globule, are secreted and form the lipoprotein particles of Table 1.2.

Excretion. The glandular epithelium, consisting of layers of secretory cells, form spherical bodies called *alveoli*. Each of these has a central lumen into which the freshly formed milk is secreted. From there, the milk can flow through small ducts into larger and still larger ones until it reaches a cavity called the *cistern*. From the cistern, the milk can be released via the teat. A cow has four teats and hence four separate mammary glands, commonly called (udder) quarters.

Excretion of the milk does not happen spontaneously. The alveoli have to contract, which can be achieved by the contraction of muscle tissue around the alveoli. Contraction is induced by the hormone oxytocin. This is released into the blood by stimulation of the teats of the animal, be it by the suckling young or by the milker. The udder is not fully emptied.

Lactation. When a calf is born, lactation — i.e., the formation and secretion of milk — starts. The first secretion greatly differs in composition from milk (see Subsection 2.7.1.5). Within a few days the milk has become normal and milk yield increases for some months, after which it declines. The yield greatly varies among cows and with the amount and the quality of the feed taken by the cow. For milch cows, milking is generally stopped after about 10 months, when yield has become quite low. The duration from parturition to leaving the cow dry is called the lactation period, and the time elapsed after parturition is the stage of lactation.

1.3 SOME PROPERTIES OF MILK

Milk as a Solution. Milk is a dilute aqueous solution and behaves accordingly. Because the dielectric constant is almost as high as that of pure water, polar substances dissolve well in milk and salts tend to dissociate (although this dissociation is not complete). The ionic strength of the solution is about 0.073 M. The pH of milk is about 6.7 at room temperature. The viscosity is low, about twice that of water, which means that milk can readily be mixed, even by convection currents resulting from small temperature fluctuations. The dissolved substances give milk an osmotic pressure of about 700 kPa (7 bar) and a freezing-point depression close to 0.53 K. The water activity is high, about 0.995. Milk density (ρ^{20}) equals about 1029 $kg \cdot m^{-3}$ at 20°C; it varies especially with fat content.

Milk as a Dispersion. Milk is also a dispersion; the particles involved are summarized in Table 1.2. This has several consequences, such as milk being white. The fat globules have a membrane, which acts as a kind of barrier between the plasma and the core lipids. The membrane also protects the globules against coalescence. The various particles can be separated from the rest.

The *fat globules* can be concentrated in a simple way by creaming, which either occurs due to gravity or — more efficiently — is induced by centrifugation.

In this way cream and skim milk are obtained. Skim milk is not identical to milk plasma, though quite similar, because it still contains some small fat globules. Cream can be churned, leading to butter and buttermilk; the latter is rather similar in composition to skim milk.

Likewise, *casein micelles* can be concentrated and separated from milk, for instance, by membrane filtration. The solution passing through the membrane is then quite similar to milk serum. If the pores in the membrane are very small, also the serum proteins are retained. When adding rennet enzyme to milk, as is done in cheese making, the casein micelles start to aggregate, forming a gel; when cutting the gel into pieces, these contract, expelling whey. Whey is also similar to milk serum but not quite, because it contains some of the fat globules and part of the κ-casein split off by the enzyme. Casein also aggregates and forms a gel when the pH of the milk is lowered to about 4.6.

Moreover, *water* can be removed from milk by evaporation. Altogether, a range of liquid milk products of various compositions can be made. Some examples are given in Table 1.5.

Flavor. The flavor of fresh milk is fairly bland. The lactose produces some sweetness and the salts some saltiness. Several small molecules present in very small quantities also contribute to flavor. The fat globules are responsible for the creaminess of whole milk.

Nutritional value. Milk is a complete food for the young calf, and it can also provide good nutrition to humans. It contains virtually all nutrients, most of these in significant quantities. However, it is poor in iron and the vitamin C content is not high. It contains no antinutritional factors, but it lacks dietary fibre.

Milk as a Substrate for Bacteria. Because it is rich in nutrients, many microorganisms, especially bacteria, can grow in milk. Not all bacteria that need sugar can grow in milk, some being unable to metabolize lactose. Milk is poor in iron, which is an essential nutrient for several bacteria, and contains some antibacterial factors, such as immunoglobulins and some enzyme systems. Moreover, milk contains too much oxygen for strictly anaerobic bacteria. Altogether, the growth of several bacteria is more or less restricted in raw milk, but several others can proliferate, especially at high ambient temperatures.

1.4 VARIABILITY

Freshly drawn milk varies in composition, structure, and properties. Even within the milk from a single milking of one cow, variation can occur. The fat globules vary in size and, to some extent, in composition, and the same applies to casein micelles.

Natural Variation. The main factors responsible for natural variation in milk are the following:

- *Genetic factors:* Breed and individual.
- *The stage of lactation:* This can have a significant effect. Especially the milk obtained within 2 or 3 d after parturition tends to have a very different composition; it is called *colostrum* or beestings.

- *Illness of the cow:* Especially severe *mastitis* (inflammation of the udder) can have a relatively large effect. The milk tends to have an increased content of somatic cells.
- *Feed:* The amount and the quality of the feed given strongly affect milk yield. However, the effect of the cow's diet on milk composition is fairly small, except for milk fat content and composition.

In a qualitative sense, cows' milk is remarkably constant in composition. Nevertheless, individual milkings show significant differences in composition. The variation is small in milk processed at the dairy, because this consists of mixtures of the milk of a large number of cows from many farms.

Other Causes. As soon as the milk leaves the udder, it becomes contaminated, for instance, with oxygen and bacteria (milk within the udder of a healthy cow tends to be sterile). Contamination with other substances can occur. The temperature of the milk generally decreases. These factors can lead to changes in milk properties. Far greater changes occur during long storage and in milk processing (see the next section).

1.5 CHANGES

Milk is not a system in equilibrium. It changes even while in the udder. This is partly because different components are formed at various sites in the mammary secretory cell and come into contact with one another after their formation. Furthermore, several changes can occur due to the milking, the subsequent lowering of the temperature, and so on. Changes may be classified as follows:

1. *Physical changes* occurring, for instance, when air is incorporated during milking: Because of this, additional dissolution of oxygen and nitrogen occurs in milk. Moreover, a new structural element is formed: air bubbles. Milk contains many surface-active substances, predominantly proteins, which can become attached to the air–water interface formed. Furthermore, by contact with the air bubbles, fat globules may become damaged, i.e., lose part of their membrane. Fat globules may cream. Creaming is most rapid at low temperature because the globules aggregate to large flocs during the so-called cold agglutination (Subsection 3.2.4). On cooling, part of the milk fat starts to crystallize, the more so at a lower temperature. But even at 0°C part of the fat remains liquid. The presence of fat crystals can strongly diminish the stability of fat globules against clumping.
2. *Chemical changes* may be caused by the presence of oxygen: Several substances may be oxidized. In particular, light may induce reactions, often leading to off-flavors. Composition of salts can vary, for example, with temperature.
3. *Biochemical changes* can occur because milk contains active enzymes: Examples are lipase, which causes lipolysis; proteinases, which cause

proteolysis; and phosphatases, which cause hydrolysis of phosphoric acid esters.

4. *Microbial changes* are often the most conspicuous: The best-known effect is production of lactic acid from lactose, causing an obvious decrease in pH. Numerous other changes, such as lipolysis and proteolysis, may result from microbial growth.

Cooling of the milk to about 4°C is generally applied to inhibit many of the changes mentioned, especially growth of microorganisms and enzyme action. In many regions, the milk is already cooled at the farm, directly after milking, in a so-called bulk tank. The milk should be kept cold during transport to the dairy and subsequent storage.

Processing. At the dairy milk is always processed. Of course, this causes changes in composition and properties of the milk, as it is intended to do. These changes can be drastic, as the following examples will show, and it can be questioned whether the resulting product can still be called milk; however, it is standard practice to do so. The most common processes applied are briefly described in the following paragraphs.

Heat treatment is virtually always applied, primarily to kill harmful bacteria. It also causes numerous chemical and other changes, the extent of which depends on temperature and duration of heating. Low pasteurization (e.g., 15 s at 74°C) is a fairly mild treatment that kills most microorganisms and inactivates some enzymes but does not cause too many other changes. High pasteurization (e.g., 15 s at 90°C, but varying widely) is more intense; all vegetative microorganisms are killed, most enzymes are inactivated, and part of the serum proteins become insoluble. Sterilization (e.g., 20 min at 118°C) is meant to kill all microorganisms, including spores; all enzymes are inactivated; numerous chemical changes, such as browning reactions, occur; and formic acid is formed. UHT (ultrahigh-temperature) heating (e.g., at 145°C for a few seconds) is meant to sterilize milk while minimizing chemical changes; even some enzymes are not inactivated fully.

Separation, usually by means of a flow-through centrifuge called a cream separator, yields skim milk and cream. The skim milk has a very low fat content, 0.05 to 0.08%. Milk skimmed after gravity creaming has a much higher fat content. Unless stated otherwise, the term *skim milk* will refer to centrifugally separated milk. By mixing skim milk and cream, milk may be standardized to a desired fat content.

Homogenization (i.e., treatment in a high-pressure homogenizer) of milk leads to a considerable reduction in fat globule size. Such milk creams very slowly but is also altered in other respects. All types of sterilized milk or, more generally, all long-life liquid milk products are homogenized in practice.

Evaporation removes water, producing milk that is more concentrated. Many properties are altered; the pH decreases, for example.

Membrane processes may be applied to remove water; this is called reverse osmosis. Ultrafiltration separates milk into a concentrate and a permeate that is rather similar to milk serum. Electrodialysis removes some inorganic salts.

Fermentation or culturing of milk, usually by lactic acid bacteria, causes considerable alteration. Part of the lactose is converted to lactic acid, causing a

TABLE 1.5
Gross Composition and Some Properties of Milk and Liquid Milk Products

Milk Product	Fresh Milk	Beverage Milk	Beverage Milk	Skim Milk[b]	Whipping Cream	Evaporated Milk	Whey (Sweet)[b]	Buttermilk (Sour)[c]	Yogurt (Plain)[d]
Heat Treatment[a]	No	LP	ST	LP	HP	ST	LP	HP	HP
Contents (% w/w)									
Fat	4.0	3.5	3.5	0.07	35	8.0	0.05	0.4	3.5
Casein	2.6	2.6	2.6	2.7	1.7	5.7	trace	2.6	2.5
Other proteins	0.70	0.70	0.70	0.66	1.15	1.5	0.81	0.9	0.8
Lactose	4.6	4.6	4.6	4.8	3.1	9.5	5.0	3.9	3.7
Salts	0.85	0.85	0.85	0.88	0.57	1.85	0.66	0.88	0.85
Water	87.1	87.5	87.5	90.6	58.4	73.1	93.3	90.2	87.5
Properties									
pH	6.7	6.7	6.6	6.7	6.7	6.2	6.6	4.6	4.4
Viscosity (mPa·s)	1.9	1.8	1.9	1.65	8.7	17	1.2	100[e]	400[e]
Density ($kg \cdot m^{-3}$)	1029	1030	1030	1035	990	1070	1025	1034	1031
Fat globule size (μm[f])	3.4	0.5	0.3	0.4	4.0	0.3	—	—	0.4

Note: Approximate examples; values at room temperature.

[a] LP = low-pasteurized, HP = high-pasteurized, ST = sterilized.
[b] Centrifuged.
[c] Undiluted.
[d] Stirred yogurt.
[e] Highly variable.
[f] d_{32} (see Subsection 3.1.4).

decrease in pH to such an extent that the casein becomes insoluble. This makes the milk much more viscous. The bacteria also produce other metabolites, the kind and concentrations of which depend on the bacterial species.

Cheese making. As mentioned, milk can be clotted by adding rennet, which contains a specific proteolytic enzyme. The enzyme transforms the casein micelles in such a way that they start to coagulate. The resulting gel can be broken into pieces; stirring the material then results in the formation of curd particles and whey. The curd contains the micellar casein and most of the fat, the liquid whey contains most of the water-soluble components of the milk and some protein split off casein by the rennet. The curd is further processed to form cheese.

Table 1.5 gives some idea of properties and composition of milk treated in various ways and of some liquid milk products, including whey. More extensive information is given in the Appendix.

Suggested Literature

Most aspects are further discussed in later chapters, and literature will be mentioned there.

A general reference for many aspects treated throughout the book: H. Roginski, J.W. Fuquay, and P.F. Fox, Eds., *Encyclopedia of Dairy Sciences,* Vols. 1–4, Academic Press, London, 2003.

A good monograph on milk synthesis, secretion, and collection, although slightly outdated in a few aspects: B.L. Larson, Ed., *Lactation,* Iowa State University Press, Ames, Iowa, 1985.

2 Milk Components

In this chapter, the properties of the various (classes of) components of milk are discussed. The emphasis is on chemical properties and reactivity, although some other aspects are included such as crystallization. Synthesis of the components and nutritional aspects are briefly mentioned. The chapter ends with a section on the natural variation in milk composition. Chapter 3 emphasizes the physical aspects of milk's structural elements.

The reader should be cautioned that people often speak of a component whereas what is being referred to is actually part of a larger molecule; for instance, 'the linoleic acid content of milk' generally refers to linoleic acid esterified in the triglycerides, rather than to the 'free' fatty acid implied.

2.1 LACTOSE

Lactose, or 0-4-D-galactopyranosyl-(1,4)-glucopyranose, is the major carbohydrate of milk. This sugar has been found in the milk of nearly all mammals and is unique to milk. Both glucose and galactose are abundant in the mammalian metabolism; lactose is only synthesized in the Golgi vesicles of the lactating cells. This occurs due to the presence of α-lactalbumin, a protein unique to milk. This protein modifies the action of the common enzyme galactosyl-transferase to catalyze the formation of lactose from uridine-diphosphate-galactose and glucose.

Cows' milk contains traces of other carbohydrates, e.g., glucose and galactose, but no polysaccharides. Furthermore, glucidic compounds such as hexosamines and N-acetyl neuraminic acid (see Figure 2.28, NANA) occur in milk, but most of these are covalently bound to proteins, especially membrane proteins, or in cerebrosides (see Table 2.8).

Lactose can be separated from milk or, in industrial practice, from whey, by letting it crystallize. Crystalline lactose is produced in large amounts, and it is mainly used in foods and in pharmaceuticals; nearly all pills contain lactose as a filling material. Lactose is also used as raw material for a range of chemical or enzymatic derivatives, such as lactitol, lactulose, and oligosaccharides.

2.1.1 Chemical Properties

Lactose is a disaccharide composed of D-glucose and D-galactose. The aldehyde group of galactose is linked to the C-4 group of glucose through a β-1, 4-glycosidic linkage (Figure 2.1). Both sugar moieties occur predominantly in the pyranose ring form. Chemical reactions of lactose involve the hemiacetal linkage between

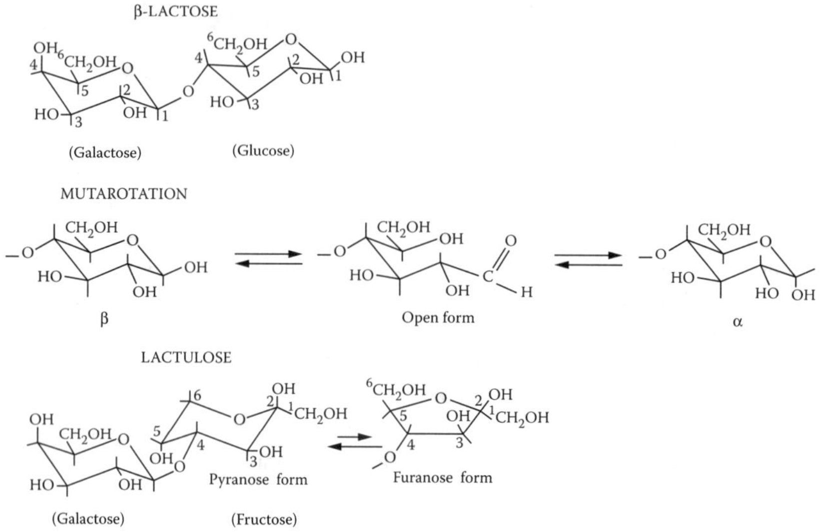

FIGURE 2.1 Chemical structure of β-lactose and of lactulose, and the mutarotation of the glucose moiety of lactose.

C^1 and C^5 of the glucose moiety, the glycosidic linkage, the hydroxyl groups, and the –C–C– bonds.

Furthermore, lactose is a *reducing sugar.* As shown in Figure 2.1, the O–C^1 bond in the glucose moiety can break, leading to an open-chain form that has an aldehyde group. It is also shown that conversion of the α-anomer into the β-anomer, and vice versa, does occur via the open-chain form. This phenomenon is called *mutarotation.* Presumably, less than 0.1% of the lactose in fresh milk is in the open-chain form. At high temperatures, and also at high pH values, this is a much higher proportion, say, between 1 and 10%. Because the aldehyde group is by far the most reactive one of lactose, this means that the reactivity of the sugar then is greatly enhanced.

Suitable reagents or enzymes can cause mild *oxidation* of lactose, whereby the aldehyde group is converted to a carboxyl group. Somewhat more vigorous oxidation ruptures the glycosidic linkage and produces carboxyl groups in the remaining sugars. Gentle *reduction* of lactose converts the aldehyde group to an alcohol group. More intense reduction cleaves the glycosidic linkage and results in the formation of alcohol groups in the remaining sugars. *Hydrolysis* of lactose by acid does not occur easily. If it occurs (high temperature and low pH), many other reactions take place as well.

Several reactions of lactose occur when milk is heated. Lactose may isomerize into *lactulose.* That means that the glucose moiety converts to a fructose moiety (Figure 2.1). Isomerization of the glucose moiety into mannose may occur as well, yielding epilactose, but the latter compound is formed in only trace amounts. In these isomerization reactions milk components are active as catalysts. (It is not fully clear as to which milk components are involved.) The quantity of lactulose in heated milk products can be used as an indicator of the intensity of the heat treatment.

Other reactions occurring during heat treatment are *caramelization* and *Maillard reactions*, which are to some extent related. The latter occur in the presence of amino groups, especially the ε-amino group of lysine residues in proteins. These reactions can lead to formation of flavor compounds and brown pigments, and to a decrease in the nutritionally available lysine. They are discussed in some detail in Subsection 7.2.3.

Sweetness: a lactose solution is approximately 0.3 times as sweet as a sucrose solution of the same concentration. In milk, the sweetness is, moreover, to some extent masked by the protein, primarily casein. Consequently, unsoured whey has a sweeter taste than milk. If the lactose in milk is hydrolyzed into glucose and galactose, the sweetness is considerably enhanced.

2.1.2 Nutritional Aspects

Lactose primarily provides the young animal with energy (about 17 kJ per gram of lactose), but it has other functions, such as giving a sweetish flavor. Lactose cannot be taken up into the blood; it must first be hydrolyzed into glucose and galactose. This occurs slowly, which prevents a sudden large increase of the glucose level of the blood after ingesting a substantial amount of milk. High blood glucose levels are considered to be detrimental. Moreover, some sugar (galactose and lactose) can reach the large intestine (colon), where it serves as a carbon source for several benign colon bacteria.

Lactose is hydrolyzed by the enzyme *lactase*, more precisely, β-galactosidase (EC 3.2.1.23), which is secreted in the small intestine. Naturally, the suckling young needs this enzyme, but after weaning, the amount of enzyme produced decreases to an insignificant level. This is not so for all humans. The estimates vary, but in at least 60% of people over 4 years old the enzyme activity is greatly reduced (to 5 to 10%), and they thus poorly metabolize lactose; these people are called *lactose mal-absorbers*. Drinking milk considerably enhances the activity of their colon flora. Roughly half of the mal-absorbers then develop significant problems, ranging from flatulence to severe diarrhea; these are called *lactose intolerant*. They cannot drink milk in significant quantities (say, 100 ml per day). The proportion of lactose mal-absorbers varies widely among regions, from, say, 10 to 90%. In populations in which milk has been part of the diet for numerous generations (people living in, or originating from, most of Europe and parts of Central and Western Asia, India, and Eastern Africa), mal-absorbers are scarce; in other populations they are very common. It has often been observed that fermented (sour) milk products hardly produce lactose intolerance, although they still contain about two thirds of the original lactose; the explanation is not fully clear. Another possibility is to treat milk with lactase: the lactose then is almost fully hydrolyzed into glucose and fructose (moreover, some oligosaccharides are formed). As mentioned, it markedly increases the sweet taste of the milk.

2.1.3 Physicochemical Aspects

2.1.3.1 Mutarotation

In solution, conversion of α- to β-lactose and vice versa occurs via the open-chain form (Figure 2.1), which is very short-lived. We have

α-lactose → β-lactose, reaction constant k_1, and
β-lactose → α-lactose, reaction constant k_2

Both are first-order reactions. We denote the equilibrium ratio [β]/[α] by R, and $R = k_1/k_2$.

The rate of the mutarotation reaction has the constant $K = k_1 + k_2$. If we dissolve, for instance, α-lactose, and if we define $x = [\alpha]/[\alpha]_{\text{equilibrium}}$, we have $\ln[R/(x-1)] = Kt$. In other words, the proportion of the mutarotation reaction that has been completed at time t is given by $1 - e^{Kt}$. The same holds for conversion of β-lactose. Examples are given in Figure 2.2A. These changes may be observed by using a polarimeter. The rotation of the plane of polarization then is found to change (to mutate) with time because α- and β-lactose differ in specific rotation. Hence, the term 'mutarotation.' Values for the specific rotation of α- and β-lactose are given in the Appendix, Table A.7.

The mutarotation rate K depends strongly on temperature (Figure 2.2A). At 20°C and pH 6.7, $K \approx 0.37\ \text{h}^{-1}$, and it increases by a factor of 3 or more per 10°C rise in temperature. At room temperature it takes many hours before mutarotation equilibrium is reached; at 70°C a few minutes. Figure 2.2B shows the great effect of pH on K. Several substances may affect the mutarotation rate. For example, the salts in milk increase the reaction rate by a factor of almost 2 as compared to the rate in water.

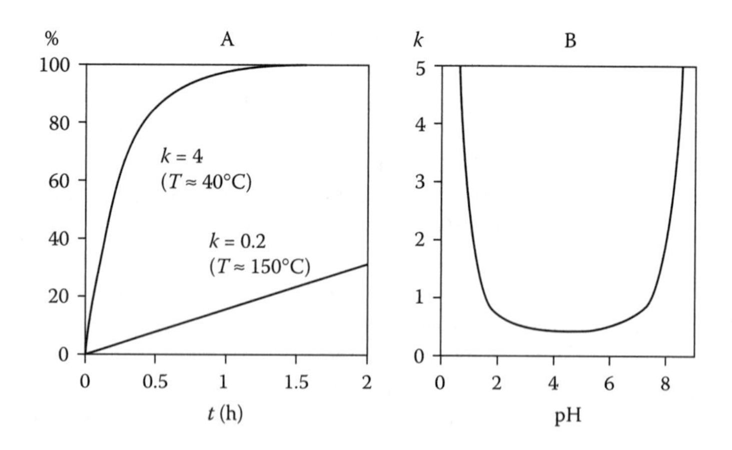

FIGURE 2.2 Mutarotation in lactose solutions. (A) Course of the reaction (% finished) against time t. (B) Mutarotation reaction constant K (h^{-1}) as a function of pH (~25°C). (Data from H.C. Troy and P.F. Sharp, *J. Dairy Sci.*, **13**, 140, 1930.)

The mutarotation equilibrium likewise depends on temperature: $R \approx 1.64 - 0.0027T$, where T is in degrees Celsius. Thus, a change in temperature causes mutarotation.

Mutarotation depends on lactose concentration. With increasing concentration, K decreases and R changes as well. K decreases considerably if other sugars such as sucrose are present in high concentration. At very high lactose concentration, i.e., in amorphous lactose as, for example, occurs in spray-dried milk powder, after equilibration $R \approx 1.25$, independent of temperature; mutarotation may still occur, but extremely slowly.

2.1.3.2 Solubility

As seen in Figure 2.3, α- and β-lactose differ considerably in solubility and in the temperature dependence of solubility. If α-lactose is brought in water, much less dissolves at the outset than later. This is because of mutarotation, α-lactose being converted to β, so the α-concentration diminishes and more α can dissolve. If β-lactose is brought in water, more dissolves at the outset than later (at least below 70°C); on mutarotation more α-lactose forms then can stay dissolved, and α-lactose starts to crystallize.

The solubility thus depends partly on the mutarotation equilibrium, the rate of dissolution on the mutarotation rate. The so-called final solubility is identical whether we dissolve α- or β-lactose. It is $R + 1$ times the initial solubility of

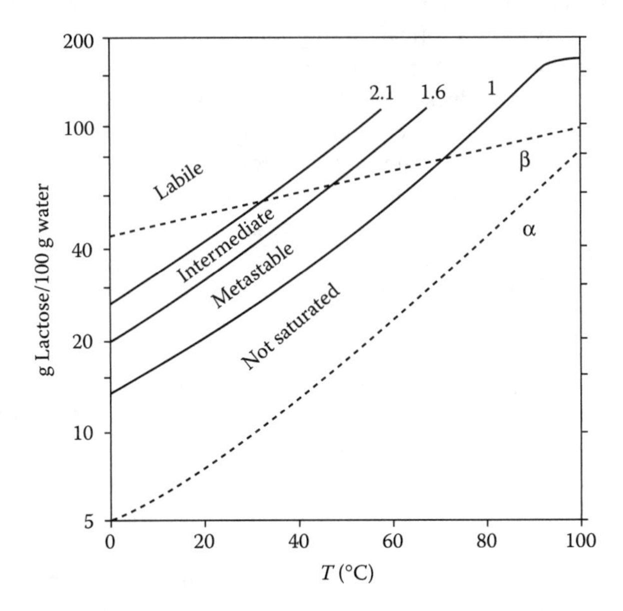

FIGURE 2.3 Solubilities of α- and β-lactose, final solubility of lactose (curve 1), and supersaturation by a factor of 1.6 and 2.1, as a function of temperature. (From P. Walstra and R. Jenness, *Dairy Chemistry and Physics,* Wiley, New York, 1984. With permission.)

α-lactose. This applies below 93.5°C because above this temperature β-lactose determines the final solubility. At low temperatures, it takes a long time to reach equilibrium.

Lactose solutions can be supersaturated easily and to a considerable extent. This is indicated roughly in Figure 2.3. At concentrations over 2.1 times the saturation concentration, spontaneous crystallization occurs rapidly, probably because of homogeneous nucleation (i.e., formation of nuclei in a pure liquid). At less than 1.6 times the saturation concentration, seeding with crystals usually is needed to induce crystallization, unless we wait a very long time; the solution is thus metastable. In the intermediate region, the occurrence of crystallization depends on several factors, such as time.

2.1.3.3 Crystal Forms

Usually, α-lactose crystallizes as a hydrate containing one molecule water of crystallization. The crystals are very hard, slightly hygroscopic, often fairly large, and dissolve slowly. The water of crystallization is very strongly bound. Above 93.5°C, anhydrous β-lactose crystallizes from an aqueous solution. β-lactose is not very hygroscopic, and it dissolves quickly; its solubility is good. Obviously, dehydrating α-hydrate is difficult. It may cause problems when determining the dry-matter content of milk and milk products; this determination implies evaporation of water at elevated temperature. Maintaining the temperature >93.5°C during the assay is of paramount importance to prevent formation of α-lactose hydrate crystals.

Amorphous lactose is formed during rapid drying, as in a spray drier. It is present in the glassy state (see Subsection 10.1.4), which means that many properties, including hardness, density, and specific heat, are similar to those of the crystalline sugar but that the packing of the molecules does not show perfect order. Amorphous lactose contains at least a few percent of water and can quickly dissolve on addition of water. But then, α-lactose hydrate may start to crystallize. If the water content of the amorphous lactose is low, say, 5%, crystallization is postponed. However, the product attracts water from moist air, and when moisture content rises to about 8%, α-lactose hydrate starts to crystallize (at room temperature). The postponed crystallization is an important factor in relation to spray-dried powders made from skim milk or whey because it leads to hard lumps in the powder; eventually, the whole mass of powder turns into one solid cake.

Several other crystal modifications of lactose may occur. Figure 2.4 gives a survey, including the transitions. In principle, it shows the methods for preparing the different modifications. In practice, it is almost impossible to obtain pure crystals of whatever form. For example, α-hydrate usually contains a little β-lactose, and vice versa.

The different forms mentioned are different crystal modifications, i.e., they have different crystal lattices. As a result, the properties are different. For instance, the stable anhydrous α-modification, also called S-lactose, is quite soluble in water. But a concentrated solution of it is unstable, which means that α-hydrate soon starts to crystallize. S-lactose is not very hygroscopic. The unstable anhydrous α

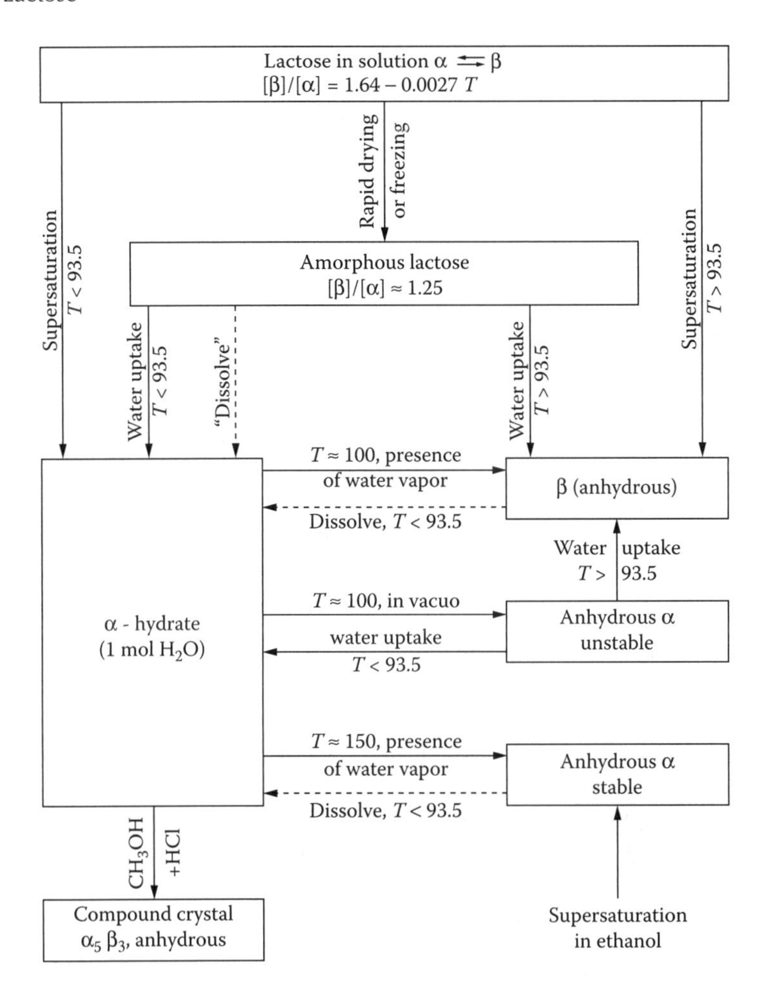

FIGURE 2.4 The different forms of lactose. T = temperature (°C).

form, however, is hygroscopic; accordingly, its transition to α-hydrate occurs easily, and the sugar dissolves faster, though not better, than α-hydrate.

2.1.3.4 Crystallization of α-Lactose Hydrate

This crystallization is of great practical importance. Because α-hydrate is poorly soluble, it may crystallize in some milk products, especially ice cream and sweetened condensed milk. Large crystals can easily be formed because both nucleation and crystal growth are slow. We usually have to add numerous tiny seed crystals to ensure the rapid formation of sufficient, hence small-sized, crystals. To prevent segregation and development of 'sandiness' in milk products, the largest

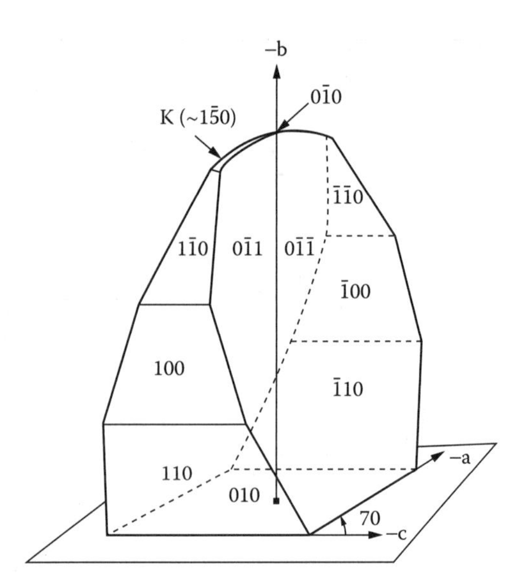

FIGURE 2.5 Common shape of the α-lactose hydrate crystal. The main axes (a, b, c) and the indices of the various faces are given. (Adapted from P. Walstra and R. Jenness, *Dairy Chemistry and Physics,* Wiley, New York, 1984.)

crystals formed should be no more than 10 μm in size. This implies that at least 10^{10} crystals per gram of crystalline lactose should be present.

α-hydrate crystals can have many geometrical forms (but the crystal lattice is always the same). The commonest shape is the 'tomahawk,' depicted in Figure 2.5. Usually, the crystal does not grow in the direction of the b axis, i.e., the crystal faces $0\bar{1}0$ and K or $1\bar{5}0$. Likewise, the $0\bar{1}1$ lateral faces do not grow at all. Consequently, the 'apex' of the crystal is also the point where the crystal started to grow. Furthermore, crystal growth is slow, far slower than may be accounted for by the combined effects of mutarotation and diffusion of α-lactose to the crystal.

Presumably, there is some difficulty for a molecule to fit into the crystal lattice. But otherwise the observations of the preceding paragraph are largely explained by inhibition by β-lactose. It appears as if β-lactose fits well into the $0\bar{1}0$ and $0\bar{1}1$ faces of the crystal lattice but then prevents any further uptake of α-lactose. Growth of other faces is inhibited as well (Figure 2.6 and Table 2.1). If very little β-lactose is present (this condition is difficult to achieve), the $0\bar{1}1$ faces grow fast, causing formation of needle crystals. Several other substances can retard crystal growth; the individual crystal faces are inhibited in different ways, which leads to variation in crystal habit. An example of a growth 'inhibitor' is riboflavin, present in milk, though at a low concentration (about 2 ppm).

One particular crystal growth inhibitor should be discussed. If α-lactose hydrate is purified by recrystallization, its rate of crystal growth decreases (Table 2.1). Moreover, the pH of the lactose solutions falls upon further recrystallization. It appears that a crystal growth inhibitor is present; it has a stronger affinity for the lactose

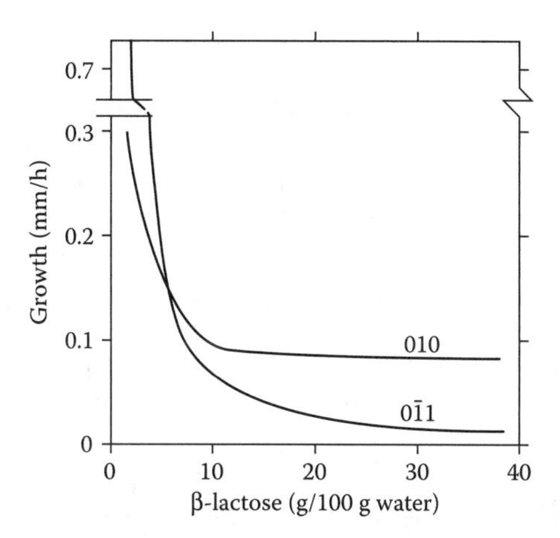

FIGURE 2.6 Effect of the concentration of β-lactose on the growth rate of some faces of the α-lactose hydrate crystal. Supersaturation of α-lactose is by 170%. (Data from A. van Kreveld, *Neth. Milk Dairy J.*, **23**, 258, 1969.

crystal than α-lactose itself. The inhibitor is a mixture of lactose monophosphates; its concentration in milk is about 15 $mg \cdot l^{-1}$. It particularly inhibits growth at low supersaturation and causes inhibition of nucleation in lactose solutions. The substance can be removed by ion exchange.

TABLE 2.1
Examples of the Rate of Growth of Some Faces of an α-Lactose Hydrate Crystal as Affected by Liquid Composition

		Growth ($\mu m \cdot h^{-1}$) of Face			
Supersaturation (%)	**Remarks**	**010**	**110**	**100**	**1$\bar{1}$0**
55	—	3.8	3.3	1.3	0.3
120	—	43	34	21	12
55	+100 ppm riboflavin	2.7	0.0	0.0	0.0
55	Own pH (~4)	3.2	2.7	1.6	0.4
55	pH 7	6.6	5.0	2.7	1.2
55	3 × recrystallized	0.2	0.7	1.3	0.5
55	Nonionic[a]	19.1	9.1	3.1	1.2
55	Nonionic + inhibitor[b]	0.0	0.0	0.9	0.5

[a] Solution passed through an anion exchanger.
[b] Lactose monophosphates.

2.2 SALTS

Milk contains inorganic and organic salts. The concept of 'salts' thus is not equivalent to 'mineral substances.' Salts are by no means equivalent to 'ash' because ashing of milk causes loss of organic acids including citrate and acetate and because organic phosphorus and sulfur are transferred to inorganic salts during ashing.

2.2.1 Composition and Distribution among the Phases

Average concentrations of salts in milk are given in Table 1.3, Table 2.2, and Table 2.4. The salt composition varies, but the various components do not vary independently of each other.

Not all of the salts are dissolved, and not all of the dissolved salts are ionized. The casein micelles contain the undissolved salts. In addition to counterions of the negatively charged casein (mainly Ca, Mg, K, and Na), the micelles contain the so-called colloidal calcium phosphate, which also contains some citrate. The colloidal phosphate is amorphous, can vary in composition, and may have ion exchange properties.

The distribution of phosphorus among the fractions is even more intricate. Details are given in Table 2.3. It is important to note that phosphatases present in

TABLE 2.2
The Most Important Salts in Milk and Their Distribution between Serum and Casein Micelles

Compound	Molar Mass (Da)	Range (mmol/kg)	Average (mg/100 g)	Fraction Present in Serum	In Micelles (mmol/g Dry Casein)
Cations					
Na	23	17–28	48	0.95	0.04
K	39.1	31–43	143	0.94	0.08
Ca	40.1	26–32	117	0.32	0.77
Mg	24.3	4–6	11	0.66	0.06
Amines	—	~1.3	—	~1	—
Anions					
Cl	35.5	22–34	110	1	—
CO_3	60	~2	10	~1?	—
SO_4	96.1	~1	10	1	—
PO_4 [a]	95	19–23	203	0.53	0.39
Citrate[b]	189	7–11	175	0.92	0.03
Carboxylic acids	—	1–4	—	~1?	—
Phosphoric esters[c]	—	2–4	—	1	—

[a] Inorganic only.
[b] $(CH_2–COO^-)–(COH–COO^-)–(CH_2–COO^-)$.
[c] Soluble.

TABLE 2.3
Approximate Distribution of Phosphorus in Milk

Type	Location	Distribution (%)	Dialyzable against Water	Dialyzable at Low pH	Soluble in TCA[a]	Extractable with Ethanol/Ether
Esterified to casein	Casein micelles[b]	22	No	No	No	No
Esterified in phospholipids	Fat globules and serum	1	No	No	No	Yes
Various esters	Serum	9	Yes	Yes	Yes	Yes
Inorganic, 'colloidal'	Casein micelles	32	No[c]	Yes	Yes	No
Inorganic, dissolved	Serum	36	Yes	Yes	Yes	No

Note: Milk contains approximately 1 g phosphorus per kg.

[a] Final concentration 12% trichloroacetic acid.
[b] A small part in the serum, especially at low temperature.
[c] Partly dialyzable against an excess of water.

TABLE 2.4
Estimated Average Composition of Milk Serum, i.e., the Salt Solution of Milk, including the Other Dissolved Substances

Cations					Anions					Neutral	
Ion	*m*	*mz*	*mz*^{2}	*a*	Ion	*m*	*mz*	*mz*^{2}	*a*	Molecule	*m*
K^+	36.3	36.3	36.3	29	Cl^-	30.9	30.9	30.9	25.0	KCl	0.7
Na^+	20.9	20.9	20.9	17	SO_4^{2-}	1.0	1.9	3.8	0.4	NaCl	0.4
Ca^{2+}	2.0	4.0	8.0	0.84	KSO_4^-	0.1	0.1	0.1	0.1	$CaSO_4$	0.1
$CaCl^+$	0.3	0.3	0.3	0.2	$Citrate^{3-}$	0.3	0.8	2.3	–	$CaHPO_4$	0.6
$CaH_2PO_4^+$	0.1	0.1	0.1	0.1	H $Citrate^{2-}$	–	0.1	0.2	–	$MgHPO_4$	0.3
Mg^{2+}	0.8	1.6	3.2	0.34	Ca $Citrate^-$	7.0	7.0	7.0	5.6	KH_2PO_4	0.2
$MgCl^+$	0.1	0.1	0.1	0.1	Mg $Citrate^-$	2.0	2.0	2.0	1.6	NaH_2PO_4	0.1
RNH_3^+	0.7	0.7	0.7	0.6	$RCOO^-$	3.0	3.0	3.0	2.4	$CaROPO_3$	0.2
H^+				$2 \cdot 10^{-4}$	HPO_4^{2-}	2.6	5.3	10.6	1.1	$MgROPO_3$	0.1
					$H_2PO_4^-$	7.5	7.5	7.5	6.1	CO_2	0.2
					$KHPO_4^-$	0.5	0.5	0.5	0.4	H_2CO_3	0.1
					$NaHPO_4^-$	0.4	0.4	0.4	0.3	Lactose	147
					$ROPO_3^{2-}$	1.6	3.2	6.4	0.7	Urea	5
					$HROPO_3^-$	0.5	0.5	0.5	0.4	Other	4
					$KROPO_3^-$	0.1	0.1	0.1	0.1		
					$NaROPO_3^-$	0.1	0.1	0.1	0.1		
					HCO_3^-	0.5	0.5	0.5	0.4		
Total	61.2	64.0	69.6		Total	58.1	63.9	75.9		Total	159

Note: pH = 6.7; *m* = concentration in mmol/l solution; *z* = valency; *a* = free-ion activity; –= <0.05.
To convert the composition to millimoles per kg of milk, multiply by 0.904; to convert to millimoles per kg water in milk, multiply by 1.045.

milk may hydrolyze phosphoric esters, causing the content of organic (esterified) phosphate to decrease and that of inorganic phosphate to increase. Milk contains sulfur, again in several forms. Not more than 10% of the sulfur in milk, amounting to about 36 mg per 100 g of milk, is present as inorganic sulfate, whereas the remainder is present in protein.

The dissolved salts affect various milk properties, e.g., protein stability. These salts are only present in the serum. Note that the solute content in the serum is approximately 1.09 times the content in milk; in the plasma it is approximately 1.04 times the content in milk (see also Section 10.1).

The composition of the salt solution of milk is best determined from a dialyzate of milk. The solution can be obtained by dialysis of water against a large excess of milk, as it is in equilibrium with the colloidal particles and dissolved proteins in milk. It does not reflect precisely the concentrations of the various dissolved salts in the water. To begin with, part of the water in milk is not available as a solvent (Section 10.1). Second, the proteins have a 'diffuse double layer,' so that they are accompanied by counterions, i.e., positively charged ions that compensate for the net negative charge on the protein. Of the cations associated with the casein micelles (Table 2.2), all of the Na and K, most of the Mg, and a far smaller portion of the Ca are present as counterions. The serum proteins and the fat globules also take along some counterions.

Trace elements are elements of which not more than a trace is found in milk. Of these, zinc is present in the highest concentration, that is, approximately 3 mg per kg of milk, whereas the other elements are present in far lower concentrations. The number of known trace metals increases with the development of increasingly sensitive methods of analysis. Trace elements are listed in the Appendix, Table A.8.

Trace elements are natural components in milk. Concentrations of some of these elements in milk can be increased by increasing their level in the feeding ration of the cow. Consequently, their concentration in milk can vary widely. For instance, Se can range from 4 to 1200 μg per kg of milk. Concentrations of other metals are not affected by cattle feed, except on shortage (e.g., Cu and Fe), or if extreme levels in the ration cause poisoning of the cow (e.g., Pb). Finally, some elements can enter the milk by contamination following milking. Contamination can considerably increase the concentrations. For instance, the natural Cu content of milk is about 20 $\mu g \cdot kg^{-1}$ (colostrum contains more); contact of milk with bronze parts in milking utensils or with copper pipes can increase its Cu content easily to 1 $mg \cdot kg^{-1}$. Fe also can readily enter the milk due to contamination.

Little is known about the distribution of the trace elements among the fractions. Part of the elements are likely to be associated with protein, e.g., some heavy metals with lactoferrin, whereas most of the other elements are dissolved. About 10% of Cu and nearly half of Fe are associated with the fat globule membrane. Zn and Sr are predominantly in the colloidal phosphate.

For the dairy manufacturer, Cu is of great importance due to its catalytic action on fat autoxidation. 'Natural Cu' in milk does not promote oxidation, or does so hardly at all, but 'added Cu' often does, even when added in minute amounts. Mn is of importance in the metabolism of some lactic acid bacteria, especially for fermentation of citrate, and in some milks the Mn content is too low for production of diacetyl by leuconostocs.

2.2.2 Properties of the Salt Solution

Only the dissolved salts are considered here, i.e., roughly the salts in the milk serum. The composition does not follow simply from Table 2.2, mainly because of the extensive association of ions. Some of the acids and bases in milk (phosphoric acid, carbonic acid, secondary amines, etc.) are not fully ionized at milk pH. But some of the salts may also be partly associated. This primarily concerns binding of Ca^{2+} and Mg^{2+} to $citrate^{3-}$ and to HPO_4^{2-}. Consequently, the concentration of Ca^{2+} ions is much lower than that of dissolved Ca because $Ca\text{-}citrate^-$ and $CaHPO_4$ are present in appreciable amounts. Also other salts, such as the chlorides of Na, K, Ca, and Mg, are not completely ionized. The approximate ion composition can be calculated from the atomic composition of the milk salt solution and the association constants. Results are given in Table 2.4.

Such calculations should be based on *activities* (a, in mol/l) rather than on concentrations (m, in mol/l). Activities govern reaction rates and thereby the state of ionization. By definition, the relation between a and m of a substance x is given by

$$a_x \equiv \gamma_x m_x \tag{2.1}$$

where γ is the activity coefficient. If the total ionic strength (I) is not too high, say, $I < 0.1$ M, the *free-ion activity coefficient* of ionic species in water of room temperature is approximately given by

$$\gamma \approx \exp(-0.8z^2 I^{1/2}) \tag{2.2}$$

where z is the valency of the ion (i) involved. I is defined by

$$I = \frac{1}{2}\Sigma m_i z_i^2 \tag{2.3}$$

Results in Table 2.4 show that in milk $I \approx 0.073$. Hence, γ for mono, di, and trivalent ions in milk would be 0.81, 0.42, and 0.14, respectively. These are approximate values. All association and dissociation constants and solubility products as listed in handbooks and the like are so-called *intrinsic* constants that refer to activities. They only apply to concentrations if $I \to 0$, where $\gamma \to 1$ and, hence, $a \to m$. Relating association constants to concentrations yields *stoichiometric* constants; in these cases, the ionic strength always should be recorded.

Neglecting the association of ions and not allowing for the activity coefficients can cause considerable deviations, especially with respect to polyvalent ions. Let us consider, for example, the solubility of calcium citrate. The solubility product of $Ca_3citrate_2$ is given as 2.3×10^{-18} $mol^5 \cdot kg^{-5}$ (Table 2.5). If one takes the total concentrations of calcium and citrate in milk serum (these amount to, say, 10 $mmol \cdot kg^{-1}$), a concentration product of 10^{-10} $mol^5 \cdot kg^{-5}$ is found, i.e., about 4×10^7 times the solubility product. In other words, milk serum would be supersaturated with respect

TABLE 2.5
Intrinsic Solubility Products at 20°C (Ion Activities in mol/kg Water)

Compound	Solubility Product	Unit
$Ca_3Citrate_2$	2.3×10^{-18}	$mol^5 \cdot kg^{-5}$
$CaHPO_4 \cdot 2H_2O$	2.6×10^{-7}	$mol^2 \cdot kg^{-2}$
$Ca_4H(PO_4)_3$	1.2×10^{-47}	$mol^8 \cdot kg^{-8}$
$Ca_3(PO_4)_2$	$\sim 10^{-29}$	$mol^5 \cdot kg^{-5}$
$Ca_5OH(PO_4)_3$	$\sim 10^{-58}$	$mol^9 \cdot kg^{-9}$
$Ca(CH_3CHOHCO_2)_2$	$\sim 10^{-4}$	$mol^3 \cdot kg^{-3}$
$MgHPO_4 \cdot 3H_2O$	1.5×10^{-6}	$mol^2 \cdot kg^{-2}$

to calcium citrate by a factor of $(4 \times 10^7)^{1/5} = 33$. But, obviously, only the Ca^{2+} and $citrate^{3-}$ ions must be considered. From Table 2.4 we see that

$$[Ca^{2+}]^3 \times [citrate^{3-}]^2 = (2.1 \times 10^{-3})^3 \times (0.3 \times 10^{-3})^2 \approx 8.3 \times 10^{-16}$$

This corresponds to a supersaturation by a factor of $360^{1/5} \approx 3.2$. The solubility product is, however, an intrinsic property. Consequently, the activities rather than the concentrations must be inserted, and the product of the ions must be multiplied by $(\gamma_{Ca^{2+}})^3 \times (\gamma_{citrate^{3-}})^2 \approx 0.42^3 \times 0.14^2 \approx 1.5 \times 10^{-3}$. Hence, the product of activities is only about half the solubility product, i.e., the concentration is 88% of the saturation concentration. Incidentally, the data used are not very precise, so the result may not be exactly correct. Milk must be about saturated with respect to calcium citrate because a little citrate is undissolved (Table 2.2).

Addition of a neutral salt such as NaCl to milk will cause some further association of Na^+ with $citrate^{3-}$ and of Cl^- with Ca^{2+}. More importantly, the ionic strength increases, as a result of which all ion activity coefficients decrease (see Equation 2.2) and, consequently, the ion activities and the activity product decrease as well. As a result, more calcium citrate can dissolve. For example, increasing the ionic strength from 0.08 M to 0.12 M would increase the solubility of $Ca_3citrate_2$ by about 35%. The general rule, thus, is that any increase of the ionic strength increases the solubility of salts.

Increasing the ionic strength also causes an increase of the dissociation of salts. Consider the equilibrium

$$Ca^{2+} + HPO_4^{2-} \rightleftarrows CaHPO_4$$

The dissociation constant is

$$K_D = a_{Ca^{2+}} \times a_{HPO_4^{2-}} / a_{CaHPO_4}$$

We may state that $a_{CaHPO_4} = [CaHPO_4]$ because in dilute mixtures the activity coefficient of electroneutral molecules is about 1. Increasing I will reduce the activity coefficients of Ca^{2+} and HPO_4^{2-}. Because K_D is constant, $[Ca^{2+}]$ and $[HPO_4^{2-}]$ will increase, whereas $[CaHPO_4]$ will decrease. In other words, the dissociation increases.

This is also true of acids in milk. The stoichiometric pK of monovalent acids is shifted downward by about 0.1 pH unit in milk, as compared to $I \to 0$. The stoichiometric pK is defined by

$$\log([\text{ion}]/[\text{acid}]) = \text{pH} - \text{p}K \quad (2.4)$$

pK values for acids in milk are approximately as follows:

Phosphoric acid	2.1,	7.1,	12.4
Citric acid	3.0,	4.5,	4.9
Carbonic acid	6.3,	10.0	
Fatty acids	4.7		

Note that the pH is an intrinsic quantity: $\text{pH} = -\log a_{H^+}$.

2.2.3 Colloidal Calcium Phosphate

Table 2.2 shows that part of the salts in milk is present in or on the casein micelles, i.e., in colloidal particles. This undissolved salt is called, in brief, colloidal, or micellar, calcium phosphate, though it includes other components, i.e., K, Na, Mg, and citrate. The total amount is about 7 g per 100 g of dry casein. Part of it is to be considered as counterions. This is because the casein is negatively charged at the milk pH and is thus associated with positive counterions. This presumably involves the K, Na, Mg, and part of the Ca in the micelles. The rest, which is mainly calcium and phosphate together with a little citrate, is present in a different state. Milk is supersaturated with respect to calcium phosphate and, accordingly, a large part of it is undissolved. Designating this part as a precipitate would be incorrect. The calcium phosphate in the casein micelles consists of small, noncrystalline regions and is, moreover, bound to the protein.

The molar ratio Ca/inorganic phosphorus in the micelles is high. Even if the part of the Ca present as counterions is subtracted, a ratio of about 1.5 remains for the calcium phosphate, i.e., as in tricalcium phosphate. That seems astonishing because we would expect a diphosphate (Ca/P = 1) at most, due to the pK_2 of phosphoric acid being about 7. Therefore, the phosphate esterified to serine residues of casein, i.e., the organic phosphate, is believed to participate in the colloidal phosphate as a result of which a ratio of ~1 would be found. However, the colloidal phosphate should not be seen as one of many known types of calcium phosphate. Moreover, it has a variable composition that depends on the ion atmosphere. For instance, as stated above, it contains citrate as well as traces of several other ions, e.g., Mg and Zn. In other words, the colloidal phosphate can be considered to have ion exchange properties.

2.2.4 Nutritional Aspects

A very important nutrient in milk is *calcium phosphate*, which is needed for growth and maintenance of bones. Accordingly, calcium phosphates are poorly soluble in water, which makes it difficult to realize high concentrations in an aqueous system like milk. Ruminants, such as the cow, have some mechanisms to overcome this problem. First, cows' milk has a relatively high casein content, and casein can accommodate large quantities of undissolved calcium phosphate by forming casein micelles, as mentioned. Second, the concentration of citrate in milk is high, and it strongly associates with calcium, which means that the calcium ion activity in milk is quite low; see Table 2.4. Third, part of the phosphate is esterified (see Table 2.3) and thereby kept in the casein micelles or in solution.

Milk and dairy products form an important source of calcium for humans; the calcium in these products is well absorbed. Moreover, the molar ratio Ca/P (including organic phosphate) equals 0.9 in milk; this is quite a high ratio compared to most foods, and it is considered to be beneficial for people who are prone to suffer from osteoporosis (loss of bone mass). Most of the calcium ingested is not absorbed: it stays in the gut as an amorphous calcium phosphate on which several soluble compounds can adsorb, which induces some beneficial effects. Gastroenteritis, as caused by some pathogens, can be significantly alleviated. Moreover, it decreases the risk of colon cancer in rats and probably also in humans.

Milk is a good source for many other minerals, including trace elements (see Appendix, Table A.8), especially zinc. Milk is quite low in iron, which helps in reducing growth of pathogenic bacteria in the gut of the young calf. (A calf has a considerable iron reserve at birth, as has a human infant.)

Some trace elements are toxic (e.g., Cd, Hg, and Pb), but they are hardly ever detected in milk in unsafe concentrations.

2.2.5 Changes in Salts

The salts of milk are in *dynamic equilibrium*: among themselves in solution, between solution and colloidal phosphate, and between solution and proteins. Changing external conditions of milk may cause alterations in equilibria. To be sure, there is no true equilibrium, especially not of calcium phosphates, but local or pseudoequilibria exist. The stress here is on the word dynamic.

Milk is saturated with respect to $CaHPO_4$ (solubility in milk serum is about 1.8 mmol/l, that of $Ca_3(PO_4)_2$ about 0.06). Furthermore, a small part of the citrate in milk is undissolved, as can be seen in Table 2.2. Milk is not saturated with respect to other salts (e.g., solubility of $MgHPO_4$ in water is approximately 12 mmol/l). Milk has a buffering capacity for some ions, primarily Ca^{2+}. This is mainly caused by the presence of undissolved colloidal phosphate, which can vary not only in quantity but also in composition, as a result of ion exchange. Imposing changes in ionic composition by altering temperature, pH, etc., has therefore a different effect on the salt solution of milk, where an exchange with the micellar salts may occur, as opposed to the salt solution of whey or ultrafiltrate.

TABLE 2.6
Effect of Various Additions to Milk on Increase (+) or Decrease (−) of Ca and Inorganic Phosphate in Various States

Substance Added	Effect on the Concentrations of		
	Ca^{2+}	Dissolved Ca	Phosphate in the Micelles
HCl	+0.2	+0.3	−0.2
NaOH	−0.1	−	+0.2
NaCl	+	+0.005	−
$CaCl_2$[a]	+0.3	+0.3	+0.4
Citric acid[a]	−0.1	+0.4	−0.2
Phosphoric acid[a]	−0.05	−0.1	+0.1
EDTA[a,b]	−	+	−

Note: As far as is known, the approximate magnitude of the change is given in moles Ca or phosphate per mole substance added, for a small addition.

[a] Plus as much NaOH as is needed to keep the pH constant.
[b] EDTA, ethylenediaminetetraacetic acid, a chelating agent.

Table 2.6 gives examples of what will happen when some substances are added to milk. Moreover, some changes may occur during storage of milk. The milk loses CO_2; the original content in the udder is roughly twice that mentioned in Table 2.4. Enzyme action on dissolved phosphoric esters causes a decrease in pH and in [Ca^{2+}], and an increase in the amount of colloidal phosphate. Lipolysis yields free fatty acids that decrease the pH and bind some Ca^{2+} ions. The calcium salts of fatty acids are poorly soluble (solubility product of fatty acid with $Ca^{2+} \approx 10^{-10}$ to 10^{-12} $mol^3 \cdot kg^{-3}$ and the pK of fatty acids ≈ 4.7).

Chelating agents such as EDTA or — more common — polyphosphates [$PO_3^{2-} - (OPO_2^-)_n - OPO_3^{2-}, n = 1, 2, 3 \ldots$] are often used in milk products, for example, concentrated milks or processed cheese, to reduce the calcium ion activity without the formation of insoluble calcium phosphates. Polyphosphate ions strongly associate with Ca^{2+} and the formed salts are well soluble. Citrates also strongly associate with Ca^{2+}, but at about neutral pH calcium citrate will become supersaturated.

2.2.5.1 Acidity

The pH may change as a result of additions, by concentrating or heating the milk and so forth. Microbial fermentation of lactose to lactic acid (Section 13.2) is of great importance. The ensuing drop in pH causes a partial dissolution of the colloidal phosphate (Figure 2.7), and a decrease of the negative charge of the proteins, which is accompanied by a decrease in the association with counterions. Moreover, a decrease in pH reduces the dissociation of weak acids, increases the [Ca^{2+}] (see addition of HCl in Table 2.6, and also Figure 2.9), and increases the ionic strength.

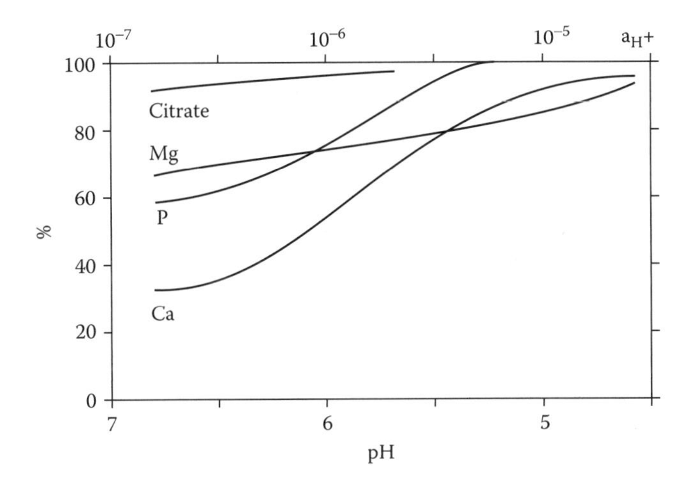

FIGURE 2.7 Approximate percentages of calcium, inorganic phosphate, magnesium, and citrate that are dissolved as a function of the pH of milk.

Several lactic acid bacteria also break down citrate, which enhances the increase in [Ca^{2+}]. On the other hand, lactate associates to some extent with Ca^{2+}, and the increase in [Ca^{2+}] upon lactic fermentation will thus be less than would follow from the drop in pH.

2.2.5.2 Temperature Treatment

We should distinguish between three types of experiments:

1. Measure the actual state in milk at various temperatures. For example, the dissociation constants are temperature dependent. Very few direct measurements with respect to alterations in milk have been made, but we may try to infer the composition of the salt solution at different temperatures from the following experiment.
2. Separate milk serum (e.g., by ultrafiltration) at various temperatures and investigate it at room temperature. It then follows that the pH of serum made at 0°C is ~0.1 unit higher, and that made at 93°C is ~0.5 unit lower, as compared to serum made at 20°C (see Figure 4.2).
3. Heat milk at various temperatures for various times, cool to room temperature, and investigate. This is the most common type of experiment.

During heating, the most important change is that dissolved calcium and phosphate become supersaturated and partly associate with the casein micelles. The additional colloidal phosphate so formed has a molar ratio Ca/P ≈ 1. The reaction is roughly

$$Ca^{2+} + H_2PO_4^- \rightarrow CaHPO_4 + H^+$$

This implies that the milk becomes more acidic. (The drop in pH partly counterbalances the insolubilization of Ca and phosphate.) The reactions are slow and

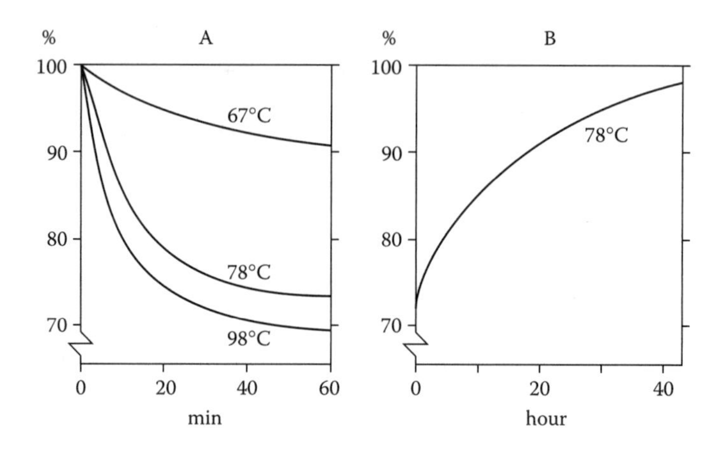

FIGURE 2.8 The amount of Ca in milk that is dissolved in percentage of the original amount. (A) effect of the time of heating at various temperatures (determined after 1 h at 20°C); (B) effect of holding at 20°C after heating for 30 min. Approximate examples. (From data by Jenness and Hilgeman, unpublished.)

occur especially in a fairly narrow temperature range. Below 60°C, changes are small, whereas above 80°C any further increase in temperature has little effect (Figure 2.8A). The reactions reverse at room temperature, though very slowly (Figure 2.8B). At low temperature, the reverse occurs: after 24 h at 3°C, dissolved Ca is increased by about 7%, dissolved phosphate by about 4%, and Ca^{2+} concentration is also increased. The magnitude of all of these changes may vary.

2.2.5.3 Concentrating

Concentrating of milk by evaporation of water causes several changes, but it should be taken into account that heating is usually involved as well. The pH drops by about 0.3 unit for 2:1 concentration (i.e., concentrating to twice the original dry-matter content) and by about 0.5 unit for 3:1 concentration. Again, the main cause of the changes is formation of additional calcium phosphate in the casein micelles. Accordingly, the Ca^{2+} concentration does not increase appreciably. The fractions of dissolved citrate, phosphate, and calcium decrease, but less than proportionally to concentration; for instance, dissolved calcium decreases from 40% to 30% for 2.5:1 concentration. This can be attributed partly to the pH decrease, partly to the increase of the ionic strength, but the changes have been insufficiently studied.

2.2.5.4 Calcium Ion Activity

The Ca^{2+} activity ($a_{Ca^{2+}}$) in milk is an important variable, especially for the stability of casein micelles. It differs markedly from the content of dissolved calcium. Because of this, direct determination of $a_{Ca^{2+}}$ is essential. This can be achieved by using a calcium ion selective electrode, in much the same way as the pH is measured. Figure 2.9 summarizes the effect of some variables. Addition of sugar, e.g., sucrose,

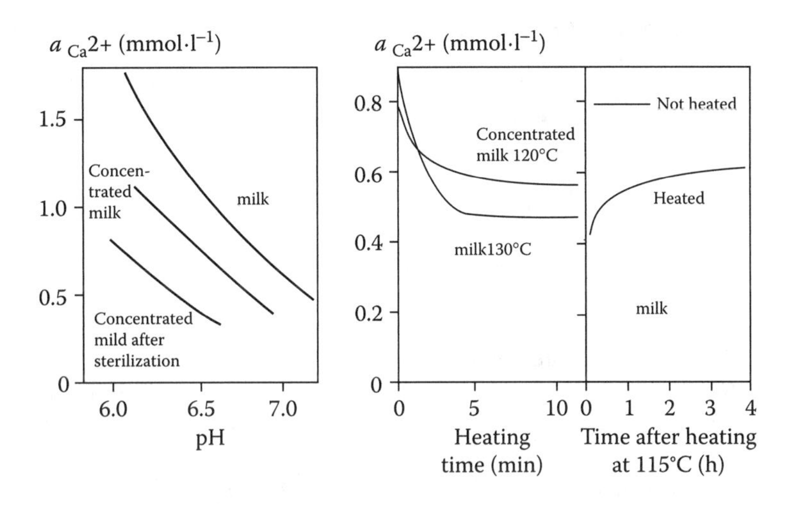

FIGURE 2.9 Effects of pH, heat treatment, concentration (by a factor of 2.6), and time after heating on calcium ion activity. All measurements at 20°C.

to milk (as is applied in the manufacture of sweetened condensed milk, ice cream mix, and other milk products) significantly increases $a_{Ca^{2+}}$ (expressed in mmol per kg of water). For example, the activity coefficient of Ca^{2+} increases from 0.40 to 0.46 by the addition of 150 g of sucrose to 1 l of milk. Different milks vary widely in $a_{Ca^{2+}}$, from 0.6 to 1.6 mmol·l^{-1}. The variation usually correlates significantly with pH, i.e., the higher the a_{H^+} (the lower the pH), the higher the $a_{Ca^{2+}}$.

2.3 LIPIDS

Lipids are esters of fatty acids and related components that are soluble in nonpolar organic solvents and insoluble, or nearly so, in water. Alternatively, the term *fat* is used. But fat is usually considered to consist largely of a mixture of triglycerides, especially when the mixture is partly solid at room temperature.

Nearly all of the fat in milk is in fat globules. It can therefore be concentrated readily by means of gravity creaming, possibly followed by churning. Products rich in fat, such as cream and butter, have a specific and often desired flavor and texture. On the other hand, milk fat is prone to deterioration, leading to serious off-flavors. The consistency of high-fat products greatly depends on the crystallization of the fat. In turn, crystallization behavior of milk fat depends on variation in fat composition and on processing and storage conditions.

In this section, composition and properties of lipids are discussed. The triacylglycerols, generally called triglycerides, make up over 98% of the milk lipids, which does not imply that the other lipids are unimportant. The various lipid classes are discussed in Subsection 2.3.2. We will start with a subsection on fatty acids, as these are essential constituents of nearly all lipids (except cholesterol), and their properties vary widely.

2.3.1 Constituent Fatty Acids

The fatty acid pattern is an important factor in determining lipid properties, such as melting range, chemical reactivity, and nutritional value. Table 2.7 gives the main fatty acids and some of their properties.

2.3.1.1 Fatty Acids Occurring

The following are the main variables among fatty acids in milk fat:

1. *Chain length.* Most fatty acids contain 4 to 18 carbon atoms; even-numbered acids are predominant.
2. *Number of double bonds,* in other words, the degree of unsaturation. It mainly determines chemical reactivity, including proneness to autoxidation.
3. *Position of double bonds.* For instance, conjugated (–CH=CH–CH= CH–) or nonconjugated (e.g., –CH=CH–CH_2–CH= CH–).
4. *Configuration of a double bond.* Each double bond can be either in the

   cis 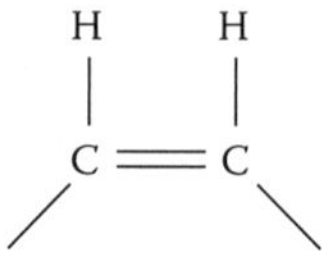 or in the *trans*

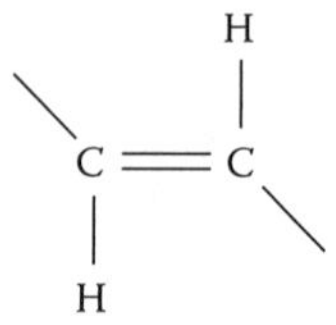

   configuration. The cis form is the common one in nature. Milk fat contains about 3 mol % trans acids, predominantly monounsaturated.
5. *Branching.* Nearly all of the fatty acids have an unbranched carbon chain. But some have a terminal –$CH(CH_3)_2$ group.
6. *Keto* or *hydroxy* group. Some fatty acids have a keto [–(C=O)–] or a hydroxy [–(HCOH)–] group.
7. A fatty *alcohol* or a fatty *aldehyde* residue. A very small proportion of the fatty acid residues is replaced by these residues.

2.3.1.2 Fatty Acid Pattern

On the basis of Table 2.7 and more detailed data, the fatty acid pattern of milk fat can be characterized as follows:

1. Its fatty acid composition is very wide. It includes fatty acids with keto or hydroxy groups, with an uneven number of C atoms, or with a branched carbon chain; the total number is some 250 different acid residues. Of these, 11 amount to over 1 mol % of the mixture of fatty acid residues.
2. It contains a relatively high proportion (15 to 20 mol %) of short-chain fatty acid residues, with 4 to 10 C atoms.

3. The proportion of saturated fatty acid residues is high; for example, 70 mol % (≈ 63% w/w).
4. Oleic acid is the most abundant of the unsaturated fatty acid residues (about 70%).
5. The other unsaturated fatty acid residues are present in a wide variety of chain length as well as in number, position, and configuration of the double bonds.

The pattern is very different from that of most vegetable oils, in which C_{18} acids predominate, with a high proportion of unsaturates — including polyunsaturates — and with a far smaller range of fatty acid residues.

2.3.1.3 Synthesis

In the rumen, the redox potential is low and extensive *hydrogenation* of double bonds occurs. This means that by far most of the fatty acids taken up from the intestinal tract are saturated (the content of polyunsaturated acids is especially small). Esterification of the resulting fatty acids with glycerol, which occurs in the mammary gland, would result in a fat of high melting range, being partly solid at physiological temperature. This cannot be tolerated by the animal. Consequently, a few mechanisms have evolved by which ruminants keep their milk fat liquid. These include:

1. In the mammary gland, short-chain fatty acids are synthesized in high quantities, and these acids give the triglycerides a relatively low melting range. The rumen flora produces β-hydroxybutyrate, which gives rise to synthesis of butyric acid (C_4). The rumen also produces much acetate, which can be added (by covalent bonds) to C_4, resulting in C_6–C_{14} acids. The C_{16} and C_{18} acids in milk fat are derived from blood triglycerides.
2. The mammary gland contains a desaturase, an enzyme that can convert stearic acid ($C_{18:0}$) into oleic acid ($C_{18:1}$), and to a far smaller extent, $C_{16:0}$ into $C_{16:1}$. Unsaturated fatty acids also give the fat a relatively low melting range.
3. The distribution of fatty acid residues over the triglycerides is uneven; the short-chain residues are predominantly in the 3-position rather than in the 2-position, as shown in Table 2.7. Also this gives a lower melting range.

These mechanisms qualitatively explain much of the fatty acid pattern described in the preceding text.

The variety of synthetic pathways may also explain the relatively strong dependence of the fatty acid pattern on feed composition. A ration containing much C_{18} acids gives a relatively unsaturated milk fat because of the extensive formation of $C_{18:1}$, whereas much C_{16} results in a low proportion of unsaturated fatty acid residues. Rations with a high proportion of concentrates yield little acetate in the rumen, hence, relatively little short-chain acids, whereas a high

TABLE 2.7
Fatty Acids in Milk Fat

	Notation[a]				Composition (in mol %) of			
Acid	x	y	Melting Point (°C)	Solubility[b] (g/l)	Neutral Glycerides[c]	Phospholipids	Free Fatty Acids[d]	Percentage in 3-Position
Saturated:					69 (57–80)	45	72	
Butyric	4	0	–8	Miscible	8.5 (7–14)	0.0	14.5	97
Caproic	6	0	–4	174	4.0 (2–7)	0.0	4.5	84
Caprylic	8	0	16	58	1.8 (1–3.5)	0.2	2	45
Capric	10	0	31	17	3.0 (1.5–5)	0.2	2	33
Lauric	12	0	44	5.6	3.6 (2.5–7)	0.5	2	26
Myristic	14	0	54	1.6	10.5 (8–15)	3	9	17
Palmitic	16	0	63	0.49	23.5 (20–32)	19	21	12
Stearic	18	0	70	0.14	10.0 (6–13)	12	13	22
Odd-numbered					2.5 (1.5–3.5)	4.5	2.5	7
Branched					1.1 (0.7–1.8)	0.7	1	
Other					0.7 (0.3–2)	5		
Monoene:					27 (18–36)	41	23	
Palmitoleic	16	1 Δ9[e]			1.4	?	1?	23
Oleic	18	1 Δ9[e]	16	0.42	21 (13–28)	38	20	32
Other					5.5	3	3.5	
Diene:					2.5 (1–4.3)	8		
Linoleic	18	2 Δ9,12[e]	–5		1.8	8	2.3	20
Other					0.7	0.2		

Polyene:					0.8 (0.4–2)	4		29
α-Linolenic	18	3 Δ9,12,15[e]	−12		0.4	2		
Other					0.4	2		
Keto					0.3	?		
Hydroxy					0.3	?		
Fatty alcohol					0.01	0.15		
Fatty aldehyde					0.02	0.01		
Unclassified						2	1	

Note: Properties, approximate average fatty acid composition of some lipid classes, and average percentage of each fatty acid residue esterified in the 3-position of the triglycerides.

[a] x = number of C atoms; y = number of double bonds; Δ refers to the position in the carbon chain: Δ9, 12, for instance, indicates that the two double bonds occur at the 9th and 12th bonds, counting from the carboxyl group.

[b] Critical micellization concentration in the presence of Na^+ ions.

[c] In parentheses is the approximate range.

[d] Free fatty acids liberated by the action of milk lipase.

[e] All *cis*.

proportion of roughage causes the opposite. The physical form of the feed also has an effect: a significant part of the unsaturated fatty acids in a concentrate consisting of relatively large and hard bits escapes hydrogenation, which can result in a milk fat with a relatively high proportion of polyunsaturates.

2.3.1.4 Reactivity

On *heating*, some of the fatty acids exhibit chemical reactions. Residues of 3-ketoacids give rise to free *methyl ketones* (R–CO–CH_3). 4- and 5-hydroxy fatty acid residues give γ- and δ-*lactones*, respectively. Especially the latter

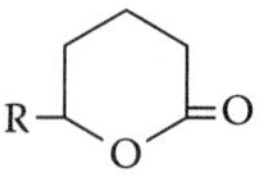

R O =O

are also present in fresh milk and are partly responsible for the characteristic flavor of milk fat. Higher quantities, which may arise from heat treatment or during long storage of dried milk, cause an atypical flavor. At still higher heating temperatures (e.g., 150°C) the position of part of the double bonds changes, and some are transformed from *cis* into *trans*.

The double bonds are subject to oxidation, forming peroxides that are further degraded: see Subsection 2.3.4. Oxidation gives rise to off-flavors. Another off-flavor is observed when lipolysis occurs, that is, the hydrolysis of ester bonds catalyzed by a lipase. It gives rise to the formation of free fatty acids and of mono- and diglycerides (see Subsection 3.2.5).

Some modifications are aimed at altering the crystallization behavior. In a physical method, the fat is *fractionated* into high and low melting portions after letting it partly crystallize (Subsection 18.5.2). The fat can be modified chemically by *interesterification* or randomizing (i.e., interchange of fatty acid residues among their positions in the triglyceride molecule) under the influence of a catalyst; it also occurs somewhat during intense heating. Another method is saturation of double bonds by *hydrogenation*, which is achieved at high temperature in the presence of H_2 and a metal catalyst. The chemically modified fats can no longer be called milk fat.

2.3.2 Lipid Classes

Table 2.8 gives the various lipid classes in milk, their concentration, and composition. Table 2.7 gives additional information about composition. Some properties of the individual lipid classes will be discussed below.

2.3.2.1 Triacylglycerols

The molecular structure is given in Figure 2.10. Triacylglycerols, called triglycerides for short, make up the bulk (generally more than 98%) of the lipids and,

TABLE 2.8
Lipids of Fresh Milk

Lipid Class	Alcohol Residue	Other Constituent	$\overline{MW}$	Fatty Acid Residues			Percentage in Milk Fat (w/w)	Percentage of the Lipid in		
				Number	$\overline{x}$	$\overline{y}$		Core of Globule	Globule Membrane	Milk Plasma
Neutral glycerides:							98.7			
Triglycerides	Glycerol		728	3	14.4	0.35	98.3	~100	+	
Diglycerides	Glycerol		536	2	14.9	0.38	0.3	90?	10?	?
Monoglycerides	Glycerol		314	1	15.0	0.36	0.03	+	+	+
Free fatty acids	—		253		15.8	0.36	0.1	60	10?	30
Phospholipids[a]:		Phospho group					0.8	0	65	35
Ph. choline (lecithin)	Glycerol	Choline	764	2	17.2	0.6	0.27			
Ph. ethanolamine[b]	Glycerol	Ethanolamine	742	2	17.9	1.0	0.26			
Ph. serine[b]	Glycerol	Serine	784	2	17.8	0.8	0.03			
Ph. inositide[c]	Glycerol	Inositol	855	2			0.04			
Sphingomyelin[d]	Sphingosine	Choline	770	1	19	0.2	0.20			
Cerebrosides[c,d]	Sphingosine	Hexose	770	1	20	0.2	0.1	0	70	30
Gangliosides[c,d]	Sphingosine	Hexose[e]	~1600	1			0.01	0	70?	30?
Sterols:							0.32	80	10	10
Cholesterol	—		387				0.30			
Cholesteryl esters	Cholesterol		642	1	16	0.4	0.02?			
Carotenoids + vitamin A	—						0.002	95?	5?	+

Note: See also Figure 2.10; approximate average values, not complete; $\overline{x}$ = average number of carbon atoms; and $\overline{y}$ = average number of double bonds.

[a] Approximately 1% is present as lysophosphatides.
[b] Phosphatidylethanolamine + Ph. serine = cephalin.
[c] Glycolipids.
[d] Sphingolipids.
[e] Also neuraminic acid.

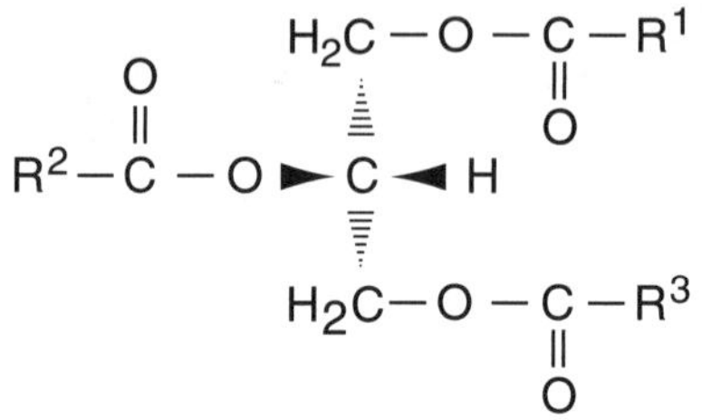

FIGURE 2.10 Triacylglycerol molecule. R denotes an aliphatic chain. For the configuration of the four bonds of the central C-atom as drawn, R^1, R^2, and R^3 denote the stereospecific numbering (*sn*) of the three acyl positions.

accordingly, largely determine the properties of milk fat. These properties vary with the fatty acid composition. Because the number of different fatty acid residues is great, the number of different triglycerides is much greater. The 11 major fatty acid residues alone would yield 11^3, or 1331, different triglycerides. Assuming that any other minor fatty acid residue would not appear more than once in a triglyceride molecule, we arrive at 10^5 different molecules at least. Moreover, there is presumably no single triglyceride species present in a concentration over 2 mol %. Clearly, milk fat shows a wide compositional range.

The distribution of fatty acid residues over the position in the triglyceride molecule is far from random. For example, butyric and caproic acid are largely in the 3- and stearic acid in the 1-position (see Table 2.7). The position of the fatty acid residues in the triglyceride molecules considerably affects the crystallization behavior of milk fat. Most other properties depend only on fatty acid composition.

Triglycerides are very apolar and not surface active. In the liquid state they act as a solvent for many other apolar substances, including sterols, carotenoids, and tocopherol. A small amount of water (about 0.15% at room temperature) dissolves in liquid milk fat. Some physical properties are given in the Appendix, Table A.4.

2.3.2.2 Di- and Monoglycerides

Some of these occur in fresh milk fat. Lipolysis increases their quantities. Diglycerides are predominantly apolar and do not differ much from triglycerides in properties. Monoglycerides, present in far smaller quantities, are somewhat polar; they are surface active and thus accumulate at an oil–water interface.

Most lipolytic enzymes, including that of milk, especially attack the 1- and the 3-position of the triglyceride molecule (Figure 2.10). This means that most monoglycerides have a fatty acid residue at the 2-position, and that most of the free fatty acids formed originate from the other positions, including the short-chain types that are predominantly in the 3-position.

2.3.2.3 Free Fatty Acids

These already occur in fresh milk and lipolysis increases their amount. Especially the shorter acids are somewhat soluble in water. In water, the acids can, of course, dissociate into ions; their p*K* is about 4.8. In milk plasma, they are thus predominantly in the ionized form (i.e., as soaps), and these are much more soluble in pure water than the pure fatty acids. Table 2.7 gives solubilities of the Na^+ soaps in pure water. In milk plasma, the concentration is always below the critical micellization concentration (see Subsection 3.1.1.3a).

Fatty acids dissolve well in oil, though only in the nonionized form. Moreover, they tend to associate into dimers, by forming hydrogen bonds:

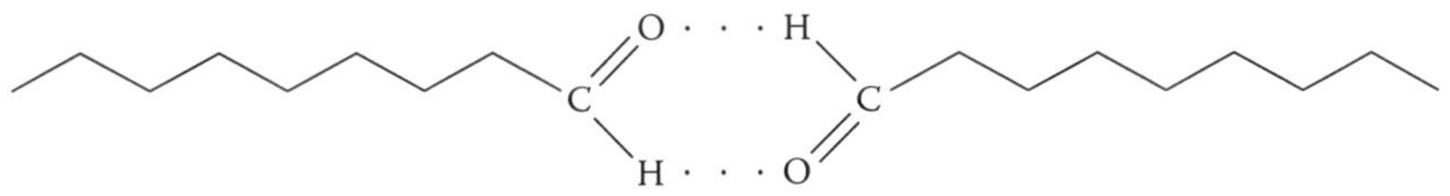

The partition of the acids over the oil and water phases is rather intricate. All in all, the shorter acids (C_4 and C_6) are predominantly in the plasma, the longer ones (from C_{14} on) in the fat. The other acids are distributed between both fractions, though more go into the fat with decreasing pH (i.e., with ionization becoming weaker). This is even more complicated because the fatty acids, especially the long-chain ones, are surface active and tend to accumulate in the oil–water interface. The distribution over the phases is of much importance because acids dissolved in the aqueous phase (in the form of soaps) — hence, the shorter acids — are responsible for the soapy-rancid flavor perceived after lipolysis.

2.3.2.4 Compound Lipids

These are also called polar lipids, because they contain charged, that is, acidic and/or basic groups. They are strongly amphipolar and are virtually insoluble in water as well as in oil. They are highly surface active and form the typical bilayers that are the basic structure of cellular membranes. In milk they are mainly present in the fat globule membrane and in the poorly defined lipoprotein particles.

Most compound lipids are *phospholipids*. The main types derive from phosphatidic acid, by attachment of various organic bases (see Figure 2.11). Sphingomyelin is different in that it is a derivative of sphingosine, as are the cerebrosides. Most of the compound lipids have an acidic and a basic group, and are thus amphipolar at neutral pH, but some (phosphatidyl serine and inositol) have a net negative charge. In milk, small quantities of phosphatidic acid and of lysophosphatides occur, which also have a negative charge. The latter form by splitting off one of the fatty acid residues catalyzed by phospholipases. These enzymes can be produced by several bacteria.

As seen in Table 2.8, the fatty acid pattern of the compound lipids differs markedly from that of the triacylglycerols. The average chain length is longer,

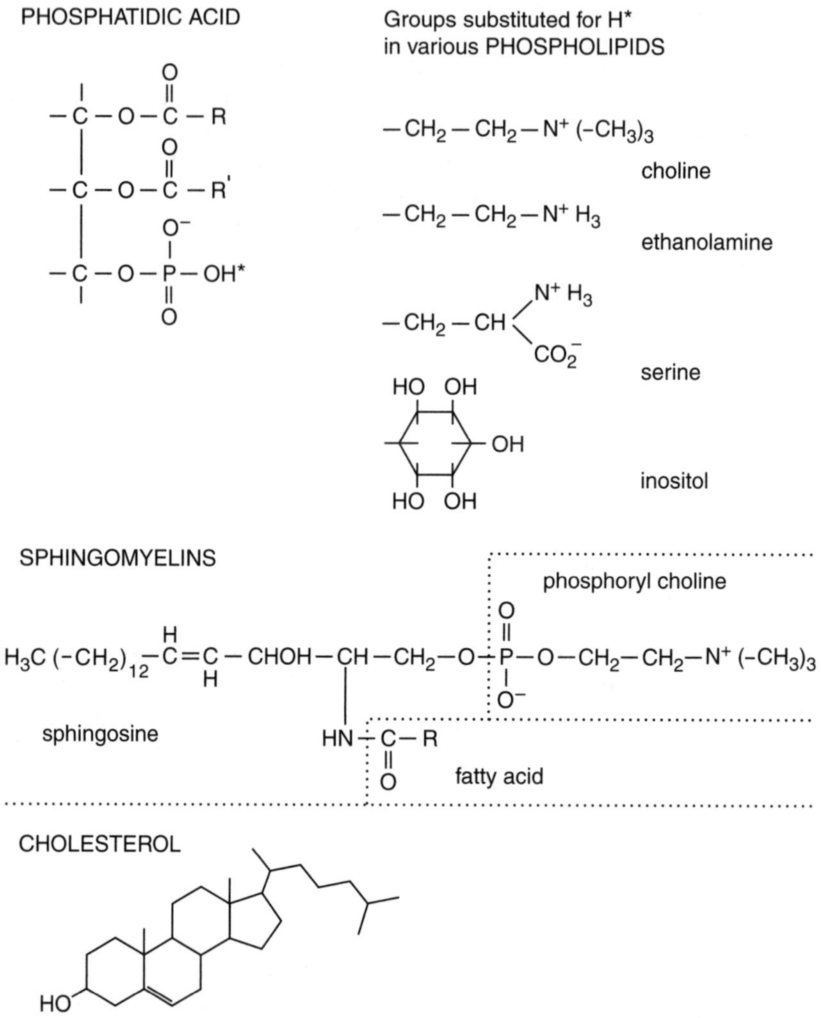

FIGURE 2.11 Molecular structure of some of the compound lipids of milk. R denotes an aliphatic chain.

especially for the sphingolipids, and the average number of double bonds is high in the glycerol-containing phospholipids. Especially phosphatidyl ethanolamine contains a high proportion of polyunsaturates. These phospholipids are highly susceptible to oxidation.

In some milk products, phospholipids are added, e.g., as a surfactant. It then often concerns soya lecithin or a fraction thereof. The composition is, however, markedly different from that of milk phospholipids. The added lipids tend to form vesicles. (A vesicle is a hollow, water-filled sphere of some 20 to 40 nm, formed by a curved phospholipid bilayer.)

TABLE 2.9
Approximate Content of Lipids in Some Milk Products

	Composition (% w/w)			
Product	**Total Fat**	**Phospholipids**	**Sterols**	**Free Fatty Acids**
Separated milk	0.06	0.015	0.002	0.002
Milk	4	0.035	0.013	0.008
Cream	10	0.065	0.03	0.017
Cream	20	0.12	0.06	0.032
Cream	40	0.21	0.11	0.06
Buttermilk from 20% cream	0.4	0.07	0.005	0.002[a]
Buttermilk from 40% cream	0.6	0.13	0.011	0.002[a]
Butter (unsalted)	82	0.25	0.21	0.12[a]
Anhydrous milk fat	> 99.8	0.00	0.25	0.15[a]

[a] Higher if the cream has been subject to lipolysis, especially after its separation.

2.3.2.5 Unsaponifiable Lipids

In milk, the unsaponifiable fraction consists largely of *cholesterol* (Figure 2.11), which is quite apolar and easily associates with phospholipids; accordingly, part of the cholesterol is in the fat globule membrane, the remainder being dissolved in the fat. A fraction of the cholesterol is esterified to a fatty acid (i.e., actually being saponifiable). Carotenoids — in milk, especially β-carotene — are responsible for the yellow color of the fat. Tocopherol, being an antioxidant and a vitamin (E), is mainly dissolved in the fat. Vitamin A, a carotenoid, and vitamin D, a sterol, are also dissolved in the fat.

2.3.2.6 The Fat in Milk Products

Because the various lipids are unevenly distributed among the physical fractions of milk (Table 2.8), the fat composition of different milk products varies. The largest differences originate from variations in the amount of material from the fat globule membranes. Examples are given in Table 2.9. Anhydrous milk fat is prepared from butter by melting it, and by separating and drying the oil layer obtained; its composition is virtually equal to the fat in the core of the milk fat globules.

2.3.3 Nutritional Aspects

The fat in milk primarily provides energy, about 37 $kJ \cdot g^{1}$. Fat is a very efficient carrier of edible energy because of its high energy content and because it does not increase the osmotic pressure of the milk; the latter is physiologically limited to about 700 kPa. Proteins (providing about 17 $kJ \cdot g^{1}$) do not greatly increase the osmotic pressure, but sugars (16 $kJ \cdot g^{1}$) do. (Incidentally, if a monosaccharide would be the milk sugar rather than lactose, milk could contain only half as much sugar.) Milk fat is well digestible, irrespective of the physical form (natural or homogenized fat globules, or butter) in which it is taken. However, long-chain saturated fatty acids, especially stearic acid, as liberated from the fat in the small intestine, may not be completely absorbed in the presence of calcium, because Ca-stearate is poorly soluble at the prevailing neutral pH value.

The milk fat has other functions, one being a solvent for some vitamins (A, D, and E). Another is providing the nutrients linoleic and linolenic acid, often called essential fatty acids (EFA). These are needed as precursors for some hormones and other essential metabolites, as well as to provide the low-melting, long-chain fatty acid residues in the phospholipids that every cell needs for making and maintaining membranes. The content of EFA in (cows') milk fat is low, although it is rarely below 2 energy percent. Deficiency symptoms are not observed in people or laboratory animals having milk fat as the only source of dietary fat.

Milk fat is often considered to be hypercholesteraemic, i.e., causing an increased blood cholesterol level, due to its high content of saturated and low content of polyunsaturated fatty acids. Consequently, milk fat intake has been incriminated as increasing the incidence of (fatal) coronary heart disease (CHD). Milk fat is indeed slightly hypercholesteraemic, at least in some individuals, but whole milk is not (rather, it is the opposite). More importantly, intervention studies do not show a significant effect of the intake of milk fat on mortality.

A high fat intake has been incriminated as enhancing the occurrence of certain types of cancer. The effect is very slight, however. On the other hand, some scientists suppose that the presence of certain conjugated linoleic acids (CLAs) in milk fat may protect against cancer; however, the significance of this effect has not been established. CLAs originate from the rumen fermentation, but the concentration of the precursors in the feed is highly variable. Hence, the quantities in milk fat vary with the feed. Values of up to about 2 mol % have been observed.

2.3.4 Autoxidation

The double bonds in a fatty acid or a fatty acid residue can oxidize. From the oxidation products obtained, several components can be formed. Some of these can be perceived in exceptionally low concentrations and thereby cause off-flavors, including tallowy, fatty, fishy, metallic, and cardboard-like. Off-flavor development can cause problems in beverage milk, sour-cream buttermilk, cream, and especially long-keeping high-fat products such as butter and whole milk powder. The complex of reactions involved is highly intricate, and there are several complicating factors.

2.3.4.1 Reactions Involved

Molecular oxygen (i.e., triplet oxygen, $^3O_2$, with two unpaired electrons in the $2p\pi$ orbital) exists in a relatively unreactive state. The following oxides are much more reactive:

1. Singlet oxygen $^1O_2$
2. Superoxide anion radical $O_2^{\cdot-}$
3. Hydroxyl radical $OH\cdot$

1. *Initiation,* e.g.
   a. Formation of singlet oxygen $^1O_2$

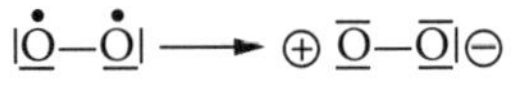

$$|\dot{\underline{O}}-\dot{\underline{O}}| \longrightarrow \oplus\overline{\underline{O}}-\overline{\underline{O}}|\ominus$$

   b. Formation of peroxide

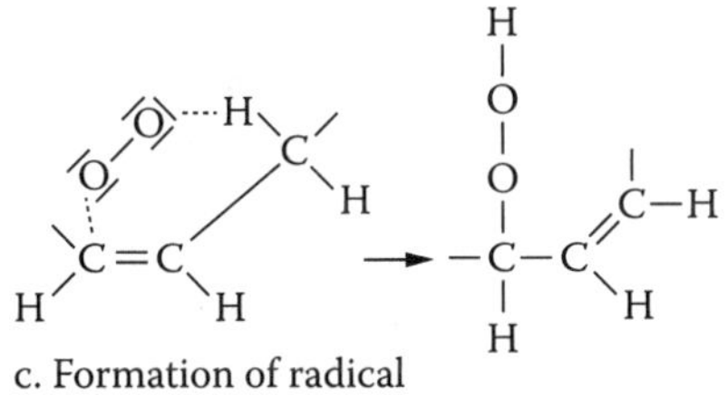

$$\text{C=C(H)(H)} \cdots \text{O=O} \cdots \text{H–C} \longrightarrow -\underset{H}{\overset{H-O-O}{C}}-C{=}C-H$$

   c. Formation of radical

$$ROOH \xrightarrow{\text{catalyst}} ROO\bullet + H\bullet$$

2. *Chain reaction*

$$ROO\bullet + RH \longrightarrow ROOH + R\bullet \quad \text{(slow)}$$
$$R\bullet + {}^3O_2 \longrightarrow ROO\bullet \quad \text{(fast)} \quad +$$
$$RH + {}^3O_2 \longrightarrow ROOH$$

3. *Termination,* e.g.

$$ROO\bullet + R\bullet \longrightarrow ROOR$$
$$R\bullet + R\bullet \longrightarrow R{-}R$$

4. *Breakdown of hydroperoxides*

$$ROOH \longrightarrow \text{unsaturated aldehydes and ketones } (C_6{-}C_{11}),$$

$$\text{e.g.}\quad R'-CH=CH-CH=CH-C(=O)H$$

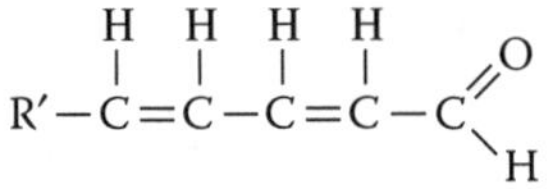

FIGURE 2.12 Approximate profile of the autoxidation reaction of unsaturated fatty acid residues.

Singlet oxygen is the principal agent initiating oxidation of fat. This is because it is highly electrophilic: one of the oxygen atoms has two paired 2pπ electrons, but the other has none in the 2pπ orbital. It can readily react with a double bond to yield a hydroperoxide while shifting the double bond (Figure 2.12, reaction 1b).

The question now is how singlet oxygen (or one of the other reactive species) can be generated. In milk, there are a few pathways for $^1O_2$ to be formed:

1. As a result of photooxidation, i.e., a reaction between triplet oxygen and riboflavin, excited by light
2. By action of peroxidase and/or xanthine oxidase
3. By means of Cu^{2+} and ascorbic acid
4. By a reaction between the superoxide anion radical and H_2O_2
5. By degradation of hydroperoxides during autoxidation of the fat

A simplified outline of fat oxidation is given in Figure 2.12. First of all, peroxide radicals should be formed (initiation). There are several pathways to achieve this. Usually the initiation is very slow until hydroperoxides are formed. In milk, the main catalyst with respect to reaction 1c in Figure 2.12, probably is Cu^{2+}. Then a chain reaction sets in (propagation). It keeps itself going; hence the name autoxidation. Now, the concentration of hydroperoxides increases significantly. Occasionally, radicals react with each other to yield stable final products (in which fatty acid residues become covalently connected), thereby terminating the reaction (termination).

The formed hydroperoxides have no flavor. But they are fairly unstable and can break down in various ways to form unsaturated ketones and aldehydes, some (especially those with the group –CH=CH–CH=CH–COH) having a very strong flavor, that is, they may have a threshold concentration as low as 10^{-3} ppm.

Many other reactions occur during fat oxidation. Moreover, the ratio of the reaction rates depends significantly on such conditions as temperature. Consequently, the reaction products can also vary in character. It implies that off-flavors developed under certain conditions (such as high temperature) do not always correlate satisfactorily with those under other conditions, such as low temperature. The rule is that a higher temperature leads to more rapid development of defects. The rate at which off-flavors develop especially depends on the extent of unsaturation. For example, the reaction rates for C_{18} acids with one, two, and three double bonds roughly are in the proportion 1:30:80.

Antioxidants can block the chain reaction or prevent the initiation. However, they are consumed in this process. Natural antioxidants are tocopherol (which reacts with the radicals) and β-carotene, which can react with singlet oxygen. Several synthetic antioxidants have been made as well. So-called synergists (for example, phospholipids dissolved in the fat) enhance the action of antioxidants. Other substances such as citrate can bind the catalyzing metal ions.

Under practical conditions, autoxidation usually takes some time to set in. Initially, the antioxidants are consumed; after this has been achieved, peroxides are first liberated and subsequently are broken down to form perceptible amounts of flavor products, as has been stated above (see Figure 2.13).

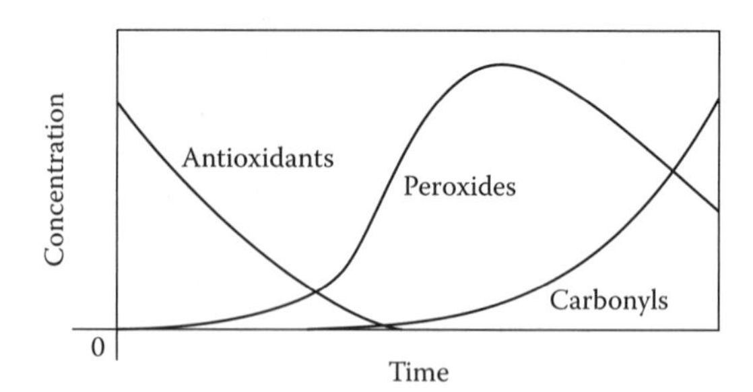

FIGURE 2.13 Relative concentrations of antioxidants, hydroperoxides, and free carbonyl components during autoxidation of a fat. Highly schematic, not to scale.

2.3.4.2 Milk Products

Autoxidation of milk fat in milk and milk products usually starts with the phospholipids of the fat globule membrane. These lipids have highly unsaturated fatty acid residues (Table 2.7 and Table 2.8). Moreover, Cu, the main catalyst, can be in the membrane of the fat globules but not in the core, which contains the triglycerides. Copper is particularly active as a catalyst if phospholipids are present; moreover, at least in milk and buttermilk, a little ascorbic acid is needed. Cu entering the milk by contamination during or after milking is much more active as a catalyst than natural Cu. There are large variations in susceptibility to autoxidation of fat among different lots of milk. In some milks, contamination with as little as 5 μg of Cu per kg suffices for an oxidized flavor to develop, whereas in others 200 $\mu g \cdot kg^{-1}$ is insufficient. Some workers assume that oxidation is spontaneous in some milks. Minute Cu contamination (e.g., 10 $\mu g \cdot kg^{-1}$) is, however, hardly avoidable in practice. Incidentally, Fe can also be active as a catalyst, but not in the presence of proteins, that is, not in milk.

Autoxidation of fat in milk can also be enhanced by exposure to light of short wavelengths. Putting a glass bottle of milk in sunlight for 10 min usually suffices to subsequently produce a distinct tallowy flavor. Other light-induced off-flavors can also develop, i.e., flavors in which lipid oxidation is not involved (Section 4.4). Riboflavin is an essential factor in light-induced autoxidation.

Early-lactation milk on the average is more prone to develop flavors caused by fat oxidation. The natural variation in this tendency may be ascribed to variations in redox potential, concentrations of tocopherol (antioxidant), ascorbic acid, and oxidases (peroxidase, xanthine oxidase). Variation in activity of superoxide dismutase may also be of importance. This enzyme catalyzes the decomposition of superoxide anion (Subsection 2.5.2), and thereby acts as an antioxidant.

Several **treatments** can affect autoxidation reactions:

1. The role of Cu in autoxidation primarily depends on its concentration in the fat globule membrane. The concentration of *natural* Cu in the membrane presumably is too low (the amount being constant at about 10 μg per 100 g of fat, even if the natural Cu concentration in the milk is high) to cause significant oxidation. Cooling of the milk (e.g., keeping at 5°C for 3 h) causes a further decrease, as nearly half of the natural Cu moves to the plasma.
2. In milk, 1 to 10% of *added* Cu goes to the membrane, the percentage increasing with the quantity added.
3. Heating of the milk causes part of the copper to move from the plasma to the membrane. Heating for 15 s at 72°C has a significant effect; 15 s at 90°C has much more. The Cu content of the membrane may increase by 15-fold. Heating only the cream results in a smaller increase because then a smaller amount of Cu per unit mass of fat globules is available. To be sure, all of this copper does not necessarily have full catalytic activity (see item 6).

4. Souring milk or cream causes 30 to 40% of added Cu to move from the plasma to the fat globule membrane. This may explain why ripened-cream butter is more prone to autoxidation than sweet-cream butter. But the oxidation reaction itself may also be faster at lower pH.
5. Heating of the cream (e.g., 15 s at 83°C) before acidification largely prevents the transport of Cu, mentioned in item 4. The explanation is uncertain, but the next point may be the cause.
6. Intensive pasteurization causes exposure of additional sulfhydryl groups and especially formation of free H_2S; most likely, the latter originates from a fat globule membrane protein. H_2S is a strong antioxidant. It dissociates to form S^{2-}, which associates with Cu^{2+}. Assuming 3 $\mu mol \cdot l^{-1}$ of H_2S to be formed, and the dissociation constant of H_2S to be 10^{-14} $mol \cdot l^{-1}$, the concentration product of CuS will be on the order of 10^{-20} $mol^2 \cdot l^{-2}$. Because the solubility product of CuS is as small as 10^{-47} $mol^2 \cdot l^{-2}$, virtually all Cu ions are removed. Presumably, Maillard products formed during heating can also act as an antioxidant. All in all, intensive heating markedly inhibits autoxidation.
7. Heat treatment can cause inactivation of superoxide dismutase (EC 1.15.1.1), which probably is an important antioxidant. Moderate heating of milk can increase the proneness to autoxidation, maybe because of Cu migration (item 3).
8. Homogenized milk is much less prone to autoxidation, even if induced by light. The change in surface layer of the fat globules must be the cause, but the explanation is unknown.
9. In general, the rate of autoxidation increases with increase in temperature ($Q_{10} \approx 2$). This also holds for many milk products. In fresh raw milk, however, oxidized flavor develops more quickly if temperature is lower. The explanation is not certain, but the decreased activity of superoxide dismutase probably is involved.
10. The rate of autoxidation reactions in dried products depends on water activity (a_w) (see Figure 10.5). Clearly, water is an antioxidant. This is a factor in milk powder where tallowy flavors develop faster if the water content, and hence a_w, is lower. At low a_w, ascorbic acid is not needed to cause autoxidation, and tocopherols do not act as inhibitors.
11. The oxygen content becomes a limiting factor if it is below about 0.8 ml O_2 per 100 ml of fat, corresponding to an oxygen pressure of 0.02 bar. Such low levels of O_2 can only be achieved in fermented milks and cheese, or in products in hermetically sealed packages, such as milk powder packaged in tins.

2.3.5 Triglyceride Crystallization

Crystallization of a fat, i.e., a mixture of various triglycerides and small quantities of other components, is a complicated phenomenon, especially for milk fat with its very broad composition. Fat crystals greatly affect consistency and mouthfeel of high-fat products, especially butter; and the physical stability, especially to partial coalescence, of milk fat globules.

2.3.5.1 Melting Range

Fatty acids have widely varying melting points. This is reflected in the melting range of the fats in which the acids have been esterified. The shorter the chain length and the greater the number of double bonds, the lower the melting point (Table 2.7). Moreover, *cis* double bonds give a lower melting point than *trans*, and the position of the double bonds also has an effect. Some minor triglycerides, such as those containing a branched fatty acid residue, will probably not crystallize at all under most conditions.

The melting point of a triglyceride molecule also depends on the distribution of the fatty acid residues over the three positions. For example, a strongly asymmetric triglyceride (e.g., PPB, where P = palmitic acid, and B = butyric acid) has a lower melting point than a symmetric one with the same fatty acid residues (e.g., PBP).

Milk fat is a mixture of many widely different triglycerides (Subsection 2.3.2) with different melting points. The multicomponent fat thus has a wide melting range, as is shown in Figure 2.14. It is seen that the melting curve depends on fat composition (compare winter and summer fat) and also on pretreatment (compare rapid and slow cooling).

Between −35 and +35°C, milk fat usually consists of liquid as well as solid fat, that is oil with various crystals. The oil, consisting of relatively low-melting triglycerides, acts as a solvent for the higher melting ones. This means that an individual triglyceride in the mixture can be fully liquid at temperatures considerably below its melting point. For example, the highest-melting triglyceride in milk fat is tristearate; its melting point is 72°C, that is, about 35°C above the final melting point of milk fat. The solubility of a single triglyceride in oil can be calculated

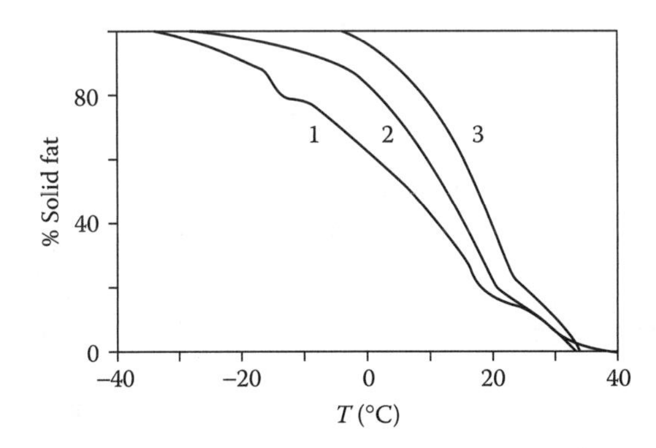

FIGURE 2.14 Melting curves of milk fat. 1. "Summer" fat, slowly cooled. 2. Same fat, rapidly cooled. 3. "Winter" fat, rapidly cooled. (Adapted from J. Hannewijk and A.J. Haighton, *Neth. Milk Dairy J.*, **11**, 304, 1957.)

from the thermodynamic theory for ideal solutions, but this is usually not so if several triglycerides crystallize at the same time, as they then affect each other's solubility. Moreover, the latter effect depends on external conditions, as will become clear later. Consequently, the melting curve of milk fat cannot be simply derived from its composition. Naturally, a higher content of high-melting triglycerides causes a higher content of solid fat at, say, room temperature.

Solidification curves depend even more strongly on external conditions than melting curves do. This is because undercooling may occur, as becomes clear below.

2.3.5.2 Nucleation

A substance cannot crystallize unless *nuclei* have formed, i.e., tiny embryonic crystals, just large enough to escape immediately dissolving again. (The solubility of a small particle increases as its radius of curvature decreases; see Subsection 3.1.1.4.) Homogeneous nucleation (i.e., the formation of nuclei in a pure liquid) often requires considerable undercooling to form nuclei within, say, a few hours. In most fats, an undercooling of about 35°C below the final melting point is needed. But nucleation is usually heterogeneous, that is, it takes place at the surfaces of very small 'contaminating particles.' Such particles are called *catalytic impurities*. As a rule, the number of impurities that catalyze nucleation significantly increases with decreasing temperature.

In milk fat, if it is present in bulk (i.e., as a continuous mass), undercooling by, say, 5°C, results in a sufficient number of catalytic impurities to induce crystallization. As soon as fat crystals have been formed, they can, in turn, act as catalytic impurities for other triglycerides; for this to happen, very small undercooling suffices. Milk fat in bulk may indeed show little hysteresis between solidification and melting curves (Figure 2.15A).

The situation may be very different if the fat has been emulsified. In bulk fat, some 10^4 catalytic impurities per gram would suffice to ensure rapid crystallization. But in milk, 1 g of fat is divided over about 10^{11} globules; in homogenized milk,

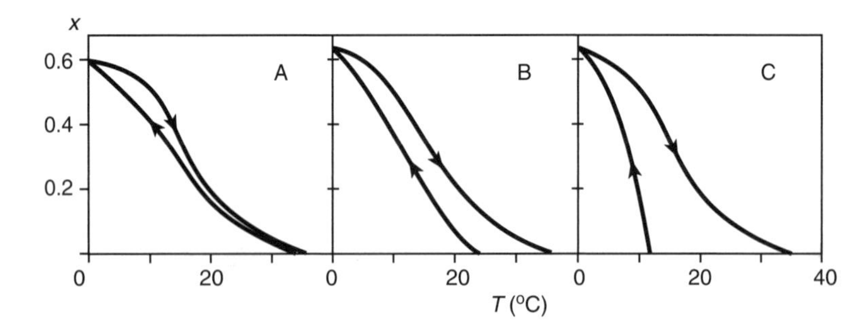

FIGURE 2.15 Examples of the proportion of fat being solid (x) after 24 h cooling to temperature T, and after warming again (after keeping it at 0°C). (A) Fat in bulk. (B) Same fat in cream. (C) Same fat in homogenized cream. (Examples adapted from P. Walstra and E.C.H. van Beresteyn, *Neth. Milk Dairy J.*, **29**, 35, 1975.)

1 g may be distributed even over $>10^{14}$ globules. In each of those, at least one nucleus must be formed. Consequently, considerable undercooling may be necessary, and significant hysteresis between solidification and melting curves occurs (Figures 2.15B and 2.15C). Naturally, the magnitude of this effect depends on globule size, i.e., the undercooling should be deeper if the globules are smaller. If there are B catalytic impurities per ml of fat and if globule volume is v ml, then

$$y_{\max} = 1 - e^{-vB}$$

where y is the proportion of the fat in globules containing one or more crystals. Variable B greatly depends on temperature. As a rule of thumb, B in milk fat doubles for each 1.75 degree lowering of temperature. In a certain temperature range, part of the fat may thus be devoid of any crystal present if the fat has been finely emulsified. In principle, each globule should eventually contain crystals. In practice, however, no further changes occur after, say, 24 h. Crystallization in finely divided fat is less and slower than in fat present in bulk. Figure 2.16 gives examples of undercooling for milk fat of a certain composition.

Presumably, micelles or crystals of monoglycerides are the main catalytic impurities in milk fat. For instance, less deep undercooling is needed with increasing lipolysis, which leads to an increased concentration of monoglycerides, and hence to higher values of B.

Another important phenomenon is *secondary nucleation*, which appears to be prominent in some triglyceride mixtures, especially in milk fat. It means that as soon as a crystal is formed from a nucleus, other crystals form rapidly in the vicinity of the first one. This happens if the undercooling is fairly deep. It implies a high concentration of nuclei, that is, a high number of crystals per unit volume.

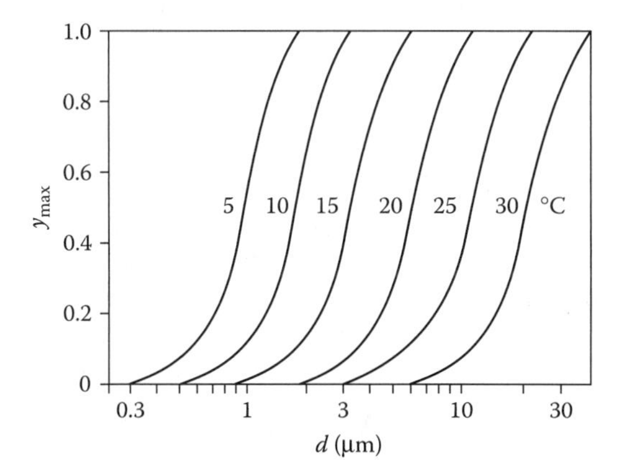

FIGURE 2.16 Examples of the proportion of milk fat present in fat globules that eventually contain solid fat ($y_{\max}$) as a function of the globule diameter (d) and temperature. (Example adapted from P. Walstra and E.C.H. van Beresteyn, *Neth. Milk Dairy J.*, **29**, 35, 1975.)

2.3.5.3 Crystal Growth

In milk fat, growth of crystals is very slow. The large, elongated, and flexible triglyceride molecules take a long time before obtaining the correct position and conformation to fit into the crystal lattice. Before a molecule actually fits in, it often diffuses away again. Moreover, there are many competing molecules, which are so similar to those in the crystal that they almost fit into its lattice. They have to diffuse out again before a properly fitting molecule can occupy a site in the crystal lattice.

To achieve reasonably fast crystallization, the supersaturation should be high, hence the undercooling deep (say, by 20° or more). This means rapid cooling to a low temperature. For the fat in milk or cream, this can readily be done. For fat in bulk it is difficult, because removal of the heat of fusion is difficult. Per gram of fat crystals formed, about 150 J of heat is released, whereas the heat capacity of milk fat is about 2.2 $J \cdot g^{-1} \cdot K^{-1}$. This means that the temperature will rise considerably, for instance by about 13°C if 20% of the fat crystallizes, thereby strongly lowering the supersaturation. Because crystallization causes the fat to obtain a more or less solid consistency, stirring of the fat to enhance heat removal is frustrated. The general result is that the fat in fat globules tends to crystallize at a much higher supersaturation than fat in bulk. The fat in globules must anyhow be undercooled strongly for nucleation to occur, as mentioned earlier.

For small samples of bulk fat, isothermal crystallization can be realized. Then at 25°C, it generally takes 1 to 2 h before half of the final amount is crystallized. The time needed is roughly halved for each 5 to 6° temperature decrease.

2.3.5.4 Polymorphism

Chain Packing. Most molecules with long aliphatic chains can crystallize in different *polymorphic modifications*, the most important ones being denoted α, β′, and β. Each modification or polymorph is characterized by its mode of chain packing, and hence by the distances (spacings) between chains. Consequently, these modifications can be identified by x-ray diffraction (from the so-called short spacings).

The modifications are also observed in triglycerides. Table 2.10 gives some properties of tristearate polymorphs. It is seen that the melting point, the heat of fusion, and the density of the crystals increase in the order $\alpha \rightarrow \beta' \rightarrow \beta$. This implies that closeness and intricacy of fit of the molecules in the lattice increase and their freedom of motion decreases. In the α-modification, the chains can, to some extent, rotate and wiggle. In the β-form, the chain packing is very tight.

The α and β′ modifications are metastable. Transitions can only take place according to the scheme

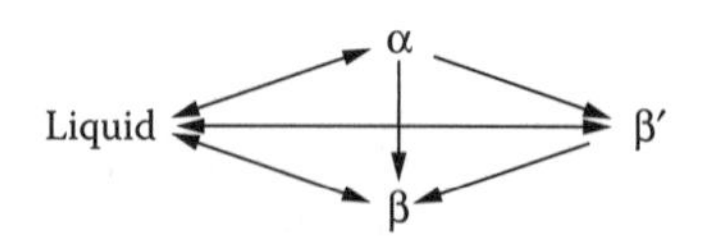

TABLE 2.10
Properties of Some Polymorphic Modifications of Tristearate

Polymorph		α	β′	β
Melting point	(°C)	55	63	73
Melting heat	(kJ/mol)	110	150	189
Melting dilatation[a]	(ml/kg)	119	(147)	167

[a] Increase in volume per unit mass upon melting of the crystals.

Other transitions cannot occur. This means that the α-modification of pure tristearate can only be formed after cooling to below 55°C; subsequent heating to, say, 60°C would then lead to melting of the α crystals, but after a while β′ or β crystals can form, which melt at a higher temperature. Such double melting points can also be observed in milk fat after rapid deep cooling and subsequent rapid warming.

Nucleation usually occurs in the α-modification. Often, transition to a stabler polymorph occurs after a while. In most fats, the α-modification has only a short lifetime, whereas β′ may persist longer. But in milk fat, α crystals can be quite persistent (this is because of formation of compound crystals; see the following text).

Molecular Packing. Within a given chain packing, i.e., within the α-, β′-, or β-polymorph, various packing modes of the triglyceride molecules are possible. The main types are illustrated in Figure 2.17. It is seen that molecular layers can span two or three times a fatty acid residue length, designated L2 and L3, respectively. Also the tilt angle between the chain and the molecular plane can be different. (In α crystals it is always 90°; in β crystals angles of 72° and 81° have been observed.) Also these packing types can be distinguished by x-ray diffraction (from the so-called long spacings). The L2-type packing generally occurs if the three fatty acid residues are saturated and of about equal length

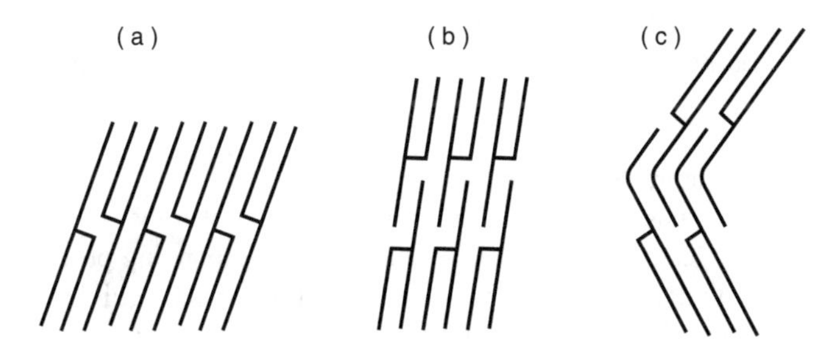

FIGURE 2.17 Various packing modes of triglyceride molecules in a crystal. See text.

(Figure 2.17a). L3 types occur especially if one of the chains is much different from the other ones, be it in chain length (Figure 2.17b), e.g., capric acid; or in having a double bond (Figure 2.17c), e.g., oleic acid.

Altogether, several different polymorphs can thus occur. Milk fat has, especially, an L3-α type and some β′ types, both L2 and L3. During storage, and certainly during slow heating, transitions to stabler polymorphs can occur. Upon heating, the α-polymorph tends to disappear at about 13°C, and two main β′ types at 20 to 23°C and at 35 to 40°C. The β-polymorph is generally not observed, or at most in a very small amount. The proportions and melting properties of the various types depend on triglyceride composition and especially on temperature history.

2.3.5.5 Compound Crystals

Two or more different components may occur together in one crystal. In some of these compound (or mixed) crystals, the components may occur in all proportions (at least within a certain compositional range); these are named *solid solutions.*

Milk fat exhibits extensive compound crystallization in the form of solid solutions, probably because of the very great number of distinct, though quite similar, triglyceride molecules. Compound crystals are formed easily and abundantly in the α-modification because the packing density is low enough to allow molecules that somewhat differ in shape to fit into one and the same lattice. In β crystals, the molecular packing is so dense that different kinds of molecule cannot fit in the same crystal lattice.

The total supersaturation with respect to molecules that can form a compound crystal is much higher than that of each of the individual triglyceride molecules separately. This produces the driving force for compound crystallization. In milk fat, compound β′ crystals may even be more stable than pure β crystals; this is probably the explanation of the virtual absence of the latter.

Above about 13°C, where α crystals cannot exist, only β′ crystals are observed. These are true solid solutions, but very different molecules cannot fit into one crystal lattice. Hence, there are a few different types, for the most part corresponding to those illustrated in Figure 2.17. In other words, the triglyceride molecules crystallize in a few different *groups*. When increasing the temperature, crystals melt, partly leading to formation of other types. Moreover, the composition of compound crystals changes with temperature. The group that has the highest final melting temperature is of a β′-L2 type, comprising triglycerides in which all fatty acid residues are saturated and have a long chain.

Compound crystals essentially are impure, that is, not fully ordered, crystals. Accordingly, they have a lower melting heat than corresponding pure crystals. In principle, compound crystals form in a local thermodynamic equilibrium, that is, within the polymorphic form present. However, even a slight temperature difference causes a difference in equilibrium composition. This means that in practice, equilibrium never is reached, and that changes in compound crystal composition, often

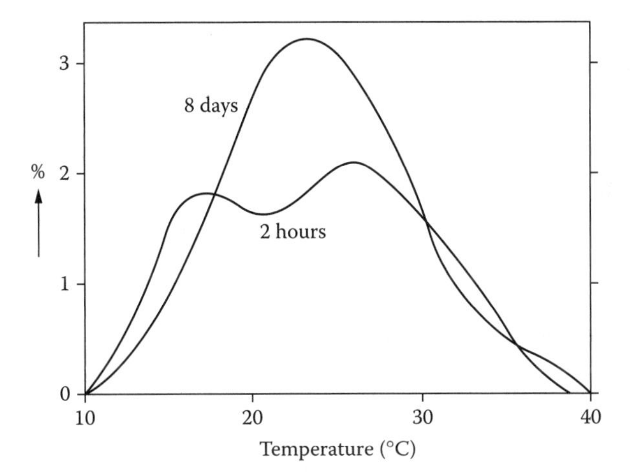

FIGURE 2.18 Differential melting curves (expressed as percentage of the fat melting per °C temperature increase) of the milk fat in a cream sample. The cream was held for 1 d at 19°C and then at 10°C for 2 h or 8 d. (From results in H. Mulder and P. Walstra, *The Milk Fat Globule,* Pudoc, Wageningen, 1974.)

accompanied by polymorphic transitions, continuously occur, although often slowly.

The theory explaining the factors that determine the proportion crystalline and the composition of compound crystals is intricate, so we will not discuss it and merely give some important consequences, which are also observed in practice:

1. Compound crystallization causes narrowing of the melting range.
2. The temperature range at which most of the fat melts depends on the temperature at which solidification took place. It is found to be slightly above the latter temperature. Consecutive solidification at two temperatures thus yields a differential melting curve with two melting maxima; see Figure 2.18.
3. Cooling in steps or slow cooling gives less solid fat than rapid and direct cooling to the final temperature. Compare curves 1 and 2 in Figure 2.14.
4. Cooling to a lower temperature before bringing to the final temperature gives more solid fat than direct cooling to the latter. This is shown in the hysteresis at low temperatures in Figure 2.15A.
5. During keeping, slow rearrangement of crystal composition occurs because equilibrium has not been reached. Below 5°C, this is a very slow process. At higher temperatures, it may still take days; see, e.g., Figure 2.18.
6. Also after a change of temperature, crystallization takes a long time before reaching equilibrium. After lowering the temperature, the amount

of solid fat may keep increasing over several days. After increasing the temperature to below the final melting point, it may take several minutes before melting stops.

In conclusion, the temperature history of the fat has a significant effect on the amount, the composition, and the stability of fat crystals. Moreover, a system such as partially crystalline milk fat is virtually never in equilibrium.

2.3.5.6 Size and Shape of the Crystals

Milk fat crystals tend to be quite *small*, the longest dimension often being below 1 μm. Figure 2.19 shows that at temperatures above about 10°C, the characteristic time for nucleation to occur (curve 2) is longer than that for crystal growth (curve 3). This suggests that relatively few crystals will be formed, which thus would become large. Nevertheless, the number of crystals formed is quite high (several per μm^3 of fat), implying that crystals remain small. This is because copious *secondary nucleation* tends to occur, as mentioned. Only at quite low supersaturation, which means temperatures above about 25°C, do larger crystals form. In milk fat globules, fat crystallization will hardly occur above 25°C: see Figure 2.16. Hence the crystals in fat globules always are small. In a bulk fat, slow temperature fluctuations cause recrystallization at very low supersaturation, which can lead to the formation of quite large crystals (up to, say, 100 μm).

The *shape* of a crystal should be distinguished from the discussed polymorphic modification. Within each polymorph, widely varying shapes can be formed. In

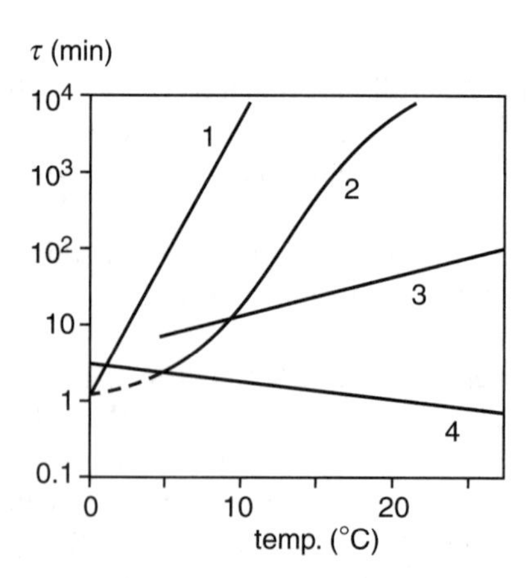

FIGURE 2.19 Characteristic times (τ) for homogeneous (curve 1) and heterogeneous nucleation (2), crystal growth (3), and for the crystals to aggregate (4), as a function of temperature (assuming very fast cooling to that temperature). Approximate results for milk fat; meant to illustrate trends.

milk fat, long and slender platelets predominate. The ratio of length:width:thickness is approximately 50:10:1. The large crystals formed at very low supersaturation tend to be spherulites, that is, spheres that are built of radially oriented branched needles.

2.3.5.7 Crystal Networks*

Fat crystals suspended in oil attract each other because of van der Waals forces and only repel each other because of hard-core repulsion (i.e., when they touch). Accordingly, they always aggregate, and this causes the formation of a *space-filling crystal network*, in which the oil is held. The network causes the fat to have a perceptible firmness, as soon as 1 or 2% of the fat has crystallized.

Figure 2.19 shows that the time needed for aggregation of the fat crystals (curve 4) is quite short. This means that soon after cooling has started, when still only relatively few crystals exist, a network is formed. This is illustrated in Figure 2.20, Frames 1 to 3. This network is fractal (Subsection 3.1.3). In the meantime, nucleation and crystal growth go on, leading to the formation of new crystal aggregates, which tend to 'fill' the pores in the existing network, Frames 4 to 5. Thereby, the network becomes more homogeneous (the fractal geometry is lost) and much stiffer. Moreover, part of the crystallizing molecules are deposited onto existing crystals, whereby they can cause *sintering* of aggregated crystals; this is illustrated in the two magnified frames of Figure 2.20. Sintering occurs readily because compound crystallization predominates. Altogether, quite a firm network with very small pores — on average about 0.2 μm — results. The fat thus is firm and does not show oiling off. (Pure triglyceride crystals, such as those in the β-polymorph, tend to be larger and will not sinter, leading to a relatively weak and coarse network.)

The firmness of the fat is greater for a larger volume fraction of solid fat (φ). Even after φ has almost reached a constant value, *setting* (that is, increase of firmness) goes on. Because true equilibrium is not reached, recrystallization occurs, leading to increased sintering. Setting is markedly enhanced by small temperature fluctuations; if it occurs at higher temperatures, however, the crystal size may slowly increase, causing the network to weaken. All these phenomena affect the firmness — hence, the spreadability of butter — and are further discussed in Subsection 18.3.2.

A space-filling network is also formed inside a fat globule, provided φ is at least about 0.1. This is important for the stability of the fat globules against partial coalescence (Subsection 3.2.2.2).

2.3.5.8 Summary

Table 2.11 summarizes the influence that various factors have on crystallization. Of course, the composition of the fat also has a considerable effect.

---

* The reader is advised to consult Section 3.1 first.

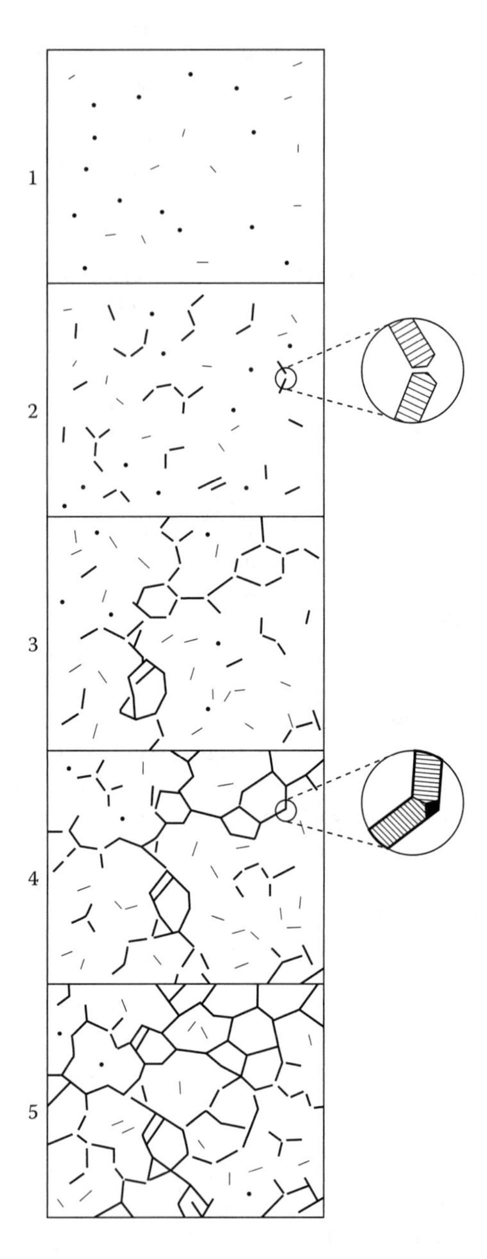

FIGURE 2.20 Various stages in the formation of a fat crystal network. Approximate scale: the edge of a frame would be some micrometers. See text. (From P. Walstra, *Physical Chemistry of Foods,* Dekker, New York, 2003. With permission.)

TABLE 2.11
Summary of Factors Affecting Crystallization of Milk Fat

Factor	Effect on		
	Melting Curve[a]	Amount of Solid Fat	Crystals
Lower crystallization temperature	Maximum at lower temperature	Greater	Smaller[c]
Faster cooling	Maximum at higher temperature	Generally greater	Smaller[c]
Cooling in steps	More than one maximum	Generally less	Often larger: spherulites[c]
Precooling to low temperature	Maximum at lower temperature	Greater	Somewhat smaller[c]
Keeping at not too low a temperature	Flatter	Generally greater	Larger
Fat in globules rather than in bulk	Higher final melting temperature[b]	Low temperature: greater; high temperature: less	Smaller
Smaller globules		Generally less	(Smaller)

[a] Described as differential curve: –d(solid fat)/dT, i.e., the amount of fat that melts within a small temperature interval, as a function of T.
[b] Because of variation in composition among fat globules.
[c] This does not apply to fat in globules.

2.4 PROTEINS

This section primarily discusses the chemical properties of milk proteins. We start with a brief summary of protein chemistry for the benefit of readers who are not quite familiar with it. Formation and properties of casein micelles are discussed in Section 3.3.

2.4.1 Chemistry of Proteins

Proteins are made up of amino acids — more precisely, L-α-aminocarboxylic acids:

$$R - \overset{\displaystyle NH_2}{\underset{\displaystyle H}{C}} - CO_2H$$

At neutral pH, the acids ionize to form $R-CH(NH_3^+)-CO_2^-$. They thus are so-called zwitterions, being at the same time acid and base. An amino group and a carboxyl group can react with each other to yield a *peptide linkage* while releasing a water molecule. The structure of the peptide linkage is intermediate between

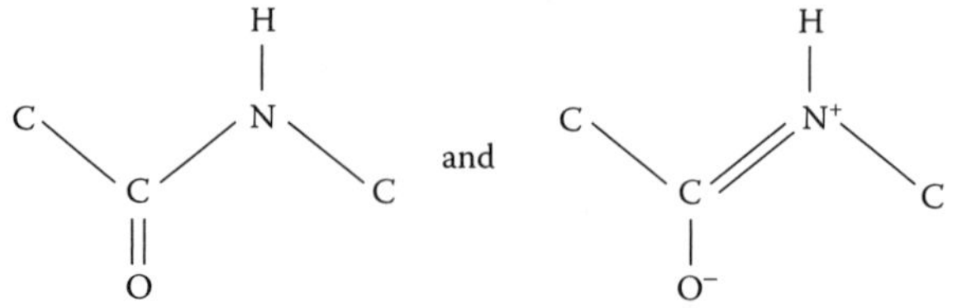

Because of this, the peptide bond has a dipole moment, and is planar and rigid; that is, it remains in the *trans* configuration. Rotation is possible to some extent about the other bonds in the peptide chain (N–CR and CR–CO; Figure 2.21), except for the N–CR bond in proline residues.

A linear peptide chain can be formed from a number of amino acids, as is shown in Figure 2.21. If such a chain is short, it is called peptide; if it is long, the term 'polypeptide' or 'protein' is applied. Most proteins contain at least 100 amino acid residues. There are 20 different natural amino acids; hence, there are 20 different kinds of side chain R. Moreover, modifications of the side-chain groups can occur (see the following text). Properties of amino acids are listed in Table 2.12.

2.4.1.1 Primary Structure

Primary structure is defined as the sequence of the different amino acid residues in the peptide chain. It is specific for every individual protein and is genetically determined.

The specific properties of a protein thus are determined by the side chains R of the amino acids in the polypeptide chain (Table 2.12). The aliphatic apolar side chains are hardly reactive. Side chains with a carboxyl group (Asp and Glu) or an amino group (Lys and Arg) ionize at neutral pH to yield $-CO_2^-$ and $-NH_3^+$, respectively, and so do terminal carboxyl and amino groups; also, His is partly protonated. The free carboxyl groups can be esterified. The free amino groups, especially those of lysine, can contribute to Maillard reactions (Subsection 7.2.3).

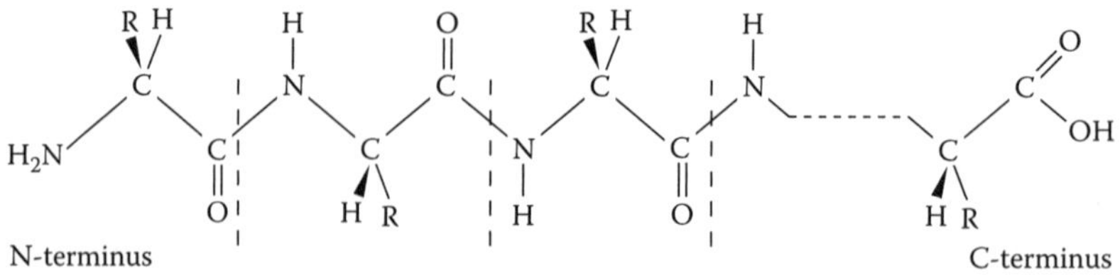

FIGURE 2.21 Peptide chain.

Posttranslational Modifications. The genetic code for a protein gives rise to a sequence of amino acids. However, several proteins within the cell undergo changes in composition after the primary structure has formed. This is called posttranslational modification. In many globular proteins –S–S– linkages are formed between two cysteine residues. Other modifications involve phosphorylation by esterification of phosphoric acid to the hydroxyl groups of Ser or Thr; this gives rise to *phosphoproteins.* Various kinds of saccharide groups can be linked to hydroxyl or amide groups, yielding *glycoproteins.* These modifications can, and often do, lead to variation in the composition within one species of protein.

The same reactions can occur after the protein has been secreted from the cell. Proteins in solution are thus prone to –S–S– bond formation or reshuffling (i.e., breaking of an –S–S– and forming a new one that involves another cysteine residue), especially at high pH and temperature. Esterification of hydroxyl groups can also occur. See further Subsection 7.2.2. Moreover, some cleavage of the peptide chain can occur by the action of proteolytic enzymes.

Genetic Variants. Most proteins occur in two or more variants of the primary structure, which thus are genetically determined. The variants are commonly denoted A, B, C, etc. If an organism, such as a cow, is homozygous for variant A, its genotype is AA, and it only produces the variant A. An organism can also be heterozygous and be of geno-type AB; it then produces variants A and B, generally in equal quantities. In most cases, two variants differ in one amino acid residue; this often leads to no significant difference in properties of the protein. In other cases, especially when the variants differ in charge, clear differences in properties, such as solubility, conformation, or conformational stability, can occur. If the protein is an enzyme, its catalytic activity may differ among variants.

For all major milk proteins, as well as for several enzymes, genetic variants have been identified; Table 2.22 gives some examples. The frequency of occurrence of a variant tends to depend on breed.

2.4.1.2 Conformation

Several intramolecular bonds can form in a protein molecule. This concerns predominantly noncovalent bonds of two main types. *Hydrogen bonds* can form between hydrogen donor groups (especially –OH and =NH) and hydrogen acceptor groups (especially =O, –O–, and =N–). *Hydrophobic interactions* occur between hydrophobic side groups, indicated in Table 2.12; the latter bonds tend to be very weak or absent at 0°C, and their strength markedly increases with increasing temperature.

Hydrogen bonds can lead to the formation of some special types of *secondary structure.* Whether this occurs depends closely on the primary structure; for example, their formation is counteracted by the presence of proline residues. The main types are the α-*helix* and the β-*strand*; such structural elements involve, say, 20 amino acid residues. A single β-strand is unstable, but a number of strands can form (by hydrogen bonding) a stable stack, named β-*sheet.*

TABLE 2.12
Properties of Amino Acid Residues

Name of Acid	Symbol	Side Chain	Reactive Group	pK^a	Charge at pH 6.6	Φ^b
Glycine	Gly	$-H$				0
Alanine	Ala	$-CH_3$				3
Valine	Val	$-CH(CH_3)CH_3$				10
Leucine	Leu	$-CH_2\ CH(CH_3)CH_3$				12
Isoleucine	Ile	$-CH(CH_3)CH_2CH_3$				12
Serine	Ser	$-CH_2OH$	Hydroxyl			–
Threonine	Thr	$-CHOHCH_3$	Hydroxyl			–
Aspartic acid	Asp	$-CH_2CO_2^-$	Carboxyl	4.0	−1	
Asparagine	Asn	$-CH_2CONH_2$	Amide			–
Glutamic acid	Glu	$-CH_2CH_2CO_2^-$	Carboxyl	4.5	−1	
Glutamine	Gln	$-CH_2CH_2CONH_2$	Amide			–
Lysine	Lys	$-(CH_2)_4NH_3^+$	ε-Amino	10.6	+1	–
Arginine	Arg	$-(CH_2)_3NHC(NH_2)_2^+$	Guanidine	12.0	+1	–
Cysteine	Cys	$-CH_2SH$	Thiol	8.5	0	?
Methionine	Met	$-CH_2CH_2SCH_3$	Thioether			?
Phenylalanine	Phe	$-CH_2-C_6H_5$ (phenyl ring)	Phenyl			9

Tyrosine	Tyr	$-CH_2-C_6H_4-OH$	Phenol	~10	0	?
Tryptophan	Trp	$-CH_2-$(indolyl)	Indole			10
Histidine	His	$-CH_2-$(imidazolium, HN, NH^+)	Imidazole	6.4	$+^1/_2$	–
Proline	Pro	(Note[c])				?
Phosphoserine	SerP	$-CH_2O-PO_3^{2-}$	Phosphoric acid	1.5, 6.5	$-1^1/_2$	–
Terminal groups	–	$-CO_2^-$	α-Carboxyl	3.6	–1	
	–	$-NH_3^+$	α-Amino	7.6	+1	

[a] Approximate intrinsic ionization constant in unfolded peptide chains.
[b] Rough estimate of the hydrophobicity in kJ per residue (– = value small or negative; ? = value uncertain).
[c] Secondary amino acid: (pyrrolidine ring, N–H)–CO_2H

Further folding leads to a *tertiary structure*. Here, hydrophobic interactions are essential; moreover, a few internal salt bridges may be formed, and some proteins have one or a few intramolecular –S–S– bridges, and hence covalent bonds. The tertiary structure leads to three main types of proteins:

- *Globular proteins:* The protein chain, including the secondary structure elements, is tightly folded into a more or less spherical shape. Most proteins are globular, and in each, the folding of the chain is specific and needed for its physiological function, such as enzyme action, transport of some essential metabolite, or immunoresponse. Most milk serum proteins are globular.
- *Fibrous proteins:* These usually contain much β-sheet, and form regular elongated structures. Such proteins function as a building material. They do not occur in milk.
- *Disordered proteins:* These generally have some secondary and tertiary structure, but it is not fixed: the structure varies between molecules and during time. This group includes the caseins.

Globular proteins can only be formed if the primary structure contains a sufficient proportion of amino acid residues with a highly *hydrophobic side group*, viz Val, Leu, Ile, Phe, and Trp: see Table 2.12. (It may be mentioned, however, that there is no full agreement among protein chemists about the degree of hydrophobicity of the various side groups.) Ideally, all the hydrophobic side groups are in the core of the folded molecule, all strongly hydrophilic ones at the outside, that is, in contact with water. In the hydrophobic core, where very little water is present, the hydrogen bonds needed for the formation of secondary structure can be formed; otherwise, the various groups on the protein will preferably form H-bonds with water, that is, be hydrated.

The segregation between hydrophilic and hydrophobic groups is by no means complete. By far most of the charged groups are, indeed, at the outside, but most of the peptide bonds — which are also polar — are in the core. Moreover, the outside generally has some hydrophobic patches, and if these are relatively large, association of molecules into larger units (doublets, quadruplets, etc.) can occur; this is called the *quaternary structure.* On the other hand, a very large globular protein molecule would have a relatively small surface area, insufficient to accommodate all the strongly hydrophilic groups. Consequently, many proteins of high molar mass have two or more separate globular *domains*, each of 100 to 200 residues, which are connected to each other by short peptide chains.

2.4.1.3 Denaturation

The conformation of a globular protein is thus stabilized by a great number of weak bonds. If the peptide chain unfolds, its conformational entropy greatly increases, and several groups, especially the peptide bonds, become hydrated. These factors make up a large amount of free energy, and the total free energy of the bonds causing a compact conformation is only slightly larger. In other words, the conformational stability is relatively small. Slight changes in conditions can therefore lead to unfolding. Such an unfolding is called *denaturation*. It transforms a native globular protein into a more or less disordered protein.

Several agents can cause denaturation of globular proteins, and some will be briefly mentioned:

- *High temperature* always leads to denaturation (increased effect of the change in conformational entropy), although the temperature needed varies; a common value is 70°C.
- *Low temperature* can cause some proteins to denature (hydrophobic bonds become quite weak or even repulsive); cooling is nearly always needed to at least −20°C. Somewhere between the two temperatures mentioned — say, at 25°C — the stability of the conformation is at maximum.
- *High pressure*, i.e., a hydrostatic pressure above, say, 200 MPa (2 kbar) can also cause denaturation (by breaking H-bonds).
- *High pH*, e.g., above 8 or 9, may cause denaturation (due to mutual repulsion between negatively charged groups: see Figure 2.23a, further on); such pH values are rarely applied in practice. Also at very low pH denaturation may occur.
- Several *reagents* added to a protein solution in high concentrations cause denaturation, often by breaking H-bonds, but these reagents are not applied in dairy manufacture. In the laboratory, urea is often used.
- Often, a *combination* of two agents can cause ready denaturation, for example, a moderate temperature increase and a moderate pH increase.

Denaturation generally is a *cooperative transition*. This means that most of the weak bonds stabilizing the native conformation break (almost) simultaneously. Presumably, breaking of only a few bonds makes the conformation quite unstable. Consequently, denaturation occurs over a small range of intensity of the denaturing agent. An example is given in Figure 2.22, where urea is the denaturing agent. It is seen that the conformation of the globular protein β-lactoglobulin strongly changes over a narrow concentration range, whereas for the disordered protein β-casein only a small gradual change occurs. Also heat denaturation occurs over a small temperature range, as discussed in Section 7.3.

In practice, some proteins do not comply with the relatively simple picture sketched. For instance, there may be conformations that are intermediate between the native and a (fully) disordered state. Such a conformation is often designated as a molten globule state. Moreover, some proteins change their conformation while retaining the properties of a globular protein upon some change in condition, for example, pH.

Denaturation leads to the loss of the native protein's functionality, for example, to act as an enzyme. In principle, however, denaturation is *reversible*: upon changing the condition to its original value, for instance by lowering the temperature again, the original conformation and the functional activity can be regained. In

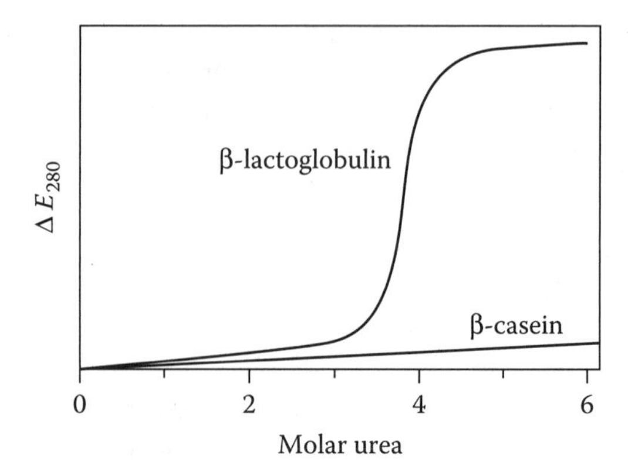

FIGURE 2.22 Change of the specific extinction at 280 nm as an indicator for denaturation by urea.

practice, such regeneration is often absent because the protein in its unfolded state is more reactive; it may thus undergo covalent bond changes, which then prevent refolding to the original (native) conformation. This can especially happen at high temperatures. See further Subsections 7.2.2 and 7.3.2.

2.4.1.4 Solubility

The three main factors determining solubility of a protein are intermolecular electrostatic and hydrophobic interactions, and molar mass. In general, a substance of a higher *molar mass* tends to be less soluble. Macromolecules often do not precipitate when the solubility limit is surpassed, but form a coacervate; this is a layer of very high concentration, but still containing a substantial amount of solvent, also within the molecules. Proteins can, in principle, form crystals upon supersaturation, but this never happens in industrial practice. Something intermediate between a precipitate and a coacervate may be formed.

Electrostatic Interactions. Proteins are salts, although with a generally large and variable *valency* (z). The valency is primarily a function of pH because pH affects the ionization of several side groups: see Table 2.12. An example of the relation is given in Figure 2.23a, which gives a (proton) titration curve for β-lactoglobulin. Such curves vary to some extent with total ionic strength. The pH at which the average net z value equals zero is named the *isoelectric point* (IEP) on the pH scale, in the present case about 5.1. (Because the charge always fluctuates around the average, some molecules have a small negative, and others a small positive charge at the IEP.) Proteins greatly vary in IEP. The IEP value also depends somewhat on ionic strength, as illustrated in Figure 2.23b by the effect of ionic strength on the pH of minimum solubility.

Some basic theory for the solubility of salts is given in Subsection 2.2.2. The main factor governing solubility is the ion activity coefficient γ; the lower its value,

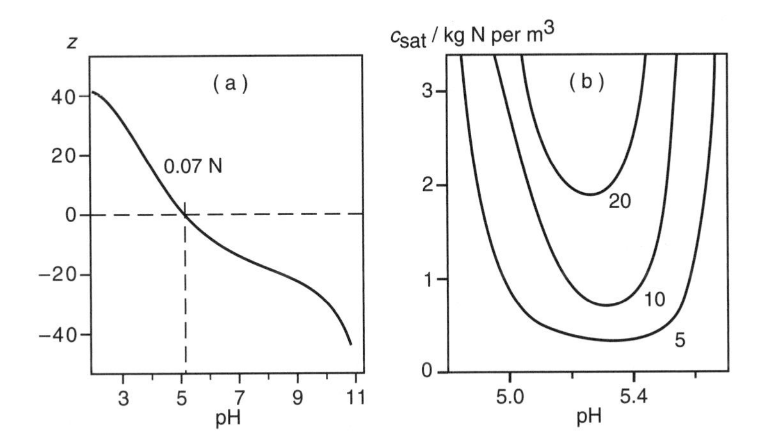

FIGURE 2.23 Some properties of β-lactoglobulin. (a) Titration curve: the average valency z as a function of pH. (b) Solubility (c_{sat}) as a function of pH for various NaCl concentrations (indicated, millimolar).

the greater the solubility. The value of γ decreases strongly with increasing z^2; hence, the solubility will be at minimum near the IEP. Moreover, the value of γ decreases with increasing ionic strength (I); hence, solubility increases with increasing I, which is called *salting in*. An example of these effects is shown in Figure 2.23b. The trend shown is generally observed, but the quantitative relations, e.g., the value of the minimum solubility, vary greatly among proteins.

Hydrophobic Interactions. If many hydrophobic side groups are exposed to the solvent, the protein tends to form intermolecular hydrophobic bonds, which can lead to precipitation, unless the electric charge is high. The solubility thus depends on the amino acid composition of the protein, and also on its conformation. Denaturation of a globular protein causes many hydrophobic side groups to become exposed, and it often leads to precipitation.

The solubility also depends on solvent properties. Several solutes can affect the strength of hydrophobic interactions, and thereby the solubility. This especially concerns salts at high concentration (a few multiples of 1 M), which lower the solubility. This is called *salting out*. The effect greatly varies among salts (e.g., sulfates being quite effective), and it is thus not the ionic strength that is determinant. Several other solutes do something similar, e.g., ethanol. If hydrophobic interactions are determinant, the solubility markedly increases by lowering the temperature close to 0°C. This is observed for casein.

Finally a few words about *hydration*. Ionized groups always are hydrated when in contact with water. This means that one or a few water molecules stay far longer near the group than they would near a noncharged group; say, 10^{-9} rather than 10^{-12} s. These water molecules may be considered *bound*. Something similar happens near a strong dipole, which means, for proteins, the peptide bonds (on average about 0.5 water molecule per peptide bond). Altogether, bound water rarely amounts to more than 0.2 g per gram of dry protein; the value hardly depends on pH.

Bound water should not be equated to the immobilized or *held* water in several protein-containing products. Held water is merely entrapped in the structure; it may involve far greater quantities (up to 100 g per gram of dry protein), and its amount tends to depend greatly on pH and other variables.

2.4.1.5 Reactivity

The reactive side groups of a *disordered protein* all are exposed to the solvent and thus can react. The various reactions are summarized in Subsection 2.4.1.1. The protein is also subject to hydrolytic cleavage of peptide bonds, if suitable enzymes are present. These enzymes come in two main types. Exopeptidases split off amino acids one by one from the polypeptide chain, either from the N-terminal or the C-terminal end. Endopeptidases and proteinases cleave somewhere in the middle of the chain. Individual enzymes are more or less specific for bonds between certain amino acid residues. Proteolysis thus yields peptides and amino acids, several of which have a distinct flavor, such as bitter. (Proteins themselves have no flavor.) Other kinds of enzymatic hydrolysis can occur — for instance, dephosphorylation of SerP.

Globular proteins are less reactive because several side groups are buried in the core. Most ionized groups are exposed, for instance, lysine, which is involved in the Maillard reaction (Subsection 7.2.3). Cysteine is rarely exposed, and it can be quite reactive at high temperature or pH. Globular proteins are also resistant to many proteolytic enzymes. This all means that these proteins become far more reactive upon denaturation.

Most of the chemical reactions occur especially at high temperature. This is discussed in some detail in Subsection 7.2.2.

2.4.2 Survey of Milk Proteins

About 95% of the nitrogen in milk is in the form of proteins. Multiplication of the nitrogen content by a *Kjeldahl factor* of 6.38 is generally accepted to give the protein content of milk and milk products, although the Kjeldahl factor varies among milk proteins (see Table 2.13), and the average factor for whole milk is presently calculated at about 6.36. More important is the presence of *nonprotein nitrogen* components, comprising about 5% of the total nitrogen in fresh milk (see Subsection 2.6.1). The values generally given for the 'protein content of milk' are thus too high by 5 to 6%, or nearly 0.2 percentage units. It would be better to speak of the crude or total protein content. The data given in this book refer to true or net protein contents, unless mentioned otherwise.

Milk proteins make up a complicated mixture from which the individual pure components can be hard to separate. This is partly because some of the proteins are closely related. Moreover, a protein species generally occurs in some compositional variants, due to posttranslational modifications and the presence of genetic variants.

TABLE 2.13
Proteins in Milk

Protein	mmol/m^3 Milk	g/kg Milk	g/100 g Protein	Molar Mass	g Protein/g N	Remarks
Casein	1120	26	78.3		6.36	IEP ≈ 4.6
α_{S1}-Casein	450	10.7	32	~23600	—	Phosphoprotein
α_{S2}-Casein	110	2.8	8.4	~25200	—	Same, contains –S–S–
β-Casein	360	8.6	26	23983	—	Phosphoprotein
κ-Casein	160	3.1	9.3	~19550	—	"Glycoprotein"
γ-Casein	40	0.8	2.4	~20500	—	Part of β-casein
Serum proteins	~320	6.3	19	—	~6.3	Soluble at IEP
β-Lactoglobulin	180	3.2	9.8	18283	6.29	Contains cysteine
α-Lactalbumin	90	1.2	3.7	14176	6.25	Part of lactose synthase
Serum albumin	6	0.4	1.2	66267	6.07	Blood protein
Proteose peptone	~40	0.8	2.4	4000–40000	~6.54	Heterogeneous
Immunoglobulins	~4	0.8	2.4	—	~6.20	Glycoproteins
IgGl, IgG2	—	0.65	1.8	~150000	—	Several types
IgA	—	0.14	0.4	~385000	—	
IgM	—	0.05	0.2	~900000	—	Part is cryoglobulin
Miscellaneous	—	0.9	2.7	—	—	
Lactoferrin	~1	0.1	—	86000	6.14	Glycoprotein, binds Fe
Transferrin	~1	0.01	—	76000	6.21	Glycoprotein, binds Fe
Membrane proteins	—	0.7	2	—	~7.1	Glycoproteins, etc.
Enzymes	—	—	—	—	—	

Note: Approximate composition. IEP = isoelectric pH.

TABLE 2.14
Some Properties of the Main Groups of Protein in Skim Milk

Property	Caseins	Globular Proteins	Proteose-Peptone
Present in	Casein micelles[a]	Serum	Both
Soluble at pH 4.6	No	Yes	Yes
Clotting by rennet[b]	Yes	No	Partly
Heat denaturation	No	Yes	No

[a] At low temperature part is in the serum.
[b] At pH 6.7.

All the same, the protein composition is well known. Table 2.13 presents an overview of the milk proteins, and Table 2.14 summarizes some practical properties of the main groups. Various chemical properties of the main milk proteins are given in Table 2.15 and amino acid compositions in the Appendix, Table A.5.

Casein is defined as the protein precipitating from milk near pH 4.6. It thus is not soluble at its isoelectric pH. Casein is not a globular protein; it associates extensively and is present in milk in large aggregates, the casein micelles, which also contain the colloidal calcium phosphate (CCP). On acidification, the CCP dissolves.

Casein is a mixture of several components (Table 2.13). According to the genetically determined primary structures, we can distinguish α_{S1}-, α_{S2}-, β-, and κ-casein, but each of these occurs in a number of variants. Most of the κ-casein molecules are glycosylated to various extents. Part of the β-casein is split by proteolytic enzymes into γ-casein and proteose peptone. The α_S- and β-caseins are phosphoproteins that have a number of phosphate groups esterified to serine residues; they precipitate with Ca^{2+} ions, but κ-casein protects them from precipitation. However, κ-casein is easily attacked by the rennet enzyme chymosin, which splits off a portion of the κ-casein molecule; it thereby loses its protective ability. As a result, the casein precipitates in the presence of Ca ions. These reactions are the basis of the clotting of milk by rennet and, thus, of cheese making. Casein altered in this way is called *paracasein* and can be obtained by means of renneting. The resulting rennet casein has a high content of calcium phosphate. (*Note:* Casein and paracasein are chemical names; acid casein and rennet casein are names of commercial products.)

Casein does not show denaturation. However, heating at temperatures above approximately 120°C causes the casein to slowly become insoluble due to chemical changes.

Serum proteins are present in a dissolved form, in the serum. They are often called whey proteins, although they are not precisely identical to the proteins of rennet whey, which also contains the peptides split off from κ-casein. The immunoglobulins in milk vary widely in concentration and composition (colostrum has

TABLE 2.15
Properties of Some Milk Proteins

Property	α_{s1}-Casein (B)	α_{s2}-Casein (A)	β-Casein (A^2)	κ-Casein (A)	β-Lactoglobulin (B)	α-Lactalbumin (B)	Serum Albumin
Molar mass	23,614	25,230	23,983	19,023[a]	18,283	14,176	66,267
Amino acid residues/molecule	199	207	209	169	162	123	582
Phosphoserine (res./mol.)	8	11	5	1	0	0	0
Cysteine (res./mol.)	0	2	0	2	5	8	35
–S–S– linkages/mol.	0	1	0	—	2	4	17
Hexoses (res./mol.)	0	0	0	~2.3[b]	0[c]	0[d]	0
Hydrophobicity[e]	25	23	29	22	29	28	24
α-Helix (approximate %)	7?	?	10?	?	11	30	46
Charged residues (mol %)	34	36	23	21	30	28	34
Net charge/residue	–0.10	–0.07	–0.06	–0.02[b]	–0.04	–0.02	–0.02
Distribution of charge	Uneven	Uneven	Very uneven	Very uneven	Even	Even	
Isoelectric pH	4.5	5.0	4.8	5.6	5.2	~4.3	4.8
Association tendency	Strong	Strong	$f(T)$[f]	Strong	Dimer	No	No
Ca^{2+} binding	++	++	+	–	–	([g])	–

[a] Exclusive of carbohydrate residues.
[b] Average.
[c] 8 in a rare variant (Dr).
[d] A small fraction of the molecules has carbohydrate residues.
[e] % hydrophobic side groups (Val, Leu, Ile, Phe, Trp).
[f] Poor below 5°C, strong (micelle formation) at 37°C.
[g] Binds 1 mol Ca^{2+} per mole; very strong bond.

a high immunoglobulin content). Except proteose peptone, all serum proteins are globular proteins. At their IEP they remain in solution, but they are heat sensitive.

Miscellaneous proteins are numerous. The membrane of the fat globule contains several of these, including various glycoproteins. Most of the many enzyme proteins present in milk are also located in the fat globule membrane. All membrane proteins also occur in the plasma, although in very small concentrations.

2.4.3 Serum Proteins

Most serum proteins typically are globular proteins: they have relatively high hydrophobicity and compactly folded peptide chains. Most contain an appreciable proportion of α-helix and β-sheet; the charge distribution is rather homogeneous (see Table 2.15). They become insoluble at pH values below 6.5 if milk is heated. No doubt this change is related to the denaturation of the proteins involved. To be sure, the reaction is quite complicated (see Subsection 7.2.2). The denaturation does not result in aggregation, but the proteins precipitate onto the casein micelles and remain dispersed. Colostrum, which has a very high content of serum proteins, gels when it is heated in a way comparable to the white of an egg.

α-lactalbumin's biological function is as coenzyme in the synthesis of lactose. The protein is a small, compactly folded, more or less spherical molecule. It does not associate, except at low ionic strength.

α-lactalbumin has a specific nonexposed binding site for a Ca ion. The Ca is strongly bound and stabilizes the protein conformation. Removal of the Ca, or lowering the pH to about 4, which also loosens the Ca ion, causes partial unfolding into a molten globule state. In this state the protein is subject to irreversible heat denaturation at relatively low temperatures. Native α-lactalbumin shows complete renaturation after heat treatment if no other proteins were present during heating.

β-lactoglobulin is the major serum protein, and its properties tend to dominate the properties of whey protein preparations, especially the reactions occurring upon heat treatment. Its solubility strongly depends on pH and ionic strength, as illustrated in Figure 2.23, but it does not precipitate on acidification of milk; the same holds true for the other serum proteins. Heat denaturation is discussed in Subsection 7.2.2.

The secondary and tertiary structures of β-lactoglobulin are known in detail. It has two –S–S– bonds and one free sulfhydryl group; the latter is not exposed in the native state. The protein is subject to a number of changes in tertiary and quaternary structure with changes in pH or temperature. In milk, it is present as a dimer (hence, MW = 36.6 kDa). Both molecules are tightly bound to each other, mainly by hydrophobic interactions. The dimer dissociates at high temperature. Below pH 5.5, β-lactoglobulin associates to form an octamer, although smaller aggregates also occur. At still lower pH values, i.e., below 3.5, there is no association. Also at pH values above 7.5, only monomers occur. This change accompanies the so-called Tanford transition at that pH, which is a change in tertiary structure that causes exposure of the sulfhydryl group; the latter then becomes quite reactive.

β-lactoglobulin occurs in three main genetic variants, A, B, and C. These associate each to a different extent, i.e., A > B > C. These variants form separate dimers; hence AB, AC, or BC do not occur. Variants A and B also differ in conformational stability.

β-lactoglobulin tends to bind some apolar molecules, which may not be strange because of its high hydrophobicity (see Table 2.15). The binding concerns, for example, retinol (vitamin A) and some fatty acids. It appears unlikely that this has technological or nutritional significance.

(Blood) serum albumin is a minor protein that presumably gains entrance to milk by 'leakage' from blood serum. It is a large molecule having three globular domains, resulting in an elongated shape, about 3 × 12 nm in size. It has 17 –S–S– bonds and one –SH group.

Immunoglobulins are antibodies synthesized in response to stimulation by specific antigens. They specifically occur in blood. Immunoglobulins are large glycoprotein molecules of heterogeneous composition, even within one subclass. This is no surprise, considering that they are formed by different secretory cells that each may produce different peptide chains. Moreover, a portion of the molecule is specifically formed to neutralize a particular antigen; this is the so-called hypervariable portion.

Various classes of immunoglobulins are distinguished, including G (gammaglobulins), A, and M (macroglobulins), which occur in milk (Figure 2.24). Each IgG molecule is a polymer of two heavy (H) and two light (L) chains; MW ≈ 150,000. The molecule has two identical reactive sites or junctures. The antigen involved, or often a small part of it, fits exactly into these sites. It is bound by means of several interactions: H-bonds, hydrophobic bonds, and electrostatic attraction. The distance between both reactive sites is flexible. This facilitates adhesion onto the antigen. The IgG immunoglobulin can exert action against many antigens and may inhibit bacterial growth. IgM consists of a pentamer of IgG-like molecules joined by a so-called J component (Figure 2.24). It is a large molecule, with a molar mass about 900,000 and a diameter of about 30 nm. IgM can be an

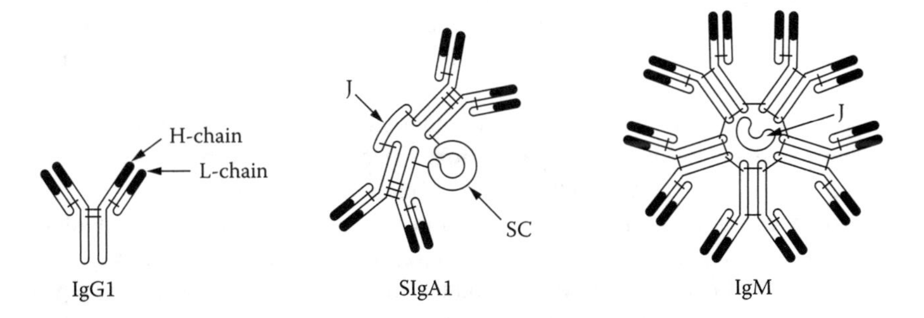

FIGURE 2.24 Schematic shape of immunoglobulins G1, secretory A1, and M. Disulfide linkages are designated by dashes. Variable portions are black. (From P. Walstra and R. Jenness, *Dairy Chemistry and Physics,* Wiley, New York, 1984. With permission.)

antibody against polysaccharide groups (as occur in the bacterial cell wall), and especially acts against 'particles,' including bacteria and viruses. It can flocculate these particles because a single IgM molecule becomes attached to two of these (the reactive sites are on the outside of the molecule). This flocculation is called *agglutination*, and the IgM concerned an *agglutinin*. The agglutination reaction is specific with respect to the antigen, but also depends on factors such as pH and ionic strength; generally, optimum pH is 5.5 to 7, and optimum ionic strength ≈0.05. Some agglutinins exhibit cryoprecipitation, i.e., they precipitate at low temperature (< 37°C, better < 15°C). In doing so they can also agglutinate other particles; this is a partly nonspecific flocculation, but for the rest similar to agglutination. The proteins involved are called cryoglobulins.

In milk, IgG (1 and 2), IgA, and IgM all are present (Table 2.13). The concentrations are highly variable. Concentrations in colostrum can be 30 to 100 times as high, whereas very little is present in late-lactation milk; there are also significant variations among individual cows. Little is known about the action of IgG (which has the highest concentration of the various classes) and IgA in milk; some propionic acid bacteria are inhibited by one or both of these. However, IgM is of great importance in milk. It includes the so-called lactenins L_1 and L_3, which are inhibitors of Gram-positive bacteria. These lactenins are agglutinins. L_3 especially acts against some strains of *Lactococcus lactis*. Their agglutinative action is highly specific; often, there are sensitive as well as nonsensitive bacteria in one strain. Naturally, the specific action will depend on the antigens (in this case, bacteria) that the cow has encountered.

IgM includes at least one cryoglobulin. The latter is involved in the flocculation of milk fat globules (Subsection 3.2.4), which is a nonspecific reaction: each cryoglobulin present flocculates fat globules of all kinds of milk. Bacteria are also agglutinated onto fat globules. Presumably, this is a specific reaction. All these reactions cause removal of bacteria from the bulk of the milk. Common agglutination will sediment the bacteria to the bottom of the vessel. If they are agglutinated onto the fat globules, they accumulate in the cream layer. As a result, metabolism and growth of the bacteria involved can be significantly inhibited.

The agglutinins are inactivated by heat treatment (Subsection 7.2.2). The inactivation reaction is coincident with the immunoglobulins becoming insoluble. Homogenization also inactivates the agglutinins, but the explanation is unclear.

The main natural function of the immunoglobulins is to immunize the calf. During the first few days after parturition, the calf can absorb intact immunoglobulins from colostrum into the blood through its gastrointestinal tract. Colostrum does contain a component (a globulin-like protein) inhibiting proteolytic enzymes — trypsin, in particular. In milk, this component is virtually absent. Moreover, chymosin, which does not attack immunoglobulins, occurs as a proteolytic enzyme in the abomasum of the newborn calf, rather than pepsin. As the calf grows older, ever more pepsin is produced.

Proteose peptone is defined as non-heat-sensitive, not precipitated at pH 4.6, and precipitated by 12% trichloroacetic acid. This fraction is quite different from the other serum proteins. Three different degradation products of β-casein (the complement of the γ-caseins) largely account for the fraction. It also contains a glycoprotein (called PP3) that is a fat globule membrane constituent, and presumably there are traces of other proteins. Clearly, at neutral pH a considerable part of the proteose peptone is present in the casein micelles, so that rennet cheese whey by no means contains all of the proteose peptone, but serum obtained upon acidification of milk does.

Lactoferrin (Table 2.13) is an inhibitor of some bacteria including *Bacillus stearothermophilus* and *Bacillus subtilis*. The inhibition is caused by removal of iron, more precisely Fe^{3+} ions, from the serum. To be sure, the lactoferrin concentration in cows' milk is low; in human milk it is far higher.

2.4.4 Casein

The properties of the caseins differ from those of most proteins (Table 2.15; Figure 2.25). Caseins are hydrophobic; they have a fairly high charge, many prolines, and few cysteine residues. They do not form anything more than short lengths of α-helix and have little tertiary structure. This does not imply that the casein molecules are random coils, though in dilute solution the chains are partly unfolded. Many hydrophobic groups are exposed, so that the molecules readily form hydrophobic bonds. The caseins thus show extensive association, both self-association and association with each other. (Association in casein micelles is discussed in Subsection 3.3.1.) The relatively high charge is needed to keep casein in solution.

Casein molecules cannot or can hardly be denatured, because they have little secondary and tertiary structure. An example is given in Figure 2.22. β-lactoglobulin, a globular protein, shows a steep conformational change at about 4-M urea, whereas β-casein changes little. Because of this, casein does not become insoluble by heating at temperatures below 100°C.

The high charge of casein is partly caused by the phosphate groups. These are for the most part esterified to serine residues; near the pH of milk they are largely ionized (Table 2.12). The groups strongly bind divalent ions like Ca^{2+}, especially at a higher pH. Figure 2.26 shows that the Ca binding parallels the content of these groups.

Several different caseins occur in milk, but their separation is not easy. Reactions that cause their precipitation from milk (acidification, renneting, and centrifugation after adding calcium) all yield a more or less complete mixture of caseins. It was only after electrophoresis came into use that resolution of the caseins was feasible, at first into the three components, α, β, and γ. Later on, α-casein could be separated into a fraction sensitive to Ca^{2+} (α_s = α-sensitive) and a Ca^{2+}-insensitive fraction, i.e., κ. Still later, further separation turned out to be necessary to obtain pure components. Currently, the complete primary structures are known. This has revealed that there are four different peptide chains — α_{s1}, α_{s2}, β, and κ, of which the molar ratio is about 11:3:10:4. Differences in phosphorylation and glycosylation, as well as some proteolysis, cause additional heterogeneity.

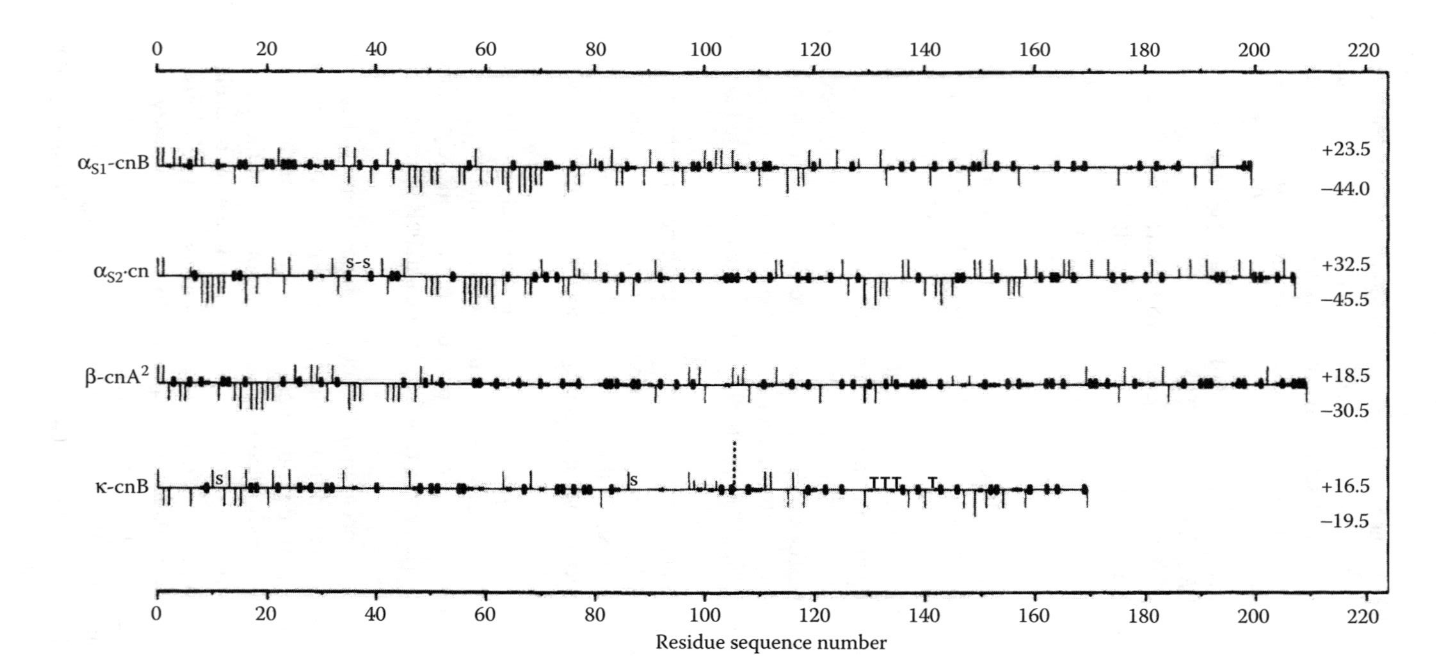

FIGURE 2.25 Peptide chain of caseins. The vertical bars stand for positive and negative charges, respectively (exclusive of those of neuraminic acid), where the long negative bars denote SerP and the short positive ones His. The crosses stand for proline residues, the black squares for hydrophobic amino acids, Ile, Leu, Phe, Trp, and Val. S indicates a cysteine residue and S–S a linkage. The broken bar indicates the point of cleavage by chymosin. Possible location of glucide residues esterified to threonine: T. For amino-acid sequences of the caseins, see Appendix, Table A.6.

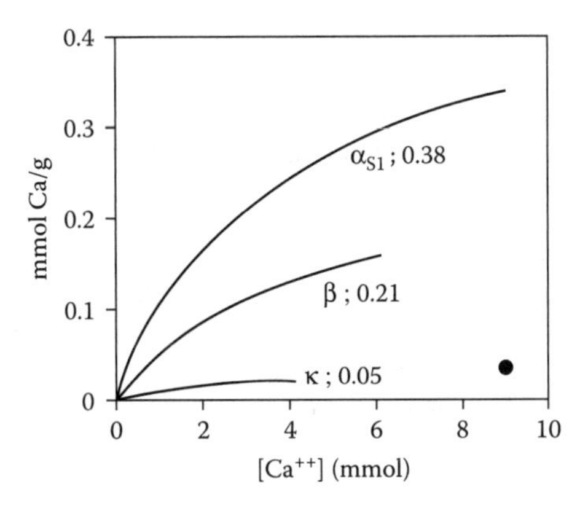

FIGURE 2.26 Binding of Ca^{2+} by caseins at pH 7.4. The ester-bound phosphate content (mmol · g^{-1} casein) is indicated. The dot refers to dephosphorylated α_{s1}-casein. (Adapted from I.R. Dickson and D.J. Perkins, *Biochem. J.*, **127**, 235, 1971.)

2.4.4.1 α_{s1}-Casein

α_{s1}-Casein has a high net negative charge and a high phosphate content. Figure 2.27 shows that α_{s1}-casein associates in two steps at pH 6.6 and 0.05 M ionic strength. Obviously, very low casein concentrations are needed to obtain nonassociated molecules. On the other hand, reducing the ionic strength, and thereby increasing the effective range of the electrostatic repulsion, decreases the

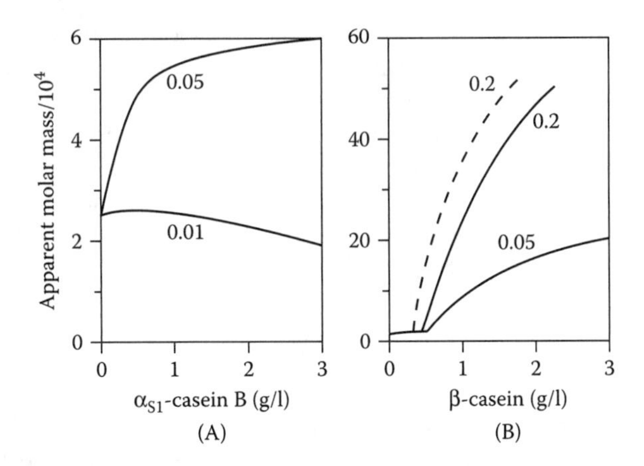

FIGURE 2.27 Association of α_{s1}- and β-casein as a function of concentration. pH 6.6 in α_{s1}-, 7.0 in β-casein. Temperature 21°C, for broken line 24°C. Ionic strength (mol · m^{-3}) indicated on curves. (Adapted from D.G. Schmidt and T.A.J. Payens, *Surface and Colloid Science*, **9**, 165, 1976.)

association. Hydrophobic interactions are also involved in the association. At higher pH, that is, greater charge, the association decreases and eventually disappears, even if casein concentration and ionic strength are high. A variant that occurs in small amounts has nine rather than eight phosphate groups.

2.4.4.2 α_{s2}-Caseins

Some variants of this protein exist. They differ in the number of ester phosphate groups, i.e., 10 to 14 per molecule. α_{s2}-Caseins contain two cysteine residues (forming an –S–S– bridge) and no carbohydrate groups. They are rather Ca^{2+} sensitive. The association pattern is similar to that of α_{s1}-casein.

2.4.4.3 β-Casein

β-Casein is the most hydrophobic casein, and it has a large number of proline residues. Moreover, as is shown in Figure 2.25, the charge is unevenly distributed. For example, dividing the molecule into two pieces, starting from the N terminus, results in the following:

Residue sequence number	1–44	45–209
Proline frequency	0.02	0.20
Charge frequency	0.57	0.13
Net charge	−15.5	+2.5
Hydrophobic residue frequency	0.25	0.30

Obviously, both parts differ considerably in properties. Thus, β-casein is somewhat like a soap molecule with a polar 'head' and a long-chain, apolar 'tail.' The association of β-casein is somewhat similar to that of a soap in that a critical micellization concentration occurs; see Figure 2.27B. The micelles (here in the strict sense used by physical chemists) comprise some 20 or 30 molecules. Note the strong dependence of the association on temperature and ionic strength. Below 5°C no association of β-casein occurs and the molecule remains unfolded. Now it behaves like a random coil. In milk, part of the β-casein goes into solution at low temperature, thereby increasing the viscosity of the milk. These changes are reversible but occur slowly (over several hours).

2.4.4.4 γ-Casein

γ-Casein is a degradation product of β-casein. It corresponds for the most part to amino acid residues 29 to 209 of the β-casein sequence, that is, the more hydrophobic portion. Accordingly, it is fairly soluble in ethanol (900 $mg \cdot l^{-1}$ in 50% ethanol; α_s-casein, for example, only 9 $mg \cdot l^{-1}$). The cleavage is caused by the enzyme plasmin (EC 3.4.21.7), present in milk. The amount of γ-caseins can vary widely, depending on the age and the keeping temperature of the milk. The split-off parts constitute most of the proteose peptones.

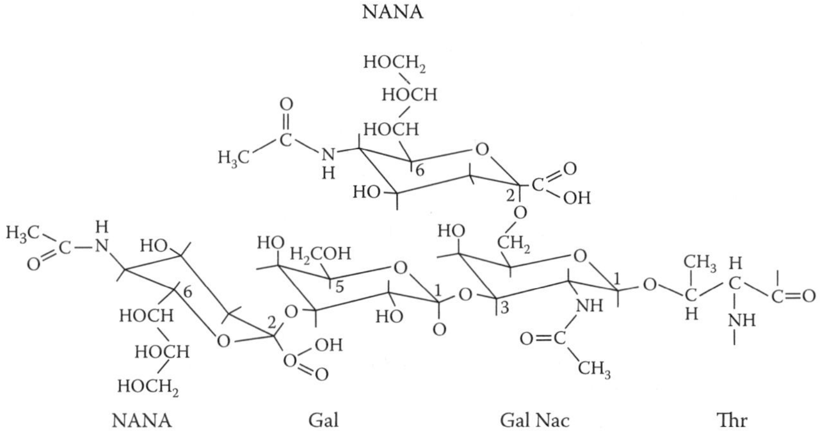

FIGURE 2.28 Example of a glucide group linked to κ-casein. Often the NANA residue at the top is lacking.

2.4.4.5 κ-Casein

κ-Casein greatly differs from the other caseins. It has two cysteine residues that form intermolecular disulfide bonds. Because of this, κ-casein occurs in milk as oligomers containing 5 to 11 monomers, with an average MW about 120 kDa. The properties of κ-casein given nearly always refer to the 'reduced' or monomeric form (the reduction breaks the –S–S– linkages).

About two thirds of the (monomeric) molecules contain a carbohydrate group, which is esterified to one of the threonines (131, 133, 135, or 142) and has galactosamine, galactose, and one or two N-acetyl neuraminic acid (NANA or *o*-sialic acid) residues; see Figure 2.28. These groups each have one or two negative charges and are hydrophilic. Some other, minor configurations occur as well. So there are obvious differences among κ-casein molecules, also because some of them have two ester phosphate groups rather than one. This so-called microheterogeneity always occurs, even within an individual milking of one cow.

The peptide bond between residues 105 and 106 is rapidly hydrolyzed by proteolytic enzymes. Note that there is a positively charged region near this site (Figure 2.25).

κ-Casein also strongly associates to yield micelles that contain over 30 molecules including protruding carbohydrate groups. The association is somewhat like that of β-casein.

2.4.5 Nutritional Aspects

In this section, we will only consider the nutritional value for people older than, say, a year. Milk as a food for babies will be discussed in Section 16.6.

The primary nutritional role of proteins is the provision of *essential amino acids*, i.e., Ile, Leu, Lys, Met, Phe, Thr, Trp, and Val (for infants His may also be essential). To this end, the protein should be well digestible and contain enough of these amino acid residues. The digestibility of milk protein is close to 100%; it is slightly less in intensely heated products. The nutritional value can be expressed in various ways; the best estimator may be the protein efficiency ratio (PER), the ratio of weight gain of a growing animal to its protein intake. The PER value of casein is ~2.5, of serum protein ~3.0, and of whole milk protein ~3.3, which is higher than that of most proteinaceous foods. The value of the mixture of casein and serum protein is higher than that of each one separately because they are complementary in their concentration of some essential amino acids: casein is relatively high in Tyr and Phe, serum protein in the sulfur-containing Cys and Met. A little milk protein may considerably enhance the PER of a diet consisting of vegetable foods.

Another nutritional function of casein is that it can bind large amounts of the important nutrients calcium and phosphate. This property naturally evolved to provide the calf with bone minerals; it is also important for human nutrition.

Proteins can have several specific nutritional effects, be it detrimental, such as proteinase inhibition or the promotion of atherosclerosis, or beneficial, such as antimicrobial activity. However, significant detrimental effects for humans ingesting milk have never been established.

Similar to the above is the presence of *bioactive peptides*, or, rather, peptide sequences; digestion in the intestine may then yield the peptides. Several such sequences have been identified in milk proteins, especially in casein. Well known are the caseinomorphins from β-casein, which have opioid properties, and would maybe induce sleep. Others inhibit the angiotensin-converting enzyme (ACE), so would reduce blood pressure. Still others, containing SerP, may aid the adsorption of divalent cations in the small intestine. All such activities have been established in studies *in vitro*, but clinical evidence for significant effects in humans has not yet been obtained.

Some individual humans exhibit *allergic responses* to one or more proteins. The symptoms vary widely, from fairly mild ones like rhinitis or diarrhea, to more serious ones like dermatitis or asthma. It is assumed that about 2% of babies can develop a cows' milk allergy, but several of these are far less sensitive at a later age. Both caseins and serum proteins can invoke allergic reactions, but β-lactoglobulin seems to be the most common agent in milk. Heat denaturation may somewhat diminish the allergenic property; substantial hydrolysis of the protein is more effective. Substitution of cows' milk by goats' milk generally does not diminish the allergic reaction.

2.5 ENZYMES

Milk contains scores of enzymes. The native or indigenous enzymes are those known to be excreted by the mammary gland. Most of these are synthesized by the secretory cells, others derive from blood, for example, plasmin. Moreover, several enzymes are present in the leukocytes, e.g., catalase. In addition, enzymes of microbial origin may be involved. The latter may be present in microorganisms, secreted by the organisms (such as proteinases and lipases), or released after lysis.

The native enzymes can be present at different locations in the milk. Many of them are associated with the fat globule membrane. This is no surprise, considering that most of the membrane originates from the apical cell membrane, which contains several enzymes. Other enzymes are in solution, i.e., dispersed in the serum, but some of these (such as lipoprotein lipase) are for the most part associated with the casein micelles.

Most of the milk enzymes seem to have no biological function in milk, even if they are present in high concentrations (e.g., ribonuclease; Table 2.16). Often, they do not significantly alter the milk. Some enzymes have an antimicrobial function or play other beneficial roles. A few of the enzymes may facilitate resorption of milk constituents into the blood if, and when, milking is stopped. It presumably concerns plasmin and lipoprotein lipase, which are not very active in fresh milk though they are present in high concentrations (Table 2.16). These, as well as some other enzymes, can cause spoilage of milk during storage. Some enzymes are used for analytical purposes. Formerly, catalase activity was estimated to detect mastitis, but the correlation is too weak. N-acetyl-β-D-glucosaminidase (EC 3.2.1.30), also called NAGase, is now considered a better marker, although again the relation with mastitis is far from perfect. Furthermore, particular enzymes are used to monitor pasteurization. Some examples of native milk enzymes and their action are discussed in Subsection 2.5.2.

2.5.1 Enzyme Activity

An enzyme catalyzes a specific reaction. The activity of an enzyme depends on many factors. The rate of catalysis is expressed as k_{cat} or turnover number of the enzyme, that is, the number of molecules of substrate converted per enzyme molecule per second. The total turnover rate V then is given by the product $k_{cat}[E]$, where [E] = enzyme concentration. According to Michaelis and Menten the initial velocity of the reaction as a function of the substrate concentration [S], for example, expressed in $mol \cdot l^{-1} \cdot s^{-1}$, is given by

$$v_i = V_{max} \cdot [S]/(K_m + [S]) \quad (2.5)$$

where V_{max} is the turnover rate for 'infinite' [S]. An example of the relationship is in Figure 2.29a. The Michaelis constant K_m is a measure of the affinity of the enzyme for its substrate (the lower the value of K_m, the greater the affinity). It depends on the type of substrate used and is thus a variable in addition to k_{cat}. K_m equals the substrate concentration when $v_i = V_{max}/2$, as shown. Equation 2.5 only applies to the initial velocity v_1 of the reaction because (1) the presence of reaction products generally decreases the net reaction rate (product inhibition) and (2) the substrate concentration decreases during the reaction.

Equation 2.5 is only valid for an 'ideal' enzyme in a dilute solution. It does not apply for so-called allosteric enzymes (which will not be discussed). For enzymes that act at a phase surface, such as lipases at an oil–water interface, other relations hold. Moreover, enzyme reactions can be diffusion controlled, e.g., in concentrated products. In principle, the reaction rate will be proportional

TABLE 2.16
Some Enzymes in Milk

Name	EC Number	Optimum		Activity[a]		Where in Milk	Inactivation[b]
		pH	Temperature (°C)	Potential	Actual		
Xanthine oxidase	1.1.3.22	~8	37	>>40	40	Fat globule membrane	7 min 73°C
Sulfhydryl oxidase	1.8.?	~7	~45	?	?	Plasma	3 min 73°C
Catalase	1.11.1.6	7	37?	?	300	Leukocytes	2 min 73°C
Lactoperoxidase	1.11.1.7	6.5	20	?	22000	Serum	10 min 73°C
Superoxide dismutase	1.15.1.1	?	37?	~2000	?	Plasma	65 min 75°C
Lipoprotein lipase	3.1.1.34	~9	33	3000	0.3	Casein micelles	30 s 73°C
Alkaline phosphatase	3.1.3.1	~9	37	500	<<500	Fat globule membrane	20 s 73°C
Ribonuclease	3.1.27.5	7.5	37	([c])	?	Serum	?
Plasmin	3.4.21.7	8	37	3	0.05	Casein micelles	40 min 73°C

[a] $\mu mol \cdot min^{-1} \cdot l^{-1}$.
[b] Heat treatment needed to reduce activity to approximately 1%.
[c] 11–25 mg enzyme per kg of milk.

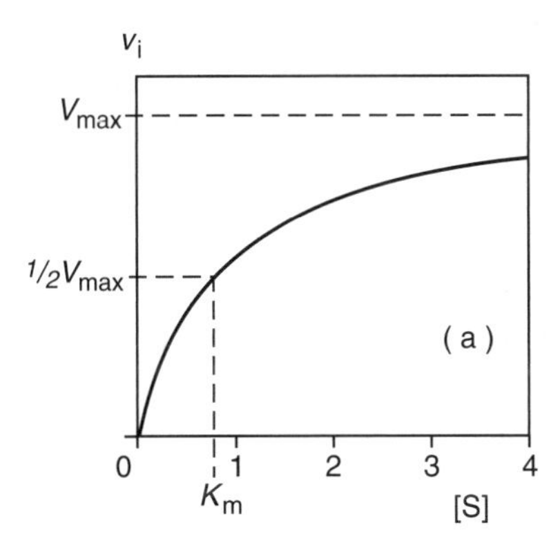

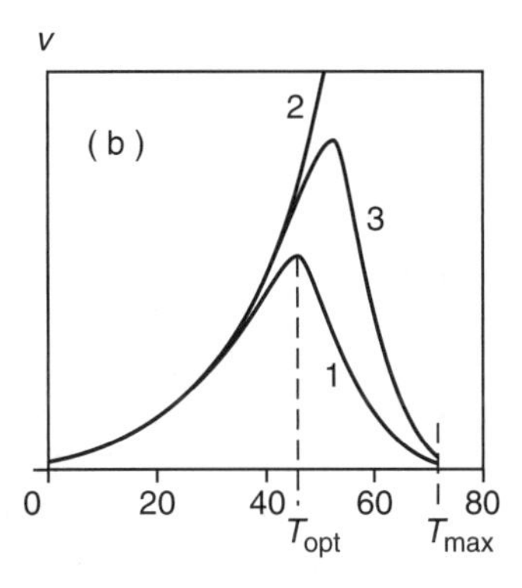

FIGURE 2.29 Enzyme kinetics. (a) Effect of substrate concentration [S] on the initial reaction rate v_i, according to Equation 2.5. (b) Effect of temperature T (°C) on the rate v of proteolysis caused by plasmin. Curve 1, results obtained during 30 min of reaction; 2, extrapolated to high T, assuming the activation energy to be constant; 3, approximate result during a few minutes. (From data by A. Metwalli et al., *Int. Dairy J.*, **8**, 47, 1998.)

to the enzyme concentration [E], but there are important exceptions (see items 3 to 6 below). As mentioned, the rate depends on the substrate concentration. For most enzymes active in milk, [S] is and remains greatly in excess, implying that v_i is close to V_{max}. (This is not the case for the action of chymosin on κ-casein during milk clotting, where all of the substrate is eventually used up.) Further variables affecting enzyme action are:

1. *Temperature:* The temperature dependence of most enzyme reactions follows Arrhenius kinetics (see Subsection 7.3.2), as shown in Figure 2.29b, curve 2. However, at high temperature the enzyme, which is a globular protein, starts to denature and thereby loses activity. (The denaturation need not be irreversible; see Subsection 7.3.3.) This results in a temperature dependence with an optimum and a maximum temperature; see curve 1. However, the denaturation is not infinitely fast, which implies that the resulting $v(T)$ curve will depend on the length of time over which the enzyme reaction is observed, as illustrated by curve 3. This, then, means that T_{opt}, and possibly T_{max}, depend on time or on heating rate.
2. *Solvent properties:* These can affect the conformation of the enzyme molecule — and thereby its activity — and its affinity for the substrate — i.e., K_m. The susceptibility of the substrate to attack may also depend on solvent properties, especially if it is a protein. The pH generally has considerable effect, and most enzymes have a minimum, an optimum, and a maximum pH; these values may also depend on temperature. Other variables are ionic strength, many enzymes being inhibited at high strength. A high concentration of other solutes, e.g., sugar, may also affect the reaction rate.

3. *Inhibitors:* In addition to the mentioned reaction products, several substances may inhibit enzyme activity, for instance, because these substances bind also to the active site of the enzyme (competitive inhibition) or because they affect the conformation of the enzyme molecule.
4. *Cofactors:* Many enzymes need a cofactor to be fully active. An example is the apoprotein needed for milk lipoprotein–lipase action (Subsection 3.2.5). The concentration of cofactors in milk often is variable.
5. *Stimulators:* There may be other stimulating substances — for instance, those inactivating an inhibitor.
6. *Compartmentalization:* This can make the substrate inaccessible. An example is the triglycerides that are screened from enzymes in milk by the fat globule membrane. Another example is adsorption of the enzyme onto particles, thereby lessening activity. Some milk enzymes, notably lipases and proteinases, are for a considerable part associated with the casein micelles, whereby their activity is decreased.
7. *Enzyme formation:* The enzyme may be present in a nonactive form, a so-called zymogen, and slowly be converted to the true enzyme. An example is plasmin, largely occurring in milk as the inactive plasminogen; see Subsection 2.5.2.5.
8. *Enzyme destruction:* This can also occur. For instance, lipoprotein lipase activity in milk slowly decreases, presumably caused by an oxidative reaction on the enzyme.

2.5.2 Some Milk Enzymes

2.5.2.1 Antibacterial Enzymes

The main representative is *lactoperoxidase* (EC 1.11.1.7). It catalyzes the reaction:

$$H_2O_2 + 2HA \rightarrow 2H_2O + 2A$$

where the substrate HA can include several compounds: aromatic amines, phenols, vitamin C, and so on. The enzyme can also catalyze oxidation of thiocyanate (SCN) by H_2O_2 to OSCN and other oxyacids, which inhibit most bacteria. If the bacteria themselves produce H_2O_2, as most lactic acid bacteria do, they are inhibited. (In milk, H_2O_2 decomposition by catalase, EC 1.11.1.6, is too weak to prevent this.) Milk is rich in lactoperoxidase (about 0.4 μM), but the thiocyanate concentration in milk varies widely because it depends on the cyanoglucoside content of the feed. Sometimes thiocyanate together with a little H_2O_2 is added to raw milk to prevent spoilage. In this way, even in a warm climate spoilage of milk can be delayed for many hours. In milk, the action of the lactoperoxidase–hydrogen peroxide–thiocyanate system (which also occurs in saliva) can be enhanced by xanthine oxidase and possibly sulfhydryl oxidase (see the following subsection), which can form H_2O_2 from some substrates.

Lysozyme (EC 3.2.1.17) is another bactericidal enzyme; it hydrolyzes polysaccharides of bacterial cell walls, eventually causing lysis of the bacteria (see also

Section 26.2). In cows' milk the lysozyme activity is very weak; in human milk it is much stronger.

2.5.2.2 Oxidoreductases

Xanthine oxidase (EC 1.1.3.22) can catalyze oxidation of various substances, by no means xanthine only. Many substances, including O_2, can be a hydrogen acceptor. The enzyme can reduce nitrate (which occurs in milk only in trace quantities) to nitrite. This property is put to use in the manufacture of some cheeses, in which nitrate is added to milk to prevent the proliferation of the detrimental butyric acid bacteria (Section 26.2). Nitrite inhibits these bacteria. If nitrate has been added to the milk, nitrite is present in sufficient amounts in cheese, though it is fairly rapidly decomposed. Cows' milk has a relatively high xanthine oxidase content. Most of the enzyme is associated with the fat globule membrane. Because of this it is only partly active, but the activity is increased by such treatments as cooling and homogenization, which may release enzyme from the membrane.

Superoxide dismutase (EC 1.15.1.1) catalyzes the dismutation (dismutation here meaning oxidation of one molecule and simultaneous reduction of another) of superoxide anion $O_2^{\bar{\cdot}}$ to hydrogen peroxide and triplet oxygen according to:

$$2O_2^{\bar{\cdot}} + 2H^+ \rightarrow H_2O_2 + {}^3O_2$$

The enzyme in milk is identical to that in blood. Its biological function is to protect cells from oxidative damage. In milk, the superoxide anion can be generated by oxidations catalyzed by xanthine oxidase and lactoperoxidase, and by photo-oxidation of riboflavin. Superoxide dismutase may inhibit oxidation of milk constituents. The enzyme is supposed to counteract autoxidation of lipids (this oxidation causes off-flavors; see Section 2.3). It is not inactivated by low pasteurization.

Sulfhydryl oxidase (EC 1.8.?) catalyzes oxidation of sulfhydryl groups to disulfides, using O_2 as electron acceptor:

$$2\,RHS + O_2 \rightarrow RSSR + H_2O_2$$

–SH groups of both high- and low-molar-mass compounds are decomposed. It may be responsible for the virtual absence of (exposed) –SH groups in raw milk. Most of the enzyme is bound to lipoprotein particles. Pasteurization inactivates the enzyme only partially. In pasteurized milk products, the enzyme may be responsible for reducing the cooked flavor caused by –SH compounds.

Lactoperoxidase (see the preceding subsection) and *catalase* are also oxidoreductases.

2.5.2.3 Phosphatases

Several phosphatases occur in milk. Best known is milk *alkaline phosphatase* (EC 3.1.3.1), which catalyzes the hydrolysis of phosphoric monoesters. Generally, determination of the activity of the enzyme by the phosphatase activity test is applied as a check of low pasteurization of milk. Inactivation of the enzyme ensures that all of the non-spore-forming pathogenic microorganisms (as far as these can grow in milk) present in the milk during heat treatment have been killed; most but not all of the lactic acid bacteria and Gram-negative rods have also been killed. Most of the enzyme is in the membranes of the fat globules. Accordingly, the phosphatase test is less sensitive when applied to skim milk.

Milk also contains an *acid phosphatase* (EC 3.1.3.2); it occurs in the serum and is quite heat resistant (see Figure 2.31). These phosphatases catalyze the hydrolysis of certain phosphoric esters in milk, but slowly.

Some phosphatases can release phosphoric acid groups esterified to casein; this especially occurs in cheese during maturation.

2.5.2.4 Lipolytic Enzymes

Several esterases, which can hydrolyze fatty acid esters, occur in milk. Some of these attack esters in solution, but the principal lipolytic enzyme of cows' milk, that is, *lipoprotein lipase* (EC 3.1.1.34), liberates fatty acids from tri- and diglycerides and is only active at the oil–water interface. It is bound largely to casein micelles. In milk, lipolysis causes a soapy, rancid off-flavor, and this is further discussed in Subsection 3.2.5.

2.5.2.5 Proteinases

In milk at least two trypsin-like endopeptidases occur. One of these is *plasmin* (EC 3.4.21.7) or alkaline milk proteinase. Most of the plasmin in milk is present as the inactive plasminogen. The enzyme is largely associated with the casein micelles. Its activity in milk varies widely, partly because of a variable ratio of plasmin to plasminogen. Usually, the activity increases with time as well as by heating, for example, pasteurization. The explanation appears to be that milk contains one or more promoters (especially urokinase) that catalyze the hydrolysis of plasminogen to yield plasmin. Moreover, milk contains at least one substance that inhibits the promoter(s). The inhibitor is inactivated by heat treatment. Leukocytes also contain a promoter, and milk with a high somatic cell count generally shows enhanced plasmin activity.

Plasmin can hydrolyze proteins to yield large degradation products and is responsible for production of γ-casein and proteose peptones from β-casein. This reaction even proceeds at 5°C, presumably because β-casein then is better accessible, as part of it dissociates from the casein micelles. The enzyme causes proteolysis in some products, e.g., in cheese. In UHT milk products (Figure 2.30) its proteolytic action causes a bitter flavor and eventually can solubilize the casein micelles; in some cases, gelation has been observed. This is because the enzyme is very heat resistant (Figure 2.31). Accordingly, appropriate UHT treatment (for example, 140°C for 15 s) should be applied to prevent such problems.

Milk also contains *acid proteinases*. These originate from somatic cells in the milk. The most active one is cathepsin D (EC 3.4.23.5), though its activity in milk is low. This will partly be due to the low optimum pH (4.0) of the enzyme.

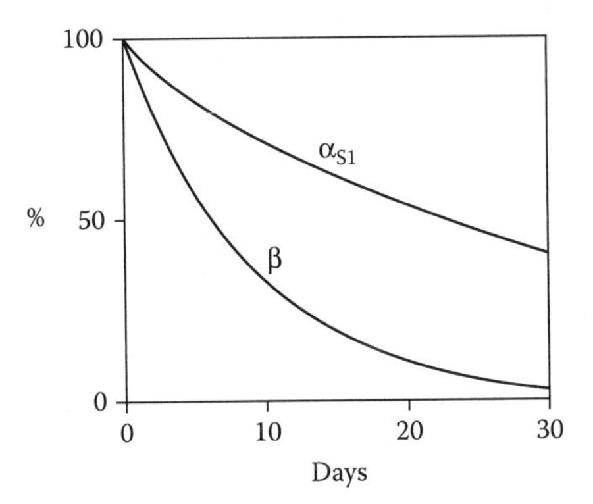

FIGURE 2.30 Plasmin action on α_{s1}- and β-casein in milk at 20°C. Casein left (in %) as a function of storage time. The milk had been heated for 5 s at 134°C. (Adapted from results by F.M. Driessen, Ph.D. thesis, Wageningen, 1983.)

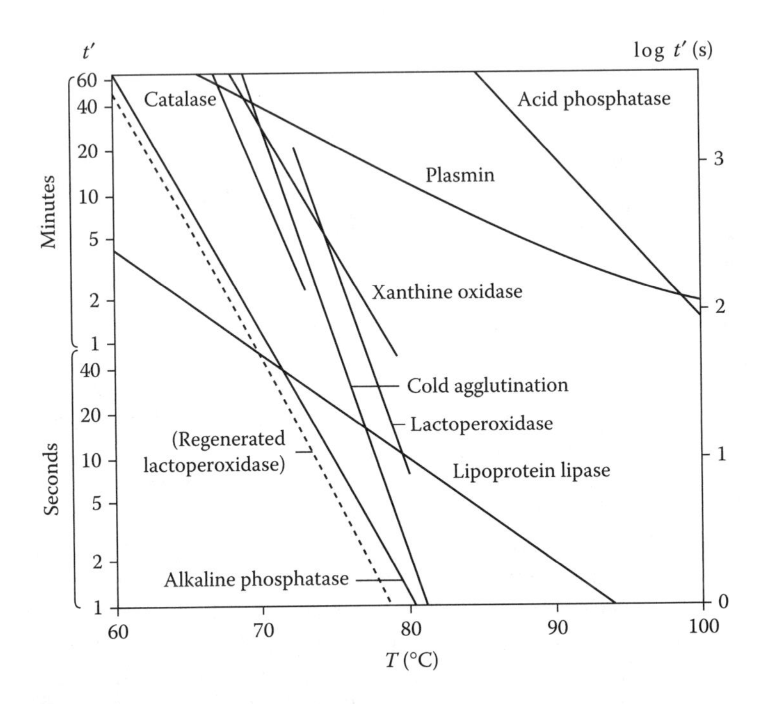

FIGURE 2.31 Time (t') and temperature (T) of heating milk needed to inactivate some enzymes (i.e., reduce the activity by about 99%) and to prevent cold agglutination. Approximate examples. (Modified after P. Walstra and R. Jenness, *Dairy Chemistry and Physics*, New York, Wiley, 1984.)

It is less heat resistant than plasmin. It is also associated with the casein micelles, and it can cause some proteolysis in cheese.

2.5.3 Inactivation

The term 'inactivation' implies that the enzyme is chemically changed to such an extent that it can no longer act as a catalyst (in some cases, the inactivation is reversible). Inactivation of some of the molecules thus results in a decreased enzyme activity. The activity can also decrease because the turnover rate is decreased, which generally applies to all molecules. This can be due to a change in any of the conditions that affect enzyme activity mentioned in Subsection 2.5.1. The two cases should be clearly distinguished.

To inactivate enzymes, heat treatment is mostly applied. Inactivation is generally due to denaturation (unfolding) of the enzyme molecule. The kinetics of the heat-induced reactions are discussed in Subsection 7.2.3. The heating time–temperature relationship for the inactivation of various enzymes in milk is given in Figure 2.31. Inactivation by heat treatment is of great importance because:

1. Enzymes that cause spoilage can be inactivated.
2. Systems that inhibit bacterial growth can be inactivated. This can be either desirable or undesirable.
3. Spoilage-inhibiting enzymes (e.g., superoxide dismutase) can be inactivated.
4. The intensity of milk pasteurization can be checked: low pasteurization through alkaline phosphatase and high pasteurization through lactoperoxidase.

The rate of inactivation by heat treatment may strongly depend on conditions such as pH and the presence or absence of substrate. Moreover, different isozymes (that is, genetic variants of one enzyme) that differ in heat stability may be involved. Often, they cause curved plots for the relationship between the log of the required heating time and the temperature.

Another complication is reactivation, which means that the heat-inactivated enzyme returns to the active form after cooling. Many enzymes do not exhibit reactivation, whereas other enzymes slowly do so. Examples are alkaline phosphatase, which becomes partly reactivated a few days after heat treatment, especially in cream (because of its high phosphatase content), and lactoperoxidase. Reactivated lactoperoxidase is also heat labile, even more than the original enzyme (Figure 2.31). Some enzymes are reactivated very rapidly after cooling. It then seems as if the heat treatment has had hardly any effect on the enzyme activity. On the other hand, heating may have been so intense that the unfolded enzyme molecule has reacted such that reactivation of the enzyme is not possible anymore. The inactivation plots then usually are curved and not very steep, and high temperatures and long times are needed. An example in Figure 2.31 is plasmin.

Alternatively, enzymes can be inactivated by oxidation processes, by very high pressure treatment, and by excessive aeration, etc. Homogenization rarely has a significant effect.

2.6 OTHER COMPONENTS

Sections 2.1 to 2.4 deal with the main components in milk (with the exception of water, which actually is the main component). Several minor components are also present, but these are not necessarily unimportant; see, for instance, Section 2.5 on enzymes. The number of compounds detected in milk is large and will certainly increase with further research and with improvements in the sensitivity of analytical methods. It is often uncertain as to whether such a component occurs as such or is chemically bound. Sometimes components alter during separation and analysis.

Most of the minor components are natural, which means that they are normally produced by the mammary gland. Others enter the milk by contamination; moreover, some natural components can (greatly) increase in concentration due to contamination. Other minor components are formed due to enzymatic or microbial action, or result from reactions occurring during processing of the milk.

2.6.1 NATURAL COMPONENTS

Most of these components stem directly from the blood or are intermediate products of the metabolic processes in the secretory cell. Some groups of components that were not discussed as yet will be considered here. They are grouped arbitrarily and several components belong to more than one category:

1. *Organic acids:* In addition to citric acid (Section 2.2) and low-molar-mass fatty acids (Section 2.3), small quantities of other organic acids (e.g., trace amounts of lactic and pyruvic acid) occur in milk serum. Bacterial action may greatly increase the concentration of such acids.
2. *Nitrogenous compounds:* On average, about 5% of total nitrogen in milk is nonprotein nitrogen (NPN). Groups of NPN compounds are listed in Table 2.17. The compounds are partly intermediate products of the protein metabolism of the animal (e.g., ammonia, urea, creatine, creatinine, and uric acid). Most of the amino acids as well as their derivatives (amines and serine phosphoric acid) are also found in trace amounts free in solution. Milk also contains small peptides. These compounds may be essential nutrients for some bacteria.
3. *Vitamins:* All known vitamins are present in milk. Table 2.18 gives an overview and also indicates the nutritional significance. It is seen that milk is a good source for most vitamins, especially vitamin A and most B vitamins. It is not a good source of vitamins C and E, but these are readily acquired from other foods in most diets. Vitamin-D content is

TABLE 2.17
Nonprotein Nitrogenous Compounds in Milk

Compound	Concentration, mg per kg of Milk	
	Nitrogen[a]	Total Mass[b]
Urea	84–280	250
Creatine	6–20	30
Creatinine	2–9	10
Uric acid	5–8	18
Orotic acid	4–30	70
Hippuric acid	4	13
Thiocyanate	0.2–4	1
Ammonia	3–14	10
α-Amino acids	29–51	280
Peptides	~30[c]	
Total NPN	230–310	

[a] Approximate range of contents reported.
[b] Average, very approximate.
[c] Greatly depends on analytical method.

also not high, but a very limited exposure of the skin to sunlight leads to sufficient synthesis by the organism, including man. The contents of some vitamins are reduced by processing or storage, which is discussed in Section 16.4.

Some vitamins play other roles in milk and milk products. Carotenes cause the yellow color of milk fat. Riboflavin is a fluorescent dye and is responsible for the yellowish color of whey. It is involved in redox reactions (Section 4.3), formation of singlet oxygen, and hence, in fat oxidation (Subsection 2.3.4), and is involved in the formation of light-induced off-flavors (Section 4.4). Also, ascorbic acid is involved in redox reactions. Tocopherols are antioxidants and reduce off-flavor formation in fats.

4. *Ribonucleic acids and their degradation products*, e.g., phosphate esters and organic bases. Furthermore, orotic acid typically occurs in milk of ruminant animals; it is a growth factor for *Lactobacillus delbrueckii* ssp. *bulgaricus*.
5. *Sulfuric acid esters:* Only indoxyl sulfate has been found in milk.
6. *Carbonyl compounds:* An example is acetone; more of it occurs if the cow suffers from ketosis. Fat-soluble aldehydes and ketones are mentioned in Table 2.7 and Subsection 2.3.4.
7. *Gases:* In milk the amount of nitrogen is about 16 $mg \cdot kg^{-1}$, that of oxygen about 6 $mg \cdot kg^{-1}$, or about 1.3% and 0.4% by volume, respectively. Milk

TABLE 2.18
Vitamins in Milk and Recommended Daily Intake (Approximate Values)

Vitamin	Chemical Name	Concentration per kg of Milk	RDI[a]	Present in/at
A	Retinol	0.7–1.3 mg RE[b]	0.4–1	Fat
B_1	Thiamine	0.5 mg	0.5–1	Serum
B_2	Riboflavin	1.8 mg	1–2	Serum
B_3	Niacin + its amide	8 mg[c]	18	Plasma
B_5	Pantothenic acid	3.5 mg	3–8	Serum
B_6	Pyridoxine, etc.	0.5 mg	1–2	Serum
	Biotin[d]	20–40 μg	100–200	Serum
	Folic acid[d]	50–60 μg	200–400	Protein[e]
B_{12}	Cobalamin	4.5 μg	1.5–2.5	Protein
C	Ascorbic acid	10–25 mg	40–70	Serum
D	Calciferols	0.1–0.8 μg	2–10[f]	Fat
E	Tocopherols	1–1.5 mg	5–10	Fat globules
K_2	Menaquinone	10–50 μg	100–1000	Fat

[a] Approximate recommended daily intake. Recommendations vary, e.g., with age; those for babies are excluded.
[b] Retinol equivalents: retinol + β-carotene/6.
[c] Niacin equivalents; includes 1/60 times tryptophan present in excess of its RDI value.
[d] B-vitamin.
[e] Specific folate-binding proteins.
[f] RDI greatly depends on exposure of the skin to sunlight.

is almost saturated with respect to air; however, it contains relatively far more carbon dioxide though in the form of bicarbonate (Section 2.2). The O_2 content of milk while in the udder is lower, i.e., about 1.5 $mg \cdot kg^{-1}$.

8. *Hormones:* Several hormones are present in trace quantities in milk. Examples are prolactin, somatotropin, and steroids.
9. *Somatic cells:* Together with all compounds they contain; see Subsection 2.7.1.

In addition to the compounds mentioned in the preceding text, milk contains numerous others. For instance, it contains about 3 mg ethanol per kg (per year, a cow produces about 20 g of ethanol, corresponding to 200 ml of wine).

2.6.2 Contaminants

In principle, the number of compounds that can enter milk by contamination is endless. There is much concern about compounds that may be harmful to the consumer because of their potential toxicity or mutagenicity. It has hardly ever happened that a compound, though harmful in principle, was found in such concentration in milk that it would produce a health hazard. Furthermore, investigations have concentrated on contaminants that cause undesirable effects during manufacture or storage of milk or milk products.

There are several pathways by which contaminants can gain entrance into milk; some are known to enter milk in more than one way:

1. *Illness of the cow:* For example, severe mastitis causes blood compounds and somatic cells to enter the milk (Subsection 2.7.1).
2. *Pharmaceuticals* (drugs) that have been administered to the cow: Antibiotics and sulfonamides are widely used; they are introduced into the udder to treat mastitis and can still be detected in milk 3 or 4 d after they have been administered. Antibiotics in milk may slow down the action of lactic acid bacteria used in the manufacture of fermented products. Several pharmaceuticals can enter the milk through the blood.
3. *Feed:* Many compounds can gain entrance into the milk through the feed, though the cow may act as a filter. Sometimes, substances are partly broken down first. The following list gives examples:
   Chlorinated hydrocarbons, such as several pesticides (DDT, aldrin, dieldrin); PCBs (polychlorinated biphenyls), which are widely used in materials; dioxins, which are potentially harmful to the consumer even in extremely low concentrations. Some of these components are toxic or carcinogenic, and occasionally too high a level (that is, higher than the standard, which normally has a safety factor of, say, 100) has been detected — for instance, in milk from cows fed large quantities of vegetables sprayed with pesticides. These substances are lipophilic and hence tend to accumulate in the fat.
   Other pesticides, herbicides, and fungicides such as phosphoric esters and carbamates. Most of these components are broken down by the cow.
   Mycotoxins may originate from molds growing on concentrates fed to cows. Particularly suspect are the harmful aflatoxins. In many countries, the feed has to comply with strict requirements.
   Heavy metals. Pb, Hg, and Cd are especially suspect, but toxic levels have virtually never been found in milk. Most heavy metals do not gain entrance into the milk, because the cow acts as a filter, unless extremely high quantities are fed.
   Radionuclides. See Subsection 2.6.3.
4. *Compounds that may enter the milk during milking and milk handling:*
   Pesticides can also gain entrance into the milk through the air, e.g., when aerosols with insecticides are used.
   Plasticizers from plastics or antioxidants from rubber (teat cup lining).
   Metals ions, especially Cu, may cause off-flavors via autoxidation of fat.

TABLE 2.19
Most Important Radionuclides that Can Occur in Milk

Radionuclide	Physical Half-Life	Biological Half-Life	Location in Milk
^{89}Sr	52 d	~50 yr	>80% in casein micelles, the rest in serum
^{90}Sr	28 yr	~50 yr	>80% in casein micelles, the rest in serum
^{131}I	8 d	~100 d	Serum (~2% in the fat)
^{137}Cs	33 yr	~30 d	Serum

Cleaning agents and disinfectants may cause off-flavors and decreased activity of starters. Small quantities of chloroform can be formed.

5. *Substances added on purpose.* Sometimes disinfectants are added to milk to arrive at a low colony count. This is, of course, adulteration. Determination of active chlorine may detect such adulteration. Addition of water is best checked by determining the freezing point.

2.6.3 Radionuclides

Radioactive isotopes of several elements are always present in milk (as in nearly all foods), but in minute quantities. It especially concerns ^{40}K. If feed or drinking water contaminated after radioactive fallout is ingested by the cow, part of the radionuclides will be secreted in the milk, despite the fact that the cow acts as a filter. For example, of the radioactive Sr ingested only a small part enters the milk, of ^{131}I much more.

Table 2.19 lists radionuclides that are the most harmful to the consumer and that may enter milk, with some particulars. ^{90}Sr is of great concern because the physical as well as the biological half-life times are long. The physical half-life refers to the period needed to reduce the radioactive emission by an isotope to half of its original level. The biological half-life refers to the time it takes to excrete half of the amount of a compound ingested by the body. The latter depends strongly on conditions. Table 2.19 refers to the half-life of the most tenaciously held pool of the element; a large part of it is often excreted more quickly. Furthermore, if appreciable accumulation of ^{90}Sr in the bone is to be prevented, it is a low ratio of ^{90}Sr to Ca that is important rather than a small quantity of ^{90}Sr taken up. ^{131}I is considered to be particularly hazardous for fairly short periods after serious fallout. Iodine accumulates in the thyroid gland. Atomic bomb testing especially leads to emission of radioactive Sr and I. Nuclear reactor accidents can cause fairly serious contamination by ^{137}Cs.

Strontium is distributed in milk in much the same way as Ca, but because $SrHPO_4$ is very poorly soluble, by far the greater part of the strontium in milk is in the colloidal phosphate. Cs behaves similar to K^+ and Na^+. Most iodine is found as dissolved iodide.

2.7 VARIABILITY

As briefly discussed in Section 1.4, fresh milk varies in composition and structure, and hence in properties. The causes are variation in genetic make up of the animal (especially species), its physiological condition (especially stage of lactation), and in environmental factors (especially feed). These aspects will be discussed here, with the emphasis on variation in milk composition.

The variation in composition is not precisely known, though numerous reports on the subject have been published. This is mainly because so many factors affect the composition, and because many variations are interdependent. Furthermore, the extent of variation in the factors affecting milk composition varies greatly with climate, management practices on the farm, breeding programs, etc. In other words, results obtained in one country or region may not apply elsewhere.

2.7.1 Sources of Variability

Species. Different mammals produce milk varying widely in composition. The available data, covering about 150 species, show that dry-matter content ranges from 8 to 65%, fat 0 to 53%, protein 1 to 19%, carbohydrates 0.1 to 10%, and ash 0.1 to 2.3%.

The only species raised specifically for milk production are hoofed animals, the most important of which (cow, zebu, buffalo, goat, and sheep) are ruminants. Table 2.20 gives an overview of the variation. For example, milk of buffalo and sheep contain much fat.

The composition of the main groups of components varies also. The main milk protein, for instance, may be casein or serum protein. Moreover, the relative proportions of the various caseins vary significantly among species, and the primary structures of the proteins as well. This often causes considerable variation in such properties as the stability and the rennetability of the casein micelles. Similar variations occur in serum protein composition. Other differences include fat composition. Goat and sheep milk fats have low contents of butyric acid residues but high contents of caproic, caprylic, and capric acid residues (capra = goat). Buffaloes' milk has comparatively large fat globules and a high colloidal phosphate content.

Breed. Often various subspecies can be distinguished within a species, but the breeds of cow are predominantly the result of selection by man. Various breeds have been obtained, according to the intended use (milk, meat, and draught power) and local conditions, such as climate, feed, terrain, and customs. This has led to a wide variability in milk yield and composition. However, the strongly directed selection over the last 100 years has decreased the variation in milk composition among typical dairy breeds. Some examples are given in Table 2.21. Variation in composition of, for instance, buffaloes' or goats' milk is much wider.

Individuals. Variation in milk composition among individual cows of one breed may be greater than that among breeds (see Figure 2.35). Differences in

TABLE 2.20
Approximate Average Composition (% w/w) of Milk of Some Milch Animals

Animal	Genus/Species	Dry Matter	Fat	Casein	Serum Protein	Carbohydrates	'Ash'
Donkey	*Equus asinus*	10.8	1.5	1.0	1.0	6.7	0.5
Horse	*Equus caballus*	11.0	1.7	1.3	1.2	6.2	0.5
Camel	*Camelus dromedarius*[a]	13.4	4.5	2.7	0.9	4.5	0.8
Reindeer	*Rangifer tarandus*	35	18.0	8.5	2.0	2.6	1.5
Cow	*Bos taurus*	12.8	3.9	2.7	0.6	4.6	0.7
Zebu[b]	*Bos indicus*	13.5	4.7	2.6	0.6	4.7	0.7
Yak	*Bos grunniens*	17.7	6.7	5.5		4.6	0.9
Buffalo[c]	*Bubalus bubalis*	17.2	7.4	3.3	0.6	4.8	0.8
Goat	*Capra hircus*	13.3	4.5	3.0	0.6	4.3	0.8
Sheep	*Ovis aries*	18.6	7.5	4.5	0.8	4.6	1.0

[a] Also *Camelus bactrianus* and crossbreeds.
[b] The zebu is often thought to be a subspecies of the cow.
[c] Also called swamp or water buffalo, or carabao.

composition in the milk of different quarters of the udder of one cow mostly are negligible, unless the cow suffers from mastitis.

Stage of Lactation. This is by far the most important physiological variable. Examples are given in Figure 2.32. These refer to the average composition of the

TABLE 2.21
Approximate Average Composition (% w/w) of the Milk of Some Breeds of Cow

Breed	Dry Matter	Fat	Crude Protein	Lactose	'Ash'
Black and white (in the Netherlands)	13.4	4.4	3.5	4.6	0.75
Black and white (other sources[a])	12.4	3.6	3.3	4.6	0.75
Brown Swiss	12.9	4.0	3.3	4.7	0.72
Jersey	15.1	5.3	4.0	4.9	0.72

[a] For example, Holsteins in the U.S. and Canada.

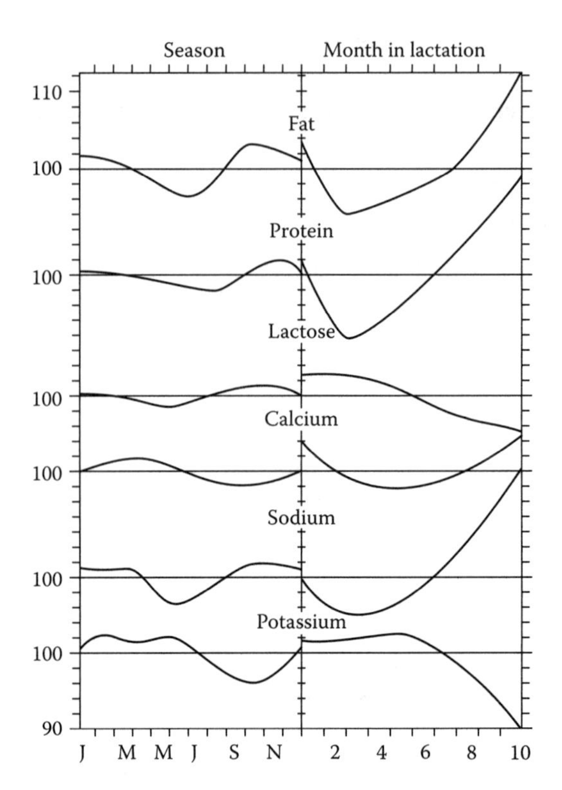

FIGURE 2.32 Content of some components in milk as a function of season and of lactation stage. The average content is put at 100%. One division corresponds to 2%. (Adapted from P. Walstra and R. Jenness, *Dairy Chemistry and Physics,* Wiley, New York, 1984.)

milk of ten cows from the same farm that calved in different seasons. Plotting the data as a function of seasons results in far smaller variations. This shows that lactation stage is the main variable, though it is difficult to separate the effect from that of other variables such as feeding regime and grazing. Phosphate content shows a trend similar to calcium; that of chloride parallels sodium. Prolonging lactation after 10 months may eventually cause milk composition to become very different.

Colostrum. Colostrum (or beestings) has a markedly different composition. An example is given in Figure 2.33, but the compositional change varies widely among cows. In colostrum, immunoglobulins make up a large proportion of the great amount of serum proteins. Immunoglobulin content in the first colostrum is on average about 7%. Colostrum gels when heated at, say, 80°C, because all of that serum protein becomes insoluble. Colostrum is also high in somatic cells, Cu and Fe. The pH can be as low as 6 and the acidity as high as 40°N.

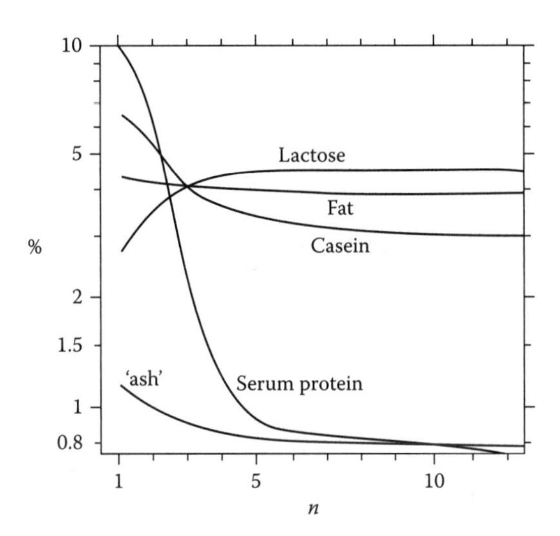

FIGURE 2.33 Example of composition (% w/w, log scale) of milk (colostrum) just after parturition. n = number of milking. Serum protein exclusive of proteose-peptone. (Adapted from P. Walstra and R. Jenness, *Dairy Chemistry and Physics,* Wiley, New York, 1984.)

Other Physiological Factors. Estrus and gestation do not have a great effect on milk composition, but they have on milk yield. Most of the milk components decrease slightly in concentration with age of the cow, and Na increases.

Mastitis. When pathogenic bacteria enter a mammary gland, a severe inflammation can result, called *mastitis*. It may cause a significant decrease in milk yield and a change in milk composition. The change in composition is comparable to, but stronger than, the change observed toward the end of a lactation period: the milk composition becomes closer to that of blood serum (see Table 1.4). The concentration of somatic cells, especially polymorphonuclear leukocytes, increases also.

The somatic cell count (number of cells per milliliter) is often taken as a measure of the severity of the mastitis and of the change in milk composition. However, the correlations are far from perfect. This is mainly because other factors also affect the cell count: for healthy cows it increases with age and with stage of lactation, and it varies significantly among individuals. But if the cell count of the milk from a quarter is over, say, $3 \cdot 10^6$, it is almost certain that this quarter suffers from severe mastitis.

It is thus difficult to establish relations between cell count and milk composition. Because of the nonlinearity of the relations, it makes also a difference whether one samples quarters, individual cows, or herds. Moreover, studies on the relations between cell count and milk composition often have been done on samples of foremilk (i.e., the first few milliliter obtained when milking a cow), which has a significantly lower cell count than the average milk. Finally, the common infrared methods to determine milk composition can be subject to considerable error for

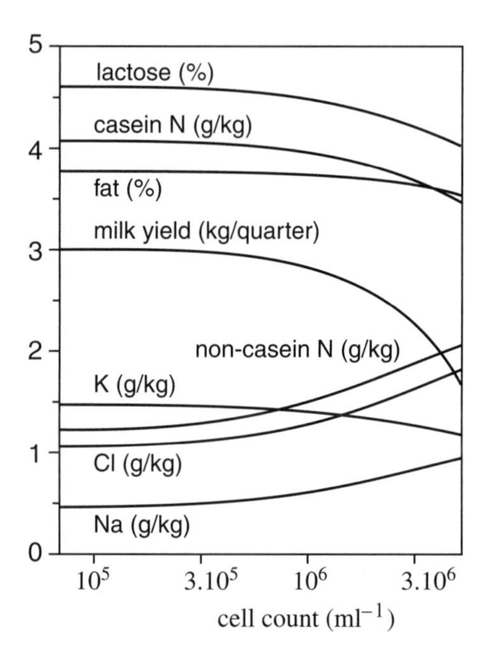

FIGURE 2.34 Approximate average changes in milk composition and milk yield (per milking) as a function of somatic cell count.

mastitic milk. In other words, reliable data are hard to come by. In Figure 2.34 we give an 'educated guess' of the average dependence of milk composition on its cell count. Other changes occur as well, especially in the activities of some enzymes. As mentioned in Subsection 2.5.2, plasmin activity is enhanced in mastitic milk.

The presence of mastitic milk rarely causes significant problems for the dairy manufacturer. Highly abnormal milk is generally not delivered to the dairy, and the mildly abnormal milk is strongly diluted by normal milk. On the other hand, mastitis can cause a significant loss to the farmer.

Feed. Environmental factors may greatly affect milk yield but have less influence on milk composition. The capability of mammals to maintain a constant composition of body fluids and cells (called homeostasis) is reflected in a qualitatively constant milk composition. Feed composition can affect the fat content of milk and especially its fat composition (Subsection 2.3.1.3). A low-protein diet causes the protein content of the milk to decrease somewhat, whereas a high-protein diet causes nonprotein N content to increase. Several minor components are strongly affected by the content in the feed (see also Subsection 2.6.2).

Other Environmental Factors. Climate has little effect on milk composition unless it is extreme, causing heat stress. All other kinds of stress, exhaustion, and housing are associated with a decrease in milk yield and usually a small increase in dry-matter content.

Milking. The shorter the time elapsed after the previous milking, the lower the milk yield and the higher the fat content will be. Hence, evening milk usually has a higher fat content than morning milk, the difference amounting to, say, 0.25% fat. During milking the fat content of milk increases (e.g., from 1 to 10%), but the difference varies markedly among cows. Incomplete milking thus can decrease the fat content of a milking, although not that of the milk on average. Short time intervals between milkings increase the susceptibility of the milk to lipolysis (Subsection 3.2.5). The somatic cell count also increases during milking.

Random Variations. Day-to-day fluctuations occur, especially in the fat content.

2.7.2 Nature of the Variation

Most of the figures and tables of Section 2.7 give examples of variation in milk composition. Among the main components, the widest variation usually occurs in fat content. Variation in protein is less and that in lactose and ash still less. This is illustrated in Figure 2.35. The composition of a main component can also vary, especially the fatty acid pattern of the milk fat and the ratio between minerals, for example, Na to K. Each individual protein is of constant composition except for genetic variants, but the ratio between them may vary somewhat; casein is relatively constant, but the proportions of the individual serum proteins are less constant, for example, immunoglobulins and serum albumin are variable. The so-called casein number, i.e., the percentage of the N that is present in casein, largely determines the yield of cheese per kilogram of milk protein and thus is an important variable.

The distribution of components among the physical fractions of milk, as well as the sizes of fat globules and casein micelles are also variable. The fat globule size varies considerably, with the volume–surface average d_{vs} ranging from 2.5 to 6 μm, which corresponds to a difference by a factor of 14 in average volume.

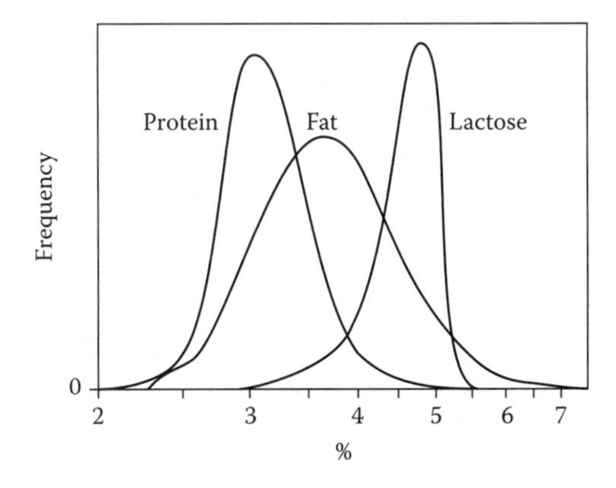

FIGURE 2.35 Frequency distributions of crude protein, fat, and lactose contents (logarithmic scale) of samples of milk from individual cows, taken throughout a year. (From P. Walstra and R. Jenness, *Dairy Chemistry and Physics,* Wiley, New York, 1984. With permission.)

Within a single milking of one cow, fat globules and casein micelles vary in composition.

All of these variations cause differences in physical properties. Examples are density (standard deviation for individual cow samples about 2 $kg \cdot m^{-3}$), titratable acidity (1 $mmol \cdot l^{-1}$), pH (0.04 unit), viscosity (5% relative), and refractive index of the fat (10^{-3} unit).

The extent of variation greatly depends on the sample population. The fat content of separate milkings of individual cows may range from 2 to 9%, but the range in fat content of milk as received at a dairy factory will be far smaller. Obviously, lots of milk that are each an average of a greater number of (hypothetical) sublots will generally show a smaller spread.

For the dairy manufacturer, differences among geographic regions (for example, due to a difference in breed of cow and farming plan) and the seasonal variation are most important. The latter variation strongly depends on calving pattern of the cows. In some regions (for example, in much of California), at all times of the year an equal number of calvings occurs. This means that any seasonal pattern in milk composition cannot be due to variation in stage of lactation. Another extreme is that all cows calve within a couple of weeks at the end of winter (for example, in much of New Zealand). This means a large variation in milk composition during the season, for by far the most part due to lactation stage; such variation can pose problems for the dairy manufacturer. In most countries, the situation is in-between, with milk production throughout the year but most cows calving somewhat before the most favorable season for milk production, be it the summer or the wet season. This would mean a moderate effect of lactation stage on seasonal variation. Figure 2.36 gives an example (for a temperate climate), and its shows that the variation is not very wide, i.e., not more than ± 8% relative. Figure 2.32 also provides information.

At a decrease of the milk yield as caused by external conditions (bad weather and stress), the fat and protein contents of the milk often tend to increase slightly.

2.7.2.1 Correlations among Variables

Correlations among variables often occur and some can qualitatively be explained. Milk is isotonic with blood, and the osmotic pressure is mainly determined by lactose and the dissolved salts. Consequently, if one of these is relatively low in concentration, the other will be high. When cows suffer from mastitis or are in an advanced stage of lactation, a larger amount of low-molar-mass blood components is leaking from the blood into the milk. The lactose content of the milk of these cows will be low because blood serum contains a greater amount of dissolved salts and far less sugar than milk does. Generally some mastitic cows are present, although not in large numbers, and the opposite (low content of dissolved salts) never occurs. As a result, the distribution of the lactose content of individual milkings will be negatively skewed (Figure 2.35). Similarly, the cell count frequency of quarters is positively skewed because only a few quarters suffer from mastitis, and hence have a high cell count.

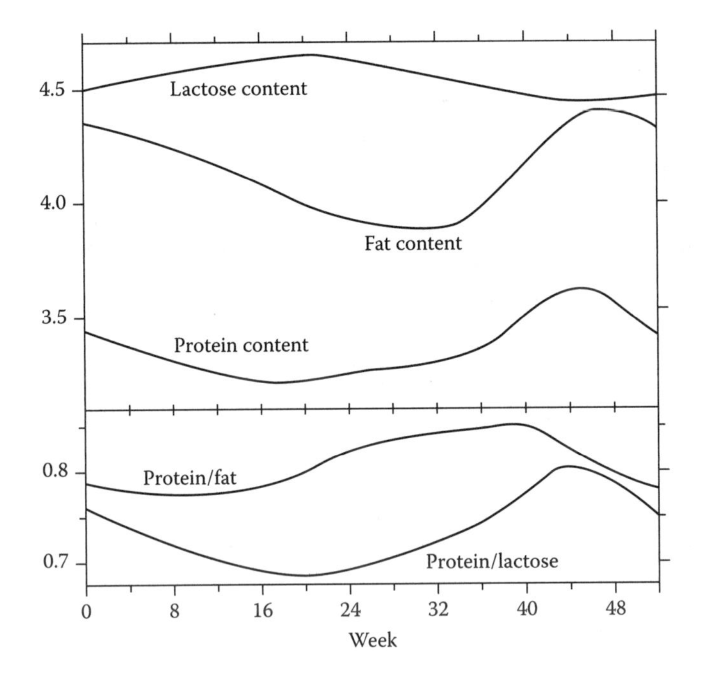

FIGURE 2.36 Examples of lactose, fat, and crude protein contents (% w/w) of milk delivered to a dairy throughout a year (January to December).

The negative correlation between Na and K concentrations (Figure 2.32) must be ascribed to the 'Na-K pump' in the plasmalemma of the lactating cell. The concentrations of undissolved salts and casein are positively correlated because the casein micelles contain colloidal calcium phosphate. Any part of a whole, of course, will be correlated with that whole, e.g., casein N with total N.

The composition of milk can also be correlated with the *genetic variant* of one of the milk proteins. Table 2.22 gives examples. It shows that cows with variant B for κ-casein produce milk with a higher protein content, of which a larger part is casein, of which in turn a larger part is κ-casein. This especially is true of homozygotic animals, that is, those with genotype BB, having only variant B in their milk. To be sure, the data refer to the milk on average, as there are, of course, other factors that affect the content. The genetic variant of κ-casein is also correlated with other contents in the milk. For example, cows with genotype AA tend to produce milk with a higher pH and a lower calcium content. All this shows that there is not always a causal connection between genetic variant and milk composition. Presumably, the loci of the genes that encode for the differences involved (genetic variant of a protein and other composition of the milk) are close together on a chromosome, so that the genes concerned may be coupled. The genetic variant of β-lactoglobulin appears to be correlated with the protein composition of milk rather than with the protein content.

TABLE 2.22
Genetic Variants of κ-casein and β-lactoglobulin, Their Frequency, and the Relation with Protein Content and Protein Composition of Milk

Genetic Variant[a]	Frequency (%)	Crude Protein (%)	Casein N/ Total N	κ-Casein N/ Casein N
κ-A	64	3.58	0.77	0.15
κ-A and B	32	3.67	0.78	0.16
κ-B	4	3.76	0.79	0.18
β-lg-A	19	3.68	0.77	0.17
β-lg-A and B	51	3.65	0.78	0.16
β-lg-B	30	3.69	0.79	0.16

[a] A cow of genotype AA produces variant A, genotype AB gives both variants.
Note: Samples of milk of 10,000 black and white cows in the Netherlands

2.7.3 Some Important Variables

Following are some examples of variability in milk products caused by variation in the milk:

1. Yields of product may vary. For example, butter yield depends on fat content of milk, cheese yield primarily on casein content, and yield of skim milk powder on solids-not-fat content.
2. Composition of most products depends on milk composition. In standardizing cheese milk the ratio of protein to fat is important, whereas for the composition of skim milk powder the protein-lactose ratio prevails (Figure 2.36). The fat content of skim milk and hence of skim milk powder depends on fat globule size (Figure 2.37 B).
3. The crystallization behavior of milk fat greatly depends on fat composition, which then affects the firmness of butter (often butter is firm in winter; Figure 2.37A). The seasonal effect especially involves feed, although considerable differences exist among cows.
4. Heat stability (Figure 2.37D) is an important factor in the manufacture of evaporated milk.
5. Fouling of heat exchangers due to deposition of protein may vary significantly with milk composition. The salt composition appears to have some effect, and this may be related to heat stability (point 4). A very high content of immunoglobulins, as in colostrum, causes severe fouling. (Moreover, milk that has turned somewhat sour due to bacterial growth causes far more fouling.)

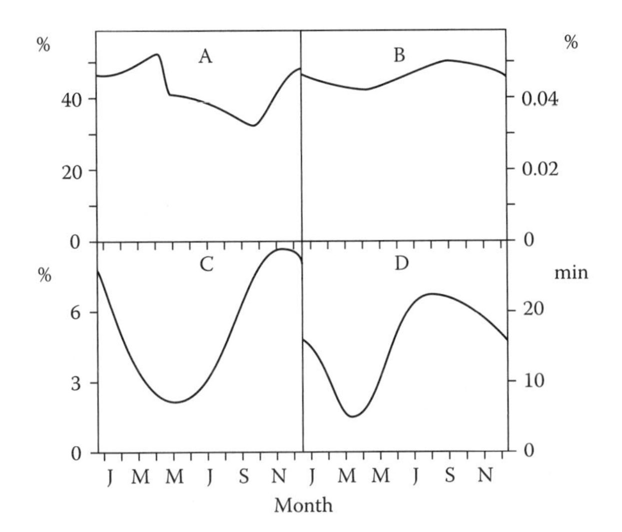

FIGURE 2.37 Examples of seasonal variation. (A) Firmness of butter (yield stress). (B) Fat content of separated milk. (C) Percentage of herd samples having a fat acidity >1.0 mEq per 100 g of fat. (D) Heat coagulation time of milk at 140°C. Approximate examples.

6. Rennetability mainly is a function of Ca^2 activity (Subsection 24.3.3).
7. Creaming of milk, especially rapid creaming due to the action of cold agglutinin, is highly variable. Agglutinin content of milk decreases markedly during lactation, but there are large differences among cows (see Subsection 3.2.4).
8. Factors that inhibit or stimulate the growth of microorganisms may vary. This concerns, for instance, the concentration of agglutinins (which probably decreases with increasing lactation stage).
9. In terms of the flavor of milk, the ratio of the contents of dissolved salts to lactose is important. The mass ratio 100[Cl]/[lactose] varies from 1.5 to 3; in extreme cases from 1.2 to 4.5. A ratio greater than 3 gives a salty taste. Proneness of milk to develop off-flavors also varies widely; lipase activity (Figure 2.37C) and autoxidation rate strongly vary among cows (variation by a factor of ten or more) and usually increase and decrease, respectively, with advancing lactation. Proneness of milk to develop "sunlight" flavor appears strongest in winter.
10. The color of milk, and especially that of butter and cheese, varies widely because of differences in the β-carotene content of the fat. This content strongly depends on feed (grass yields yellow and hay yields white fat), but also on the ability of the cow to convert β-carotene to vitamin A. This ability varies widely among individual cows. Jersey cows give yellow-, even orange-colored milk fat, whereas buffalo, sheep, and goat give virtually colorless milk fat.

Suggested Literature

More extensive treatment of many aspects discussed in this chapter: P. Walstra and R. Jenness, *Dairy Chemistry and Physics,* Wiley, New York, 1984, although parts are now somewhat outdated.

Extensive information on the chemistry and some physical and nutritional aspects of milk components: P.F. Fox, Ed., *Advanced Dairy Chemistry,* in three volumes: *Proteins,* 3rd ed., Kluwer Academic, New York, 2003; *Lipids,* 2nd ed., Chapman and Hall, London, 1995; *Lactose, Water, Salts and Vitamins,* 2nd ed., Chapman and Hall, London, 1997. Volume 1 also includes indigenous milk enzymes.

Another reference book: N.P. Wong, R. Jenness, M. Keeney, and E.H. Marth, Eds., *Fundamentals of Dairy Chemistry,* 3rd ed., Van Nostrand Reinholt, New York, 1988. It also contains a chapter on Nutritive value of dairy foods.

Nucleation, growth and network formation of fat crystals, including rheological properties of plastic fats: P. Walstra, *Physical Chemistry of Foods,* Dekker, New York, 2003.

Basic aspects of the chemistry of food components (also found in texts on organic chemistry and biochemistry): O.R. Fennema, Ed., *Food Chemistry,* 3rd ed., Dekker, New York, 1996.

Contaminants of milk: Monograph on residues and contaminants in milk and milk products, International Dairy Federation, Special Issue. Brussels, 1991.

3 Colloidal Particles of Milk

A glance back at Figure 1.1 and Table 1.2 shows that colloidal aspects must be of great importance. Milk is a dispersion and contains many particles of colloidal dimensions (say, between 10 nm and 100 μm in diameter), especially fat globules and casein micelles; together they make up 12 to 15% of the milk by volume. Moreover, small gas (especially air) bubbles may be incorporated during some processes.

The presence of these particles has several consequences:

- Some substances are present in separate particles. This means that interactions with compounds in the serum have to occur via the surface of the particles, which may act as a barrier. Moreover, compounds in one fat globule generally cannot reach another globule.
- The particles can be subject to various instabilities. The most important ones are illustrated in Figure 3.1. Most of these instabilities eventually lead to an inhomogeneous product with greatly altered properties.
- Several physical properties depend on the state of dispersion; for instance, turbidity and, hence, color and viscosity. If aggregation of particles occurs, this will further affect viscosity, and often a gel will eventually result.

Most of these points will be discussed in this chapter, with some emphasis on the second.

3.1 BASIC ASPECTS

In this section we briefly discuss some rudiments of colloid and surface science for the benefit of readers who are insufficiently acquainted with these aspects. The matter discussed will also be useful in some of the other chapters.

Two kinds of colloids can be distinguished:

- *Lyophobic colloids*. They are, in principle, unstable and all of the physical instabilities depicted in Figure 3.1 will take place, although the rate of change may be extremely small in some conditions. Lyophobic particles constitute a true phase. They have a phase surface onto which substances can adsorb. In milk, the fat globules, and any gas bubbles present, are of the lyophobic type. Fat crystals in oil also can be considered as lyophobic colloids.

- *Lyophilic colloids.* These are, in principle, stable and can form spontaneously. They have no phase surface and can only undergo physical changes if conditions change, e.g., pH. In milk, the term refers to the casein micelles. (*Note:* Casein micelles are not chemically stable: the chemical composition changes with time, which may eventually lead to physical instability, although the rate may be extremely slow.)

The terms lyophobic and lyophilic imply 'solvent-hating' and 'solvent-loving,' respectively. For particles in milk serum, where the continuous phase is aqueous, lyophobic and lyophilic thus imply that the colloidal particles are hydrophobic and hydrophilic, respectively.

Surface science is about the properties of phase surfaces. It concerns forces that act in the direction of the surface. These forces are greatly modified by the adsorption of substances onto the surface.

Colloid science is about the interaction forces acting between particles, which determine, for instance, whether particles will aggregate or not. These forces act in the direction perpendicular to the surface. In the case of a lyophobic surface, adsorption of substances can greatly modify the forces.

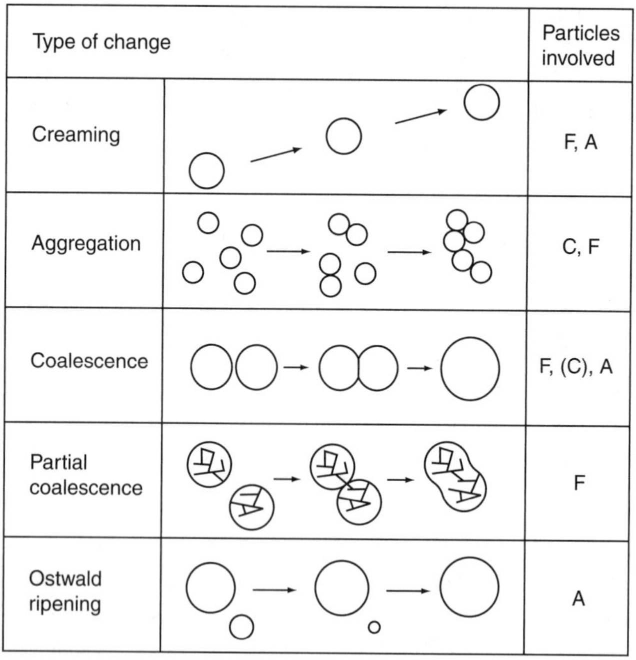

Type of change		Particles involved
Creaming		F, A
Aggregation		C, F
Coalescence		F, (C), A
Partial coalescence		F
Ostwald ripening		A

FIGURE 3.1 Illustration of the various changes that can occur with colloidal particles. A is air bubbles (diameter, e.g., 50 μm), C casein micelles (e.g., 0.1 μm), and F fat globules (e.g., 3 μm). The solid lines in the fat globules (partial coalescence) denote fat crystals.

It may further be noted that creaming, or, more generally, sedimentation, is not due to colloidal interaction, but due to an external force caused by gravity or centrifugation. Also, lyophilic particles, including casein micelles, are subject to sedimentation, provided that the centrifugal acceleration is large enough. Finally, all of the instabilities, except Ostwald ripening, are arrested if the particles are immobilized, which can be realized by giving the liquid a yield stress (see Subsection 4.7.1).

3.1.1 Surface Phenomena

Various types of interfaces can exist between two phases, the main ones being gas–solid, gas–liquid, liquid–solid, and liquid–liquid. If one of the phases is a gas (mostly air), one usually speaks of a *surface*; in the other cases the term would be *interface*, but these words are often considered to be interchangeable. More important is the distinction between a solid interface, where one of the phases is a solid, and a fluid interface between two fluids (gas–liquid or liquid–liquid). A solid interface is rigid; a fluid interface can be deformed.

3.1.1.1 Surface Tension

An interface between two phases contains an excess of free energy, which is proportional to the interfacial area. Consequently, the interface will try to become as small as possible, to minimize the interfacial free energy. This then means that one has to apply an external force to enlarge the interfacial area. The reaction force in the interface is attractive and acts in the direction of the interface. If the interface is fluid, the force can be measured, and the force per unit length is called the surface or interfacial tension: symbol γ, units $N{\cdot}m^{-1}$. (γ_{OW} is the tension between oil and water, γ_{AS} between air and a solid, etc.; cf. Figure 3.4.). A solid also has a surface tension, but it cannot be measured.

The magnitude of γ depends on the composition of the two phases. Some examples are given in Table 3.1. The interfacial tension also depends on temperature, and it nearly always decreases with increasing temperature.

3.1.1.2 Adsorption

Some molecules in a solution that is in contact with a phase surface can accumulate at this surface, forming a monolayer. This is called *adsorption* (to be distinguished from absorption, when a substance is taken up in a material). A substance that does adsorb is called a *surfactant*. It adsorbs because that gives a lower surface free energy — hence, a lower surface tension. Examples are in Figure 3.2a. It is seen that the decrease in γ depends on the surfactant concentration left in solution after equilibrium has been reached. The lower the value of c_{eq} at which a given decrease in γ is obtained, the higher the *surface activity* of the surfactant. Substances in a gas phase, such as water in air, can also adsorb onto a surface, and the same relations apply.

TABLE 3.1
Values of the Interfacial Tension (γ) of Some Systems

Between Phases	γ
Water–air, 0°C	76
Water–air, 25°C	72
Water–air, 60°C	66
Na laurate[a]–air	43
Protein solution–air	~50
Oil[b]–air	35
Oil–water	30
Protein solution–oil	~10
Ice–water, 0°C	25
Fat crystal[c]–water	31
Milk fat globule–milk serum	~1.5[d]
Fat crystal–oil	4

Note: Approximate values in $mN \cdot m^{-1}$ at 25°C, unless stated otherwise. The values involving a solid interface are rough estimates.

[a] 0.02 *M* aqueous solution.
[b] Pure triglyceride oil.
[c] Pure triglyceride crystal.
[d] Measured values range from 0.9 to 2.5.

An important variable is the *surface load*, Γ, i.e., the amount (in moles or in mass units) of adsorbed material per unit surface area. For $\Gamma = 0$, $\gamma = \gamma_0$, the value for a clean interface. At a relatively high surfactant concentration, Γ reaches a plateau value, i.e., a packed monolayer; this corresponds with the concentration

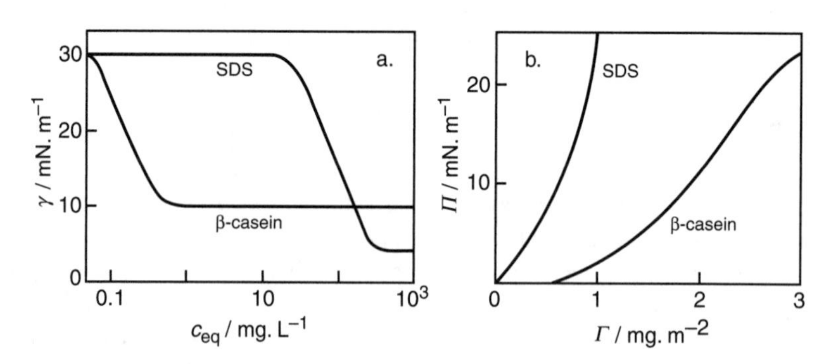

FIGURE 3.2 Adsorption of β-casein and SDS (sodium dodecyl sulfate) at an oil–water interface. (a) Interfacial tension γ as a function of equilibrium surfactant concentration c_{eq}. (b) Relation between surface pressure $\Pi = \gamma_0 - \gamma$ and surface load Γ. Approximate results.

at which γ reaches a plateau value. The magnitude of $\Gamma_{plateau}$ varies among surfactants — for the most part, between 1 and 4 $mg{\cdot}m^{-2}$. The relation between Γ and surfactant concentration is called an *adsorption isotherm.*

Each surfactant has at equilibrium (and at a given temperature) a fixed relation between the magnitude of Γ, and the decrease of γ; the latter is called the *surface pressure* $\Pi = \gamma_0 - \gamma$. Examples of the relation are in Figure 3.2b. Values for Π at $\Gamma_{plateau}$ can be derived from Table 3.1: compare, e.g., γ for water–air and Na laurate solution–air, yielding $\Pi = 19$ $mN{\cdot}m^{-1}$. The maximum value of Π varies among surfactants; for many surfactants (though not for all) the value is roughly the same for air–water and oil–water interfaces.

The *rate of adsorption* of a surfactant depends primarily on its concentration. It will often be transported to a surface by diffusion. If its concentration is c and the surface load to be obtained Γ, a layer adjacent to the surface of thickness Γ/c will suffice to provide the surfactant. Application of a simple equation, i.e., $L^2 = D\, t_{0.5}$, where D is diffusion coefficient and $t_{0.5}$ the time needed to halve a concentration difference over a distance L, now leads to

$$t_{0.5} = \Gamma^2/Dc^2 \tag{3.1}$$

In aqueous solutions D is generally on the order of 10^{-10} $m^2{\cdot}s^{-1}$. In milk, the most abundant surfactant is the serum proteins, at a concentration of about 6 $kg{\cdot}m^{-3}$, giving Γ values of about 3 $mg{\cdot}m^{-2}$. Application of Equation 3.1 then leads to $t_{0.5} \approx 2.5$ ms. Adsorption will be complete in 10 or 20 times $t_{0.5}$, i.e., well within 1 s. If the surfactant concentration is lower, adsorption will take (much) longer, but then stirring will markedly enhance adsorption rate. In other words, adsorption will nearly always be fast in practice.

3.1.1.3 Surfactants

Surfactants come in two main types, small-molecule amphiphilic compounds, hereafter called amphiphiles for short, and polymers, especially proteins.

1. *Amphiphiles.* These substances are surface active because they are amphiphilic. The molecules have a hydrophobic tail, generally of a fatty acid, and a hydrophilic head group. Most amphiphiles are poorly soluble in water, but they tend to form micelles above a certain concentration, the critical micellization concentration (CMC). At the CMC also $\gamma_{plateau}$ is reached. The longer and the more saturated the fatty acid chain, the lower the CMC, the higher the surface activity, and the lower the HLB number. The latter variable is the *hydrophile-lipophile balance* of the molecules. The surfactant has equal solubility in oil and water at an HLB value arbitrarily set at 7; at lower values the surfactant is more soluble in oil than in water, and vice versa.

   Some important amphiphiles are monoglycerides and Spans (i.e., fatty acid sorbitan esters), which have low to medium HLB-values.

Tweens are derived from Spans by attaching a few polyoxyethylene chains to the sorbitan moiety (see Figure 3.3); they have higher HLB numbers. Besides these neutral or nonionic surfactants, ionic ones exist. These include sodium soaps, lactic acid esters, and SDS (sodium dodecyl sulfate); they are anionic and have high HLB numbers. Phospholipids are also ionic surfactants (most are zwitterionic), but they have a very low solubility in water. Except SDS, all of these surfactants may be used in foods. Some are used in ice cream (Section 17.3).

2. *Proteins*. These are also amphiphilic molecules, but the main cause for their high surface activity is the large size. They tend to change conformation upon adsorption, and globular proteins often become denatured to some extent. For instance, most enzymes become irreversibly inactivated after adsorption onto an oil–water interface. The conformational change takes a much longer time than needed for adsorption, ranging, e.g., from 10 s (β-casein) to 15 min (β-lactoglobulin). Consequently, it takes some time before an equilibrium γ value is reached. Figure 3.3 illustrates how molecules can adsorb onto an interface. Micellar casein and serum proteins are adsorbed onto fat globules during homogenization; see Section 9.5.

Figure 3.2 shows considerable differences between the surface properties of a protein and an amphiphile, and these differences hold in general in a qualitative sense. The protein is much more surface active than the amphiphile, although even for the latter the concentration at which plateau values are reached is as low as 0.02%; for the protein it is 0.00005%. A protein will hardly desorb from an interface when the solution is diluted. On the other hand, the amphiphile can give a lower γ value than a protein. This implies that a protein will be displaced from an interface by an amphiphile if the latter's concentration is high enough. Finally, as shown in Figure 3.2b, the surface load of a protein needs to be higher than that of an amphiphile to obtain a significant decrease of γ.

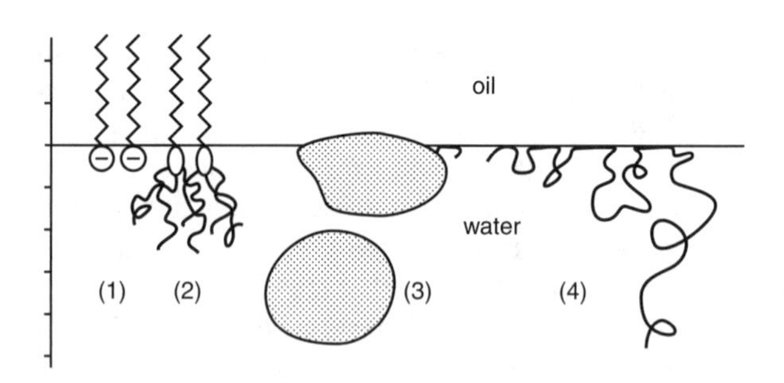

FIGURE 3.3 Mode of adsorption of some surfactants at an oil–water interface. At left is a scale of nanometers. (1) a soap, (2) a Tween, and (3) a small globular protein; for comparison, a molecule in solution is shown: (4) β-casein. Highly schematic.

3.1.1.4 Curved Interfaces

Consider a spherical drop or bubble. The surface tension tries to minimize the surface area, so tries to compress the drop. This means that the pressure inside the drop is increased. The magnitude of this *Laplace pressure* is given (for a sphere) by

$$p_{\mathrm{La}} = 4\gamma / d \tag{3.2}$$

where d is the sphere diameter.

Quite in general, the pressure at the concave side of a curved surface is higher than that at the convex side, the more so for a stronger curvature (smaller d) and a higher γ value. Hence, it is quite difficult to deform a small drop. This is an essential aspect in the breakup of fat globules during homogenization (Section 9.3). For a small protein-covered emulsion drop of $d = 1$ μm and $\gamma = 10$ mN·m^{-1}, the Laplace pressure will be $4 \cdot 10^4$ Pa, or 0.4 bar.

Another consequence is that the solubility (s) of a material inside a small particle is enhanced relative to the solubility at a flat interface (s_∞). This is given by the *Kelvin equation*, which reads

$$\frac{s(r)}{s_\infty} = \exp\left(\frac{2\gamma V_{\mathrm{D}}}{rRT}\right) \tag{3.3}$$

where r is radius, V_{D} the molar volume of the disperse phase (in m^3 per mole), and $RT \approx 2.5$ kJ·mol^{-1}. The Kelvin equation holds for solids, liquids, and gases, but the increased solubility is most readily envisaged for a gas bubble: inside the bubble the pressure is increased, and the solubility of a gas is proportional to its pressure. To give an example, for a small air bubble of 10 μm diameter in milk the solubility will be increased by about 20%.

The consequence of the dependence of solubility on curvature is *Ostwald ripening*, illustrated in Figure 3.1. The material in small particles gradually diffuses through the continuous medium toward larger ones. The latter grow and the former disappear. This is especially prominent in foams (where the phenomenon is often called disproportionation) because air is well soluble in water. Significant Ostwald ripening may occur in minutes. In oil-in-water emulsions it is not observed: the solubility of triglyceride oil in water is negligible, hence the rate of oil diffusion is negligible. Water-in-oil emulsions can show Ostwald ripening, as water has a small, but significant, solubility in oil; see Appendix, Table A.4.

By the use of protein as a surfactant, the process can be retarded. As a bubble (or a drop) decreases in size, the value of Γ increases; this causes γ to decrease, as shown in Figure 3.2b; this causes a decrease in Laplace pressure; and this retards Ostwald ripening. However, the excess surfactant in the surface layer will desorb, and the original (equilibrium) γ value will be restored. The latter will happen rapidly if the surfactant is an amphiphile, but proteins desorb sluggishly. In this way proteins can slow down the ripening process.

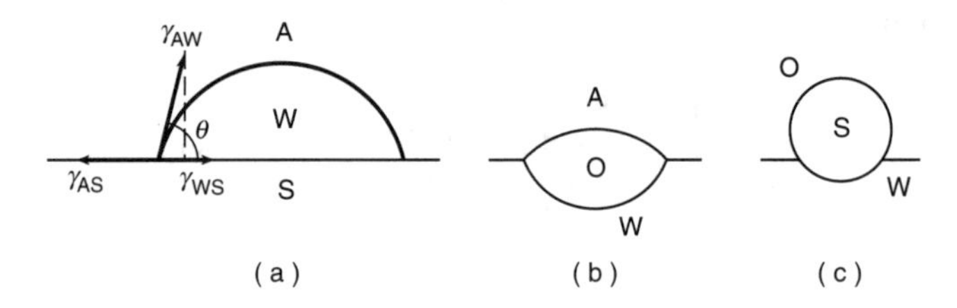

FIGURE 3.4 Contact angles (θ). Examples of three-phase systems: A = air, O = oil, S = solid, and W = water. In (a) the relation $\gamma_{AS} = \gamma_{WS} + \gamma_{AW}\cos\theta$ holds.

3.1.1.5 Contact Angles

Where three phases meet, the angles that the interfaces make with each other are determined by the three interfacial tensions. This is illustrated in Figure 3.4. If one phase is a solid, the geometry is characterized by one contact angle, which is by convention measured in the densest fluid phase. This angle is then given by the *Young equation*. For the case depicted in Figure 3.4a it reads

$$\cos\theta = \frac{\gamma_{AS} - \gamma_{WS}}{\gamma_{AW}} \tag{3.4}$$

If $(\gamma_{AS} - \gamma_{WS}) > \gamma_{AW}$ the equation would yield $\cos\theta > 1$, which is clearly impossible. In this case the water will spontaneously spread over the solid because that situation gives the lowest interfacial free energy. In the case of Figure 3.4b, the oil will spread over the water surface if $(\gamma_{AO} + \gamma_{OW}) < \gamma_{AW}$. See further Subsection 3.2.3.

For the case of Figure 3.4a, the condition $(\gamma_{AS} + \gamma_{AW}) < \gamma_{WS}$ would imply $\theta >$ 180°, which implies that the liquid cannot wet the solid. A similar argument applies in the case depicted in frame (c): for $\theta >$ 180°, the solid particle will not become attached to the interface and stay in the oil phase. Fat crystals tend to adsorb at an oil–water interface, about as depicted in the figure, with $\theta \approx$ 150° (as measured in the aqueous phase). Fat globules can adsorb onto an A-W interface. Particles larger than, say, 20 nm, then, are very tenaciously held at the interface. The contact angle also is an important variable in wetting phenomena — for instance, in cleaning (removal of fat from a solid surface) and in the dispersion of powders (see Subsection 20.4.5).

Under some conditions, casein micelles can become adsorbed onto oil–water and air–water interfaces, but they cannot be considered solid particles and have no phase boundary. Hence, the reasoning given here does not apply. See further Section 9.5.

3.1.1.6 Functions of Surfactants

The presence of surfactants in a system can have several different effects. Hence, surfactants are applied to fulfill a number of functions. Some changes leading to important effects are briefly discussed in the following text:

1. *Laplace pressure*. Addition of a surfactant generally lowers surface tension; hence, the Laplace pressure. This facilitates the deformation of fluid particles and thereby the formation of small particles during emulsion and foam formation.
2. *Ostwald ripening*. The rate of Ostwald ripening is decreased because of the lowering of interfacial tension and, in the case of proteins as surfactant, because of the mentioned reluctance of proteins to become desorbed.
3. *Contact angle*. Surfactants affect contact angles and thereby the adhesion of particles at an interface and wetting phenomena, including the propensity of a cleaning liquid to remove dirt from a solid surface. The contact angle can also affect the occurrence or rate of partial coalescence (Subsection 3.2.2.2).
4. *Surface tension gradients*. The surface tension at a clean surface is the same everywhere. If a surfactant is present, the surface tension can, in principle, vary from place to place. This will happen when liquid flows along the surface: it will sweep surfactant molecules downstream, as depicted in Figure 3.5a, which causes a surface tension gradient to be formed. This gradient ($d\gamma/dx$, where x is distance) exerts a stress on the liquid. The flowing liquid exerts a stress on the surface of magnitude viscosity × shear rate. If the surface tension gradient can be large enough, these stresses will become equal and opposite, which implies that the surface is arrested: it does not move in a tangential direction. This may be the most important property of surfactants in technology. It is all that makes the formation of foams and emulsions possible. Without surfactant, the liquid will flow very rapidly away from the gap between two bubbles (or two drops), as depicted in Figure 3.5b, because the A/W surface cannot offer any resistance; this will generally

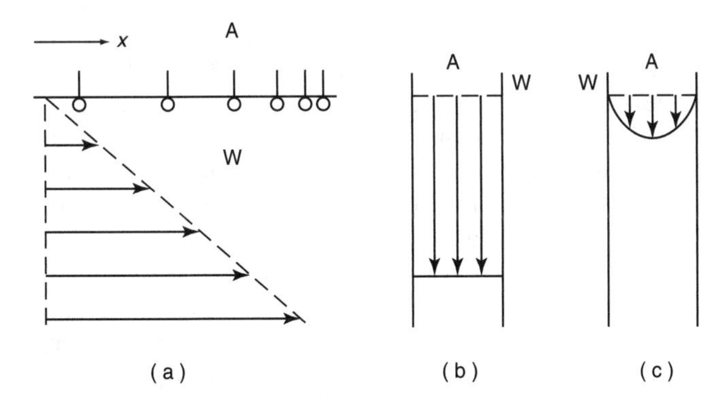

FIGURE 3.5 Surface tension gradients at an A/W surface induced by flow. (a) Shear flow along the surface inducing a gradient of surfactant molecules. (b) Drainage of water from a vertical film in the absence of surfactant. (c), Same, in the presence of surfactant.

cause immediate coalescence of the bubbles. If a surfactant is present, the drainage will be very much slowed down, as depicted in Frame (c).

5. *Colloidal interaction.* The adsorption of a surfactant onto particles can greatly affect — in most cases increase — the repulsive colloidal interaction forces between these particles (as will be discussed in Subsection 3.1.2), and thereby determine whether they will aggregate or not. A main function of surfactants is to prevent particle aggregation.
6. *Coalescence.* Another main function of surfactants is to prevent coalescence of emulsion droplets and foam bubbles. In most situations, this function depends for a considerable part, though not exclusively, on the colloidal repulsion provided. Surfactants greatly vary in their effectivity to prevent coalescence. Some may even promote it under some conditions. For instance, small-molecule surfactants can displace proteins from an interface and may then induce instability. This is applied in the making of ice cream (Section 17.3).
7. *Micellization.* As mentioned, many small-molecule surfactants form micelles in water above the CMC. These micelles can take up, i.e., harbor in their interior, lipid molecules. This is an important aspect of detergency.

3.1.2 Colloidal Interactions

Colloid scientists consider the free energy V needed to bring two particles from infinite separation distance to some close distance h. If the energy involved is positive, we have net repulsion between the particles, if it is negative, net attraction. The interaction free energy is usually given in units of kT (about $4 \cdot 10^{-21}$ J) because that is a measure of the average kinetic energy of particles. For $V < kT$, the interaction energy often can be neglected.

To illustrate the consequences of the interactions, we will consider curve 2 in Figure 3.6. Coming from a large mutual distance, the particles will probably arrive near C, which is called a secondary minimum in the curve. In the example, the interaction free energy in the minimum is about −3 times kT, which means that the particles have a tendency to stay together at that distance; i.e., they are aggregated. Occasionally, the particles in an aggregate may diffuse away from each other, and the chance that they do so is smaller as the secondary minimum is deeper. They may also diffuse further toward each other, but to come very close, they have to pass over the maximum near B. In the figure, the height of the maximum is about 10 kT, which means that the probability to do so is small (e.g., once in 1000 encounters between two particles). If so, they arrive in the primary minimum A, and this is often deep enough to prevent disaggregation. At very small separation — say, $h < 0.5$ nm — there is always strong repulsion between particles (*hard-core repulsion*). Notice also that the distance over which the colloidal interactions are important is quite small, rarely above 20 nm, which is a small percentage of the particle diameter.

The shape of the interaction curve will thus determine whether the particles will aggregate and whether the aggregation is reversible. In several situations,

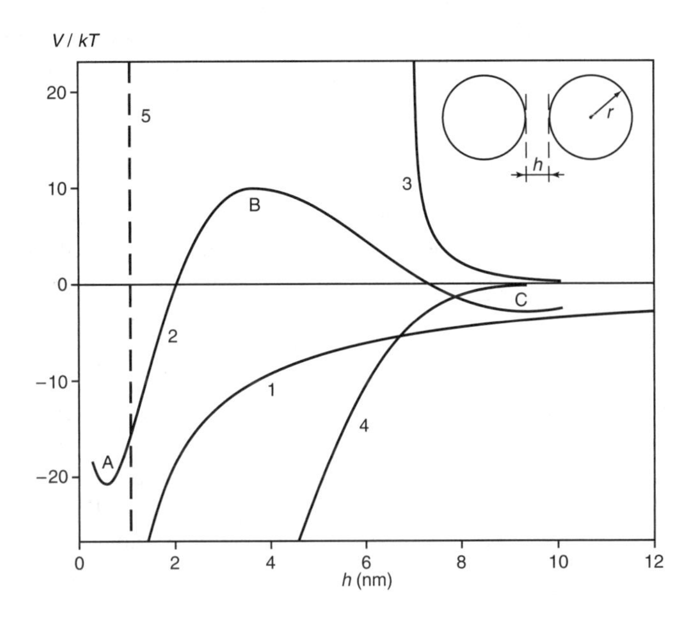

FIGURE 3.6 Examples of the interaction free energy V between two particles as a function of their separation distance h. Curve (1) van der Waals attraction; (2) DLVO-interaction; (3) steric repulsion; (4) depletion interaction. See the text for further explanation. The broken line 5 gives the thickness of the electrical double layer ($1/\kappa$) in milk. The insert shows the geometry considered.

the curve can be calculated from theory and measurable parameters. In the classical DLVO theory of colloid stability (named after Deryagin, Landau, Verwey, and Overbeek), van der Waals attraction and electrostatic repulsion are considered. Moreover, other interactions come into play. For the colloidal particles in milk and milk products, such a calculation is generally not possible with sufficient accuracy. Consequently, we will merely give some qualitative relations. This may help the reader in understanding the factors of importance in colloidal stability.

3.1.2.1 Van der Waals Attraction

The attractive van der Waals forces act between all molecules and particles. For two equal particles of radius r dispersed in another medium, the attractive free energy is given by

$$V_{\mathrm{vdW}} = -\frac{Ar}{12h} \tag{3.5}$$

provided $h << r$. The Hamaker constant A depends on the material of the particles and of the medium. For milk fat globules in milk plasma, $A \approx 0.75\ kT$. Curve 1 in Figure 3.6 gives an example. The aggregation of fat crystals in an oil, as discussed in Subsection 2.3.5.7, is caused by van der Waals attraction; there is no repulsive force in this case, except hard-core repulsion.

3.1.2.2 Electrostatic Repulsion

Particles in an aqueous medium nearly always bear an electric charge. This leads to an electric potential at the surface, ψ_0. In milk products, the absolute values of the *surface potential* are generally below 25 mV. Examples are given in Figure 3.7: it is seen that these potentials are negative at physiological pH and become zero at the isoelectric pH. (Actually, the figure gives electrokinetic or zeta potentials, but these are often considered to be representative for ψ_0.)

Due to the negative surface charge, *counterions* (in this case, cations) accumulate near the surface and *co-ions* (anions) tend to stay away. At some distance from the surface the charge is neutralized and the potential reduced to zero. The layer over which this occurs is called the *electric double layer.* (Particles moving through the liquid take counterions with them, but water molecules can freely diffuse in and out of the double layer.) The potential decay with distance h from the surface is given by

$$\theta = \psi_0 \, e^{-\kappa h} \tag{3.6a}$$

and the electric shielding parameter κ is given by

$$1/\kappa \approx 0.30/\sqrt{I} \tag{3.6b}$$

where $1/\kappa$ is in nanometers and I in moles per liter. See Subsection 2.2.2, Equation 2.3, for the ionic strength I. $1/\kappa$ is called the Debye length or the nominal *thickness of the*

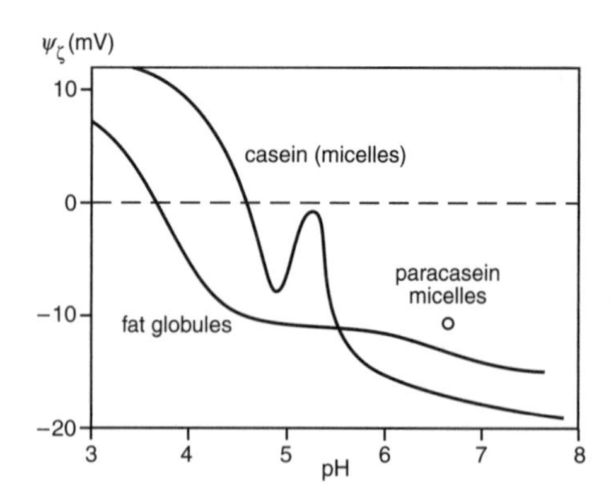

FIGURE 3.7 Zeta potential (ψ_ζ) of particles in milk at various pH. Approximate results at room temperature. (Adapted from P. Walstra and R. Jenness, *Dairy Chemistry and Physics*, Wiley, New York, 1984.)

electric double layer. For $h = 1/\kappa$, the potential is reduced to $e^{-1} = 0.37$ times ψ_0. At $h = 3/\kappa$, the reduction is to $e^{-3} = 0.05$, which generally means that the potential is negligible, at least in milk. For values of

$I =$	1	10	100	1000	millimolar
$1/\kappa \approx$	10	3	1	0.3	nm

In milk $I \approx 73$ mmol/l (Subsection 2.2.2), hence $1/\kappa \approx 1.1$ nm.

When charged particles come close to each other, their double layers overlap, which leads to *repulsion.* We will not give the full theory but merely an approximate dependence of repulsion on surface potential and distance:

$$V_{\mathrm{el}} \propto \psi_0^2 e^{-\kappa h} \tag{3.7}$$

An example of the sum $V_{\mathrm{vdW}} + V_{\mathrm{el}}$ as a function of h is given in Figure 3.6, curve 2. If the surface potential is decreased (e.g., by decreasing the pH) or the ionic strength increased, the maximum at B will become smaller or even disappear, and the particles would aggregate. For the opposite changes, the maximum would increase and the secondary minimum at C disappear, providing stability to aggregation. In practice, pH and ionic strength are thus the important variables.

3.1.2.3 Steric Repulsion

When several polymer chains protrude from a surface, forming a *hairy layer,* repulsion can occur if two particles come close enough for these layers to overlap. The polymer concentration then is locally increased, causing an increased osmotic pressure. To minimize the osmotic pressure, solvent is sucked into the gap, which drives the particles apart. Hence, a repulsive force is acting, which can be very strong if the polymer chain density in the layer is high, the layer is several nanometers thick, and the solvent quality is good. (In a 'good' solvent, solvent molecules preferentially closely surround the solute molecules, or — in the present case — the protruding chains.) An example of steric repulsion by a fairly thin layer is shown in Figure 3.6, curve 3, and it is seen that the repulsive force very steeply increases with decreasing interparticle distance.

In milk products, the polymer chain is likely to be (part of) a protein. Such chains nearly always are charged, and if the charge is large enough, this implies a good solvent quality. The repulsion can also involve an electrostatic component, and one may speak of *electrosteric repulsion.*

3.1.2.4 Depletion Interaction

Besides polymer chains protruding from a surface, polymer molecules in solution can affect colloidal interaction. Consider a liquid dispersion, e.g., an emulsion, which also contains some dissolved polymer, say, xanthan. The random-coil xanthan molecules, which may have a radius of about 30 nm, cannot come closer to the surface of the emulsion droplets than about that radius; hence, a layer of liquid is depleted of polymer or, in other words, it is not available as a solvent for the polymer. This is analogous to the situation depicted in Figure 10.2, albeit at a different scale. The osmotic pressure of the polymer solution is thus increased. If now the droplets aggregate, the depletion layers overlap, more solvent comes available for the xanthan, and the osmotic pressure decreases. There is thus a driving force for aggregation. An example is given in Figure 3.6, curve 4. If the effect is strong enough, as in the figure, aggregation will occur.

The effect can occur in milk when a suitable polysaccharide is added in sufficient concentration. The casein micelles cause depletion interaction between the fat globules, but the effect is too weak to cause their aggregation.

3.1.2.5 Other Interactions

Especially if the particles are covered by a hairy layer, other kinds of attractive interactions may occur. This is because the steric repulsion does not prevent contact between protruding chains. If the surfaces or the hairs contain distinct groups of positive and negative charges, electrostatic attraction may cause aggregation. Even if all charges are negative, bridging may occur due to the addition of a positively charged polymer or, in some cases, by divalent cations, e.g., Ca^{2+}. Proteinaceous particles may also aggregate during heat treatment because of chemical cross-links formed between exposed side groups of the proteins (see Subsection 7.2.2). By and large, protein-covered emulsion droplets aggregate under conditions (e.g., pH) where the protein itself is insoluble.

3.1.3 Aggregation

Aggregation occurs if particles being close together remain in that state for a much longer time than they would do in the absence of attractive forces between them. The terms flocculation and coagulation are also used, in which the former especially refers to weak, i.e., reversible, aggregation and the latter to irreversible aggregation.

Aggregation is of importance because the aggregates have properties that differ from those of the individual particles, such as faster sedimentation or causing an increase of viscosity. Aggregation is often needed for drops to coalesce, as drops generally must stay close together for a rather long time for coalescence to occur. Moreover, aggregation of particles can lead to gel formation.

3.1.3.1 Aggregation Rate

Particles that move randomly by Brownian motion will frequently encounter each other and may then aggregate. If so, the rate of this *perikinetic aggregation*, which is the decrease in the number concentration of particles N with time, i.e., $-dN/dt$, is given by

$$J_{\text{peri}} = \frac{4kT\,N^2}{3\eta W} \tag{3.8}$$

where kT has its normal meaning, η is the viscosity of the continuous liquid, and W is the stability factor. If nothing hinders aggregation $W = 1$. W, then, is the factor by which the aggregation is slowed down, and the main reason for $W > 1$ is colloidal repulsion, as discussed in Subsection 3.1.2.

The time needed for the value of N to be halved is given by

$$t_{0.5} = \frac{\pi d^3 \eta W}{8kT\varphi} \tag{3.9}$$

where φ is the volume fraction of the particles and d their diameter. Assuming $W = 1$, this results for the casein micelles in milk in $t_{0.5} \approx 0.2$ s, and for the fat globules, 3 min.

Particles can also encounter each other due to agitation or, more precisely, by a velocity gradient Ψ (expressed in s^{-1}) in the liquid. The rate of such *orthokinetic aggregation* then is given by

$$J_{\text{ortho}} = \frac{4\varphi N \Psi}{\pi W} \tag{3.10}$$

Assuming the value of W to be the same in both cases (which is not always true), the ratio between ortho and perikinetic aggregation will be

$$\frac{J_{\text{ortho}}}{J_{\text{peri}}} = \frac{d^3 \eta \psi}{2kT} \tag{3.11}$$

which reduces for particles in water at room temperature to $0.12\ d^3\Psi$, if d is expressed in μm. This means that a very small velocity gradient of 1 s^{-1} (as may occur due to free convection) will already cause the fat globules in milk to aggregate by a factor 10 faster than in perikinetic aggregation. For casein micelles, orthokinetic aggregation will nearly always be negligible.

3.1.3.2 Fractal Aggregation

In perikinetic aggregation, particles form aggregates, which will aggregate with other aggregates to form larger aggregates, and so on. It turns out that large aggregates have a very open structure, as depicted in Figure 3.8. The following relation is observed to hold, albeit with some statistical variation:

$$N_p = (R/a)^D \tag{3.12}$$

where N_p is the number of particles in the aggregate, R is aggregate radius (see Figure 3.8) and a is particle radius. The exponent D is called the *fractal dimensionality*, and its value is always smaller than 3; it is generally between 1.7 and 2.5. This means that a large aggregate will be more open (tenuous, rarefied) than

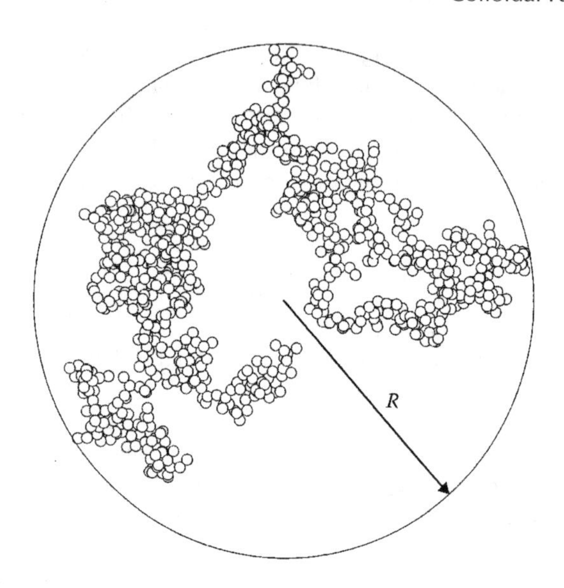

FIGURE 3.8 Side view (projection) of a fractal aggregate of 1000 particles, fractal dimensionality = 1.8. (Courtesy of J.H.J. van Opheusden.)

a smaller one. The maximum number of particles that can be packed in a sphere of radius R is given by

$$N_{\mathrm{m}} = (R/a)^3 \tag{3.13}$$

For the volume fraction of particles in an aggregate we now have

$$\varphi_{\mathrm{A}} = N_{\mathrm{p}}/N_{\mathrm{m}} = (R/a)^{D-3} \tag{3.14}$$

Because the exponent is negative, a larger value of R will lead to a smaller value of φ_{A}, and so to a more open aggregate.

Gel formation. This relation has an important consequence. As aggregation goes on, the total volume of the aggregates increases (while the number of aggregates decreases), until they fill the whole system and form a space-filling network, i.e., a gel. This will happen when the average φ_{A} becomes equal to the original volume fraction of particles in the system φ. The aggregate radius at the moment of gelation is given by

$$R_{\mathrm{g}} = a\varphi^{1/(D-3)} \tag{3.15a}$$

and the time needed to form a gel by

$$t_{\mathrm{g}} = t_{0.5}\ \varphi^{D/(D-3)} \tag{3.15b}$$

where $t_{0.5}$ is given by Equation 3.9.

Such a 'fractal gel' is characterized by tortuous strands of particles (see Figure 3.8), alternated with thicker nodes, leaving pores of various size. The largest pores in the gel have a radius comparable to R_{g}. For example, if $a = 0.5\ \mu\mathrm{m}$, $\varphi = 0.1$

and $D = 2.0$, we obtain $R_g = 5\,\mu m$; therefore, a maximum pore size of some 10 μm. R_g will be larger for a larger value of a, a smaller φ, and a larger D.

In the case just mentioned, the aggregation time would be given by $0.1^{-2} = 100$ times $t_{0.5}$. Assuming that casein micelles would aggregate unhindered ($W = 1$), it would only take 20 s; actually, W is much larger than unity, in this case. The gelation time will be shorter for a smaller value of a, a smaller D, and a larger φ.

Some complications. The theory as given here is an oversimplification. All of the equations should contain proportionality constants; these are generally unknown but they are always close to unity. Several variables affect the magnitude of D. A gel is only formed if aggregation can proceed unhindered, which means in the absence of any agitation. Moreover, aggregates may sediment before a gel is formed when the particles are relatively large and differ significantly in density from the medium.

Generally, the fractal structures are subject to change. Short-term rearrangement can occur in just-formed small aggregates; they then tend to become quite compact. These compact aggregates then form fractal structures and eventually a gel, but now an effective a value has to be inserted. For instance, $a_{eff} = 3a$. It implies that R_g becomes proportionally larger. Long-term rearrangements can occur after a gel has been formed. This will be discussed in relation to syneresis of rennet milk gels (Subsection 24.4.4).

Fractal aggregation can occur with casein micelles and fat globules, and with fat crystals in oil, as discussed in various sections in this book.

3.1.4 Size Distributions

It may be clear from the discussion so far that several properties of dispersions, especially physical stability, can depend on the size of the particles. As the particles within a dispersion generally vary considerably in size, the size distribution also has to be considered. Therefore, some basic aspects of these distributions will briefly be discussed for spherical particles.

Often the particle size range is divided into size classes of width Δd. If, in size class i, N_i is the number of particles with diameter d_i (or, more precisely, with diameters between $d_i + \frac{1}{2}\Delta d$ and $d_i - \frac{1}{2}\Delta d$) per unit volume of liquid, the number frequency is given by $N_i/\Delta d$ (dimension [length^{-4}]) as a function of d. The volume frequency, which represents the amount of disperse material present in particles as a function of d, is given as $\pi N_i d_i^3/6\Delta d$ (dimension [length^{-1}]). We may also express the volume frequency as the percentage of the total volume of dispersed material per unit class width (which is almost equal to the mass frequency). See the examples in Figure 3.9 for milk fat globules.

Table 3.2 gives some parameters of size distributions that are useful for various properties. (The parameter S_n is merely meant to facilitate the calculations.) The first characteristic of a distribution is its *average*, and several types of average can be given. It depends on the property considered as to which type is needed. The number average diameter $\bar{d}$ generally is not very useful; we are often more interested in the amount of material. Moreover, the total number of particles S_0 is

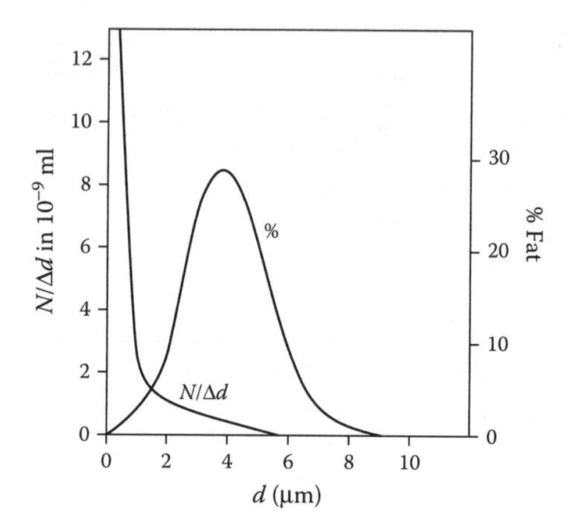

FIGURE 3.9 Average size frequency distribution of the fat globules in milk of Friesian cows (d_{vs} = 3.4 μm, fat content 3.9%). Number ($N/\Delta d$) and volume frequency (% of the fat) per μm class width against globule diameter. (Adapted from P. Walstra, *Neth. Milk Dairy J.*, **23**, 99, 1969.)

often difficult to obtain because of the relatively large number of very small particles. The average diameter type d_{vs} is more useful. It relates the volume of dispersed material to its surface area, and it is needed, for instance, to calculate the quantity of surfactant that is required to cover the total surface area. The sedimentation parameter H is discussed in Subsection 3.2.4. The mean free distance x is the average linear distance a particle can move before touching another.

A size distribution cannot be represented by its mere average. The width of the distribution is equally important, and the *relative width* is more useful than the absolute one. Relative width can be defined as c_s, the standard deviation in particle size divided by d_{vs}. If need be, the shape of the distribution also can be given.

TABLE 3.2
Parameters of Size Frequency Distributions

Parameter	Symbol	Given by	Dimension
*n*th moment of distribution	S_n	$\Sigma N_i \cdot d_i^n$	$[L^{n-3}]$
Number average *d*	$\bar{d}$	S_1/S_0	[L]
Volume/surface average *d*	d_{vs}	S_3/S_2	[L]
Sedimentation parameter	H	S_5/S_3	$[L^2]$
Relative distribution width	c_s	$(S_2S_4/S_3^2 - 1)^{1/2}$	—
Volume fraction of particles	φ	$\pi S_3/6$	—
Specific surface area of particles	A	$\pi S_2 = 6\varphi/d_{vs}$	$[L^{-1}]$
Mean free distance	x	$0.225d_{vs}(0.74/\varphi - 1)$	[L]

Note: $N_i/\Delta d$ = the number frequency; d = particle diameter; and L = length.

3.2 FAT GLOBULES

Nearly all of the fat in milk is in separate small globules; about 0.025% lipid material is in the plasma. Milk is thus an oil-in-water emulsion. The physicochemical aspects of this emulsion are essential, especially when considering the changes that occur during storage and processing of milk and milk products.

3.2.1 PROPERTIES

3.2.1.1 Size Distribution

The milk fat globules vary in diameter from about 0.1 to 15 μm, as illustrated in Figure 3.9. It is seen that milk contains very many small globules, which comprise only a small fraction of the total fat. Globules smaller than 1 μm make up about 75% of the number of globules, about 2% of the globular fat, and about 7% of the fat globule surface area. Freshly drawn milk also contains a few globules of 10 to at most 15 μm; presumably, these have formed by coalescence of smaller globules during flow in the glandular ducts.

An emulsion is characterized by its droplet size distribution and the volume fraction of droplets φ. The fat content in mass per cents equal $100\varphi\rho_{fat}/\rho_{product}$, which equals about 90φ in milk at 20°C. Fat content and globule size determine such parameters as the specific surface area ($A = 6\varphi/d_{vs}$) and the mean free distance (x). The effect of fat content on the latter is illustrated in Figure 3.10.

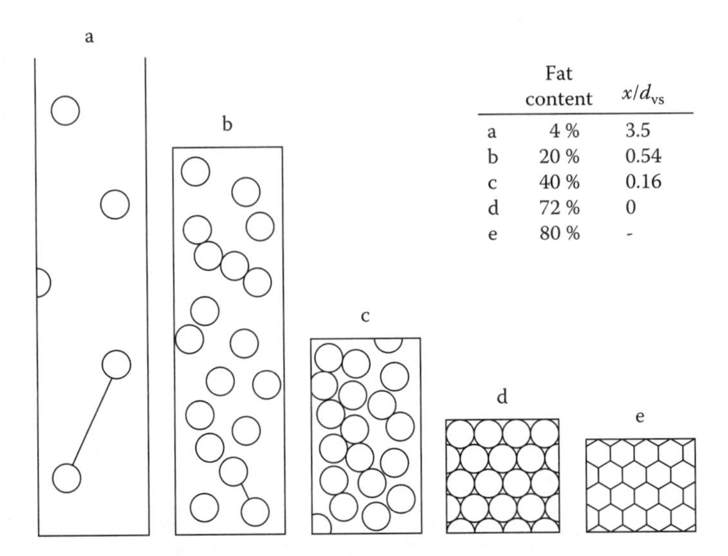

FIGURE 3.10 Distance between fat globules in milk and cream. Remember that a true representation cannot be given in a two-dimensional diagram. The line segments indicate the mean free distance x. (Adapted from H. Mulder and P. Walstra, *The Milk Fat Globule*, Pudoc, Wageningen, 1974.)

TABLE 3.3
Size Distribution of Fat Globules in Milk and Milk Products

Product	Fat Content (% w/w)	d_{vs} (μm)	c_s	A(cm²/ml Product)	Mean Free Distance x(μm)	x/d_{vs}
Milk, black and white	3.8	3.4	0.45	750	12.5	3.7
Milk, Jerseys	5.2	4.5	0.45	770	12.0	2.7
Homogenized milk	3.2	0.6	0.85	3,500	2.5	4.5
UHT milk	3.2	0.3	0.85	7,200	1.3	4.5
Half-and-half	10	3.5	0.45	1,900	4.5	1.3
Light cream	18	1	0.95	12,000	0.5	0.54
Whipping cream	35	4	0.5	6,000	0.6	0.16
Evaporated milk	8	0.4	0.85	13,000	0.6	1.6
Sweetened condensed milk	8	1	0.9	5,300	1.1	1.1
Ice cream mix	12	0.6	0.9	13,000	0.6	1.0
Separated milk	0.03[a]	1.2	0.5	17	600	500

Note: Approximate examples.

[a] Globular fat only.

Volume-surface average diameter d_{vs} is on average about 3.4 μm for milk of several breeds of cows and the relative distribution width c_s about 0.4. On average, 1 g of fat is divided among about $4 \cdot 10^{11}$ globules, of which 10^{11} globules are >1 μm in diameter. Diameter d_{vs} varies among breeds (Jerseys have d_{vs} 4.5 μm), with stage of lactation (d_{vs} decreases on average from about 4.3 to 2.9 μm), etc. But the shape of the size distribution is almost constant. This implies that different size distributions can be made to coincide merely by altering the scales. Of course, the size distribution can be altered by treatment, especially by homogenization (Chapter 9). Quantitative data are given in Table 3.3.

3.2.1.2 Surface Layers

Each fat globule of milk is surrounded by a surface layer or *membrane*. The layer functions to prevent the fat globules from coalescence. Its composition is completely different from either milk fat or milk plasma and is like that of a cell membrane, from which the fat globule membrane largely derives.

Table 3.4 gives the approximate composition of natural milk fat globule membrane, at least the main components. The phospholipids and cerebrosides have compositions similar to those given in Table 2.7 and Table 2.8; they have many unsaturated fatty acid residues. The composition of the proteins of the membrane is intricate; there are at least 10 major species and several minor compounds. They are predominantly glycoproteins specific for membranes and include butyrophilin, which appears to be specific for milk fat globules. Another

TABLE 3.4
Estimated Average Composition of the Membranes of Milk Fat Globules

Component	mg Per 100 g Fat Globules	mg Per m^2 Fat Surface	Percentage of Membrane Material[a]
Protein	1800	9.0	70
Phospholipids	650	3.2	25
Cerebrosides	80	0.4	3
Cholesterol	40	0.2	2
Monoglycerides	+	+	?
Water	+	+	—
Carotenoids + Vitamin A	0.04	2×10^{-4}	0.0
Fe	0.3	1.5×10^{-3}	0.0
Cu	0.01	5×10^{-5}	0.0
Total	>2570	>12.8	100

Note: Incomplete; several other components occur in trace quantities.
[a] Dry matter

component is proteose peptone 3, which is also present in milk serum. Several of the membrane proteins are enzymes. Alkaline phosphatase and xanthine oxidase are, for the most part, in the membrane and make up a considerable part of its protein.

Nearly all lipids in the membrane are polar lipids, especially phospholipids. It has often been said that the membrane contains a significant amount of high-melting triglycerides, but that is very unlikely. (Preparations obtained from milk that are assumed to represent the membrane material can readily become contaminated with triglyceride crystals). Traces of monoglycerides and free fatty acids are present.

Hypothetical structures of the membrane have been given in several publications, but such pictures are very uncertain because membrane structure is poorly understood. The original structure of the membrane is presumably a phospholipid monolayer adsorbed from the cytoplasm, surrounded by a layer of proteins and, on top of this, a lipid bilayer interspersed with proteins, some of which protrude into the milk plasma. However, much of this structure is lost during and after milk secretion, and the membrane shows considerable variation from place to place. The average layer thickness is some 15 nm but varies from about 10 to 20 nm. The globules have a negative charge: zeta potential in freshly drawn milk is about −12 mV (see Figure 3.7). The interfacial tension is about 1.5 $mN \cdot m^{-1}$.

Material that is typical of the fat globule membrane is also found in skim milk, in which it often amounts to about one third of the total amount of membrane material present in milk (see also Table 1.3 and Table 2.8). Not all of this originates from the fat globules but part does because treatment of milk causes the fat globules to lose membrane material. Coalescence of fat globules (Subsection 3.2.2) causes a decrease of their surface area, which leads to release of membrane material. Fat globules also lose part of their membrane if they come into contact with air (Subsection 3.2.3). Cooling leads to a migration of membrane material to milk plasma (the change is irreversible); about 20% of the phospholipids as well as some protein, xanthine oxidase, and Cu are released. By contrast, cooling causes adsorption of other proteins (cryoglobulins) onto the fat globules; but this process is reversible (Subsection 3.2.4).

Releasing part of the membrane from the surface of a fat globule causes surface-active substances (mainly protein) to adsorb from the plasma onto the denuded fat–water interface. This may happen when air is beaten in (Subsection 3.2.3). Alternatively, increasing the fat surface area by reducing average globule size creates an uncovered interface, which subsequently acquires a coat of plasma proteins. This especially happens during homogenization (Chapter 9). Table 9.4 compares various properties of natural, homogenized, and recombined fat globules.

3.2.1.3 Crystallization

Crystallization of fat in fat globules differs from that of fat in bulk (Subsection 2.3.5). Undercooling must be deeper to induce crystallization. The crystals in a fat globule cannot grow larger than the globule diameter. The arrangement of the crystals may also be different from that of fat in bulk. If there are sufficient crystals in a globule, they can aggregate into a space-filling network that provides a certain rigidity to the globule. Sometimes, especially after churning, crystals tend to be sited in the oil–water interface (see Subsection 3.1.1.5) and orient tangentially: this causes a bright layer in polarized light microscopy. As crystallization proceeds (e.g., by cooling), the tangentially oriented crystals may grow into a solid layer.

Crystallization of fat in the globules is of great importance for their stability (Subsection 3.2.2). Globules containing crystals cannot be called droplets, and they can generally not show coalescence, but partial coalescence.

3.2.1.4 Variation among Fat Globules

Differences between individual fat globules refer especially to their size. Differences in size are associated with variations in composition. For example, the phospholipid content of fat globules will be inversely proportional to d_{vs}. However, globules of the same size also show variations in composition, especially with respect to triglycerides. There are considerable variations in composition among globules in one milking of one cow; for example, the final triglyceride melting point of the globules in such milk can vary by up to 10°C. Membrane composition of individual fat globules can also vary, but quantitative data are not available.

3.2.2 Emulsion Stability

In many cases a stable emulsion is desired; the emulsion should not change on standing or during treatment. But moderate instability is desired during some treatments (e.g., in the whipping of cream and freezing of ice cream); sometimes the emulsion should break, as in churning.

3.2.2.1 Types of Instability

Various kinds of instability can be distinguished, as illustrated in Figure 3.1. These instabilities can to some extent occur simultaneously. Moreover, the changes can influence each other. All changes that cause an increase in particle size promote creaming, and thereby demixing. Aggregation thus enhances creaming, but the creaming also enhances aggregation. Aggregation of fat globules rarely leads to formation of a particle gel because the aggregates soon start creaming before a gel can be formed. But a cream layer can be like a particle gel (see Subsection 3.1.3.2). If aggregation occurs in a cream of, say, over 25% fat, it may proceed so fast that a gel results before significant creaming has occurred. Coalescence can only occur if the droplets are close, be it due to aggregation or creaming. If coalescence goes on, it will lead to phase separation: an oil layer on top of an emulsion layer of decreased oil content. This rarely occurs in milk products (see the following).

Fat globules can *aggregate* in various ways. Three types of aggregates are illustrated in Figure 3.11. We have given these types arbitrary names (others may use different names):

1. In *floccules*, attractive forces between globules are weak, and stirring disrupts the floccules. Flocculation does not happen normally with milk fat globules because electrostatic and steric repulsion prevents it. Milk fat globules do not flocculate even at their isoelectric pH, which is approximately 3.7, as shown in Figure 3.7. Some of the glycoproteins in the membrane cause sufficient steric repulsion. Agglutination in raw milk, which refers to spontaneous flocculation at cold temperatures, is discussed in Subsection 3.2.4.2.
2. In *clusters*, two globules share part of the membrane material, generally micellar casein. Examples are so-called homogenization clusters (Section 9.7) and heat-coagulated fat globules (Subsection 7.2.4). Clusters usually cannot be disrupted by stirring.
3. In *granules*, fat touches fat. Aggregation to granules can only occur if the fat globules contain a network of fat crystals, giving the globules a certain rigidity. Granules usually cannot be disrupted.

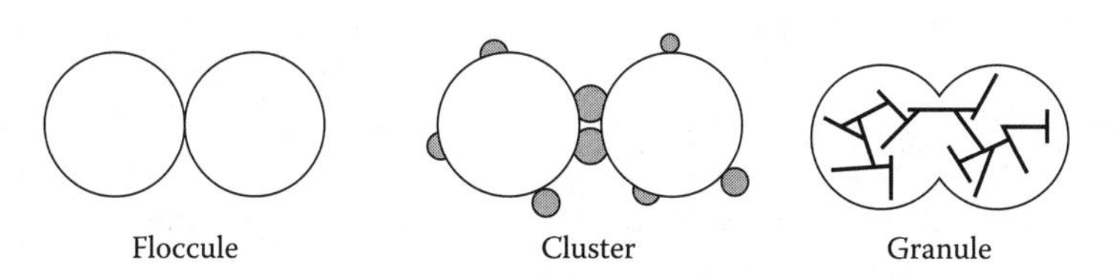

FIGURE 3.11 Different types of aggregates of fat globules. Gray dots denote (parts of) casein micelles; heavy lines denote fat crystals. Highly schematic, not to scale.

3.2.2.2 Partial Coalescence

If two emulsion droplets are close together and the thickness of the film between them has been reduced to a few nanometers, the film can rupture. The droplets will then fuse or coalesce into one droplet. But milk fat globules have a 'hairy' surface layer that causes sufficient steric and electrostatic repulsion between them to prevent such a close approach, hence, prevent coalescence. This is thus a rare event in milk products (unless the droplets are pressed together); moreover, it can only occur if the fat is liquid.

However, a milk fat globule often contains fat crystals, and if the solid fat content is some 10% or higher, the crystals tend to form a space-filling network in the globule (Subsection 3.1.3.2). One or a few of these crystals may slightly protrude from the globule surface into the surrounding liquid. If so, a protruding crystal may pierce the film between two approaching globules, which often leads to the onset of coalescence. However, because of the crystal network in the globules, full coalescence is impossible. But oil contact between the globules can be made and then a granule is formed. This process is called partial coalescence. It will readily occur in liquid milk products if the fat is partly solid when the product is subject to (vigorous) flow.

Partial coalescence (formation of granules, clumping) differs from true coalescence in its *effects*. The globules are not transformed into larger spherical globules, but into large, irregularly shaped granules (sometimes called butter grains) or even into a space-filling network, as in whipped cream. Increasing the temperature and thereby melting the fat crystals will cause coalescence of the formed granules into (large) droplets. Consequently, the oil–water interface decreases and substances of the fat globule membrane (e.g., phospholipids) are released into the plasma. During partial coalescence release of membrane components may also occur, although to a lesser degree.

The following are factors affecting the *rate* of partial coalescence:

1. Stirring or, more precisely, bringing about a velocity gradient or shear rate in the liquid has a considerable effect (Figure 3.12.a). This is typical of partial coalescence because stirring usually has little effect on coalescence if the globules are liquid. The velocity gradient increases the rate of encounters of the fat globules, causes encountering globules to roll around each other (enabling a protruding crystal to pierce the film between the globules), and can bring them closer together (enabling crystals that protrude less far to pierce the film).
2. Beating in of air is, again, a kind of stirring; moreover, it enhances clumping in another way (see Subsection 3.2.3).
3. Fat content has considerable effect (Figure 3.12.b). This is because partial coalescence follows second-order kinetics.
4. The proportion of solid fat is crucial (see Figure 3.12.c). If there are insufficient crystals to form a space-filling network, partial coalescence will not occur. If the globules contain too much solid fat (e.g., after

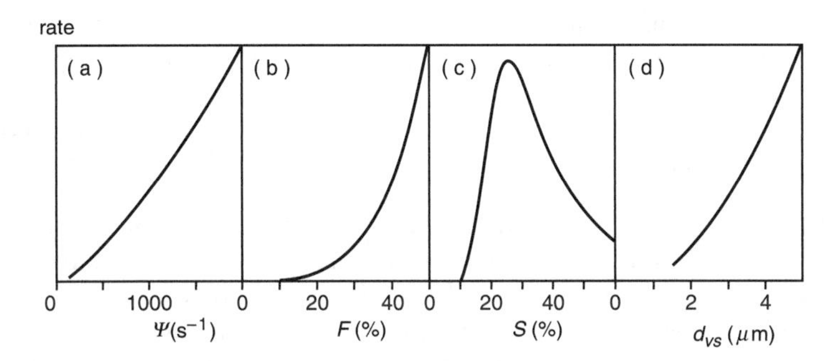

FIGURE 3.12 Rate of partial coalescence (or churning) of cream subjected to laminar flow as a function of (a) velocity gradient or shear rate Ψ; (b) fat content F, (c) fraction of the fat that is solid S, and (d) average globule size d_{vs}. Approximate results showing trends.

keeping them for some hours at < 5°C), the residual liquid fat is retained in the pores of the crystal network, leaving no sticking agent to hold the globules together.

5. The smaller the globules, the stabler they are (see Figure 3.12d). Larger globules have larger crystals, and therefore the probability of a crystal sticking far enough out of the globules to pierce the film between two globules is increased. For larger globules, fewer aggregation events are needed to form visible granules. The effect of globule size is considerable. For example, homogenized cream cannot be churned. The globule size often is the main cause of variation in coalescence stability if different emulsions are compared.
6. The surface layers of the fat globules are also essential. Natural fat globules are reasonably stable. But if globules of a similar size have a surface layer of protein, as formed by homogenization or recombination (Section 9.5), they are far more stable. Displacing part of that proteinaceous surface layer by adding surfactants such as monoglycerides or Tweens markedly decreases the stability to coalescence. Such surfactants therefore are commonly applied in ice cream mix in which the globules should aggregate during the beating process.
7. Temperature fluctuations can have a considerable effect. For instance, keeping cream of 25% fat or more at 5°C for some time, then warming to about 30°C (e.g., 30 min) and cooling it again — not very rapidly — produces a strong increase in its viscosity, and it may even gel. This process is called *rebodying*. The explanation is that the size of the fat crystals has been greatly increased. For this to happen, warming to a temperature at which not all but most of the crystals melt is of paramount importance. Rebodying is caused by partial coalescence. It only occurs without stirring if the fat content is so high that the fat globules are very close together. It may also occur in a cream layer formed on high-pasteurized, unhomogenized milk.

Size, orientation, and network formation of the crystals in the globules will have a significant effect on the tendency to form granules. An example is the above-mentioned rebodying.

3.2.2.3 'Free Fat'

The quotation marks in the heading signify that the authors consider the term free fat to be misleading. It is used to indicate that the fat globules in the product have been subject to instability, but the type of instability is generally left unclear. Sometimes it is defined as 'uncovered fat,' but that is certainly a misnomer. An uncovered oil–plasma interface can be momentarily formed, e.g., by interaction with air bubbles, as discussed in the next section. Or globules may be disrupted into smaller ones, whereby the fat surface area is enlarged; this does not occur during normal processing, except in a homogenizer. In any case, protein will immediately adsorb onto the clean interface, and this will generally take less time than 0.1 s (Subsection 3.1.1.2 and Section 9.5).

Various methods for the determination of 'free fat' have been published, but it remains unclear what the results would mean. Moreover, different tests often give very different results. A change of the fat globules can be due to (partial) coalescence, and then an increase in particle size has occurred, which can generally be estimated. Loss of fat globule membrane material can occur due to coalescence or due to contact with air bubbles; this leads to an increased phospholipid content in the plasma, which can also be estimated. Altogether, it is much better to avoid the confusing term free fat.

3.2.3 Interactions with Air Bubbles

Skim milk foams readily, especially at low temperatures. The foam is subject to drainage, Ostwald ripening, and coalescence, and has a lifetime of about an hour. At temperatures around 40°C, the lifetime is longer, say 5 h. Milk fat globules depress foaming. If only 1% of whole milk is added to centrifugally separated milk, this diminishes the foaming capacity to less than half. Presumably, the fat globules can induce rupture of the foam lamellae, hence, coalescence of bubbles; in particular, bubbles in the upper layer of a foam coalesce with the air above.

Figure 3.13 shows what may happen if an emulsion droplet approaches an air–water interface; see also Subsection 3.1.1.5. Important variables are the spreading pressure p_s and the presence of adsorption layers onto the air–water and oil–water interfaces. From the data given in Table 3.1 we calculate for p_s in the absence of adsorbed layers 72 – (30 + 35) = 7, and when protein is adsorbed about 50 – (10 + 35) = 5 $mN \cdot m^{-1}$, i.e., positive values. This means that *spreading* of oil from the drop over the air–water interface will occur. The spreading rate is on the order of 2 mm/s.

In most cases both the droplet and the air–water interface will bear an adsorbed layer (e.g., of protein), and this causes colloidal repulsion. The oil now cannot make contact with the interface, as depicted in the third row of Figure 3.13. If the air–water interface is temporarily clean, contact will be made and spreading occurs.

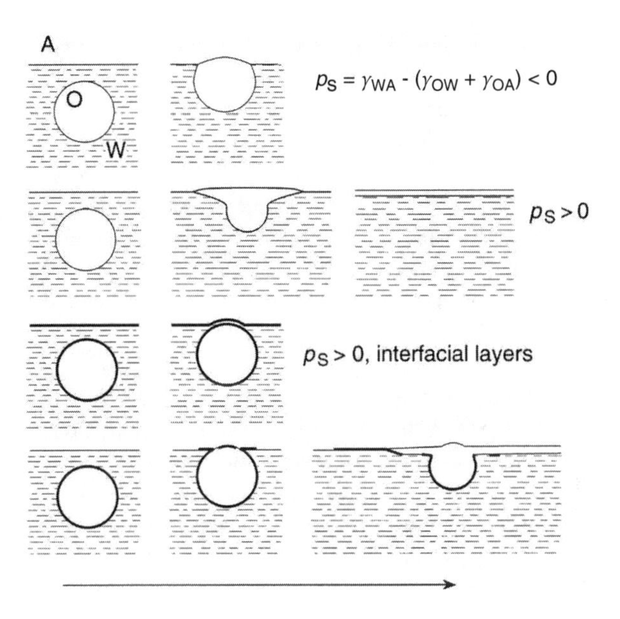

FIGURE 3.13 Interactions between an oil droplet (O) and the air–water (AW) interface, as a function of the spreading pressure (p_s; γ = interfacial tension) and of the presence of interfacial adsorption layers (heavy lines). The bottom row refers to a milk fat globule contacting a newly formed air–plasma interface. Schematic, not to scale. (After H. Mulder and P. Walstra, *The Milk Fat Globule,* Pudoc, Wageningen, 1974.)

For milk fat globules the situation is more complicated. If the fat is partially solid, a crystal may possibly be able to pierce the thin film between globule and the air–water interface, thereby establishing contact between air and oil. Initially, material of the fat globule membrane will spread; given time and space, this is followed by oil. If it occurs in a foam lamella, the spreading will drag the continuous liquid with it, which leads to local thinning, and possibly rupture, of the lamella. In other words, the fat globules can induce bubble coalescence. If the fat is, for the greater part, solid, e.g., at 5°C, spreading of oil can hardly occur, but the fat globules will nevertheless become attached to the air bubbles. In this way, fat globules can be accumulated in a foam formed of milk or cream by bubbling or agitation; such a process is called *flotation.*

Rapid beating in of air in milk or cream (as in churning or whipping) causes new air–water interface to be continually formed, and fat will spread over the interface. If the fat is fully liquid, the subsequent breaking up of air bubbles covered with fat causes disruption of the fat. Churning warm milk or cream thus yields smaller fat globules. If the globules also contain solid fat, they become attached to the air bubbles. As the air surface area diminishes (because air bubbles coalesce), the attached fat globules are driven nearer to each other; the liquid fat spread over the air bubble surface, readily causes the globules to form granules (see also Figure 18.4). Furthermore, the liquid fat makes a foam less stable; in other words, the lifetime of the air bubbles is short. Further aggregation of granules yields butter grains, in which a phase inversion has apparently taken place; i.e., oil is the continuous phase. But the grains still contain fat globules and moisture droplets. Concentrating and then working the grains removes excessive moisture and reduces the moisture droplets in size. In this way, butter is obtained (Subsection 18.2).

If there is a very little liquid fat during beating in of air and if, moreover, the fat content is fairly high, structures of fat clumps are formed. However, churning does not occur; the structures entrap the air bubbles, so that whipped cream is obtained. Similar processes occur during the freezing and whipping of ice cream mix.

3.2.4 Creaming

Because of the difference in density between milk plasma and fat globules, the globules tend to rise. This property is of great importance because it causes (undesirable) creaming during storage, and it enables milk to be separated into cream and skim milk. Creaming is much enhanced if the fat globules have been aggregated into floccules or clusters.

3.2.4.1 High-Pasteurized Milk

In this milk, creaming of single fat globules may occur. The velocity v_s of a rising globule is usually obtained from Stokes's equation:

$$v_s = -a\,(\rho_p - \rho_f)\,d^2/18\eta_p \qquad (3.16)$$

where ρ_p is density of plasma, ρ_f is density of fat globules, η_p is viscosity of plasma (not of the milk); a is acceleration, i.e., g (9.81 $m \cdot s^{-2}$) if creaming is due to gravity. For Stokes's law to hold, several conditions must be fulfilled, but the equation is quite useful to predict trends.

In practice, globule size and temperature mainly determine the extent of creaming in high-pasteurized milk. In first approximation, the creaming rate is proportional to the sedimentation parameter H (Table 3.2). H varies among lots of milk but processing of milk, especially homogenization, is the main variable. Data with respect to H are given in Table 9.1 and Table 9.3. Temperature affects the factor $(\rho_p - \rho_f)/\eta_p$ (Appendix, Table A.10; Figure 8.2B).

Clusters of fat globules cream much faster than single globules. Clustering may be caused by homogenization (Section 9.7) or by intense heating (sterilization). The latter clustering can occur in evaporated milk in which it causes undesirable creaming.

3.2.4.2 Raw Milk

Creaming in cold raw milk is usually determined by the flocculation of the globules by 'agglutinin,' i.e., a complex of cryoglobulins (predominantly immunoglobulin M) and lipoproteins; see also Subsection 2.4.3. In cold agglutination

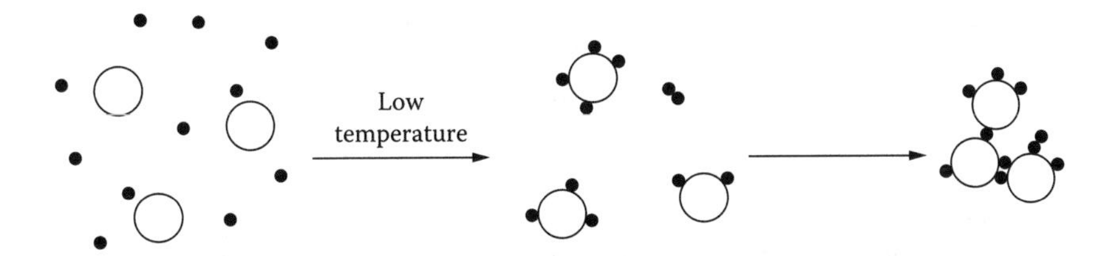

FIGURE 3.14 Adsorption of agglutinin (black dots, greatly exaggerated in size) onto milk fat globules and the ensuing flocculation of the globules. Highly schematic.

the following events may be envisioned (Figure 3.14), though the actual mechanism of creaming probably is more complicated:

1. In the cold, cryoglobulins precipitate onto all kinds of particles, especially fat globules (at a rate about 10^{-3} times that predicted by the theory for fast aggregation).
2. The cryoglobulin-covered fat globules aggregate into fairly large floccules.
3. Large floccules rise rapidly.
4. Large floccules overtake smaller ones and single fat globules, thereby enhancing aggregation and rising still faster. In this way, a cream layer forms rapidly, even in a deep vessel, such as a large milk tank.

The following are the main variables affecting natural creaming in raw milk (Figure 3.15):

1. *Temperature*. No agglutination occurs at 37°C. The colder the milk, the quicker the creaming is.

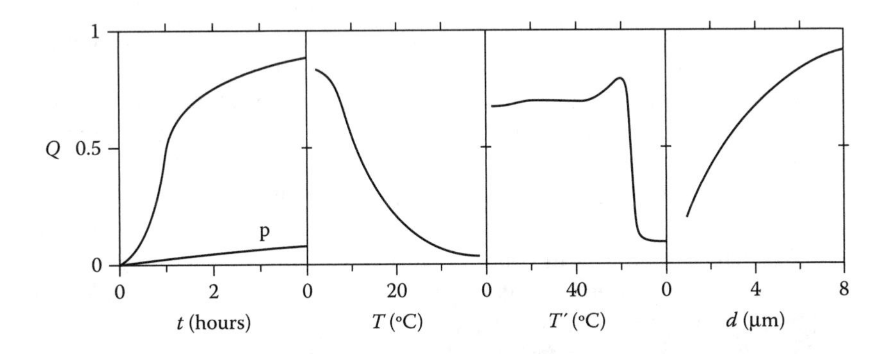

FIGURE 3.15 Gravity creaming in raw milk. Effect of creaming time (t), creaming temperature (T), temperature of pretreatment (T', during 30 min), and globule size (d) on fat fraction creamed (Q). p = high-pasteurized milk. Approximate examples. (From P. Walstra and R. Jenness, *Dairy Chemistry and Physics,* Wiley, New York, 1984. With permission.)

2. *Concentration of agglutinin* varies among cows and with lactation stage, i.e., high in colostrum, negligible in late-lactation milk.
3. *Fat globule size*, partly because smaller globules have a relatively larger surface area and, consequently, need more agglutinin.
4. *Fat content.* The higher the fat content of milk, the faster the creaming is because formation of floccules is quicker.
5. *Agitating milk* for a while at low temperature seriously impairs creaming. A possible explanation is that aggregates of agglutinin are formed causing a decrease of the number of active particles.
6. *Warming* such milk largely restores the ability to flocculate because it dissociates agglutinin.
7. *Heat treatment* of milk can inactivate agglutinin. The effect precisely parallels the insolubilization (by denaturation) of the immunoglobulins. Heating for 20 s at 71°C has no effect, 73°C causes 25% less creaming, and 78°C often leads to complete inactivation (Figure 2.31).
8. *Homogenization* inactivates agglutinin, also if it takes place without the fat globules being present. The phenomenon is not well understood.

3.2.4.3 Cream and Skim Milk

Cold agglutination in raw or low-pasteurized milk leads to a deep cream layer of loosely packed floccules, containing much plasma; the layer may contain 20% to 25% fat. It contains many of the bacteria and somatic cells from the milk. The cream can readily be redispersed throughout the milk. In low-pasteurized cream of 20% fat or higher, creaming will hardly occur: the globules aggregate into a fractal gel that fills the whole volume. However, gravity tends to compact this gel and a layer of plasma may form at the bottom, slowly increasing in height.

A different cream layer forms on high-heated milk. This is a thin layer of a high fat content (40 to 50%), but the closest possible packing of globules (about 70% fat) is not attained. When the fat starts crystallizing in the cream layer, this may promote partial coalescence of the fat globules, and so does stirring the layer. Consequently, it may then be impossible to redisperse the cream throughout the milk. By centrifugal action, cream with a much higher fat content, i.e., 80% fat or more ('plastic' cream), can be obtained so that globules are deformed (Figure 3.10).

Skim milk obtained after natural creaming rarely has a fat content of less than 0.5% (though a content as low as 0.1% may be achieved under ideal conditions); the skim milk still contains fairly large globules. Centrifugally separated milk has a far lower fat content, made up by the smallest fat globules and about 0.025% 'nonglobular fat.'

Most of the agglutinin is found in the cream if raw or low-pasteurized milk is separated when cold, e.g., 5°C. This can cause considerable agglutination in the cream. Most of the agglutinin is in the skim milk if the milk is separated when warm, say, >35°C. Consequently, a mixture of cold-separated skim milk and of cream obtained by warm separation displays hardly any cold agglutination.

3.2.5 Lipolysis

In milk, lipolysis (i.e., enzymatic hydrolysis of triacyl glycerols) causes free fatty acids to be formed, and this may give the milk a soapy-rancid taste. Several enzymes can be responsible for lipolysis, but here we consider the main lipolytic enzyme of milk, i.e., lipoprotein lipase (Subsection 2.5.2.4). In blood, this enzyme liberates fatty acids from lipoproteins and chylomicrons. A cofactor, in the present case the apoprotein part of certain lipoproteins, is necessary for the enzyme to act at the oil–water interface. Optimum temperature of the enzyme is about 33°C, optimum pH about 8.5. Milk contains 10 to 20 nmol of the enzyme per liter. Under optimal conditions, k_{cat} is greater than 3000 s^{-1} and would suffice to make milk rancid in 10 s, but this never happens.

There are several factors that slow down or enhance lipolysis in raw milk. The situation with respect to lipolysis is highly intricate and may be as follows:

1. Ionic strength, ionic composition, and pH of milk are suboptimal.
2. Most of the enzyme is bound to the casein micelles. This diminishes its activity considerably. Factors 1 and 2 would cause k_{cat} in milk to be reduced to some 100 s^{-1}, which can induce perceptible rancidity in milk in 5 min. In exceptional cases, this indeed may happen (see the following text).
3. The natural fat globule membrane protects the inside of fat globules against enzyme attack. The cause is the very low interfacial tension between globule and plasma, mostly < 2 $mN \cdot m^{-1}$. Because the enzyme alone cannot produce such a low interfacial tension, it cannot penetrate the membrane to adsorb onto the fat. (A component can only adsorb onto an interface if it reduces the interfacial tension.)
4. Together with certain lipoproteins from blood (the apoproteins alone probably do not suffice), the lipase can adsorb onto the fat. Addition of blood serum to milk can cause fast lipolysis.
5. Milk contains one or more enzyme inhibitors (besides the casein micelles) that presumably counteract the stimulating lipoproteins. Probably proteose peptone component 3 is the most important inhibitor.
6. Product inhibition occurs, although not because of the free fatty acids formed. Probably long-chain saturated monoglycerides, if formed in fairly high concentration, act as inhibitor.
7. Temperature affects the partitioning of enzyme activity (enzyme, stimulator, and/or inhibitor) among the fractions. At low temperature more enzyme activity is associated with the fat globules. This may explain why in raw milk the optimum temperature for lipolysis generally is about 15°C.
8. The activity of the enzyme slowly decreases in raw milk, by about 10% per day at 5°C, and faster at higher temperatures.

Most raw milks scarcely become rancid on standing, although the milk of some cows does. Increased fat acidity especially occurs when milk yield becomes

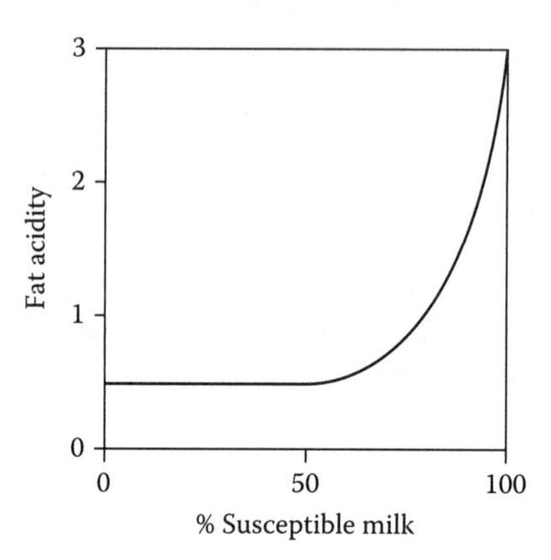

FIGURE 3.16 Acidity of milk fat (as mmol/100 g) of milk 'susceptible' to lipolysis, 'normal' milk, and mixtures thereof. The milks were kept for 24 h at 4°C. (Adapted from unpublished results of A. Jellema.)

low (< 3 kg/milking), i.e., at the very end of lactation. As is illustrated in Figure 2.37C, this can cause seasonal variation in degree of lipolysis. Increasing the number of milkings also leads to increased lipolysis. All of these conditions enhance 'leakage' of lipoproteins from the blood into the milk. Milks of some cows are highly susceptible to lipolysis, leading to 'spontaneous lipolysis.' But addition of 'normal' to 'susceptible' milk considerably slows down lipolysis, as is shown in Figure 3.16. This is because inhibitors are present; consequently mixed milk, if treated properly, rarely becomes rancid.

Lipolysis can, however, readily be induced in milk. Removing the natural membrane from the fat globules by intense beating in of air (Subsection 3.2.3), or increasing the fat–plasma interface by homogenization, leads to a globule surface layer of plasma proteins (Section 9.5), which causes an interfacial tension of about 10 $mN \cdot m^{-1}$. As a result, lipoprotein lipase can penetrate the membrane. Homogenization of raw milk thus causes very rapid lipolysis; see Figure 9.10.

Lipolysis can also be induced by cooling raw milk to 5°C, warming it to 30°C, and cooling it again, although there is wide variation among milk samples. The explanation is uncertain.

3.3 CASEIN MICELLES

The fact that casein in milk is not present in solution but in micelles has important consequences for the properties of milk. To a large extent the casein micelles determine the physical stability of milk products during heat treatment, concentrating, and storage. Their behavior is essential in the first stages of cheese making. The micelles largely determine the rheological properties of sour and concentrated milk products. The interaction of casein micelles with oil–water interfaces is of importance with respect to properties of homogenized milk products.

3.3.1 Description

Almost all casein in fresh uncooled milk is present in roughly spherical particles, mostly 40 to 300 nm in diameter. On average, the particles comprise approximately 10^4 casein molecules. These casein micelles also contain inorganic matter, mainly calcium phosphate, about 8 g per 100 g of casein (Table 2.2). They con-tain also small quantities of some other proteins, such as part of the proteose peptone and certain enzymes. The micelles are voluminous, holding more water than dry matter. They have a negative charge.

3.3.1.1 Structure

There is still disagreement among workers on the supramolecular structure of casein micelles, although several aspects of it are common ground. The following are some observations that are relevant for such a model:

- Electron micrographs show roughly spherical particles, average size on the order of 0.1 μm, with a marked spread in size. The particles are not quite spherical: they have a bumpy surface, and most of them are slightly anisometric (axial ratio 1 to 1.4).
- As discussed before, the various caseins tend to associate (Subsection 2.4.4) into small aggregates (4 to 25 molecules, according to casein species). In a solution of whole casein in (simulated) milk ultrafiltrate, association occurs as well; the small aggregates are of mixed composition. If calcium and phosphate ions are slowly added at constant pH to such a solution, it becomes white, and shows particles that look just like native casein micelles.
- In the lactating cell (Figure 1.2), it is seen that most of the Golgi vesicles contain many small particles, about 15 nm in size, whereas in other vesicles these particles have apparently aggregated into casein micelles.
- From determination of casein composition of micelles of various diameter, it has been calculated that the core of a micelle consists of roughly equal amounts of α_s- and β-casein, with very little κ-casein, whereas the outer layer appears to consist of about equal amounts of κ- and α_s-casein, with very little β-casein. The κ-casein concentration is nicely proportional to the specific surface area of the micelles.
- Nearly all of the κ-casein is present in the form of polymers of 2 to 9 (average about 6) molecules, linked to each other by –S–S– bridges.
- The hydrodynamic voluminosity of casein micelles is considerable, about 4 ml per gram of dry casein, implying that the casein micelles must contain a lot of water.

From these and other observations and considerations, the model illustrated in Figure 3.17 has resulted. An important element is that the micelle is built of mixed-composition *submicelles*, each 12 to 15 nm in size, and containing some

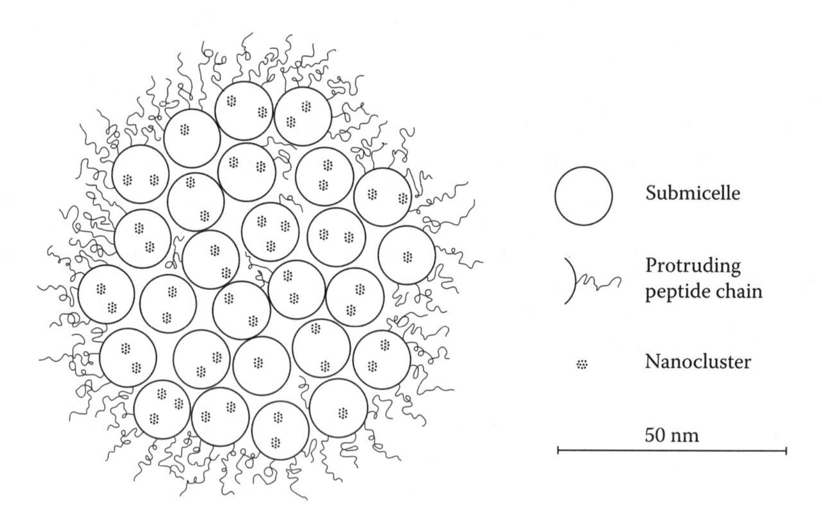

FIGURE 3.17 Cross section through a tentative model of a casein micelle.

20 to 25 casein molecules. Some of these submicelles would contain one or two polymers of κ-casein and are at the outside of the micelles. The other submicelles would contain no or very little κ-casein.

This model agrees with most of the properties of casein micelles. However, some workers are of the opinion that the existence of submicelles is unlikely, and assume a more homogeneous structure. In our opinion, this point is as yet undecided. However, several characteristics are generally accepted. These include:

- The existence of a 'hairy layer,' consisting of the C-terminal end (about 75 amino acid residues) of κ-casein. (The cross-links between the molecules are in the N-terminal part: see Figure 2.25.) The hairs are quite hydrophilic and are negatively charged; they also contain the carbohydrate moieties of κ-casein. The hydrodynamic thickness of the layer is about 7 nm. The layer is essential in providing colloidal stability.
- The presence of so-called nanoclusters of calcium phosphate of about 3-nm diameter. These contain the inorganic phosphate and much of the calcium in the micelles (the CCP), but also the organic phosphate of the SerP residues, and probably some glutaminic acid residues. In other words, a nanocluster is not pure calcium phosphate but also contains protein moieties.
- The forces keeping the structural elements of a micelle together are, at least at physiological conditions, hydrophobic bonds between protein groups and cross-links between peptide chains by the nanoclusters. Probably, ionic bonds are also involved.
- NMR studies have shown that the protein molecules in a casein micelle are almost fully immobile, except for the hairs, which will show

continuous Brownian motion. This concerns immobility at very short timescales (on the order of nanoseconds). At longer timescales, molecules can move in and out of a micelle.

As mentioned, the micelles are fairly voluminous, but experimental results on their voluminosity vary widely. Presumably, the results involved may be interpreted such that the voluminosity of the micelles without the hairy layer is some 2 to 2.5 ml per gram of dry casein. Dry casein would occupy about 0.7 ml/g, so that the rest of that volume is water, partly in and partly between submicelles. Taking the hairy outer layer into consideration, micelle voluminosity is about 4 ml/g of casein. Hence, the total volume fraction of casein micelles in milk approximates 0.1. Small solute molecules can penetrate the micelles, but large ones such as serum proteins can do so hardly if at all. Obviously, the liquid in the micelles is not water only but neither is it identical to milk serum (see Subsection 10.1.1).

3.3.1.2 Variability

The micelles are not all the same. They show a distinct size distribution, a typical example being given in Figure 3.18. Note the great number of very small particles. These appear to be loose submicelles, which make up only a small part of the total casein (often called 'soluble casein'). Authors do not fully agree about the size distribution. In early-lactation milk, a small number of very large micelles are found, diameter up to 600 nm. We estimate the volume-surface average diameter (d_{vs}), excluding the loose submicelles, to be 120 nm. Further data are given in Table 3.5. Two kinds of variation should be distinguished: (a) the variation between the micelles of one milking of one cow and (b) the variation between different lots of milk, from different cows, etc. Different cows produce milk with a different size distribution. This is roughly represented in Table 3.6.

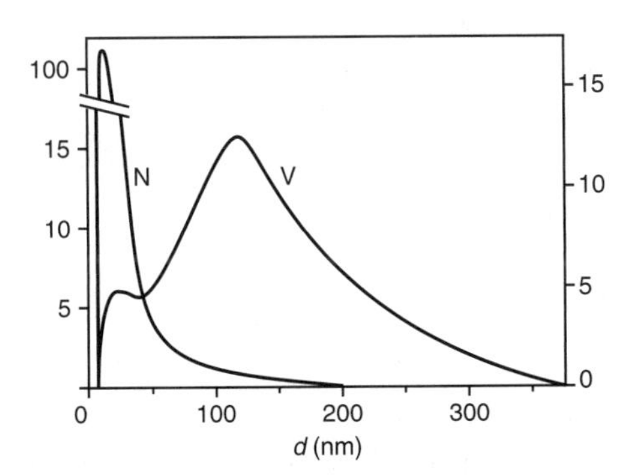

FIGURE 3.18 Estimate of the average size frequency distribution of casein micelles. Number (N, left-hand ordinate) and volume frequency (V, right-hand ordinate), both in percentage of the total per 20-nm class width, against micelle diameter (d).

TABLE 3.5
Size Distribution of Casein Micelles

Parameter	Full Distribution	Without Particles <20 nm
Number per μm^3 milk	600	110
Volume fraction, φ	0.06[a]	—
Number average diameter, $\bar{d}$ (nm)	25	75
Volume-surface average diameter, d_{vs} (nm)	90	120
Width[b]	0.45	—
Mean free distance, x (nm)	—	210
Surface area (m^2/ml milk)[c]	4	3
Same of submicelles (m^2/ml milk)	20	—

Note: Approximate example.

[a] Including the hairy layer: about 0.11.
[b] Standard deviation/average of the volume distribution.
[c] Excluding the hairy layer and assuming a spherical shape.

Table 3.6 shows that the protein composition is also variable. In particular, the proportion of κ-casein varies, since it largely determines the casein micelle size. The voluminosity of the micelles also varies. It will markedly increase with decreasing micelle size because of the constant thickness (approximately 7 nm) of the hairy layer and the low protein content in that layer. The content and composition of the colloidal calcium phosphate may vary, too. Large micelles probably have a higher CCP content.

TABLE 3.6
The Extent of Variation in the Properties of Casein Micelles

Property	Variation	
	Within Milk	Among Milks
Particle size	+ + +	+
Voluminosity	+ +[a]	+ +[a]
Protein composition	+ +[a,b]	+[a,b]
Amount of colloidal phosphate	+	+
Composition of colloidal phosphate	?	+

Note: Approximate results of the variability between the micelles of one milking and between different milks, respectively.

[a] Is partly linked with micelle size.
[b] Mainly because of variable κ-casein content.

3.3.2 Changes

During storage of milk the casein micelles alter, though slowly. This is because there is no thermodynamic equilibrium between the micelles and their surroundings. The main change probably is proteolysis of β-casein into γ-casein and proteose peptone by plasmin (Subsection 2.5.2.5); part of the proteose peptone enters the serum. Even from a purely physicochemical point of view, the micelles are not stable. The main cause is that the colloidal phosphate is not in the stablest form. The phosphate thus will (usually slowly) be converted to stabler phosphates (octa calcium phosphate or hydroxyapatite, depending on conditions), associated with the casein in another way, or in the form of a precipitate that is separate from the micelles.

Furthermore, the casein micelles will alter during changes in the external conditions, especially temperature and pH. Some of these alterations are reversible, whereas others are not or partly so.

3.3.2.1 Dynamic Equilibria

A casein micelle and its surroundings keep exchanging components. The principal exchanges are represented schematically in Figure 3.19. The exchanges can be considered dynamic equilibria, though they may be pseudo rather than true equilibria.

Part of the mineral compounds exchanges the fastest. The counterions, present as free ions in the electrical double layer, would exchange very rapidly. Some of the components of the colloidal phosphate, including Ca, phosphate, and citrate, also exchange fairly rapidly (relaxation time, e.g., 1 h), whereas the remaining part appears to be strongly bound. Casein can diffuse in and out of each micelle, presumably mainly in the form of submicelles, and the (average) equilibrium situation depends on such factors as temperature, pH, and $a_{Ca^{2+}}$. Casein micelles can be broken up to smaller units by mechanical forces, e.g., by very intensive

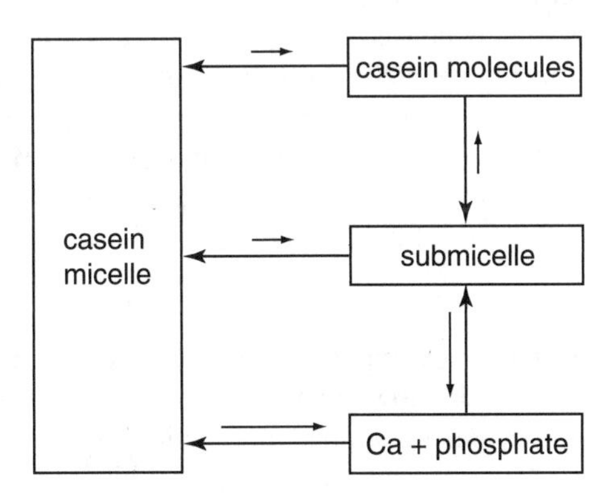

FIGURE 3.19 Outline of the main dynamic equilibria between casein micelles and serum.

homogenization; the formed fragments then rapidly reaggregate (relaxation time a few minutes) into the original size distribution.

All of this means that a model of the casein micelle, as depicted in Figure 3.17, would represent an instantaneous picture, a 'snapshot' as it were, with an exposure time of about 10^{-12} s. The protruding hairs would show the fastest motion. Small changes in micelle shape would take less than 1 s. At a timescale of 1 min, some submicelles would leave the micelle and new ones would be incorporated. Diffusion of a submicelle from one micelle to a neighboring micelle would take a few milliseconds.

Probably, free casein molecules also occur in milk, though they comprise only a tiny part of the casein at physiological conditions. At low temperatures, their amount increases considerably (see the following text).

The relative rate of most of the exchanges strongly decreases with decreasing temperature. Although the processes may be reversible to the extent that compounds that have left the micelles can return to them (e.g., when increasing temperature again), it remains to be seen as to whether this is a complete return to the native micelle structure.

Furthermore, the voluminosity of the micelles and their electrostatic charge may vary, usually as a result of the above-mentioned changes.

3.3.2.2 Low Temperature

Figure 3.20 illustrates some changes that occur by lowering the temperature. Dissolution of a considerable part of the β-casein occurs. But it is important to note that essentially a part of the dissolved β-casein is included in loose submicelles (experimentally differentiating between the two states of β-casein is difficult). Undoubtedly, the main cause of the dissolution of β-casein is that the hydrophobic bonds, which are predominantly responsible for its binding, are much weaker at low temperature. It therefore is not surprising that other caseins will dissolve as well, although to a lesser extent (α_s-caseins least); see also Figure 3.21D. Proteolytic enzymes can far better attack casein in a dissolved state. Because of this, for example, β-casein is fairly rapidly converted by plasmin at low temperature.

As is illustrated in Figure 3.20, the voluminosity of the micelles increases markedly at low temperature. This increase should be partly ascribed to the formation of another category of hairs. In addition to some β-casein molecules going into solution, others may be loosened so that β-casein chains now may protrude from the core surface of the micelles. This hairy layer is discussed further in Subsection 3.3.3.1. The voluminosity of the core of the micelles probably increases also.

Disintegration of micelles upon cooling may also be due to dissolution of a part of the CCP. Figure 3.20 shows that the association of Ca^{2+} ions with α_{s1}-casein decreases with decreasing temperature, and the same is true of the other caseins. The loss of CCP presumably causes a weaker binding of individual casein molecules in the micelles.

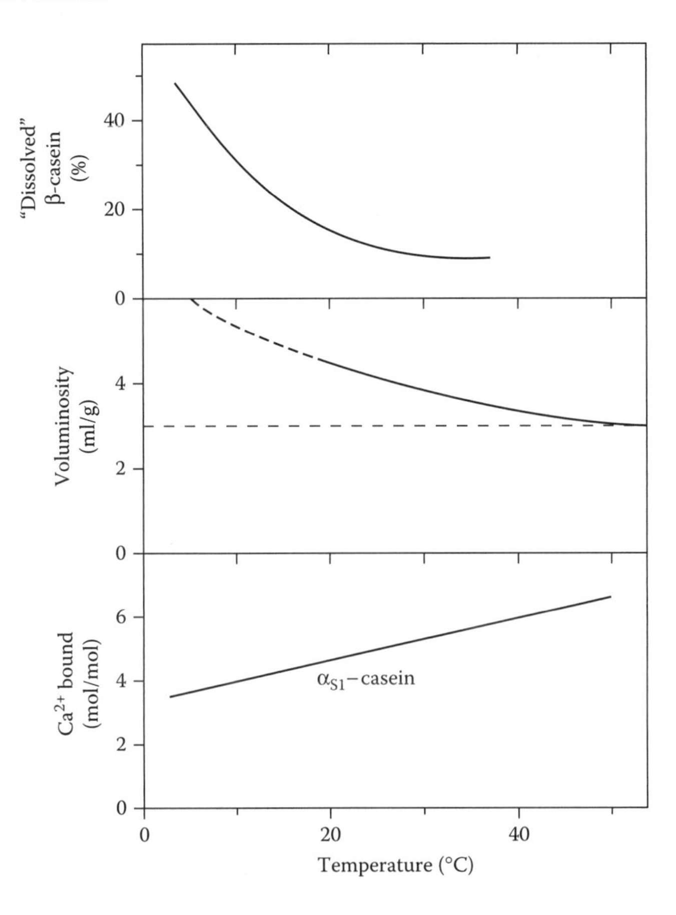

FIGURE 3.20 Influence of the temperature (after keeping milk for about 24 h) on some properties of casein micelles. Percentage of the β-casein not associated with the micelles. Voluminosity (in ml/g dry casein) from intrinsic viscosity; below 20°C (broken line) the values are probably overestimated. Binding of Ca^{2+} ions to α_{s1}-casein. Results at physiological pH.

The above-mentioned changes of the micelles cause the milk to obtain other properties. For example, its viscosity increases significantly. The colloidal stability of the casein micelles is definitely greater; consequently, the milk shows poor rennetability during cheese making. All of these changes do not occur immediately on cooling, but they take some 24 h at 4°C before being more or less completed. On subsequent heating of the milk, β-casein returns to the micelles, and the amount of colloidal phosphate increases again. These changes occur slowly at, say, 30°C. Heating the milk briefly to 50°C and then cooling it to 30°C reestablishes its original properties, at least as far as rennetability and viscosity are concerned. As mentioned, it is questionable whether the casein micelles have become identical to the original micelles.

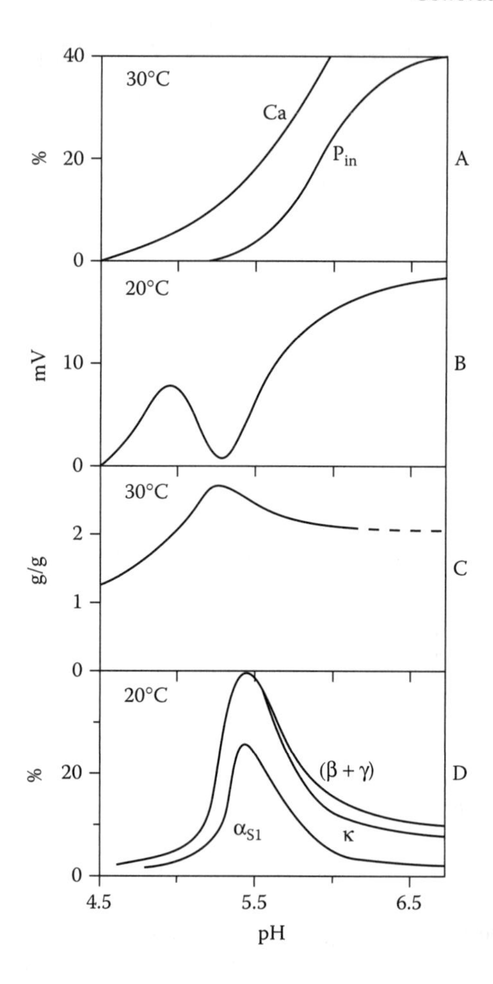

FIGURE 3.21 Properties of casein micelles as a function of pH. (A) Percentage of calcium and inorganic phosphorus in the micelles. (B) Negative electrokinetic or zeta potential. (C) Amount of water per gram of dry casein in micelles separated by centrifugation. (D) Percentage of the different caseins that cannot be separated by high-speed centrifugation (in milk).

3.3.2.3 High Temperature

On increasing the temperature, there is an initial continuation of the trends shown in Figure 3.20. The micelles shrink somewhat and the amount of colloidal phosphate increases. Figure 2.8A shows that the latter change occurs slowly. Moreover, the additional colloidal phosphate may not have the same properties as the natural phosphate.

At temperatures above 70°C the casein molecules become more flexible, as if part of the micelle structure melts. At still higher temperatures (above 100°C), dissolution of part of the κ-casein occurs. The extent closely depends on pH; no dissolution occurs below pH 6.2 (measured at room temperature), but there is almost complete dissolution at pH 7.2. The explanation is uncertain. At least part of the effect must somehow be due to the increased effect of entropy at high temperature. But other factors, such as the absence of serine phosphate in the part of the κ-casein chain that is inside the micelle, may also be involved. Because of this, κ-casein is weakly or not bound to the colloidal phosphate. However, micelle-like particles remain at high temperatures, even at 140°C.

If serum proteins are present, as is always the case in milk, another essential change occurs at high temperature. Serum proteins become largely associated with the casein micelles during their heat denaturation, and they largely become bound to the micelle surface. The association should at least partly be ascribed to formation of –S–S– linkages. An example is the association of β-lactoglobulin with κ-casein. Most of these associations are irreversible on cooling.

3.3.2.4 Acidity

Figure 3.21 illustrates some of the changes that result from a pH decrease. The colloidal phosphate goes into solution, with the dissolution being completed at pH 5.25. Removal of all of the calcium requires a still lower pH, i.e., until below the isoeletric pH of casein. Figure 3.21 further shows that the absolute value of the zeta potential decreases by decreasing the pH. This is because of increasing association of hydrogen ions with the acid and basic groups of the protein and also because of an increasing calcium ion activity, as calcium ions also associate with acid groups. In other words, Ca substitutes calcium phosphate to a certain extent. On further decrease of the pH, the negative charge of casein increases, due to dissociation of the calcium ions from the micelles, and eventually decreases again, due to association with H^+ ions. At still lower pH, casein becomes positively charged. Furthermore, lowering the pH leads at first to swelling of the particles and eventually to considerable shrinkage (Figure 3.21C). When the pH is lowered to, say, 5.3, a large part of the caseins goes into 'solution'; more so with increasing hydrophobicity of the casein concerned. Figure 3.21D refers to a temperature of 20°C; at higher temperatures the effect is much smaller, at lower temperatures greater (though then part of the 'dissolved' casein is included in loose submicelles).

The average particle size changes little in the pH region considered. All the same, particles are quite different at high and low pH values. At physiological pH, it is primarily the colloidal calcium phosphate that keeps the micelles intact. When the pH is lowered, the phosphate dissolves, resulting in increasingly weaker bonds. Consequently, swelling of the micelles occurs, along with dissolution of part of the casein. At low pH, internal salt bridges between positive and negative groups on the protein keep the molecules together. Obviously, the total attraction is strongest near the isoelectric pH of casein, i.e., near pH 4.6. We may conclude that the number and/or the strength of the sum of all kinds of bonds is weakest near pH 5.25; this optimum pH depends somewhat on temperature.

Increasing the pH of milk causes swelling of the micelles and their eventual disintegration. Presumably, the colloidal phosphate passes into another state.

3.3.2.5 Disintegration

Weakening of the bonds between the submicelles or those between protein molecules in the submicelles can lead to disintegration of the micelles. The former may be due to dissolution of the colloidal phosphate at constant pH, e.g., by adding an excess of a Ca binder like citrate, EDTA, or oxalate. Examples of the effect of various additions are given in Table 2.6. The second type of disintegration occurs by addition of reagents like sodium dodecyl sulfate or large quantities of urea, which break hydrogen bonds and/or hydrophobic interactions. Reagents that break –S–S– linkages do not disintegrate the micelles, but it is not known if less rigorous changes occur.

3.3.3 Colloidal Stability

Casein micelles are lyophilic colloids, which implies that they are physically stable. Sure enough, the micelles do not aggregate under physiological conditions. However, most of the particles are large enough to aggregate owing to van der Waals attraction; see Equation 3.5, where the Hamaker constant A presumably will be about 0.25 kT. Hence, there must be repulsive forces that prevent aggregation. The conditions of milk can be altered in such a way that the micelles do aggregate, for example, by adding considerable quantities of calcium chloride or ethanol or by applying a very high temperature.

In all such cases, casein micelles are much less stable than is dissolved casein. This must be due to the much higher entropy (S) of the free molecules as compared to the micelles. Aggregation can only occur if it results in a lower free energy (G), which, in turn, is due to the decrease of enthalpy (H), i.e., bond energy. However, aggregation causes of decrease of entropy, and this increases the free energy (because $\Delta G = \Delta H - T\Delta S$), and thereby counteracts aggregation. But most of the entropy of the casein has already been lost in the formation of the micelles, so that the additional decrease in entropy on aggregation of the micelles is almost negligible.

3.3.3.1 Cause of Stability

From the discussion in Subsection 3.1.2, combined with the model of the casein micelle in Figure 3.17, it will be clear that the *hairy layer* of κ-casein chains provides colloidal stability by means of steric and electrostatic repulsion. (It can be calculated from the DLVO theory that electrostatic repulsion will generally not suffice.) Aggregation can thus occur if (1) the hairs are removed, as by enzymatic cleavage caused by milk clotting enzymes; or (2) if the hairs collapse. The latter may occur if the hairs lose most of their effective net charge (by lowering of pH or by strongly increasing ionic strength) or by significantly decreasing their solvent quality.

However, this explanation may not always be sufficient, since in some kinds of aggregation, e.g., during intense heat treatment, the hairy layer appears to remain present. Moreover, electrostatic interactions seem to affect the instability. This has led us to the following tentative explanation.

When two micelles closely approach due to Brownian motion, the hairy layers of the micelles may overlap. Only by overlapping can steric repulsion occur. The hydrodynamic thickness of each of the layers is approximately 7 nm. At least some of the hairs, therefore, protrude over a greater distance from the core surface of the micelles. This implies that repulsion will start to occur at a distance of some 20 nm. At this distance any electrostatic repulsion resulting from charged groups on the micelle surface would be negligible because of the small thickness of the electrical double layer in milk. However, not all of the charge is on the core surface of the micelles. A considerable part of it is on the hairs, as is schematically shown in Figure 3.22a. (The situation is more complicated than depicted as there are more charges on each hair, including a few positive ones.) It implies that electrostatic repulsion extends over a distance much farther from the micelle surface. Though not large enough in itself to prevent aggregation, electrostatic repulsion can affect the distance of closest approach of the micelles. Variation in the electrostatic charge has therefore an effect on the extent of interpenetration of the hairy layers. This is diagrammatically shown in Figure 3.22b.

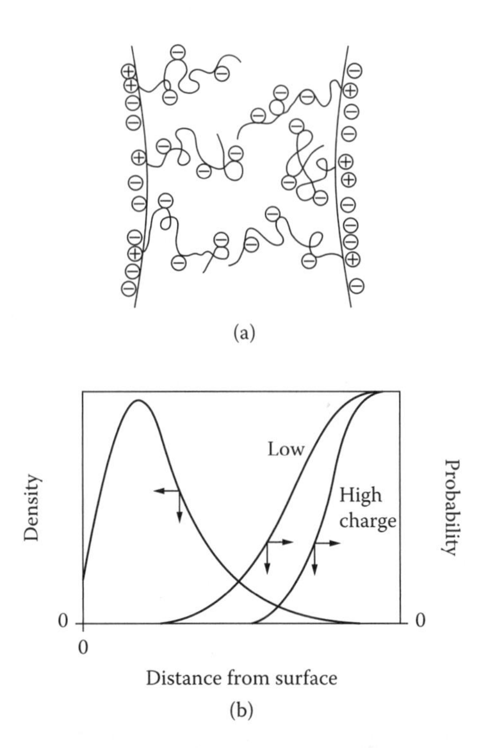

FIGURE 3.22 Hypothetical picture of interactions between two casein micelles. (a) Overlapping hairy layers and electrostatic charges. (b) Average density of material of the protruding peptide chains of the left micelle as a function of the distance from the core surface of that micelle, as well as the relative probability of finding a segment of a hair of the right micelle in the hairy layer of the left one, for a low and a high net charge of the hairs, respectively.

Hairs of both micelles will frequently touch one another if the hairy layers overlap during an encounter. Cross-linking between reactive groups on the chains may occur if such groups happen to touch. The probability that linkages form presumably increases with increasing interpenetration depth of the hairy layers. Such linkages may be salt bridges, especially –Ca– bridges between negatively charged groups; at high temperature, covalent bonds between amino acid residues can be formed (Table 7.2). When it concerns noncovalent bonds that are very short-lived, like –Ca– bridges, numerous bonds presumably must be simultaneously formed between two micelles for the contact to be lasting.

This brings us to a general point. Two aggregated casein micelles are not linked by a mere bond, but by a considerable number of bonds that jointly form a *junction*. (The same can hold for aggregated fat globules). The bonds in a junction need not be all of the same nature. Moreover, more, and possibly other, bonds may gradually form after a junction has been established, thereby strengthening the junction.

3.3.3.2 Causes of Instability

Native casein micelles are very stable to aggregation. Applying Equation 3.9 for perikinetic aggregation and inserting values given in Table 3.5, we obtain for the time needed to halve the number of micelles $t_{0.5} \approx 0.003 \cdot W$ seconds. Because no significant change in the micelle size distribution is observed after a year ($3 \cdot 10^7$ s) in (sterilized) skim milk, it follows that the stability factor W would be about $3 \cdot 10^7 / 0.003 = 10^{10}$ or higher.

However, a change in environment may lead to aggregation of the micelles. The main causes for aggregation are listed in Table 3.7. Generally, the change in

TABLE 3.7
Various Causes for the Aggregation of Casein Micelles

Cause	Micelles Changed?	Aggregation Reversible?	Aggregation at Low Temperature?
Long storage (age gelation)	Yes	No	No
At air–water interface	Spreading	No	No
High temperature (heat coagulation)	Chemically	No	—
Acid to pH ≈ 4.6	No CCP left	(Yes)[a]	No
Ethanol	Presumably	No	?
Renneting	κ-Casein split	No	No
Excess Ca^{2+}	More CCP	Yes	?
Freezing plus thawing	Presumably	(Yes)[b]	—
Addition of some polymers	No	Mostly	Yes

CCP = colloidal calcium phosphate.

[a] At neutral pH, the aggregates dissolve again but the natural micelles do not reappear.

[b] Partly, depending on conditions.

conditions results in changes in composition and structure of the micelles before aggregation occurs. Often, it is even questionable whether the resulting particles can still be called casein micelles. The aggregation appears to be irreversible in most cases. The various cases will be briefly discussed.

Age thickening and gelation mainly occur in evaporated and sweetened condensed milk. The explanation is unclear. Electron microscopy reveals that the micelles in these products become much less smooth and increasingly show protrusions; see Figure 18.7. This change causes an increase of the viscosity of the product and eventual formation of a space-filling network; hence, a gel.

Beating in of air in milk causes adsorption of casein micelles onto the air bubbles formed. The micelles can partly spread over the bubble surface. After the air itself has dissolved, the adsorbed material remains behind as a kind of bag (deflated balloon) in which the micelles can be recognized. This so-called ghost membrane is very stable, which implies that the casein micelles remain aggregated. The molecular explanation of the phenomenon is uncertain; the partly spread micelles can conceivably touch each other at sites devoid of hairs, leading to their fusion.

Heat coagulation of milk is preceded by substantial changes in chemical composition. It is discussed in Subsection 7.2.4.

Acidification causes a number of changes. Calcium and inorganic phosphate gradually dissolve and the net negative electric charge of the casein micelles decreases, including that of the hairy layer, which causes the layer to collapse. The casein itself is insoluble near its isoelectric pH (about 4.6). Altogether, this results in aggregation. Even a small decrease of the pH leads to a decreased charge, partly because of increased calcium ion activity; this diminishes colloidal stability.

Addition of *ethanol* lowers the solvent quality for the hairs of κ-casein. This causes the hairy layer to collapse and the electrosteric repulsion to diminish or even to change in attraction. The latter effect is enhanced by a decrease of the dielectric constant, resulting in a reduced electrostatic repulsion. Moreover, the colloidal phosphate passes into another, unknown state which causes aggregation of the 'micelles' to be irreversible. The lower the pH of the milk, the smaller the ethanol concentration needed to cause coagulation. This principle may be applied to quickly detect slight sourness in milk. In the so-called alcohol stability test, milk and ethanol are mixed in fixed proportion. If visible flocculation occurs, the milk is taken to be sour.

Renneting causes removal of the κ-casein hairs. It is discussed in Section 24.3.

An *excess of calcium ions* enhances the possibilities of –Ca– bridge formation. Moreover, it decreases the charge of the micelles and increases the supersaturation with respect to calcium phosphate in the milk serum. The latter would cause formation of additional colloidal phosphate (presumably, the serine phosphate residue of the protruding part of κ-casein can take part in it), which would cause fusion of micelles. In other words, if much CCP can be deposited, the steric repulsion is overcome, leading to aggregation. Occasionally, this seems to occur in fresh milk that spontaneously curdles just after milking. This so-called Utrecht milk abnormality is related to a low citrate content and an ensuing high calcium ion activity in the milk.

Freezing of milk leads to highly increased salt concentrations in the remaining nonfrozen solution. The situation is comparable to that in the preceding paragraph, but there also is true salting out. Slow changes occurring in the CCP cause the aggregated micelles to be not fully redispersible after thawing (see also Section 11.2).

The last mentioned cause for aggregation — *addition of polymers* — is of a completely different nature. Some long-chain polymer molecules can adsorb onto casein micelles. If the polymer tends to form a network, the micelles are incorporated into it. For example, a weak gel is obtained by adding a little κ-carrageenan to milk. Higher concentrations of suitable polymers may cause a kind of coprecipitation of polymer and casein. Presumably, the micelles remain virtually unaltered. Therefore, the aggregation concerned is reversible.

3.3.3.3 Effect of Temperature

The lower the temperature, the greater the colloidal stability of casein micelles. Table 3.7 shows that most aggregation reactions do not occur at low temperature, e.g., 5°C. That does not imply that no strong bonds between micelles can exist at low temperature. After their aggregation at, for instance, 30°C by renneting or acidification of milk, the formed gel cannot be dispersed again by lowering the temperature to, say 5°C; it even becomes firmer, which implies that the number or the strength of the bonds between the micelles increases (see Figure 3.23b). In other words, the casein micelles fail to aggregate at low temperature because of a high activation free energy for aggregation. This may be explained as follows. Figure 3.20 shows that lowering the temperature increases the voluminosity of

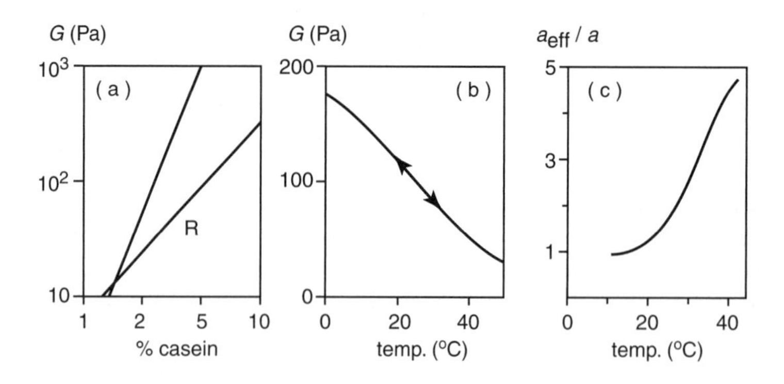

FIGURE 3.23 Properties of acid skim milk gels; in frame (a) also a rennet gel (R) is considered. (a) Elastic shear modulus G as a function of casein concentration (log–log plot). (b) Modulus as a function of measurement temperature; the temperature of making and aging the gel was 30°C. (c) The effective radius of the building blocks of the gel a_{eff} relative to the particle radius a, as a function of gelation temperature. Meant to illustrate trends.

the micelles. This can partly be ascribed to formation of a thicker hairy layer or of a layer of another composition; β-casein chains especially may protrude from the micelle surface at low temperature. Because of this, steric repulsion would be stronger, and repulsion remains after removal of the hairs of κ-casein, as occurs on renneting. Apart from that, swelling of the micelles results in a smaller van der Waals attraction.

3.3.4 Gel Formation and Properties

When the casein micelles in skim milk quiescently aggregate, a fractal gel is formed, as described in Subsection 3.1.3.2. If they aggregate in milk, the fat globules are entrapped in the pores of the gel. These processes occur during renneting and slow acidification of milk: long before the micelles start to aggregate, any agitation due to addition of rennet or starter has subsided. Also during heat coagulation of concentrated milk, fractal aggregation can occur.

In this section we will briefly discuss rennet and acid skim milk gels, with some emphasis on the latter, as rennet gels are amply discussed in relation to cheese making (Subsection 24.3.4). It may be noticed that in these cases the particles in the gel are not proper casein micelles. Rennet gels are made of paracasein micelles: these have lost their κ-casein hairs. The bonds between particles probably are for the most part –Ca– bridges. At low pH, the micelles have lost the colloidal calcium phosphate, and keep their integrity by internal salt bridges and hydrophobic interactions. The same bonds are presumably formed between the particles.

Gels can be characterized by various parameters. Important is the gel structure, especially the coarseness of the network, which can be given as the *pore size* (distribution). The rheological properties are also essential. One mostly determines the elastic *shear modulus* or stiffness, i.e., the ratio of the shear stress applied over the resulting strain (relative deformation). This parameter is measured at very small strain. At larger strain, the proportionality between stress and strain is lost, and when the *fracture stress* or strength of the material is reached, the test piece breaks; the *fracture strain* is also a relevant variable. All of these parameters can vary widely and are important in relation to the usage properties of the gel. The above discussion is an oversimplification: the rheological parameters also depend on other factors, especially on the rate of deformation.

In Table 3.8 and Figure 3.23, some data on properties of skim milk gels are given. From the results in the table, both types do not seem to be greatly different, though the moduli differ by a factor three. It should be realized that the value of the modulus depends on several factors: (1) the number of junctions, which depends, in turn, on particle concentration and the geometry of the network; (2) number and strength of the bonds in a junction; and (3) the modulus of the particles (casein micelles are deformable).

It is seen that for an acid gel the fracture stress is much larger and the deformation at fracture much smaller than for a rennet gel. It will be partly due to the particles being stiffer at pH 4.6 than at 6.6; this also affects the gel modulus. Another difference is that rennet gels show considerable long-term structural

TABLE 3.8
Some Properties of Gels Made by Renneting or by Slow Acidification of Skim Milk at 30°C

Property	Rennet Gel	Acid Gel
pH	6.65	4.6
Fractal dimensionality	2.25	2.35
Elastic shear modulus (Pa)	30	100
Fracture stress (Pa)	10	100
Fracture strain	~3	~1
Size of largest pores (μm)	~10	~18
Occurrence of syneresis	Yes	Virtually not

Note: Approximate examples.

rearrangement (i.e., after a gel has formed), which leads to stretching of the initially tortuous stands in a fractal gel. This must be the main explanation of the difference in slope of the two lines in Figure 3.23a. Another consequence of long-term rearrangement is the occurrence of *syneresis* (spontaneous expulsion of whey), which does occur in rennet gels, but hardly in acid gels; at 30°C, the syneresis rates differ by more than factor 50. Acid gels made and kept at 40°C do show syneresis.

Another type of rearrangement is fusion of micelles. Directly after two micelles have aggregated, they form a small junction, which would thus include a small number of bonds. The size of the junction increases with time; hence, the number of bonds involved increases. This is a relatively rapid process in rennet gels, and the value of the modulus in the table (30 Pa), which is obtained 1 h after addition of rennets, increases to about 100 Pa after some hours. The modulus of the acid gel will increase to about 120 Pa on storage for some days.

Short-term rearrangement of particles (before a gel is formed) leads to an increase in the size of the building blocks of the gel a_{eff}. This is illustrated in Figure 3.23c for acid gels. Below 20°C, no such rearrangement occurs, but it is extensive at high temperature. The main result of practical importance is that the pore sizes of the gel increase. Short-term rearrangement also occurs in rennet gels, though somewhat less (see Table 3.8, size of largest pores).

Another temperature effect is shown in Figure 3.23b. Once a gel is made and aged, it materially increases in modulus upon lowering the temperature, and vice versa; these changes are reversible. The formation temperature has a similar effect: a lower modulus at a higher temperature, but time effects may confuse the relations. Moreover, gels cannot be made within a reasonable time at low temperature (see Subsection 3.3.3.3).

Suggested Literature

Nearly all basic aspects discussed in this chapter: P. Walstra, *Physical Chemistry of Foods,* Dekker, New York, 2003.

Aspects of milk fat globules are discussed in: P.F. Fox, Ed., *Advanced Dairy Chemistry, Vol. 2: Lipids,* 2nd ed., Chapman and Hall, London, 1995; Chapter 3 (fat globule synthesis), Chapter 4 (physicochemical aspects), and Chapter 7 (lipolysis) are especially relevant.

Far more comprehensive, but partly out of date: H. Mulder and P. Walstra, *The Milk Fat Globule,* Pudoc, Wageningen, 1974.

Lipolysis of milk is discussed in an IDF report: Flavour impairment of milk and milk products due to lipolysis, international Dairy Federation, Document 118, Brussels, 1980.

Casein micelles are discussed in: P.F. Fox, Ed., *Advanced Dairy Chemistry, Vol. 1: Proteins,* 3rd ed., Kluwer Academic, New York, 2003, especially in Chapter 5 (casein micelles) and Chapter 18 to Chapter 22 (stability); Chapter 27 discusses interfacial properties of milk proteins.

The proceedings of a symposium on "Casein Micelle Structure" have been published in: *J. Dairy Sci.,* **81** (11), 2873–3028, 1998, and those of an (earlier) symposium on "Caseins and caseinates: Structures, Interactions, Networks" in: *Int. Dairy J.,* **9** (3–6), 161–418, 1999.

4 Milk Properties

Several chemical and physical properties of milk follow from its composition and structure, discussed in Chapter 2 and Chapter 3. These properties can affect the processing of milk and the quality of liquid milk products. Some important aspects are discussed in this chapter.

4.1 SOLUTION PROPERTIES

Milk is a good *solvent* for many compounds, including most salts, sugars, and proteins. This is largely due to its high relative dielectric constant, about 80 at 20°C, which is almost identical to that of water.

The total *ionic strength* of milk, discussed in Subsection 2.2.2, is on average 0.073 molar; in individual cow samples it varies for the most part between 0.067 and 0.080. The ionic strength affects the thickness of the electric double layer of Debye length $1/\kappa$, discussed in Subsection 3.1.2.2. Its average value in milk or milk serum is about 1.1 nm. Consequently, electrostatic interactions in milk will usually be of little importance over distances greater than about 2.5 nm. Divalent counterions (Ca^{2+}, Mg^{2+}) are roughly 60 times as active as Na^+ and K^+ in causing flocculation of colloidal particles. Though present in only small amounts, Ca^{2+} and Mg^{2+} are thus of great importance. Ionized calcium has a large effect on the properties of casein and casein micelles. Consequently, the calcium ion activity $a_{Ca^{2+}}$ (Subsection 2.2.2) is an essential parameter for milk and milk products. Variation in citrate concentration mainly determines the natural variation in $a_{Ca^{2+}}$.

The *electric conductivity* of milk is about 0.5 $A{\cdot}V^{-1}{\cdot}m^{-1}$ (variation ~0.4 to 0.55) at 25°C. This roughly corresponds to the conductivity of 0.25% (w/w) NaCl in water.

Freezing-point depression, osmotic pressure, –ln(water activity), etc., are called the *colligative properties*. Nonionic solutes as well as ions determine the magnitude of these properties. A freezing-point depression of 0.53 K (which corresponds to that of a solution of 0.90% of NaCl) is calculated from the total concentration of dissolved substances (0.28 mol/l solution), the molal freezing-point depression of water (1.86 K for 1 mol in 1 kg of water), the average osmotic coefficient (~0.93 for ionic species and 1.00 for neutral molecules), and from the fact that 1 l of milk serum contains about 950 g of water. The calculated value corresponds satisfactorily to the observed freezing-point depression of about 0.53 K, which varies among milk samples from ~0.515 to ~0.55 K.

The freezing-point depression of milk is very constant (relative standard deviation between individual milkings about 1%) inasmuch as it is proportional to its osmotic pressure, which is essentially equal to that of blood, which in turn is kept almost constant. The osmotic pressure of milk is about 700 kPa (7 bar) at 20°C. It follows from Table 2.4, that lactose accounts for over half of that value, complemented by K^+, Na^+, and Cl^-. Furthermore, a boiling point elevation of milk of 0.15 K is calculated from the molar concentration, and a water activity of 0.995.

Upon *acidification* of milk, the salt solution changes considerably, primarily because the colloidal calcium phosphate goes into solution (see Figure 2.7) but also because of altered dissociation of various salts. It has several consequences. All of the colligative properties increase in magnitude. For instance, the freezing-point depression increases by about 2 mK for each mmol of acid added per liter. That means that at pH = 4.6, the freezing point will be about −0.63°C. The electric conductivity will increase by about 4 $mA \cdot V^{-1} \cdot m^{-1}$ for each mmol of acid per liter, which means an increase of about 0.2 $A \cdot V^{-1} \cdot m^{-1}$ at pH 4.6. The total ionic strength will increase from 75 to about 130 $mmol \cdot l^{-1}$ upon lowering the pH to 4.6.

4.2 ACIDITY

The acidity of milk is usually expressed as pH. Several conditions and processes can affect the pH value. Examples of titration curves of milk and of the buffering index derived from it are given in Figure 4.1; d B/d pH is the molar amount of HCl or NaOH needed to change the pH by one unit. It is also seen, in Figure 4.1c, that titration results can markedly depend on the pH history. Upon adding HCl, micellar calcium phosphate goes into solution, and upon increasing the pH again, the calcium phosphate generally does not return to the casein micelles, at least not at the same pH and not in the same composition (see Subsection 2.2.5). This

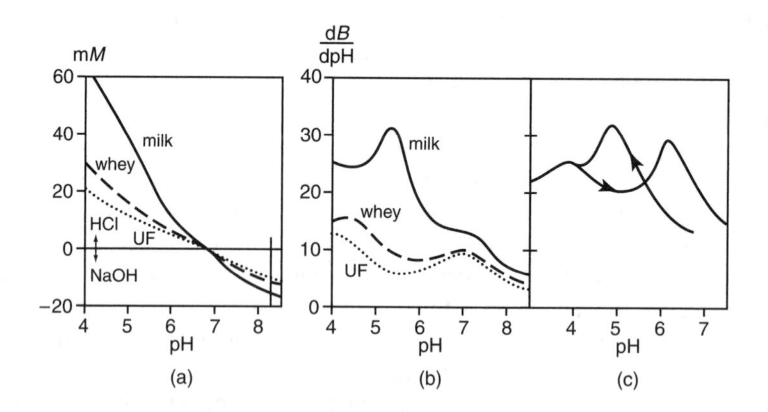

FIGURE 4.1 Titration curves (a) and buffering index (b and c) of milk, sweet whey, and milk ultrafiltrate, all expressed in mmol per liter. In (c) milk is titrated at first to pH 3 and subsequently to high pH. Approximate examples. (Figures [a] and [b] are derived from data in P. Walstra and R. Jenness, *Dairy Chemistry and Physics,* Wiley, New York, 1984, and [c] is adapted from results by J.A. Lucey et al., *Milchwiss.,* **48**, 268, 1993.)

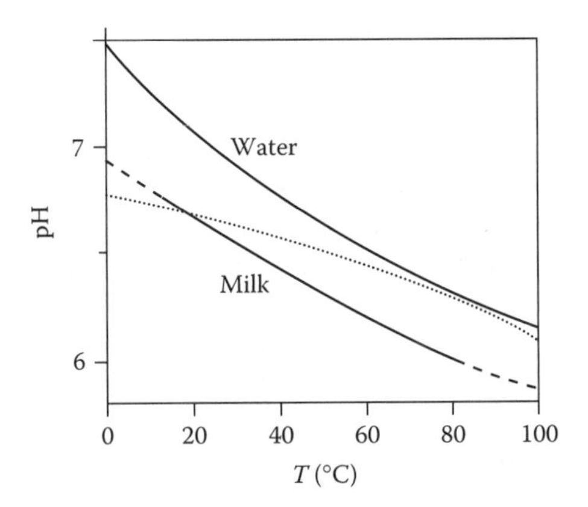

FIGURE 4.2 The pH of milk and water as a function of temperature. The dotted line refers to the pH at 20°C of milk serum obtained after ultrafiltration at the indicated temperature. Approximate values. (Adapted from P. Walstra and R. Jenness, *Dairy Chemistry and Physics,* Wiley, New York, 1984.)

qualitatively explains the hysteresis observed. For comparable reasons, the titration values can depend on the rate of addition (i.e., mmol · s^{-1}) of the titrant.

Figure 4.1a and Figure 4.1b also show results for whey and ultrafiltration permeate obtained from the same milk. These results indicate that the phosphates and the proteins primarily determine the buffering in the range given. p*K* values for acids in milk are given in Subsection 2.2.2, and the values for ionizable groups of protein in Table 2.12.

The pH of milk depends strongly on temperature. It is seen in Figure 4.2 that the pH of water also decreases with increasing temperature, and with it the neutral pH (that is, the pH value at which the activities of H^+ and OH^- are equal). The dissociation of most of the ionizable groups depends on temperature but the dependence varies widely. Influences of heat treatment, cooling, and some other processes on pH are discussed in Subsection 2.2.5. The acidity of milk can also decrease somewhat by heat treatment due to loss of CO_2, but at high temperatures (>100°C) acidity slowly increases due to the formation of acids (Figure 7.1).

In practice *titratable acidity* is often used for milk and milk products. It is defined as the buffer capacity between its own pH and pH ≈ 8.3; it is usually expressed in °N = mmol NaOH per liter of milk or milk product. The average contributions of milk components to the acidity are approximately:

2.2 °N per % casein, thus on average	~5.7 °N
1.4 °N per % serum protein	~0.9 °N
0.1 °N per mM colloidal inorganic phosphate	~1.0 °N
0.7 °N per mM dissolved inorganic phosphate	~7.8 °N
1.5–2 °N for other compounds	~1.7 °N
Total average	~17 °N

In most fresh milk samples the titratable acidity ranges from 14 to 21°N, (average about 17 °N). It tends to be high at the onset of lactation, say, 3 °N above the level reached later. The pH of most samples of milk is 6.6 to 6.8; average 6.7 at 20°C. This implies that the H^+ ion activity ranges from 0.16 to 0.25 $\mu mol \cdot l^{-1}$. Titratable acidity and pH show a weak negative correlation.

The titratable acidity of cream is lower than that of milk because the fat globules hardly contribute to the acidity. Naturally, the buffering capacity of milk serum, and hence whey, is lower than that of milk, as Figure 4.1 shows. Lipolysis (enzymatic production of free fatty acids from triglycerides) causes the titratable acidity to increase appreciably, especially in high-fat cream. Hydrolysis of esters (especially phosphoric esters) by enzymes can also decrease the pH.

When acid is added, or produced by bacteria, the titratable acidity increases proportionally, while the pH decreases (Figure 4.1). Lactic acid is largely dissociated (as most of the other organic acids are) until the pH falls below ~5.5. Titration down to this pH takes about equimolar quantities of lactic acid or of HCl (see also Figure 13.4). Formation of 0.1% lactic acid (MW = 90) increases the titratable acidity by $\rho^{20}/90 = 11.4$°N. Thus, titration provides an easy method to monitor the amount of lactic acid formed. Note that rather than the lactic acid, it is the reduced dissociation of acid groups and the enhanced dissociation of basic groups that are titrated. Essentially, the pH is a more meaningful parameter for characterizing milk acidity than titratable acidity; pH determines the conformation of proteins, the activity of enzymes, the dissociation of acids present in milk, etc. Undissociated acids cause an acid taste and can inhibit the activity of microorganisms.

4.3 REDOX POTENTIAL

The oxidation–reduction or redox potential (E_h) of a redox system at 25°C is given by

$$E_h = E_0 + 0.059 n^{-1} \log [Ox]/[Red] \qquad \text{(volts)} \qquad (4.1)$$

where E_0 = standard redox potential (a characteristic of each system that is dependent on temperature and especially on pH); n = number of electrons per molecule involved in the oxidation–reduction reaction; [Ox] and [Red] = molar concentrations of the compound concerned in the oxidized and reduced forms, respectively. Equation 4.1 only holds for reversible reactions. For $n = 1$, an increase of E_h by 0.1 V thus corresponds with an increase of the relative concentration of the oxidized form, e.g., from 50% ($E_h = E_0$) to 98%.

Table 4.1 shows some redox systems occurring in milk. The standard potential mentioned is not the only determinant because the concentration of each redox system present also determines to what extent a system of a different standard potential can be oxidized or reduced. Moreover, the concentration determines the sensitivity of E_h to additions such as oxidants — in other words, the poising capacity (which is comparable to buffering capacity).

TABLE 4.1
Standard Redox Potentials (E_0) of Some Redox Systems Important for Milk and Their Concentration in Fresh Milk (T = 25°C)

Redox System	n[a]	E_0 at pH 6.7 (V)	Concentration ($\mu Eq \cdot L^{-1}$)
Fe^{2+}/Fe^{3+}	1	+0.77	~ 4[b]
Cu^{+}/Cu^{2+}	1	+0.15	< 0.5
(Dehydro)ascorbate	2	+0.07	180–310[c]
Riboflavin	2	−0.20	4–14
Lactate/pyruvate	1	−0.16	([d])
Methylene blue	2	+0.02	11[e]

[a] Number of electrons transferred per molecule.
[b] Probably only partly reversible.
[c] In pasteurized milk usually less than 50% of this concentration.
[d] In fresh milk irreversible. Action and concentration depend on bacteria.
[e] Concentration in the methylene blue reduction test.

In fresh oxygen-free milk, $E_h \approx +0.05$ V and is mainly dependent on ascorbate. On holding the milk, ascorbic acid shows reactions according to

```
 O=C────┐              O=C────┐              O=C—OH
   |    |                |    |                |
HO—C    |     −2H      O=C    |      H2O     O=C
   ||   O    ⇌           |    O     ──→        |
HO—C    |              O=C    |              O=C
   |    |                |    |                |
 H—C────┘              H—C────┘              H—C–OH
   |                     |                     |
HO—C—H                HO—C—H                HO—C—H
   |                     |                     |
  CH2OH                 CH2OH                 CH2OH
L-ascorbic acid    Dehydro-L-ascorbic acid   Diketo-L-gulonic acid
```

The latter reaction is irreversible but only occurs if both riboflavin and O_2 are present; light acts as a catalyst. Riboflavin itself is not susceptible to O_2 but is highly light-sensitive.

In actual practice, however, fresh milk always contains O_2 and, accordingly, E_h is higher, i.e., +0.2 to +0.3 V. Eventually, only dehydroascorbate is left, which can be subsequently hydrolyzed. Upon heating of milk, free sulfhydryl groups (Subsection 7.2.2.1) are formed; these can cause a decrease of the E_h by about 0.05 V. The cysteine–cystine system itself does not contribute to the E_h because it is not reversible at neutral pH.

Bacterial action, e.g., lactic acid fermentation, removes O_2 from milk and produces reductants. As a result the redox potential decreases steeply, ultimately to –0.1 to –0.2 V, depending upon the bacterial species. In such cases, methylene blue, if added to milk, will be converted to the colorless reduced form. One takes advantage of this change of color when applying the methylene blue reduction test for estimating the number of lactose-fermenting bacteria in milk.

4.4 FLAVOR

Flavor perception is highly complex. Traditionally, two chemical senses are distinguished: taste and odor, which are sensitive to specific water-soluble and volatile compounds, respectively. Such compounds can be identified and characterized. It has long been known that a combination of taste and odor leads to the perceived flavor. Actually, flavor perception involves other stimuli as well, in a manner only partly understood. Mouthfeel and physical structure of the food can have significant effects.

The minimum concentration at which a flavor compound can be perceived, generally called the *threshold value*, varies widely among compounds, for the most part ranging between 10^3 and 10^{-4} mg per kg. This means that some compounds give a perceptible flavor even if present in minute quantities. The threshold value of a compound can also depend on the material in which it is present. For example, reported thresholds for methyl ethyl ketone are 3 mg/kg in water, 80 mg/kg in milk, and 20 mg/kg in oil. The difference between water and milk may be partly due to the ketone associating with the milk proteins, whereby its effective concentration (more precisely, its thermodynamic activity) is reduced. Most volatile flavor compounds are hydrophobic and readily associate with proteins, substantially decreasing flavor perception. It is also clear that substances that cause a strong flavor mask the perception of weak flavors. Moreover, a substance giving a pleasant flavor when present at low concentration often causes a quite unpleasant sensation if the food contains much larger amounts. However, all this depends on the individual observer because perception is, by its nature, purely subjective. The sensitivity for a given compound varies significantly among individuals. Moreover, flavor perception is significantly dependent on one's experience with various foods and on the conditions during flavor assessment.

Altogether, results given about the flavor of a food should be interpreted with care. Nevertheless, a great deal of important information has been obtained, especially about various off-flavors, i.e., identification of the compounds responsible and of the cause of their presence. Another important observation is that in nearly all cases quite a number of compounds contribute to the flavor sensation, especially if the observer considers the food to taste good; in other words, a 'flavor component balance' is often desirable.

The main nonvolatile flavor compounds in *fresh milk* are lactose and dissolved salts, which cause a sweet or salty taste, respectively. The sweetness caused by lactose is decreased by the dissolved salts and by the micellar casein in milk (Subsection 2.1.1). Nevertheless, the sweet taste generally prevails. The salty taste is prevalent if the Cl^-/lactose ratio is high, as in mastitic or late-lactation milk. Some volatile compounds, especially dimethyl sulfide, and also diacetyl, 2-methylbutanol, and some aldehydes are responsible for the characteristic flavor of fresh raw milk. However, skim milk and whole milk differ considerably in flavor. The fat globules are largely responsible for the 'creamy' or 'rich' flavor (although milk with a high fat-free dry-matter content also has enhanced 'richness'). It is not fully clear what causes the 'creaminess'; some fat-soluble compounds contri-bute, especially δ-lactones (see Chapter 2, subsection 2.3.1.4), and also free capric and lauric acid, indole (1-benzopyrrole), and skatole (3-methylindole); trace amounts of 4-*cis*-heptenal (resulting from autoxidation) can also contribute. It is often assumed that compounds from the fat globule membrane contribute to the creaminess, but the evidence is weak. It is quite likely that the physical presence of the fat globules plays an important part because creaminess can also be enhanced by other small spherical particles. Naturally, cream tastes much creamier than milk.

All in all, the flavor of milk is generally considered mild, especially when cold, which implies that few people dislike it. On the other hand, the mildness means that flavor defects are readily noticed.

Fresh milk may have *off-flavors* originating from the cow's feed. The compounds responsible for this enter the milk through the cow or from the air, and sometimes via both pathways. Examples are clover and garlic flavors. If the cow suffers from ketosis, for instance because its feed is deficient in protein, increased concentrations of ketones (especially acetone) are found in the milk. Consequently, the milk exhibits a typical 'cowy' flavor. Vacuum heating may remove part of such volatile flavor compounds if they are hydrophilic. Removal of the many fat-soluble compounds is more difficult. Hay feeding leads to the presence of coumarin (1-2-benzopyrone) in milk; this flavor compound is not always considered undesirable.

Microbial spoilage of milk may produce flavor defects; the various defects are referred to as acid, 'unclean,' fruity or ester, malty or burnt, phenolic, bitter, rancid, etc. Enzymatic spoilage includes lipolysis (Subsection 3.2.5). The resulting free fatty acids of 4 to 12 C atoms are responsible for a soapy-rancid flavor. Several proteolytic enzymes, including the native plasmin (Subsection 2.5.2.5), can attack casein, giving rise to bitter peptides.

Autoxidation of fat, as caused by catalytic action of Cu, can lead to 'oxidative rancidity' (Subsection 2.3.4). Several carbonyl compounds contribute to the off-flavor, especially some polyunsaturated aldehydes. The resulting flavor in milk is generally called 'tallowy'; in butter one speaks of 'oily' or, if the flavor is quite pronounced, 'fishy.' In milk a 'cardboard' flavor may also occur, which results from autoxidation of phospholipids; it can also be observed in skim milk. The phospholipids in the plasma appear to be oxidized readily. In sour-cream buttermilk, which is rich in phospholipids, this may lead to a 'metallic' flavor if the defect is weak and to a sharp (pungent) flavor if it is strong.

Flavor defects in milk can also be induced by *light*. Like Cu, light may catalyze fat autoxidation. Direct sunlight applied for 10 min or diffused natural light for a somewhat longer time often suffices. A tallowy flavor develops, not immediately, but only after some time. Light from fluorescent lamps especially can induce the cardboard flavor. Exposure of milk to light can also lead to the development of a 'sunlight flavor.' In the presence of riboflavin (vitamin B_2), oxidation of free methionine can occur, yielding methional (CH_3–S–CH_2–CH_2–CHO); moreover, free –SH compounds can be formed from protein-associated sulfur-containing amino acids. Presumably, these compounds jointly cause the sunlight flavor.

Moderate *heat treatment* of milk (say, 75°C for 20 s) causes the characteristic raw milk flavor to weaken so that a fairly flat flavor results. More intense heat treatment, e.g., 80 to 100°C for 20 s, results in a 'cooked' flavor, caused mainly by H_2S. This compound primarily derives from a protein of the fat globule membrane, which is the reason why cream is much more sensitive to the development of a cooked flavor than milk. (H_2S is also a strong inhibitor of Cu-catalyzed fat autoxidation; see Subsection 2.3.4.)

More intense heat treatment, e.g., 115°C for 10 to 15 min, leads to a 'sterilization' flavor, which can be especially strong in concentrated milk. Several flavor compounds result from the degradation of lactose, partly via the Maillard reaction (see Subsection 7.2.3); the main compounds causing off-flavor are maltol, isomaltol, and certain furanones. Other components that can be responsible are formed by the breakdown of fatty acids. Aliphatic methyl ketones are formed from decarboxylation of β-keto fatty acids, and lactones from γ- and δ-hydroxy fatty acids. The typical flavor of UHT milk is mainly due to H_2S as well as the mentioned ketones and lactones.

4.5 DENSITY

Mass density or volumic mass is the mass per unit of volume. It is expressed in $kg{\cdot}m^{-3}$ (SI units) or $g{\cdot}ml^{-1}$ (c.g.s. units; $1\ g{\cdot}ml^{-1} = 1000\ kg{\cdot}m^{-3}$.) The symbol is ρ or d. The density should be distinguished from the specific gravity (s.g.), i.e., the weight of a volume of the substance divided by the weight of an equal volume of water. Thus, s.g. $(T_1/T_2) = \rho\ (T_1)/\rho_{water}\ (T_2)$. Because $\rho_{water}(4°C) \approx 1\ g{\cdot}ml^{-1}$, s.g. $(T_1/T_2) \approx \rho(T)$ in $g{\cdot}ml^{-1}$ if $T = T_1$ and $T_2 = 4°C$. For example, $\rho^{20} \approx$ s.g. (20/4). Accordingly, unlike s.g., ρ depends strongly on T. Numerical values are given in the Appendix, Table A.10.

The density of a mixture of components, such as milk, depends on its composition and can be derived from

$$\frac{1}{\rho} = \sum\left(\frac{m_X}{\rho_X}\right) \tag{4.2}$$

where m_x is the mass fraction of component x, and ρ_x its apparent density in the mixture. A value of 998.2 $kg{\cdot}m^{-3}$ can be taken for ρ^{20} of water, about 918 $kg{\cdot}m^{-3}$ for fat, about 1400 $kg{\cdot}m^{-3}$ for protein, about 1780 $kg{\cdot}m^{-3}$ for lactose, and about 1850 $kg{\cdot}m^{-3}$ for the residual components (≈ 'ash' + 0.3%) of milk. All of these are apparent densities in aqueous solution, and not the densities of the components in a dry state (except for fat). This is because dissolution, especially that of low-molar-mass components, causes contraction. Generally, contraction increases with concentration, though less than proportionally.

The density of milk is rather variable. On average, ρ^{20} of fresh whole milk is about 1029 kg·m^{-3} provided that the fat is fully liquid. The latter can be realized by heating the milk to, say, 40°C and subsequently cooling it to 20°C just before determining the density; the fat then remains liquid (i.e., undercooled) for some time (see Subsection 2.3.5.2). Crystallization of fat causes the density to increase; for instance, the equilibrium density of whole milk at 10°C is about 1031 kg·m^{-3}. Further data are given in the Appendix, Table A.10.

The density of milk increases with increasing content of solids-not-fat (snf) and decreases with increasing fat content. From the density and the fat content (F), the dry-matter content (D) of whole milk can be approximately calculated. Taking $\rho_{\text{fat}} = 917$, $\rho_{\text{snf}} = 1622$ and $\rho_{\text{water}} = 998$ kg·m^{-3}, leads to

$$D = 1.23F + \frac{260(\rho^{20} - 998)}{\rho^{20}} \pm 0.25 \quad \%(\text{w/w}) \tag{4.3}$$

The standard deviation refers to the variation between individual cow samples.

4.6 OPTICAL PROPERTIES

The *refractive index*, n, of a transparent liquid is defined as the ratio of the velocity of light in air to that in this liquid. It depends on the wavelength of the light and decreases with increasing temperature. It is generally determined at a wavelength of 589.3 nm and at 20°C.

The refractive index of milk (about 1.338) is determined by that of water (1.3330) and the dissolved substances. Particles larger than about 0.1 μm do not contribute to n. Consequently, fat globules, air bubbles, or lactose crystals do not contribute to the refractive index of milk and milk products, though they may hamper the determination of n by the turbidity they cause. Casein micelles, though many are larger than 0.1 μm, do contribute to n because they are inhomogeneous (consisting of far smaller building blocks) and have no sharp boundary.

The difference between n of water and an aqueous solution is given by

$$n(\text{solution}) - n(\text{water}) = \Delta n \approx \rho \Sigma m_i r_i \tag{4.4}$$

where ρ is the mass density of the solution, m the mass fraction of a solute, and r is its specific refraction increment. The equation is not precise because the contribution of the various components (i) is not always additive. Values for r (for 589.3 nm wavelength and 20°C) are in ml/g (which implies that ρ has to be in g/ml):

Casein micelles, 0.207 (expressed per gram of casein)
Serum proteins, 0.187
Lactose, 0.140
Other dissolved milk components, ~0.170
Sucrose, 0.141

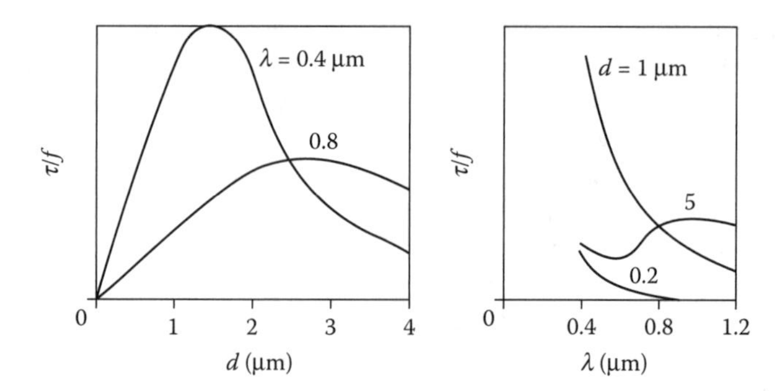

FIGURE 4.3 Light scattering by milk fat globules. Total turbidity (τ) per percentage of fat (f) as a function of globule diameter (d) and wavelength (λ). Globules should be of uniform size and the dispersion highly dilute.

With these data, the refractive index of milk products can be calculated. When concentrating milk or another liquid by evaporation, $\Delta n/\rho$ increases proportionally with the concentration factor. Since n can be determined easily, rapidly, and accurately (standard deviation 10^{-4} or less), it is a useful parameter to check changes in composition such as snf content.

Light scattering is caused by particles whose refractive index differs from that of the surrounding medium. For instance, $n_{\text{fat globule}}/n_{\text{plasma}} \approx 1.084$. Figure 4.3 gives some examples of light scattering as caused by fat globules. Because the fat globules in a sample of milk vary considerably in size, mean curves should be calculated. From results as presented in Figure 4.3, particle size or fat content can be derived through turbidity measurements on strongly diluted samples. The casein micelles also scatter light, though far less than fat globules do. This is because they are smaller in size and inhomogeneous; furthermore, the difference in refractive index with the solution is somewhat smaller.

Absorption: In the ultraviolet plasma as well as fat (double bonds) strongly absorb light, especially for wavelengths < 300 nm. In the near-infrared, numerous strong water absorption bands occur, but there are also some absorption bands that can be used to estimate the contents of fat, protein, and lactose in milk.

The *color* of an opaque material generally is the resultant of the scattering and absorption of visible light. Skim milk shows very little absorption, and its color is bluish because the small casein micelles scatter blue light (short wavelength) more strongly than red light (see Figure 4.3 for particles of 0.2 μm). Milk serum and whey look slightly yellowish due to the presence of riboflavin. This substance absorbs light of short wavelength, but its concentration in milk is only about 1.5 mg per liter: too low to bestow color to the milk. Milk scatters more light due to the presence of fat globules, and this scattering is not strongly dependent on wavelength. Hence milk is not blueish. Moreover, milk fat has a yellow color, primarily due to the presence of β-carotene. At a wavelength of 0.46 μm milk fat may give an absorbency of about 0.3 in a 1-mm optical cell. However, the carotene content of milk varies widely.

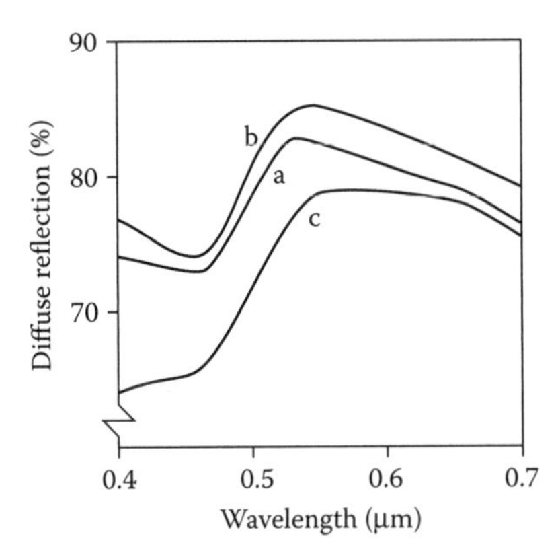

FIGURE 4.4 Diffuse reflection spectra of homogenized milk. (a) Unheated; (b) heated for 30 s at 130°C; (c) heated for 40 min at 115°C. The intensity is given as a percentage of the reflection of a purely white reference material. (Courtesy of H. Radema, Netherlands Institute for Dairy Research.)

Quantitative results can be provided by diffuse reflection spectra. When a sample of milk is illuminated with white light, it penetrates the milk, is scattered by the particles present and can lose intensity by absorption. Most of the light leaves the milk again, and its intensity can be measured as a function of the wavelength (λ). Figure 4.4 gives some examples: The unheated homogenized milk shows a spectrum where the reflection is quite high and not very dependent on λ, which means that the sample looks white (even whiter than unhomogenized milk). There is a weak dip in the curve near $\lambda = 0.46$ μm, undoubtedly due to the carotene in the fat globules; this results in a slightly creamy color. A moderately intense heat treatment causes the reflection to increase somewhat at all wavelengths, presumably due to serum protein denaturation and aggregation, which increases the number of scattering particles. This means that the milk has become a little whiter. Intense heat treatment causes the reflection to be significantly diminished at low wavelengths, resulting in a brownish color. This must be due to the formation of pigments by Maillard reactions (see Subsection 7.2.3).

4.7 VISCOSITY

This section is about viscous properties of milk and liquid milk products. Some basics of rheology will be briefly discussed for readers insufficiently familiar with it.

4.7.1 Some Fluid Rheology

A liquid can be made to flow by applying a stress to it. The stress σ equals force over area, units N/m^2 = Pa. The flow is characterized by a velocity gradient (or shear rate) $\mathrm{d}v/\mathrm{d}x$ (unit reciprocal seconds), meaning that the flow velocity v changes

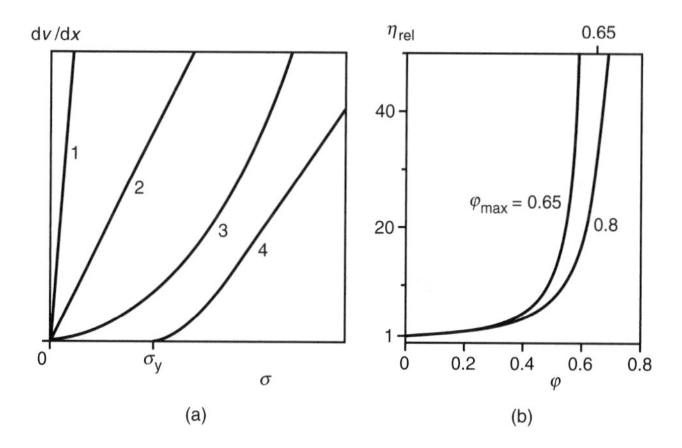

FIGURE 4.5 Viscosity. (a) Examples of flow curves: velocity gradient (dv/dx) as a function of (shear) stress (σ) applied. (b) Relative viscosity (η_{rel}) as a function of volume fraction (φ) calculated for two values of φ_{max}.

in a direction x, often perpendicular to the direction of flow. The relation between the two properties is given by

$$\sigma = \eta \frac{dv}{dx} \tag{4.5}$$

where the proportionality factor η (in Pa·s) is called the *viscosity*. This is illustrated in Figure 4.5a, curve 1 and curve 2, where the viscosity of 1 is smaller than that of 2.

Viscosity is a material property. For water at 20°C η = 1 mPa·s, and it markedly decreases with increasing temperature (see Appendix, Table A.10). The viscosity increases if other substances are added to water, be it small molecules (e.g., sugar), polymers (e.g., xanthan), or particles (e.g., fat globules). According to Einstein the relation is

$$\eta = \eta_s\,(1 + 2.5\varphi) \tag{4.6}$$

if the volume fraction φ is very small, e.g., <0.01. η_s is the viscosity of the solvent, in this case water, and the ratio η/η_s is called the relative viscosity η_{rel}. Notice that particle or molecule size is not a variable.

For higher φ values, more complicated relations are needed. For hard spherical particles, the Krieger-Dougherty equation is often used, which reads

$$\eta = \eta_s(1 - \varphi/\varphi_{max})^{-2.5\varphi_{max}} \tag{4.7}$$

where φ_{max} is the maximum volume fraction attainable; at $\varphi = \varphi_{max}$, the viscosity would increase to infinity. For monodisperse spheres, φ_{max} is about 0.7; for particles varying in size, its value can be significantly higher, e.g., 0.8 or even more, because small particles (e.g., casein micelles) may fit into the pores left between larger particles (e.g., fat globules). Some calculated examples are shown in Figure 4.5b. It is seen that addition of an amount of disperse material at initially low φ gives a much smaller increase in viscosity than the same amount added at high φ.

If more types of particles are present, various types of η_{rel} can be taken, depending upon what is considered the solvent. For instance, for milk and cream η_s could be that of milk plasma, of serum (i.e., without casein micelles), or of ultrafiltrate (i.e., also without soluble proteins). It should further be noted that the value of φ to be taken is an effective one that can be substantially higher than the net value. If the particles are anisometric, $\varphi_{eff} > \varphi_{net}$, the more so for stronger anisometry. To be sure, Equation 4.6 then does not hold anymore, but it can be modified to fit the results, and the trends remain the same. Also, for particles with a rough surface (see, e.g., Figure 9.7), φ is effectively increased. The value of φ_{eff} can be quite high for particle aggregates because these enclose a lot of, more or less immobilized, solvent (Figure 3.8). For polymeric substances the relationships are different and Equation 4.7 does not apply. Suffice it to say that random coil polymers give a high increase in viscosity per unit mass, the more so for a higher degree of polymerization (longer chains).

So far, we have considered Newtonian liquids, which are defined as those for which the viscosity does not depend on the shear stress or the velocity gradient. All one-component liquids behave in this manner. However, many other liquids are *non-Newtonian*. An example is given by curve 3 in Figure 4.5a: here the viscosity (stress over shear rate) decreases with increasing stress (or with increasing shear rate). Such a liquid is called shear (rate) thinning, and it has an apparent viscosity η_a, depending on the value of the shear stress or the shear rate (velocity gradient). This is further illustrated by curve 2 in Figure 4.6.

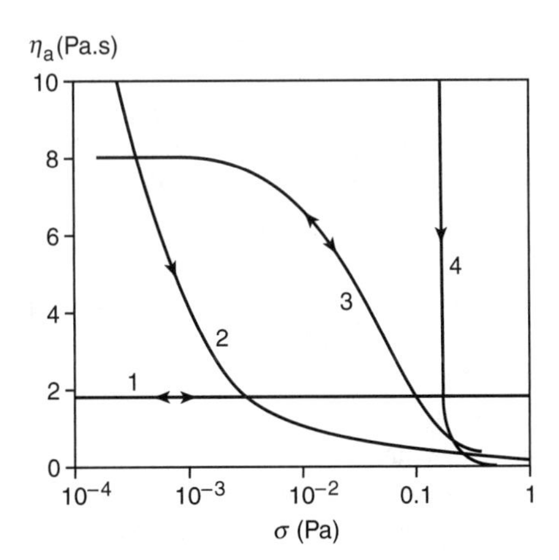

FIGURE 4.6 Apparent viscosity (η_a) as a function of shear stress (σ) applied. Curve 1: Newtonian behavior. Curves 2 and 3: shear rate thinning. Curve 4: material with a yield stress. The arrows indicate in what direction the stress is changed.

Another situation is given by curve 4 in Figure 4.5a. In this case, the material does not flow if a weak stress is applied. It will be deformed, but the deformation is elastic, which means that the material will return to its original shape upon release of the stress. Upon increasing the stress to a value called the *yield stress* σ_y, the material starts to flow, and upon further increase of the stress, the behavior becomes shear rate thinning.

Figure 4.6 (curve 1) shows a highly viscous Newtonian liquid (pure glycerol, $\eta = 1730\ \eta_{water}$) and three examples of non-Newtonian flow behavior. Curve 2 can be explained by weak attractive forces between quite small particles. The curve is not immediately reversible: upon increasing the stress more bonds between particles are broken and it takes some time for these bonds to re-form (see Subsection 3.1.3.1). This curve represents a sample of (aged) evaporated milk. Curve 3 is a typical example of a polysaccharide solution: 0.25% of xanthan, a thickening agent used in some milk products. It is seen to be Newtonian at very low stress and strongly shear rate thinning at higher stresses; for $\sigma > 10$ Pa it is again Newtonian but at a much lower viscosity. This flow behavior is fully reversible. The explanation is intricate. Curve 4 represents a liquid with a yield stress. Here, small particles or a mixture of polysaccharides form a (weak) network (hence, a gel) which is ruptured at the yield stress. Again, the behavior is not immediately reversible. If it is reversible it will take, say, 15 minutes for the weak gel to be restored. Curves like this are obtained when a small quantity, 10 to 50 mg/l, of κ-carrageenan is added to milk. Carrageenan and casein (micelles) jointly form a weak network.

Figure 4.6 shows a range of very small stresses. One pascal is the stress exerted under gravity by a vertical 'column' of water of 0.1-mm height. If a liquid has a yield stress of 1 Pa, its existence will not be noticed in practice: if the liquid is poured out of a bottle, the stress acting will always be far larger than 1 Pa. Nevertheless, very small stresses may occur and be of importance, especially during sedimentation (creaming or settling) of small particles. The stress exerted under gravity (acceleration g) by a particle of diameter d is roughly given by $dg\,\Delta\rho$, $\Delta\rho$ being the density difference. For a fat globule of 1 μm in milk, the stress would thus be of the order of 10^{-3} Pa, and the apparent viscosity needed to calculate the creaming rate (see Subsection 3.2.4.1) would have to be taken at that value. A glance at Figure 4.6 shows that η_a then can be very much higher than is measured in common rheometers, where the smallest stress applied is often about 1 Pa. It is possible that the liquid has a yield stress above 10^{-3} Pa, preventing sedimentation. (It should be mentioned, however, that σ_y can in practice be much smaller, say, by a factor of 10, at a long timescale.) Another point to be noted is that predicting creaming rate under gravity from the results of a centrifuge test may give false results because the stress exerted by a globule in the centrifuge may be by 10^3 or 10^4 times higher than that under gravity.

Shear rate thinning behavior is by no means restricted to low shear stresses, and a yield stress can also be quite high. We will see some examples for a number of dairy products.

4.7.2 Liquid Milk Products

For most products mentioned here, Equation 4.7 applies reasonably well, with $\varphi_{max} \approx 0.8$, because the particles involved tend to be quite polydisperse. For natural fat globules, the volume is ~1.12 ml/g of fat, for casein micelles ~4 ml/g of casein, and for serum proteins ~1.5 ml/g. The viscosity of 'particle-free milk', i.e., milk ultrafiltrate, is about 1.17 times that of water.

The influence of fat globules is shown in Figure 4.7A. Generally, the behavior of the liquid is Newtonian up to $\varphi_{fat} \approx 0.4$, unless d$v$/d$x$ is quite low. In raw cream or milk of low temperature, cold agglutination of fat globules occurs (see Subsection 3.2.4.2). It causes η to increase and to become shear rate thinning. Figure 4.7C gives examples; the liquid is shear rate thinning till dv/d$x \approx 10^3$ s^{-1}. It is also seen that even at 55°C, when cold agglutination is absent, a little shear rate thinning occurs; presumably, this is due to a slight depletion aggregation of the fat globules caused by casein micelles.

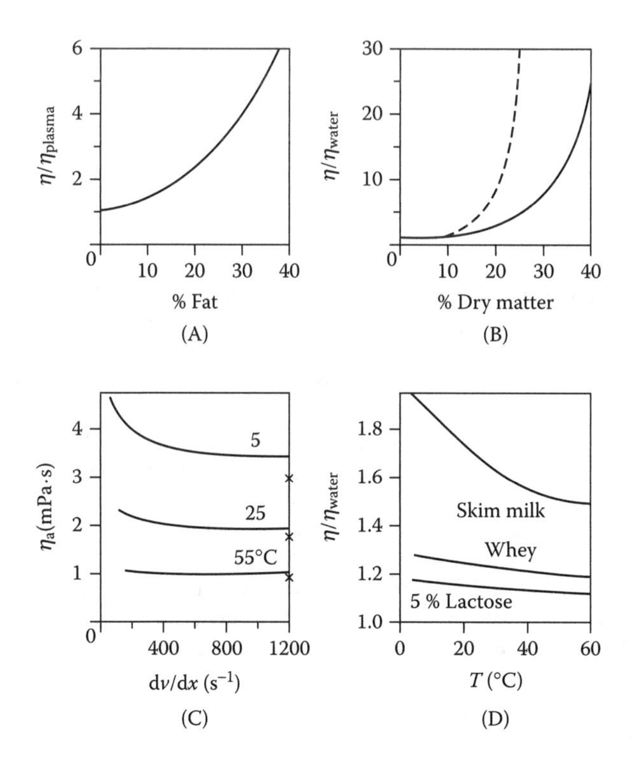

FIGURE 4.7 Influence of some variables on viscosity (η). (A) Milk and cream at 40°C. (B) Skim milk and concentrated skim milk at 20°C; the broken line refers to ultrafiltered skim milk. (C) Raw milk of 5% fat; dv/dx = shear rate; η_a = apparent viscosity; crosses indicate $2 \times \eta_{water}$. (D) Influence of measuring temperature (T).

The casein micelles contribute significantly to the value of η (Figure 4.7D) because of their high voluminosity (Subsection 3.3.1.1). Compare also both curves in Figure 4.7B. In the UF-concentrated skim milk, the highly voluminous casein micelles make up a far greater part of the dry mass — for a given percentage of dry matter — than in the case of evaporated skim milk.

The viscosity significantly decreases with increasing temperature, for a considerable part due to the temperature dependence of the viscosity of water. In addition it can be seen in Figure 4.7D that the contribution of the micelles also depends on the temperature. At low temperatures, the voluminosity of the micelles is markedly increased and part of the β-casein becomes dissociated from the micelles. Consequently, the viscosity increases steeply. Heat treatment of skim milk to such a degree that the serum proteins become insoluble causes an increase in viscosity by about 10%. This may be explained by the increase of the voluminosity of the serum proteins.

Increasing the pH of milk also increases its viscosity, presumably owing to additional swelling of casein micelles. Slightly decreasing the pH usually leads to a small decrease of η. A more drastic pH decrease causes η to increase due to the aggregation of casein. Homogenization of milk has little effect on η, but homogenization of cream may considerably enhance apparent viscosity (see Section 9.7 and Subsection 17.1.4).

Further data are given in Table A.10.

Suggested Literature

Several physical properties of milk: J.W. Sherbon, in N.P. Wong, Ed., *Fundamentals of Dairy Chemistry,* 3rd ed., Van Nostrand Reinhold, New York, 1988, Chapter 8; H. Singh, O.J. McCarthy, and J.A. Lucey, in P.F. Fox, Ed., *Advanced Dairy Chemistry,* Vol. 3, *Lactose, Water, Salts and Vitamins,* 2nd ed., Chapman & Hall, London, 1997, Chapter 11.

The same book contains a chapter on flavor: P.L.H. McSweeney, H.E. Nursten, and G. Urbach, Flavours and off-flavours in milk and dairy products, Chapter 10.

More specific on the flavor of milk: H.T. Badings, in H. Maarse, Ed., *Volatile Components in Foods and Beverages,* Dekker, New York, 1991, Chapter 4.

More detailed information on solution properties, optical properties and fluid rheology: P. Walstra, *Physical Chemistry of Foods,* Dekker, New York, 2003.

The principles of fluid rheology: H.A. Barnes, J.F. Hutton, and K. Walters, *An Introduction to Rheology,* Elsevier, Amsterdam, 1989.

5 Microbiology of Milk

Milk must be of good hygienic quality. This is essential in terms of public health, the quality of the products made from milk, and the suitability of milk for processing. Components that are foreign to milk but enter the milk via the udder or during or after milking, as well as any changes occurring in the milk, are often detrimental to its quality. These matters are the subject of milk hygiene. Microbial, chemical, and physical hygiene may be distinguished. Thus, microorganisms may produce a health hazard (food infection or food poisoning) or spoil the milk, e.g., because they turn it sour during storage. Light-induced off-flavors, fat oxidation, and fat hydrolysis result from chemical or enzymic transformations. Furthermore, compounds that are potentially harmful to the consumer, such as antibiotics, disinfectants, pesticides, and heavy metals, may enter the milk.

In this chapter, microbiological aspects of milk hygiene are discussed. Chemical contamination of milk is covered in Subsection 2.6.2.

5.1 GENERAL ASPECTS

Milk is a good source of nutrients and edible energy, not only for mammals but for numerous microorganisms, which thus can grow in milk. These microorganisms are primarily bacteria, but some molds and yeasts can also grow in milk. In this section, some general aspects of these microorganisms, their analysis, growth, and inhibition in milk will be discussed.

5.1.1 MICROORGANISMS

Microorganisms are living creatures that are not visible with the naked eye. They must not be thought of as being sharply differentiated from the world of macroscopic life, but their small size lets them disappear beyond the limits of our unaided vision. On looking into the microscope, the extended world of algae, protozoa, yeasts, bacteria, and viruses becomes brightly illuminated. Usually, in the macroscopic world, the properties of plants distinguish them clearly from animals. But when one cannot clearly see the creatures, it is much more difficult to place them in any kingdom. Basically, among the lower and more primitive forms, plants and animals are becoming similar. Many microorganisms classified as plants possess some animal characteristics and vice versa. Therefore, a separate kingdom for microorganisms has been suggested: the protists. This kingdom includes organisms that differ from animals and plants through their limited morphological differentiation and that are often unicellular. On the basis of their cell structure, two different groups can be distinguished. The higher protists have a cell structure similar to that of animals and plants; they belong to the eukaryotes. The eukaryotic cell possesses a real nucleus, surrounded by a membrane, and organelles, such as mitochondria or chloroplasts (similar to plants). Algae, protozoa, yeasts, and molds belong to this group. The lower protists comprise bacteria and cyanobacteria; they are prokaryotes and differ considerably from eukaryotes. The DNA (genetic material) of the prokaryotic cell is not present in a nucleus but lies free in the cytoplasm, and organelles are absent.

The viruses, as noncellular particles, are different from all other organisms. They are not self-reproducible and require living cells for their multiplication.

Because bacteria are the principle microorganisms associated with milk and dairy products, the following subsection (5.1.2) contains some further basic information on their properties. Subsection 5.1.3 considers yeasts and molds. Bacterial viruses, the bacteriophages, will be discussed in Chapter 13.

5.1.2 Bacteria

Bacteria include the smallest living organisms known. Their small size is typical and has important consequences not only for the morphology, but also for the activity and flexibility of the metabolism of bacteria. The diameter of most of bacteria is less than 1 μm, and their average volume is 1 μm^3. Many bacteria are rod-shaped, others are spherical, and some are spirally curved. Reproduction in bacteria typically occurs by binary division and is universally independent of sexual events; it is asexual. Sexual events can occur in bacteria, but reproduction is not associated with them. Cells of certain unicellular bacteria tend to adhere to each other following cell division. The cells accumulate in characteristic arrays that reflect the planes in which cell divisions occur. Thus, more or less regular clumps, sheets, or chains of cells are formed (Figure 5.1).

In a few groups of bacteria, the cells may enter a dormant stage. They form *endospores*, usually one per cell, by differentiation of a portion of the protoplast into a typically thick-walled, relatively dehydrated unit that is released from the remainder of the cell when the cell lyses. This property is typical for the genera *Bacillus* and *Clostridium.*

An important taxonomic characteristic for bacteria is the *Gram stain reaction.* This reaction is determined by microscopic examination of cells that have been successively stained with the basic dye crystal violet, treated with an iodine solution,

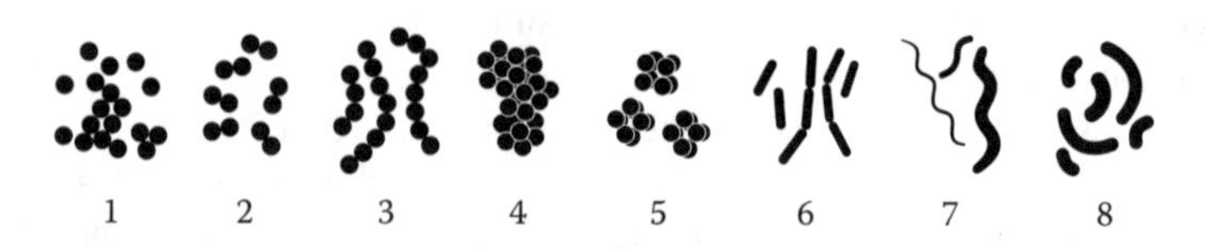

FIGURE 5.1 Forms of unicellular bacteria: (1) Micrococci, (2) Diplococci, (3) Streptococci, (4) Staphylococci, (5) Sarcina, (6) Rods, (7) Spirilla, (8) Vibrios, Pseudomonads. (From H.G. Schegel, *Algemeine Mikrobiologie,* Thieme Verlag, Stuttgart, 1985. With permission.)

and rinsed with alcohol. Gram-positive cells retain the violet stain, whereas Gram-negative cells are decolorized by alcohol. The final step in the procedure is application of a red stain (usually safranin) so that Gram-negative cells can be visualized. The Gram stain reaction is correlated with other properties of the bacteria, such as the structure of their cell envelope.

The bacterial *cell envelope* consists of the cytoplasmic membrane and the cell wall. The cytoplasmic membrane has common characteristics for all bacteria and has a bilayer structure (unit membrane) as found in animal and plant cells. Its presence is essential to the maintenance of an osmotic barrier against the external environment and segregates the protoplasmic activity into a discrete cellular unit, the protoplast. The cytoplasmic membrane contains the enzyme systems of the electron transport chain and oxidative phosphorylation, systems for the active transport of solutes and excretion of waste products as well as the synthetic apparatus necessary for the production of external layers. The cytoplasmic membrane is surrounded by the cell wall, which is usually rigid and, indeed, is responsible for the shape of the bacterium. Bacterial cell walls contain peptidoglycan, a unique polymer composed of repeating units of hexosamine-muramic acid and a small peptide. This polymer forms a sacculus around the bacterial protoplast and protects it against bursting when under considerable osmotic pressure due to solutes, and it holds the cell content within the fragile cytoplasmic membrane. Because peptidoglycan is unique in bacteria and essential for their survival, several successfully applied antibiotics, such as penicillin, act specifically on steps in the biosynthesis of this polymer. Examination of high-resolution electron micrographs of sections through Gram-negative and Gram-positive bacteria reveals gross morphological differences. The cytoplasmic membrane of Gram-positive bacteria is encased in a thick peptidoglycan layer and, in some cases, in an extracellular capsule (see the following text). The envelope of Gram-negative bacteria is more complex (Figure 5.2). The cell wall petidoglycan appears to be less substantial and not as closely associated with the cytoplasmic membrane as the equivalent structure in Gram-positive bacteria. The most notable feature of the Gram-negative envelope

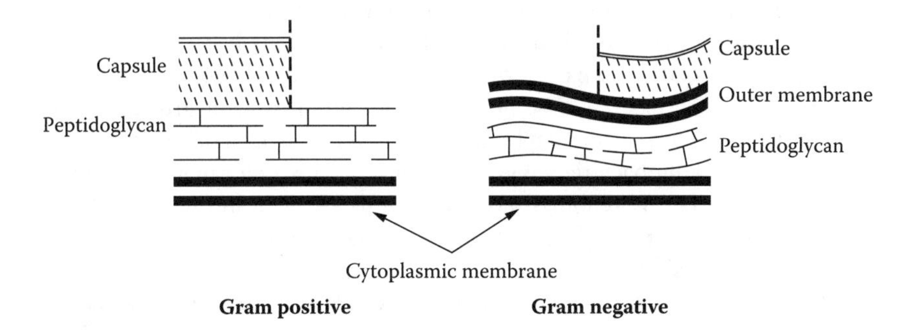

FIGURE 5.2 Diagrammatic representation of the cell envelope of Gram-positive and Gram-negative bacteria. (Redrawn from S.M. Hammond, P.A. Lambert, and A.N. Rycroft, *The Bacterial Cell Surface,* Croom Helm, Australia, 1984.)

is the presence of a second membrane outside the peptidoglycan, the outer membrane. Also, this membrane consists of lipid/protein bilayers; however, beyond this superficial resemblance, few structural or physiological similarities between the cytoplasmic and outer membranes exist, reflecting the differing roles of the membranes in the life of the bacterium. The outer membrane of Gram-negative bacteria has no role in electron transport and has extremely limited enzymic activity. It is distinguished by the presence of certain proteins (porins) and lipopolysaccharide, which are both unique to the envelope of Gram-negative bacteria. The outer membrane constitutes a barrier, making the surface of these bacteria less permeable than that of Gram-positive bacteria for a wide variety of molecules, e.g., certain antibiotics. The lipopolysaccharide in the outer membrane is important to the survival and the growth of the bacterium in the animal host. It aids resistance to the emulsifying action of bile salts and to host defense mechanisms such as phagocytosis.

The surfaces of many bacteria are covered with an extracellular substance collectively termed the *slime material* or *glycocalix*. When present, this viscous polymer may be organized neatly around the outside of the bacterial cell wall as a capsule. Frequently, the polymer is not smoothly distributed about the cell and occurs in more disorganized configurations, and is then generally referred to as slime layer. The slime layer or capsule appears to be composed of a secretory polysaccharide or polypeptide substance that is distinct from the cell wall of the bacterium. Most bacteria synthesize slime polymer, especially under natural conditions, and it has been suggested that the polymer may contribute to the survival of the cell. Some lactic acid bacteria produce *exopolysaccharides* during growth in milk, which may affect the structure of the fermented product. Encapsulation may also be responsible for the invasiveness of certain pathogenic bacteria in establishing infections and may assist these organisms in escaping the cellular immune responses of the host.

Several bacteria are *motile*, and locomotion in these bacteria is due to the activity of flagella, of which there may be one to several per cell. The distribution of flagella on the cell is used as a primary taxonomic criterion among bacteria. Two general patterns occur: polar and peritrichous (lateral) distribution. Other mechanisms for locomotion than by flagella occur in some bacteria.

All the morphological characteristics mentioned above, together with several physiological properties (e.g., the need for oxygen during their growth: aerobic and anaerobic bacteria), are important to classify the bacteria. In the binary nomenclature system, bacteria are given a genus and a species name. The genera are grouped in a family (ending in *aceae*) and these again in orders (ending in *ales*). In Table 5.1, some examples of the major families of bacteria that contain representatives possibly associated with milk are given.

Even if a complete taxonomic classification of bacteria seems feasible, one has to realize that these organisms are very variable and flexible. Although they do not show many morphological differences, their physiological properties, on the contrary, are extremely variable. This variability and, at the same time, flexibility is supported by their broad biosynthetic capacity and the simple structure of their

TABLE 5.1
Some Genera of Bacteria Possibly Associated with Milk

Family	Genus	Morphology	Motility	Gram Reaction	Oxygen Requirement
Micrococcaceae	*Micrococcus*	Coccus	–	+	Aerobic
	Staphylococcus	Clump of cocci	–	+	Aerobic
Lactobacillaceae	*Lactococcus*	Diplococcus	–	+	Aerotolerant
	Streptococcus	Chain of cocci	–	+	Aerotolerant
	Lactobacillus	Rod, chain	–	+	Aerotolerant
Bacillaceae	*Bacillus*	Rod, spores	–	+	Aerobic
	Clostridium	Rod, spores	–	+	Anaerobic
Enterobacteriaceae	*Escherichia*	Rod	+ (Peritrichous)	–	Facultative aerobic
	Salmonella	Rod	+ (Peritrichous)	–	Facultative aerobic
Pseudomonadaceae	*Pseudomonas*	Rod	+ (Polar)	–	Aerobic

genetic material. As a reaction to changes in the environment, bacteria may adapt themselves physiologically and genetically, giving rise to variants of bacteria that differ in properties from the original ones. These adaptations may be due to different levels of expression of certain genes, thus influencing protein (enzyme) synthesis, or due to mutation of certain genes. In both instances, a bacterium may emerge that is better equipped to combat the changed circumstances. In the course of time, many bacterial variants, subspecies, and strains have been discovered. Recent advances in the genetic analysis of bacteria have helped to discriminate between them. For example, the genus *Lactococcus* harbors the well-known species *lactis*, which consists of two subspecies, *lactis* and *cremoris*. The first subspecies has a variant that is famous for its capacity to convert citrate into diacetyl. From this bacterium, *Lactococcus lactis* ssp. *lactis* biovar. *diacetylactis*, several strains are known. Another example is the occurrence of several strains in the Gram-negative bacterium *Escherichia coli.* Some of these strains are saprophytic and others enteropathogenic. The multiplicity of bacterial types and their classification are readably presented in the literature. The systems of classification are based on an international code of nomenclature of bacteria, and *Bergey's Manual* (8th edition, 1974) is still widely used in this respect.

5.1.3 Yeasts and Molds

Yeasts and molds are eukaryotic microorganisms with a relatively large size and evident nuclear structures. These properties clearly distinguish them from bacteria. They grow readily on artificial culture media, but in methods of reproduction they are more complex than bacteria. Because of their enzymatic activities, many common species of yeasts and molds are of great importance in medicine and industry. Some cause fermentations that yield valuable substances such as ethyl and isopropyl alcohols and acetone. Others cause damage through decay of many kinds of organic matter. Some molds and yeasts are causes of diseases in plants, animals, and human beings. Other molds such as *Penicillium*, from which penicillin is produced, assumed enormous importance as the source of antibiotics. In the dairy industry, specific molds and yeasts are essential for the ripening of certain cheese types (Chapter 25 and Chapter 27). Most molds are able to grow in situations in which yeasts and bacteria cannot survive because of high osmotic pressure, acidity, or low water content. Molds are characteristically strict aerobes, whereas yeasts can grow under aerobic as well as anaerobic conditions.

Differentiating between yeasts and molds is sometimes difficult because the transition in form and manner of reproduction, from one group to another, is so subtle that it is hardly possible to draw sharp lines of demarcation. Ordinarily we think of yeasts as permanently microscopic organisms with ovoid or elliptical cells, each cell living as a complete individual. Molds typically form greatly elongated, branched hair-like filaments or hyphae, which grow to form microscopic, tangled masses called mycelia. A difficulty in the classification arises from the fact that while some of these fungi form more or less definite mycelia under certain conditions of growth and nutrition, typical yeast-like cells are formed under other conditions.

Yeasts cells are on average larger than bacterial cells. Most oval yeasts range around 5 μm in diameter by 8 μm in length, yielding a volume of around 100 μm^3, which is approximately 100 times that of bacteria. The diameter of the filaments of most species of molds is relatively large, ranging from 5 μm to 50 μm or even more (often visible with the naked eye). The length of filaments may be enormous. An entire mold mycelium may cover several square centimeters. Cottony and powdery growth of mycelia in black, green, yellowish, or white colors may be seen on jam, old bread, fruit, or cheese.

Yeasts and molds reproduce both asexually and sexually. Asexual reproduction in yeasts goes mostly via budding of large mature cells giving rise to one or more daughter cells, which are at first much smaller and which may remain attached to the parent cell till the division is complete. Several well-defined forms of asexual reproduction are known to occur in molds. Molds with septate hyphae often form a number of rather closely spaced divisions, resulting in the separation of a number of short, more or less ovoid cells, which are called arthrospores or oidia. The milk fungus *Geotrichum candidum* exhibits this type of asexual reproduction. Other molds form sporangiospores by repeated asexual nuclear division within globular envelopes called sporangia. When the sporangium ruptures, the sporangiospores are released. The well-known mold genus *Mucor* forms this type of spores. Conidiospores are formed at the free end of fertile hyphae by several molds. In stead of being enclosed in a sporangium, they are free, sometimes being produced in long chains. The size, form, and arrangement of the often branched hyphae (conidiophores), which form and support conidiospores, as well as the color of the conidia, are distinctive of the different genera and species. Molds in the genera *Aspergillus* and *Penicillium* form conidiospores. In several varieties of yeasts and molds also, sexual reproduction has been observed. In this type of reproduction two cells, generally called gametes, fuse and form characteristic bodies containing several spores with nuclear material from both gametes.

5.1.4 Enumeration of Microorganisms

Analytical tests for bacterial counts are routinely done to characterize the microbial pollution in milk samples. The so-called total bacterial count is typically determined by standard plate counting, which measures all bacteria able to form colonies on a nutrient agar medium within 48 h under aerobic conditions at 32°C. The samples are diluted so that plates are obtained that show only about 30 to 300 colonies. The numbers of colonies multiplied by the dilution gives the concentration of bacterial cells in the original sample. This concentration is referred to as colony-forming units (CFUs). Because several bacteria tend to remain attached to each other after division, the colony count may be much smaller than the actual numbers of living cells present. This is especially true for *Lactococcus* and *Streptococcus* species, as well as for some species of *Bacillus* and *Lactobacillus*. For the application of the (diluted) samples, several alternative but less commonly applied techniques exist, including loop count and membrane filter count.

The method of serial dilution can also be used by inoculating 1-ml quantities of decimal dilutions of the sample in triplicate or in fivefold into series of tubes with broth. From the number of tubes with visible growth after incubation, the most probable number of bacteria present in the undiluted sample can be calculated using mathematical tables. This method can also be used for the enumeration of specific groups of bacteria by adding special indicator substances to the broth. For example, it is standard practice to add lactose and to note the highest dilution of sample in which lactose fermenters (acid and gas) are found.

A notable rapid method for the determination of total bacterial counts in milk, for example with the 'Bactoscan,' utilizes fluorescent staining. In this technique, somatic cells are chemically degraded and, with other particles, separated from bacterial cells by centrifugation. Bacterial cells are then stained with acridine orange and irradiated with blue light, which causes orange light pulses to be emitted from live bacteria. These pulses are detected by a photodetector fitted to the objective of an epifluorescence microscope, beneath which the stained bacteria are channeled. Differences in acridine orange intercalation into cell DNA cause dead cells to emit green light, whereas live cells emit orange light, thus ensuring that the Bactoscan only counts individual live bacterial cells. Calibration of the Bactoscan using reference standards allows the counts found to be translated into CFUs. The Bactoscan method is unique in that it counts individual cells rather than CFUs, leading to values of bacterial counts that may be significantly higher than corresponding CFU values, particularly in the presence of bacteria that form clusters or chains.

Although total bacterial cell counts are useful to ensure that milk meets specific regulations, they are of less utility for identification of the source of bacterial contamination and for assessing risks to milk quality posed by a particular bacterial population. Selective tests that detect and quantify a specific type or group of bacteria can prove to be more useful. The identity of dominant organisms in a given population can often suggest a possible contamination source or route and can help to assess bacterial threats to milk quality and safety. Numerous selective and differential tests have been developed to determine the presence or absence of specific types of bacteria in milk. Procedures have been established to detect and quantify, among others, psychrophilic bacteria, thermophilic organisms (terms defined below), proteolytic bacteria, coliforms, enterococci, lactic acid bacteria, or aerobic bacterial spores. Also, tests for counting yeasts and molds are available.

Any analytical technique to characterize the bacterial population present in milk has its limitations. No one test can detect all bacteria and give a complete picture of the microbial population, not even the nonselective tests designed to determine total bacterial numbers. Also, knowledge of the total microbial mass present is generally more informative than the number of microorganisms. Therefore, microbial analysis must involve deciding which tests will provide the most useful information about the microbial population of the particular product or situation examined.

5.1.5 Growth

Bacteria multiply by division. Every cell division yields two new bacterial cells. The multiplication is a geometrical progression $2^0 \rightarrow 2^1 \rightarrow 2^2 \rightarrow 2^3 \cdots \rightarrow 2^n$. If a growing bacterial culture contains N_0 cells per milliliter, the bacterial count N after n divisions is

$$N = N_0 \cdot 2^n \tag{5.1}$$

or

$$\log N = \log N_0 + n \log 2 = \log N_0 + 0.3\,n \tag{5.1a}$$

The time needed for a full cell division thus determines the growth rate. It is called the generation time g, and can be derived from the number of divisions occurring during a certain time t:

$$g = t/n \tag{5.2}$$

Consequently, in well-defined conditions the count N after an incubation time t can be calculated from Equation 5.1a and Equation 5.2, if N_0 and g are known:

$$\log N = \log N_0 + 0.3\,t/g \tag{5.3}$$

Generation time g depends on several factors. In milk, the bacterium species (or strain) and the temperature are of special importance. Other factors involved are pH, oxygen pressure, and concentrations of inhibitors and nutrients, which are all fairly constant in raw milk.

Growth of bacteria thus means an increase in the number present. Current methods for determination generally give the number of CFUs per milliliter. Because this number does not always reflect the actual number of cells present (Subsection 5.1.4), determination of the biomass of bacteria present would be preferable in some cases in order to estimate the real growth rate. The same holds true for yeasts and molds also.

The preceding equations apply to the exponential-growth phase of the bacteria (sometimes called logarithmic or log phase). Figure 5.3 illustrates the various growth phases that can be distinguished. During the lag phase, bacteria do not multiply, primarily because their enzyme system needs adaptation, enabling them to metabolize the nutrients in the medium. The duration of the lag phase closely depends on the physiological state of the bacteria, the temperature, and the properties of the medium. During the exponential phase, the growth rate remains at a maximum until the stationary phase is approached. In the latter phase, some growth still occurs, together with dying off. The decrease in the growth rate is usually caused by the action of inhibitors formed by the bacteria themselves, or by a lack of available nutrients. Eventually, the stationary phase turns into the dying-off phase, during which the count decreases. These two phases are of special importance for the quality and the keeping quality of milk. In fermented milk products also, these phases are essential.

Temperature has a significant effect on bacterial growth. Lowering the temperature retards the rate of nearly all processes in the cell, thereby slowing down growth and decreasing fermentation rate (e.g., acid production). Moreover, it

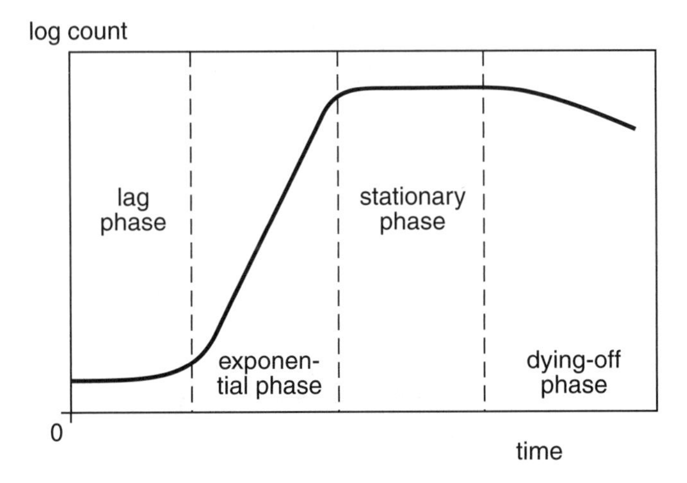

FIGURE 5.3 Growth curve of a bacterial culture.

extends the duration of effectiveness of some of the natural bacterial inhibitors in milk. Furthermore, many bacteria coming from a medium such as dung or teat surface and entering a substrate such as milk must adapt themselves to the new medium, hence the lag phase. At a lower temperature, the lag phase thus will last longer. The extent to which a lowering of the temperature affects bacterial growth depends on the type of organisms present.

For convenience, bacteria are grouped into ranges of *temperature* preference for biological activity. These groups include the psychrophiles, which thrive at low temperatures (0 to 20°C), the mesophiles, which have growth optima in the range between 20 and 45°C, and the thermophiles, which prefer temperatures of 45 to 60°C for optimal growth. The preferred temperature range is specific for each species of bacterium and, frequently, the range is surprisingly narrow, especially for organisms that occupy unique ecological habitats such as boiling springs or the mammalian intestinal tract. For example, the response of the bacterial growth of *E. coli* to changes in temperature is depicted in Figure 5.4. The minimum, maximum, and optimum temperatures of a bacterium are referred to as cardinal temperatures. Note that the curve in Figure 5.4 is typically asymmetric, with a steep decline of the growth above the optimum temperature. Two other terms have been introduced in food microbiology to characterize the temperature preference of bacteria prevailing in the food environment. These are the psychrotrophs, a group of Gram-negative and Gram-positive rods that may grow in foods such as milk at refrigerator temperatures, although they have an optimum temperature between 20 and 30°C. The other group consists of thermoduric bacteria, which are often Gram-positive and are able to grow up to temperatures around 50°C.

Table 5.2 gives some examples of the effect of temperature on generation time. It shows that lactic acid bacteria will not spoil cold-stored milk, and that at 30°C pseudomonads grow more slowly than other bacteria. The temperature dependence of the growth rate has consequences for the keeping quality of milk, as is shown

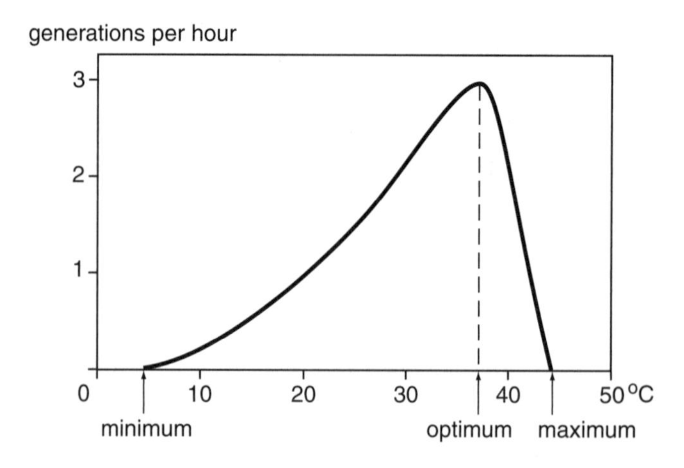

FIGURE 5.4 The response of the growth of *Escherichia coli* to changes in temperature.

TABLE 5.2
Generation Time (h) of Some Groups of Bacteria[a] in Milk

Temperature (°C)	5	15	30
Lactic acid bacteria	>20	2.1	0.5
Pseudomonads	4	1.9	0.7
Coliforms	8	1.7	0.45
Heat-resistant streptococci	>20	3.5	0.5
Aerobic sporeformers	18	1.9	0.45

[a] Within these groups of bacteria, generation time varies widely among species and strains. The values mentioned are approximately true for the fastest-growing representatives at the given temperatures. The lag phase is not included in the figures given.

in Table 5.3. At high storage temperatures, milk has poor keeping quality, even if its initial count is low; it should be processed within a few hours after production.

Figure 5.5 shows that a low initial count and a low storage temperature are essential. Whether milk is kept at a low or at a higher temperature, a lower initial count always means that it takes more time for the milk to spoil. Clearly, the combination of low initial count and low storage temperature is to be preferred. It is important to note that for raw milk to be kept for several days, the type of contamination may be of greater importance than the total count. For example, contamination of milk by 10^5 ml^{-1} mastitis bacteria has less effect on the keeping quality at low temperature than contamination by 10^3 ml^{-1} psychrotrophs. These aspects are further illustrated in Figure 5.6.

TABLE 5.3
Approximate Example of the Effect of the Keeping Temperature of Raw Milk on Its Count after 24 h, and on Its Keeping Quality (Initial Count 2.3×10^3 ml^{-1})

Milk Held at (°C)	Count after 24 h (ml^{-1})	Keeping Quality[a] (h)
4	2.5×10^3	>100
10	1.2×10^4	89
15	1.3×10^5	35
20	4.5×10^6	19
30	1.4×10^9	11

[a] Keeping quality is defined here as the storage time during which milk remains suitable for processing (count not exceeding 1×10^6 ml^{-1}).

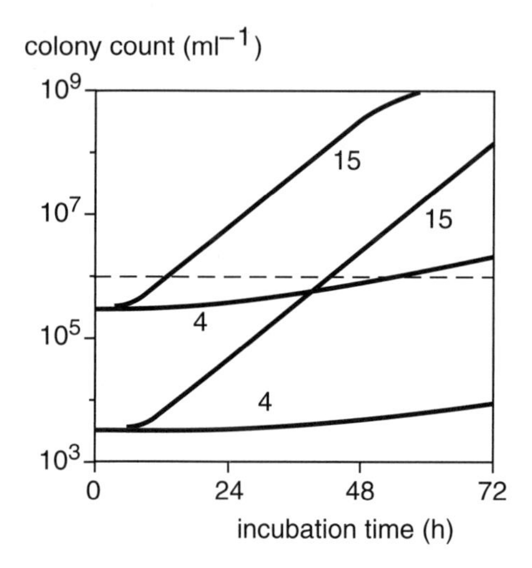

FIGURE 5.5 Change of colony count during the storage of milk of two initial counts at two temperatures (4 and 15°C). Approximate examples. The broken line marks the critical number of counts in which spoilage generally becomes manifest.

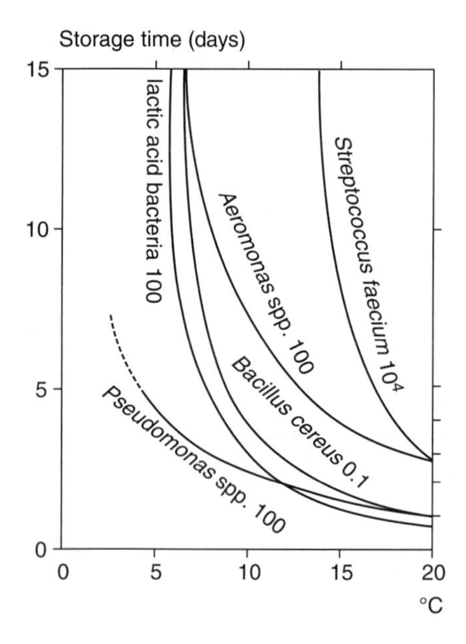

FIGURE 5.6 Time needed to reach a count of 10^6 ml^{-1} bacteria when keeping raw milk at various temperatures. The bacterium considered and its assumed initial count (in ml^{-1}) are indicated. These are approximate examples meant to illustrate trends.

Growth of microorganisms in raw milk is generally undesirable. Of all treatments known to reduce the growth of microorganisms, only lowering of the temperature is generally feasible for raw milk. Heat treatment kills bacteria (Subsection 7.3.3). Milk contains some natural growth inhibitors (Subsection 5.1.6), but addition of a bacterial inhibitor to milk generally is not allowed by the public health authorities. Such inhibitors may pose a health hazard or cause off-flavor development. But in some tropical countries, high temperatures and poor hygienic standards prevail; in order to transport milk in good condition to where and when it is needed for processing, addition of a bacterial inhibitor seems unavoidable. The use of hydrogen peroxide can under certain conditions be tolerated. Better preservation may result from activation of the lactoperoxidase–thiocyanate–H_2O_2 system (Subsection 5.1.6).

5.1.6 Milk as a Substrate for Microorganisms

This subsection discusses the effect of the properties, especially composition, of raw milk and treated milk on the growth and metabolic action of microbes. Ecological aspects are briefly mentioned in Subsection 5.3.1.

Milk contains such a wide range of nutrients, including all of the vitamins, that numerous species of bacteria find sufficient raw material in it for fermentation and growth. The result is well known: Raw milk spoils rapidly at room temperature. But because the bacteria that can grow in milk may have very different properties, we should be cautious in applying general rules. For some bacteria, lactose is not a suitable energy source. Others rely on free amino acids as a nitrogen source, and fresh milk contains only tiny amounts of amino acids. Consequently, such bacteria often start to grow after other bacteria have hydrolyzed proteins, thus providing suitable nutrients. Another example is the production of CO_2 by some lactic streptococci, which stimulates growth of some lactobacilli. (On the other hand, some Gram-negative bacteria are inhibited by CO_2.) Some bacteria need specific trace components for growth or fermentation that are missing or present in insufficient concentrations.

Some conditions in milk may be unfavorable for growth of some bacteria. A possible lack of nutrients has been mentioned already. Water activity and ionic strength of milk are never limiting, and pH is so only for a few organisms. But redox potential and O_2 pressure are mostly such that strictly anaerobic bacteria cannot grow. The growth of aerobic bacteria also depends on location; in a cream layer, O_2 pressure may be much higher than near the bottom of a deep vessel of milk. At room temperature, bacterial action usually lowers both O_2 level and pH. For very few microorganisms, conditions in milk are so unfavorable as to kill them.

Milk contains natural *inhibitors*. Some bacteria do not grow in milk despite the presence of sufficient nutrients and suitable conditions. A mere delay (or lag phase) in growth after addition of the bacteria to milk is not proof of inhibition; the bacteria may not be adapted to milk (i.e., they have to change their enzyme system before they can use the nutrients available).

An important class of inhibitors is the *immunoglobulins* (Subsection 2.4.3), which are antibodies against specific antigens, often bacteria. Thus they are specific for the species and strains of bacteria encountered by the cow, and other bacteria are not inhibited. Mixed milk may contain immunoglobulins active against a wide variety of bacteria, but the concentration is usually low. Although examples of inhibition by IgG and IgA (probably in cooperation with 'complement') in milk are known, the agglutinative action of IgM is most conspicuous. There is an agglutinin acting against strains of *Streptococcus pyogenes*, another against strains of *Lactococcus lactis* ssp. *lactis* and *cremoris*; generally, *Bacillus cereus* also exhibits agglutination in raw or low-pasteurized milk.

Agglutination means that bacteria meet each other owing to their Brownian motion and are kept together in (large) floccules by the sticking action of the agglutinin. The floccules can become so large that they sediment, removing the bacteria from most of the milk. Bacteria also can become attached by agglutinins to fat globules if the temperature is low. The cold agglutination and rapid creaming of the fat globules (Subsection 3.2.4) then cause rapid removal of bacteria from most of the milk. The organisms are not killed but are restricted in growth because of the depletion of nutrients and the accumulation of inhibiting metabolites in the sedimented layer of floccules. Those concentrated in the cream layer also may be inhibited by the higher O_2 pressure. If agglutination is hindered mechanically, e.g., when milk is renneted directly after adding a starter (the bacteria now become enclosed in the meshes of the paracaseinate network), inhibition is insignificant. The content of agglutinins in milk is highly variable; usually, colostrum has relatively high concentrations.

Some nonspecific bacterial inhibitors are lysozyme and lactoferrin, but their concentrations in cows' milk are so low as to have little effect; human milk contains much higher levels. Lysozyme is an enzyme (EC 3.2.1.17) that hydrolyzes the structural polymer peptidoglycan of the bacterial cell wall, in particular, splitting the linkage between the N-acetyl muramic acid and N-acetyl glucosamine units; this may cause lysis of the bacteria. Lactoferrin binds Fe, thus reducing the activity of Fe^{2+} ions, which are needed by several bacteria. Possibly some fragments of lactoferrin produced by proteolysis exhibit other antimicrobial activities.

The most important nonspecific inhibitor of milk is the *lactoperoxidase–thiocyanate–*H_2O_2 system, which is also quite active in saliva. The milk enzyme lactoperoxidase (Subsection 2.5.2.1) as such does not cause inhibition, but it catalyzes the oxidation of NCS^- by H_2O_2, and the intermediates are powerful bacteria killers, especially $SCNO^-$. Milk contains far more than sufficient peroxidase, up to 0.4 μM, to activate the amount of NCS^- present. The NCS^- content is variable (it depends on the feed, as it is mainly derived from glucosides in species of *Brassica* and *Raphanus*) and is mostly in the range 0.02 to 0.25 mM; currently, low values are more likely in many areas. If the concentration is indeed 0.25 mM, NCS^- is, in conjunction with peroxidase, active against many bacteria that have no catalase and thus produce H_2O_2 (e.g., all lactic acid bacteria, although some of them have enzyme systems that metabolize H_2O_2, so that they are not inhibited or are less so). Fresh milk contains no H_2O_2, but if a little is added, say 0.25 mM (i.e., far too little to be active as such against bacteria or to cause oxidative defects), the system also inhibits catalase-positive organisms, similar to most Gram-negative bacteria. If sufficient NCS^- is also present (either natural or added), an effective preservative results, and even in heavily contaminated milk, bacterial growth may be prevented for about 24 h at 15°C. Addition of some glucose oxidase (EC 1.1.3.4/5) also induces formation of H_2O_2, and it may be that the natural milk enzyme xanthine oxidase (EC 1.1.3.22) can do the same under certain conditions. The natural content of catalase (EC 1.11.1.6) in milk does not interfere with the system, even when it is high (mostly because of a high somatic cell count). Nevertheless, the inhibitory effect in milk is quite variable, largely because of the variation in thiocyanate content.

Inhibitors may enter milk by contamination. This is undesirable, not only because fermentation may be impaired but because these substances may be detrimental to the consumers' health. Antibiotics such as penicillin may be present because they have been infused into the udder to control mastitis, and they can be found in the milk up to about 3 days after infusion. In particular, some lactic acid bacteria are sensitive to these antibiotics. Disinfectants used to treat milking or processing equipment may contaminate milk easily and then inhibit or even kill bacteria. In some countries, H_2O_2 may be added to raw milk as a preservative (10 to 15 mM); it is removed by addition of catalase before processing the milk.

Treatment of milk may profoundly alter its suitability as a substrate for bacteria. The most important is heat treatment (Chapter 7), which kills bacteria and may activate sporulation but also alters the milk. Inhibitors are inactivated; this pertains to the immunoglobulins (if heated for, say, 20 s at 76°C) and to the lactoperoxidase system (e.g., 20 s at 85°C). Consequently, pasteurization may considerably stimulate growth of bacteria (that have entered the milk afterward); the higher the heating intensity, the greater the growth, up to around 20 s at 85°C. Very intense heating may lead to formation of stimulants, such as formic acid for certain lactobacilli (see Subsection 7.3.3).

Homogenization appears to inactivate the agglutinins, but it usually is applied to milk that has been heated to such an extent that the agglutinins are inactivated anyway.

Fermentation by lactic acid bacteria causes the formation of lactic acid, and this is an effective inhibitor for many bacteria if it is undissociated. Its pK is about 3.95, which implies that the inhibition is stronger for a lower pH. Hardly any bacteria can grow in milk brought to a pH of <4.5 by lactic acid, but some yeasts and molds can. Bacteria also can produce other inhibiting substances, such as acetic acid and antibiotics. Some strains of *Lactococcus lactis* ssp. *lactis* produce the powerful antibiotic nisin.

Fermentation implies a drastic change in composition. Other similar changes also may be effective in inhibiting bacteria. Conditions can be made strictly anaerobic. At very low water activity, bacteria do not grow (Section 10.1), as in milk powder. Often a combination of lack of suitable nutrients, unfavorable conditions, and inhibiting agents prevents growth; this is especially true in many types of cheese (no sugar, low redox potential, not too high temperature, high salt, low pH, sufficient lactic acid).

5.2 UNDESIRABLE MICROORGANISMS

The presence of several species of microorganisms in raw milk is undesirable, either because the organisms can be pathogenic, or because their growth results in undesirable transformations in the milk.

Pathogenic microorganisms that enter milk can be pathogenic for humans or animals. Human pathogens are usually classified into those causing *food infection* and those causing *food poisoning*.

Food infection implies that the food, e.g., milk, acts as a carrier for the microorganism, which enters the human body through milk. So a person can become ill, often not until a day or so after drinking the milk. In food infection, fairly small numbers of microorganism may suffice to cause illness, according to the pathogen involved. Almost any pathogenic bacterium can occasionally be present in milk in very small numbers, but if it does not grow in milk, it is very unlikely to cause illness.

In food poisoning, the microorganism forms a toxin in the food (or such a toxin contaminates the food by another route). The consumer rapidly falls ill. Large numbers of the pathogenic microorganism are usually needed to cause food poisoning. The amount of toxin produced should be large enough to produce symptoms. Unlike food infection, food poisoning does not imply that the pathogenic organism is still in the food. Some toxins are more heat resistant than the toxin-producing microorganism itself, e.g., *Staphylococcus* spp.

Nonpathogenic microorganisms by themselves would not impair milk quality. But the organisms require nutrients, which are obtained by producing enzymes that hydrolyze lactose, protein, fat, or other substances in the milk, in order to yield compounds suitable for their growth. These conversions cause the milk to develop off-flavors and to be less suitable for processing into retail milk and milk products, e.g., because of a decreased heat stability of the milk. Furthermore, most heating processes applied in dairy processing do not destroy all microorganisms or all microbial enzymes.

Some of the pathogenic and spoilage microorganisms will now be discussed briefly.

5.2.1 Pathogenic Microorganisms

Some pathogens important for milk and milk products will be discussed in this subsection (see also Section 25.8). The pathogens generally do not grow very well in milk, contrary to spoilage microorganisms (Subsection 5.2.2). Milk may merely act as a carrier of pathogens. In many countries, the hygienic condition of milk with respect to these microorganisms is satisfactory. Occasionally, however, food infections caused by consumption of raw milk occur, especially in tropical and subtropical countries, where the risk of infection may be greater. Generally, drinking of raw milk is inadvisable. Table 5.4 gives a survey of pathogenic microorganisms associated with milk and milk products.

Enterobacteriaceae occur widely in nature, e.g., in dung and polluted water. Coliforms (coli-like) belong to this family and are among the normal inhabitants

TABLE 5.4
Groups of Human Microbial Pathogens Possibly Occurring in Milk and Milk Products

Organism	Disease
Enterobacteriaceae	
Escherichia coli[a]	Gastroenteritis
Salmonella	Gastroenteritis, typhoid fever
Shigella	Gastroenteritis
Yersinia enterocolitica[b]	Gastroenteritis
Other Gram-negative bacteria	
Aeromonas hydrophila[b]	Gastroenteritis
Brucella abortus	Brucellosis (abortion)
Campylobacter jejuni	Gastroenteritis
Gram-positive spore formers	
Bacillus cereus[a,b]	Intestinal intoxication
Bacillus anthracis	Anthrax
Clostridium perfringens	Gastroenteritis
Clostridium botulinum	Botulism
Gram-positive cocci	
Staphylococcus aureus[a]	Emetic intoxication
Streptococcus agalactiae[a]	Sore throat
Streptococcus pyogenes	Scarlet fever, sore throat
Miscellaneous Gram-positive bacteria	
Mycobacterium tuberculosis	Tuberculosis
Mycobacterium bovis	Tuberculosis
Mycobacterium paratuberculosis	Johne's disease (only ruminants)
Corynebacterium spp.	Diphtheria
Listeria monocytogenes[b]	Listeriosis
Spirochetes	
Leptospira interrogans	Leptospirosis
Rickettsia	
Coxiella burnetii	Q fever
Viruses	
Enterovirus, rotavirus	Enteric infection
Fungi	
Molds	Mycotoxicoses
Protozoa	
Entamoeba histolytica	Amoebiasis
Cryptosporidium muris	Cryptosporidiosis
Toxoplasma gondii	Toxoplasmosis

[a] Grows well in milk.

[b] Psychrotrophic strains known.

of the animal and human intestines. They include *Escherchia coli*, a Gram-negative rod, which is facultatively anaerobic and ferments lactose. Some strains of this bacterium are enteropathogenic and among them is the well-known *E. coli* O157:H7, which can cause hemorrhagic colitis. *Salmonella* and *Shigella* also belong to this family and can cause intestinal disorders. *Yersinia enterocolitica* is a psychrotrophic member of this family and may cause enteric infection that mimics appendicitis. Low pasteurization is adequate to kill the bacteria of this family. Milk and milk products are hardly ever involved in food infection or poisoning by these bacteria.

Among the other Gram-negative bacteria are some known potential pathogens. *Aeromonas hydrophila* is an opportunistic pathogen found in feces of healthy animals. This bacterium is psychrotrophic and can grow in media with 5% salt. *Brucella abortus* has been found in milk and is a pathogen to man and animal. It is the most important cause of brucellosis. *Campylobacter jejuni* belongs to the family of Spirillaceae and can occur in the intestinal tract of many animals. *C. jejuni* is a common cause of gastroenteritis, diarrhea and abdominal cramps being the main features of the disease. In all these cases, milk is generally contaminated with these bacteria by dung. Because they are very heat sensitive, these bacteria will not survive low pasteurization. They also die rapidly in cheese, partly because of the low pH. So far, a few outbreaks have been reported as occurring due to raw milk.

The Gram-positive spore-forming bacteria belong to the family Bacillaceae. The two most important genera are *Bacillus* and *Clostridium*. Their spores make them resistant to heat and other destructive agents. *Bacillus cereus* is a common soil bacterium that is often found in milk. Some strains are psychrotrophic and can grow at 7°C. *B. cereus* is capable of producing enterotoxins that cause food poisoning. However, large numbers are needed before milk becomes toxic. At that point, the milk will be obviously spoiled (awful flavor and sweet curdling), so that it will be neither consumed nor processed. Therefore, the health hazard will be limited. Some starchy milk products pose a larger risk because spoilage is more difficult to observe. Heating to a temperature above 100°C for at least 5 min kills the spores of *B. cereus*, which belongs to the least heat-resistant bacilli. Another well-known pathogenic bacillus is *B. anthracis*, which causes anthrax primarily in animals. Because this illness is practically absent in dairy herds, milk is not likely to be a carrier of these germs, which may gain entrance to the human body through lesions in the skin.

Of the pathogens among the strict anaerobic clostridia, *Clostridium perfringens* produces a toxin during sporulation in the digestive tract. Although spores of this bacterium are often present in raw milk through contamination via soil, dust, or dung, milk or milk products are hardly ever the cause of food poisoning by this organism. This is due to the fact that these bacteria are outnumbered by other bacteria in raw milk, and large numbers of vegetative cells of the bacterium are required to cause illness. Because babies are more susceptible to this organism than adults, milk intended for manufacture of baby formulas must be heated sufficiently. Sterilization as applied in the dairy factories kills *C. perfringens*.

Another bacillus, *C. botulinum*, occurs sometimes in soil and surface water. It causes botulism, which results from an exceedingly poisonous toxin that attacks the nervous system and is formed during its growth in food products. Milk and milk products are never the cause of botulism, though *C. botulinum* can occur in milk. Milk is too aerobic to allow growth of this organism. Most cheese is anaerobic and has a low redox potential, but contains no carbon source suitable for this organism, nor for *C. perfringens.* Industrial sterilization as used for milk products such as sterilized milk or evaporated milk kills any *C. botulinum* present.

Staphylococcus aureus is one of the Gram-positive cocci, which often occur in the udder of a cow with mastitis. The bacterium is also abundant in humans. Some strains can form a heat-stable toxin and cause inflammation (ulcers). Large numbers of the bacterium are required to form the toxin. Its growth can be slowed down by lowering the temperature of milk, or by lowering the pH and formation of antagonistic components by lactic acid bacteria, as happens during cheese manufacture. Low pasteurization kills *S. aureus*. All these factors limit the frequency of food poisoning by *S. aureus* through milk and milk products, despite the fairly frequent presence of the organism in raw milk. *Streptococcus agalactiae* belongs to the group B streptococci and is, besides *S. aureus*, an important agent of bovine mastitis. It thus occurs in raw milk and is a potential cause of disease in humans. *S. pyogenes* is pathogenic to humans and some animals. It resists phagocytosis and produces an erythrogenic toxin that causes scarlet fever. Both streptococci are killed by low pasteurization.

Mycobacterium tuberculosis may gain access to milk from infected animals through milk secretion and fecal contamination, or from milkers and other environmental sources. The organism does not multiply in milk but survives in unpasteurized milk and milk products. It is pathogenic to humans, as is the bovine type, *M. bovis.* Among the non-spore-forming pathogenic organisms, *M. tuberculosis* is the most heat-resistant. It is killed by low pasteurization of milk, e.g., 15 s at 72°C. Beverage milk should be pasteurized to inactivate alkaline phosphatase to the extent that it is no longer detectable (Subsection 16.1.1). Inactivation of this enzyme ensures the killing of *M. tuberculosis*. The related *M. paratuberculosis* causes infection in ruminants, which is commonly known as Johne's disease. Pathogenicity of this bacterium to humans is not yet established.

Coryneform (club-shaped) bacteria have been isolated from raw milk, especially in cases in which mastitis occurred. *Corynebacterium pyogenes* is one of the causes of mastitis and a potential cause of diphtheria-like syndrome in humans. *Listeria monocytogenes* is often found in nature. It can be pathogenic to humans, and it causes abortion and meningitis in its severe form. In animals it causes mastitis and abortion. A few cases of contamination through milk are known. The organism is aerobic and can grow at temperatures as low as 5°C. It is killed by the usual pasteurization.

The genus *Leptospira* consists of Gram-negative flexible helicoidal rods and belongs to the order Spirachaetales. *Leptospira interrogans* causes leptospirosis in animals and humans. The kidney is the natural habitat, and contamination via urine is the main route of infection. *Coxiella burnetii* belongs to the family

Rickettsia and causes Q fever in humans. It can occur in cows, goats, and sheep, and may be carried by ticks. The organism can cause mastitis, but animals are often carriers without becoming ill. Although the bacterium is rather heat-resistant, it is killed by low pasteurization.

Viruses are intracellular parasites and do not multiply in milk, but some of them may survive for long periods. Viruses are infective at very low doses, and most food-borne viruses cause gastroenteritis. Some may originate from the cow and enter the milk by fecal contamination, or from polluted water. They are inactivated by pasteurization of the milk. This is also true for the virus causing food and mouth disease, which is specific to cloven-hoofed animals.

Different genera of molds, such as *Aspergillus*, *Penicillium*, and *Fusarium* can produce mycotoxins in milk and milk products. These mycotoxins are toxic, carcinogenic, emetic, or mutagenic. Some aspergilli produce a toxin when growing on cheese.

Entamoeba histolytica is a protozoan that causes amoebiasis, the third most common cause of death by parasites in the world. Transmission occurs via ingestion of cysts in contaminated foods or water. *Cryptosporidium muris* causes cryptosporidiosis, with diarrhea as the primary symptom of infection. Cows can shed oocysts of *C. muris* in their feces, which may lead to milk contamination. Drinking of contaminated water is the most common vehicle for transmission of this disease. *Toxoplasma gondii* causes a parasitic infection in humans and many warm-blooded animals, especially dairy goats. Infection can occur by ingesting foods or water contaminated with oocysts. Contaminated milk is considered a potential source of human toxoplasmosis.

Molds and protozoa are generally killed by pasteurization.

5.2.2 Spoilage Microorganisms

Milk is a suitable culture medium for many microorganisms, and an attempt to discuss them all would be beyond the scope of this book. It is sufficient to mention some groups of bacteria, often consisting of several genera, that are responsible for a certain type of deterioration or that are typical of the source of contamination or the treatment of milk.

5.2.2.1 Lactic Acid Bacteria

These mainly produce lactic acid from carbohydrates such as lactose. They are widespread and include the genera *Lactococcus* and *Lactobacillus*. *Lactococcus lactis* sspp. *lactis* and *cremoris* grow rapidly in milk, especially above 20°C. So milk mostly turns sour if kept uncooled, and aggregation of casein particles may occur. Before the milk is considered truly sour, it has become unfit for processing, mainly because of the loss of heat stability. The mesophilic lactic acid bacteria are killed by low pasteurization (e.g., 15 s at 72°C) and largely even by thermalization (e.g., 15 s at 65°C). Low pasteurization does not kill thermophilic lactic acid bacteria such as *Streptococcus thermophilus*.

The dairy manufacturer exploits lactic acid bacteria in making fermented milk products, including yogurt, cheese, and butter. Carefully selected strains of bacteria are grown under controlled conditions (Chapter 13).

5.2.2.2 Coliforms

These belong to the Enterobacteriaceae family and are widespread, e.g., in the digestive tract. They include *Escherichia coli* and *Klebsiella aerogenes*, but several other genera and species are involved. They grow rapidly in milk, especially above 20°C, and attack proteins and lactose, as a result of which gas (CO_2 and H_2) is formed and the flavor of the milk becomes unclean.

Low pasteurization kills the coliforms to virtually the same extent as *Mycobacterium tuberculosis* (Subsection 5.2.1). This, as well as the fact that the organisms occur widely, has led to their use as indicator organisms. If coliforms are absent, the heated product has been heated sufficiently and has most likely not been recontaminated, and so pathogenic microorganisms, apart from heat-resistant ones, will most likely be absent.

5.2.2.3 Psychrotrophs

These are also designated pseudomonads or Gram-negative rods, occur widely, and include the genera *Pseudomonas*, *Achromobacter*, *Flavobacterium*, and *Alcaligenes*. Psychrotrophs grow readily at low temperatures (<15°C); in milk they proliferate even at a temperature as low as 4°C. Their optimum temperature is far higher, ranging from 20 to 30°C. Most of these organisms produce proteases and lipases, and thus attack protein and fat, causing putrid and soapy-rancid off-flavors. They do not lower the pH. Unlike the bacteria themselves, the enzymes produced can be highly resistant to heat and may cause off-flavors and alter physicochemical properties even in stored UHT milk. For example, they can hydrolyze protein, as a result of which the milk becomes bitter and eventually becomes more or less transparent. More than 5×10^5 psychrotrophs per ml of original milk can be harmful. In low-pasteurized and in raw milk, flavor defects do not occur until numbers are over 10^7 ml^{-1}, because of the brief storage time and low storage temperature of the milk. In tropical areas, where storage time is generally longer and storage temperature of the milk higher, serious infection via contaminated water may give rise to very high counts of Gram-negative rods. They cause severe spoilage of milk in a very short time without acidification, sometimes accompanied with sweet curdling.

5.2.2.4 Thermoduric Bacteria

Some bacteria, including *Microbacterium lacticum*, thermophilic streptococci, and certain *Micrococcus* species, do not form spores but the vegetative cells survive low pasteurization. Heating above, say, 80°C for 20 s kills them. The organisms are chiefly encountered in places where other bacteria die due to high prevailing temperatures, e.g., the high temperature used during cleaning of milking units. These bacteria are not very active in cold-stored milk, but they are undesirable

TABLE 5.5
Some Microorganisms Associated with Possible Spoilage of Milk and Milk Products

Organism	Source	Growth in Milk	Heat Resistance	Spoilage
Spore formers				
Bacillus cereus	Feed, dung, soil, dust	++	+[a]	Sweet curdling, bitty cream in pasteurized milk and cream
Bacillus subtilis	Feed, dung, soil, dust	++	+	Spoil sterilized milk
Bacillus stearothermophilus	Feed, soil	++	+	Spoil evaporated milk
Clostridium tyrobutyricum	Soil, silage, dung	–	+	Late blowing in cheese
Coliforms				
Escherichia coli	Feces, milking utensils, contaminated water	++	–	Spoil milk and cheese
Klebsiella aerogenes		++	–	Spoil milk
Lactic acid bacteria				
Lactobacillus species	Milking utensils, parlor	++	–	Sour milk
Lactococcus lactis	Milking utensils, parlor	++	–	Sour milk
Streptococcus thermophilus	Milking utensils, parlor	++	+	Sour milk
Psychrotrophs				
e.g., *Pseudomonas*	Milking utensils, cold-stored milk	++	–	Hydrolyze protein and fat in cold-stored milk
Thermoduric bacteria				
e.g., *Micrococcus* species	Milking utensils	+	+	Can grow in pasteurized products
Yeasts	Dust, milking utensils	+/–	–	Spoil cheese, butter, sweetened condensed milk
Molds	Dust, dirty surfaces, feed	+/–	–	Spoil cheese, butter, sweetened condensed milk

[a] + = Survive low pasteurization.

because they are still present in the heated product and can grow out if conditions are favorable, especially at high ambient temperatures.

5.2.2.5 Spores of Bacteria

The genera *Bacillus* (aerobic or facultatively anaerobic) and *Clostridium* (strictly anaerobic) can form spores. Most of these survive fairly intense heat treatment. They originate especially from soil, dust, and dung, and also from cattle feed. Some important species and their effects are as follows:

B. cereus can spoil pasteurized milk by causing sweet curdling, a bad off-flavor, or clumps of fat globules. It is not very heat-resistant. Organisms may grow at temperatures down to about 7°C.

B. subtilis and *B. stearothermophilus* are sufficiently heat-resistant to spoil sterilized milk if insufficiently heated.

C. tyrobutyricum belongs to the butyric acid bacteria and can cause late blowing, which is a serious defect in such cheeses as Gouda or Emmentaler. Formation of gas, including H_2, leads to large holes and cracks. Production of butyric acid from the lactic acid in the cheese causes an awful flavor.

The vegetative cells of spore-forming bacteria and the spores of yeasts and molds generally are not very heat resistant.

A survey of the potential spoilage microorganisms in milk and milk products is presented in Table 5.5.

5.3 SOURCES OF CONTAMINATION

This section starts with some general remarks about ecological aspects. Subsequently, the sources of microorganisms in the environment of the cow are described, and then the way in which they contaminate milk. The measures to be taken to avoid, or at least reduce, such contamination are discussed in Section 5.4.

5.3.1 Microbial Ecology

Microorganisms have inhabited our planet since the early stages of evolution. Selective pressures in the course of time have resulted in the development of a broad diversity of microbes by mutation and adaptation. It is, therefore, not surprising that today microorganisms, especially bacteria, are everywhere. Whether a given type of microorganism is present in a certain ecosystem is determined by the biotic (presence and abundance of other organisms) and abiotic components (chemical composition, temperature, etc.) of the system. Each microorganism finds its ecological niche, i.e., an environment in which it can find nutrients, generate energy, grow, and compete with other organisms and tolerate adverse conditions. This implies an interaction between microorganisms and environment: the latter determines which organisms can proliferate, but the organism in turn alters the environment, affecting its suitability for other organisms. For example, lactic acid bacteria growing in milk produce lactic acid from lactose and decrease the redox potential, thereby eventually precluding growth of most other bacteria (and themselves); some yeasts, however, can tolerate these conditions.

Raw milk left in contact with the external environment is essentially an open ecosystem, and it may contain a wide range of bacterial species. In a temperate climate, and in the absence of cooling machines, lactic acid bacteria tend to predominate in any place where milk is kept: they can grow fast and outcompete most other organisms. Because of the large-scale introduction of cooling tanks on the farm, this has changed dramatically: it is psychrotrophs that predominate now. In tropical countries, still other types of bacteria may prevail.

Many milk products are closed or controlled ecosystems, and the microbial changes occurring greatly depend on the particular contamination by bacteria that occurred. Not only are species and numbers of bacteria important, but so is their physiological state, which especially depends on the growth phase (Figure 5.3), and the possible presence of bacteriophages (Section 13.3). The effect of the environment is usually different for growth, fermentation, and, for certain species, germination of spores. Generally, conditions permitting growth are more restricted than those for catabolism (fermentation). For instance, several lactic acid bacteria do not grow to any extent near 5°C (see Figure 5.6), but, other conditions being favorable, they may go on producing lactic acid from lactose. Conditions inducing germination of spores are usually quite restricted.

5.3.2 Microorganisms Present in the Udder

A distinction should be made between healthy and unhealthy cows, although it may not be clear-cut, especially for some types of mastitis.

5.3.2.1 Healthy Cows

In most cows, no microorganisms are present in the milk in the alveoli, ducts, cistern, and teat cistern, but they are present in the teat canal and the sphincter of the teat, mainly non-heat-resistant *Micrococcus* and *Staphylococcus* spp. and *Corynebacterium bovis*. Sometimes other bacteria are also involved. During milking, these bacteria enter the milk. Directly after milking, their number varies widely among cows, from hardly any to about 15,000 ml^{-1}; the colony count of aseptically drawn milk of healthy cows is usually low, often <100 ml^{-1}. At 5°C the bacteria hardly grow and, after low pasteurization, these organisms can often not be detected. Obviously, microbially high-grade milk can be collected from healthy cows.

The cow has several defense mechanisms to keep microorganisms away from the udder:

- The sphincter of the teat has some circular muscles that can keep the opening closed.
- Bacteriostatic and bactericidal agents present in the keratin material of the teat canal and in the milk itself, and the leukocytes in the milk.
- The ‘rinsing effect’ due to discharge of the milk.

5.3.2.2 Unhealthy Cows

When a cow is ill due to microbial infection, the organisms involved can enter the milk. In the case of mastitis, pathogenic organisms are already present in the udder and, thereby, in the milk. Because of this, mastitic milk usually has a high count. Some of these mastitic organisms, including certain streptococci, *Staphylococcus aureus*, and certain strains of *Escherichia coli*, are also pathogenic to humans.

If organs other than the udder are inflamed, pathogens may directly enter the milk through the body, especially if the cow is also mastitic. Naturally, the organisms can also enter the milk through, for instance, dung or urine (see Subsection 5.3.3.1). Among such organisms that are pathogenic to humans are *Leptospira interrogans*, *Mycobacterium tuberculosis*, *Campylobacter jejuni*, *Listeria monocytogenes*, *Bacillus anthracis* (causes anthrax), and *Brucella abortus* (causes an illness resembling Malta fever in humans). Obviously, it is essential to exclude milk of diseased animals from being processed, and to heat the milk in order to kill any pathogens.

5.3.3 Contamination during and after Milking

The hygienic measures taken during and after (mechanical) milking essentially determine what foreign microorganisms enter the milk, including human pathogens. This applies also to their numbers. The count of properly drawn mixed milk from healthy cows is about 10,000 ml^{-1}, sometimes even less. If, however, the level of hygiene during milking is poor, freshly drawn mixed milk can have a much higher count, up to one million ml^{-1}. Potential sources of contamination of milk, together with the characteristic microorganisms involved, will now be discussed.

5.3.3.1 The Cow

During milking, microorganisms can enter the milk from the skin of the teats, which often are contaminated by dung, soil, or dust. Flakes of skin, hairs, and dirt from the feet and flanks can also enter the milk. Several types of microorganisms can contaminate the milk, including coliforms, fecal streptococci, other intestinal bacteria, bacterial spores (mostly *Clostridium* spp.), yeasts, and molds. Some of these microorganisms are human pathogens.

Appropriate housing and care of the cows is an essential measure to promote clean udders. As a result, dry treatment, including removal of loose dirt, suffices during milking. Such dry treatment, moreover, causes less leakage of milk against the teats. Fewer bacteria then become detached from the teat skin. Dirty udders have to be cleaned thoroughly before milking. However, the complete removal of bacteria is impossible.

5.3.3.2 Soil, Dung, and Dust

All these contaminants can reach the milk and thereby increase counts. Moreover, spores of bacteria, yeasts, and molds also occur in air. Well known is *B. subtilis*, originating from hay dust. The spores can enter the milk through air sucked in during mechanical milking, or fall directly into it during milking in open pails.

The cleanliness of the milking parlor and the restfulness of the cows during milking are among the factors determining contamination of the milk.

5.3.3.3 The Feed

Feed often contains large numbers of microorganisms. Feed can sometimes fall directly into the milk but, more significantly, certain microorganisms in the feed survive passage through the digestive tract and subsequently enter the milk through dung; it includes some human pathogens. Spore-forming bacteria, including *Bacillus cereus*, *B. subtilis*, and *Clostridium tyrobutyricum*, which can spoil milk and milk products, are especially involved. Large numbers of *C. tyrobutyricum* occur in silage of inferior quality. The bacterial spores survive low pasteurization of cheese milk, to which a more intense heat treatment cannot be applied, and may cause "late blowing" in some types of cheese (Section 26.2). Accordingly, high-quality silage is of paramount importance, and contamination of the milk, e.g., by dung, should be rigorously combated. In some regions, the use of silage is strictly prohibited, e.g., in areas of Switzerland and northern Italy where Emmentaler and Parmesan cheese, respectively, are made. Incidentally, when the cows suffer from diarrhea (caused, for instance, by feeding too much concentrate), the contamination of the milk by dung is increased.

5.3.3.4 Milking Unit

Contact infection poses the largest threat of contamination to almost all foods, including milk. Poorly cleaned and disinfected milking equipment can contain large numbers of microorganisms. Because these organisms generally originate from milk, they will grow rapidly and can decrease quality. Residual milk often contains about 10^9 ml^{-1} of bacteria, and even 1 ml of such milk entering 100 l of milk during the following milking would increase the count by 10,000 ml^{-1}.

The methods of cleaning and disinfection applied largely determine the species of the contaminating organisms. If high temperatures are used and cleaning and disinfection of milking utensils are unsatisfactory, the main species will be heat-resistant, including micrococci, *Microbacterium lacticum*, some streptococci, and spore-forming bacteria. If, on the other hand, ambient temperatures are used, lactic acid bacteria, e.g., *Lactococcus lactis*, pseudomonads, and coliforms will mainly be involved. Use of milking equipment that can be adequately cleaned and disinfected is thus paramount. Small cracks in worn-out rubber units and 'dead ends' in the equipment that are insufficiently rinsed should be avoided.

5.3.3.5 Water Used

Tap water may be of good quality. Any private water supply must be examined at intervals. Surface water can contain many microorganisms, including human pathogens, and it must therefore on no account be used for cleaning and rinsing. Gram-negative rods such as *Pseudomonas*, *Achromobacter*, *Flavobacterium*, and

TABLE 5.6
Contribution of Some Sources of Contamination to the Colony Count of Raw Milk

Source of Contamination	Estimate of the Contribution to the Count (ml^{-1})
Udder of a healthy cow	Up to several thousand
Udder of a mastitic cow	Up to several million
Skin of cow	A hundred up to several thousand
Milking parlor (soil, dung, dust, air)	Up to a thousand
Feed	Up to a thousand
Milking unit	A thousand up to several million
Water for cleaning, rinsing	Up to several thousand
Good milker	Generally negligible

Note: Approximate examples.

Alcaligenes spp., most of which are psychrotrophic, often occur in contaminated water (also in dung, soil, and poorly cleaned utensils). Especially in the tropics, water may have very high counts.

5.3.3.6 The Milker

Milkers influence many of the preceding factors and thereby the microbiological quality of the milk. They can also contaminate the milk directly, e.g., with the hands. If they suffer from microbial infections, they might directly contaminate the milk with pathogens.

Table 5.6 gives an overview of the contribution of some sources of contamination to the count of milk.

5.4 HYGIENIC MEASURES

In discussing measures that would result in satisfactory bacteriological milk quality, contamination by undesirable bacteria should be distinguished from growth of the bacteria in milk. Butyric acid bacteria, for example, cannot grow in milk, but the presence of more than 1 spore/ml of milk is undesirable in the production of some types of cheese. Psychrotrophic bacteria, however, grow rapidly in milk, and contamination by 10^2 to 10^3 ml^{-1} during milking is hard to avoid. Counts smaller than 10^5 ml^{-1} are harmless. Hygienic measures should aim at suppressing pathogens and inhibiting spoilage organisms. These subjects will now be discussed.

5.4.1 Protection of the Consumer against Pathogenic Microorganisms

The following are the main reasons why contamination of raw milk by pathogens and growth of these organisms in milk during storage should be avoided as much as possible:

1. During microbial growth in raw milk, toxins may be formed. Some toxins are fairly heat-resistant.
2. Some pathogens survive heat treatments such as pasteurization. Fortunately, this is exceptional. The higher the count in the raw milk, the greater the number of organisms that may survive heat treatment. This is of importance if the heat treatment applied leaves only a narrow margin for error.
3. The heavier the contamination of raw milk by pathogens, the greater the risk of recontamination of the heated milk.

Contamination of raw milk by pathogens can never be ruled out. Milk intended for liquid consumption or for transformation into milk products is therefore often required by law to be heated to such an extent that the common pathogens are killed; this implies at least low pasteurization.

The dosage, or numbers of infecting organisms, is an important factor in establishing infection. Yet it also involves other factors, such as port of entrance and virulence of the parasite. In general, we may say that under ordinary circumstances the larger the dose of infective microorganisms, the greater the chance that an infection will result. However, certain qualifications are necessary. For example, very large numbers of some microorganisms may be present in certain situations without causing any problems. The intestine contains thousands of billions of deadly bacteria at all times, yet if only a dozen or so of some of these are introduced into the peritoneum or injected into the brain, they can quickly set up a fatal infection. Similarly, one might swallow three or four typhoid bacilli without ill effects, yet a dosage of several hundred might overcome local resistance and cause typhoid fever. With some organisms, a single cell or particle is invariably sufficient to cause an infection. Obviously, much depends on the virulence of the particular organism involved and on the resistance of the tissues that it contacts, as well as on the dosage. In case of intoxication via the intake of food, the concentration of pathogens in the food is generally high, or had been high before in order to produce a sufficient dose of toxin.

Measures taken to prevent growth of spoilage organisms also stop growth of pathogenic bacteria that can produce heat-resistant toxins. Pasteurized milk is, therefore, among the safest food products of animal origin.

5.4.2 Measures against Spoilage Organisms

A low level of contamination by microorganisms is the first aim. To achieve this, the sources of contamination should be known. Some contamination occurs before milking, especially in housing (clean cows) and fodder production (butyric acid bacteria). Cleaning and disinfection of the milking equipment is essential (Chapter 14). It is specifically meant to remove and kill bacteria. Bacteria originating from inadequately cleaned equipment usually have no lag phase and can grow rapidly in milk (Figure 5.5).

Cooling is the main means of slowing down the growth of bacteria in milk. The maximum storage time of milk closely depends on the storage temperature. Satisfactory operation of refrigerated milk tanks on the farm is essential. However, cooling of milk kills no bacteria, and it cannot remedy unsatisfactory hygiene.

In dairy factories, the raw milk received often is not simply stored before processing, but is thermalized and then cooled to below 4°C. Thermalization is a mild heat treatment, e.g., 15 s at 65°C. It kills nearly all psychrotrophic bacteria, which are not at all heat-resistant. In this way, growth of these bacteria to harmful numbers during cold storage of the milk in the factory is prevented, as is the formation of heat-resistant enzymes (lipases and proteinases). Thermalization kills some of the other bacteria too, including many lactic acid bacteria.

Suggested Literature

A useful textbook on general microbiology is: H.G. Schlegel, *General Microbiology,* Cambridge University Press, 1993.

An advanced and comprehensive textbook on food microbiology, including aspects of milk, is: D.A.A. Mossel, J.E.L. Corry, and C.B. Struijk, *Essentials of the Microbiology of Foods,* Wiley, Chichester, 1995.

Microbiology of milk (and some milk products) is treated in: R.K. Robinson, Ed., *Dairy Microbiology,* Vol. 1, *Microbiology of Milk,* and Vol. 2, *Microbiology of Milk Products,* 2nd ed., Elsevier, London, 1990.

Applied aspects of microorganisms in milk can be found in: E.H. Marth and J.L. Steele Eds., *Applied Dairy Microbiology,* 2nd ed., Dekker, New York, Basel, 2001.

Encyclopedic information, also on various microorganisms in milk, are in: H. Roginski, J.W. Fuquay, and P.F. Fox, Eds., *Encyclopedia of Dairy Sciences,* Academic Press, San Diego, 2002.

Microbiology of raw milk is discussed in the IDF report: Factors influencing the bacteriological quality of raw milk, International Dairy Federation, Document 120, Brussels, 1980.

Methods for quality assessment of raw milk are given in: *Bulletin of the International Dairy Federation No. 256,* Brussels, 1990.

Monographs on spore-forming bacteria occurring in milk are contained in: *Bulletin of the International Diary Federation No. 357*, Brussels, 2000.

Part II

Processes

6 General Aspects of Processing

Before coming to specific processes, we will discuss some general considerations about milk processing and quality assurance.

6.1 INTRODUCTION

Milk is a raw material in the manufacture of several food products. These products are predominantly made in dairy factories (or dairies, for short). Their mode of operation is dominated by the properties of the raw material. Some typical *characteristics* of the dairy industry are as follows:

1. Milk is a liquid, and it is homogeneous (or it can readily be made homogeneous). This implies that transport and storage are relatively simple and it greatly facilitates the application of continuous processes.
2. Milk properties vary according to source, season, and storage conditions, and during keeping. This may imply that processes have to be adapted to the variation in properties.
3. Milk is highly perishable and the same is true of many intermediates between raw milk and the final product. This requires strict control of hygiene and storage conditions.
4. Raw milk may contain pathogenic bacteria, and some of these can thrive in milk. This also requires strict control of hygiene and the application of stabilization processes.
5. Generally, raw milk is delivered to the dairy throughout the year, but in varying quantities (in some regions there is even no delivery during part of the year). Because the milk must be processed within a few days at the most, this generally implies that the processing capacity of a dairy cannot be fully used during most of the year.
6. Milk contains several components, and it can be separated in fractions in various ways, e.g., in cream and skim milk, in powder and water, or in curd and whey. Moreover, several physical transformations and fermentations can be applied. This means that a wide variety of products can be made.
7. Relatively small amounts of raw material (besides milk) are needed for the manufacture of most milk products, but consumption of water and energy may be high.

8. One and the same unit operation can often be applied in the manufacture of a range of products. This includes heat treatment, cooling, cream separation, and homogenization.

Nearly all process steps, or *unit operations*, that are applied in food processing are applied in the dairy. They can be grouped as follows:

1. Transfer of momentum: pumping, flow.
2. Heat transfer: heating and cooling.
3. Mixing/comminution: stirring, atomization, homogenization, and recombination. The last two can also be considered physical transformations.
4. Phase separation: skimming, separating milk powder from drying air, part of the churning process, etc.
5. Molecular separation: evaporation, drying, membrane processes, and crystallization (of water, lactose, and milk fat).
6. Physical transformation: gel formation (as due to renneting or acidification of milk), important elements of butter-making, making of ice cream, etc.
7. Microbial and enzymatic transformation: production of fermented products, cheese manufacture and ripening.
8. Stabilization: pasteurization, sterilization, cooling, and freezing. At least one of these operations is virtually always applied. Most stabilization processes are also, or even primarily, aimed at ensuring food safety.

In some cases, general knowledge of food process engineering may suffice to apply these unit operations. Some operations that are essential in dairy manufacturing are, however, not treated or are hardly treated in texts on food engineering. Moreover, the process affects the material — which is why it is applied — but also the material affects the process, of which numerous examples are given in the ensuing chapters. Often, such mutual interactions are intricate, specific, and of practical importance. Consequently, a thorough knowledge of the physics, chemistry, and microbiology of milk and its components is needed to understand the changes occurring, both intended and undesired, in the material during processing.

Objectives. In the development of processes for the manufacture of food products, several constraints have to be taken into account. These include availability of materials, machinery, skilled staff, and specific knowledge, as well as legal conditions. However, the objectives of the production process are of paramount importance. The ensuing requirements can be grouped as follows:

1. *Safety of the product for the consumer.* The health of the consumer can be threatened by pathogenic bacteria (or their toxins) and by toxic or carcinogenic substances. The first of these nearly always provides by far the most serious hazard. These aspects are generally discussed in

Section 6.3, and more specifically in Chapter 5 and the chapters in Part III (Products).

2. *Quality of the product.* Apart from product safety, which may be considered a quality aspect, this generally involves:
   - Nutritional value.
   - Eating quality: taste, odor, and mouthfeel.
   - Appearance: color and texture.
   - Usage properties, e.g., spreadability of butter, whippability of cream, and dispersibility of milk powder; and, in general, ease of handling.
   - Keeping quality or shelf life, i.e., the length of time a product can be kept before it significantly decreases in quality or may have become a health hazard.
   - Emotional values: a wide range of aspects, greatly varying among consumers. To be sure, most of the aspects just mentioned may also be subject to emotional considerations by the consumer.

   The quality requirements vary widely among products, and even if they are the same (e.g., the shelf life), different measures may be needed to meet them.
3. *Quality of the process.* The process should be safe and convenient for the staff involved as well as for other people in the vicinity. It should not cause environmental problems, such as pollution, or excessive depletion of exhaustible resources (e.g., energy and water).
4. *Expenses.* Often, the necessity to maintain the processing costs within limits is overriding. Concerns may include the price of raw materials (including packaging), use of energy, equipment expenditure, and labor intensity, etc. Also the flexibility and complexity of the process, with the ensuing probability of making mistakes (resulting in poor quality or even the need to discard products), may affect production costs. The same is true of the costs of storage.

It may be added that the objectives are manifold and often mutually conflicting. This means that process optimization may be far from easy.

6.2 PRESERVATION METHODS

The manufacture of milk products virtually always involves some form of preservation, which means taking measures to prevent, or at least postpone, deterioration. Most technologists primarily think of deterioration caused by microorganisms, but it can also involve enzymatic, chemical, or physical changes. The measures needed to prevent the latter two are highly specific and will not be considered in this section, but we will include enzymatic causes. It may be added that deterioration and becoming unsafe are often not correlated. A product that is obviously deteriorated (has turned bad) can be, and often is, perfectly safe; whereas a product that still looks and tastes quite good may sometimes contain a dangerous level of pathogenic bacteria or toxin.

To counteract *microbial action*, one can (1) kill the microorganisms, (2) physically remove them, (3) inhibit their growth (although this will not always prevent metabolic action by the enzyme systems of the organism), and (4) prevent (re)contamination with microbes. (Methods 1 and 2 can also be used to eliminate pathogens.) There is considerable variation in the resistance of bacteria to preservation agents; moreover, bacterial spores are much more resistant than the corresponding vegetative cells. Yeasts and molds do not tend to be very resistant. The resistance of viruses is variable. Moreover, the resistance of an organism to a preservation agent depends on environmental conditions, such as pH, ionic strength, and temperature.

Bacteria, yeasts, and molds tend to die off under conditions in which they cannot grow, albeit slowly; even after some years viable organisms can be found. Spores can survive under harsh conditions for a very long time.

To counteract *enzymatic action* one can irreversibly inactivate the enzyme; the resistance substantially varies among enzymes and with environmental conditions. One can also reduce the specific activity of an enzyme by changing the environment.

Several preservation methods can be applied, and all have specific advantages and disadvantages. We will briefly discuss the more important methods.

- *Heat treatment.* This is generally the method of choice for liquid products. It is active against microbes and enzymes. The method is convenient, flexible, well-studied, and fairly inexpensive. The disadvantage is that undesirable chemical reactions occur, especially at high heating intensity, for instance, causing off-flavors. The method is extensively discussed in Chapter 7.
- *Pressure treatment.* The hydrostatic pressure applied must be high, well over 100 MPa (1 kbar). The high pressure leads to unfolding of globular proteins and thereby to killing of microbes and inactivation of some enzymes. For example, to reduce the number of vegetative bacteria by a factor of 10^5 or 10^6, a pressure of about 250 MPa must be applied for 20 min, or 500 MPa for 10 s. Spores, as well as most enzymes, are far more resistant. The great advantage of this method is that undesirable chemical reactions hardly occur. A disadvantage is that in milk, the casein micelles tend to dissociate irreversibly, leading to a significantly changed product. Moreover, the method is expensive because the process is discontinuous and can only be applied to relatively small volumes at a time. It is not applied for milk products.
- *Irradiation.* This can be ionizing radiation, e.g., β- or γ-rays emitted by radioactive materials, or ultraviolet light. The former needs to be of high intensity to kill bacteria, and especially spores. This causes off-flavors. Moreover, there is considerable public opposition against the use of radioactive materials. The method is only used for some

condiments and for sterilizing surfaces. UV light kills microbes, but it can only penetrate clear liquids. It is occasionally used for water sterilization and also for surface decontamination. Neither of these methods is suitable for inactivation of enzymes.

- *PEF*, i.e. short Pulses of a high Electric Field. Such pulses can kill microbes, presumably, by damaging the cell membrane. The smaller the cell dimensions, the higher the intensities needed, and spores are very difficult to kill. Enzymes are generally not inactivated. The method is not used for milk products.
- *Removal of microbes*. This can have the obvious advantage that no chemical reactions occur. On the other hand, enzymes are not inactivated, and complete removal of microbes cannot generally be achieved. A suitable and fairly inexpensive method is microfiltration (see Subsection 12.1.1). Because of the very small pore size needed, say 0.5 μm, fat globules and some of the casein micelles are also removed. This makes the method impractical for many milk products. It is used for water, cheese brine, and also, in combination with other unit operations, for liquid milk (see Subsection 16.1.3). Another method is bactofugation, i.e., the removal of bacteria and especially spores in a centrifugal separator. It operates at about 70°C, and part of the casein is also separated; see Section 8.2. This method is applied to milk for some special applications.
- *Drying*. At high concentrations of water-soluble substances, most microorganisms stop growing, presumably because the contents of the cell become highly concentrated. There is considerable variation among microbes and with the nature of the solute, but a concentration corresponding to a water activity below 0.65 suffices in most milk products. Dried milk is thus free from microbial growth (see Subsection 10.1.5). To stop enzyme action, lower water activities are generally needed, below 0.2 or even less.
- *Freezing*. This causes freeze concentration and acts much like evaporation or drying (see Section 11.2); moreover, the temperature is so low that microbial or enzymatic action is anyway very sluggish.
- *Mild preservatives*. This means high concentrations of salt (e.g., cheese), acid (e.g., fermented milks), or sugar (e.g., sweetened condensed milk). Acids act in their undissociated form, and thus need a pH low enough to greatly decrease their dissociation to be effective. Acids and salts can substantially decrease enzyme action; sugars generally do not. The preservatives naturally affect product properties, especially flavor.
- *Disinfectants* should never be added to milk or milk products. They may be harmful and give off-flavors. They are used for disinfection of surfaces (see Section 14.3).
- *Specific inhibitors* for bacteria present in milk are discussed in Subsection 5.1.6. The inhibitors can be boosted (e.g., by adding H_2O_2 and thiocyanate for the peroxidase system) or additional quantities can be added (e.g., lysozyme). Specific inhibitors (e.g., antibiotics) are also

made by several bacteria used in producing fermented milk. Some antibiotics are added to specific products, e.g., nisin. Increased concentrations of carbon dioxide (e.g., by 20 mM) inhibit growth of some Gram-negative bacteria. Specific inhibitors for enzymes also exist, but these are not used in industrial practice.

- *Reduce the concentration of essential compounds.* A well-known example is exclusion of oxygen, often due to its consumption by bacteria, which inhibits or fully prevents the growth of molds and aerobic bacteria. Many types of cheese are devoid of sugars, again due to bacterial action, preventing several other bacteria from growing.
- *Prevent (re)contamination.* This needs strict hygienic measures (Sections 5.4 and 14.3) combined with rigorous control and adequate packaging (Chapter 15).

Finally, preservation can often be achieved by a combination of (several) measures. The prime example is cheese (Section 23.3); fermented milks and ice cream are also in this category.

6.3 QUALITY ASSURANCE

Quality assurance is of paramount importance in all food manufacture and handling. It involves a coherent system of activities that assures (guarantees) that the products meet a set of defined quality marks. Specific aspects will be discussed throughout the rest of the book. By way of example, Section 20.3 gives a detailed description of controlling hygienic quality in the manufacture of milk powder. Some general considerations are given in the following subsections.

6.3.1 Concepts

Quality can be defined in various ways. A well-known definition (by J.M. Juran) is: "Quality is fitness for use." This needs some elaboration. A product or a service is fit for use if it meets the expectations of the user. However, it is far from easy to establish what these expectations are. This is because the expectations vary, often widely, among consumers, and generally depend on conditions under which a product is purchased or used. Moreover, several quality marks are highly subjective, and it is difficult to translate these into measurable product attributes. High quality does not merely mean that the product complies with legal requirements or preconceived ideas of the manufacturer. Marketing specialists and technologists should cooperate in establishing the desired quality marks.

Food technologists then play a key role in translating quality aspects into defined criteria and in developing methods for determining whether and to what extent a criterion is met. In Section 6.1, a list of quality aspects is given. For some of these, the value can be estimated by more or less objective methods (e.g., safety, shelf life, and dispersibility), and others can only be assessed by consumer panels (e.g., flavor). Often, attempts are made to establish correlations between objective criteria and consumer opinions, e.g., between acidity or diacetyl content of fermented milks and their flavor, or between a rheological parameter and the subjectively observed spreadability of butter.

To ensure that high quality is obtained, it does not suffice to define criteria and then to inspect whether they are met. Quality must be controlled (enforced) and is thus a management function. The current approach is a system of integrated or *total quality management*. It involves integration in three directions:

1. Throughout the product chain, i.e., from the farm to the consumer. It may even have to start before the farm, for instance in the design of milking machines or in the specifications for concentrates fed to the cows. Distribution of the products also involves several steps that can bear on product quality.
2. For the product in the widest sense, including service. This would involve the way in which the product reaches the consumer and the information given about the product.
3. Throughout the organization, i.e., at all hierarchical levels and in all departments.

The quality concept has to be built in from the beginning: in the definition of the product and its positioning in the market, in the development of the manufacturing process, in the design of equipment (e.g., 'cleanability'), in the specifications for raw materials, in the planning of logistics for distributing the product, etc. In other words, quality begins with the *design*: can good products be made by the planned procedures? The next question is whether the desired quality can be *reproduced*: does every item produced comply with the set quality criteria? For the latter, a control system has to be installed. However, the general rule should be that prevention is better than cure.

Of greatest concern is the *safety* of the product for the consumer. Milk may contain several types of pathogenic bacteria. Because their presence is largely determined by chance, and because a single bacterial cell can in principle be dangerous (because some pathogens can grow in milk), safety cannot be assured by selecting and inspecting samples. In practice, it is almost never possible to check every unit of the product. Consequently, other measures must be taken, such as

1. Treating the raw milk in such a way that all the pathogenic bacteria that can be present and harmful are killed.
2. Prevention of recontamination of intermediates and end product. This requires strictly enforced hygienic measures and packaging.
3. Transformation of the material into a product in which pathogens cannot grow; a good example is fermented milk. Preferably, any pathogens present will die off.

A combination of the three treatments will give the least chance for 'accidents' to happen. However, all these measures, especially the third one, cannot always be taken. Consequently, a rigorous inspection and control system must be established.

Health hazards due to toxic or carcinogenic concentrations of substances in the product are very rare in milk products. Most contaminations with hazardous substances are restricted to one farm, and the dilution achieved by mixing the milk with that from several other farms generally causes the concentration in the final product to be far below the toxic level. Sampling of the final product will generally establish the contamination, and sampling of all road tankers, and subsequently of individual deliveries, will then readily locate the source of contamination.

6.3.2 Hazard Analysis/Critical Control Points (HACCP)

HACCP is a method to establish, for an existing production process, the control measures that are essential to ensure the safety of the product. The same method can be applied for other quality characteristics, but the emphasis generally is on safety. HACCP should be applied separately to every manufacturing process in operation; this means a separate system for every product or group of closely related products. The main features of the method are what the name says: make an analysis of the potential hazards, identify critical points in the process, and establish criteria for control. A *critical point* is defined as one that must be controlled to ensure safety or good quality. There may be several other points (process steps or product properties) that can be, but need not be, controlled because control of another, critical, point will also detect the imperfection.

HACCP is also a control system applied after the analysis has been made. It involves corrective measures where needed, e.g., via feedback or control loops that adjust process variables if needed; a simple example is the adjustment of heating temperature. An HACCP study may reveal that the process should be changed to allow efficient control.

Instruction manuals give more details, as shown in Table 6.1. The product and its use must be described in detail. The manufacturing process is carefully described in a flow diagram, including the control points for adjusting the process (e.g., temperature, flow rate, mixing intensity, and rate at which a component is added). Each step in the process is analyzed for its potential hazards, and these are evaluated and quantified. It is then analyzed as to what measures can be taken to minimize the hazard, and it is finally decided, on the basis of systematic criteria, whether this is to be made a critical control point. If so, a monitoring scheme is devised, an essential point of which is the monitoring frequency. Too frequent monitoring is unnecessarily expensive and tends to demotivate the operators; monitoring too seldom may lead to an unacceptable hazard. This procedure is applied to every process step, leading to a complete HACCP system. Furthermore, a corrective action plan should be developed, i.e., what measures should be taken when some critical parameter is observed to be outside the limits set. The system should regularly be evaluated and verified during its application, and modified when needed.

TABLE 6.1
Procedure for Developing a Control System for the Manufacture of a Food Product, According to the European Hygienic Design Group

Stage	Action
1	Define terms of reference
2	Select the HACCP team
3	Describe the product
4	Identify intended use of product
5	Construct a flow diagram
6	On-site verification of flow diagram
7	List all hazards with each process step and list all measures which will control the hazards
8	Apply HACCP decision tree to each process step in order to identify CCPs
9	Establish target level(s) and tolerance for each CCP
10	Establish a monitoring system for each CCP
11	Establish a corrective action plan
12	Establish record keeping and documentation
13	Verification
14	Review the HACCP plan

Note: CCP = critical control point; HACCP = hazard analysis/critical control points.

An essential aspect is that HACCP systems cannot be copied. Each manufacturer has his own particulars in the process applied and in outside conditions and constraints. Moreover, the development and repeated evaluation of the system by the people involved in its application is a prerequisite for its success. Consequently, this book will not give prescriptions for HACCP systems for particular products, although potential hazards and critical points in a process will often be indicated.

6.3.3 Quality Assurance of Raw Milk

As mentioned, total quality assurance involves the full chain from the production of raw milk to the consumption of dairy products. Obtaining high-quality raw milk is a matter of particular and lasting concern for a dairy. This is because in milk production, so many steps and aspects play a role and so many individual producers are involved. Figure 6.1 illustrates the relationships between a dairy and the outside world. It follows that measures for quality assurance also have to be taken in the distribution of dairy products.

Raw milk quality has several aspects, the most important being gross composition and hygienic quality. The former can readily be assessed by determination of, say, fat and protein contents in random samples; the price of the raw milk

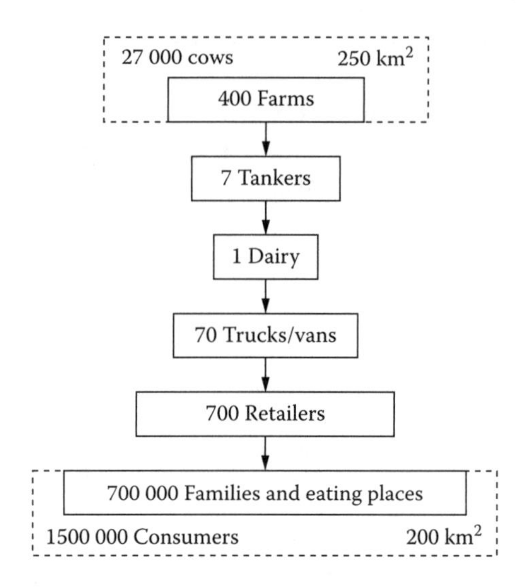

FIGURE 6.1 Relations between a dairy, its producers of raw milk, and its consumers of liquid milk products. Total quantity of milk processed would be $2 \cdot 10^8$ kg per annum. Hypothetical and simplified example. (Courtesy of M.G. van den Berg.)

is determined by its composition. Assurance of the hygienic quality poses more problems, mainly because mistakes leading to poor quality can readily be made and because sampling and analysis of every lot delivered would often be too expensive. The actual measures to be taken will vary greatly with local conditions, problems, and regulations, but in any case, milk must be sampled and analyzed on a regular basis. The success of a quality assurance system would further depend on a number of conditions:

1. The farmer should be knowledgeable about hazards and remedies, and should be committed to delivering high-quality milk. This means that training and information should be provided.
2. The farmer should be financially rewarded for producing milk of good hygienic quality, and penalized for delivering milk that is adulterated or potentially harmful.
3. If the farmer encounters difficulties in producing high-quality milk, help should be provided in establishing the cause and in finding remedies.
4. If the farmer suspects that his milk accidentally has become of poor quality, e.g., because the cooling system has failed, he should have the option of reporting this to the dairy. The milk can then be collected separately, and any financial penalty for poor-quality milk can be restricted to the quantity of that lot.

Another problem is that raw materials for concentrates (cattle feed) are sometimes contaminated with substances that may reach the milk, e.g., aflatoxins.

The best solution seems to be that the dairy industry and the cattle feed industry come to agreement on how to curtail such problems.

It goes without saying that milk of unacceptable hygienic quality should always be rejected. On the other hand, quality criteria for raw milk should not be more stringent than is necessary to make safe and good-quality milk products.

6.4 MILK STORAGE AND TRANSPORT

Milk storage and transport operations are aimed at having good-quality milk available where and when needed for processing. The milk should not be contaminated by microorganisms, chemicals, water, or any other substance. Obviously, the costs involved in storage and transport should be kept low, which implies that, for example, loss of milk should be minimized. Simple and effective cleaning of all the equipment involved should be possible. Furthermore, a satisfactory record of actual losses is desirable; most manufacturers determine the mass and fat balance on a daily basis.

Transport and storage refer to raw milk as well as to intermediate products.

6.4.1 Milk Collection and Reception

Milk may be supplied to the dairy in milk cans (churns) or by a tanker after it has been cold-stored at the farm (tank milk).

During transport, *milk in cans* usually has a temperature of >10°C, but may vary between 0 and 40°C according to the climate. Consequently, bacterial growth often occurs between milking and the milk's arrival at the dairy, as this may take as long as a day. The extent of bacterial growth depends primarily on the level of hygiene during milking, the temperature, and the storage period (see Subsection 5.1.5). Spoilage of the milk is mainly by mesophilic bacteria and usually involves lactic acid fermentation; however, heavy contamination with polluted water (mainly pseudomonads) may cause a nonsouring spoilage. On reception at the dairy plant, milk is cooled to < 6°C, which helps to more or less stabilize its bacteriological quality for at most 2 d.

Tank milk has been kept at low temperature, but for a longer time. It mainly contains psychrotrophs and consequently requires treatment different from milk in cans. Among the advantages of tank milk over milk in cans are the cheaper transport costs (if the collection routes are not too long) and a regular supply of good-quality milk, provided that the temperature of the milk at the farm and during transport is satisfactorily controlled.

On reception, the quantity of milk is recorded first. At the dairy, milk in cans is weighed by a platform balance. The quantity of tank milk is determined by metering the intake line of the milk tanker. Milk volume is then converted to weight.

Collected milk ought to be routinely examined to identify poor-quality milk supplies. A simple, rapid examination of the sensory properties would include odor, appearance, and temperature. In addition, the intake pipe of the milk tanker will be equipped with a continuously recording thermometer and a pH meter that will switch off the intake pump if the values recorded exceed a predetermined level. Incidentally, an off-flavor is more easily detected in the warmer milk in cans than in tank milk, and souring of milk can be detected easier than the growth of psychrotrophs. In addition to this simple inspection, the milk can be tested at the laboratory of the dairy for the presence of antibiotics, as well as its freezing-point depression, acidity, and bacterial count.

It is advisable that the reception of milk in cans at the dairy occur as soon as possible after milking. This implies twice-a-day milk collection. Often this is not practical, and the evening milking is cooled by mains or well water. Once-a-day collection may, however, seriously impair the milk quality in cans. Tank milk should be refrigerated to < 4°C. After 4 or 5 d storage, substantial growth of psychrotrophs may have occurred (Subsection 6.4.2). Consequently, tank milk can normally be kept on the farm for 3 d, i.e., six milkings, and stored for another day at the dairy before processing.

Milk can be contaminated during transport if the tanker was inadequately cleaned. Milk tankers can contaminate milk with high numbers of psychrotrophs. This means that rigorous cleaning of the tanker and routine monitoring are essential. Furthermore, the temperature of the milk during transport must be kept low, i.e., < 5°C.

Paying strict attention to the measures mentioned in the preceding text will ensure a satisfactory quality of the raw material supplied. A small quantity of milk of somewhat inferior quality will have little effect due to its dilution in the large storage tanks of the dairy. However, milk supplies of poor quality should preferentially be eliminated.

6.4.2 Milk Storage

Variations in composition, properties, and quality of the raw milk directly affect the manufacturing processes as well as the composition and quality of the final products, and are therefore undesirable. Some variation is inevitable, but mixing many deliveries in large storage tanks, containing, for example, 300,000 kg of milk, results in only a small variation among lots of milk within 1 or 2 d.

6.4.2.1 Bacterial Growth

The duration for which raw milk can be kept in storage tanks is mainly determined by the growth of psychrotrophs. Prior to processing, bacterial numbers greater than $5 \cdot 10^5$ ml^{-1} in milk imply a risk that psychrotrophs have produced heat-stable enzymes, i.e., bacterial lipases and proteinases, which may impair the quality of the final product. It is important to note that a high count, originating from mixing a small quantity of milk containing many psychrotrophs with a large quantity of milk of a low count, is more harmful than a similar count resulting from limited growth in the whole lot. This is because extracellular enzymes are predominantly produced at the end of the exponential-growth phase. Examples of the growth of psychrotrophic and other bacteria in milk during storage on the farm and at the dairy are given in Figure 6.2. Initially, during storage on the farm, the total count

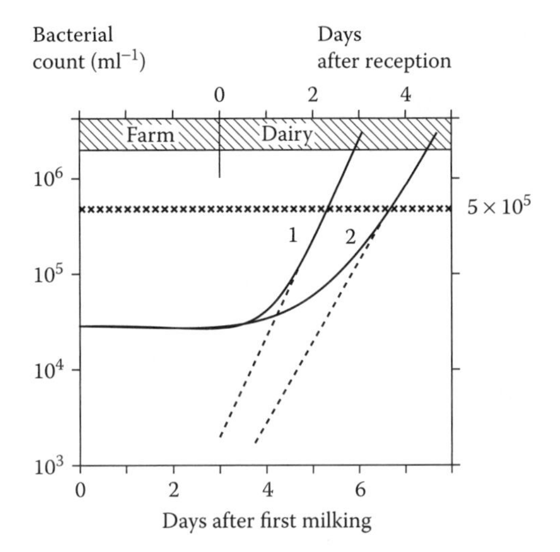

FIGURE 6.2 The growth of psychrotrophic and other contaminating bacteria of different generation times (g) in tank milk at 4°C. Calculated examples. Total count (—), number of psychrotrophs (---), desirable upper limit for processing (xxx). 1. Fast-growing contaminating bacteria, $g = 6$ h; 2. Contaminating bacteria of $g = 8$ h.

remains almost constant and only starts to increase after 4 or 5 d. The delay in the growth of psychrotrophs to high numbers is often thought to be due to an extended lag phase at low temperatures. However, a very low initial contamination with fairly fast-growing bacteria, i.e., < 10 ml^{-1}, e.g., from an improperly disinfected bulk tank, may also be responsible for the delay, as illustrated in Figure 6.2.

Depending on the age of the milk supplied to the dairy, it can be stored for another 1 or 2 d without further treatment. All milk supplies should, however, be cooled to < 4°C because the temperature of milk may have increased during transport from the farm to the dairy and the generation time of bacteria is markedly shorter at high temperatures (see Table 5.2 and Figure 5.5). Often, the dairy is unable to process all milk supplies within 4 d of milking. Consequently, measures must be taken to keep the raw milk for a longer time. Pasteurization (72°C for 15 s) is not desirable, because it will be done later on and pasteurizing twice may impair the quality of the finished products. A more moderate heat treatment (e.g., 65°C for 15 s, called thermalization) reduces the number of psychrotrophs considerably, leaving most milk enzymes and agglutinins intact (Section 7.3). After thermalization, the milk can be kept for another 3 or 4 d at 6 to 7°C without substantial increase in the bacterial count, provided there is no recontamination by psychrotrophs. Milk should be thermalized as soon as possible after arrival at the dairy. Thermalization is a far better method for controlling the quality of dairy products than merely cooling the raw milk, but it also is more expensive. Because many bacteria survive thermalization, considerable bacterial growth can occur at 30 to 40°C in the regeneration section of the heat exchanger (Subsection 7.4.4).

TABLE 6.2
Examples of Standards for (Pooled) Milk before Processing

Quality Mark	Standard	Absolute Limit	Unit
Acidity	17	≤18	mmol/l
Count (raw)	100; 95% < 250	< 500	μl^{-1}
Count (thermalized)	50; 95% < 100	< 250	μl^{-1}
Heat-resistant bacteria	5; 95% < 10	< 25	μl^{-1}
Bacillus cereus	0.1; 95% < 0.2	≤ 1	ml^{-1}
Fat acidity	0.6; 95% < 0.8	≤ 0.9	mmol/100 g
Freezing-point depression	520–525	> 515	mK
Antibiotics	Not detectable		
Disinfectants	Not detectable		

Therefore, it may be necessary to clean it after operating for 4 to 6 h. The quality of thermalized milk may still be threatened by the presence of any psychrotrophs that are fairly heat resistant, e.g., *Alcaligenes tolerans.*

Usually, the quality of milk is examined after it arrives at the dairy. It is advisable to test the milk again just before processing. Standards for milk quality before processing are given in Table 6.2.

6.4.2.2 Enzyme Activity

Lipase activity is usually the main problem in fresh milk (Subsection 3.2.5), although other milk enzymes, e.g., proteases and phosphatases, also cause changes. Therefore, extensive temperature fluctuations, in the range of 5 to 30°C, and damage to fat globules (see the following text) should be avoided.

6.4.2.3 Chemical Changes

Exposure to light should be avoided because it results in off-flavors (Section 4.4). Contamination with rinsing water (causes dilution), disinfectants (causes oxidation), and especially with Cu (catalyzes lipid oxidation) should be avoided.

6.4.2.4 Physical Changes

The following are the main physical changes that can occur during storage:

1. Raw or thermalized milk stored at low temperature creams rapidly (Subsection 3.2.4.2). Formation of a cream layer can be avoided by regular stirring of the milk, e.g., stirring for a few minutes every hour. This is often done by aeration; the air supplied should be sterile, for obvious reasons, and the air bubbles fairly large, as otherwise too many fat globules would adsorb onto the bubbles (see point 2).

2. Damage to fat globules is mainly caused by air incorporation and by temperature fluctuations that allow some fat to melt and crystallize. These events can lead to increased lipolysis, to disruption of fat globules if the fat is liquid, and to clumping of fat globules if the fat is partly solid (10 to 30°C).
3. At low temperatures, part of the casein, primarily β-casein, dissolves from the micelles to end up in the serum. This dissolution is a slow process and reaches equilibrium after approximately 24 h (Subsection 3.3.2.2). The dissolution of some casein increases the viscosity of the plasma by approximately 10% and reduces the rennetability of the milk. The reduced rennetability may be partly due to a changed calcium ion activity (Subsection 2.2.5.4). Temporarily heating milk to ~50°C or higher almost fully restores the original rennetability of the milk.

6.4.3 Transport of Milk in the Dairy

To move the milk about, a dairy needs an intricate system of pipelines, pumps, and valves, as well as controlling units. The system should be flexible while excluding such errors as milk running off or unintentional mixing of different products. To save on pumping costs, gravity is often used. Centrifugal pumps are preferentially used for milk because they keep turning without great problems if the milk cannot be discharged; they are not suitable for viscous products.

Following are some specific problems:

1. *Milk loss:* This concerns residues in pipes and equipment after processing, spillage, and mixing of milk with different products or with water when valves are switched over. Ensuring a satisfactory discharge of the milk, avoiding 'dead ends' in pipes, and minimizing the surface area wetted by milk, are all obvious measures to reduce loss. Minimizing the diameter (D) of pipes can reduce the amount of mixing that occurs between milk and water; the volume of the mixing region is proportional to $D^{2.55}$. Milk diluted with water may be evaporated, used to dissolve skim milk powder, or used as cattle feed. Proper operation reduces the cost due to milk losses to approximately 1% of the total cost of the raw material.
2. *Damage to milk:* Air incorporation may damage milk fat globules. Excessive shear rates and intense turbulence during transport may cause clumping, i.e., formation of visible lumps of fat, especially in cream. In transporting cream, it is hence advisable to avoid narrow and long pipes, as well as obstacles (e.g., sharp bends) in the pipeline system. The cream should not be transported at temperatures between 10 and 40°C. Furthermore, the viscosity of products like yogurt and custard can be markedly reduced by high deformation rates occurring during transport (irreversible breakdown of structure).

3. *Bacterial growth:* During transport, contamination of milk by bacteria can readily occur. Balance tanks are often situated before various kinds of processing equipment to ensure a constant milk-flow rate. If the temperature in such a balance tank is high enough for bacterial growth, the tank tends to act as a continuous fermentor, allowing growth of bacteria in the milk to be processed. To prevent such problems, two balance tanks are put in parallel, so that either of them can be used; this allows cleaning of one tank whilst the other is in operation, without interrupting the processing. Leaving raw milk for some time in non-insulated pipelines favors bacterial growth. Obviously, all such situations should be avoided.

6.5 STANDARDIZING

Standardization of the composition of a milk product is needed because it is legally required or because manufacturers set a standard for their product. It mostly concerns the fat content, often also the dry-matter content (or the degree of concentration), sometimes the protein content, or still another component.

From an economic point of view, continuous standardization is desirable; turbidity or density measurements can be applied for fat content and density or refractive index for dry-matter content. Infrared-reflection measurement is also used, e.g., to determine the water content of milk powder. In continuous standardization, the (amplified) measuring signal may control the position of a regulating valve, e.g., a valve in a cream line or in a steam-supply pipe; in this way the desired content can be adjusted. To achieve this, the relations between turbidity and fat content, between density and dry-matter content, etc., in the original milk must be known. This is because these relations are not always the same. The adjustment is often difficult because great fluctuations can easily occur when the adjustments are being made. Therefore, a double adjustment is often employed, based on measurement of both volume flows and a concentration-dependent variable.

After the standardization, performed tentatively or by means of continuous determination, the desirable content will have to be checked. This implies that it may be necessary to make an adjustment by the addition of cream, skim milk, water, etc. Any bacterial or other contamination should be rigorously avoided. The added compound should have been treated (especially with respect to heating) in a way similar to that of the product itself.

Standardization is always subject to inaccuracy because the results of the methods of determination and the measuring or weighing of the components have a certain inaccuracy. This is also true of determination by the supervising authority. Therefore, a certain margin should be left, e.g., twice the standard deviation. In some cases, for example, with respect to the fat content of beverage milk, a deviation of ±0.05% fat may be permitted, whereas the average value over a prolonged period should deviate by no more than 0.01% fat from the accepted standard value.

Standardization of products (e.g., beverage milk) with respect to protein content is generally not allowed. All the same, the nutritive value and the cost price of the milk greatly depend on the (variable) protein content. Technically, standardization is possible by applying ultrafiltration.

Suggested Literature

There are several text and reference books about dairy technology (processing and products), but most of these are very elementary. Some interesting aspects are discussed in: R.K. Robinson, Ed., *Modern Dairy Technology,* Vol. 1, *Advances in Milk Processing,* and Vol. 2, *Advances in Milk Products,* 2nd ed., Elsevier, London, 1993.

Technical information, both of a general nature and as applied in the manufacture of various milk products: A.Y. Tamime and B.A. Law, Eds., *Mechanization and Automation in Dairy Technology,* Sheffield Academic Press, Sheffield, U.K., 2001.

A heavily illustrated book, useful in providing information about equipment and technical processes: *Dairy Processing Handbook,* 2nd ed., text by G. Bylund, Tetra Pak Processing Systems, Lund, Sweden, 2003, but the treatment of dairy science and technology is weak and not up-to-date.

Several new developments: G. Smit, Ed., *Dairy Processing: Improving Quality,* Woodhead Publishing, Cambridge, 2003.

A recent review on new preservation techniques for milk products: F. Devliegere, L. Vermeiren, J. Debevere, *Int. Dairy. J.*, 14, 273, 2004.

A general text on HACCP: M.D. Pearson and D.A. Gorlett, *HACCP: Principles and Applications,* AVI, New York, 1992.

See also: S. Leaper, ed., *HACCP: A Practical Guide,* Technical Manual 38, Camden Food and Drink Research Association, Camden, U.K., 1992.

7 Heat Treatment

The manufacture of virtually all milk and dairy products involves heat treatment. Such treatment is mainly aimed at killing microorganisms and inactivating enzymes, or at achieving some other, mainly chemical, changes. The results greatly depend on the intensity of the treatment, i.e., the combination of temperature and duration of heating. It is useful to distinguish between irreversible and reversible changes. The latter are often involved when milk is brought to a higher temperature to facilitate some reaction or process, such as renneting of cheese milk, growth of starter organisms, water evaporation or centrifugal separation, etc.

Heat treatment may also cause undesirable changes, although desirability may depend on the kind of product made and on its intended use. Examples are browning, development of a cooked flavor, loss of nutritional quality, inactivation of bacterial inhibitors, and impairment of rennetability. This means that heat treatment should be carefully optimized.

After defining the objectives of heat treatment, the various chemical and physical reactions occurring at high temperatures will be covered. This will be followed by a discussion on the kinetics of the changes occurring. Finally, more practical aspects of heat treatment will be reviewed. For the benefit of readers not well acquainted with the fundamentals of heat transfer, some aspects are briefly given in Appendix A.11.

7.1 OBJECTIVES

The main reasons for heat treatment of milk are the following:

1. *Warranting the safety of the consumer:* It specifically concerns killing of pathogens like *Mycobacterium tuberculosis, Coxiella burnetii, Staphylococcus aureus, Salmonella* species, *Listeria monocytogenes*, and *Campylobacter jejuni*. It also concerns potentially pathogenic bacteria that may accidentally enter the milk. A fairly moderate heat treatment kills all of these organisms. Highly heat-resistant pathogens either do not occur in milk (e.g., *Bacillus anthracis*), or become readily overgrown with other bacteria (e.g., *Clostridium perfringens*), or cannot grow at all in milk (e.g., *Clostridium botulinum*), or are pathogenic only at such high numbers (e.g., *Bacillus cereus*) that impending spoilage of the milk is detected long before these high counts are reached. To be sure, some toxins (especially from staphylococci) can withstand moderate heat treatment.
2. *Increasing the keeping quality:* It primarily concerns killing of spoilage organisms and of their spores if present. Inactivation of enzymes, native

to milk or excreted by microorganisms, is also essential. Chemical deterioration by autoxidation of lipids (Subsection 2.3.4) can be limited by intense heat treatment. Rapid creaming can be avoided by inactivating agglutinin (Subsection 3.2.4).

3. *Establishing specific product properties:* Examples are (1) heating the milk before evaporation to increase the coagulation stability of evaporated milk during its sterilization (Subsection 19.1.4), (2) inactivating bacterial inhibitors such as immunoglobulins and the lactoperoxidase-CNS-H_2O_2 system (see also Subsection 7.3.4) to enhance the growth of starter bacteria, (3) obtaining a satisfactory consistency of yogurt (Section 22.3), and (4) coagulating serum proteins together with casein during acidification of milk (Chapter 21).

7.2 CHANGES CAUSED BY HEATING

7.2.1 Overview of Changes

Changes in the composition of milk caused by an increase in temperature may be reversible or irreversible. Here we are mainly interested in the irreversible or slowly reversible reactions; such changes scarcely occur at heat treatments of lower intensity than low pasteurization. All the same, reversible reactions must be taken into account because they determine the state of the milk at increased temperatures, i.e., the conditions in the milk at which the irreversible changes take place. Reversible changes include the mutarotation equilibrium of lactose (Subsection 2.1.3.1) and changes in ionic equilibriums, including pH (see, e.g., Subsection 2.2.5, and Section 4.2).

Numerous changes that occur on heating are discussed in this book. Here we give a brief survey of the main changes. The list is by no means complete. Moreover, several changes are interdependent, and the various changes may occur at very different heating intensities.

7.2.1.1 Chemical and Physical Changes

Possible chemical and physical changes caused by heat treatment include:

1. Gases, including CO_2, are partly removed (if they are allowed to escape from the heating equipment). Loss of O_2 is important for the rate of oxidation reactions during heating, and for the growth rate of some bacteria. The loss of gases is reversible, but uptake from the air may take a long time.
2. The amount of colloidal phosphate increases and the [Ca^{2+}] decreases (Figure 2.8). Again, the changes are reversible, though slowly (~24 h).
3. Lactose isomerizes and partly degrades to yield, for instance, lactulose and organic acids (Subsection 7.2.3).
4. Phosphoric esters, those of casein in particular, are hydrolyzed (Subsection 7.2.2.3). Phospholipids are also split. Consequently, the amount of inorganic phosphate increases.

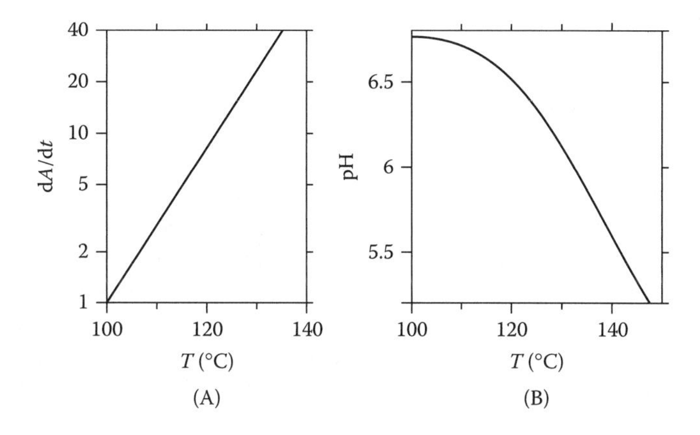

FIGURE 7.1 Acid production in milk during heat treatment as a function of the temperature T. Approximate results. (A) Titratable acidity produced, in $mEq \cdot l^{-1} \cdot h^{-1}$; (B) pH at room temperature after 30 min of heating. (Adapted from P. Walstra and R. Jenness, *Dairy Chemistry and Physics,* Wiley, New York, 1984.)

5. The pH of the milk decreases, and the titratable acidity increases, mainly due to changes 2, 3, and 4 (see Figure 7.1). All of these changes depend somewhat on the prevailing conditions. See also Subsection 2.2.5.1.
6. Most of the serum proteins are denatured and thereby rendered insoluble (see Subsection 7.2.2.2).
7. Part of the serum protein (especially of β-lactoglobulin) becomes covalently bound to κ-casein and to some proteins of the fat globule membrane.
8. Enzymes (Section 2.5) are inactivated. See Figure 2.31, Figure 7.9, and Subsection 7.3.3.
9. Reactions between protein and lactose occur, Maillard reactions in particular (Subsection 7.2.3). This involves loss of available lysine.
10. Free sulfhydryl groups are formed. This causes, for instance, a decrease of the redox potential (Section 4.3).
11. Other reactions involving proteins occur. See Subsection 7.2.2.
12. Casein micelles become aggregated. Aggregation may eventually lead to coagulation (Subsection 7.2.4).
13. Several changes occur in the fat globule membrane, e.g., in its Cu content.
14. Acylglycerols are hydrolyzed and interesterified (Subsection 2.3.1).
15. Lactones and methyl ketones are formed from the fat (Subsection 2.3.1).
16. Some vitamins are degraded.

7.2.1.2 Consequences

Usually, the main effect of heat treatment is the far slower rate of deterioration caused by microbial and enzymatic action. The most important other effects are:

a. *Bacterial growth rate* of the organisms surviving, or added after heat treatment, can be greatly affected, generally increased. This is mainly because bacterial *inhibitors* are inactivated. Immunoglobulins are denatured at relatively low intensity: see the curve for cold agglutination in Figure 7.9D. *Bacillus cereus* is especially sensitive to IgM (agglutinin). The lactoperoxidase system, especially affects lactic acid bacteria. It is inactivated due to denaturation of lactoperoxidase (see Figure 7.9A). Lactoferrin, especially affects *Bacillus stearothermophilus*; it needs conventional heat sterilization to be inactivated. *Bacteriophages* can be inactivated, depending on the heating intensity (see Subsection 13.3.4), and this is especially important for lactic fermentations. Some *stimulants* can also be formed, e.g., formic acid, which enhances growth of lactic acid bacteria, especially thermophilic ones; its formation needs intense heat treatment (see Subsection 7.2.3).
b. *Nutritive value* decreases, at least for some nutrients, due to changes 16 and 9, and maybe 11, in the list above. Examples are given in Table 16.3.
c. The *flavor* (Section 4.4) changes appreciably, mainly due to 9, 10, 11, and 15.
d. *Color* may change. Heating milk at first makes it a little whiter, maybe via change 2. On increasing the heating intensity, the color becomes brown due to 9. See Section 4.6, Figure 4.4.
e. *Viscosity* may increase slightly due to 6 and much more due to 12 (if it happens). The latter change especially occurs when concentrated milk is sterilized.
f. *Heat coagulation* in evaporated milk (Subsection 19.1.5) is markedly decreased when the milk is heated so that most of the serum protein is denatured before concentrating.
g. *Age gelation* in sweetened condensed milk (Subsection 19.2.2) is also reduced when the milk is intensely heated before concentrating.
h. The *rennetability* of milk and the rate of *syneresis* of the rennet gel decrease (Subsection 24.3.6), for the most part due to 7.
i. *Creaming* tendency of the milk decreases (Subsection 3.1.4), mainly caused by 6 (denaturation of IgM).
j. The proneness to *autoxidation* is affected in several ways (Section 2.4), mainly due to 13, 10, and 8.
k. The composition of the *surface layers* of the fat globules formed during homogenization or recombination is affected by the intensity of heating before homogenization, mainly because of change 6. This affects some product properties; for example, the tendency to form homogenization clusters (Section 9.7) is increased.

7.2.2 Reactions of Proteins

Several reactions of side chain groups (and possibly of terminal groups) of proteins can occur at high temperature. Table 7.1 gives examples. Many of these reactions, i.e., 5, 6, 9, 10, 11, and especially 12, can form cross-links within or between peptide chains; cross-linking reactions may reduce the solubility of the

TABLE 7.1
Possible Reactions of Side Chain Groups of Amino Acid Residues Linked in the Peptide Chain (|) of Proteins at High Temperature

No.	Reactants		Products
1.	⊢ $CH_2–CONH_2 + H_2O$ Asparagine	⟶	⊢ $CH_2–COO^- + NH_4^+$ Aspartic acid
2.	⊢ $(CH_2)_2–CONH_2 + H_2O$ Glutamine	⟶	⊢ $(CH_2)_2–COO^- + NH_4^+$ Glutamic acid
3.	⊢ $CH_2–O–PO_3^{2-} + H_2O$ Phosphoserine	⟶	⊢ $CH_2–OH + HPO_4^{2-}$ Serine
4.	⊢ $CH_2–SH + OH^-$ Cysteine	⇄	⊢ $CH_2–S^- + H_2O$
5.	⊢ $CH_2–S–S–CH_2$ ⊣ + ⊢ $CH_2–S^-$	⇄	⊢ $CH_2–S^-$ + ⊢ $CH_2–S–S–CH_2$ ⊣
6.	⊢ $CH_2–S^- + {}^-S–CH_2$ ⊣ Cysteine	⟶	⊢ $CH_2–S–S–CH_2$ ⊣ + 2 ⊖ Cystine
7.	⊢ $CH_2–S^-$ Cysteine	⟶	⊨ CH_2 + HS^- Dehydroalanine
8.	⊢ $CH_2–O–PO_3^{2-}$ Phosphoserine	⟶	⊨ CH_2 + HPO_4^{2-} Dehydroalanine
9.	⊨ CH_2 + $HS–CH_2$ ⊣ Dehydroalanine Cysteine	⟶	⊢ $CH_2–S–CH_2$ ⊣ Lanthionine
10.	⊢ $(CH_2)_4–NH_3^+ + H_2C$ ⫤ $+ OH^-$ Lysine Dehydroalanine	⟶	⊢ $(CH_2)_4–NH–CH_2$ ⊣ $+ H_2O$ Lysinoalanine
11.	⊢ $CH_2–(C_3H_3N)–NH^+ + H_2C$ ⫤ $+ OH^-$ Histidine Dehydroalanine	⟶	⊢ $CH_2–(C_3H_3N)–N–CH_2$ ⊣ $+ H_2O$ Histidinoalanine
12.[a]	⊢ $CH_2–COOH + H_2N–(CH_2)_4$ ⊣ Aspartic acid Lysine	⟶	⊢ $CH_2–CO–NH–(CH_2)_4$ ⊣ $+ H_2O$ Isopeptide
13.[b]	⊢ $(CH_2)_4–NH_2 + C_6H_{12}O_6$ Lysine Glucose	⟶	⊢ $(CH_2)_4–NH–C_6H_{11}O_5 + H_2O$ Amadori product

[a] Reaction also occurs with glutaminic acid residues.
[b] First step in the Maillard reaction with glucose or another reducing sugar. See Figure 7.4.

protein. The rate of most of the reactions and their equilibrium states are poorly known. Besides reactions 4 to 6, which may occur upon denaturation of the protein, most of the reactions involved require high temperatures (sterilization). Because casein contains phosphoserine, dehydroalanine can be formed (reaction 8); it also results from reaction 7. Dephosphorylation by hydrolysis (reaction 3) occurs faster than by β-elimination (reaction 8). All in all, in milk, most of the reactions considered in Table 7.1 may occur, though only very small amounts of, for instance, lysino-alanine (reaction 10) are formed unless the pH is very high; lysinoalanine might be toxic because its ingestion can cause changes in the kidney of rats (though not observed in humans). Deamidation of glutamine (reaction 2) is far slower than that of asparagine (reaction 1).

As stated above, a high temperature is needed for most of the reactions to occur. It is not always necessary for the reaction itself, but it is needed for the unfolding of the peptide chain (denaturation) whereby the groups involved become exposed and available for reactions. Most reactions proceed faster at higher pH but not reaction 3 and reaction 11.

7.2.2.1 Reactions of Thiol (Sulfhydryl or –SH) Groups

The –SH group of cysteine is very reactive in the ionized form (reaction 4). In the peptide chain its pK is about 9.5 at 25°C. This means that at pH 6.1, 6.4, and 6.7, on average 0.04%, 0.08%, and 0.16% of the group, respectively, is dissociated. Consequently, the rate of reaction 4 will strongly depend on pH. Of course, before this reaction can occur the peptide chain must be unfolded, unless the thiol group is on the outside of the native molecule, which is exceptional. Heat treatment of milk such that denaturation of serum proteins occurs, therefore, results in a considerable increase in the number of reactive thiol groups. Upon such heating, it is mainly reaction 5 that takes place, thus shifting the position of the –S–S– linkages involved; it should be noted that the reaction can even occur at low pH (4.5) and low temperature (20°C), although quite slowly (over a few days). The disulfide interchange may considerably affect the conformation of a protein molecule. Reaction 6 closely depends on the redox potential because this cross-linking reaction occurs by oxidation.

Formation of H_2S (reaction 7, but presumably other reactions as well) causes a cooked or even 'gassy' flavor to develop in milk. As a rule, no more than, say, 1% of the thiol groups react. The formed dehydroalanine residue readily reacts, according to reactions 9, 10, and 11.

Table 2.15 shows in which of the milk proteins –S–S– and –SH groups occur; the immunoglobulins also contain much cystine. The main contributor of –S–S– and –SH groups probably is β-lactoglobulin, due to its high concentration in milk and its free thiol group. Upon heating, the thiol group becomes very reactive and because of that, irreversible changes occur in the molecule. As far as the formation of H_2S is concerned, a part of the fat globule membrane protein is by far the most active agent, at least 10 times more active than β-lactoglobulin; heating of skim milk hardly produces any H_2S, whereas heating of cream produces far more.

7.2.2.2 Denaturation of Serum Proteins

Globular proteins can be subject to denaturation, as discussed in Subsection 2.4.1.3. They exhibit unfolding of their peptide chains at a high temperature, say above 70°C, although marked variation in the temperature needed is observed among proteins. As mentioned, reactions occurring in or between side groups in the peptide chain at the prevailing temperature may then prevent refolding of the peptide chain into its original, i.e., native, conformation. In other words, the protein remains denatured. As a result, most proteins lose their biological activity, e.g., as an enzyme or as an antibody. Generally, they also become less soluble. Otherwise, the changes may be mild and the nutritive value is rarely impaired.

These changes occur with the globular serum proteins of milk, namely, β-lactoglobulin, α-lactalbumin, serum albumin, and the immunoglobulins (as well as most of the minor serum proteins). The proteose-peptones, just like the caseins, are not denaturable. Heat denaturation of β-lactoglobulin has been studied in some detail. At high temperatures, the free thiol group becomes exposed and it reacts with one of the –S–S– groups (Table 7.1, reaction 5), often of another molecule, whereby both molecules become bonded, forming a dimer. In the same way, trimers and tetramers, etc., are also formed but the aggregates may remain fairly small and soluble. Depending greatly on prevalent conditions, especially pH, but also ionic composition and temperature, further aggregation may now occur, resulting in large insoluble particles; at higher concentrations a gel may be formed. Much the same happens with the other proteins, although some of them lack free thiol groups. This implies that either the reaction scheme is different or that a free thiol group of another protein (β-lactoglobulin, bovine serum albumin, or some immunoglobulin) is involved. When heating a solution of serum proteins, e.g., whey, at various pH, it is observed that they only become insoluble at low pH (see Figure 7.2A). Also, heating at neutral pH and acidification after cooling leads to insolubility. Insolubility is enhanced at a high Ca^{2+} activity.

If *milk* is heated, covalent binding to other proteins can occur via –S–S– bridges, depending on the pH and temperature during heat treatment. This involves, of course, bonds between various serum proteins but also bonds with proteins of the fat globule membrane and with κ- and α_{s1}-caseins. Some serum protein thus becomes associated with the fat globules.

The interaction with casein (micelles) is more complicated. Heat denaturation at pH 6.9 (as measured at room temperature), does much the same as in whey: small aggregates of serum protein are formed, e.g., 60 nm in size. These aggregates remain 'dissolved,' and they may or may not contain some κ-casein. At pH values of 6.5 or less, all denatured serum protein associates with the casein micelles. At pH 6.7, about 30% is in dissolved aggregates, the rest is associated with the micelles. The association involves –S–S– linkages. The casein micelles do not become 'covered' by serum protein: there is not enough for full coverage and, moreover, the association is for a substantial part in the form of serum protein aggregates. The micelles thus obtain a 'bumpy' surface, and the effective voluminosity of the

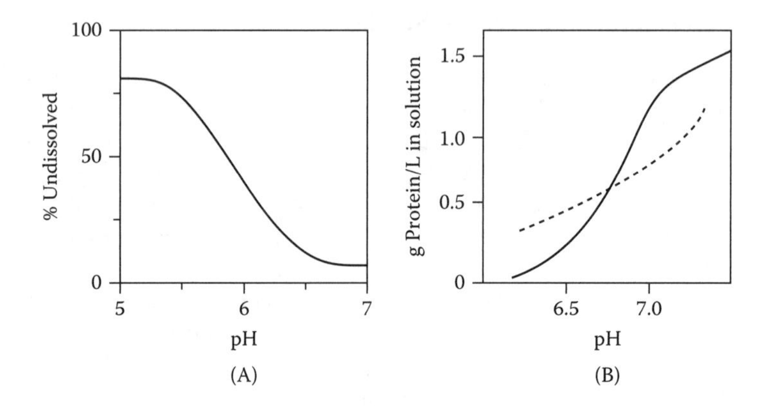

FIGURE 7.2 Influence of pH on the effects of heating on proteins. (A) Percentage of the proteins that become precipitated after heating whey for 10 min at 80°C. (B) Amount of protein that remains in solution, i.e., not associated with the casein micelles, after heating milk (—) or serum protein free milk (----) at 140°C.

overall milk protein is increased, leading to increased viscosity. When milk is heated at physiological or higher pH and then acidified to pH 6.5 or lower after cooling, all of the denatured serum protein becomes associated with the micelles.

The changes mentioned occur at temperatures between about 70°C and 90°C. Above about 120°C a considerable amount of the casein, predominantly κ-casein, leaves the micelles at high pH (see Figure 7.2B). The κ-casein, whether in solution or in the micelles, reacts with denatured serum protein, which modifies the pH dependence of the dissociation of κ-casein, as shown in the figure. This is of considerable importance for the heat stability of (concentrated) milk (Subsection 7.2.4).

Kinetics of the denaturation reaction are given in Figure 7.3, and it can be seen that the various serum proteins differ in heat sensitivity (see also Figure 7.9E and Subsection 7.3.2).

7.2.2.3 Degradation

Upon heating at high temperature, cleavage of various parts of the molecules may occur. Such cleavage has been mainly demonstrated to occur in casein. An example is dephosphorylation of caseinate. Furthermore, severe heat treatment cleaves peptide chains, yielding soluble peptides. N-acetyl neuraminic acid and possibly other carbohydrates may be cleaved from casein as well. For example, in a study on heating caseinate solutions, a treatment of 20 min at 120°C caused 2.7%, 0.9%, and 1.1% of the nitrogen and 9.5%, 7.5%, and 14.4% of the organic phosphorus of α_s-, β-, and κ-casein, respectively, to become 'soluble' (i.e., not precipitated with the casein at pH 4.6). After treatment at 135°C for 1 h, all of the organic phosphorous and 15% of the nitrogen became 'soluble'.

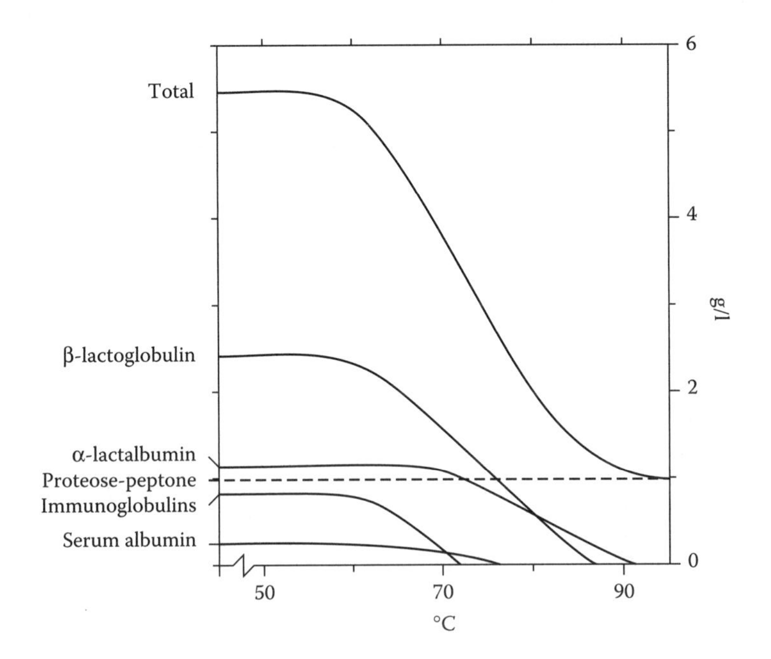

FIGURE 7.3 Effect of heating milk for 30 min at various temperatures on quantity of serum proteins that remain dissolved after cooling and acidification to pH 4.6. (Mainly adapted from B.L. Larson and G.D. Rolleri, *J. Dairy Sci.,* **38**, 351, 1955.)

7.2.3 Reactions of Lactose

During heating of milk, lactose undergoes reactions that have important consequences for the milk: changes in flavor, color, nutritive value, and pH may result. As is discussed in Subsection 2.1.1, lactose is a reducing sugar that reacts with amino groups (in milk, supplied mainly by lysine residues) in the Maillard reaction. Besides, lactose may isomerize into other sugars; it may also degrade into galactose and degradation products of glucose, including various organic acids. Figure 7.4 schematically shows the main reactions. Formulas of several of the reaction products are:

Acetol	CH_3–CO–CH_2OH
Methylglyoxal	CH_3–CO–CHO
Formaldehyde	HCOH
Formic acid	HCOOH
Acetic acid	CH_3–COOH
Pyruvic acid	CH_3–CO–COOH
Hydroxymethyl furfural	H_2COH O COH

ISOMERIZATION AND SUGAR DEGRADATION REACTIONS

Epilactose
⇅
Lactose → Galactose + C_6
⇅
Lactulose → Galactose + C_6
↘ Galactose + C_5 + Formic acid

Galactose → Tagatose → C_5 + Formic acid
↘ C_5 + Formic acid

MAILLARD REACTIONS

Initial

Lactose + lysine–R → Lactulosyl-lysine–R

Galactose + lysine–R → Tagatosyl-lysine–R

Intermediate

Lactulosyl-lysine–R → Lysine–R + Galactose + C_6
↘ Lysine–R + Galactose + C_5 + Formic acid

Tagatosyl-lysine–R → Lysine–R + C_n (n = 1–6)

Advanced

C_n + Lysine–R → Melanoidins

C_n + Arginine–R → Melanoidins

FIGURE 7.4 Simplified scheme of reactions occurring with lactose during the heating of milk at sterilization temperature. R stands for a peptide chain, C_n for an organic compound containing n carbon atoms.

Furfuryl alcohol

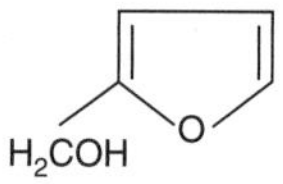

H_2COH

Maltol

O
OH
O
CH_3

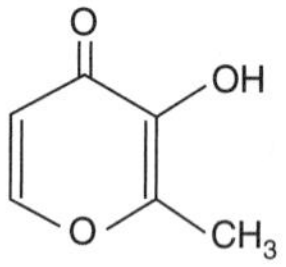

The *isomerization* and direct *degradation* reactions thus proceed in the absence of amino groups, although amino groups supposedly catalyze the isomerization reaction. Lactulose (a disaccharide of galactose and fructose; see Figure 2.1) is formed in fairly large quantities, from 0.3 $g{\cdot}l^{-1}$ to over 1 $g{\cdot}l^{-1}$ (3 mM), in sterilized milk. Epilactose (a disaccharide of galactose and mannose) is only formed in trace amounts. In principle, all of these isomerization reactions are reversible. Furthermore, the fructose moiety of lactulose may be split into formic acid and a C_5 compound or be changed into another C_6 compound, whereas the galactose moiety is left unchanged. The latter can also be degraded, either directly or after isomerization, to the ketose sugar tagatose. Among the C_5 compounds detected are furfural, furfuryl alcohol, deoxyribose, and 3-deoxypentulose. The latter two are unstable compounds. Formation of C_6 compounds includes trace quantities of hydroxymethyl furfural (HMF) in addition to other, unidentified products. These reactions are irreversible. The formic acid formed is primarily responsible for the increased acidity of heated milk (see Figure 7.1).

It is only in a later stage of the heating of milk that the *Maillard reaction* manifests itself in the changes in flavor and color. In its initial stage, it is a reaction between lactose and a free $-NH_2$ group, generally of a lysine residue. Through a number of steps, the more or less stable intermediate product lactulosyl-lysine-R is formed. (This is an Amadori product that is converted into furosine upon hydrolysis in 6 M HCl; the resulting amount of furosine can be readily determined.) From the fructose moiety of lactulosyl-lysine, reactive intermediates can be formed; galactose is left and the lysine residue is released again. (Actually, lysine thus acts as a catalyst.) Galactose, in turn, also participates in the Maillard reaction. The amount of HMF produced is on the order of some tens of $\mu mol{\cdot}l^{-1}$, i.e., far less than lactulose. Also, other low-molar-mass reaction products are formed from the intermediate lactulosyl-lysine, including acetol, methylglyoxal, maltol, and several other aldehydes and ketones. To be sure, all of these compounds may be formed in fairly small amounts but they are important with reference to flavor development (especially maltol) and because of their reactivity. Figure 7.5 gives examples of the quantities of reactants formed.

At a later stage of the Maillard reaction, polymerization reactions of amino compounds with reactive sugar intermediates occur. The latter substances also polymerize without the amino compounds being present. The polymers formed are called melanoidins and they are covalently linked to the proteins. They cause the milk to have a brown color. The molecular structure of the melanoidins is quite complex and has not been sufficiently elucidated. Advanced Maillard reactions also cause cross-linking of proteins.

Of all the reactions occurring, the Maillard reaction in particular is an intricate one, and much about it is unclear. Nevertheless, color development can now be modelled reasonably well as a function of time and temperature. Phosphate catalyzes the Maillard reaction. The various reactions that occur affect one another. For example, the drop in pH, mainly resulting from formic acid production, decreases the rate of isomerization as well as that of the Maillard reactions. The reactions proceed faster at a higher temperature (Q_{10} is 2 to 3) but the temperature dependence of the reactions is probably different for each of them. Once the Maillard reaction gets started, it proceeds at an appreciable rate even at a lower temperature, as can be noticed during storage of evaporated milk.

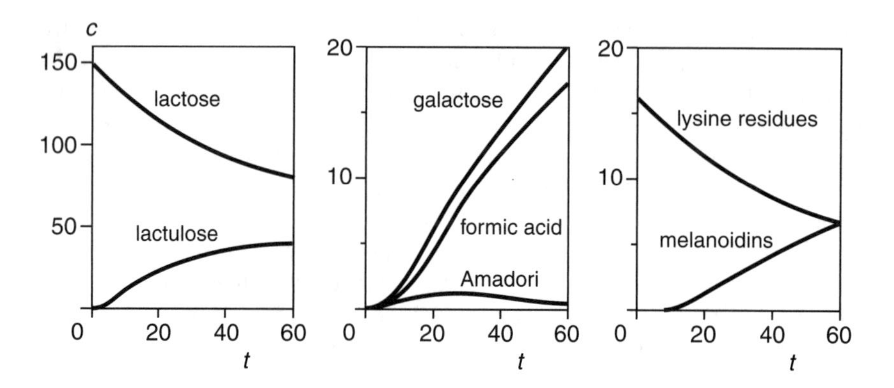

FIGURE 7.5 Concentration (c, mmol·l^{-1}) of lactose, lactulose, galactose, formic acid, Amadori products, unaltered lysine residues, and melanoidins, as a function of time (t, min) of heat treatment at 120°C, of a solution of 150 mM lactose and 3% Na-caseinate in a 0.1 M phosphate buffer of pH 6.8. Approximate results. (From a study by C.M.J. Brands and M.A.J.S. van Boekel.)

Of course, the composition of milk and milk products also affects the reactions, not only because of variations in the concentration of reactants but also because of the possible presence of components that are active as catalysts. Therefore, it is hard to predict the effects resulting from a change in milk composition.

Summarizing, it may be stated that in not too intensely heated milk, lactose will be decomposed predominantly by isomerization reactions, whereas a smaller part will be degraded through Maillard reactions. During very intense heating, and also in the later stages of the reactions, the Maillard reaction plays an important part. The resulting manifestations are changes in flavor, development of a brown color by formation of melanoidins, and a certain loss in nutritive value caused by lysine being rendered unavailable for absorption in the gut.

7.2.4 Heat Coagulation

Casein does not show heat denaturation as suffered by globular proteins. However, at a very intense heat treatment it can aggregate under certain conditions, especially if in micellar form (general aspects of the casein micelle stability are discussed in Subsection 3.3.3). Under practical conditions, the reaction can manifest itself as coagulation during sterilization. The coagulation may become visible when large aggregates have emerged or by the formation of a gel. The time needed for this to occur is called the *heat coagulation time* (HCT).

7.2.4.1 Milk

The heat coagulation of milk is an intricate phenomenon. This is because several interactions and conditions play a role. The most important variable is pH.

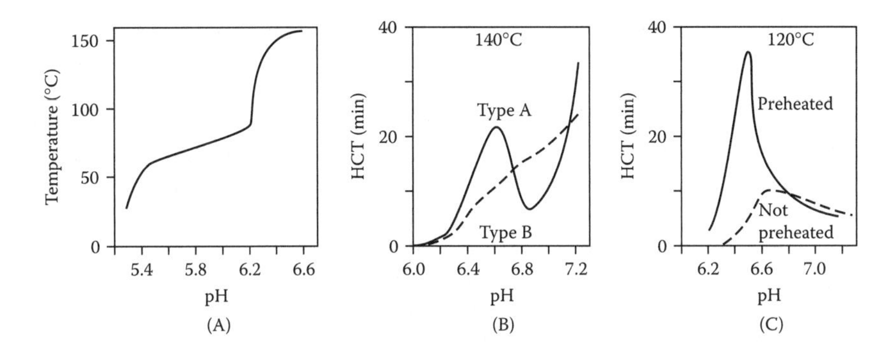

FIGURE 7.6 Heat coagulation of milk as a function of the initial pH. (A) Temperatures at which coagulation starts at fairly rapid warming of the milk (approximate results). (B) Heat coagulation time at 140°C of two different samples of fresh milk. (C) HCT at 120°C of evaporated skim milk, with or without preheating of the milk before concentration.

The initial pH of milk considerably affects HCT, i.e., the lower the pH, the lower the temperature at which coagulation occurs (Figure 7.6A). By and large, at constant temperature the rate of coagulation increases with decreasing pH. But often a local minimum in the heat coagulation time occurs near pH 6.8 to 6.9 (called the pessimum pH) and a local maximum near pH 6.6 (called the optimum pH). Milk that behaves like this is referred to as milk of type A; if it does not, it is milk of type B, as shown in Figure 7.6B. Type-A milk is by far the most common. There is considerable variation in heat stability among different lots of milk. These are not completely understood; in many regions a seasonal effect is observed (see, e.g., Figure 2.37D), partly caused by variation in the natural urea level.

Apart from what happens near the pessimum pH, coagulation only occurs when the pH of the milk has become low, about pH < 6.2 (the pH is lowered during heating; see Figure 7.1). Nevertheless, the aggregation generally is irreversible, that is, the aggregates formed cannot be redispersed by increasing the pH. It thus appears that the aggregates are held together by chemical (covalent) cross-links. All the same, colloidal interaction forces are essential because the micelles have, in any case, to come close enough before cross-linking can occur. The following are factors that determine the colloidal interaction:

1. κ-Casein (provides steric and electrostatic repulsion)
2. pH (affects electrostatic repulsion)
3. Ca^{2+} activity (Ca^{2+} may form salt bridges; affects electrostatic repulsion)

See also Subsection 3.3.3.1, especially Figure 3.22.

The pH decrease that occurs during the heating of milk is an essential factor in the heat coagulation of milk. The initial decrease in pH is mainly caused by 'precipitation' of calcium phosphate (Subsection 2.2.5) and the further decreases by production of formic acid from lactose (Subsection 7.2.3). The rate of pH decrease largely determines the rate of coagulation. The influence of a number of factors on heat coagulation is reflected in the rate of pH decrease. Coagulation generally occurs after milk pH falls to a value below 6.2, so a higher initial pH would require a longer time to reach a sufficiently low level for heat coagulation to occur. As stated earlier, heat coagulation is, however, not simply the same as coagulation by acid. Obviously, additional reactions play a part.

The following model may explain most observations on heat coagulation of milk. There are two different reactions that may cause coagulation. The first is *colloidal aggregation*, in which Ca^{2+} ions play a crucial role, presumably via Ca bridging. The coagulates formed can be dissolved by adding Ca-chelating agents (unless a fairly long heating time was needed for coagulation to occur). The reaction is second order. Its rate is not strongly dependent on temperature. It depends very much on Ca^{2+} activity. For a lower pH, $a_{Ca^{2+}}$ of milk is higher. Heating itself causes two effects — a lower pH, and a lower $a_{Ca^{2+}}$ at the same pH (Figure 2.9). This means that during the heating of milk $a_{Ca^{2+}}$ does not greatly alter because both effects roughly compensate each other.

The second reaction is a *chemical cross-linking*, although the cross-links involved have not been identified. (The problem is that at high temperatures several types of cross-links are formed, inside casein micelles as well as between micelles that had already formed aggregates.) The reaction is much faster at higher temperatures ($Q_{10} \approx 3$) and it very much increases in rate as the pH decreases. This means that the chemical reaction often overtakes the colloidal one, at least during heating of untreated milk: The first reaction proceeds at a slow rate, until the pH has reached a value, generally about 6.2 as measured at room temperature, when the second reaction becomes quite fast. The HCT is then largely determined by the rate of acid production during heating.

Another essential point is the *depletion of* κ-*casein* from the micelles, making them less 'hairy.' This may be explained by the results in Figure 7.2B, where we will first consider the curve in the absence of serum proteins. At high temperatures, more protein is outside the micelles at higher pH values; this is for the most part κ-casein. There is an equilibrium between κ-casein in the micelles and in solution, which shifts toward the solution with increasing pH. It is also seen that the change is more pronounced in the presence of serum proteins, occurring over a narrower pH range. As mentioned earlier (Subsection 7.2.2.2), β-lactoglobulin and κ-casein react at high temperatures. At high pH (> 6.7), this occurs for the most part in the solution, lowering the concentration of κ-casein in solution, and thereby disturbing the partition between the κ-casein in solution and in the micelles. The result is that more κ-casein leaves the micelles. At a low pH (< 6.7) the opposite will happen. Consequently, the state of depletion of the casein micelles is very strongly dependent on pH. Depleted micelles will be much less stable to aggregation than those with κ-casein hairs. Once fully depleted, the micelles will be more stable at higher pH because that implies a higher negative charge; this is also true of micelles that are not depleted of κ-casein.

The relations discussed lead to the model given in Figure 7.7. At fairly high initial pH, the micelles become depleted of κ-casein and the colloidal reaction is

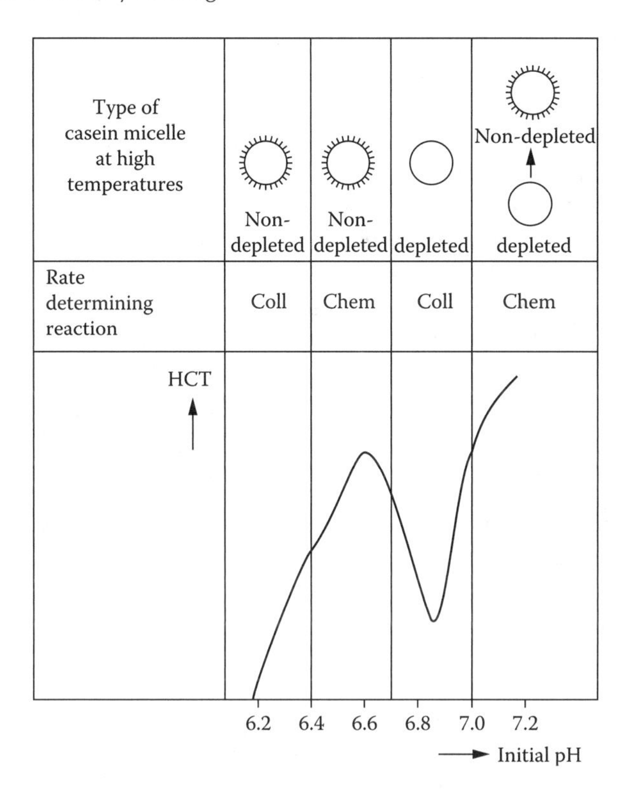

FIGURE 7.7 Model for the effect of initial pH on the type of casein micelle emerging at high temperature and thereby on the heat coagulation time (HCT) of milk. coll = colloidal aggregation, chem = chemical reaction.

rate determining for coagulation. If the initial pH is higher still, the micelles are sufficiently stable (high electric charge, very low $a_{Ca^{2+}}$) to allow the pH to decrease until they become hairy again; presumably, the complex formed by β-lactoglobulin and κ-casein precipitates onto the micelles, providing steric repulsion. Now the pH is low enough for the chemical cross-linking to become rate determining, as is also the case if the initial pH is around 6.6. At very low pH the Ca^{2+} activity is so high that the colloidal reaction is the fastest one.

7.2.4.2 Concentrated Milk

In practice, heat coagulation of milk is rarely a problem but *concentrated milk* (e.g., evaporated milk) may coagulate during sterilization. It is far less stable than untreated milk, as seen by comparing parts B and C of Figure 7.6; note the difference in temperature. This lower stability is primarily due to the higher concentration of casein: A second-order reaction proceeds faster at higher concentrations (provided that the rate constant remains the same). Moreover, other conditions have also changed by concentrating the milk. Although for the most part the same mechanisms act during heat treatment, there are important differences in the consequences between milk and concentrated milk.

To begin with, the heat stability of concentrated milk is considerably increased in the acid pH range if the original milk had been preheated (preheating has little, if any, effect on the heat stability of plain milk). This is explained as follows: In nonpreheated concentrated milk the serum proteins are in the native state. During warming to 120°C, the serum proteins become denatured and in the acid pH range they strongly aggregate. Due to the high concentration of the serum proteins (which have also been concentrated) a gel is formed. In other words, the casein micelles become incorporated in a serum protein gel. At higher pH, the denatured serum proteins remain dissolved and no serum protein gel is formed. In concentrated milk made from preheated milk, the serum proteins have already been denatured and have become associated with the casein micelles. During preheating of the nonevaporated milk, formation of a gel is not possible because the serum protein concentration is too low, and in the concentrated milk it will not occur because the serum proteins have already been denatured.

Furthermore, in concentrated milk the increase in stability from pH 6.2 to pH 6.5 is, as in milk, ascribed to the decreasing Ca^{2+} activity. The decrease of the stability at pH > 6.6 is, again as in milk, caused by dissociation of κ-casein, as a result of which depleted micelles remain, which are susceptible to Ca. If the pH increases to >7.0, the stability does not increase again as happens in milk (Figure 7.6B). This is due to the increased salt concentration; rising salt concentration in unconcentrated milk has virtually the same effect on heat stability at high pH. Presumably, during heating so much calcium phosphate associates with the micelles as to make them very unstable. This is somewhat comparable to the Ca^{2+}-induced coagulation of depleted micelles in unconcentrated milk, but the reaction with calcium phosphate is much faster.

There is another complication, which is that for the same reaction rate the coagulation time can be very different under different conditions. This is best illustrated for concentrated skim milk, as is done in Figure 7.8, which compares heat coagulation at a pH near the optimum (about 6.5) with that at a lower and a higher value; cf. Figure 7.6C. Near pH 6.3 the micelles aggregate to form open clusters, which soon fill the whole volume, whereby a gel is formed; this is discussed as fractal aggregation in Subsection 3.1.3.2. In agreement with the open structure of the aggregates the turbidity does not increase greatly but the (apparent) viscosity does: the effective volume fraction of the material greatly increases. At pH 6.8 the depleted micelles tend to fuse into larger ones upon aggregation. This leads to a large increase in turbidity but not to a higher viscosity: the volume fraction does not alter. Only in the final stages do the aggregates become of irregular shape and form a gel. At pH 6.3 about 10 micelles can form an aggregate of critical size for gelation, whereas the almost isometric aggregates formed at pH 6.8 would be made up of about 1000 micelles. Nevertheless, the HCT is the same, which means that the aggregation reaction rate ($-dN/dt$, where N is the particle number) is much greater at the higher pH. At pH 6.5, near the maximum HCT, the reaction proceeds for the most part as at pH 6.3, although some fusion

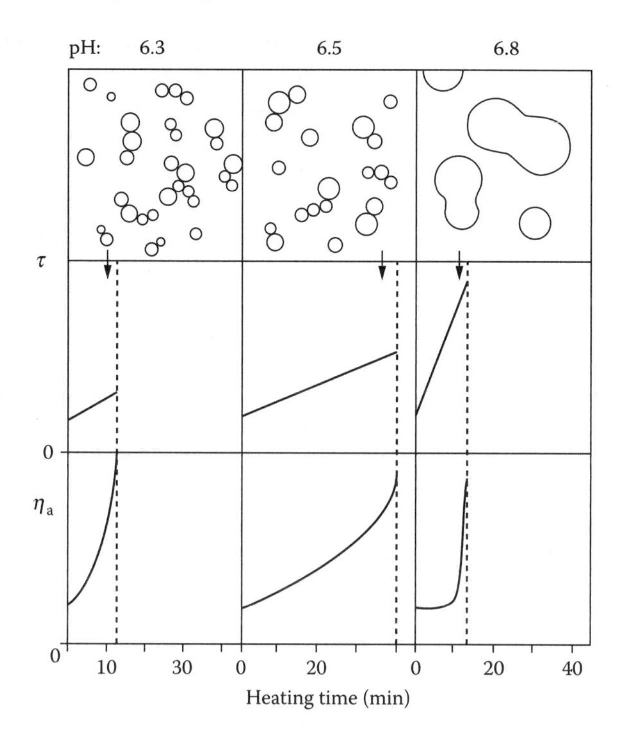

FIGURE 7.8 Heat coagulation at 120°C of concentrated skim milk at various initial pH. The upper row shows the appearance of the casein micelles (derived from electron micrographs) at the moments indicated by arrows, i.e., shortly before heat coagulation. The HCT is indicated by a vertical broken line. The second row gives the turbidity (τ) as a function of heating time, the lowest row the apparent viscosity (η_a). τ and η_a were determined in situ, i.e., at 120°C. Approximate results. (Adapted from results obtained by J.A. Nieuwenhuijse et al., *Neth. Milk Dairy J.*, **45**, 193–224, 1991.)

of micelles occurs. However, the reaction is slower because of the higher pH, presumably due to stronger electrostatic repulsion.

7.2.4.3 Conclusions

Most of the effects on heat coagulation of milk can be explained by using the model described above. For example, we have:

1. *Protein composition:* Its effect specifically concerns the ratio between κ-casein and β-lactoglobulin. The larger the amount of β-lactoglobulin, the higher the maximum HCT and the lower the minimum. This is explained by β-lactoglobulin enhancing the dissociation of κ-casein at pH > 6.7, which results in formation of more strongly depleted micelles. The higher maximum HCT at pH 6.6 may result from an increased association of β-lactoglobulin with the micelles, which may enhance colloidal repulsion.

2. *Urea content:* The higher the urea content, the stabler the milk toward heat coagulation, at least near the optimum pH. This can partly be explained by the fact that urea slows down the pH decrease but there are also other effects. Incidentally, urea does not affect the heat stability of concentrated milk (unless urea concentration is very high).
3. *Salt composition:* Its main influence is through the calcium and phosphate contents. The addition of a certain salt to milk can strongly disturb all salt equilibriums involved (Subsection 2.2.5). Addition of calcium and phosphate to milk, up to the concentrations that are found in concentrated milk, causes its heat stability at pH > 6.8 to be equal to that of concentrated milk, i.e., zero.
4. *Fat:* Fat in itself does not affect heat stability. This is no surprise, considering that heat coagulation is a coagulation of casein micelles. It is different if casein enters the fat globule–plasma interface, as occurs in homogenization (see Figure 9.7). This makes the fat globules behave like large casein micelles and they coagulate along with the micelles. As a result, in products like cream and concentrated milk, homogenization tends to cause a lower heat stability.

7.3 HEATING INTENSITY

The intensity of a heating process follows from the duration (t') of heating and the temperature (T). Figure 7.9 gives several examples. The effects of a certain combination of t' and T will differ, because they depend on the reaction considered, e.g., inactivation of a certain enzyme or formation of Maillard products. Some reactions occur fairly quickly at relatively low temperatures, whereas others need a much higher temperature before they can have an appreciable effect. The dependence of the reaction rate on temperature varies widely among reactions, which explains why at a certain combination of t' and T (e.g., 15 min at 110°C) reaction A may have advanced further than reaction B, whereas it is just the opposite at another combination, e.g., 10 s at 140°C. (See also Figure 7.10)

7.3.1 Processes of Different Intensity

In classifying heating processes on the basis of their intensity, special attention is usually paid to the killing of microorganisms and to the inactivation of enzymes. The following are customary processes.

1. *Thermalization:* This is a heat treatment of lower intensity than low pasteurization, usually 20 s at 60 to 69°C. The purpose is to kill bacteria, especially psychrotrophs, as several of these produce heat-resistant lipases and proteinases that may eventually cause deterioration of milk products. Except for the killing of many vegetative microorganisms and the partial inactivation of some enzymes, thermalization causes almost no irreversible changes in the milk.

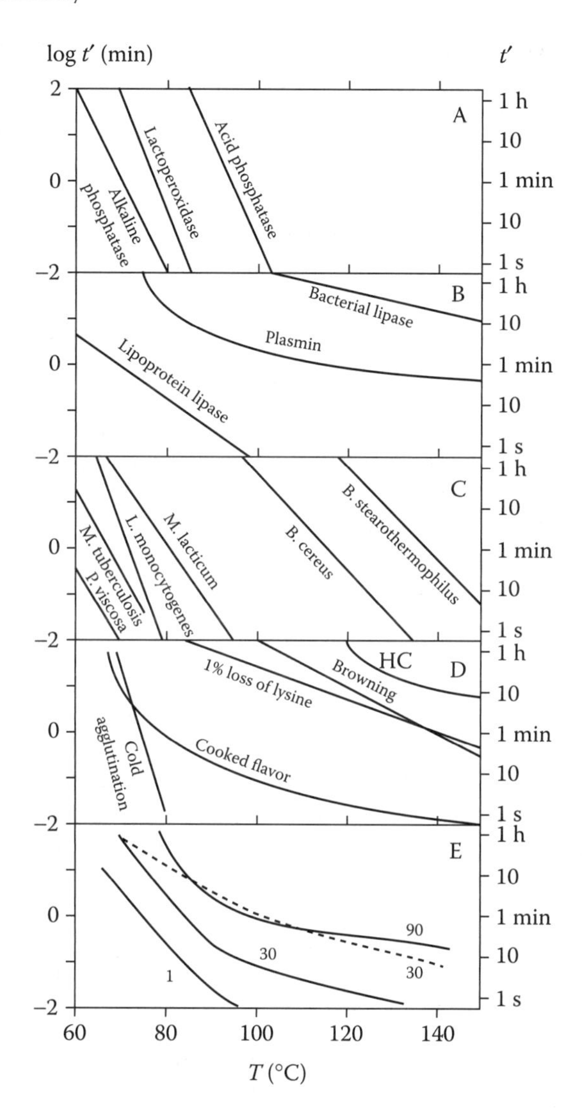

FIGURE 7.9 Combinations of temperature (T) and time (t') of heat treatment of milk that cause (A, B) inactivation (reduction of activity to about 1%) of some milk enzymes and a bacterial lipase; (C) the killing (reduction of the count to 10^{-6}) of strains of the bacteria *Pseudomonas viscosa, Mycobacterium tuberculosis, Listeria monocytogenes,* and *Microbacterium lacticum,* and of spores (10^{-4}) of *Bacillus cereus* and *B. stearothermophilus;* (D) visible heat coagulation (HC), a certain degree of browning, decrease in available lysine by 1%, a distinct cooked flavor and inactivation of cold agglutination; (E) insolubilization of 1%, 30%, and 90% of the β-lactoglobulin, and of 30% of the α-lactalbumin (----). Approximate results.

2. *Low pasteurization:* This is a heat treatment of such intensity that the enzyme alkaline phosphatase (EC 3.1.3.1) of milk is inactivated. It may be realized by heating for 30 min at 63°C or for 15 sec at 72°C. Almost all pathogens that can be present in milk are killed; it specifically concerns *Mycobacterium tuberculosis*, a relatively heat-resistant organism that formerly was among the most dangerous pathogens. All yeasts and molds and most, but not all, vegetative bacteria are killed. Some species of *Microbacterium* that grow slowly in milk are not killed (Figure 7.9C). Furthermore, some enzymes are inactivated but by no means all of them. The flavor of milk is hardly altered, little or no serum protein is denatured, and cold agglutination and bacteriostatic properties remain virtually intact. A more intense heat treatment is, however, often applied (e.g., 20 s at 75°C; see Subsection 16.1.1). This causes, for instance, denaturation of immunoglobulins (hence, decrease in cold agglutination and bacteriostatic activity) and sometimes a perceptible change in the flavor of milk.
3. *High pasteurization:* This heat treatment is such that the activity of the enzyme lactoperoxidase (EC 1.11.1.7) is destroyed, for which 20 s at 85°C suffices. However, higher temperatures, up to 100°C, are sometimes applied. Virtually all vegetative microorganisms are killed but not bacterial spores. Most enzymes are inactivated, but milk proteinase (plasmin) and some bacterial proteinases and lipases are not or incompletely inactivated. Most of the bacteriostatic properties of milk are destroyed. Denaturation of part of the serum proteins occurs. A distinct cooked flavor develops; a gassy flavor develops in cream. There are no significant changes in nutritive value, with the exception of loss of vitamin C. The stability of the product with regard to autoxidation of fat is increased. Except for protein denaturation, irreversible chemical reactions occur only to a limited extent.
4. *Sterilization:* This heat treatment is meant to kill all microorganisms, including the bacterial spores. To that end, 30 min at 110°C (in-bottle sterilization), 30 sec at 130°C, or 1 s at 145°C usually suffices. The latter two are examples of so-called UHT (ultra-high-temperature, short time) treatment. Some other effects of each of these heat treatments are different. Heating for 30 min at 110°C inactivates all milk enzymes, but not all bacterial lipases and proteinases are fully inactivated; it causes extensive Maillard reactions, leading to browning, formation of a sterilized milk flavor, and some loss of available lysine; it reduces the content of some vitamins; causes considerable changes in the proteins including casein; and decreases the pH of the milk by about 0.2 unit. Upon heating for 1 s at 145°C chemical reactions hardly occur, most serum proteins remain unchanged, and only a weak cooked flavor develops. It does not inactivate all enzymes, e.g., plasmin is hardly affected and some bacterial lipases and proteinases not at all, and therefore such a short heat treatment is rarely applied.

5. *Preheating:* This may mean anything from very mild to quite intense heating. It mostly concerns heating intensities anywhere between low pasteurization and sterilization.

7.3.2 Kinetic Aspects

As is discussed in Subsection 7.2.1, at high temperatures numerous chemical reactions occur in milk. The rate of the reactions and the temperature dependence of the rate are variable. Some aspects of the reactions will now be discussed, especially those of importance to the heat denaturation of globular proteins. The latter group of reactions is of great importance because of the consequences involved, including insolubilization of serum proteins, inactivation of enzymes and of immunoglobulins, and killing of bacteria and their spores. Denaturation is briefly discussed in Subsection 2.4.1.3, and Subsection 7.2.2.

In the present discussion it concerns irreversible changes; but denaturation as such (i.e., the unfolding of the peptide chain) is reversible. The unfolding caused by high temperatures exposes reactive side groups that they can react. This often leads to irreversible changes as discussed in Subsection 7.2.2. The rate at which insolubilization or inactivation occurs is often determined by the rate of denaturation.

7.3.2.1 Basic Equations

A *first-order reaction* equation is usually used for denaturation of protein, inactivation of enzymes, and killing of bacteria and spores. We thus have

$$-\mathrm{d}c/\mathrm{d}t = Kc \qquad (7.1)$$

where c is concentration, t is time, and K is the rate constant. Often, but not always (see below), this means of calculation appears to be a suitable approximation. Integration of Equation 7.1 yields

$$\ln\,(c_0/c) = Kt \qquad (7.2a)$$

or

$$c = c_0\,e^{-Kt} \qquad (7.2b)$$

where c_0 is the original concentration.

Parameter t' is used for referring to the duration of heat treatment needed to secure a certain effect, e.g., 99% inactivation of an enzyme or conversion of 1% of the lactose present into lactulose. We thus have

$$t' = \ln\,(c_0/c')/K \qquad (7.3)$$

A particular value of t' is that for which $c' = c_0/10$; this decimal reduction time D is given by

$$D = (\ln 10)/K \approx 2.3/K \tag{7.4}$$

By using Equation 7.3 and Equation 7.4, a given t' or D can be converted to values for other desirable heating intensities: the number of decimal reductions is proportional to t'.

A reaction equation referring to *zero-order* kinetics:

$$\mathrm{d}c/\mathrm{d}t = K \tag{7.5}$$

is often used if it concerns formation of a substance that is initially absent. This may be allowed for initial reaction steps, e.g., in the Maillard reaction, upto a few percent of the reducing sugar reacting with lysine residues. The value of K now depends on the initial concentration of these reactants. Integration of Equation 7.5 yields

$$c = Kt + c_0 \tag{7.6}$$

where often the 'blank' value $c_0 \approx 0$. The time needed to arrive at a certain amount of conversion (c') is proportional to c'.

The *temperature dependence* of a reaction is mostly assumed to follow an Arrhenius relationship. We thus have

$$K(T) = K_0 \exp(-E_a/RT) \tag{7.7}$$

where T = absolute temperature, K_0 = the assumed rate constant if E_a approaches zero, E_a = the so-called molar activation energy (in $\mathrm{J \cdot mol^{-1}}$), and R = the gas constant (8.314 $\mathrm{J \cdot mol^{-1} \cdot K^{-1}}$).

[*Note:* It is more correct not to use the activation energy E_a (according to Arrhenius), but the activation free energy $\Delta G^{\ddagger}$ (according to the 'activated-complex' theory). Because $G = H - TS$, this leads to a temperature dependence of the reaction rate of the form

$$K(T) \propto \exp(-\Delta H^{\ddagger}/RT) \exp(\Delta S^{\ddagger}/R) \tag{7.8}$$

where $\Delta H^{\ddagger}$ is the activation enthalpy (which is almost equal to E_a) and $\Delta S^{\ddagger}$ the activation entropy. For most reactions $\Delta S^{\ddagger}$ is small but not for denaturation of proteins, because the unfolding of the peptide chain causes a large increase in entropy.]

The temperature dependence is sometimes expressed as the Z value, i.e., the temperature rise needed to increase the reaction rate by a factor of 10. Consequently,

$$K(T + Z)/K(T) \equiv 10 \tag{7.9}$$

Or it is expressed as Q_{10}, which is defined by

$$Q_{10} \equiv K(T + 10)/K(T) \tag{7.10}$$

In practice we often make curves as shown in Figure 7.9, where log t' is plotted against the temperature. Usually, fairly straight lines are obtained. Such curves are very informative, e.g., the optimum combination of temperature and duration of heating may readily be obtained from them. After all, on the one hand it is desired to kill the bacteria involved and to inactivate enzymes, whereas on the other hand undesirable changes, such as formation of color and flavor substances, should be restricted. The required standards can often be approached because most of the desirable changes depend far more strongly on the temperature than most of the undesirable changes do; this is illustrated in Table 7.2. In all of these cases, it should be known to what extent the given t' values correspond to inactivation, killing, etc.; such statements as "time needed for inactivation of peroxidase" are inadequate. An example is given in Figure 7.10. It refers to the killing of spores of *Bacillus subtilis* (a potential spoilage organism in sterilized milk) and to the formation of lactulose. To be sure, lactulose as such, present in milk in small quantities, is of little importance. Determination of its concentration in milk is commonly applied to monitor the extent to which undesirable changes in flavor, color, and nutritive value have occurred due to heat treatment.

To determine the desirable heating intensity, the initial concentration of the substances in the milk should also be known. Natural substances in milk, including enzymes, often do not vary greatly in concentration, so a fixed concentration may be assumed. But the content of bacteria or of the enzymes excreted by them may vary by some orders of magnitude.

TABLE 7.2
Typical Examples of the Temperature Dependence of Some Reactions

Type of Reaction	Activation Energy[a] ($kJ \cdot mol^{-1}$)	Q_{10} at 100°C
Many chemical reactions	80–125	2–3
Most enzyme-catalyzed reactions	35–55	1.6–2.2[b]
Autoxidation of lipids	40–100	1.7–3.8[b]
Maillard reactions	100–180	2.4–5
Heat denaturation of proteins	150–500	4–75
Enzyme inactivation	450	50
Killing of vegetative bacteria	200–500	6–75
Killing of spores	250–330	9–18

[a] Often an apparent or average activation energy because it mostly concerns a number of consecutive reactions.
[b] At 25°C.

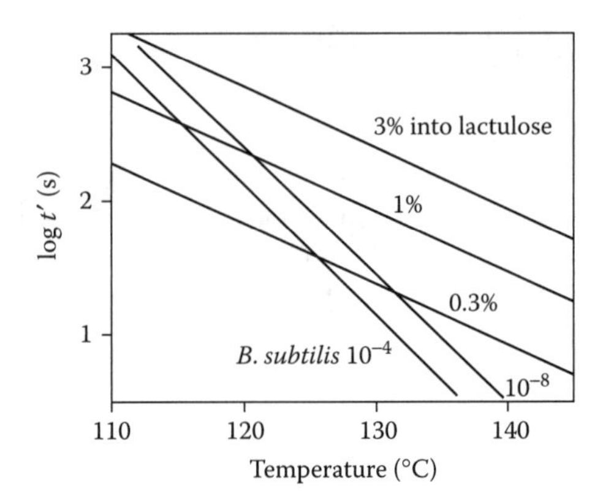

FIGURE 7.10 The time needed (t') at various temperatures to convert certain percentages of lactose into lactulose, and to obtain a certain extent of killing of *Bacillus subtilis* spores.

7.3.2.2 Complications

The relations given above may not be very precise. Possible complications are as follows:

1. A certain change observed is not the result of one reaction but of several, e.g., consecutive reactions. This is often the case in a complex material such as milk. If under all conditions the same reaction is the rate-determining one, the relations may hold, but otherwise they may not. A good example is the heat inactivation of the milk enzyme plasmin, as shown in Figure 7.9B. The plot is markedly curved, which is related to the fact that the protein molecule first has to be unfolded before a second reaction can cause irreversible inactivation. The unfolding of the peptide chain (denaturation proper) is a strongly temperature-dependent reaction, whereas the second reaction generally has a far smaller Q_{10}. In other words, at low heating temperatures the former reaction will be the rate-determining one and at high temperature the second reaction, if this is relatively slow. Such relations as well as more intricate relations often occur.

   A further complication may be that the different reactions are of different order. As an example, the insolubilization of serum protein by heat treatment is shown in Figure 7.9E. Before the protein is rendered insoluble it has to be denatured so that it can subsequently aggregate. The aggregation reaction usually follows second-order kinetics, hence the nonlinear relation. This also implies that the curves given in Figure 7.9E do not apply to milk with other protein concentrations (e.g., in evaporated milk).

2. Apart from what has been mentioned in item 1, even for a single reaction the activation enthalpy (and also the activation entropy mentioned earlier) may not be constant. There is no general reason why activation enthalpy and entropy should be independent of temperature. In actual practice, however, such independence is usually observed if the temperature range considered is not too wide, though there are exceptions. Especially for protein denaturation, $\Delta H^{\ddagger}$ and $\Delta S^{\ddagger}$ depend rather closely on temperature. It is also of importance that conditions can alter during the heat treatment. For example, at high temperatures acid will gradually be produced in milk, thereby lowering the pH (Figure 7.1). Usually, the redox potential will also be reduced (Section 4.3), partly depending on the extent to which oxygen is removed during the heating process. Whether there is an effect of all of these changes, and to what extent, will depend on the type of reaction involved. Table 7.4 includes some examples of the influence of pH on the killing of bacteria during heating.
3. According to Equation 7.7, log K should be plotted against $1/T$ to obtain a straight line. Because t' is inversely proportional to K (see Equation 7.3), we can also plot log t' against $1/T$, but plotting log t' against T (or the temperature in °C), as is commonly done, will lead to a curved plot (provided that the data points are sufficiently accurate). An example is shown in Figure 7.11. If the temperature range is comparatively small (say, < 20 K), fitting of a straight line to the results may be acceptable, but extrapolation to other temperatures is usually not allowed. This can readily be seen when the relationship of Z with the temperature is derived:

$$Z = 2.303\ RT^2/E_a \tag{7.11}$$

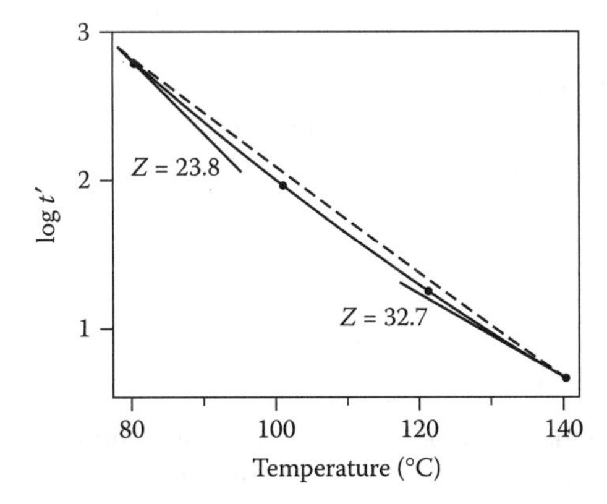

FIGURE 7.11 Example of a relation between log t' and the temperature as calculated from Equation 7.7 for an activation energy of 100 $kJ \cdot mol^{-1}$.

Z thus greatly depends on temperature, as is clearly shown in Figure 7.11 where Z changes by more than 5% per 10 K in the temperature range considered. We also have

$$Q_{10} \approx \exp\,(10\;E_a/RT^2) \tag{7.12}$$

corresponding to a change of about 5% per 10 K.

4. In the customary heating processes, the liquid (e.g., milk) is warmed to a temperature of x°C, held at that temperature for y s, and then cooled. Assuming that these nominal values are determinant, in other words that the heating is for y s at x°C, may lead to serious errors.

   In the first place, there is a spread (statistical variation) in the *residence time* — i.e., the time during which a volume element of the product is in the heating equipment. This is especially important in relation to the killing of pathogenic bacteria or bacterial spores. In most heat exchangers (the commonly used type of equipment), the spread in residence time is small, but it is generally not negligible in the so-called holding section (generally a tube), where the liquid is kept at a fixed temperature. This means that one often takes a holding tube that is 10 to 20% longer than would be needed if the average residence time in the tube were to apply for every bacterial cell present. Moreover, the temperature in the holding section may not be fixed but decrease somewhat from entrance to exit due to radiative loss of heat; this especially occurs in small-scale equipment.

   On the other hand, the reactions (including killing of bacteria, etc.) occurring *during warming and cooling* of the liquid cannot usually be neglected. For a first-order reaction, the overall effect of a reaction is given by $\int K(T)\,\mathrm{d}t$. When T, and thereby K, are constant, the integral simply yields $t'K(T)$. But if the temperature is a function of time the process is more complicated. Usually, numerical or graphical integration is needed. The result can most readily be described as an effective duration of heating t_{eff}, that is, the time during which the product should be held at the nominal temperature (assuming the times needed for warming and cooling to be negligible) to arrive at the same effect. We thus have

$$t_{\mathrm{eff}} = \int_0^\infty K(t)\,\mathrm{d}t / K_T \tag{7.13}$$

   where K_T is the first-order rate constant at the nominal temperature. Of course, the result depends on the warming and cooling profiles and, thereby, on the type of apparatus used — but also on the activation energy of the reaction. The weaker the temperature dependence of the reaction, the larger is the difference between the nominal and the effective duration

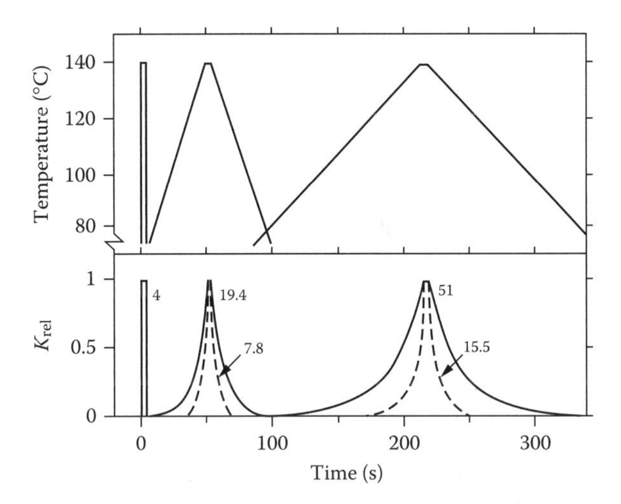

FIGURE 7.12 Schematic examples of temperature profiles during a heating process (nominally 4 s at 140°C) and of the ensuing rate constant compared to that at 140°C (K_{rel}) for an activation energy of 110 kJ · mol^{-1} (—) and 330 kJ · mol^{-1} (----). The figures refer to the effective duration of heating (s).

of heating. Figure 7.12 gives some examples of the relative reaction rate as a function of heating time for two warming and cooling profiles and two values of activation energy. The surface area below the curves represents the effective duration of heating at the nominal temperature. Note that the reaction starts to clearly occur at 120°C for E_a = 330 kJ·mol^{-1} (or Z = 10 K at 140°C), which is a common value for the killing of bacterial spores; for E_a = 110 (Z = 30), a typical value for a chemical reaction, such is already the case at 80°C. Note also that the differences between nominal and effective t' can be considerable.

The subject discussed in item 4 is also of importance in optimizing heating processes. On the one hand, such a change as the killing of spores is desirable, and, to achieve this, a certain minimum effect is required. On the other hand, changes that cause a decrease in product quality should be minimized. The best compromise is found when all of the particles of milk undergo precisely the necessary combination of time and temperature. This means that the warming and cooling times should be as brief as possible, the temperature in the holder as constant as possible, and the residence time of the milk in the apparatus should vary as little as possible.

5. In so-called direct UHT heating (see Subsection 7.4.2), steam is injected into the milk (or vice versa) to heat it, e.g., from 70°C to 140°C. The steam condenses and the water thus added is removed by evaporation at reduced pressure, thereby cooling the milk. During the heat treatment, however, the milk is diluted with water. This would cause

bimolecular reactions to proceed more slowly, as the reaction rate is proportional to the product of the concentrations of both reactants. This may apply, for instance, to the Maillard reaction. Reactions causing inactivation of enzymes and killing of microorganisms are probably not affected by the dilution. The heat given up by condensation of the steam would be, for instance, 2.1 kJ per gram of steam and the amount of heat taken up by the milk about 280 J per gram of milk. This implies addition of about 0.13 g of water per gram of milk, or dilution by a factor of 0.88. This, then, would mean that a bimolecular reaction would proceed more slowly by a factor of 0.88^2, or 0.78.

7.3.3 Inactivation of Enzymes

Heat inactivation of most enzymes follows first-order kinetics as occurs during denaturation of globular proteins. The inactivation is strongly temperature dependent, Q_{10} usually being at least 50. A D value of 1 min is usually reached between 60°C and 90°C. But milk contains some enzymes that can cause spoilage and that show wide variations in susceptibility to heat inactivation (Table 7.3).

TABLE 7.3
Heat Inactivation of Some Enzymes in Milk

Enzyme	EC Number	Temperature (°C)	D (s)	Q_{10}
Milk enzymes				
Alkaline phosphatase	3.1.3.1	70	33	60
Lipoprotein lipase	3.1.1.34	70	20	13
Xanthine oxidase	1.1.3.22	80	17	46
Lactoperoxidase	1.11.1.7	80	4	230
Superoxide dismutase	1.15.1.1	80	345	150
Catalase	1.11.1.6	80	2	180
Plasmin	3.4.21.7	80	360	3.3
Plasmin	3.4.21.7	120	30	1.5[b]
Acid phosphatase	3.1.3.2	100	45	10.5
Extracellular bacterial enzymes[a]				
Lipase *Pseudomonas fluorescens*		130	500	1.3[b]
Lipase *Pseudomonas* sp.		130	700	2.4
Lipase *Alcaligenes viscolactis*		70	30	2.6
Proteinase *Pseudomonas fluorescens*		130	630	2.1
Proteinase *Pseudomonas* sp.		130	160	1.9
Proteinase *Achromobacter* sp.		130	510	2.1
Chymosin	3.4.23.4	60	25	70

[a] The results may vary widely among strains and may also depend on the conditions during growth.
[b] Valid only over a narrow temperature range.

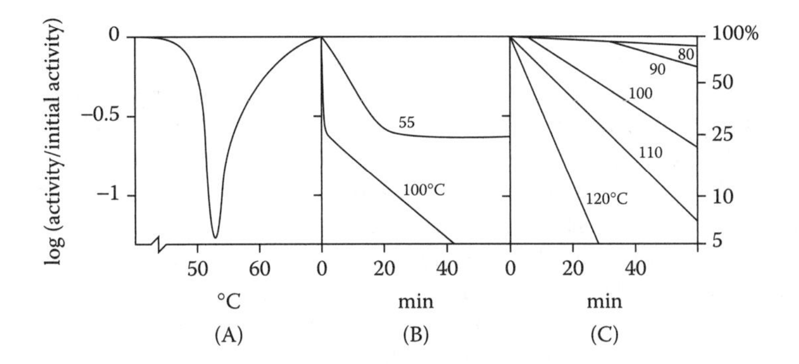

FIGURE 7.13 Heat inactivation (expressed as residual activity) of some bacterial enzymes in milk. (A) Proteinase of a *Pseudomonas fluorescens;* heating for 30 min at the temperature indicated. (B) Lipase of the same bacterium during heating at two different temperatures. (C) Proteinase of an *Achromobacter* sp. during heating at various temperatures.

After the actual denaturation of the enzyme molecule, at least one ensuing reaction is needed to prevent renaturation of the enzyme from occurring on cooling. (Renaturation would mean preserving the enzyme activity.) Various enzymes are inactivated at far higher heating intensities than mentioned above, and they also show a lower Q_{10}; see, for instance, Figure 7.9B. The high heating intensity then is needed for the ensuing reactions to proceed because these enzymes generally are in a denatured state at a temperature of, say, 80°C. The nonlinear relation between log t' and temperature that then often occurs is discussed in Subsection 7.3.2.

Yet other deviations may occur. Figure 7.13A gives an example of so-called low-temperature inactivation. The protease molecules "consume" each other, i.e., at a temperature at which part of the molecules are in a denatured state, the molecules in the native state can proteolytically hydrolyze the former, thereby inactivating them. (Most native globular proteins are quite resistant to proteolysis.) The denaturation itself appears to occur at a fairly low temperature and to have a high Q_{10}; this does not hold for the heat inactivation (which in this case proceeds roughly as in Figure 7.13C). Naturally, the proteinase can also hydrolyze other proteins if these are in an unfolded state. This would be the explanation for the unexpected shape of the curves in Figure 7.13B, i.e., at fairly low temperatures native undenatured proteinase can attack the lipase that is already denatured. In actual practice the extent to which low-temperature inactivation occurs will, therefore, greatly depend on the warming and cooling rates applied.

The relationship is again different in Figure 7.13C. Probably, three ensuing reactions occur, the first two being reversible:

1. $A \rightleftarrows B$, rate constants K_1, K_{-1}
2. $B \rightleftarrows C$, rate constants K_2, K_{-2}
3. $C \rightarrow D$, rate constant K_3

where A is the native state of the enzyme molecule, B an unfolded state, and C an intermediate state. Only the D state is irreversible. K_1 is typical of denaturation, K_2 becomes essential above 80°C, and K_3 above about 95°C (Figure 7.13C). K_{-1} often is fairly large. The magnitude of K_{-2} is of little importance, unless it is very small. In the latter case, the enzyme may exhibit a slow reactivation after cooling. Alkaline phosphatase as well as lactoperoxidase can show a slight reactivation after the heated milk has been kept cool for several days.

Finally, when plotting log activity against time a straight line is not always obtained, probably because the enzyme occurs in two or more forms that are mostly genetic variants. These isozymes may show different inactivation kinetics.

Some enzymes will now be briefly discussed.

Lipoprotein lipase (EC 3.1.1.34) of milk deviates somewhat from the norm because the Q_{10} for heat inactivation is fairly small, i.e., about 10 at 75°C ($Z \approx 10$ K).

Plasmin (EC 3.4.21.7) is very heat resistant, as Figure 7.9B shows. Above 110°C, the inactivation rate increases only slightly with increasing temperature. Even at 140°C, the milk should be heated for at least 15 s to prevent the occurrence of proteolysis during storage. The inactivation of plasmin is due to a complex set of reactions. Unfolding of the enzyme starts at 50 to 55°C, but is reversible. The reaction leading to irreversibility involves formation of –S–S– bonds with the plasmin. Sources of free –SH groups can be (1) groups exposed upon unfolding of plasmin molecules themselves, whereby unfolded molecules can react with each other (aggregate), although their concentration is very small and, hence, the inactivation quite slow; (2) free cysteine, although the concentration in milk is very small; and (3) denatured β-lactoglobulin, which is formed at a substantial rate above 75°C (Figure 7.9). However, the presence of casein greatly decreases the rate of inactivation of plasmin, possibly because of the formation of –S–S– bridges between β-lactoglobulin and κ-casein, causing a substantial decrease in the amount of –SH groups. Another complication is the presence of plasminogen, which is also very heat resistant, but which can be slowly converted into plasmin by urokinase, a plasmin activator. Milk also contains an inhibitor of urokinase and the inhibitor is presumably inactivated by low pasteurization. Anyway, low pasteurization leads to an increase of plasmin activity. See also Subsection 2.5.2.5.

Bacterial lipases, especially lipases secreted into the milk by some Gram-negative rods, may be very heat resistant (Table 7.3).

Bacterial proteinases, especially extracellular endoproteinases of Gram-negative rods, can also be very heat resistant (Table 7.3). Often, one inactivation reaction with a small Q_{10} (roughly 2) is found, but in other cases two reactions can be distinguished (see Figure 7.13C).

As a consequence of the incomplete inactivation of lipases, lipolysis may cause a rancid flavor. Residual milk proteinase especially attacks β- and α_{s2}-caseins. As a result, a bitter flavor may develop, and skim milk may finally become more or less transparent. Residual bacterial proteinases mainly attack κ-casein. The consequences of this may be bitter flavor development, gel formation, and wheying off.

The only measure available to avoid the action of the milk enzymes is an adequate heat treatment. Most of the bacterial enzymes mentioned are insufficiently inactivated by heat treatment because of their great heat resistance. Therefore, the only alternative is to prevent the growth of the bacteria involved.

7.3.4 Thermobacteriology

The kinetics of the killing, or irreversible inactivation, of bacteria and other microorganisms during heat treatment has been an intensely studied subject. For a long time, a simple approach based on first-order kinetics, as is valid for many denaturation reactions, was generally applied. However, an increasing number of exceptions have been observed and the 'established' rules of thermobacteriology have been strongly criticized. The conventional approach will be discussed first, and then the complications that compel the application of more sophisticated kinetics.

The main cause of the deviating response of many bacteria to heat treatment is that they are living creatures. This has several consequences (see Section 5.1). They show statistical variation in properties, including heat resistance. Their thermal inactivation is not a one-step process, like the unfolding of a globular protein, but can occur more gradually. Bacteria can also adapt their physiology to prevailing conditions; in the case of sporeformers, the formation and germination of spores can be affected.

7.3.4.1 The Conventional Approach

When the killing of bacteria or their spores is considered, Equation 7.2 is usually written as

$$\log N = \log N_0 - \frac{t}{D} \qquad (7.14)$$

where N is the bacterial count, generally the number of colony-forming units (CFU) per ml. This is illustrated in Figure 7.14, which applies to the simple case of the one bacterial species present, and where we will first consider curve (a). We can read from the graph that here $D = 2.5$ min. Heating for that time thus would reduce N to 10% of N_0, heating for $2D$ min reduces it to 1%, for $3D$ min to 0.1%, and so on. In curve (a), 5.7 min leads to a reduction of the count to 1 organism per ml. Of course, a longer time is needed to achieve the same count if N_0 is higher, e.g., 12.1 min for curve (b). At a higher temperature a shorter time is needed, for example, as for curve (c), where $D = 45$ s. If a greater reduction in bacterial count is desired, for instance, because the initial count is higher, it is generally preferable to increase the heating temperature somewhat, rather than increase the heating time. This has been discussed earlier in relation to Figure 7.10.

Microorganisms vary greatly in heat resistance. Generally, the characteristic parameters given are D and Z (number of degrees [K] by which the heating

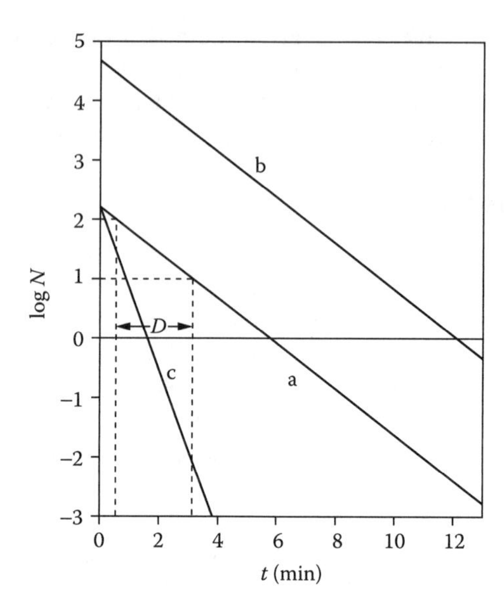

FIGURE 7.14 Examples of the reduction in bacterial count N (in ml^{-1}) as a function of time t during heating of a liquid containing one bacterial species. For curve (b) the initial count N_0 is higher than for curve (a), for curve (c) the heating temperature is higher.

temperature should be raised to reduce D by a factor of 10). Examples in Table 7.4 show that both these parameters vary widely at a given temperature, especially D. In modeling sterilization processes, it is sometimes assumed that $Z = 10$ K in all cases for the killing of spores but that assumption is by no means correct. It also appears that there can be significant variations within a species. In other words, different strains of one species can have different heat resistances.

Furthermore, the conditions during heating can affect D and Z. For example, compare (in Table 7.4) milk and whey, or the same liquid at different pH values. Ionic strength, pH, osmotic pressure, and redox potential of the medium, and also the presence and concentration of specific solutes, can all markedly affect D and Z values; the explanation is often uncertain. In most cases, the heat resistance of an organism increases with the dry-matter content of the medium, whereas the dependence of the inactivation rate on temperature decreases; examples are given in Table 7.5, which shows that these effects can be very large.

In principle, the equations discussed and the data given for D and Z should allow calculation of the treatment (combination of heating time and temperature) needed to reduce the number of specific bacteria (or their spores) to a level that is considered acceptable. The following conditions should be met:

1. The calculations should be applied to single species. The method breaks down for a total count, because D and Z values vary greatly among different bacteria, leading to curved plots of log N vs. t, and of log t' vs. $1/T$.

TABLE 7.4
Examples of Thermal Inactivation Data of Bacteria: Conventional Approach

Microorganisms	Heating Medium	Temperature (°C)	*D* (min)	*Z* (K)
Psychrotrophs				
Pseudomonas fragi	Milk	49	7–9	10–12
Pseudomonas fragi	Skim milk	49	8–10	10–12
Pseudomonas fragi	Whey, pH 6.6	49	32	
Pseudomonas fragi	Whey, pH 4.6	49	4–6	10.9
Pseudomonas viscosa	Milk	49	1.5–2.5	4.9–7.9
Pseudomonas viscosa	Whey, pH 6.6	49	3.9	
Pseudomonas viscosa	Whey, pH 4.6	49	0.5	
Pseudomonas fluorescens	Buffer	60	3.2	7.5
Listeria monocytogenes	Milk	72	0.02–0.05	6.8
Other non-spore-forming bacteria				
Salmonella (6 spp.)	Skim milk	63	0.06–0.1	4.0–5.2
Campylobacter jejuni	Skim milk	55	0.7–1.0	6–8
Enterococcus faecalis	Skim milk	63	3.5	
Enterococcus faecium	Skim milk	63	10.3	
Enterococcus durans	Skim milk	63	7.5	
Enterococcus bovis	Skim milk	63	2.6	
Escherichia coli	Skim milk	63	0.13	4.6
Escherichia coli	Whey, pH 4.6	63	0.26	6.7
Streptococcus sp., group D	Skim milk	63	2.6	
Lactococcus lactis ssp. *lactis*	Whey, pH 4.6	63	0.32	7.3
Lactococcus lactis ssp. *cremoris*	Whey, pH 4.6	63	0.036	6.7
Lactobacillus spp.	Milk	65	0.5–2.0	
Microbacterium flavum	Skim milk	65	2.0	
Microbacterium lacticum	Milk	84	2.5–7.5	
Mycobacterium tuberculosis ssp. *bovis*	Milk	64	0.1	5.0
Mycobacterium avium ssp. *paratuberculosis*	Milk	70	0.06	
Yersinia enterocolitica	Milk	58	1.6	4.3
Spore-forming bacteria				
Bacillus cereus, spores	Milk	121	0.04	9.4–9.7
Bacillus cereus, vegetative	Water or 2 M sucrose	70	0.013–0.016	6.6
Bacillus cereus, germinating spore	Water	70	0.35	6.5
Bacillus cereus, germinating spore	2 M sucrose	70	39	
Bacillus subtilis, spore	Milk	121	0.03–0.5	10.7
Bacillus subtilis, vegetative	Water	55	1.0–5.6	5.0–5.2
Bacillus coagulans, spore	Milk	121	0.6–4	4.6
Bacillus stearothermophilus, spore	Skim milk	121	2.5–4	9–11

(*Continued*)

TABLE 7.4
Examples of Thermal Inactivation Data of Bacteria: Conventional Approach (Continued)

Microorganisms	Heating Medium	Temperature (°C)	*D* (min)	*Z* (K)
Spore-forming bacteria (Continued)				
Bacillus sporothermodurans, spore	Skim milk	121	2–3.5	13–14
Clostridium sporogenes, spore	Milk, pH 7.0	121	1.7	
Clostridium botulinum, spore	Milk, pH 7.0	121	0.2	
Clostridium tyrobutyricum, spore	Milk	110	0.5	15
Clostridium perfringens, spore	Water	70	8–25	7–8
Other microorganisms				
Aspergillus sp., conidia	Buffer, pH 4.5	55	2	3.5–4
Aspergillus sp., ascospores	Buffer, pH 4.5	75	2	6–8
Saccharomyces cerevisiae, vegetative	Buffer	60	1	5.0
Saccharomyces cerevisiae, ascospores	Buffer	60	10	5.0
Foot-and-mouth disease virus	Milk	63	0.2	10–12

2. The initial count N_0 should be known or should be estimated with sufficient accuracy.
3. The effective holding time t_{eff} in the heat exchanger used should be known. See especially Figure 7.12 and the discussion regarding it. As is shown, t_{eff} can markedly depend on the Z value of the bacterium.
4. The D value should be known for the medium in which the heat treatment will occur and for strain(s) that are representative of the bacteria occurring in the liquid, i.e., milk or a milk product. To establish

TABLE 7.5
Influence of the Dry-Matter Content (Skim Milk, Evaporated Skim Milk, and Skim Milk Powder) on the Killing of Some Bacteria, as Caused by Heating

Bacterium	Temperature (°C)	Dry Matter (%)	*D* (s)	*Z* (K)
Staphylococcus sp.	70	9	70	5.2
	70	93	1800	11.6
Serratia marcescens	50	9	56	4.0
	50	93	1090	13.0
Escherichia coli	63	10	8	4.6
	63	20	15	4.9
	63	30	75	6.3
	63	40	200	7.9

thermal death rates of a bacterial strain with sufficient accuracy, relatively high initial counts are needed; to that end, the isolated bacterium must be grown to high numbers in a suitable culture medium. It should be realized, however, that bacteria cultured in a laboratory have adapted their physiology to the conditions during growth, which may significantly differ from the conditions (growth phase, chemical composition, etc.) in the product. The bacteria must thus be allowed to adapt to the conditions that prevail in practice, before estimating D values. These problems are even greater for bacterial spores.

5. The same considerations apply for the estimated Z values. Moreover, Z must be constant over the temperature range considered.
6. The dying off must follow first-order kinetics.

Conditions 5 and 6, in particular, are not met for several bacteria, as is discussed in the following subsection.

7.3.4.2 Complications

The underlying assumption for applying first-order kinetics to the thermal death curves is that the probability that a bacterium will die in the next second (or some such small time interval) is constant at a constant temperature. In practice, this means that the probability is constant during the holding time. This is expressed in Equation 7.1, whence follows an exponential decrease of the bacterial count: Equation 7.2. In reality, however, deviations are common. The probability of dying per unit time interval may increase in the course of heat treatment (K increases), or it may decrease. In such cases, the decrease in the number of viable bacteria with time is not exponential.

To better describe thermal death curves, the *Weibull equation* is often suggested. It reads

$$N = N_0 \exp - (t/\alpha)^{\beta} \quad (7.15a)$$

or

$$\log (N/N_0) = - 0.434\ (t/\alpha)^{\beta} \quad (7.15b)$$

Here α is a characteristic time, and the exponent $\beta > 0$. If β equals unity, we have first-order kinetics with $\alpha = 1/K = 0.434\ D$. If $\beta \neq 1$, the relations are different. A decimal reduction time can be defined, however the time t_n to achieve n decimal reductions is not given by $n \cdot D$, but by

$$t_n = \alpha\ (2.3\ n)^{1/\beta} \quad (7.16)$$

Figure 7.15(a) shows how the inactivation depends on the values of α and β. The latter parameter determines the shape of the curve: for $\beta < 1$, the line curves

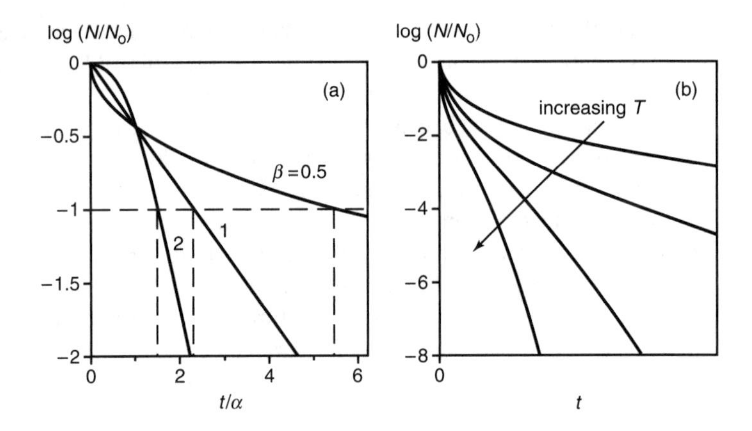

FIGURE 7.15 Hypothetical thermal death curves for bacteria. (a) Results according to the Weibull equation, for 3 values of β (indicated); dotted lines relate to one decimal reduction. (b) Approximate results as can be obtained in some cases at various temperatures. [Adapted from van Boekel (a), and Peleg (b); see Suggested Literature.]

upward, which means that the inactivation probability decreases — or that the heat resistance of the bacteria increases — during heat treatment. For $\beta > 1$, the line curves downward, implying that the resistance decreases. For $t/\alpha = 1$, the curves cross, at $\log (N/N_0) = -0.434$.

Some important *causes for deviation* from first-order kinetics are:

1. Bacteria of one species, or even of one strain, may show variation in heat sensitivity. This means that the most heat sensitive ones will, on average, die off faster than the less sensitive ones. This will result in $\beta < 1$ (upward curvature).
2. Bacteria can to some extent adapt their physiology to conditions of stress. At a relatively high temperature for growth, adaptation of the organism may result in a decrease in its heat sensitivity, hence in $\beta < 1$. Because the adaptation takes time, this result can only be expected at temperatures where heat inactivation is slow.
3. During heat treatment, bacteria can become 'heat-damaged,' i.e., lose viability without immediately dying. Due to accumulated damage their heat sensitivity increases. This leads to $\beta > 1$, especially at high temperatures, as is often observed in practice. A combination of causes 2 and 3 will lead to a change in the shape of the inactivation curves when the heating temperature is varied, for instance, as shown in Figure 7.15(b). The resulting curves can be of sigmoid shape, and then cannot be described by the Weibull equation; a three-parameter equation would be needed.
4. Several bacteria occur in clumps: staphylococci, streptococci, streptobacilli, etc. Moreover, bacteria can be present in aggregates, for

instance, in dirt particles. This implies that the number of CFU is smaller than the number of bacteria. Effectively, CFU = N/N_C, where N_C is the average number of bacteria in a clump. When the medium is at a temperature at which bacteria die off, all of the bacteria in a clump have to die before it is not a CFU anymore. This means that it takes, on average, a longer time to reduce the number of clumps than the number of non-clumped bacteria. Especially if N_C is large, and values as high as 10^4 may occur, the difference may be considerable. After some time, only small clumps and single viable bacteria are left, for which inactivation is relatively faster. In principle, the value of β as applied to CFUs will be larger than unity but after some time the slope of the curve may decrease again. The result can also be affected by breakup of clumps or aggregates into smaller ones, owing to forces exerted by the flowing liquid in the heat exchanger.

5. Of the bacteria that form spores, both vegetative cells and spores are generally present. If the product is heated at a temperature where the inactivation of the spores is slow, it is generally observed that the inactivation rate is at first high as the poorly heat-stable vegetative cells are killed, and then much slower. This means a strong upward curvature of the thermal death curve. The situation is even more complicated because heat treatment at a moderate temperature tends to induce germination of spores. For example, in milk contaminated with *Bacillus cereus* that is heated for 30 min at 70°C, only the spores survive (see Table 7.4). But the spores rapidly germinate. If the milk is cooled after heating and then directly heated again for 30 min at 70°C, the count can decrease by 1 to 2 decades. This must be due to the induced germination of spores and the subsequent killing of the resulting vegetative cells.
6. During prolonged heating at sterilization temperatures (say, 120°C) the composition of the liquid may alter, e.g., by formation of formic acid or due to loss of oxygen (depending on the construction of the heat exchanger). This may either lead to increasing or to decreasing heat sensitivity of the bacteria.

It may be noticed that these causes for deviation from first-order kinetics are for the most part very different from those encountered in the heat inactivation of enzymes.

Some results for bacteria growing in milk or a milk product are given in Table 7.6; a few results for other liquids are included to demonstrate possible effects of the heating medium and temperature. It is seen that deviation from first-order kinetics can be very large, β being much larger or much smaller than unity. For some of the organisms studied, e.g., *P. mephitica* and *S. aureus* S6, the deviation is so large that giving D and Z values would be misleading. It is also seen that the values of α and β can vary greatly among strains of the same species.

Several results have now been obtained on thermal death curves, mainly of non-spore-forming organisms (including some yeasts) in nondairy liquids. In the

TABLE 7.6
Thermal Inactivation of Some Bacteria

Species	Heating Medium	T (°C)	α (min)	β
Pseudomonas viscosa 3	Skim milk	48	5.7	2.6
Pseudomonas mephitica	Skim milk	48	6.7	3.0
Pseudomonas sp.	Egg	49.5	79	1.1
		52.5	16	0.8
		56.8	1.3	0.5
Salmonella enteritides 775W	Milk	62.5	0.01	0.4
Listeria monocytogenes	Milk	60	0.5	0.6
Staphylococcus aureus S1	Buffer	55	0.03	0.6
	Whey	55	0.4	0.8
Staphylococcus aureus S6	Milk	50	0.024	0.3
		55	0.009	0.3
		62.5	0.001	0.2
Bacillus cereus spores	Water	105	0.65	1.7

Note: Data obtained by fitting to the Weibull equation (see Equation 7.15). Approximate results from various sources.

great majority of cases, the values of β differed significantly from unity, $\beta > 1$ being more common than $\beta < 1$. In some cases the β value depended significantly on temperature. Moreover, sigmoid thermal death curves of various shapes have occasionally been observed.

It may finally be noted that the derivation of an effective heating time as given in Equation 7.13 breaks down if the β value of the thermal death curve substantially differs from unity. In such cases, a more elaborate mathematical treatment will be needed.

7.3.4.3 Practical Consequences

It will be clear from the preceding discussion that prediction of the surviving number of bacteria or their spores is fraught with difficulties. In many cases, extrapolation of an apparently linear thermal death curve to lower values of N will lead to considerable error. The same holds for application of Z values that have been obtained at relatively low temperatures. In principle, the Weibull equation would give better results but it often is difficult to determine the desired α and β values with sufficient accuracy. Consequently, manufacturers tend to remain on the safe side (see below) and practical experience has shown that 'accidents' almost never occur then. On the other hand, it implies that an overly intensive heat treatment is often applied, which may be detrimental to product quality, e.g., flavor or nutritional value. Moreover, if a hitherto unstudied pathogenic species or strain is detected, it will be very useful to precisely determine the parameters for its heat inactivation and use Weibull equations if necessary.

In practice, one often works with a more or less fixed number of the *sterilizing effect*, defined as

$$S \equiv \log N_0 - \log N \tag{7.17}$$

which is then supposed to give reliable results by applying tabulated D and Z values. The reasoning tends to be: the highest number of the most heat resistant spores in the raw milk is, say, 10^2 ml^{-1}; we do not want to have a greater chance than, say, 1 in 10^5 that a 1-liter bottle of sterilized milk can show bacterial growth; hence, the average number of surviving spores should be 10^{-5} l^{-1} or 10^{-8} ml^{-1}; hence, $S = 10$ would be needed.

It should further be noted that N never becomes zero: there is always a chance, however small, that a bacterium has survived in a given volume of milk. However, it is extremely difficult, if not impossible, to establish that 1 in 10^5 bottles is not sterile. Hence, trial sterilization can be done on milk to which, say, 10^7 spores have been added per milliliter. It will be clear from the above that this involves additional uncertainties. To remain on the safe side the manufacturer thus takes $S = 11$, which is probably too high, because in most cases $\beta > 1$ at high temperatures, which means that linear extrapolation of the thermal death curve will lead to an overestimation of the surviving number of organisms. If it nevertheless turns out that spoilage occasionally occurs, $S = 12$ is tried.

The killing of all pathogens possibly present must be very carefully controlled. Sterilization of milk kills all pathogens: some saprophytic sporeformers that have to be killed are far more heat resistant than any pathogen. Low pasteurization also kills all pathogens that can become harmful but the margin is smaller and close control is needed.

7.4 METHODS OF HEATING

Heating (and cooling; see Chapter 11) of liquids can be done in several different ways, and with various kinds of machinery.

7.4.1 CONSIDERATIONS

Prerequisites for a heating process may be defined as follows:

1. The desirable time–temperature relationship should be practicable. It also involves such aspects as controllability and reliability, and uniformity of heating. In establishing the result of the treatment (e.g., the sterilizing effect), the times needed for warming and cooling, and the spread in residence time, should be accounted for (see Subsection 7.3.2).

2. No undesirable changes should occur in the product, such as absorption of extraneous matter (including Cu, Sn, and plasticizers), loss of compounds (e.g., water), disruption or coalescence of fat globules, coagulation of protein, etc. Sometimes, excessive growth of thermophilic bacteria can occur in a pasteurizer.
3. The expenses should be low. They depend on the price, the lifetime, and the maintenance and operating costs of the machinery. Of much concern is the amount of energy needed for heating and cooling, which may be kept low by regeneration of heat and cold, respectively (Subsection 7.4.3). Furthermore, the extent of fouling (see Section 14.1) plays a role. Rapid fouling causes the heat transfer and the rate of flow to diminish. As a result, consumption of energy increases significantly. This necessitates frequent cleaning and, hence, brief operating times.
4. The method of working should fit into the planning. For example, insertion of operations such as centrifugation or homogenization in the process line may be desirable, as are good possibilities to adjust heating temperature, heating time, and flow capacity.

Furthermore, a particular apparatus or processing scheme is selected on the basis of:

1. The desired combination of time and temperature: Heating for 30 min at 68°C requires machinery that differs from that required for heat treatment of 1 s at 145°C.
2. Properties of the liquid: The main factor involved is the heat-transfer rate which, in turn, depends on the thermal conductivity and especially on the viscosity: highly viscous products show poor heat exchange (see Appendix A.11). Apart from that, the tendency to exhibit fouling is of importance.
3. Requirements for the prevention of recontamination and for the proper execution of ensuing process steps. This applies especially to packaging.

Another factor in the selection may be the effect of the method of heating on the air content, especially the O_2 content of the milk. After all, the O_2 content affects the possibilities for growth of several bacteria. For example, *Bacillus* species need some O_2; lactic acid bacteria are slowed down at high O_2 pressure, which is of special importance for lactic acid fermentations. The O_2 content of long-life milk products may affect the development of off-flavor by fat autoxidation. Holder pasteurization causes significant deaeration, but air can be reabsorbed during cooling. Heating in a heat exchanger does not affect the O_2 content, unless a special deaerator (e.g., a flash cooler) is connected as is done in direct UHT treatment (Subsection 7.4.2). The extent of deaeration during autoclaving depends on the type of sealing applied. For example, bottles fitted with crown corks lose most of the air but sealed cans lose none.

7.4.2 Equipment

Liquids can be heated and cooled in a batch process, in a heat exchanger, or in a packaged form. Moreover, direct mixing with steam can be combined with heat exchange. Highly viscous liquids can be heat-treated in a scraped-surface heat exchanger.

Originally, batch processing was in general use for pasteurizing beverage milk. It is the so-called *holder pasteurization*, e.g., 30 min at 63°C. The method is still in use in the manufacture of starters, whipping cream, and other small-scale products. Usually, the vats involved are fitted with an agitator; steam or hot water circulates through a double jacket, followed by cold water. Among the advantages of holder pasteurization are the simplicity, flexibility, and satisfactory temperature control (little fluctuation in temperature occurs unless highly viscous liquids are treated). A drawback is that the warming and cooling times are long (excessive for large containers). Furthermore, regeneration of heat is not well possible and connection to continuous processes is awkward.

Currently, flow-through heaters or heat exchangers are commonly used. Hot water or condensing steam constitutes the heating medium. Sometimes, vacuum steam heating is applied to minimize the difference in temperature with the liquid to be heated.

In *plate heat exchangers*, a large heating surface is assembled in a confined space and on a small floor area. The heating agent and the incoming liquid are present in thin layers and are separated by a thin wall, i.e., a plate. Because of the large heating surface per unit volume of liquid that is to be heated, the difference between the temperature of the heating agent and the temperature of the liquid to be heated can be small, e.g., 2°C when milk is heated from 65°C to 75°C. This may be an advantage for some heat-sensitive products, where fouling of the heat exchanger is greater for a higher wall temperature. Furthermore, warming and cooling proceed rapidly in plate heat exchangers.

Another advantage is that the energy consumption (for heating and cooling) can be relatively small because heat can be regenerated. The principle is shown in Figure 7.16. When the milk enters the heat exchanger, it is warmed by milk that has already been heated and this, in turn, is at the same time cooled by the milk coming in. The latter is subsequently heated further by hot water (or steam). It may then flow through a holder section to achieve a sufficient heating time. After being cooled by the incoming milk, it is cooled further by means of cold water (or another cooling agent). Note that the liquids being heated and cooled are always in counterflow and that the temperature difference between the two remains constant.

A plate heat exchanger is made up of various sections connected in series, including a regeneration section, a heating section, a holding section (may also be a tube), and cooling sections. Each section consists of a great number of plates, being partly connected in parallel and partly in series. In this way, the liquid is properly distributed among the plates and arrives at a speed that is high enough to reduce fouling. Figure 7.17 gives an example of the constructional setup of a

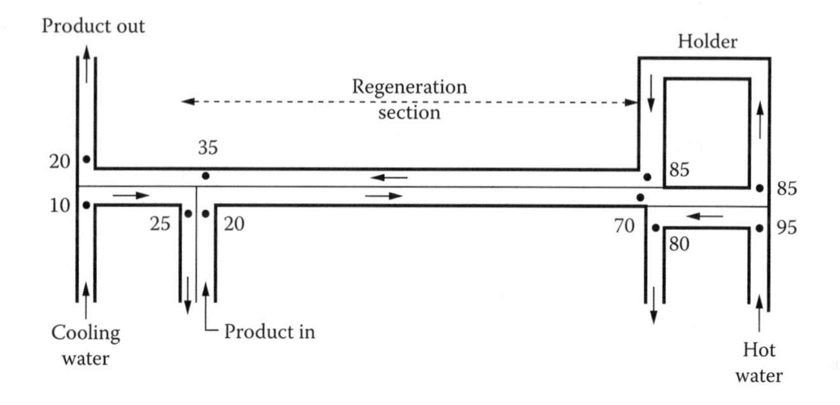

FIGURE 7.16 Simplified scheme of a heat exchanger for heating and cooling of liquids, showing the principle of regeneration. The numbers indicate temperatures (°C) and are merely an example. Thick lines denote insulating walls; thin lines are walls allowing rapid heat transfer.

plate heat exchanger and of the path the liquids take through it. The plates are shaped in such a way as to greatly enhance turbulence in the liquid. This enhances heat transfer and diminishes fouling.

Plate heat exchangers have some disadvantages. The main one is that leakage can occur, for instance, due to an imperfect or worn-out rubber gasket between two plates. If as yet unpasteurized milk in the regeneration section leaks into the milk that has passed the holder section, contamination with undesirable bacteria may occur. To prevent this, the pressure on the milk after the holder must be

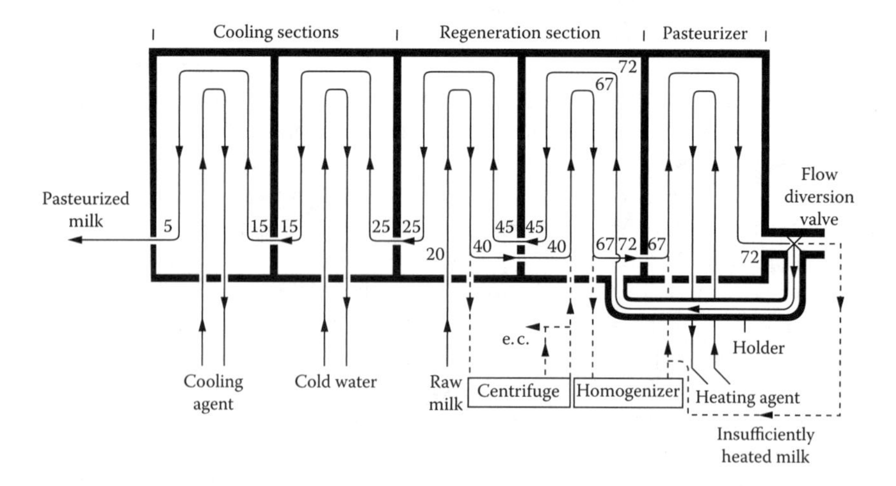

FIGURE 7.17 Example of a pasteurization process in a plate heat exchanger. Simplified diagram. e.c. = excess cream; temperatures of milk in °C.

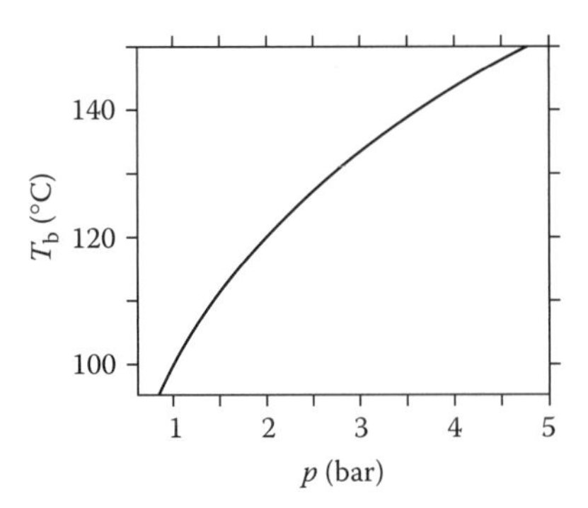

FIGURE 7.18 Boiling temperature (T_b) of water as a function of pressure (p).

higher than that on the milk before the holder; this requires an additional pump. Another problem is that the pressure on the milk cannot be very high. There is only a very small distance between plates and this precludes the heat treatment of highly viscous liquids because a high pressure would be needed to overcome the high flow resistance. It is also difficult to heat milk to well above 100°C because it needs an overpressure to prevent boiling (see Figure 7.18). Finally, liquids containing small particles, say > 50 μm, and liquids that cause heavy fouling also give problems.

In the remaining part of this subsection we will primarily discuss *sterilization*. When a flow-through process is applied, some additional measures must be taken. In the first place, the equipment should not only be clean prior to operation but also sterile. This concerns the heat exchanger and all downstream equipment like pipelines, valves, balance tanks, packaging machine, etc. The equipment must therefore be sterilized; this is generally achieved by operating it with water, which may take about half an hour. Moreover, the plant must be designed in such a way that recontamination can be rigorously prevented. Finally, the packaging material must be sterilized: see Chapter 15 under 'aseptic packaging.'

Tubular heat exchangers generally have a smaller heating surface per unit volume of liquid to be heated than plate heat exchangers. Accordingly, the difference in temperature between the heating agent and the incoming liquid will generally be greater.

To restrict fouling and enhance heat transfer, high flow rates are used, which necessitates high pressures. But this causes no problems because tubes are much stronger than plates; some tubular heat exchangers even have no sealing gaskets but have (spirally bent) concentric tubes. Tubular heat exchangers can readily be applied to obtain very high temperatures (e.g., 150°C). Accordingly, they are excellently fit for indirect UHT treatment. Like a plate heat exchanger, a tubular heat exchanger can be built of regeneration, heating, holding, and cooling sections.

In modern heat exchangers, the milk may be in counterflow with water throughout the apparatus. The water is kept circulating and is heated by means

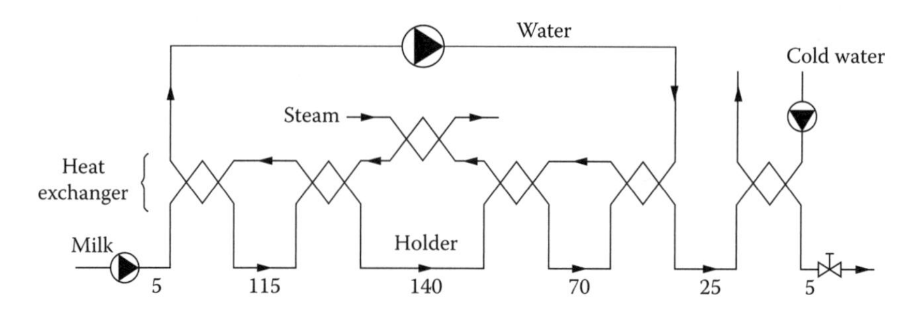

FIGURE 7.19 Example of the flow diagram of a heat exchanger in which the milk is heated and cooled by water only. The numbers refer to the milk temperature in °C.

of indirect steam heating immediately before it heats the milk to the maximum temperature desired. In such cases, the flow rates on either side of a separating wall need not be the same, implying that the temperature difference is not constant. Figure 7.19 shows an example. Often, up to about 90% heat regeneration is achieved. This method of working has advantages with respect to temperature control, rapid and even heat transfer, and saving of energy.

In *UHT treatment with direct heating* there is no wall between the heating agent and the liquid to be heated but the heating agent (steam) is injected into the liquid or the other way around. Thereby an almost instantaneous heating to the desired temperature (e.g., from 80°C to 145°C in 0.1 s) occurs, provided that the steam can immediately condense. For this to happen, finely dispersed steam and a significant back pressure in the liquid are needed (see Figure 7.18). The incoming liquid becomes diluted with the condensing steam. Of course, the steam has to be of high purity. After maintaining the liquid at the desired temperature for a few seconds, it is discharged in a vessel at reduced pressure. Here almost instantaneous evaporation of water occurs, causing very rapid cooling. The amount of water that evaporates should equal the amount of steam that had been absorbed.

Steam injection heating causes disruption of fat globules and some coagulation of protein. Homogenization (aseptic) at high pressure redisperses the coagulum. Without homogenization, the product gives an inhomogeneous astringent or nonsmooth sensation in the mouth, and a sediment tends to form on storage.

The heating section (up to, e.g., 80°C) that precedes the steam injection, and the cooling section that is connected after the 'flash cooling' (starting from, e.g., 80°C), may be plate heat exchangers. Figure 7.20 shows that the holding time of the milk above, say, 80°C can be very short, insufficient to inactivate plasmin. Therefore, holding times are often extended or the warming that precedes the steam injection is made slower and extended to a higher temperature. Also, in indirect UHT treatment, long warming times are sometimes applied to obtain a fuller inactivation of plasmin. In this way, the difference between traditionally sterilized and UHT milk is diminished.

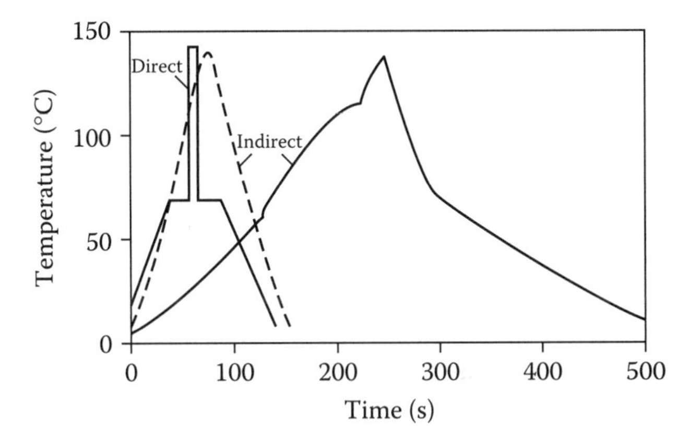

FIGURE 7.20 Temperature of milk vs. time during heat treatment. One example refers to direct and two examples to indirect UHT heating.

The so-called vacreator is quite similar to a direct UHT heater but in this apparatus the liquid is heated to pasteurization temperatures. The essential detail is evaporative cooling in vacuum, aimed at the removal of volatile flavor compounds. Such apparatus is occasionally used for pasteurizing cream for butter manufacture. Disadvantages include the considerable damage (coalescence and disruption) to fat globules and the limited heat regeneration.

A variant of steam injection is infusion. In this method the milk is sprayed into a steam flow.

Autoclaving. Heating a liquid in a hermetically sealed container (rarely larger than a liter) has the distinct advantage that recontamination of the heated liquid by microorganisms can be readily prevented. That is the reason why this method of working is primarily applied in sterilization. However, it suffers from serious disadvantages. The long warming and cooling times and the large temperature differences inside a can or bottle result in undesirable changes like browning and sterilized milk flavor. Agitating or revolving the containers during the heat treatment may restrict these changes because it considerably enhances heat transfer and evenness of temperature. However, its technical operation is difficult because the heating by steam must occur under pressure, at least when sterilization is applied. Heating to 120°C corresponds to a gauge pressure of 1 bar (see Figure 7.18).

The easiest method of heating is batchwise in a steam closet or autoclave. But that implies low capacity, and high energy costs (regeneration is impossible), and is very laborious. Because of this, continuous sterilizers are commonly applied. For bottles, it mostly concerns a hydrostatic sterilizer in which the bottles pass twice through a water seal, about 10 m high because of the 1-bar gauge pressure (Figure 7.21). In some machines, the bottles undergo a rocking motion, whereby heat transfer is speeded up. For cans, the so-called 'cooker and cooler'

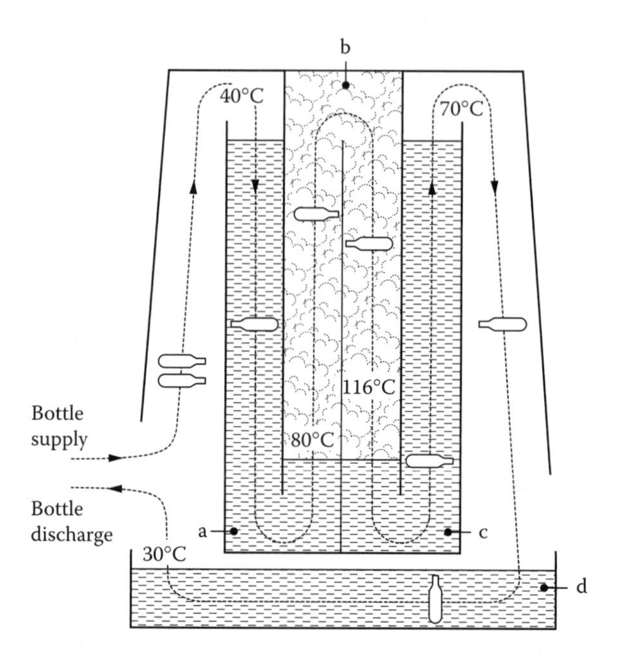

FIGURE 7.21 Diagram of a continuous in-bottle sterilizer. a = Water seal in the bottle entrance and preheating section; b = steam space with adjustable pressure; c = water seal in the bottle discharge and first cooling section; d = second cooling section. Indicated temperatures refer to that in the bottles. The water is partly circulated.

with rotary air locks is usually used. Continuous sterilizers allow considerable heat regeneration.

7.4.3 Heat Regeneration

Regeneration is the regaining of heat and the saving of cooling energy in the combined processes of heating and cooling (see Figure 7.16). The *regenerating effect* is the heat absorbed in the regeneration section as a percentage of the total heat absorption. If the specific heat is the same for all liquids at all temperatures, the regenerating effect R can simply be calculated from

$$R = \frac{T_{reg} - T_{inl}}{T_{max} - T_{inl}} \tag{7.18}$$

where T_{reg} is the temperature after heating in the regeneration section, T_{inl} that at the inlet of the heat exchanger, and T_{max} the maximum (i.e., pasteurization) temperature of the milk.

In principle, close to 100% regeneration can be realized, but this would need a very large heating surface and virtually no heat loss to the environment, which would require very costly measures. There is an optimum at which the additional savings on heating and cooling energy and the debit on additional plates, etc. balance each other. It often amounts to about 90% regeneration.

A high proportion of regeneration also has disadvantages. To begin with, it takes a long time (up to 1 h) for the heat exchanger to arrive at the desired temperature when starting the process. This not only plays a part during the startup of the heating process (which is done with water) but also in the cleaning of the exchanger. (It might become tempting to make shift with cleaning at a lower temperature!) Therefore, a special heat exchanger for heating up of the heating plant is sometimes connected. A second disadvantage is that the heating up of the milk tends to last a long time. In the simplest case, such as shown in Figure 7.17, the temperature difference in the regeneration section between the milk that is warming and the milk that is cooling down, ΔT, is constant; it is equal to the difference between pasteurization temperature and temperature after regeneration. A greater heat regeneration then leads to a smaller ΔT. ΔT is proportional to $(1 - R)$ if R is the fraction of heat regenerated. At a smaller ΔT, the heating surface should be larger and hence the holding time in the regeneration section longer. Clearly, the warming time of the milk is inversely proportional to $(1 - R)$. Consequently, increasing the regenerating effect from 70% to 90% results in a threefold increase in warming time. Figure 7.20 gives an example of a very long warming time, intentionally selected to regenerate much heat. Comparing this example with Figure 7.12 (which has roughly the same hypothetical course of the temperature) shows the strong effect of great heat regeneration on the effective heating time, especially for reactions of a low Q_{10}.

7.4.4 Control

It goes without saying that the heated product must be safe and of satisfactory quality, and this necessitates rigorous control. That may be achieved by control of the properties of the final products. But provisions against risks should also be made during heat treatment, nowadays commonly done using HACCP procedures. The following are the main risks:

1. The *heating intensity* may be insufficient because the steam supply, and so the heating temperature, may fluctuate or because of a sudden increase in fouling. Often, a pasteurizing plant has a so-called automatic flow diversion valve: The milk flows back to the supply pipe if the pasteurization temperature falls below a preset limiting value (see also Figure 7.17). Alternatively, an automatic 'pump stop' may be applied. Moreover, the heating temperature should be recorded continuously. The risk of too brief a heating time will be slight: the volumes in the heat exchanger (especially the holder) are fixed, and it is almost impossible that the milk pump would suddenly run faster.
2. *Recontamination* is a factor. Raw or insufficiently heat-treated milk may gain entrance into the heated milk, for example, because of a leaky

heat exchanger or because of a mistake made in connecting the pipes. Naturally, contamination can occur when milk passes through a machine or pipe that is not absolutely clean. Recontamination should be rigorously avoided in UHT treatment because it is usually combined with aseptic packaging (Chapter 15); it especially concerns the homogenizer. One bacterium per 1000 l of milk may cause unacceptable spoilage. However, in pasteurized and thermalized milks relatively small recontaminations can also have considerable effects.

3. *Growth of bacteria* may occur in heating equipment, e.g., in a batch pasteurizer. Especially in a vessel such as a balance tank, through which milk flows while it maintains a relatively high temperature for some time, organisms like *Bacillus stearothermophilus* (maximum growth temperature ranging from 65°C to 75°C) and *B. coagulans* (maximum growth temperature range: 55°C to 60°C) can grow. As a rule, the counts of these bacteria in raw milk are very low. Accordingly, the contamination will become perceptible only after many hours. Obviously, occasional cleaning and disinfection of the machinery can overcome these problems.

   After having been in use for hours, a pasteurizer may contain growing bacteria in the regeneration section. The bacteria involved survive the milk pasteurization and may occasionally colonize on the metal surface of plates or tubes. Bacteria in the so formed 'microcolonies' can grow rapidly, so that significantly increased counts in the pasteurized milk may be found after, say, 10 h of continuous use of the apparatus. The microorganism involved is usually *Streptococcus thermophilus* (maximum growth temperature about 53°C), but *Enterococcus durans* (maximum ~52°C), and *E. faecalis* (maximum ~47°C) may also cause problems. Timely cleaning is the obvious remedy.

Suggested Literature

Principles of heat transfer are discussed in most texts in food engineering, for example: R.P. Singh and D.R. Heldman, *Introduction to Food Engineering,* Academic Press, Orlando, FL, 1984, which gives a fairly elementary general discussion.

Heating processes used for milk: H. Burton, *Ultra-High-Temperature Processing of Milk and Milk Products,* Elsevier, London, 1988.

Effects of heat treatment on milk: P.F. Fox, Ed., *Heat-Induced Changes in Milk,* 2nd ed., International Dairy Federation, Brussels, 1995.

New methods for characterization of thermal death curves of microorganisms: M. Peleg, *Food Res. Intern.,* 32, 271–278, 1999; M.A.J.S. van Boekel, *Int. J. Food Microb.,* 74, 139–159, 2002.

8 Centrifugation

Centrifugation is usually applied to separate fat globules in the form of cream or to separate solid particles from milk and other liquids.

8.1 CREAM SEPARATION

Centrifugal cream separation serves to make cream and skim milk, to obtain some cream from whey or sweet-cream buttermilk, and to standardize milk and milk products to a desired fat content. It is applied in the industrial manufacture of nearly all dairy products.

As compared to natural creaming (see Subsection 3.2.4), centrifugal separation is far quicker (hence, also more hygienic) and far more complete. This is achieved by (1) making it a flow-through process; (2) causing the fat globules to move very much faster by means of a high centrifugal acceleration; and (3) by greatly limiting the distance over which the fat globules have to move. The latter is achieved by dividing the room in which creaming occurs into very thin compartments.

The *principle* is illustrated in Figure 8.1. The milk enters the machine along the central axis and flows into the revolving bowl through three or more openings in the bowl. It then enters a stack of conical disks with matching openings and the flow is divided over the numerous slits between disks. The centrifugal force drives the fat globules in each slit towards the lower disk, from where they move upwards and inwards in the form of cream. The skim milk, i.e., milk plasma containing some small fat globules that have escaped separation, moves outwards. Both streams then move up and remain separated by the 'cream disk', before being discharged from the centrifuge.

The machines employ different kinds of discharge of cream and skim milk. The one depicted in Figure 8.1 is called a semiopen or half-hermetic separator. The liquids have a high kinetic energy due to the fast spinning of the bowl (generally, at least 5000 r.p.m.) and are forced into centripetal pumps, also called paring disks, that cause discharge under pressure. Thus, the revolving and the static part of the machinery are separate. In a hermetic separator, the discharge pipes are connected to the revolving parts by seals that contain a flexible packing ring that can withstand the pressure of the liquid; the milk inlet is also connected in this way. The separator is thus part of a closed system, further consisting of pump, pipelines, heat exchanger, etc. The capacity of a separator is, e.g., 20,000 l of milk per hour.

Separation efficiency is generally the main concern in the manufacture of skim milk powder, in skimming cheese whey, etc. Factors affecting the proportion

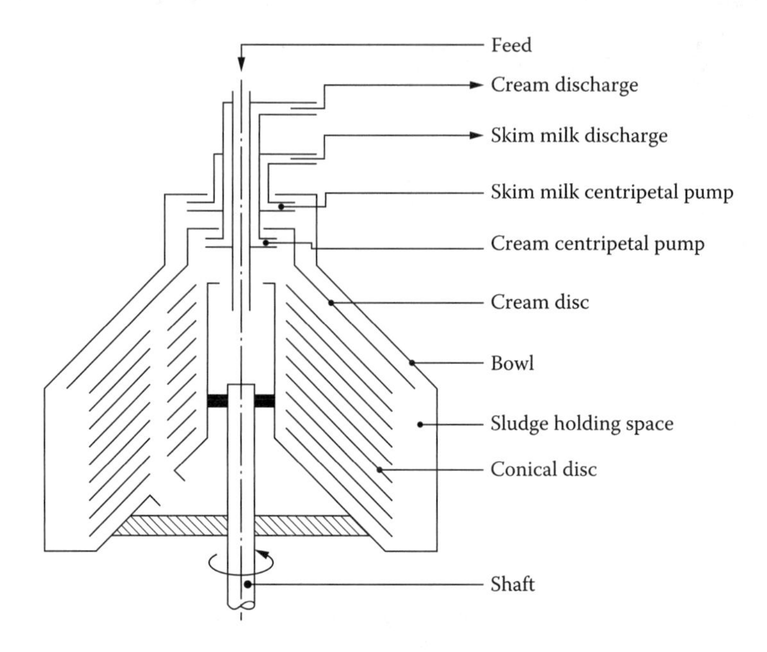

FIGURE 8.1 The basic principle of a so-called semiopen milk separator. The (revolving) bowl and the (nonrevolving) machinery for supply and discharge are shown. In reality, the bowl contains a far greater number of discs.

of the globular fat escaping separation follow partly from the Stoke's equation (Equation 3.16) for the velocity v of a fat globule relative to the surrounding liquid. For centrifugal sedimentation of fat globules, the velocity can be written as

$$v = \frac{R\omega^2(\rho_p - \rho_f)d^2}{18\eta_p} \tag{8.1}$$

where R is the effective radius of the centrifuge (i.e., the radial distance between the fat globule considered and the spinning axis), ω is the revolution rate in radians per second (which equals π/30 times the number of revolutions per minute), ρ is the density, and η is the viscosity; the subscripts p and f refer to milk plasma and fat globules, respectively. Values for density and viscosity for a range of temperatures are in Appendix A.10.

The following are the main factors determining the creaming efficiency:

1. *The centrifugal acceleration $R\omega^2$:* This is usually about 6000 g, where g is the acceleration due to gravity.
2. *The distance over which the fat globules must move:* Figure 8.1 illustrates that discs divide the room in the separator into a great number of spacings. The separation therefore occurs over only about 0.5 mm.

3. *Time available for separation:* This results from the volume and the geometry of the part of the centrifuge in which separation occurs and from the flow rate.
4. *The size distribution of the fat globules:* The critical diameter of fat globules that are just recovered by centrifugation is about 0.7 μm. This can be deduced from the constructional details of the separator, the operational conditions, and the properties of milk, by using Equation 8.1. Accordingly, the amount of fat present in small fat globules is of paramount importance. This is shown in Figure 8.2A. In addition, some nonglobular fat is present in milk (about 0.025%). All in all, at 45°C, a fat content in the separated milk of 0.04 to 0.05% can usually be obtained.
5. *Temperature:* Above all, temperature affects η_p, and also ρ_f, ρ_p, and, slightly, *d*. These variables can be lumped in an efficiency factor, i.e., the velocity calculated from Equation 8.1 divided by the velocity at 20°C. The influence of the separation temperature on this factor and on the fat content of the separated milk is shown in Figure 8.2B. If milk is to be separated at a low temperature, say at 4°C, a specially constructed separator (cold-milk separator) may be used, where the fat content of the resulting skim milk usually amounts to 0.07 to 0.1%.
6. *Proper operation of the separator:* This implies no vibrations, no leakage, etc. The construction considerably affects the result because it determines the variation in holdup time and in effective radius (see Figure 8.2C). Another aspect is the disruption of fat globules into smaller ones in the separator, which decreases separation efficiency. This can especially occur at higher temperatures, as illustrated in Figure 8.2B; the effect greatly depends on constructional details of the machine.

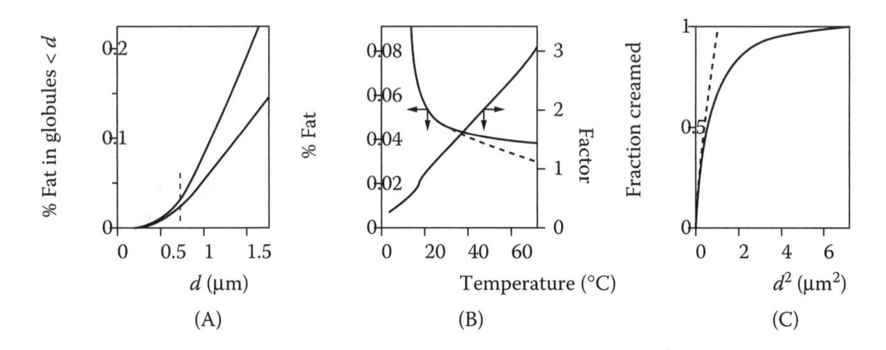

FIGURE 8.2 Centrifugal separation. (A) Amount of fat present in globules smaller than *d* in diameter; the curves represent approximately the extremes found in milks from individual cows. (B) Influence of the separation temperature on the efficiency factor (see Section 8.1) and on the fat content of separated milk; the broken line is expected if no disruption of fat globules would occur. (C) The fraction of the fat not left in the skim milk as a function of the square of the diameter of the fat globules; the broken line is expected if the operational conditions are the same for each fat globule. Approximate examples.

If a separator is only used for standardization, separation efficiency is not of major concern. More important variables then are a high capacity, even at the cost of a higher fat content of the skim milk, and a long operation time.

The latter is limited by the deposition of solid ('dirt') particles on the disks, which can locally obstruct the flow through the bowl. Deposit formation strongly depends on the concentration and type of particles in the milk; at least part of the particles can be removed by microfiltration (see Subsection 12.1.1). At conditions in which the casein micelles are unstable, e.g., at a decreased pH or at a high calcium ion activity, caseinate can be deposited. Most of the dirt, and possibly some caseinate, will be deposited as sludge in the holding space shown in Figure 8.1. At low operation temperatures, fatty deposits resulting from partial coalescence of fat globules may from at other places. See Subsection 3.2.2.2 for variables affecting this process. The extent and locations of deposit formation also depend on constructional details of the separator.

Another important variable is the proportion of the milk that is discharged as cream, which determines, together with the milk fat content, the fat content of the cream. This is achieved by the use of a throttling valve in the cream discharge line. When using a semiopen separator, there is a limit to the fat content that can be reached, say 35%. In a hermetic separator, far higher cream fat contents can be obtained. To obtain 'plastic cream,' i.e., fat content over about 65%, a specially designed separator is usually needed.

8.2 REMOVAL OF PARTICLES

Particles that have a density larger than that of milk plasma can also be removed by centrifugation. It concerns dirt particles, somatic cells, and even microorganisms. The rate of removal strongly depends on temperature because the value of η_p greatly decreases with increasing temperature. Moreover, in raw or thermalized milk, somatic cells and several bacteria participate in the cold agglutination of fat globules (see Subsections 3.2.4.2 and 3.2.4.3). This then means that most of these cells are removed with the cream, more being removed the further the temperature falls below 35°C.

Removal of dirt particles and somatic cells is a subsidiary result of centrifugal cream separation as usually applied, say at 40°C. In a traditional separator, sludge is collected in a holding space, and it is removed after stopping the operation. In the current machines, the deposit can be removed through small valves in the outer wall of the bowl; the valves can be opened and flushed with water at intervals without interrupting the separation process.

Some centrifuges, sometimes called *clarifiers*, are specifically built for the removal of solid particles. The liquid enters the bowl at the periphery to move between the disks to the center, from where it is discharged (there are thus no inlet holes in the disks). This improves separation efficiency by giving the particles a longer time to sediment. Often, the dirt is continuously removed from the bowl through a number of small holes. Clarifiers are rarely used in the dairy except for one specific purpose, which is described in the following text.

This concerns *bactofugation*, which is usually applied to remove spores from products that are low-pasteurized. This may involve removal of spores of *Bacillus cereus* from beverage milk or of *Clostridium tyrobutyricum* and related species from cheese milk. Spores are quite small, for the most part 1 to 1.5 μm, but the density difference with plasma is larger than that of bacteria, and at separation temperatures of 60 to 65°C, a substantial proportion can be removed, generally 90 to 95%. By using two bactofuges in series, a reduction by over 99% can generally be attained.

The centrifuges used are hermetic. Two types are currently used: One is like a normal clarifier, with either continuous or intermittent discharge of sludge. The sludge makes up less than 0.2% of the milk. The other type is more like a cream separator, in that it has two outlets at the top, one for the cleaned milk and the other for the portion (about 3%) that contains the particles including spores. It also has an enhanced content of casein micelles. Generally, the spore-rich liquid is sterilized and returned to the cleaned milk. If this is done with the sludge, it should first be diluted by adding milk. Heat treatment is generally for a few seconds at 130°C by steam infusion. Cooling is achieved by the immediate admixture of cleaned milk. Afterwards, the milk is low-pasteurized.

It should be noted that bactofugation is by no means equivalent to sterilization: the product still contains heat-resistant bacteria and a small number of spores. The process may serve to give pasteurized beverage milk a longer shelf life or to obtain cheese that is not subject to the defect of late blowing (Section 26.2), without inducing unfavorable changes in the milk.

Suggested Literature

More detailed information about centrifugal separation (in Chapter 8 of): H. Mulder and P. Walstra, *The Milk Fat Globule,* Pudoc, Wageningen, 1974.

9 Homogenization

Homogenization of milk causes disruption of milk fat globules into smaller ones. The milk fat–plasma interface is thereby considerably enlarged, usually by a factor of 5 to 10. The new interface is covered with milk protein, predominantly micellar casein.

9.1 OBJECTIVES

Homogenization is applied for any of the following reasons:

1. *Counteracting creaming:* To achieve this, the size of the fat globules should be greatly reduced. A cream layer in the product may be a nuisance for the user, especially if the package is nontransparent.
2. *Improving stability toward partial coalescence:* The increased stability of homogenized fat globules is caused by the reduced diameter and by the acquired surface layer of the fat globules. Moreover, partial coalescence especially occurs in a cream layer, and such a layer forms much more slowly in homogenized products. All in all, prevention of partial coalescence usually is the most important purpose of homogenization; a cream layer *per se* is not very inconvenient, because it can readily be redispersed in the milk.
3. *Creating desirable rheological properties:* Formation of homogenization clusters (Section 9.7) can greatly increase the viscosity of a product such as cream. Homogenized and subsequently soured milk (e.g., yogurt) has a higher viscosity than unhomogenized milk. This is because the fat globules that are now partly covered with casein participate in the aggregation of the casein micelles.
4. *Recombining milk products:* Recombination is discussed in Section 16.3. At one stage of the process, butter oil must be emulsified in a liquid such as reconstituted skim milk. A homogenizer, however, is not an emulsifying machine. Therefore, the mixture should first be preemulsified, for example, by vigorous stirring; the formed coarse emulsion is subsequently homogenized.

9.2 OPERATION OF THE HOMOGENIZER

Homogenizers of the common type consist of a high-pressure pump that forces the liquid through a narrow opening, the so-called homogenizer valve. Figure 9.1A gives a flowchart; for the moment, we will leave the second stage aside. The principle of operation of the valve is illustrated in Figure 9.1B. The valve has been dimensioned in such a way that the pressure in the valve (p_2) equals about zero at a reasonable homogenizing pressure ($p_{\text{hom}} = p_1$), such as $p_{\text{hom}} > 3$ MPa. Actually, p_2 tends to become negative, which implies that the liquid can start boiling; in other words, cavitation can occur. (Cavitation is the formation and sudden collapse of vapor bubbles caused by pressure fluctuations.)

During homogenization, the liquid upstream of the valve has a high potential energy. On entering the valve, this energy is converted to kinetic energy (according to the rule of Bernoulli). The high liquid velocity in the very narrow opening in the valve leads to very intense turbulence; the kinetic energy of the liquid is now dissipated, that is, converted to heat (thermal energy). Only a very small part of the kinetic energy, generally less than 0.1%, is used for globule disruption, that is, for conversion into interfacial energy. The net amount of energy dissipated per unit volume (in $\text{J}\cdot\text{m}^{-3}$) numerically equals p_{hom} (in $\text{N}\cdot\text{m}^{-2}$). If the specific heat of the liquid is c_p, the temperature rise as caused by homogenization will be p_{hom}/c_p. For milk, $c_p \approx 4 \times 10^6\ \text{J}\cdot\text{m}^{-3}\cdot\text{K}^{-1}$; thus $\Delta T \approx p_{\text{hom}}/4$ if p_{hom} is expressed in MPa, and T is expressed in K.

The passage time of the liquid through the valve is very short, generally less than 1 ms. As a result, the average power density $\bar{\varepsilon}$ (which is the energy dissipated per unit volume and per unit time in $\text{J}\cdot\text{m}^{-3}\cdot\text{s}^{-1}$ or $\text{W}\cdot\text{m}^{-3}$) is extremely high, that is, 10^{10} to $10^{12}\ \text{W}\cdot\text{m}^{-3}$ (see Figure 9.1B). Such high power densities lead to very intense turbulence. This implies that the flow pattern shows very small eddies in which high liquid velocity gradients occur. These eddies thereby cause pressure fluctuations, which can disrupt particles, especially droplets (see Section 9.3).

Figure 9.2 schematically depicts the simplest homogenizer valve in use. The pressure is adjusted by letting the control spring press down the valve with an appropriate force. In large machines hydraulic pressure rather than mechanical pressure is generally applied.

Sometimes an impact ring is present. At that point, the liquid velocity is still high, say, 50 $\text{m}\cdot\text{s}^{-1}$ (180 $\text{km}\cdot\text{h}^{-1}$). Solid particles can hit the ring and thereby be disrupted, if the particle density is significantly higher than the liquid density. This is a much-used method for reducing the size of cocoa particles in chocolate milk.

Most homogenizer valves are more complicated than the one represented in Figure 9.2. Usually, they are more or less conically shaped. More sophisticated surface reliefs occur as well. To prevent uneven wear, the valve in some homogenizers is rotated relative to the valve seat.

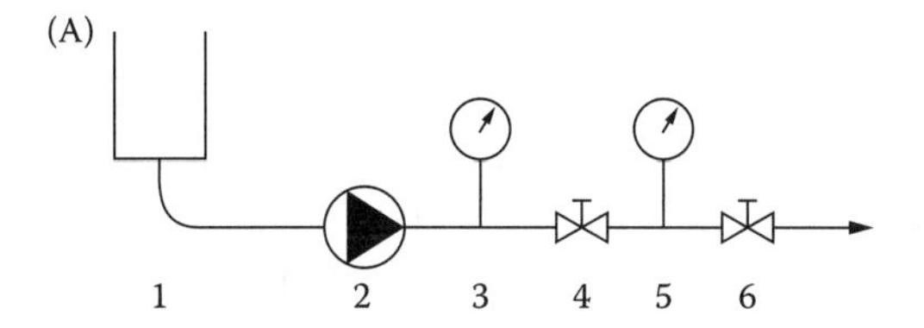

(B)

L

p_1, v_1 p_2, v_2 p_3

Bernoulli: $$p_1 + (1/2)\rho v_1^2 = p_2 + (1/2)\rho v_2^2 \quad (9.1)$$

$$p_1 = p_{\text{hom}} \quad p_2 \approx 0 \quad p_3 \approx 1 \text{ bar}$$

$$v_1 \ll v_2$$

Hence $$p_{\text{hom}} = (1/2)\rho v_2^2 \quad (9.2)$$

Power density $$\varepsilon = p_{\text{hom}}/t_p$$

$$t_p \approx L/v_2 \propto L/(p_{\text{hom}})^{1/2} \quad (9.3)$$

Hence $$\varepsilon \propto (p_{\text{hom}})^{3/2} \quad (9.4)$$

Example: $p_{\text{hom}} = 200 \text{ bar} = 2.10^7 \text{ Pa}$

hence $v_2 = 200 \text{ m.s}^{-1}$

Suppose $L_{\text{eff}} = 0.01 \text{ m}$

then $t_p = 5.10^{-5} \text{ s}$

and $\varepsilon = 4.10^{11} \text{ W.m}^{-3}$

FIGURE 9.1 Operation of the high-pressure homogenizer. (A) Flow diagram: 1, tank; 2, high-pressure pump; 3, 5, manometer; 4, homogenizer valve; 6, valve of second stage; 5 and 6 are not always present. (B) Principle of the homogenizer valve (schematic example): p = pressure; ρ = mass density; v = liquid velocity; t_p = passage time; L = passage length.

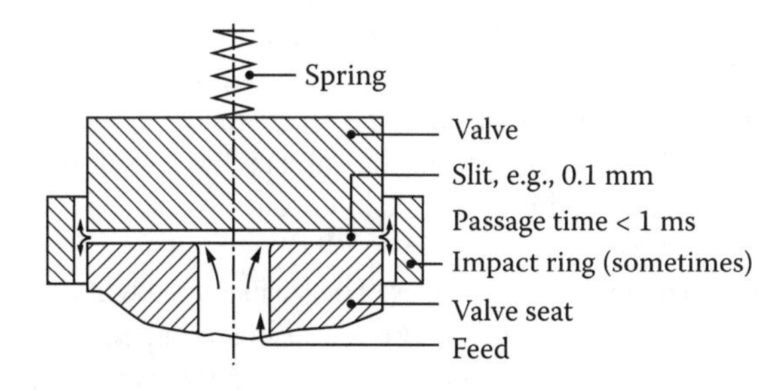

FIGURE 9.2 Cross section of a flat homogenizer valve (near to scale, but the spacing between valve and seat is much smaller than drawn).

9.3 EFFECTS OF TURBULENCE

The flow in the slit of a homogenizer valve is nearly always intensely turbulent. The Reynolds number ($\mathrm{Re} = 2\,h\,v\,\rho/\eta$, where h is slit width and v the average linear flow velocity in the valve) is, say, 40,000 in a large homogenizer. In smaller machines, Re is smaller but still sufficient to produce turbulence.

An intensely turbulent flow is characterized by the presence of eddies (whirls, vortices) of various sizes that constantly form and disappear. From the Kolmogorov theory of isotropic turbulence, the local root-mean-square flow velocity $v(x)$ in the smaller eddies can be estimated; it depends on the value of the power density ε and on the local distance x. According to the Bernoulli equation (see Figure 9.1), this will then lead to local pressure fluctuations

$$\Delta p(x) \approx \rho \langle v^2(x) \rangle, \tag{9.5}$$

Due to these fluctuations, oil drops can be deformed and possibly be disrupted into smaller ones. A drop resists deformation because of its Laplace pressure ($p_{\mathrm{La}} = 4\,\gamma/d$, as discussed in Subsection 3.1.1.4), but if the local pressure difference is larger than the Laplace pressure of the drop, its disruption generally results. It then follows from the theory that the maximum size that a drop can have in the turbulent flow (larger ones will be disrupted) is given by

$$d_{\mathrm{max}} \approx \varepsilon^{-0.4} \gamma^{0.6} \rho^{-0.2} \tag{9.6}$$

This is a very simple relation that also holds for average drop size. In milk, the value of the interfacial tension γ between oil and plasma will be about 15 or 20 $\mathrm{mN{\cdot}m^{-1}}$ at the moment that disruption occurs. The value of ε is very high in a high-pressure homogenizer because the large amount of kinetic energy available is dissipated in a very short time (see Figure 9.1).

(*Note:* The disruption of drops is thus not due to shear forces (viscous forces), as is often stated, but to inertial forces. On the other hand, the theory used is to some extent an oversimplification. Especially if Re is not very high and the drops broken up are small, viscous forces play a part, and then the resulting droplet diameter is also somewhat dependent on the viscosity of the continuous phase.)

During homogenization, several processes occur simultaneously, as depicted in Figure 9.3. Droplets are deformed (line 1) and possibly disrupted (line 1). Consequently, the total surface area is increased and additional surfactant (i.e., protein in milk) has to adsorb onto the drops (line 2). Newly formed drops will frequently collide with each other, which may result in recoalescence if the protein surface load is still small (line 3). If sufficient protein has already adsorbed, the collision may have no effect (line 4). Disruption and collision occur several, e.g., 50, times during passage of the valve. Disruption occurs in steps because a deformed drop will rarely break up into more than a few smaller ones. The surface load will thus several times decrease and increase again. (Lines 5 and 6 are discussed in Section 9.7.)

The rates of the various processes depend on a number of variables, such as power density, surfactant concentration, droplet size, and volume fraction, and each has a characteristic time τ. If, for instance, τ_{a} (adsorption) is greater than τ_{d} (deformation),

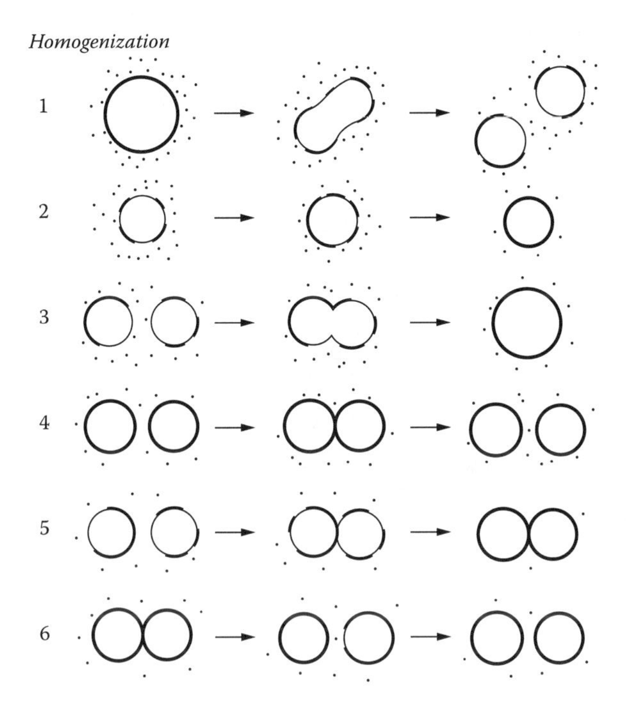

FIGURE 9.3 Processes occurring in an oil-in-water emulsion during homogenization. The droplets are depicted by thin lines, the surface-active material (e.g., protein) by thick lines and dots (schematic and not to scale).

the droplets may have a low surface load and hence a higher interfacial tension during deformation and breakup, causing a less efficient breakup. If τ_a is greater than τ_e, that is, the average time between two encounters (collisions) of a drop with some other drop, the chance that recoalescence occurs is enhanced. Both conditions will thus result in a relatively large average droplet size.

The Kolmogorov theory gives equations for the τ values, and from these we calculate for milk and cream homogenized at 60°C and 20 MPa, the following approximate times:

	Milk (4% Fat)	**Cream (25% Fat)**
Adsorption time (τ_a, in μs)	0.3	1
Encounter time (τ_e, in μs)	0.2	0.02
Deformation time (τ_d, in μs)	0.5	0.5

For milk, all of these times are approximately the same, but for cream, they are clearly not. Furthermore, all of the times mentioned are very short, that is, a factor of about 100 shorter than the passage time through the homogenizer valve.

TABLE 9.1
Average Effect of Homogenization on Milk with 4% Fat

Homogenization Pressure (MPa)	0	5	10	20	40
Temperature rise (K)	0	1.2	2.5	5	10
Number of fat globules (μm^{-3})	0.015	2.8	6.9	16	40
d_{vs} (μm)	3.3	0.72	0.47	0.31	0.21
d_{max} (μm)[a]	9	3.1	2.3	1.6	1.1
A ($m^2 \cdot ml^{-1}$ milk)	0.08	0.37	0.56	0.85	1.3
c_s	0.44	0.89	0.85	0.83	0.82
H (μm^2)	20	2.2	0.87	0.36	0.16

Note: Approximate results. See Subsection 3.1.4 for definition of variables.

[a] 99% of the fat is in globules of size less than d_{max}.

Table 9.1 and Figure 9.4 give examples of the effects created by homogenization. Because the conditions, hence ε, vary from place to place in the homogenizer valve, there will be considerable spread in d_{max}. Accordingly, the resulting size distribution is fairly wide, as shown in Figure 9.4. Repeated homogenization

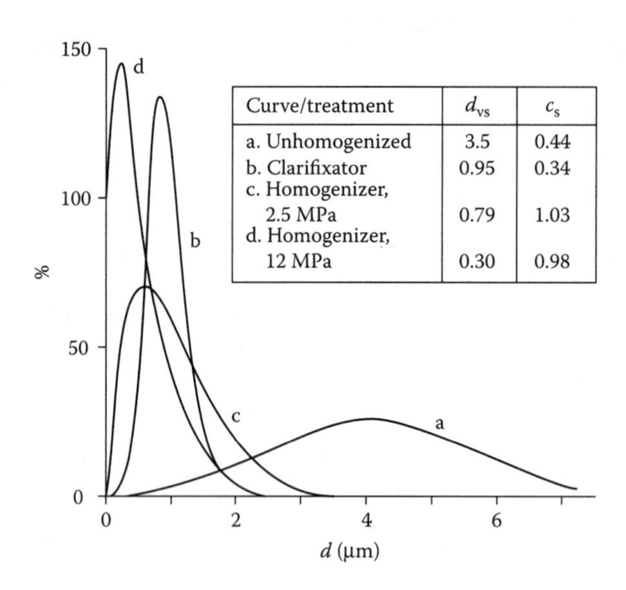

FIGURE 9.4 Some examples of the size distribution of the fat globules in homogenized and unhomogenized milk. Volume frequency (percentage of the fat per micrometer class width) as a function of globule diameter. (Adapted from H. Mulder and P. Walstra, *The Milk Fat Globule,* Pudoc, Wageningen, 1974.)

reduces the average diameter d_{vs} as well as the distribution width c_s. (Particle size distributions are described in Subsection 3.1.4.) The distribution also becomes wider if significant recoalescence of droplets occurs.

9.4 FACTORS AFFECTING FAT GLOBULE SIZE

The main factors affecting fat globule size are:

1. *Type of homogenizer,* especially construction of the homogenizer valve: For the same homogenization pressure p, the passage time t_p will differ and so will $\bar{\varepsilon} = p/\bar{t}_p$. Moreover, the spread in conditions will differ. All in all, considerable variation in results is observed. Figure 9.5 gives examples.
2. *Homogenizing pressure:* Equation 9.4 shows that $\bar{\varepsilon}$ is proportional to $p^{1.5}$. From $d_{max} \propto \varepsilon^{-0.4}$, (Equation 9.6) and from the observation that the shape of the size distribution of the fat globules depends only a little on the pressure, it can be inferred that

$$\log d_{vs} = \text{constant} - 0.6 \log p \tag{9.7}$$

and that

$$\log H = \text{constant} - 1.2 \log p \tag{9.8}$$

where the sedimentation parameter H is an average diameter squared (see also Figure 9.5). These are useful equations because from the result obtained at one pressure, the effect at other pressures can be predicted. The constant depends on the homogenizer and other parameters. Table 9.1 gives examples of results.

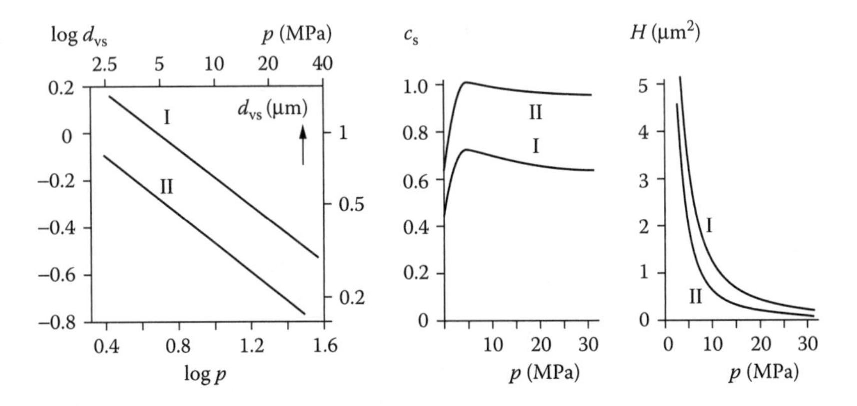

FIGURE 9.5 Effect of homogenization pressure (p) on average fat globule size (d_{vs}), relative distribution width (c_s), and sedimentation parameter (H) for two homogenizers (I and II). (From H. Mulder and P. Walstra, *The Milk Fat Globule,* Pudoc, Wageningen, 1974.)

3. *Two-stage homogenization:* The milk first passes the usual homogenizer valve, due to which the pressure is reduced, for instance, from 20 to 5 MPa (the minimum pressure *inside* the valve equals zero). Through the second homogenizer valve the pressure is reduced to about 1 bar (0.1 MPa) (see Figure 9.1). There is no significant homogenization in the second valve slit. Accordingly, the second-stage influence on the fat globule size is small. In other words, one-stage homogenization at 20 MPa leads to a result similar to that in two stages at 20 and 5 MPa. The result is worse if the pressure drop in the second stage increases to more than about 30% of the total pressure drop. The purpose of two-stage homogenization is another one (see Section 9.7).
4. *Fat content and ratio of amount of surfactant (usually protein) to that of fat:* If sufficient protein is not available to cover the newly formed fat surface, the average diameter of the fat globules (d_{vs}) and the relative distribution width (c_s) will be larger (see Figure 9.6). This results, at least qualitatively, from the theory given in Section 9.3. In cream, the time needed for formation of adsorption layers is longer than in milk, whereas the average time between encounters of one droplet with another is much shorter. As a result, in cream, far more recoalescence of newly formed droplets can occur. Figure 9.6 also shows that the simple relation of Equation 9.7 no longer holds if the fat content surpasses, say, 10% (see also Section 9.7).
5. *Type of surfactant:* When a small-molecule surfactant is added, such as Tween 20 or sodium dodecyl sulfate, the effective interfacial tension becomes lower and smaller globules result.
6. *Temperature:* Homogenization is usually done at temperatures between 40 and 75°C. Homogenization is poor if the temperature is

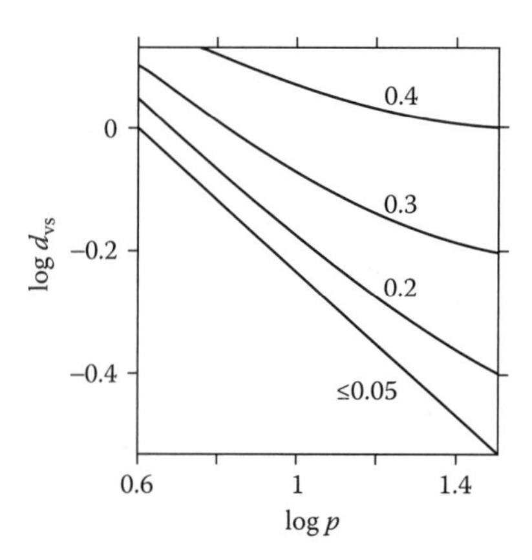

FIGURE 9.6 Effect of homogenization pressure (p, in MPa) on average fat globule diameter d_{vs} (μm) of milk and cream. The fat content, expressed as volume fraction, is near the curves. (Adapted from P. Walstra and G. Hof, unpublished.)

so low that part of the fat is crystalline (see Figure 9.11). Further increase of the temperature still has a small effect, presumably because the viscosity of the oil decreases somewhat.

7. *Proper operation of the homogenizer:* Pressure fluctuations (caused by leaking valves, etc.), a worn homogenizer valve, and air inclusion may have adverse effects. Air inclusion and wear of the homogenizer valve should be rigorously avoided. If the liquid contains solid particles such as dust or cocoa, the valve may quickly wear out, resulting in unsatisfactory homogenization.

The effect of the homogenization should be checked regularly. The average fat globule size may be derived from specific turbidity measurements at a long wavelength (such as 1 μm) after the milk has been diluted and the casein micelles dissolved. In this way, the homogenizing effect can be evaluated rapidly and simply. In principle, continuous determination is possible. In actual practice, however, an accelerated creaming test is usually done. A certain quantity of milk is centrifuged, and the fat content of the resulting skim milk determined.

9.5 SURFACE LAYERS

When butter oil is emulsified in skim milk and subsequently homogenized (as in the fabrication of recombined milk), a surface layer forms, which consists of milk proteins. Figure 9.7 illustrates the composition and structure of such a layer. When milk is homogenized, the natural membrane on the globules is disrupted but remains adsorbed. The surface layer of most globules in homogenized milk

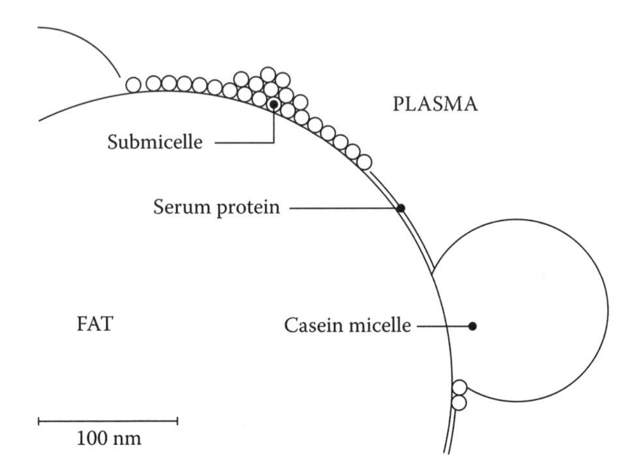

FIGURE 9.7 New surface layer of fat globules formed during homogenization (highly schematic). (From P. Walstra, The milk fat globule: natural and synthetic, *Proceedings XXth Intl. Dairy Cong.*, 1978, 75ST.)

thus contains patches of the native membrane, making up 10 to 30% of the surface area, as well as larger patches of the newly formed surface layer.

The newly formed surface layer consists of micellar casein for the greater part and also contains some milk serum protein. Some of the micelles on the surface appear hardly altered upon adsorption, but most micelles have to some extent spread over the oil–water interface. With electron microscopy, one can observe micelle fragments as well as a thinner layer (10 to 15 nm), which may consist of casein submicelles. The spreading occurs very fast, with characteristic times less than 1 μs.

It can be derived from Kolmogorov theory that the flux of (protein) particles toward the fat globules during homogenization greatly depends on the radius of the particles r_p. The relation between the rate of increase of surface load Γ and protein concentration c_p then is given by

$$d\Gamma/dt \propto c_p r(1 + r_p/r)^3 \qquad (9.9)$$

where r is the droplet radius. The factor in parenthesis is highly variable. For casein micelles, r_p ranges from 20 to 150 nm; for serum proteins, $r_p \approx 2$ nm; for the fat globules, r ranges from about 100 to 400 nm. That means that the cube of this factor may vary between 1.06 and ~8, and that it is largest for small droplets encountering large casein micelles; it virtually equals unity for fat globules with serum proteins. This has the following important consequences:

1. Casein micelles are preferentially adsorbed. Casein makes up about 80% of the protein in milk plasma but about 93% of the protein in the new surface layers. Because the 'casein layer' is thicker than the serum protein layer, casein covers approximately 75% of the surface area of the fat globules. These results apply for homogenization at about 10 MPa.
2. Large micelles are adsorbed preferentially over small ones.
3. The differences in adsorption among protein (particles) are largest for smaller fat globules because then $r \approx r_p$ for some micelles. That explains why smaller fat globules usually have a thicker protein layer than large globules (see Figure 9.8C). When homogenization pressure is very high (say, greater than 30 MPa), the surface layers obtained are virtually devoid of serum protein.

A rough average for Γ in homogenized milk is 10 $mg \cdot m^{-2}$. The following are also factors that affect Γ:

1. *Homogenization temperature:* This is explained by the casein micelles spreading more rapidly over the fat–water interface at, say, 70°C than at 40°C, which causes Γ to decrease (Figure 9.8A).
2. *Preheating*, for example, for 20 min at 80°C: this causes serum proteins to associate with the casein micelles. Consequently, Γ increases (see

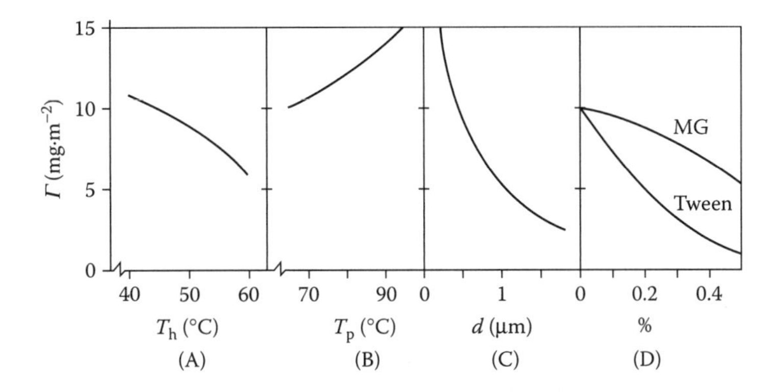

FIGURE 9.8 Effect of certain process variables on the protein surface load Γ of the fat globules in milk (or cream). Homogenization pressure is about 10 MPa. Approximate examples. (A) Homogenization temperature (T_p was 70°C). (B) Pasteurization temperature, heating during 10 min (T_h was 40°C). (C) Fat globule diameter. It concerns globules in one sample of homogenized milk. (D) Addition of a water-soluble (Tween) or an oil-soluble (monoglycerides, MG) surfactant to 12% fat cream before homogenizing.

Figure 9.8B) because formation of a thin local layer of serum proteins is no longer possible.

3. *Casein micelles size*: In evaporated milk the casein micelles may have coalesced into larger ones. This also causes Γ to increase.

In some cream products, small-molecule surfactants such as Tweens or monoglycerides are sometimes added before homogenization. These then adsorb onto the disrupted globules and thereby lower the protein surface load, as shown in Figure 9.8D. This is because these surfactants generally decrease the interfacial tension more than proteins do; hence, they are preferentially adsorbed. Some surfactants, especially Tweens, can also displace proteins when added after homogenization. The displacement of proteins tends to substantially affect various kinds of physical instability of the globules (see the following section). As mentioned, addition before homogenization also results in smaller globules.

9.6 COLLOIDAL STABILITY

Plasma protein, predominantly casein, covers a large part (up to about 90%, and in recombined milks 100%) of the surface area of homogenized fat globules. This makes the globules behave to some extent like large casein micelles. Any reaction that causes casein micelles to aggregate, such as renneting, souring, or heating at very high temperatures, will also cause the homogenized fat globules to aggregate. Moreover, aggregation will take place more quickly because homogenization has increased the apparent casein concentration (i.e., the casein concentration effective

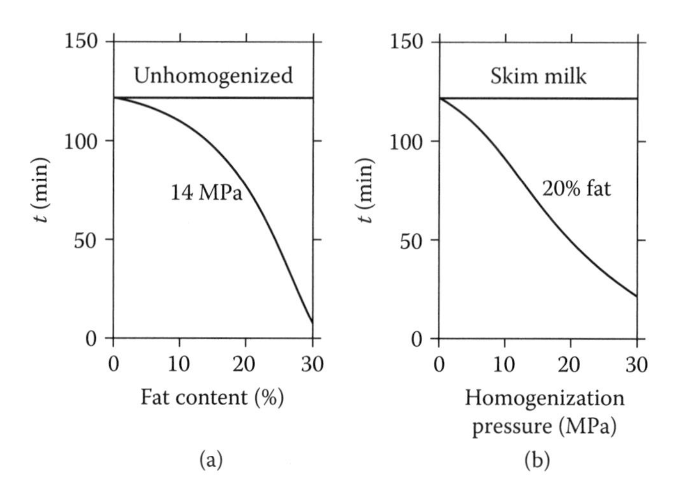

FIGURE 9.9 Time (t) needed for coagulation of milk or cream at 120°C as affected by homogenization. (Adapted from B.H. Webb and G.E. Holm, *J. Dairy Sci.,* **11**, 243, 1928.)

in aggregating). This is discussed in Subsection 3.1.3. The effect is stronger when fat content and homogenizing pressure are higher (see Figure 9.9).

The following are some important consequences of the altered surface layer:

1. Heat stability is decreased, as illustrated in Figure 9.9. This is further discussed in Subsection 17.1.2. A related aspect is the influence on the feathering of cream in coffee (Subsection 17.1.3). Addition of small-molecule surfactants before homogenization may increase stability to heat coagulation (see also Figure 9.8D).
2. Homogenized milk fat globules generally show a high stability toward partial coalescence because they are so small. Moreover, surface layers that consist fully of plasma protein (as in recombined milk) yield a very high stability to the globules. The addition of surfactants (that displace protein at the oil–water interface, see Figure 9.8D) can considerably decrease the stability to (partial) coalescence. This plays a part in a product such as ice cream (Section 17.3).
3. Sour products (yogurt and sour cream) and cheese made of homogenized milk (or cream) have different rheological properties from those of unhomogenized milk. This is caused by the fat globules becoming part of the casein network. An example is given in Figure 22.5.

9.7 HOMOGENIZATION CLUSTERS

The homogenization of cream usually causes its viscosity to be very much increased, as shown in Table 9.2. Microscopic examination shows large agglomerates of fat globules rather than single globules in the homogenized cream. These

TABLE 9.2
Effect of One- and Two-Stage Homogenization on Formation of Homogenization Clusters in Cream with 20% Fat

Homogenization[a]	Pressure (MPa)	η^b	Clusters
Not	0	1	—
One stage	7	8.9	+ +
One stage	21	30.1	+ + + +
Two stage	21/7	4.5	+

[a] Homogenization temperature = 65°C.
[b] Apparent viscosity of homogenized relative to unhomogenized cream.

Source: F.J. Doan, *J. Dairy Sci.,* **12**, 211, 1929.

so-called homogenization clusters contain very many fat globules, up to about 10^5. Because the clusters contain interstitial liquid, the effective volume fraction of particles in the cream is increased, and hence also its (apparent) viscosity. Adding casein micelle–dissolving agents can disperse the clusters. In other words, the fat globules in the cluster are interconnected by casein micelles.

The formation of homogenization clusters can be explained as follows. During homogenization, when a partly denuded fat globule collides with another globule that has been covered with casein micelles, such a micelle can also reach the surface of the former globule. As a result, both fat globules are connected by a bridge and form a homogenization cluster (see Figure 9.3, line 5; and Figure 3.11). The cluster will immediately be broken up again by turbulent eddies (Figure 9.3, line 6). If, however, too little protein is available to fully cover the newly formed fat surface, clusters are formed from the partly denuded fat globules just outside the valve slit of the homogenizer, where the power density is too low to disrupt the clusters again.

In other words, clustering may occur if c_p (= protein concentration, in $\text{kg}\cdot\text{m}^{-3}$) is less than $\Gamma\,\Delta A$ (Γ = surface load that would be obtained at low φ, in $\text{kg}\cdot\text{m}^{-2}$; ΔA = increase in surface area, in $\text{m}^2\cdot\text{m}^{-3}$); $\Delta A \approx 6\varphi/d_{vs}$ (φ = volume fraction of fat). Clearly, the following conditions promote formation of homogenization clusters:

1. High fat content.
2. Low protein content.
3. High homogenizing pressure.
4. A relatively high surface load of protein, promoted by a low homogenization temperature (less rapid spreading of casein micelles), intense preheating (little serum protein available for adsorption) and, subsequently, a high homogenizing pressure (see Figure 9.8).

Under practical conditions, clustering due to homogenization does not occur in a cream with less than 9% fat, whereas it always does in a cream with more than

18% fat. At intermediate fat contents, clustering strongly depends on pressure and temperature of homogenization (see also Subsection 17.1.4).

Clusters can be disrupted again to a large extent (but not fully) in a two-stage homogenizer (Table 9.2). In the second stage, the turbulent intensity is too low to disrupt fat globules and hence to form new clusters, whereas existing clusters are disrupted; this is accompanied by some coalescence. Two-stage homogenization of high-fat cream (e.g., 30% fat) insufficiently breaks up homogenization clusters.

9.8 CREAMING

An important purpose of homogenization usually is to slow down creaming and thereby prevent partial coalescence. This is primarily achieved by reducing the size of the fat globules. Accordingly, the factors affecting this reduction greatly affect the creaming rate (see Figure 9.5 and Figure 9.11). The influence of the fat globule size on creaming rate is given by parameter H:

$$H = \Sigma\, n_i d_i^5 / \Sigma\, n_i d_i^3 \qquad (9.10)$$

where n and d are the number and size of the globules, respectively. The largest globules especially contribute to H. That explains why the width of the size distribution can considerably affect H. Different homogenizers lead to different results (Figure 9.5).

The Stokes velocity (Equation 3.16) can be used for calculating the initial rate of cream rising in a product. Table 9.3 gives some examples. The relationship is:

$$q = 47 \cdot 10^{-5} (\rho_p - \rho_f)\, H / \eta_p h \qquad (9.11)$$

where q is the percentage of the fat in the milk that reaches the cream layer per day, h is the height of the container in cm, H is expressed in μm^2, and ρ and η in SI units. But the creaming rate often is slower because (1) the velocity of the globules is decreased due to the presence of other globules, the more so the higher the fat content, (2) the protein layer covering the fat globules after homogenization increases the globule density, and (3) the value of H below the cream layer gradually decreases during creaming because the largest globules reach the cream layer first. Table 9.3 also gives the estimated creaming rate if the latter factors are taken into account. Naturally, the extent of creaming also greatly depends on temperature (cf. Figure 8.2B) and on possible disturbances caused by stirring or convection currents.

It goes without saying that the creaming will be much faster when the fat globules are aggregated. The following are causes of aggregation:

1. *Cold agglutination:* Fat globules in most homogenized products cannot flocculate in the cold, because the cold agglutinin can be inactivated

TABLE 9.3
Initial Creaming Rate in Some Milk Products[a]

Product	$(\rho_p - \rho_f)/\eta_p$ (ks.m^{-2})	H (μm^2)	h (cm)	q 'Stokes' (% per day)	q Corrected[b] (% per day)
Pasteurized milk[c]	50	10–50	20	12–60	8–40
Homogenized milk	50	0.8–1.5	20	0.8–1.7	0.5–1.1
UHT milk	40	~0.4	20	~0.4	~0.2
Evaporated milk	10	~0.4	5	~0.4	~0.1
Sweetened condensed milk	0.2–0.07	4[d]–50	5	0.1–0.3	0.04–0.12

[a] Approximate values of the percentage of the fat reaching the cream layer per day (q). Creaming of unclustered globules at 20°C in a bottle or tin of height h.
[b] For the influence of fat content and protein layers.
[c] Unhomogenized.
[d] Slightly homogenized.

by heat treatment (Figure 7.9D) and by homogenization (Section 9.9). See Section 9.10 for partial homogenization.

2. *Homogenization clusters:* These are rarely formed except during homogenizing of cream.
3. *Heating at high temperature (sterilization):* This can cause small clusters of homogenized fat globules to be formed, especially in evaporated milk. This is an initial heat coagulation.

9.9 OTHER EFFECTS OF HOMOGENIZATION

Homogenizing milk that contains lipase strongly enhances lipolysis (Subsection 3.2.5). Raw milk turns rancid within a few minutes after homogenization (see Figure 9.10A). This can be explained by the capability of lipoprotein lipase to penetrate the membrane formed by homogenization but not the natural membrane. Accordingly, raw milk homogenization should be avoided, or the milk should be pasteurized immediately after homogenization in such a way that the lipase is inactivated. Homogenization is often done before pasteurization because in the homogenizer, the milk may readily be contaminated by bacteria. Furthermore, mixing of homogenized milk with raw milk should be prevented, again to avoid lipolysis (see Figure 9.10B).

Homogenization of milk has several other effects:

1. The color becomes whiter (Section 4.6).
2. The tendency to foam increases somewhat.

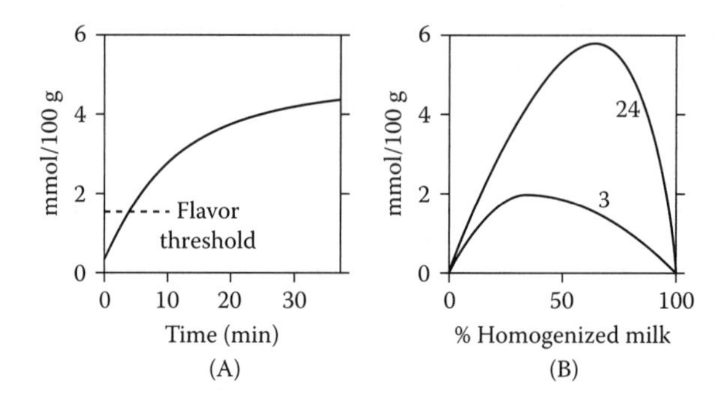

FIGURE 9.10 Influence of homogenization on lipolysis, expressed as acidity of the fat (in mmol per 100 g of fat). (A) Raw milk homogenized at 37°C; fat acidity as a function of time after homogenization. (B) Mixtures of raw milk and homogenized pasteurized milk; fat acidity after 3 h and 24 h at 15°C. (Adapted from A. Jellema, unpublished.)

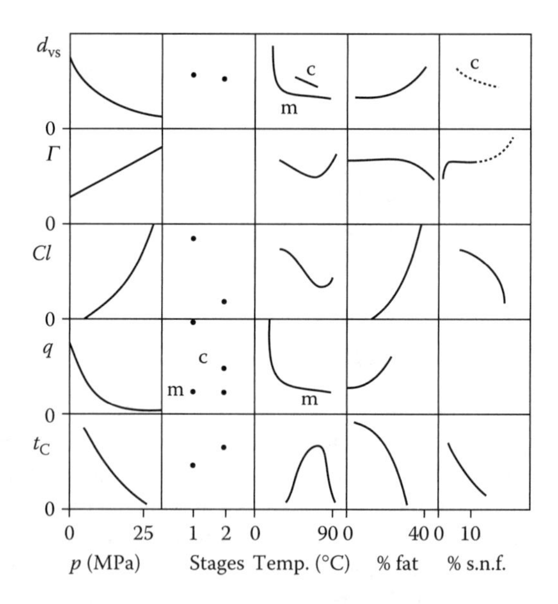

FIGURE 9.11 Effects of conditions during homogenization (p = homogenization pressure) on product properties (m: only for milk, c: cream). d_{vs} = volume/surface average diameter of the fat globules; Γ = protein surface load; Cl = homogenization clusters in cream; q = creaming rate; t_c = time needed for heat coagulation of cream. Highly schematic. Broken lines are less certain. Homogenization at, for instance, 70°C after preheating at high temperature, often leads to a result similar to that of direct homogenization at 90°C.

3. The proneness to fat autoxidation, and hence to the formation of ensuing off-flavors, is reduced (Subsection 2.3.4).
4. The fat globules lose their ability to be agglutinated upon cooling (Subsection 3.2.4.2). This is caused by inactivation of the cold agglutinin rather than by changing the fat globules; homogenization at very low pressure (1 MPa) suffices. Agglutinins (IgM) for bacteria (e.g., *Lactococcus lactis* spp.) can be inactivated also; to achieve this, a higher pressure is required, such as 10 MPa.

Figure 9.11 summarizes the effects of process and product variables on some properties of homogenized products, and Table 9.4 compares properties of the various types of fat globules.

9.10 OTHER WAYS OF WORKING

Partial homogenization is used to save on energy and machinery. (A large homogenizer is a very expensive machine and consumes much energy.) The milk is separated into skim milk and cream, and the cream is homogenized and mixed with the separated milk. The following two aspects should be considered here:

1. The agglutinin in the skim milk is not inactivated by the homogenization. Accordingly, when low pasteurization is applied, cold agglutination may occur to some extent, which enhances creaming.
2. If the fat content of the cream is too high (e.g., more than 10%), homogenization clusters are formed, resulting in rapid creaming. But Γ can be kept fairly small by making the homogenization temperature quite high, say 70°C (see Figure 9.8A). As a result, even in a cream with 14% or 15% fat, clustering is prevented when applying two-stage homogenization.

Several other types of machinery have been devised for homogenizing milk, with little success. Sometimes, the 'clarifixator' is applied. In this milk separator, the cream collides with obstacles protruding from the cream centripetal pump (paring disk) and is thereby subjected to intense turbulence, which causes a reduction of the fat globule size. Subsequently, the cream is returned to the milk stream. After the fat globules have been sufficiently reduced in size, they escape separation and are discharged in the separated-milk line, which thus actually delivers homogenized milk rather than skim milk. The resulting size distribution of the fat globules is very narrow, but the average size is not very small (see Figure 9.4).

TABLE 9.4
Comparison of the Properties of the Fat Globules in Nonhomogenized (N), Homogenized (H), and Recombined (R) Milk

Properties	N	H	R
Size distribution			
Average diameter d_{vs} (μm)	3–5	0.2–1	= H
Surface area (m^2/g fat)	1.5–2.2	7–34	= H
Fat crystallization			
Undercooling needed (K)	~20	25–33	≈ H
Surface layers			
Main component	'Lipoprotein'	Both	Plasma protein
Natural membrane (% of area)	~100	5–30	~0
Changes caused by processing	Considerable	< N	< H
Changes in dispersity			
Cold agglutination	Yes	<< N	= H
Aggregation at low pH	No	Yes	Yes
Heat coagulation	No	Yes	Yes
Coalescence	Possible	<< N	< H
Partial coalescence	Readily	< N	≈ H
Disruption of globules	Possible	<< N	≈ H
Off-flavor formation			
Lipolysis	Little	>> N	> H
Autoxidation	Prone	<< N	< H ?
Cooked flavor (H_2S, etc.)	Prone	< N	Hardly

Suggested Literature

Homogenization (comprehensively treated in Chapter 9 and Chapter 10): H. Mulder and P. Walstra, *The Milk Fat Globule,* Pudoc, Wageningen, 1974.

Fundamental aspects of emulsification (Chapter 2 of both the following books): P. Walstra, Formation of emulsions, in P. Becher, Ed., *Encyclopedia of Emulsion Technology,* Vol. 1, Dekker, New York, 1983; P. Walstra and P. Smulders, Emulsion formation, in B.P. Binks, Ed., *Modern Aspects of Emulsion Science,* Royal Soc. Chem., Cambridge, 1998.

Effects of milk homogenization: P. Walstra, *Neth. Milk Dairy J.,* **29**, 279–294, 1975.

10 Concentration Processes

10.1 GENERAL ASPECTS

Milk, skim milk, whey, and other milk products can be concentrated, i.e., part of the water can be removed. The main purpose is to diminish the volume and to enhance the keeping quality.

Water can be removed from milk by evaporation; in addition to water, volatile substances, especially dissolved gases, are removed as well. Evaporation is usually done under reduced pressure — hence, decreased temperature — to prevent damage caused by heating. Water can also be removed by reverse osmosis, i.e., high pressure is applied to a solution to pass its water through a suitable membrane (see Section 12.3). Water as well as part (1% to 20%, depending on conditions) of some low-molar-mass substances pass through the membrane. A different method of concentrating is by freezing (see Section 11.2); the more ice crystals are formed, the higher the dry-matter content in the remaining liquid. Removal of water to such a low level that the product becomes solid-like is called drying. Drying is achieved by vaporization of water, usually from concentrated milk.

10.1.1 Concentration of Solutes

The degree of concentrating can be defined as the concentration factor Q, i.e., the ratio of dry-matter content of the concentrated product (D) to that in the original material (D_1). Consequently, the mass of the concentrated product is $1/Q$ times the mass before concentrating. The concentration of solutes in water increases more than proportionally with Q. Taking the concentration of a substance relative to the amount of water (e.g., in grams per 100 g of water), its increase is given by

$$Q^* = Q(1 - D_1)/(1 - QD_1) = D(1 - D_1)/D_1(1 - D) \qquad (10.1)$$

where D and D_1 are expressed as mass fraction. Figure 10.1 illustrates the approximate relationships between Q, Q^*, and D for milk, skim milk, and whey. Q^* becomes very large when D approaches 100%.

During concentrating, some substances can become supersaturated and may precipitate after crystallization. Milk is already saturated with respect to calcium phosphate. As a result of concentrating, the amount of phosphate associated with

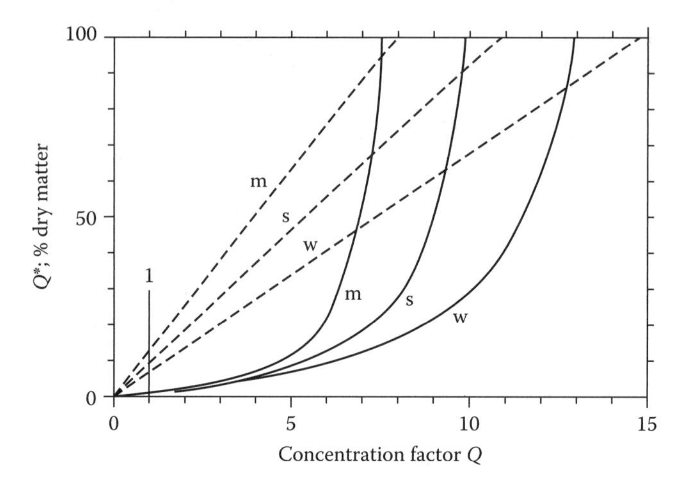

FIGURE 10.1 Percentage of dry matter (---) and relative increase in concentration of substances relative to water (Q^*) (—), as a function of the concentration factor Q for milk (m), skim milk (s), and whey (w) of average composition.

the casein micelles increases. Lactose becomes saturated in milk at room temperature when $Q \approx 2.8$, but it easily becomes supersaturated. Probably, lactose will not crystallize at all when milk is concentrated rapidly (as in a spray drier) to a low water content.

Also for substances that remain fully in solution, the thermodynamic activity will not always be proportional to Q^*. For most nonionic species, the activity coefficient increases with decreasing water content, which implies that their solubility decreases. For ions, the increasing ionic strength causes a decrease of the activity coefficient. Because of this, the ionization and solubility of salts increase (see Subsection 2.2.2).

One of the causes for the activity of a solute to become higher than corresponds to Q^* is that part of the water is not available as a solvent, which may be designated *nonsolvent water.* A small part of the water is so strongly bound (as water of crystallization or water in the interior of a globular protein molecule) that it cannot be available as a solvent. But negative adsorption of solutes at a surface present (e.g., that of proteins) may be more important. The specific surface area of milk proteins is large, e.g., 10^3 $m^2 \cdot g^{-1}$ for casein (submicelles) in milk, so that the effect may be considerable. Several solutes, especially sugars, exhibit negative adsorption. Figure 10.2 shows that the phenomenon may be interpreted as being caused by steric exclusion, i.e., solute molecules larger than water molecules must stay further away from the surface of protein particles (envisioning the location of solutes to be in their center of gravity), thus leaving a layer of water that is devoid of solute. This explains why the amount of nonsolvent water generally increases with the molecular size of the solute, as illustrated in Figure 10.2.

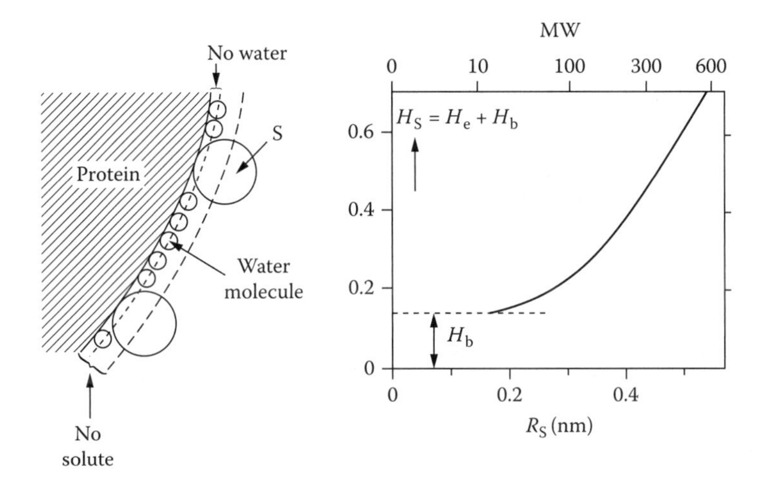

FIGURE 10.2 Steric exclusion of solute molecules (S) at a surface, e.g., of a protein particle. Left: diagrammatic explanation. Right: water not available as a solvent for S (H_s, in g per gram of protein), consisting of excluded (H_e) and bound water (H_b) as a function of radius R_s of the solute molecule. The scale of the molar mass of the solute is approximate. (Adapted from P. Walstra and R. Jenness, *Dairy Chemistry and Physics,* Wiley, New York, 1984.)

Knowledge of the amount of nonsolvent water for a particular solute allows conversion of the concentrations in milk into concentrations in plasma, serum, and other products. Formulas are given in Table 10.1. Usually, the effect is fairly small for small molecules, but it may be considerable for larger ones. For example, a content of 46 g lactose per kg of milk corresponds to 54.1 g lactose per kg of available water. An amount of 6 g serum proteins per kg milk yields 7.0 g proteins per kg serum, but taking $h = 0$ would lead to 6.45 g serum proteins per kg serum; hence, an error of about 8%.

The above holds only if there is no true (positive) adsorption onto fat globules or casein micelles. Some proteins (lipase, plasmin) do adsorb onto the latter. Some immunoglobulins can adsorb onto fat globules (see Subsection 3.2.4.2). For various serum proteins, h ranges from 0 to 3.4 (the latter figure refers to β-lactoglobulin).

Nonsolvent water should thus not be considered as being bound. Only a small part of it will actually be bound to certain groups, especially charged groups and dipoles; this concerns, for proteins, about 0.1 to 0.2 g water per gram of protein. Neither should bound water be confused with held water or imbibition water, which is mechanically entrapped. For instance, casein micelles may contain as much as 3 g of held water per gram of casein (see Subsection 3.3.1), but this water diffuses freely in and out of the micelles. If milk is renneted and forms a gel built of paracasein (see Section 24.3), all of the water is entrapped (held) in the gel, i.e., about 40 g of water per gram of paracasein, whereas nonsolvent water for lactose still is about 0.55 g per gram.

TABLE 10.1
Conversion of Concentrations of Solutes

If ρ_m = density of milk, g/ml.
ρ_p = density of plasma, g/ml.
ρ_s = density of serum, g/ml.
ρ_w = density of water, g/ml.
f = fat content as mass fraction, g/g.
d = dry-matter content as mass fraction, g/g.
p = protein content as mass fraction, g/g.
c = casein content as mass fraction, g/g.
x = content of solute in milk or concentrated milk, in units (g, mol, ml) per kg.

The content then is

In milk	x	kg^{-1}	$x\rho_m$	l^{-1}
In plasma	$x/(1 - 1.02f)$	kg^{-1}	$x\rho_p/(1 - 1.02f)$	l^{-1}
In serum	$x/(1 - 1.02f - 1.08c - hc)$	kg^{-1}	$x\rho_s/(1 - 1.02f - 1.08c - hc)$	l^{-1}
Relative to water	$x/(1 - d - hp)$	kg^{-1}	$x\rho_w/(1 - d - hp)$	l^{-1}

Where
1.02 = factor to convert fat content to content of fat globules.
1.08 = factor to convert casein content to content of dry casein micelles (factor actually ranging from 1.06 to 1.10).
h = factor (in g water/g protein) referring to the amount of nonsolvent water; h varies with molar mass of solute. For small molecules (e.g., CO_2), $h \approx 0.15$; for lactose, $h \approx 0.55$; for serum proteins, average $h \approx 2.6$.

10.1.2 Water Activity

If the water content of a product decreases, its water activity (a_w) also decreases. Water activity is expressed as a fraction. In pure water $a_w = 1$; in a system without water, $a_w = 0$. For ideal solutions, $a_w = m_w$, where m_w is the mole fraction of water in the solution, i.e.:

$$m_w = \frac{\text{moles of water}}{\text{moles of water + moles of solutes}} \tag{10.2}$$

The following explain why, in most cases, $a_w < m_w$:

1. Dissociation of compounds, e.g., salts, causes a higher molar concentration of solute species.
2. The effective concentration of solutes is considerably increased if their occupied volume is large. This is an essential factor for substances that have a high molar mass as compared to water.
3. Part of the water may not be available as a solvent (see Subsection 10.1.1).

Altogether, it means that in milk products the relation deviates from $a_w = m_w$, especially when the milk is highly concentrated.

The water activity of a product can be measured because a_w equals the relative humidity of air in equilibrium with the product. Accordingly, it can be determined by establishing the relative humidity at which the product neither absorbs nor releases water. The following are some examples of a_w values of milk products.

Milk	0.995
Evaporated milk	0.986
Ice cream mix	0.97
Sweetened condensed milk	0.83
Skim milk powder, 4.5% water	0.2
Skim milk powder, 3% water	0.1
Skim milk powder, 1.5% water	0.02
Cheese	0.94–0.98

It should further be noted that a_w does not depend on fat content. Milk and the cream and skim milk obtained from it by centrifugal separation all have precisely the same a_w value.

Any product has its characteristic relations between a_w and equilibrium water content for every temperature. Such a relation is called a water vapor pressure isotherm or *sorption isotherm*. An example is given in Figure 10.3a. At increasing temperature and constant water content, a_w increases (see Figure 10.19). The a_w of most liquid milk products appears to be fairly high.

Often, sorption isotherms show hysteresis, i.e., it makes a difference whether the curves are obtained by successively decreasing (desorption) or increasing a_w (resorption). This is because equilibrium is not reached, certainly not at very low water contents. It is very difficult to remove the last water in a product, except

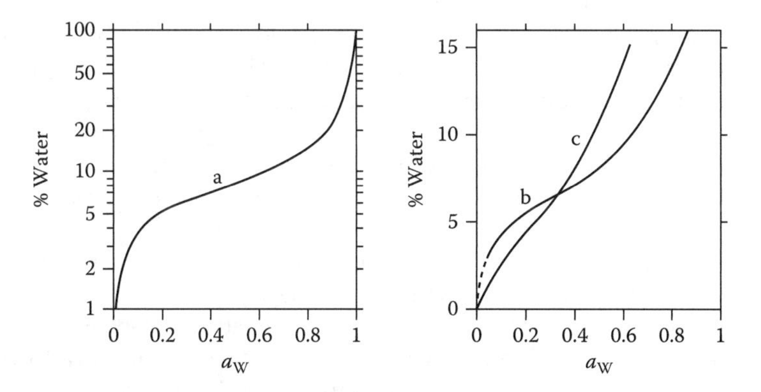

FIGURE 10.3 Water sorption isotherms of concentrated skim milk. (a) Approximate isotherm. (b) Desorption isotherm of skim milk powder with lactose in the crystalline state. (c) Resorption isotherm of skim milk powder with lactose in the amorphous state.

at high temperature, where other changes (reactions) will occur as well. This implies that, essentially, the water activity involved is unknown because a_w refers to a condition of equilibrium.

In milk, there is a complication caused by lactose. If milk is dried slowly, α-lactose hydrate will crystallize. The water of crystallization present is very difficult to remove and, hence, it does not contribute to the a_w value estimated. This implies that the point 0:0 of the sorption isotherm is not covered (Figure 10.3b). However, spray drying of milk often yields a powder with amorphous lactose. At similar water contents (i.e., water inclusive of water of crystallization), the latter powder results in a higher a_w (Figure 10.3c). This is because in the spray-dried powder almost all of the water is available as a solvent. If the powder takes up water, its a_w increases less strongly, and the isotherms b and c (Figure 10.3) intersect. Eventually, α-lactose hydrate crystallizes. Because of this, at constant a_w (= air humidity), powder c will then lose water. Incidentally, in most of the methods applied for determining water content in dried milk and dried whey, the greater part of the water of crystallization is included in the 'dry matter.'

10.1.3 Changes Caused by Concentrating

Apart from the increase of most of the solute concentrations, removal of water from milk causes numerous changes in properties, which often are approximately proportional to Q^*. The changes also depend on other conditions, such as heat treatment and homogenization before concentrating. Some important changes in properties are as follows:

1. The *water activity* decreases: Examples are given in Figure 10.3.
2. The *hygroscopicity* increases: Usually, a (dry) product is called hygroscopic if a small increase of a_w causes a considerable increase in water content. Obviously, this mainly concerns milk powder with a very low water content (Figure 10.3).
3. The *salt equilibria* change: The Ca^{2+} activity is increased only slightly because calcium phosphate, which is saturated in milk, turns into an undissolved state (see Subsection 2.2.5). Because of the latter change, the pH decreases by about 0.3 and 0.5 unit for $Q = 2$ and $Q = 3$, respectively. For $Q = 2.5$, the fraction of Ca that is in solution has decreased from about 0.4 to 0.3. As the water content decreases, association of ionic species increases and also ionic groups of proteins are neutralized.
4. The *conformation of proteins* changes: This occurs because ionic strength (hence, thickness of the electrical double layer), pH, and other salt equilibria all change. If milk is highly concentrated, the solvent quality decreases also. All in all, the tendency of the protein molecules to associate and to attain a compact conformation is increased. Coalescence of casein micelles causes them to increase in size. The increase is smaller if the milk has been intensely preheated, presumably because β-lactoglobulin and other serum proteins have become associated with casein.

5. Several *physicochemical properties* change: Osmotic pressure, freezing-point depression, boiling point elevation, electrical conductivity, density, and refractive index all increase, and heat conductivity decreases.
6. *Rheological properties* are affected: The viscosity increases (Figure 4.7) and the liquid becomes non-Newtonian (viscoelastic and shear rate thinning) and finally solid-like (say, at $Q > 9$ for skim milk). This is all highly dependent on temperature.
7. *Diffusion coefficients* decrease: At low water content the effect is very strong. The diffusion coefficient of water decreases from approximately 10^{-9} m$^2 \cdot$s^{-1} in milk to 10^{-16} m$^2 \cdot$s^{-1} in skim milk powder with a small percentage of water (see Figure 10.13.)

10.1.4 The Glassy State

Any substance that can crystallize can, in principle, also attain a glassy state. That means that the substance appears solid and brittle but is amorphous and not crystalline. The mass density is slightly lower than in the crystalline form. In a sense, the material is a liquid, although an extremely viscous one, the viscosity being over 10^{12} Pa·s (i.e., 10^{15} times that of water). Upon increasing the temperature of a glassy material, it will show a transition to a viscous state at what is called the *glass transition temperature*, T_g. Above T_g, the viscosity very strongly decreases with temperature, as illustrated in Figure 10.4a. The value of T_g is not invariable, as it somewhat depends on the compositional and temperature history of the sample. The glass transition temperature is always much below the melting temperature of the material.

To transform a pure liquid substance into a glassy state, it has to be cooled extremely fast. For instance, water has $T_g \approx -137$°C, and it must be cooled at a

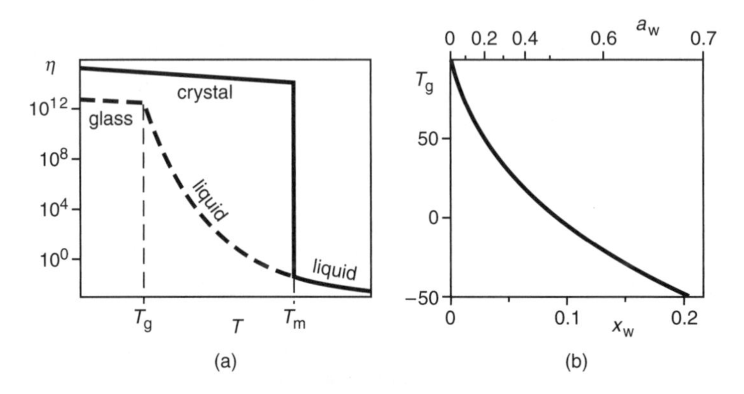

FIGURE 10.4 Glass transitions. (a) Approximate relation between the viscosity η (Pa·s) of a material and temperature T, under equilibrium conditions (solid line), and when a glass is formed (broken line). T_g means glass transition and T_m melting point. (b) Relation between the glass transition temperature (°C) and the mass fraction of water x_w of lactose-water mixtures; the water activity (a_w) at T_g is also given.

rate of at least 10^5 K·s^{-1} to obtain a glass. For a substance — say, sugar, in solution — a glassy state can also be obtained by removing water very fast; the liquid then becomes so viscous that crystallization will be too slow to occur before a glass is obtained. Such a high viscosity is reached more readily if the solution contains also other solutes in substantial quantities, especially polymeric substances. If concentrated milk or whey is spray-dried, it can attain a glassy state.

Figure 10.4b gives values of T_g for lactose, as a function of water content and water activity. Pure lactose has T_g = 101°C, and T_m (of the α-hydrate) = 214°C. It is seen that also a lactose-water mixture can form a glass, but that T_g decreases with increasing water content; water thus acts as a 'plasticizer.' Lactose dominates the glass transition of most concentrated liquid milk products. The latter have a glass transition temperature that is nearly the same as that of a lactose-water mixture of the same water activity.

A glassy material is hard and brittle. It is also quite stable to physical and chemical changes, provided that the temperature remains below T_g. It has even been assumed that diffusion of molecules in a glass is effectively zero, since a viscosity of 10^{12} Pa·s corresponds for most molecules to a diffusion coefficient D of less than 10^{-24} m^2·s^{-1}. Much faster diffusion has been observed, however. An extremely small value of D is indeed obtained for the molecules responsible for the glassy state, i.e., lactose in most milk products, but smaller molecules such as water and oxygen can still diffuse at a perceptible rate. Hence, some chemical reactions still can occur, although at a slow rate.

At temperatures above the T_g of the material, crystallization can occur. For lactose, this is seen to be reached at 30°C if it contains 10% water by mass. This corresponds to a water activity of about 0.30, which corresponds, in turn, to a water content of 5% by mass in skim milk powder (Figure 10.3, curve c). This agrees with the onset of crystallization of α-lactose hydrate in dried milks as observed in practice. Other phenomena that can occur at a few degrees above T_g are the material (say, powder particles) becoming sticky and readily deformable. Most important may be that several chemical reactions, like Maillard browning, will occur much faster, as discussed in the following section.

See Section 11.2 for the special glass transition in freeze concentrated mixtures.

10.1.5 Reaction Rates

The changes in properties caused by increasing Q^* can thus affect the rates of chemical reactions and of physical changes to a considerable extent. In most cases, a reduction in rate occurs (Figure 10.5), and often it is specifically for that reason that concentrated products, especially dried ones, are made.

Usually, the main cause of the slower reaction rate after considerable concentration is the *decreased diffusion coefficient*. Clearly, reacting molecules must collide before they can react, and the probability of collision is proportional to the diffusion coefficient; if diffusivity is sufficiently small, it becomes rate limiting for bimolecular reactions. The slower rate of physical processes such as crystallization is also ascribed to a decreased diffusivity. The larger the molecules,

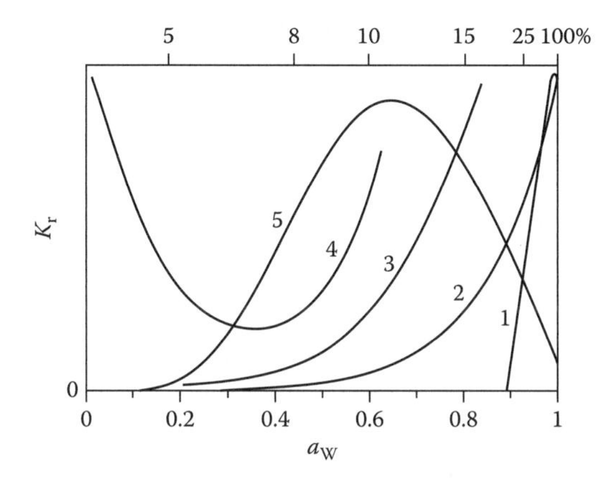

FIGURE 10.5 Relative reaction rate (K_r) of various reactions plotted against the water activity (a_w) of (concentrated) skim milk (powder). The upper abscissa scale gives the water content (% w/w). (1) Growth of *Staphylococcus aureus*; (2) oxidative degradation of ascorbic acid; (3) enzyme action (e.g., lipase); (4) lipid autoxidation; (5) Maillard reaction (nonezymatic browning). Only meant to illustrate trends. (Adapted from P. Walstra and R. Jenness, *Dairy Chemistry and Physics,* Wiley, New York, 1984.)

the stronger is the decrease of the diffusion coefficient. In milk powders with a low water content the effective diffusion coefficients may be very small because the material is in a glassy state.

It is fairly customary to plot reaction rates against water activity. This practice suggests a_w to be the rate-determining factor but often that is not true. If water is a reactant (e.g., as in hydrolysis), the reaction rate depends also on the water activity itself. However, the reaction rate then would decrease by a factor of 2 when going from a_w of 1 to a_w of 0.5, whereas the diffusion coefficient may decrease by a factor of 10^3.

Because of an increase in the concentration of reactants, the rate of bimolecular reactions at first increases due to removal of water. On further increase of Q^*, the reaction rate often decreases again; this decrease would be caused by reduced diffusivity. A good example is the Maillard reaction (Figure 10.5, curve 5). The irreversible loss of solubility of milk protein in milk powders, and the rate of gelation of concentrated milks show a trend similar to curve 5. On removal of water from milk, it thus is advisable to pass the level of approximately 10% water in the product as quickly as possible.

Autoxidation of lipids follows quite a different pattern (Figure 10.5, curve 4). The reaction rate is high for low a_w. Possible causes are that water lowers the lifetime of free radicals, slows down the decompostion of hydroperoxides, and lowers the catalytic activity of metal ions, such as Cu^{2+}.

Heat denaturation of globular proteins and, consequently, inactivation of enzymes and killing of microorganisms (see Section 7.3) greatly depend on water content. An example is given in Figure 10.6 (alkaline phosphatase). Both $\Delta H^{\ddagger}$

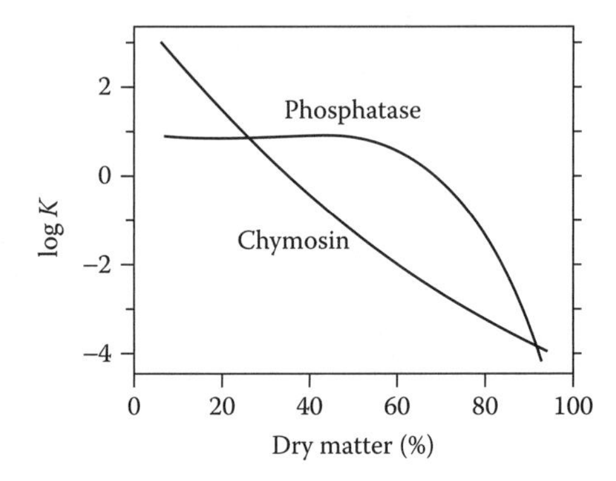

FIGURE 10.6 Reaction constants (K, in s^{-1}) for inactivation of alkaline phosphatase in concentrated skim milk and of chymosin in concentrated whey at 80°C, as a function of water content. Approximate examples. (Adapted from A.L.H. Daemen, *Neth. Milk Dairy J.*, **35**, 133, 1981.)

(activation enthalpy) and $\Delta S^{\ddagger}$ (activation entropy) for denaturation usually decrease with decreasing water content (cf. Equation 7.8). This implies that the dependence of denaturation or inactivation rate on temperature also decreases with decreasing water content. It may also occur that removal of water increases the concentration of a reactant or catalyst for heat inactivation; this is presumably the case for chymosin in whey (Figure 10.6), because at, say, 40% dry matter, a_w and diffusivity are not greatly lowered.

Growth of microorganisms can strongly depend on water content of a food. Several authors relate the possibilities of growth to the water activity, and the lowest values at which organisms can grow are given as, for instance,

Bacteria, 0.98–0.90,	halophilic bacteria down to ~0.75
Yeasts, about 0.9,	osmopholic yeasts down to ~0.6
Molds, 0.92–0.80,	xerophilic molds down to ~0.6

A simple explanation is that a low value of a_w implies a high value of the osmotic pressure Π, the relation being $\Pi = -135 \cdot 10^6 \ln a_w$ (in pascals) at room temperature. The organism is unable to tolerate high Π values as it will cause water to be drawn from the cell, damaging its metabolic system, generally because the internal concentration of harmful substances becomes high. However, microorganisms have acquired various mechanisms to neutralize the effect of a high Π value, and the effectivity of these vary greatly among organisms:

1. A fairly small difference in Π between the cell and the environment can often be tolerated. Bacterial spores can tolerate a considerable difference in Π.
2. Some solutes, e.g., most alcohols, can pass the cell membrane unhindered; hence, these do not cause an osmotic pressure difference. If such

a solute is, moreover, compatible, which means that it is not harmful to the cell at moderate concentration, the organism can survive and possibly grow.

3. The organisms may produce and accumulate low-molar-mass compatible substances that keep the internal Π value high; it often concerns specific amino acids.
4. Specific harmful substances, e.g., lactic acid, may be to some extent removed from the cell.

Consequently, it often makes much difference what solute is involved; in other words, the lowest value of a_w tolerated strongly depends on the composition of the medium. For instance, glycerol is generally tolerated to a much higher molar concentration, hence a lower a_w value, than ethanol. A halophilic bacterium can tolerate a high concentration of NaCl, but not of sugars. And an osmophilic yeast can tolerate a high concentration of sugar, but not of salts. Moreover, other factors can affect the lowest a_w tolerated, such as temperature and the concentration of growth inhibitors, such as acids, and of nutrients. These concentrations are, of course, increased upon water removal. For molds growing on the surface of a solid food (say, cheese) it may indeed be the relative humidity, i.e., a_w, that is the prime variable determining whether growth can occur.

10.2 EVAPORATING

Evaporation of products like milk, skim milk, and whey is applied:

1. To make such concentrated products as evaporated milk, sweetened condensed milk, and concentrated yogurt
2. As a process step in the manufacture of dry milk products, considering that the removal of water by evaporation requires far less energy than by drying (see Table 10.2)
3. To produce lactose (α-lactose hydrate) from whey or whey permeate (Subsection 12.2.1) by means of crystallization

Important aspects of the evaporation of milk and milk products are discussed in Section 7.2, and Section 10.1. Alternatives to evaporation are reverse osmosis (Section 12.3) and freeze concentration (Section 11.2).

Evaporation is always done under *reduced pressure*, primarily to allow boiling at a lower temperature and thus prevent damage due to heating. Figure 10.7 shows the vapor pressure as a function of temperature, i.e., the relation between pressure and boiling temperature of pure water. This relationship disregards elevation of the boiling point due to dissolved substances, which is, however, fairly small: for milk, 0.17 K; for evaporated skim milk, up to about 2 K; and for evaporated whey and sweetened condensed milk, up to slightly more than 3 K. Moreover, evaporation under vacuum facilitates evaporation in several stages; see the following text.

TABLE 10.2
Heat of Vaporization of Water and Examples of Energy Requirement in Some Processes to Remove Water

Heat of vaporization of water at 100°C	2255
Heat of vaporization of water at 40°C	2405
Sorption heat for removal of water from skim milk up to about 60% dry matter	~5
Evaporation, 3 stages	~800[a] (0.35 kg steam)
Evaporation, 6 stages, with thermal vapor recompression	~230[a] (0.1 kg steam)
Evaporation, 1 stage, with mechanical vapor recompression	~115
Roller drying	~2500[a] (1.1 kg steam)
Spray drying	~4500[a] (2.0 kg steam)
Reverse osmosis	20–35

Note: All data are in kJ or kg steam per kg of water removed.

[a] Excluding mechanical energy (pumps, etc.).

The construction of an evaporation unit, i.e., the part of the machinery in which the actual evaporation of water occurs, is variable. The oldest type is called a *circulation evaporator.* The liquid is in a vessel in which it boils under vacuum and circulates. To make it a continuous operation, a bundle of pipes is introduced. Outside the pipes is the heating medium; inside is the boiling liquid (e.g., milk), and the vapor bubbles formed drive the liquid upwards. A great disadvantage is that the height of the liquid results in a hydrostatic pressure, whereby the boiling point is elevated. For example, at a pressure in the vessel of 12 kPa, corresponding

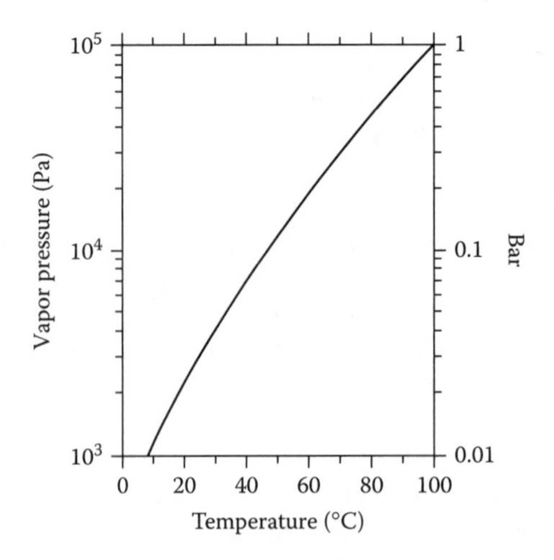

FIGURE 10.7 Vapor pressure of water as a function of temperature.

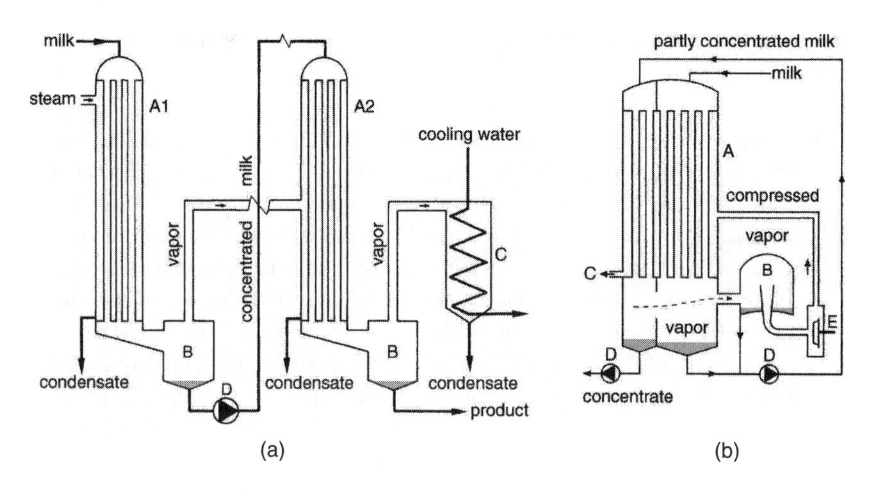

FIGURE 10.8 Diagrams showing the main operations in evaporating. A = evaporation unit; B = vapor separator; C = condenser; D = pump; E = mechanical vapor recompressor. In (a) the principle of a multiple-effect evaporator is given; the temperatures in A2 are lower than those in A1. In (b) a single-effect evaporator with mechanical vapor recompression is shown. (Based on diagrams provided by Carlisle Process Systems, Gorredijk.)

to a boiling point of about 50°C, a hydrostatic pressure corresponding to a 1-m water column, will elevate the boiling point to about 65°C. This is especially undesirable in multieffect evaporators; see below. Moreover, this type of evaporator shows heavy fouling. An advantage of the circulation evaporator is that it allows the evaporation of very viscous liquids.

Nowadays, nearly all evaporation units are of the *falling film* type (see Figure 10.8). The liquid flows by gravity along the inside surfaces of a number of pipes as a thin film; the heating medium is at the outside. The liquid passes a pipe only once and drops from the pipe in a concentrated form. A difficulty is that the pipes should be evenly and completely wetted by the liquid to avoid loss of heating surface and excessive fouling. Special devices (that, e.g., spray the milk over the top plate in the unit) ensure an equal distribution of liquid over the pipes. To prevent part of the heated surface to run dry, the volume flow rate in a pipe must be high. The problem is greatest in the lower part of a pipe, where evaporation has reduced the amount of liquid and increased its viscosity. To avoid problems, an evaporation unit can be divided into sections, as indicated in Figure 10.8b; this allows a higher film thickness for the more concentrated liquid by letting it flow through a smaller number of pipes. Falling film evaporators can also be constructed with stacks of plates, rather than bundles of pipes.

The heat needed to cause evaporation is in most cases provided by steam (generally at reduced pressure). In principle, nearly all of the heat can be regained by using the vapor produced by evaporation as the heating medium for evaporating more water. Part of the heat is lost by radiation to the environment, and the sorption heat involved (see Table 10.2) is also lost. Some systems used to regenerate heat follow:

Multiple effects: The principle is illustrated in Figure 10.8a. In the first evaporation unit, steam is introduced and part of the water is evaporated from the milk. The steam condenses, and vapor is separated from the concentrated milk. The vapor is used to evaporate water from the concentrated milk in the second effect, etc. The flows of milk and vapor are cocurrent. The number of effects varies from 3 to 7. The vapor coming from the last effect is condensed in a special condenser; the temperature of the water in the condenser determines the boiling temperature in the last effect. The boiling temperatures in the other effects are determined by the pressure drop of the vapor when being transported to the next effect. The temperature difference ΔT between the condensing vapor, and the boiling liquid will be smaller for a larger number of effects N. A higher N implies greater saving of steam but needs a larger heating surface; hence, a bigger and more expensive plant, more loss of heat by radiation, and higher cleaning costs. Especially in the last effect, the heat-transfer rate may become very small: here, the temperature is lowest, and the concentrate viscosity is very high.

Thermal vapor recompression (TVR): Some steam of atmospheric pressure can be injected (by means of a venturi) into the vapor originating from an evaporation unit, whereby the vapor is compressed and attains a higher temperature. TVR can be used in a single-effect as well as in a multiple-effect evaporator. In the latter case, a higher ΔT can be reached and the arrangement of the various effects is more flexible: the flows of vapor and milk can be in part countercurrent. Altogether, a smaller plant results and lower steam consumption (see Table 10.2).

Mechanical vapor recompression (MVR): This is illustrated in Figure 10.8b. It works with one effect. A special pump is used to recompress the vapor resulting from evaporation, whereby its temperature is raised, and it is recirculated as the heating medium. The evaporator needs only one effect, although the milk flow is generally divided into sections, as depicted in the figure. The number of sections generally is 3 to 5. A condenser is not needed. The total energy consumption is very low (see Table 10.2). Another advantage is that the evaporating liquid can be kept at a constant temperature, generally 50 to 55°C for milk and skim milk, and about 60°C for whey. ΔT is generally 4 to 5°C. The average residence time of the liquid in the evaporation unit is about 10 min. The highest dry-matter content to be reached is generally 40 to 45%.

Evaporation plants include various other apparatus: preheaters, coolers, heat regenerators, a vacuum pump to remove dissolved air, a steam inlet needed to start the operation, pumps for cleaning liquids, etc. Various configurations are used, depending on the kind of liquid that is concentrated and the degree of

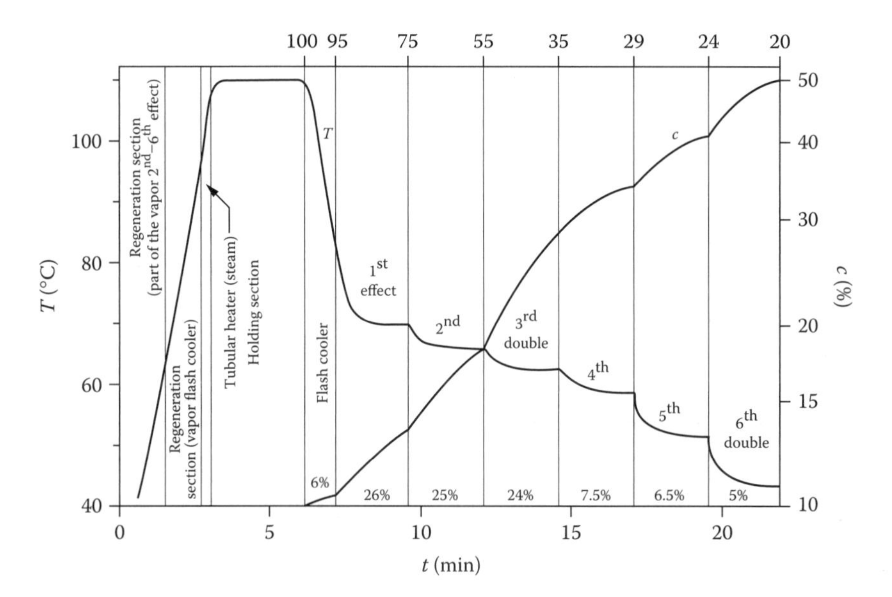

FIGURE 10.9 Example of the course of temperature (T) and dry-matter content (c) of skim milk as a function of time (t) in a six-effect evaporator with preheating. The scale for the dry-matter content is logarithmic. The milk is preheated with heat exchangers that successively use exhaust vapor of the evaporators, vapor of the cooler, and live steam. Furthermore, part of the vapor of the third effect is compressed with steam (TVR) and led to the first effect. For the rest, vapor and concentrate are cocurrent. A part of the water is evaporated in a vacuum cooler by flash evaporation. The percentages mentioned indicate the proportion of the water evaporated in the effect concerned (89% of all water being eventually removed). The numbers on top represent the residual liquid (in mass percent) after the various effects. In the third and the sixth effect, the evaporation unit is divided into two sections (as in Figure 10.8b). (From data provided by Carlisle Process Systems, Gorredijk.)

concentration desired. Figure 10.9 gives an example of the evolution of product temperature and dry-matter content in a multiple-effect evaporator. Currently, TVR is generally included. Most of the new plants built are based on MVR; it often concerns a large MVR unit, followed by a small single-effect TVR unit, to reach and fine-tune the final dry-matter content.

When manufacturing powder, the amount of water removed in the evaporator should be as large as possible. The limit is generally set by the high viscosity of the concentrate. A high viscosity retards the flow rate near the heating surface and, thereby, the heat transfer. In a falling film evaporator in which the flow rate is fairly high, the viscosity should not exceed about 0.1 Pa·s. At a low temperature, highly concentrated milk and skim milk would exceed this limit. Therefore, partial countercurrent flow is sometimes applied in multiple-effect evaporators. The concentrate passes through the last effects in the reverse order, the highest concentrated milk being evaporated not at the lowest temperature

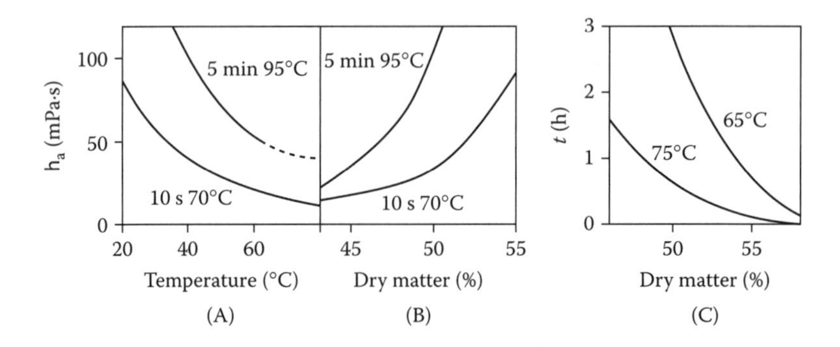

FIGURE 10.10 Apparent viscosity (η_a) of skim milk concentrate of various dry-matter content. (A) Effect of temperature, parameter is preheating, 48% dry matter. (B) Effect of the concentration, same parameter, measuring temperature 50°C. (C) Time needed (t) to cause gelation of the concentrate at two temperatures as a function of concentration. Approximate results. (Adapted from T.H.M. Snoeren et al., *Neth. Milk Dairy J.*, **38**, 43–53, 1984.)

but at a somewhat higher one, which leads to a lower viscosity. Obviously, this is only feasible if vapor recompression is applied. When making sweetened condensed milk, a viscosity of over 0.1 Pa · s is eventually reached. In the last effect a fairly large conventional circulation evaporator is then often applied, in which the milk is partially recirculated to prevent the heating surface from partly running dry.

The viscosity of the concentrate, therefore, is an important parameter in the evaporating process (as it is in the spray-drying process, where it affects the droplet size in the spray; see Subsection 10.4.2). Several factors affect the viscosity; (see Section 4.7). For the same dry-matter content, the viscosity decreases in the order skim milk > milk > whey > whey permeate. Results for skim milk are in Figure 10.10. The viscosity increases more than proportionally with the dry-matter content. The relatively strong increase upon increasing dry-matter content is explained by the particle volume fraction in the liquid already being very high (see Equation 4.7, for the case that φ approximates φ_{max}). The concentrated milk is markedly shear rate thinning, and its viscosity is an apparent one, η_a. At a shear rate of, say, 100 s^{-1}, η_a is about twice the value at 2000 s^{-1}. Preheating of the milk increases η_a considerably if the dry-matter content is high. This may be explained by the serum proteins greatly increasing in voluminosity due to denaturation. The influence of the temperature on η_a is hard to estimate because η_a rapidly increases with time at high temperature and high dry-matter content, a process called *age thickening*. Eventually this leads to gelation (see Figure 10.10C).

The degree of concentration is usually checked by means of the density ρ or the refractive index n. These parameters can be determined continuously in the concentrate flow. This enables automatic control of the evaporating process by adjusting the steam or the milk supply. This is far from easy, given the prolonged holdup time and the great number of process steps.

Product properties to be considered include the following:

1. Age thickening at high concentration and at high temperature.
2. Highly evaporated milk is susceptible to Maillard reactions (Figure 10.5).
3. Fouling occurs readily if the product is highly concentrated, the temperature is high, the temperature difference across the wall is high, and the flow rate of the liquid is slow. Preheating may significantly diminish fouling at high temperature (see Section 14.1). The construction of the equipment greatly affects the rate of fouling and the ease of cleaning. Cleaning expenses increase with the heating area of the equipment, and therefore with the number of effects.
4. Some bacteria can grow at fairly high temperatures, which mostly means in the last effects. Thermophilic bacteria are involved such as *Streptococcus thermophilus* and *Bacillus stearothermophilus*; the latter may even survive sterilization. This implies that processing must be done hygienically and that the plant must be cleaned and disinfected after not more than 20 h of continuous operation. The spread in holdup time is also of importance (see also Section 20.3).
5. Foaming mainly occurs with skim milk, at a fairly low temperature. The machinery should be adjusted to overcome this. A falling film evaporator presents few problems.
6. Disruption of fat globules especially occurs in falling film evaporators. For instance, d_{vs} may decrease from 3.8 to 2.4 μm by evaporating milk up to 50% dry matter. Usually, this is not a problem because the milk is homogenized, anyway.

It should be realized that different products allow different degrees of evaporation and that the same concentration factor affects the concentration of dissolved constituents differently. Table 10.3 gives some examples: Lactose will not

TABLE 10.3
Approximate Composition of Liquids Evaporated up to the Maximum Degree Possible

Liquid	% Dry Matter	Q[a]	Q^*	Saturation of Lactose[b]
Milk	50	4	7	1.05
Skim milk	55	6	12	1.85
Sweet whey	64	9.5	25	3.77

[a] Q = concentration factor; Q^* = concentration relative to the water content (see Section 10.1).

[b] At 40°C and on the assumption that evaporation does not alter the activity coefficient of lactose. In fact, the coefficient increases substantially, thereby increasing the actual supersaturation (especially in sweet whey).

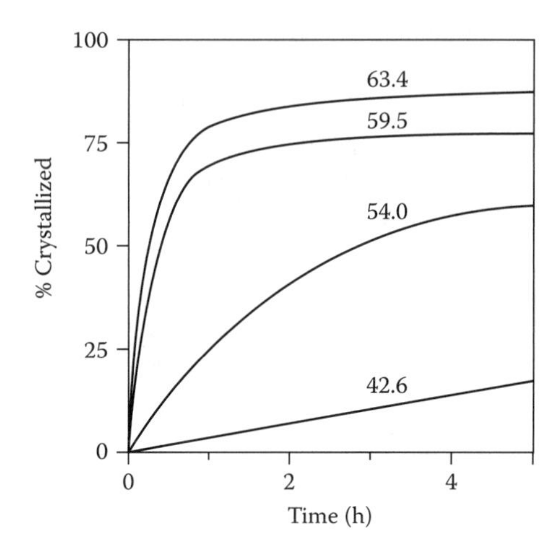

FIGURE 10.11 Crystallization of lactose in concentrated whey (parameter is % dry matter) as a function of the time after cooling to 20°C. In the range of 15 to 40°C, the crystallization rate depends little on the temperature. (Adapted from results by K. Roetman, Ph.D. thesis, Wageningen Agricultural University, 1982.)

crystallize in highly concentrated milk, whereas it may do so readily in highly concentrated skim milk. Concentrated whey with its high Q^* may show considerable fouling of the evaporator equipment due to supersaturated salts precipitating on the heating surface. This drawback can largely be overcome by keeping the partly evaporated whey outside the equipment for some time (say, 2 h) before it is further concentrated. The salts then are allowed to crystallize in the bulk and lactose crystallizes at the same time.

Highly concentrated whey shows substantial crystallization of lactose, as is needed in the manufacture of α-lactose hydrate (see Figure 10.11). Crystallization starts already in the evaporator, but it need not interfere with the evaporation process.

10.3 DRYING: GENERAL ASPECTS

10.3.1 OBJECTIVES

Drying is usually applied to make a durable product that is easy to handle and, after reconstitution with water, is very similar in properties to the original material. The resulting powders are generally in the glassy state (Subsection 10.1.4). Drying is applied to products like milk, skim milk, whey, infant formulas, cream, ice cream mix, and protein concentrates, all of which have a high water content. Removal of water is expensive, especially with respect to energy (see Table 10.2). Furthermore, driers are expensive. Therefore, the material is often concentrated to a fairly low water content by evaporation (Section 10.2) or by reverse osmosis (Section 12.3) before drying.

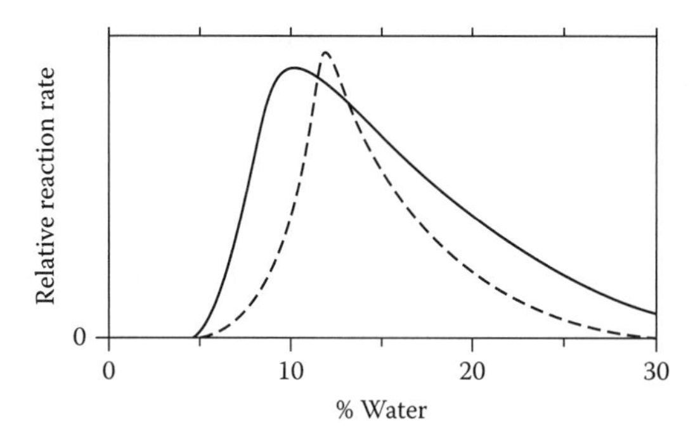

FIGURE 10.12 Rate of Maillard reactions (—) and of protein becoming insoluble (---) in concentrated skim milk at high temperature (say, 80°C) as a function of water content. The curve for insoluble protein depends on several conditions, such as preheating. Approximate examples.

The main technological problem is to prevent the drying product from undergoing undesirable changes. The rate of many reactions greatly depends on the water content; an example is given in Figure 10.12 (see also Section 10.1). It mainly concerns reactions that render protein insoluble since these strongly depend on temperature: $Q_{10} \approx 4.5$ (see also Figure 10.10C). At 80°C, about half of the protein present in concentrated skim milk with 13% water becomes insoluble in 10 s. Thus it is advantageous to pass the interval from, say, 20% to 8% water rapidly and at a moderate temperature. However, the effective diffusion coefficient of water and, consequently, the drying rate significantly decrease with decreasing water content and with decreasing temperature (Figure 10.13). The

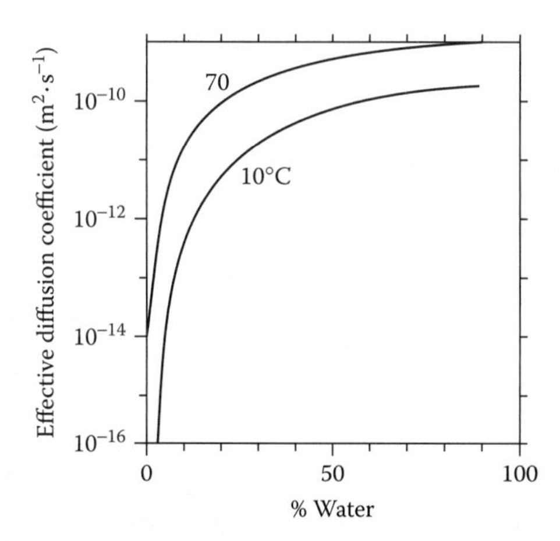

FIGURE 10.13 Effective diffusion coefficient of water in drying skim milk as a function of water content, at two temperatures. (From results by P.J.A.M. Kerkhof, Ph.D. thesis, Eindhoven Technical University, 1975.)

following calculation may be illustrative. Consider the drying of concentrated skim milk in a thin layer of thickness $x = 1$ mm on a solid support. Figure 10.13 shows that the diffusion coefficient (D) of water at 70°C is on average about $4 \cdot 10^{-11}$ $m^2 \cdot s^{-1}$ when the water content is to be halved, from 20% to 10%. According to the relation $x^2 = Dt_{0.5}$, the time required for this would be 25 ks, i.e., about 7 h. The liquid therefore will have to be atomized very finely if its drying is to be fast. Alternatively, the drying can be carried out at low temperature but that usually takes a very long time.

10.3.2 Drying Methods

There are several methods for drying liquids. The dairy manufacturer uses only a few of them.

10.3.2.1 Drum (Roller) Drying

A thin film — on the order of 0.1 mm — of milk, skim milk, etc., is dried on a large rotary metal cylinder or drum that is steam-heated internally. Often, two drums are set up side by side at a very small distance apart. The water evaporates within a few seconds, which is possible due to the high drying temperature (>100°C). The dried film is scraped off from the drum by means of a steel knife, collected, and ground. Considerable product damage due to heating occurs, mainly because scraping off is always imperfect and, accordingly, a part of the milk is repeatedly wetted and dried. The quality of the powder can be improved by using a vacuum roller drier, in which the milk is dried at a lower temperature, but this method is expensive. Nowadays the roller drying process is little used.

10.3.2.2 Foam Drying

Under pressure, air or nitrogen is injected into the concentrate, and the mixture obtained is heated in a vacuum. Many gas cells are formed in the concentrate, which soon turns into a spongy mass that can subsequently be dried fairly quickly. The process can be carried out batchwise (concentrate in shallow trays) or continuously on a conveyer belt. The dried cake is ground to a voluminous, easily soluble powder. The powder quality can be excellent due to the low drying temperature applied. The process is expensive and is only applied for special products such as some infant formulas. An advantage of the method is that it can be applied to inhomogeneous products.

10.3.2.3 Freeze Drying

A thin layer of the liquid is frozen, whereupon the ice is sublimated under a high vacuum. A voluminous cake is left (the space of the ice crystals is now occupied by holes) and is subsequently ground. A batch processing or a continuous operation in a high-vacuum belt drier can be applied. The method is expensive. Damage due

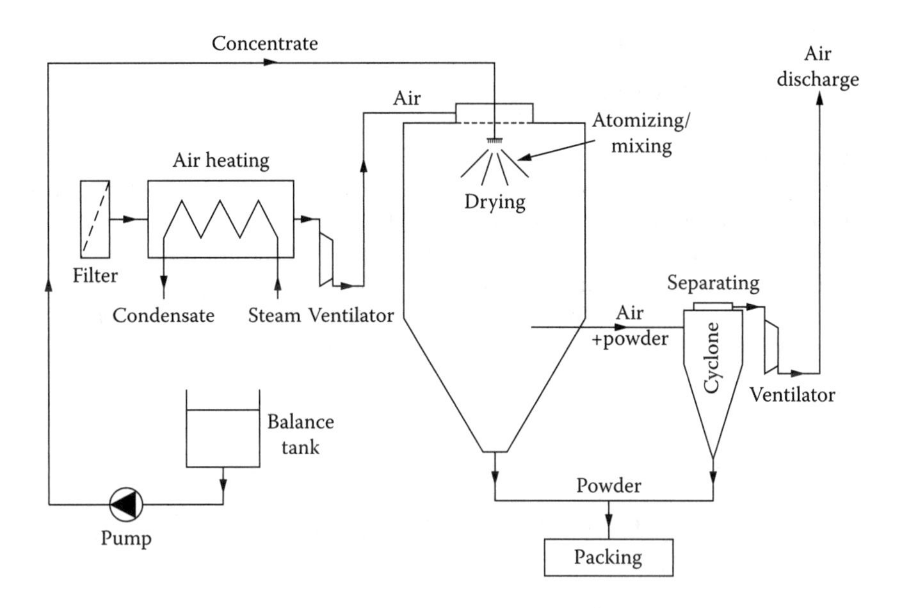

FIGURE 10.14 Simplified diagram of an example of the spray-drying process.

to heating does not occur but that is also true of spray drying if skillfully performed. A drawback is that the fat globules are subject to partial coalescence; this causes freeze-dried whole milk powder to show segregation after its reconstitution. Freeze drying is suitable for processing in small quantities and is applied in the drying of lactic starters, etc.

10.3.2.4 Spray Drying

This is the common method. There are several variants, but the following are essential process steps that are always involved (see Figure 10.14):

1. *Air heating:* The air is heated to about 200°C and leaves the drier at, say, 100°C.
2. *Atomizing the concentrate in the air:* This produces such small droplets that they will dry very quickly, with either a spinning disk or a pressure nozzle.
3. *Mixing hot air and atomized liquid:* Drying occurs correspondingly. Air and liquid usually enter the drying chamber cocurrently and are mixed so intensely that the air cools very rapidly. Consequently, the larger part of the drying process occurs at temperatures not much over the outlet temperature.
4. *Separating powder and consumed drying air:* Cyclones are commonly used.

10.3.2.5 Final Drying

In the drying of a liquid several stages can be distinguished, e.g., a stage in which the liquid turns into a more or less solid mass and a stage in which the solid mass obtained decreases further in water content (final drying). In milk products, a solid-like material is obtained at a water content near 8% (the product obtained is no longer sticky and appears to be dry), whereas a powder with, say, 3% water is desired. Traditionally, one process step included both drying stages, though in freeze drying the temperature must be raised during the final drying to complete it within a reasonable time. In spray drying, advantage is often taken of separating the final drying, which is generally achieved in a *fluid bed*, apart from the main process. This so-called two-stage drying is discussed in Subsection 10.4.5.

10.4 SPRAY DRYING

In this section, spray drying will be discussed in more detail.

10.4.1 Drier Configuration

Figure 10.14 gives one, considerably oversimplified, example of a drier layout. Actually a wide range of configurations is applied, according to type of raw material used, product specifications, and local possibilities (e.g., resources). Naturally, minimization of operating expenses is desired; this is not simple when the drier is to be used for a range of different products. The main variables are following:

Heating of the air: Generally heat exchangers are used. The classical medium is steam, but a pressure over 9 bar, hence a temperature over 175°C, is hard to reach. Consequently, hot gas is now often used, obtained by burning natural gas, predominantly methane. Another possibility is hot oil. Direct electric heating of the air can also be applied. Direct burning of natural gas in the drying air is very economical, but is undesirable (and generally illegal) as it causes some contamination of the powder with nitrogen oxides.

Atomization and air inlet: Figure 10.14 shows a spinning disk used for atomization of the liquid. The drops leave the disc in a radial and horizontal direction. The drying chamber must be wide to prevent droplets reaching, and thereby fouling, the wall of the chamber. The air inlet is generally tangentially, causing a spiral-like downward motion of the drying air. Currently, spraying nozzles are more commonly applied. Except for very small driers, several nozzles, arranged in one or more clusters, are installed. The drops leave the nozzles in a roughly downward direction. The air inlet is generally in the center of a cluster, also directed downward. The drying chamber can have a smaller diameter and often has a greater height, as compared to driers with disk atomization. The shape of the chamber is designed in such a way that the mixing of air and droplets and the course of drying of the droplets are (presumed to be) optimal.

Powder-air separation: This is primarily achieved with cyclones, arranged in various ways. In a drier configuration as depicted in Figure 10.14, where the air strongly rotates (as seen from above), the lower (conical) part of the drying chamber also acts as a cyclone, and most of the powder is discharged below. The powder often is cooled before packaging. So-called fines, i.e., the smallest powder particles present, are returned to the drier, often close to the region of atomization. The air has to be cleaned before it is returned to the atmosphere. Commonly bag-shaped cloth filters are used. Another option is wet washing: the air stream is led through a falling spray of water, which is recirculated. The outlet air is still hot and part of the heat can be transferred to the inlet air in a heat exchanger.

Aggregation of powder particles: This can be achieved by nozzle spraying when the nozzles are arranged in such a way that the sprays overlap. Another method is to return the fines in a region where the drops are still fairly liquid. A third way is rewetting in a fluid bed. In all these situations, sticky powder particles are caused to collide with each other.

Second-stage drying can be achieved in various configurations (see Subsection 10.4.5).

10.4.2 Atomization

Atomization is aimed at forming droplets fine enough to dry quickly, but not so fine as to escape with the outlet air after having been dried. Moreover, a very fine powder has undesirable properties because it is hard to dissolve.

In *disk atomization,* the disk spins very fast, i.e., at 200 to 300 revolutions per second. There are several types of disks but, essentially, the liquid falls on a disk and is flung away at a high speed, e.g., 100 $m \cdot s^{-1}$. Among the advantages of disk atomization are the following:

1. The droplets formed are relatively small.
2. The disk does not readily become clogged. For example, precrystallized concentrated whey can be atomized.
3. Disk atomization is still practicable at high viscosity; highly evaporated milk can thus be processed.

A drawback is that many vacuoles are formed in the particles (see Subsection 10.4.2.2); furthermore, the droplets are flung away perpendicularly to the axis of the disk and, accordingly, the chamber has to be wide to prevent the droplets from reaching the wall. Roughly speaking, the distance covered by the droplets in a horizontal radial direction is at least 10^4 times the droplet diameter.

With *pressure nozzles*, usually, the liquid is forced through a small opening at high pressure (up to 30 MPa) after it has been given a rotary motion. Advantages of the nozzle are its simple construction, the possibility to adjust the angle of the cone-shaped spray of the atomized liquid (thereby allowing a relatively small diameter of the drier), and a low vacuole content in the powder particles. When drying milk, the fat globules are disrupted into much smaller ones, about as they would be in a homogenizer; after all, the pressures applied are comparable.

Nozzles cannot be used for liquids that contain solid particles because they readily become clogged. The maximum liquid viscosity tolerated is somewhat lower than for a spinning disk. A given nozzle can only be used at a narrow range in capacity.

10.4.2.1 Droplet Size Distribution

Determination of the size distribution of the formed droplets is difficult. Usually, the produced powder is taken as a basis, but that involves several uncertainties: (1) the droplets often shrink very unevenly, (2) they may contain vacuoles, and (3) the powder particles may become agglomerated. None of the methods available for particle-size determination is fully reliable. Consequently, results as shown in Figure 10.15 are not quite correct.

In *disk atomization*, the average droplet diameter roughly follows:

$$d_{vs} \approx \text{constant } (Q\eta/\rho N^2 R)^{0.25} \qquad (10.3)$$

where Q is feed capacity (in $m^3 \cdot s^{-1}$), η is viscosity and ρ is density of the atomized liquid, N is number of revolutions per second of the disk, and R is disk diameter. (Note that the concentrate shows non-Newtonian behavior and is characterized by an apparent viscosity that depends on the velocity gradient. During atomizing, the velocity gradients will be high.) The constant depends strongly on the constructional details of the disk. The average droplet size will be larger at a higher dry-matter content and at a lower temperature, as both affect the viscosity. Near 60°C, d_{vs} is roughly proportional to $T^{-0.33}$ (T in °C). In Figure 10.15B, the decrease in

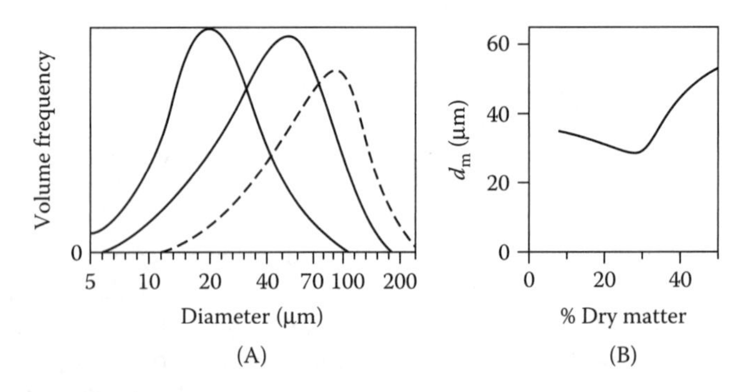

FIGURE 10.15 Particle size of skim milk powder, obtained by applying disk (—) or pressure nozzle (---) atomization. (A) Examples of the volume frequency distribution. (B) Influence of the dry-matter content of the concentrate on the median diameter.

particle size with increasing dry-matter content to about 30% follows from formation of a high vacuole volume in the powder particles if a weakly concentrated milk is atomized (see the following subsection).

In *nozzle atomization*, d_{vs} roughly follows

$$d_{vs} \approx \text{constant } (Q\eta/p)^{0.33} \tag{10.4}$$

where p = pressure in the liquid before the pressure nozzle. The constant depends strongly on the construction of the nozzle. p and Q cannot be varied greatly (otherwise the nozzle does not work at all). At a high η the size distribution becomes fairly wide.

To obtain smaller drops, the concentrate is often heated to decrease its viscosity. However, this should be done immediately before atomization because η increases rapidly (within a minute) at a high temperature, especially for concentrated skim milk.

10.4.2.2 Vacuoles

During the atomizing of a liquid, some air is always trapped in the droplets. This generally concerns some 10 to 100 air bubbles per droplet when a disk is used, whereas the number is far less when applying a nozzle, often 0 or 1 air cell per droplet (see also Figure 10.17). During the drying of the droplets, water vapor enters the air bubbles, causing them to expand. This is because the water vapor can more easily diffuse to the vacuoles than across the external layer of the drying droplets, which has already been concentrated and has become rigid. This explains why the vacuoles are only partly filled with air (see Figure 10.16A). Raising the drying

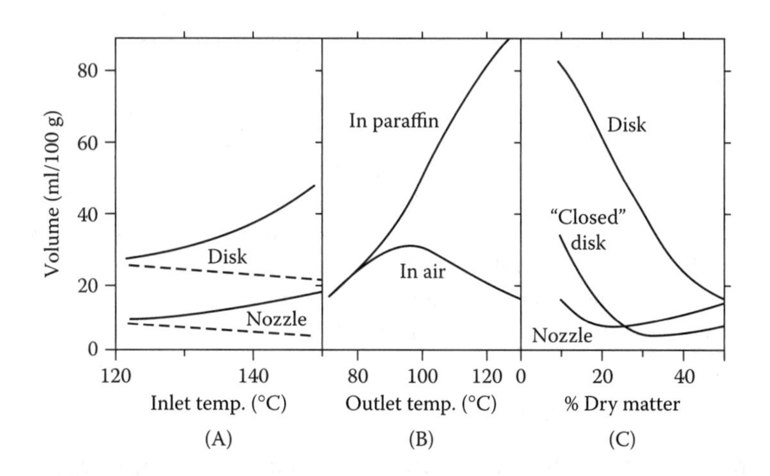

FIGURE 10.16 Volume of vacuoles (—) and of retained air (---) in powder obtained by spray-drying evaporated skim milk; determined shortly after drying. (A) Effect of the inlet temperature of the air. (B) Effect of the outlet temperature; disk atomizer. The vacuole volume was determined in paraffin oil or in air (see text). (C) Effect of the dry-matter content of the concentrate and of the construction of the atomizer.

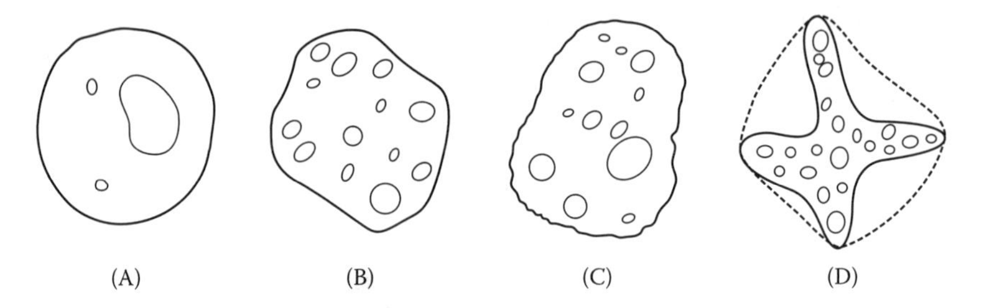

FIGURE 10.17 Cross sections of powder particles. Highly schematic; only circumference and vacuoles are indicated. Obtained by spray drying of (A) evaporated whole milk, nozzle; (B) evaporated whole milk, disk; (C) evaporated skim milk, disk; (D) skim milk, disk. The broken line shows the outer projection of the particle.

temperature increasingly expands the vacuoles and enlarges the vacuole volume. (Note that inlet and outlet temperature are both correlated to the drying temperature, with the outlet temperature generally closest; see Subsection 10.4.3.) Cracks form in the powder particles at high drying temperature. They cause the vacuoles to come into contact with the surrounding air. When the specific volume of such powder particles is determined in air, a low vacuole volume is found, whereas in paraffin oil a high value is observed (Figure 10.16B). This is because the oil penetrates the vacuoles very slowly, i.e., after many hours. Furthermore, the vacuole volume greatly depends on the dry-matter content of the concentrate (Figure 10.16C). This should largely, but not exclusively, be ascribed to the influence of the dry-matter content on the viscosity. A lower viscosity is part of the reason for a higher vacuole volume at a higher atomizing temperature (which is related to the inlet temperature).

Before, as well as during, droplet formation, air can be trapped in the droplets, especially when disk atomization is applied. The former mechanism can virtually be excluded by adapting the construction of the disk (Figure 10.16C). Entrapment of air during droplet formation can largely be overcome by substituting the air around the spray nozzle or the disk with steam. The trapped steam bubbles condense, so that few, if any, vacuoles can form. Bathing the atomizer in steam increases the temperature of the liquid during atomization and may thereby cause an increase of the insolubility of the powder. Figure 10.17 gives examples of the morphology of powder particles with vacuoles.

10.4.3 Change of State of the Drying Air

Properties of moist air can be shown in a state diagram. In analyzing drying processes, a useful diagram is the one according to Mollier. Figure 10.18 gives an example of a partial Mollier diagram. On the horizontal axis the water content is plotted in kg of water per kg of *dry* air, on the vertical axis the temperature (T) in °C as well as the enthalpy (h) per unit of mass. The lines for T and h are not horizontal. h is defined as the amount of heat needed to bring 1 kg of dry air + x kg of water from 0 to T°C,

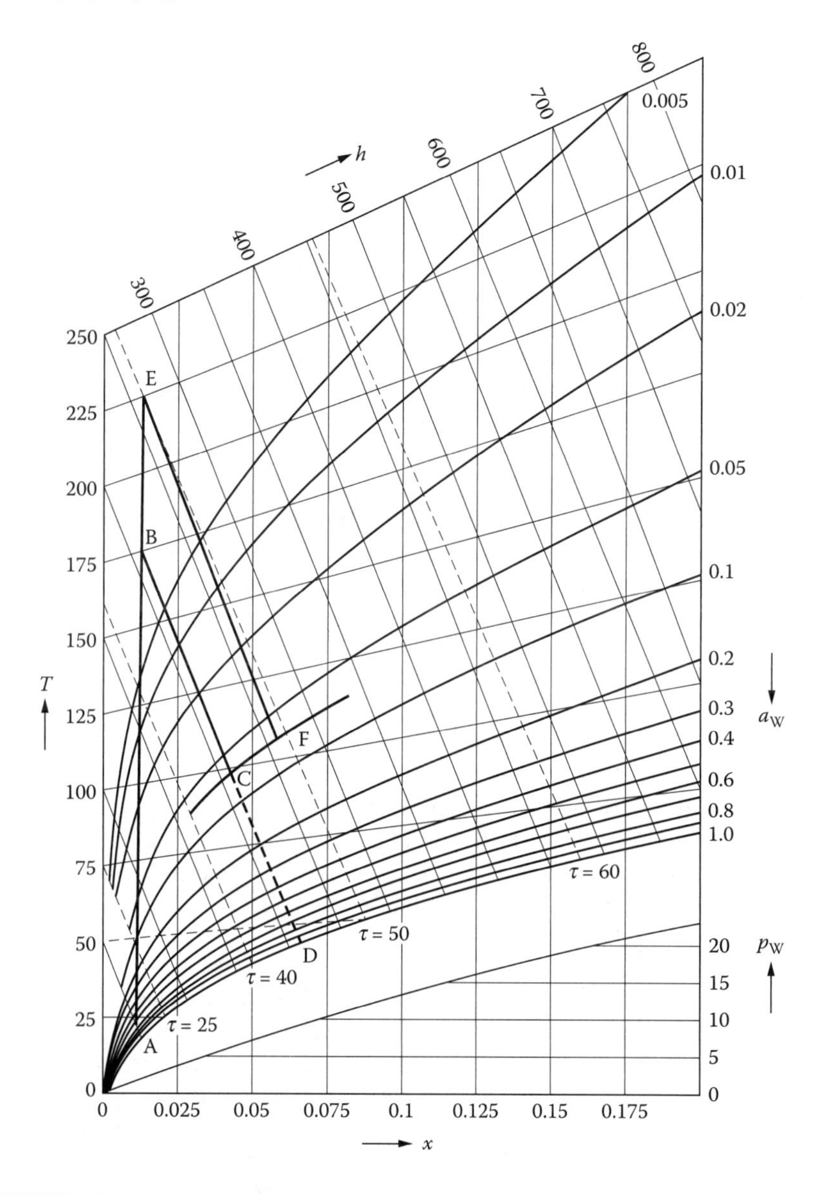

FIGURE 10.18 Partial Mollier diagram of moist air at a pressure of 10^5 Pa (= 1 bar ≈0.987 atm). x = water content in kg/kg dry air; T = temperature of the air (°C); h = enthalpy of the moist air in kJ per kg dry air; p_w = absolute water vapor pressure (kPa); a_w = water activity; τ = wet-bulb temperature (broken lines).

which includes the heat needed for vaporization of the water; h, therefore, is expressed per kg of dry air, but it is the enthalpy of the moist air, including that of the water vapor. By definition, $h = 0$ for dry air of 0°C and for water of 0°C. The diagram has been constructed in such a way that adding water vapor of 0°C to dry air of T°C corresponds to following a horizontal line from the Y axis. Starting from the Y axis, the lines of constant T therefore rise a little as these correspond to adding water vapor of T°C to dry air of T°C. The slope is $1.93T$, where the factor 1.93 is the specific heat of water vapor at constant pressure (in $kJ \cdot kg^{-1} \cdot K^{-1}$). The calibrated scale on the Y axis holds for h as well as for T because $h = T$ (if expressed in $kJ \cdot kg^{-1}$) for $x = 0$, as the specific heat of dry air at constant pressure happens to be precisely 1 $kJ \cdot kg^{-1} \cdot K^{-1}$. The lines of constant h run parallel with each other and slope down sharply. This implies that at constant T and increasing x, h increases significantly, which ensues from the definition of h, i.e., h includes the heat of vaporization of the water (2500 $kJ \cdot kg^{-1}$ at 0°C).

The properties of the air are fully determined by a point in the Mollier diagram (if the atmospheric pressure remains unchanged). Some other kinds of lines can be plotted, e.g., of constant density or constant volume. In Figure 10.18 lines of constant wet-bulb temperature τ are drawn; τ is the temperature of a water surface that shows rapid water vaporization into the air. The τ lines almost follow the lines of constant h, but not precisely. Knowledge of the quantity τ is important because the driving force for vaporization of water from a drying droplet will be proportional to the temperature difference between the droplet considered and the hot air, therefore to $T - \tau$, at least if the water activity of the drying droplet does not deviate to much from 1. Naturally, $T = \tau$ for air saturated with water vapor.

The diagram also gives lines of constant relative humidity or water activity a_w. All of these are curved plots. The importance of the quantity a_w is that it shows the water activity of the drying product after equilibrium between air and product would have been attained. Furthermore, the line for $a_w = 1$ gives the lower limit of the diagram. At the bottom of the figure there is a line that represents the absolute vapor pressure of water (p_w) at constant atmospheric pressure (1 bar); p_w is independent of the temperature (assuming, of course, that $p_w \leq$ the saturation vapor pressure, therefore, $a_w \leq 1$).

The Mollier diagram can be used in making calculations on the drying process. Consider, for example, air as characterized by point A in Figure 10.18, where $T = 20$°C and $a_w = 0.7$. The air is heated to 175°C. Because x remains constant (0.010 $kJ \cdot kg^{-1}$), we reach point B; a_w decreases sharply to about 0.002, and h increases from 45 to 203 $kJ \cdot kg^{-1}$, which means that 158 kJ per kg of dry air (=134 $kJ \cdot m^{-3}$) has been supplied. Atomizing a liquid, e.g., concentrated skim milk, will change the conditions of the air along the line BD.

The temperature to which the drying air may be cooled primarily depends on the corresponding water activity. Ideally, the desorption isotherm of the drying product should be taken as a basis (Figure 10.19). At 70°C, the curve for $a_w = 0.25$ crosses BD in Figure 10.18. This a_w corresponds to a water content in skim milk powder of, say, 2.5% (Figure 10.19) which, undoubtedly, is sufficiently low. Consequently, the outlet temperature might be adjusted to 70°C or even somewhat lower. In the above reasoning it has, however, been implicitly assumed that equilibrium between drying air and powder is established, but this is by no means true (Subsection 10.4.4). In actual practice, drying is, e.g., continued to yield $a_w \approx 0.07$;

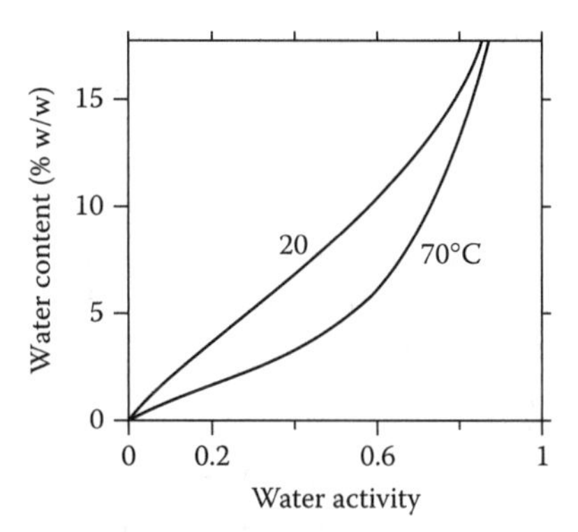

FIGURE 10.19 Approximate desorption isotherms of skim milk with noncrystallized lactose at 20°C and 70°C.

for example, this is point C on the line BD (Figure 10.18), which corresponds to 95°C. Then, if equilibrium would be attained, the water content of the powder should be 1% or even lower, whereas it actually is about 3%.

The Mollier diagram also shows that in our example the wet-bulb temperature is about 45°C. Initially, the temperature difference between air and droplet may be 175°C – 45°C = 130°C at the most, and is at least about 95°C – 45°C = 50°C. But the actual situation is more complicated; the temperature regime in the drying droplets is discussed in more detail in Subsection 10.4.4.

In our example, the water content of the air increases from 0.010 (in point B) to 0.041 (in point C) kg per kg of dry air during the drying process. Atomizing a skim milk concentrate with 54% dry matter and drying it to reach 97% dry matter requires [(100 – 54) – (54/97)3]/100(0.041 – 0.010) = 14.3 kg of dry air per kg concentrate. This corresponds to 12.1 m^3 cold air per kg concentrate, as the density of air with a_w = 0.7 is about 1.18 $kg{\cdot}m^{-3}$ at 1 bar and 20°C.

The efficiency of the heat expenditure can be expressed and calculated as follows. The heat input per kg of dry air is $(T_i - T_0)c_p$, where T_i is inlet temperature of hot air, T_0 is outside temperature, and c_p is specific heat at constant pressure. The heat output is $(T_e - T_i)c_p$, where T_e = outlet temperature of the consumed air. Because c_p of air hardly depends on its water content, the efficiency can be defined as $(T_i - T_e)/(T_i - T_0)$. In the present case, it would be (175 – 95)/(175 – 20), corresponding to 52%. The amount of heat consumed per amount of vaporized water is $(h_i - h_0)/(x_e - x_i)$ or (203 – 45)/(0.041 – 0.010) = 5097 $kJ{\cdot}kg^{-1}$, which roughly corresponds to 2.35 kg steam per kg of vaporized water. The efficiency thus is not high.

Heating the air from 20°C to 225°C, i.e., from A to E in Figure 10.18, implies drying to point F if the same a_w of the air should be reached. This means drying to 106°C; hence, to a higher outlet temperature. The average wet-bulb temperature this time is about 50°C. The efficiency amounts to (225 – 106)/(225 – 20), or 58% and 4480 kJ, or 2.08 kg steam is needed per kg water vaporized. In other words, the higher the inlet temperature, the higher the efficiency. Of course, there is an upper limit with respect to the inlet temperature, partly because of damage to the product caused by heating (Subsection 10.4.4). Moreover, the powder may catch fire in a drying chamber if it stays for a long time at a high temperature (this concerns powder deposited anywhere in the machinery). Ignition may already occur at 140°C; at 220°C, the time needed for spontaneous ignition is about 5 min.

The Mollier diagram can also be used to study the effects of varying temperature or water content of the outside air, the effect of reuse of air (e.g., by mixing it with fresh air), etc.

Summarizing, the Mollier diagram can be used to evaluate the efficiency of a drying process and to predict what will happen if the conditions in the drier are altered. It does not take into account the effect of variation in droplet size distribution. Some general relations are:

- When T_i is increased, the value of T_e should also be increased, though by a much smaller amount than T_i, to remain at a constant water content in the powder. This is done by keeping the a_w value of the outlet air constant.
- If the percentage of the water in the powder has to be increased, the a_w value of the outlet air should be increased; hence, T_e should be lowered. To what extent, depends on the relation between a_w and water content of the powder.
- If the dry-matter content of the concentrate entering the dryer is increased or its temperature decreased, the atomization results in larger droplets, primarily by the increased viscosity. To keep the water content of the powder the same, the a_w value of the outlet air should be lower; hence, T_e should be higher.

It is common practice to control the water content of the powder, which varies due to small fluctuations in the evaporating and drying process, by regulating the concentrate supply in such a way that T_e is kept constant. If T_e increases, the concentrate flux is increased, and vice versa. The points just mentioned indicate that this is not fully correct. More sophisticated methods (computer programs) have been developed to improve control of the drying process.

10.4.4 Changes of State of the Drying Droplets

Atomizing pure water in a drying chamber in the usual way causes the water droplets to reach the wet-bulb temperature and to vaporize within 0.1 s at this temperature. The presence of dry matter in the droplets, however, makes an enormous difference. Figure 10.13 shows the diffusion coefficient of water to substantially decrease with increasing dry-matter content (e.g., from 10^{-9} to 10^{-13} $m^2 \cdot s^{-1}$). Accordingly, the vaporization is significantly slowed down.

In the drying droplet, the thermal diffusivity remains at approximately 10^{-7} $m^2 \cdot s^{-1}$. This implies that in most droplets temperature equalization occurs in less than about 10 ms. In other words, the temperature is virtually constant throughout a droplet, though not at the very beginning of drying.

10.4.4.1 Drying Stages

Initially, the droplet has a very high velocity relative to the drying air. Therefore, there is a first stage during which circulation of liquid in the droplet occurs; this circulation greatly enhances transfer of heat and mass. For a droplet of 50 μm diameter, this stage lasts, say, 2 ms. In this time, the droplet covers a distance of about 10 cm and loses a small percentage of its water. Its velocity compared to the air decreases to the extent that the formed surface tension gradient of the drop surface arrests internal circulation of liquid. But in the second drying stage the difference in velocity between drop and air is still great enough to accelerate water transport. The transport in the droplet occurs by diffusion but in the air by convection. After about 25 ms the relative velocity of the droplet has decreased so far that the water transport has become essentially equal to that from a stationary droplet. Relative to the air, the droplet then has covered a distance of a few decimeters and has lost about 30% of its original water. In the third stage, lasting at least a few seconds, the droplet loses the rest of the water by diffusion.

10.4.4.2 Drying Curve

Assuming for the moment that drying air and droplet remain in equilibrium with each other, the droplet attains the wet-bulb temperature and maintains that temperature until virtually all of the water present has been vaporized. If so, the droplet temperature rises only because the increasing concentration of dry matter eventually leads to a significant elevation of the boiling point (this is equivalent to a decrease in a_w of the drying droplet). The dried droplet finally reaches the outlet temperature of the consumed air. Figure 10.20 shows this to apply reasonably well for a water droplet. But the temperature curve completely changes if dry matter is involved. A droplet of a highly concentrated liquid does not even maintain the wet-bulb temperature for some time. Note in Figure 10.20 that the time needed to arrive at a certain stage in the drying process is proportional to the square of the initial droplet diameter.

Figure 10.20 refers to drying in stationary air of constant temperature and humidity. The situation in a spray drier is very different. During drying the air temperature decreases and the humidity of the drying air increases. Moreover, the droplets vary in size and the smallest ones will dry fastest. Some calculated results are given in Figure 10.21 for a drop size distribution of common width. Drops of three sizes are taken as examples. It concerns two extremes in the mixing of air and drops. In the one case, the two flows are precisely cocurrent. It is seen that, nevertheless, the smaller drops would become far hotter than the larger ones; they very rapidly lose most water, causing their wet-bulb temperature to increase equally fast. The other case is that of perfect mixing. This would imply that the

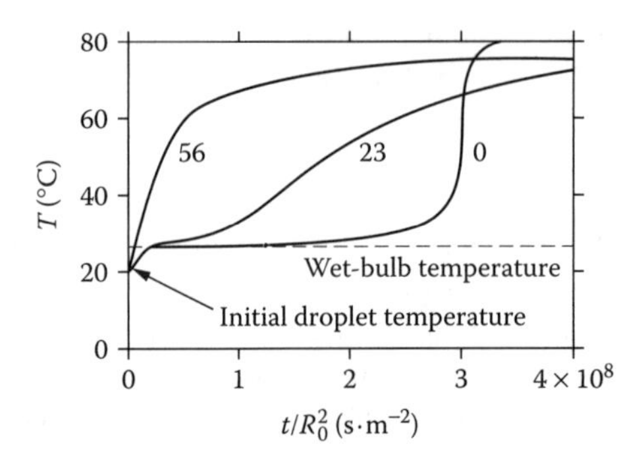

FIGURE 10.20 Temperature (T) of a stationary droplet that enters an excess of dry air of 80°C as a function of the reduced time (t = time after introducing the droplet; R_o = original radius of droplet). The figures near the curves refer to the content (%) of dry matter in the original droplet. (Measured by J. van der Lijn, Ph.D. thesis, Wageningen Agricultural University, 1976.)

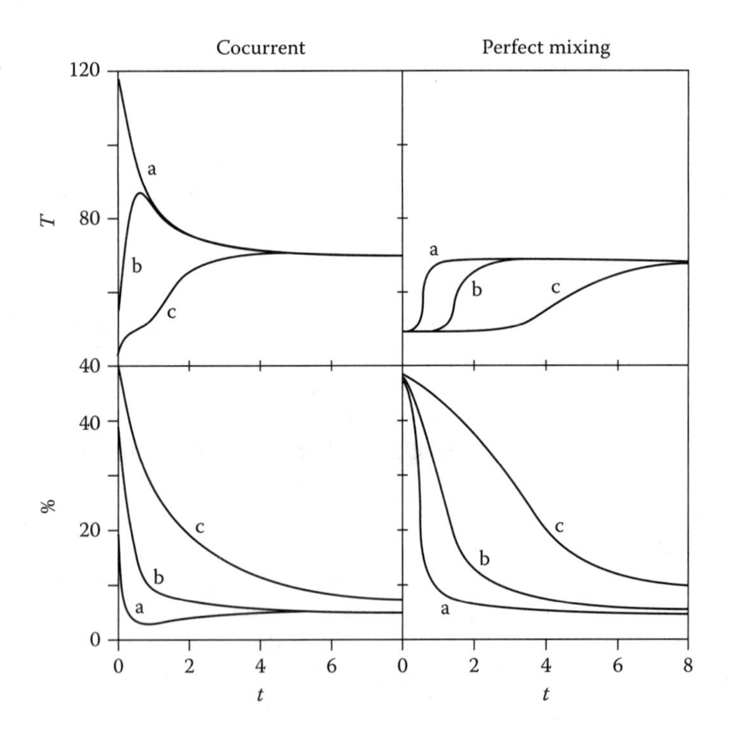

FIGURE 10.21 Calculated drying curves for droplets in a spray drier. Temperature (T, °C) and water content (%) of drops of various size as a function of drying time (t, s). Drop size: (a) 60, (b) 100, (c) 180 μm. Conditions: cocurrent flow of drops and air, or perfect mixing; T_i = 220, T_e = 70°C; nozzle atomization; initial dry-matter content 50%; log-normal drop size distribution (geometric standard deviation 0.6). (Courtesy of J. Straatsma, NIZO Food Research, Ede.)

air immediately has the outlet temperature, and the drying drops can never become hotter. Consequently, drying is slower, although the smaller drops still dry much faster than the larger ones.

In practice, the mixing is always between the two extremes. For driers with a spinning disk, the situation tends to be fairly close to perfect mixing. In most driers with nozzles, it tends to be closer to cocurrent flows. In all situations, the drying time may very by a factor of, say, 100 between the smallest and the largest drops. This is of great importance for fouling of the drying chamber; the largest drops have the greatest chance of hitting the wall and of being insufficiently dry to prevent sticking to the wall.

Another factor that affects drying rate is the presence of vacuoles in the drying drops (which has not been taken into account in the calculations of the figure). It leads to significantly faster drying.

10.4.4.3 Concentration Gradients

What causes the rapid decrease in drying rate of the droplets after the water content is reduced to, say, 15%? Figure 10.22 shows that a strong concentration gradient forms rapidly during drying. The higher the drying temperature, the stronger the effect. (This explains why stronger gradients occur for cocurrent drying.) Not surprisingly, a dry outer layer, i.e., a kind of rind, is formed; because of this, the water transport is slowed down considerably. The temperature can rise significantly in the dry outer layer because a dried material assumes the air temperature, not the wet-bulb temperature. In other words, the decrease in temperature near the surface of the droplet (caused by consumption of the heat of vaporization) becomes far smaller because the vaporization of water is slower.

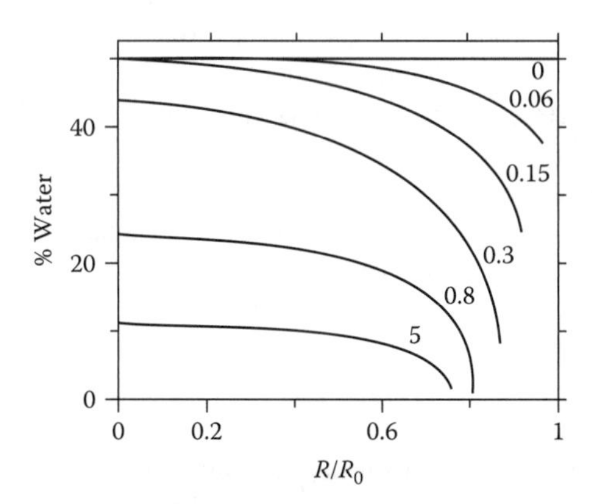

FIGURE 10.22 Water content of a drying droplet as a function of the distance R from the center of the droplet; R_0 = radius of original droplet. Parameter is drying time (s) for $R_0 = 25$ μm. Inlet temperature 175°C; outlet temperature 70°C; perfect mixing. (Examples calculated by J. van der Lijn, Ph.D. thesis, Wageningen Agricultural University, 1976.)

Because temperature equalization happens very quickly, the whole drying droplet increases in temperature.

Naturally, concentration gradients as shown in Figure 10.22 are not of a lasting nature. Let us consider a droplet that has been dried for 5 s; subsequently, it is separated and isolated from the surroundings. The water will become evenly distributed by diffusion and reach about 6% throughout the droplet. Attaining an equilibrium condition takes considerable time because the effective diffusion coefficient D of water is of the order of 10^{-13} $m^2 \cdot s^{-1}$ (Figure 10.13). From $x^2 = Dt_{0.5}$, and considering that the distance x that must be covered approximates 10^{-5} m, we derive that the time needed to halve a concentration difference equals about 10^3 s. All in all, it will take at least 1 h for the concentration gradients to become so small as to be almost negligible.

The relatively dry outer layer of the droplet soon becomes firm and eventually glassy. This causes the droplet to resist further shrinkage. The droplet can react by forming vacuoles (see Subsection 10.4.2.2) or by becoming dimpled (see Figure 10.17D). Especially at a low water content of the particles, so-called hair cracks may be formed.

10.4.4.4 Aroma Retention

Besides water, the drying droplets lose other volatile components, including flavor compounds (aroma). In many cases, however, the loss of flavor compounds is far less than expected, in spite of their volatility. This is because the effective diffusion coefficient of most flavor components in the relatively dry outer layer of the droplet decreases far stronger with decreasing water content than the diffusion coefficient of water does (due to the greater molar mass); the difference can amount to some orders of magnitude. Not surprisingly, the aroma retention (retaining flavor compounds during drying) increases with droplet size (in larger droplets the outer layer from which the flavor components get lost has a relatively smaller volume) and with drying temperature (at a higher temperature a solid rind forms more rapidly). Formation of vacuoles diminishes aroma retention, especially if (hair) cracks develop in the particles and the vacuoles come into contact with the surrounding air. In the experiment mentioned in Figure 10.16B, the loss of volatile compounds was almost proportional to the difference between the vacuole volume as determined in paraffin oil and that in air; air can penetrate the cracks in the particles during the measurements, whereas the viscous paraffin oil cannot.

10.4.4.5 Damage Caused by Heating

High drying temperatures can result in undesirable changes in the dried product. Generally, it is only after the powder has been dissolved again that the changes involved are noticed.

To understand the relations between temperature and dry-matter content of the drops as a function of time on the one hand, and the resulting extent of a reaction occurring on the other hand, Figure 10.18 and Figure 10.21 give important

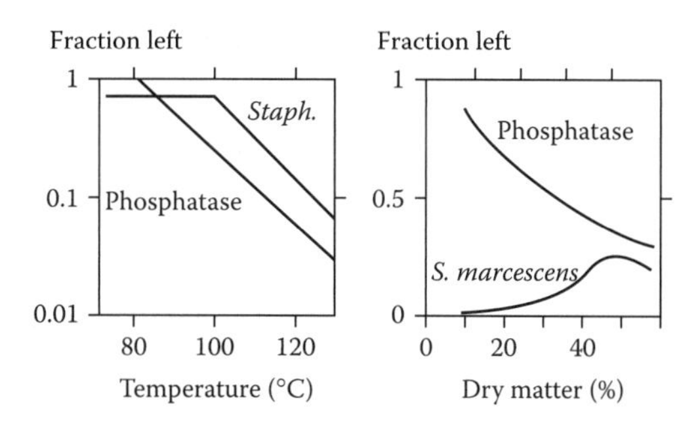

FIGURE 10.23 Examples of the effect of the outlet temperature of the drying air, and of the dry-matter content of the concentrate during atomization, on the inactivation of the enzyme alkaline phosphatase and the killing of a *Staphylococcus* species and of (the very heat-sensitive) *Serratia marcescens*. Approximate results.

information. One conclusion is immediately clear from the latter figure: perfect mixing gives far lower drop temperatures, hence, slower reactions, than cocurrent flows.

Three quite different categories of undesirable changes can be distinguished:

Heat denaturation and killing of microbes: This is discussed in Section 7.3, and Subsection 10.1.5. An important aspect is that the reaction rate is highly dependent on temperature, but that the reaction is much slower and less dependent on temperature at a low water content; see especially Table 7.5. Some results for the inactivation of phosphatase are given in Figure 10.23. It is seen that a higher T_e — hence, a higher average drying temperature — gives more inactivation. The same is true for a higher dry-matter content of the liquid. The main explanation is that this goes along with a higher viscosity and hence, on average, larger drops. Therefore, a longer heating time is needed at a dry-matter content in which the inactivation rate still is appreciable. Remember that the largest drops contain the greatest amount of material.

Much the same is true of the killing of bacteria, except for the effect of dry-matter content. The initial increase of survival of *S. marcescens* with increasing dry-matter content is presumably due to a substantial decrease in heat sensitivity of the organism.

When spray-drying a starter culture, survival of bacteria is of paramount importance. To achieve this, the drier should exhibit near-perfect mixing of air and drops, and the drops formed should be small and T_e relatively low. The latter implies that T_i also must be low, as otherwise the water content of the powder will remain too high. Moreover, the powder should immediately be cooled. Often, a relatively large proportion of an inert material, generally maltodextrin, is added to the liquid before drying, which lowers the temperature sensitivity of the bacteria. In this way, survival rates over 80% can be achieved.

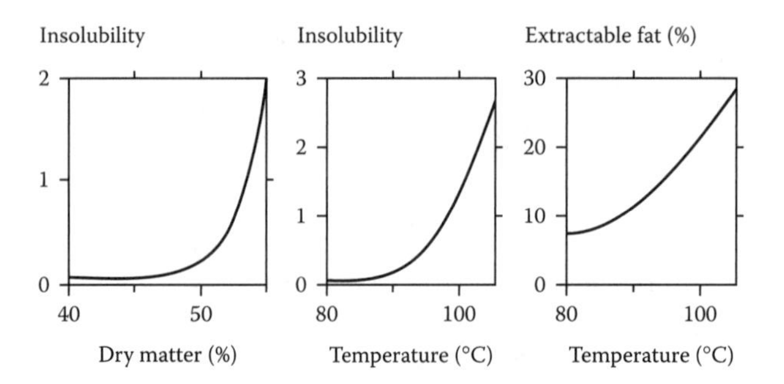

FIGURE 10.24 Effect of the dry-matter content of the concentrate during atomization, and of the air outlet temperature, on the insolubility index (ADMI method) of the resulting skim milk powder. Effect of outlet temperature on the percentage of the fat extracted from the resulting whole milk powder. Approximate examples.

Insolubility: Part of the protein may be rendered insoluble during the drying process; it is due to heat coagulation (Subsection 7.2.4). Insolubility is further discussed in Subsection 20.4.5.2: the powder contains particles that do not dissolve in water, but the amount is a tiny fraction of the powder. Figure 10.24 shows that insolubility increases with increasing T_e and increasing dry-matter content of the milk. It also greatly depends on drier design. Presumably, heat coagulation mainly occurs in some (large) drops or powder particles that recirculate in the drier and become wetted again. A cumulation of high temperatures and high dry-matter content for a relatively long time then causes the problem. Modern driers tend to give very small insolubility figures.

Formation of hair cracks: These can form at high drying temperatures because the outer rind of a drying droplet soon reaches a glassy state; the pressure gradients developing in the particle then cause these very thin cracks to form. (This is also discussed in relation to Figure 10.16B.) If it concerns whole milk powder, part of the fat can now be extracted from the powder by solvents like chloroform or light petroleum. A result is shown in Figure 10.24. The extractable fat is often called free fat, but that is a misleading term; see Subsection 20.4.2.

10.4.5 Two-Stage Drying

As stated above, spray drying is relatively expensive, e.g., with respect to energy; furthermore, the capital outlay for driers is high. A better efficiency can, in principle, be achieved by increasing the concentration factor of the milk before atomization and by applying a higher air inlet temperature, but these measures can lead to heat damage of the product. Alternatively, the powder can be separated from the air before it is completely dry, while additional drying occurs outside the drying chamber. In this way, the outlet temperature of the air can be lower, allowing the inlet temperature of the air to be higher without increased heat damage occurring. Moreover, a larger quantity of concentrate can be dried per unit time.

The powder may be discharged from the drying chamber after it has become so dry as to have lost its stickiness. The problem of stickiness is less than expected because of the concentration gradient formed in the powder particles. Consider, for example, the curve for 0.8 s in Figure 10.22. In the center of the particle the water content is 24%. It is on average about 13% but only about 2% at the periphery. Presumably, these particles would still be slightly sticky because (1) stickiness (= the tendency to stick to the machinery) considerably increases with temperature; and (2) upon removal from the drying air the outside of the powder particles rapidly increases in water content due to internal exchange of water. Moreover, larger particles will be 'wetter,' hence, more sticky. But a powder with an average water content of about 8% can readily be discharged by means of cyclones.

The final drying is often achieved in a *fluid bed drier.* A layer of powder deposited on a perforated plate can in principle be fluidized by blowing air through the layer from below. In such a fluidized bed the powder layer is expanded, containing a high volume fraction of air; the mixture can flow, almost like a liquid, if the perforated plate is slightly tilted. The particles in the bed are in a constant erratic motion, which enhances drying rate. Conditions for fluidization are (1) that the particles are larger than about 20 μm (but smaller than a few mm) and that their size distribution is not very wide; and (2) that the air flow is evenly distributed over the bed and has a suitable velocity, e.g., about 0.3 m/s for most spray-dried powders. Generally, the size distribution of milk powder particles is too wide: if all particles are to be fluidized, including the largest ones, the air velocity must be so high that the smallest particles are blown out of the bed. To overcome this problem, the machine is made to vibrate, which allows fluidization at a lower air velocity. Such a fluid bed drier can then be attached to a spray drier (by a flexible pipe), e.g., as depicted in Figure 10.25a.

In a spray drier the air inlet temperature is high; the holdup time of the powder is short, say, a few seconds. In a fluid bed drier the air inlet temperature is relatively low (e.g., 130°C), little air is consumed, and the residence time of the powder is much longer, i.e., several minutes. Because of this, a fluid bed drier is much more suitable for the final stages of drying. For example, in a comparison between traditional and two-stage drying, using the same spray drier, the same skim milk concentrate with 48% dry matter, dried to the same water content of 3.5%, may yield the following:

Number of Stages	**1**	**2**
Inlet air temperature (°C)	200	250
Outlet air temperature, chamber (°C)	94	87
a_w outlet air, chamber	0.09	0.17
Total heat consumption (kJ/kg of water)	4330	3610
Capacity (kg of powder/h)	1300	2040

The efficiency of the heat expenditure thus is better (by 17%) and the capacity greater (by 57%); against this is the capital outlay for the fluid bed drier. The

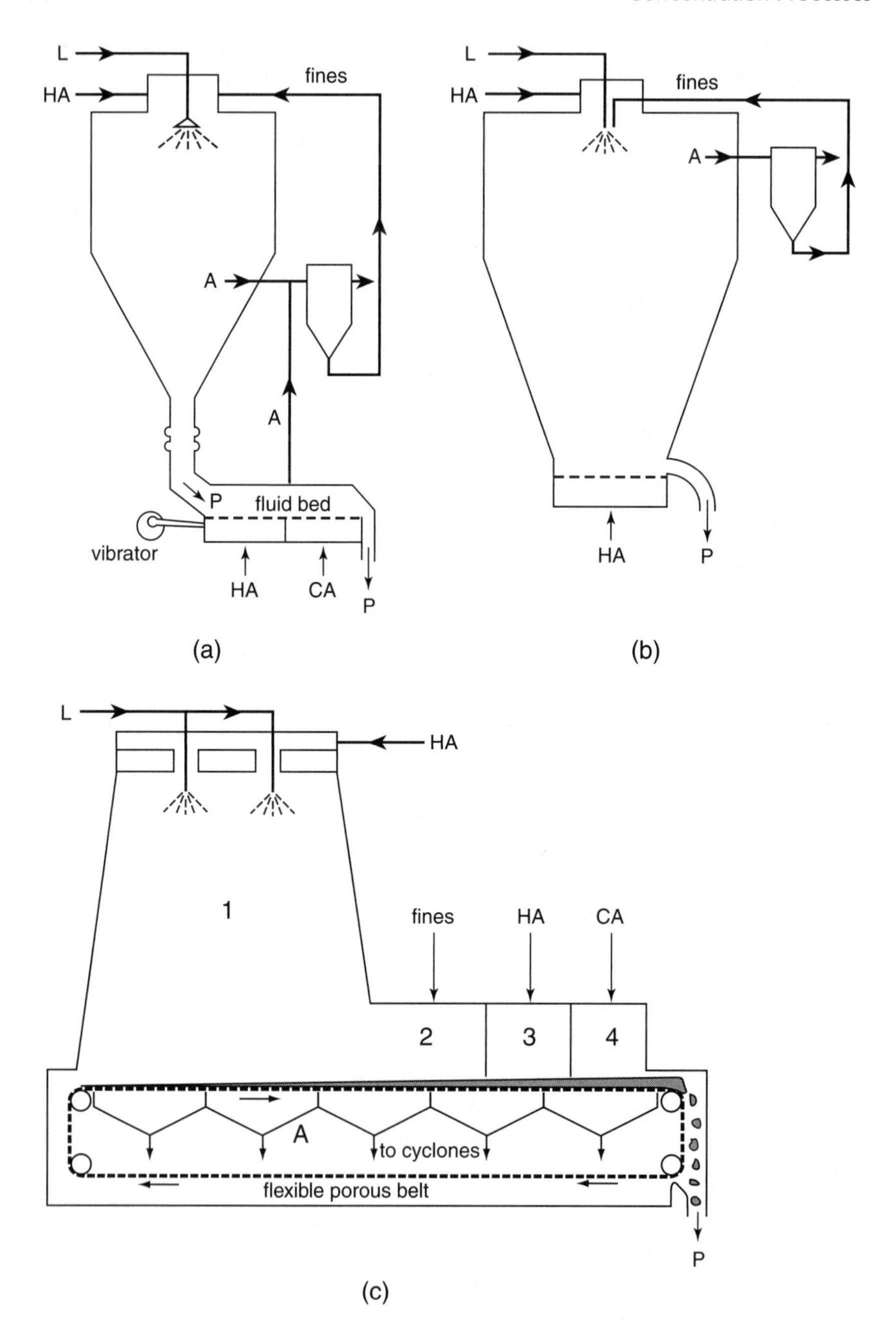

FIGURE 10.25 Two-stage drying. Highly simplified schemes of some drier configurations. (a) Classical spray drier with fluid bed attached. (b) Fluid bed in the bottom end of a spray drier. (c) Filter mat drier. L = liquid (feed); A = air; HA = hot air; CA = cold air; P = powder. (Further details in text.)

additional drying consumes only 5% of the heat. The quality of the powder (insolubility index) is certainly not poorer but, generally, better.

A fluid bed offers additional opportunities. For instance, it is quite simple to add a cooling section. The bed can also be used for agglomerating purposes. The main incentive for agglomeration is that a fine powder poorly disperses in cold water (Subsection 20.4.5.1). Therefore, often an attempt is made to produce a coarse-grained powder. In the fluid bed the powder particles collide intensely with each other. As a result, they agglomerate if they are sufficiently sticky, i.e., have a high enough water content at their periphery. Hence, agglomeration is enhanced by blowing steam into the powder (this is called rewetting, which is mostly applied when producing skim milk powder). The air velocity in the fluid bed may be adjusted in such a way that the smallest powder particles (which have already become very dry and therefore show poor agglomeration) escape separation. The latter particles are fed back to the drying chamber, gain entrance to the atomized liquid, and become agglomerated with the drying droplets (especially applied for whole milk powder).

Two-stage drying can also be effected in a modified spray drier chamber, as illustrated in Figure 10.25b. In the bottom end of the chamber, a fluid bed is realized. It need not be vibrated as the smallest particles are blown toward the atomization region where they agglomerate with the drying drops. The air circulates in the vertical direction and is removed near the top end of the chamber. A great variety of configurations has been developed: for instance, a combination of the type depicted in (b) with a fluid bed attached as in (a).

Another method of two-stage drying is carried out in the *filter mat drier*, illustrated in Figure 10.25c. The first stage is conventional spray drying. Atomization is by nozzles, and the flows of air and drop spray are cocurrent. The partly dried drops fall on a moving perforated belt. The spent air is removed through the powder bed formed on the belt and is sent to cyclones. The fines removed by the latter are added to the second section of the machine, where it agglomerates with the powder on the belt. In the third section, hot air is blown through the bed for final drying. The dry material then reaches the fourth section, where it is cooled. The powder particles become strongly aggregated and form a porous cake, which falls from the belt in large lumps. These are gingerly ground and the resulting powder is packaged.

The filter mat drier allows a greater part of the water to be removed in the second stage because the powder can hit the belt when still being sticky. The latter also makes this type of drier suitable to handle very sticky materials; cream powders, in particular, are made in filter mat driers.

Suggested Literature

General aspects of water content and activity, and the effects on food properties and stability: P. Walstra, *Physical Chemistry of Foods,* Dekker, New York, 2003; O.R. Fennema, Chapter 2 in O.R. Fennema, Ed., *Food Chemistry,* 3rd ed., Dekker, New York, 1996.

Principles of concentration and drying of foods: M. Karel and D.B. Lund, *Physical Principles of Food Preservation,* 2nd ed., Dekker, New York, 2003. (See especially Chapter 9 and Chapter 10.)

Concentrating and drying are discussed in several books on chemical and food process engineering, for example: D.R. Heldman and D.B. Lund, Eds., *Handbook of Food Engineering,* Dekker, New York, 1992.

General aspects of spray drying: K. Masters, *Spray Drying Handbook,* 5th ed., Longman, Harlow, 1991.

11 Cooling and Freezing

A food is generally cooled or frozen to retard spoilage. Fresh milk is routinely cooled to about 5°C in many regions of the world. For liquid milk products cooling may extend the shelf life by days or weeks. Cooling of milk is also applied for specific purposes, such as inducing fat crystallization or enhancing the creaming tendency. Freezing is mostly done to make a specific product, such as ice cream, and occasionally as a concentration process. Freezing is sometimes used to substantially retard deterioration of fresh milk or liquid milk products.

11.1 COOLING

Cooling of milk causes several changes, the most important ones being:

1. The growth of most microorganisms is much slower, if not stopped, and so are the changes induced in milk by their metabolism (Section 5.4).
2. Nearly all chemical and enzymic reactions are retarded.
3. Autoxidation of lipids, whether induced by light or Cu^{2+}, is enhanced (Subsection 2.3.4.2, item 9 in list), presumably because the activity of the enzyme superoxide dismutase is decreased.
4. Changes in solubility and association of salts occur. The amount of micellar calcium phosphate decreases (Subsection 2.2.5), and the pH increases (Figure 4.2).
5. The casein micelles attain a higher voluminosity and part of the casein, especially β-casein, goes into solution (Figure 3.20). This results in an increased viscosity (Figure 4.7D) and an enhanced susceptibility to attack by plasmin.
6. The fat globule membrane loses some components (Subsection 3.2.1.2), and its structure is altered. These changes are irreversible.
7. Cold agglutination of fat globules occurs (Subsection 3.2.4.2), e.g., enhancing creaming rate.
8. The triglycerides in the fat globules will partly crystallize (Subsection 3.2.1.3).

Much of what has been said about heat treatment in Chapter 7 also applies to cooling. In principle, the same equipment is used. Heat transfer will be slower because the higher viscosity of the liquid at lower temperatures causes the Reynolds number to be smaller (see Appendix A.11). This especially causes problems in high-fat cream; in flowing cream, partial coalescence (clumping) of fat globules can occur as a result of high velocity gradients, and this tends to increase the viscosity of the cream even further. As a result, the coefficient of total heat transfer k_h (cf. Appendix A.11) can fall below 100 W·m^{-2}·K^{-1}. The clumping of the fat globules is also undesirable in regard to product properties. Moreover, the resistance to flow through a plate apparatus becomes excessive, and another type of heat exchanger is needed for high-fat cream. This means that cooling is relatively expensive, especially when chilled water (iced water) or chilled has to be used rather than well water with a temperature of, say, 11°C. A larger temperature difference between the brine and the incoming product usually does not greatly enhance the heat-transfer rate because it results in local freezing of the product.

Cooling of packaged products can also take a long time, especially if it involves viscous products. For the most part, air cooling is used. It is better to package the product after cooling, if possible.

11.2 FREEZING

When milk is cooled, it starts to freeze at about −0.54°C, if no undercooling occurs. Concentrated milk naturally has a lower freezing point. Pure ice is formed, and the remaining milk thus becomes concentrated, thereby further decreasing the freezing point. The lower the temperature, the higher the proportion of the water that freezes (Figure 11.1). All the consequences of concentrating mentioned in Subsection 10.1.3, apply. If there is equilibrium among water, ice, and moist air, the water activity is a function of temperature only; this is because the vapor pressure of ice, and therefore its a_w, is a function of temperature only (for constant pressure). The relation is given in Figure 11.1.

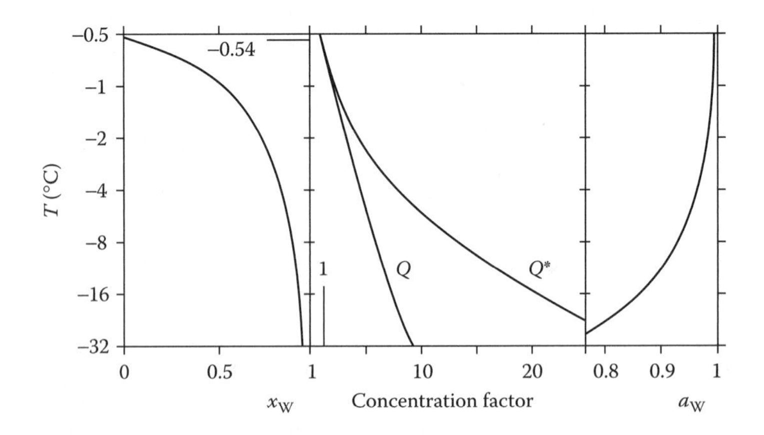

FIGURE 11.1 Freezing of skim milk. Mass fraction of the water frozen (x_w), concentration factor, and water activity (a_w) as a function of temperature (T). Approximate examples at equilibrium.

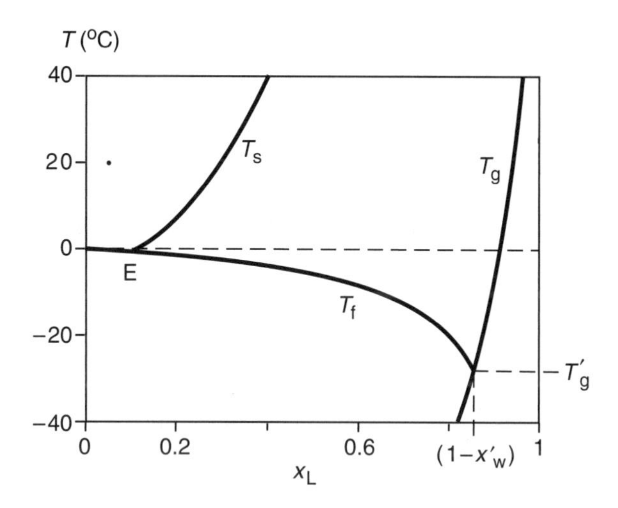

FIGURE 11.2 Partial state diagram of aqueous lactose solutions. T is temperature and x_L is mass fraction of lactose. T_s gives the solubility of lactose (for equilibrium between α- and β-lactose), T_f is freezing temperature, T_g is glass transition temperature, and E denotes the eutectic point.

The change in composition of the remaining solution is considerable. At −8°C, for example, Q^* is approximately 13, corresponding to about 60% dry matter when skim milk is frozen. Making an ultrafiltrate of the remaining liquid (at −8°C), it has a pH (at room temperature) of about 1 unit lower than that of an ultrafiltrate of the skim milk before freezing. The ionic strength may be almost tenfold higher, the calcium ion activity is significantly increased, and the quantity of calcium phosphate in the casein micelles is also increased, despite the decrease in pH.

Upon freeze concentration, the remaining liquid can attain a *glassy state* (see Subsection 10.1.4). This is illustrated in the state diagram for lactose solutions depicted in Figure 11.2. When a solution of, say, 5% lactose is cooled to below 0°C, ice will form. The liquid will thus be concentrated and the temperature decreases, as given by curve T_f; note that the values of x_L now refer to the lactose solution. At about −0.75°C and 12% lactose, the eutectic point is reached, and if equilibrium would be attained, lactose then starts to crystallize. However, lactose exhibits rather slow nucleation and crystal growth (see Subsection 2.1.3), and upon rapid cooling its crystallization will fail to occur. Freeze concentration, then, will proceed according to curve T_f, although its course can slightly depend on cooling rate. The viscosity of the liquid will markedly increase, whereby lactose crystallization becomes ever less likely, and finally the curve for T_g is reached. Now, the liquid has changed into a glassy state and further ice crystallization is prevented. This *special glass transition* at maximum freeze concentration is characterized by a special glass transition temperature T'_g and unfrozen water content x'_W. For lactose these values are approximately −28°C and 15%, respectively. (*Note*: The description given in the preceding text is, to some extent, an oversimplification because of the mutarotation of lactose and of its crystallization as a monohydrate.)

When a liquid milk product, say skim milk, is frozen, lactose becomes saturated at about −2°C and $Q = 2$, but it will not start to crystallize until Q is at least 3. Upon fast freezing, a glassy state can be attained. This transition is dominated by the behavior of lactose. Values of T_g' between −23°C and −28°C have been reported, the lower values being more likely. The unfrozen water content x_W' is approximately 4%. The product is very stable when kept below −28°C, although some lipid autoxidation may still occur. Above T_g', diffusion rates considerably increase, lactose may slowly crystallize, and further changes can occur. These phenomena are especially important in ice cream.

When lactose does not crystallize, it acts as a *cryoprotectant*. Because the equilibrium water activity is determined by temperature only, less water will freeze. Consequently, the other substances, notably salts, are concentrated less; hence, ionic strength and Ca^{2+} activity increase less. When lactose does crystallize and the frozen milk is thawed, aggregation of casein micelles is often observed, presumably due to salting out and to deposition of calcium phosphate in the micelles. When the aggregation is not too strong, it can often be undone by stirring at low temperatures (e.g., 5°C), but frozen milk that has been stored for a long time at, say, −18°C shows irreversible aggregation upon thawing. If lactose crystallization has been prevented by freezing rapidly, aggregation does not occur.

Freezing and thawing of whole milk, and especially of cream, generally causes partial coalescence (clumping) of fat globules because of the growing ice crystals pressing the globules together. It can be prevented by prior homogenization and rapid freezing.

In practice, rapid freezing is achieved in a scraped-surface heat exchanger. The liquid flows through a tube that is deeply cooled at the outside, causing ice to form at the inner surface. The ice is immediately removed by a rotating stirrer that is provided with knives that scrape over the surface. In this way, a mixture of concentrated milk product and ice crystals results. By removing the crystals, concentrated milk can be obtained. This process of freeze concentration is, however, much more expensive and more limited in possibilities than removing water by evaporation; it is occasionally used to concentrate a liquid containing aroma compounds because these may become lost during concentration by evaporation.

Suggested Literature

A fairly recent review on freezing of dairy products: H.D. Goff and M.E. Sahagian, Chapter 8, page 299, in L.E. Jeremiah, Ed., *Freezing Effects on Food Quality,* Dekker, New York, 1996.

Basic aspects (see especially Chapter 16): P. Walstra, *Physical Chemistry of Foods,* Dekker, New York, 2003.

12 Membrane Processes

Membrane processes are applied to separate a liquid into two liquids of different composition. The primary aim may be to remove a substance such as water, which amounts to concentrating milk or whey, etc. Other components to remove are salts or bacteria, for example. Another aim is to accumulate a (group of) components such as protein. Membrane separations are increasingly applied in dairy processing, particularly to whey. They often replace older separation processes, but they also give rise to new or improved products. See Subsection 10.1.1 for general aspects of concentrating.

12.1 GENERAL ASPECTS

In the application of a membrane process, a liquid is enclosed in a system confined by a semipermeable membrane. Some of its components can pass the membrane; some cannot. This is illustrated in Figure 12.1 for a flow-through process. The liquid passing the membrane is called *permeate*; the retained liquid is the *retentate* (or concentrate). The achieved separation of components primarily depends on the structure and the composition of the membrane.

In the case of membrane *filtration*, the driving force for separation is a hydrostatic pressure difference over the membrane. This is called the transmembrane pressure; it is generally realized by a pressure pump in the feed line and a throttle valve in the retentate line. In dialysis, the driving force is a concentration difference; in electrodialysis it is a difference in electrical potential.

12.1.1 Types of Processes

Various membrane processes can be distinguished. These include the following. (See also Figure 12.2.)

Microfiltration (MF) is intermediate between regular filtration and ultrafiltration. The membrane has pores that are fairly wide, i.e., > 0.2 μm, and the pressure difference to be applied is small. The method can be applied, for example, to remove small particles and microorganisms from cheese brine or waste water, and it is also suitable to remove bacteria from skim milk (Subsection 16.1.3). The amount of retentate resulting tends to be quite small. Moreover, it can be used for the standardization of milk to fat content.

Ultrafiltration (UF) effectively separates macromolecules (such as dissolved proteins) from the solution; moreover, any particles present, such as casein micelles, fat globules, somatic cells or bacteria, are also retained. The pore width of the membranes ranges from 3 to 300 nm. A common application is to accumulate protein.

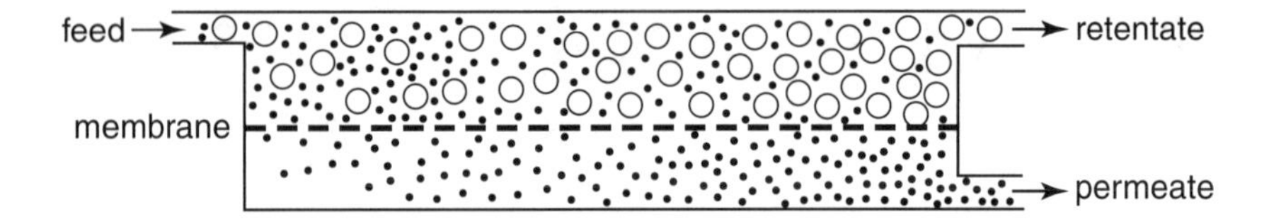

FIGURE 12.1 Highly simplified scheme of membrane filtration. The hydrostatic pressure at the retentate side is higher than that at the permeate side of the membrane.

The process is often applied to whey, and also to skim milk. It can cause appreciable changes in the composition of milk products. Cheese milk may be concentrated to such an extent that the composition approaches that of curd; the retentate can then be renneted to directly yield curd. Milk can be standardized to protein content by either adding a UF permeate or a UF retentate of skim milk.

Ultrafiltration is an almost unique process; gel filtration can provide comparable results but is hardly applied on an industrial scale, and dialysis is typically a laboratory method.

Nanofiltration (NF) is primarily used for partial desalting of whey or UF permeate; the process also causes considerable concentration of the whey (removal of water). NF is an alternative for electrodialysis. NF membranes are similar to those used in reverse osmosis and have no pores; however, the pressures applied are much lower than in RO.

Reverse osmosis (RO) is applied to remove water. The separation principle is based on the solubility of water, and the very poor solubility of most other components, in the membrane. The process causes a large difference in osmotic pressure between retentate and permeate, and high transmembrane pressures are needed. Reverse osmosis can be an attractive alternative to evaporation because it consumes less energy (Table 10.2). The process is primarily applied to whey. Disadvantages

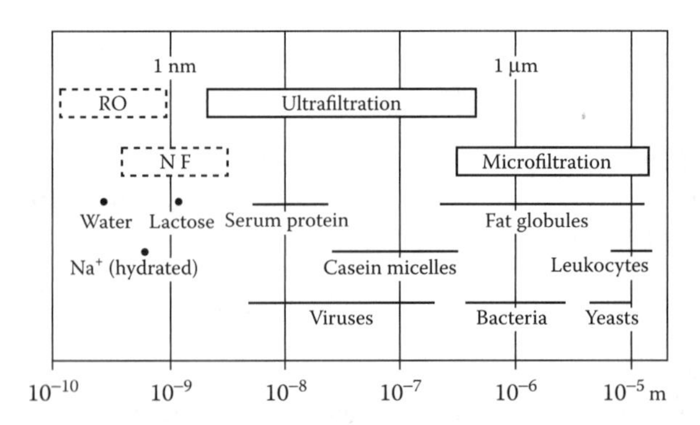

FIGURE 12.2 Approximate particle sizes for which separation by means of membrane filtration can be applied. Fundamentally, reverse osmosis (RO) and nanofiltration (NF) do not separate on a particle-size basis. The size of some molecules and particles in milk is also indicated.

may be that the liquid cannot be as highly concentrated as by evaporation and that the permeate is by no means pure water.

Electrodialysis (ED) is aimed at removing ions, for instance, in the preparation of dietary products or as a step during the manufacture of a purified protein concentrate. Passing milk or whey over ion exchange columns with suitable resins also removes ions.

12.1.2 Efficiency

The efficiency of a membrane process is primarily determined by (1) its *selectivity*, that is, to what extent are the various components to be concentrated retained and have others passed through the membrane; and (2) the *permeate flux*, for example, expressed in kg permeate per m^2 of membrane surface per unit time.

Reflection. The separation of components in a membrane process is never perfect. We define the reflection R^* of solute x as

$$R^* \equiv \frac{q_{\mathrm{w}} - (q_{\mathrm{x}}/c^*)}{q_{\mathrm{w}}} = 1 - \frac{q_{\mathrm{x}}}{q_{\mathrm{w}} c^*} \tag{12.1}$$

Here q_{w} (kg·m^{-2}·s^{-1}) is the flux of solvent, i.e., water, through the membrane and q_{x}, that of solute x. The concentration of x at the pressure side of the membrane is c^*, expressed in kg of solute per kg of water. The concentration of water, c_{w}^* then equals 1.

Ideally, $R^* = 1$ for species to be retained (for example, proteins) and $R^* = 0$ for other species (such as small molecules). In reality, most species have R^* values between 0 and 1. The factors determining the value of R^* will be separately discussed for UF and RO. The main variable is structure or composition of the membrane. The selectivity is primarily determined by the R^* values of the various components of the feed liquid.

Retention. In practice one is interested in the retention of a solute, given by

$$R = 1 - \frac{c_{\mathrm{p}}}{c_{\mathrm{f}}} \tag{12.2}$$

where c_{p} and c_{f} give the concentration (e.g., in kg·m^{-3}) in the permeate and the feed liquid, respectively. At the onset of the process, when the concentration factor Q virtually equals 1, $R = R^*$, but as Q increases, the composition of the permeate (c_{p}) changes, and thereby R. R^* gives a state (flux ratio) at one moment; it can in principle remain constant during the process. R refers to the whole process, and it will always change with time; it can be calculated from known R^* values by integration over the process time.

Diafiltration. If the aim of a membrane separation is the accumulation of a component, say protein, the retentate will always contain a considerable proportion of the species that can pass the membrane, generally the smaller molecules. The retentate must be liquid, and it will thus contain a substantial amount of water; as a first approximation, the concentration of the small molecules relative

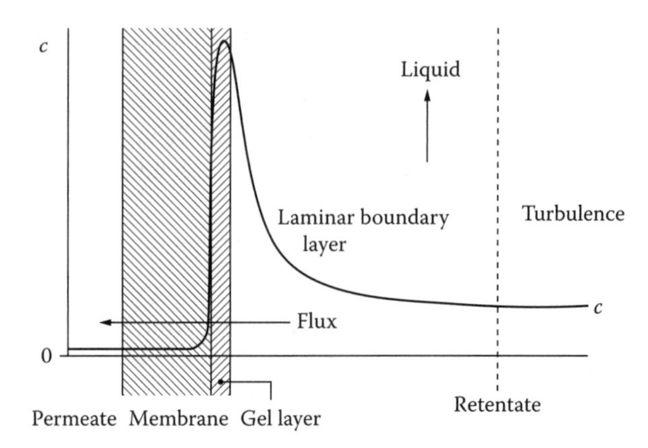

FIGURE 12.3 Concentration gradient and gel layer on a membrane along which the concentrating liquid flows. Schematic. c = concentration of a solute. Usually only a thin layer of the membrane on the retentate side is effective as a filter. The average liquid flow rate along the membrane is, for example, 10^6 times the flow rate across the membrane (the permeate flux).

to water (c^*) will be the same as in the original liquid. To obtain a purer concentrate, the retentate is diluted with water and concentrated again; this is called diafiltration. When applying diafiltration, the retention obtained changes even more strongly with time than in the case discussed earlier.

Permeate Flux. In the simplest case, the flux will primarily depend on the membrane type and be proportional to the transmembrane pressure and inversely proportional to membrane thickness. However, in the first experiments on ultrafiltration, a liquid was brought under pressure in a closed vessel of which part consisted of a (mechanically supported) semipermeable membrane. The flux was very small; moreover, the flux decreased soon to about zero, the higher the pressure, the sooner. The explanation is given with reference to Figure 12.3.

Near the membrane, a strong *concentration gradient* of the component to be retained is formed, often much stronger than depicted in the figure. This in itself will decrease the flux because the local water concentration decreases and the viscosity increases, but another consequence is often that the solubility of the retained component is surpassed, leading to deposition of a gel layer on the membrane. This very much decreases the flux, the more so as the layer becomes thicker.

The solution to this problem is *cross-flow* filtration, which is currently applied in all industrial membrane filtration processes. It is also depicted in Figure 12.3. The feed liquid is pumped at a high velocity along the membrane, resulting in turbulent flow, thereby greatly decreasing the possibility of a concentration gradient forming, except in the laminar boundary layer; hence, the gradient remains relatively small and a gel layer can only be formed after considerable concentration. A prerequisite is that the membrane be quite flat and smooth at the retentate side; otherwise, a gel layer will be formed in dents and crevices in the membrane. (*Note:* The formation of a concentration gradient and a gel layer is generally called

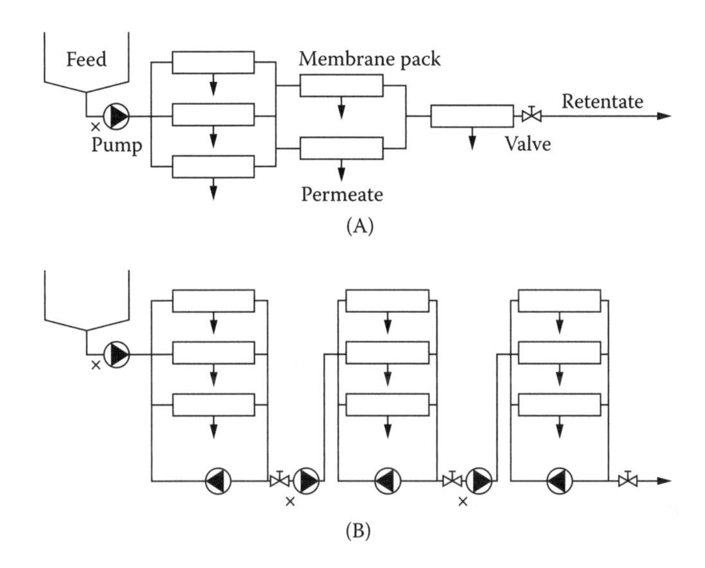

FIGURE 12.4 Arrangements for membrane filtration. Schematic. (A) Simple flow. (B) Flow in several stages with recirculation. Pumps marked with × can resist considerable back pressure; the other pumps are for circulation.

'concentration polarization,' however, polarization — as can occur at an electrode surface — is not involved.) Other factors affecting the flux will be separately discussed for UF and RO.

12.1.3 Technical Operation

A simple cross-flow system is depicted in Figure 12.4A. It can only be used if the degree of concentration is small. At increasing concentration, the viscosity of the retentate markedly increases; hence, a greater part of the transmembrane pressure is lost due to friction in the liquid. This means a decrease of the overall flux.

Consequently, multistage circulation is currently applied; a simplified scheme is in Figure 12.4B. The separation of the plant into sections (hence, the process into stages) considerably decreases the loss of transmembrane pressure. Furthermore, configurations have been devised to keep the pressure reasonably constant through each stage.

Nevertheless, the viscosity of the retentate increases during concentrating, and this is the limiting factor for the flux, also because of gel layer formation. Therefore, high operation temperatures are favorable. Another measure to keep the viscosity low is skimming of the milk before membrane filtration; if required, high-fat cream can be added after concentration.

It may be further remarked that microorganisms can grow in nearly all dairy liquids subjected to, or resulting from, membrane processing; the organisms are accumulated in the retentate. This implies that the process cannot be continued for a long time, unless the temperature is kept either quite low or quite high.

TABLE 12.1
Various Types of Filtration Units: Characteristics

Flat membrane (on supporting plates)
- Relatively small volume per square meter of membrane
- Great pressure loss due to flow resistance
- Can hardly be inspected without dismantling
- Dismantling (e.g., to replace a membrane) easily causes leaky membranes

Spiral-wound flat membrane (pile of membranes alternated with flexible supports wound around a tube)
- Relatively small volume per square meter of membrane
- Low flow resistance
- Hard to inspect
- Leakage requires replacing a complete cartridge

Tubular membrane (in hollow supporting tubes)
- Large volume per square meter of membrane
- Low flow resistance
- Can easily be cleaned, inspected, and replaced
- Suited for ceramic membranes

Membranes. Most membranes are made of polymers. Those for UF are generally made of polyether-sulfone or of polyamide. The latter material, in various compositions, is used for NF and RO membranes. Such a membrane must be very thin and would be quite fragile; consequently, it is attached (by chemical cross-linking) to a far thicker porous support layer. The currently used membranes can withstand high temperatures. A process temperature of, say, 55°C is applied, which means (1) a low liquid viscosity, hence a relatively high permeate flux, (2) virtually no bacterial growth, and (3) very little protein denaturation or other chemical changes. The membranes are also reasonably resistant to cleaning with acid and alkali (Section 14.2).

Ceramic membranes cannot be made with small pores of equal diameter and are only used in microfiltration. They are robust and can withstand high temperatures and aggressive cleaning agents.

Supporting structures for the membrane and the shape of the retentate space can vary widely. Currently, three main types are applied; some properties are given in Table 12.1. Plates with flat membranes have been much used, but they are generally replaced now by spiral-wound membranes. Tubular systems are now exclusively used for ceramic membranes.

12.2 ULTRAFILTRATION

12.2.1 Composition of the Retentate

An ultrafiltration membrane is a filter with very narrow pores (see Figure 12.2) through which most molecules and ions can pass, whereas macromolecules and particles are retained. In principle, water activity, ionic strength, and pH are equal on either side of the membrane. In the retentate, protein accumulates and its properties, including conformation, remain essentially unaltered. The ratio between, for instance, protein and sugar in the retentate changes considerably. Consequently, a retentate of skim milk has a composition completely different from that of evaporated skim milk and, as a result, has different properties: it exhibits weaker Maillard reactions during heating, is much more heat stable at an identical protein content, and has a higher viscosity at identical dry-matter content (Figure 4.7B). Similar differences are observed for whey.

The separation during ultrafiltration is not perfect. The reflection R^* (see Equation 12.1) is a function of molecular size, and that function depends on the type of membrane involved (see Figure 12.5a). R^* changes gradually with molar mass, partly because of a spread in pore width in a membrane, and differences in spread explain the differences in the slope of the curves. Even for identical pore widths, however, R^* will gradually change with molar mass; see curve 1. This is because the pores in the membrane exert a mechanical sieve action on the movement of even small molecules. The nearer the molecular size is to the pore width in the membrane, the greater the resistance. This is illustrated in Figure 12.5b. If the pores cause the same resistance to the solute as to the water molecules, $R^* = 0$.

R^* not only depends on the type of membrane, but for small molecules it also increases to some extent with the transmembrane pressure Δp (see Equation 12.5); the pressure generally is between 0.1 and 0.5 MPa. Often, R^* also depends on the presence and thickness of a gel layer. A further complication is that activities in the solution rather than concentrations are the relevant variables in Equation 12.1. At high concentrations the difference may be considerable, partly because part of the water is not available as a solvent.

In the ultrafiltration of skim milk and whey, low-molar-mass proteins or peptides are not fully retained ($R^* < 1$). Usually, for lactose R^* varies from 0.02 to 0.15, and for citrate from 0.01 to 0.10. Anions in milk serum are, on average, larger than cations. The ensuing difference in flux has to be counterbalanced by a flux of hydroxyl ions, which causes the pH of the permeate to be 0.04 to 0.10

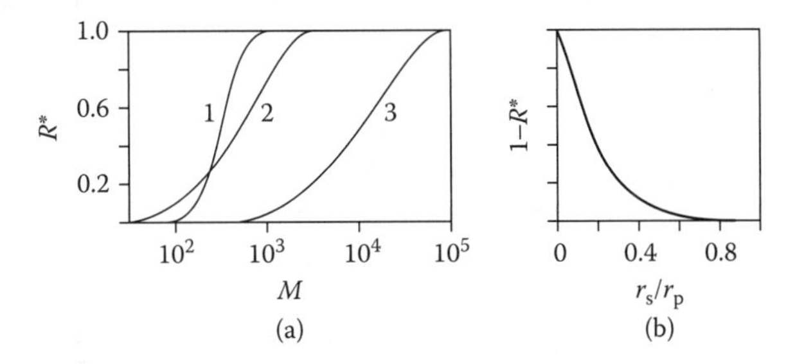

FIGURE 12.5 Dependence of the reflection R^* of ultrafiltration membranes on molecule or particle size. (a) Approximate examples of the dependence on molar mass of the solute for three membranes; in membrane 1 the pores have equal diameter. (b) The relative passage velocity $(1 - R^*)$ of a sphere of radius r_s through a cylindrical pore of radius r_p, as a function of their ratio; approximate calculated result.

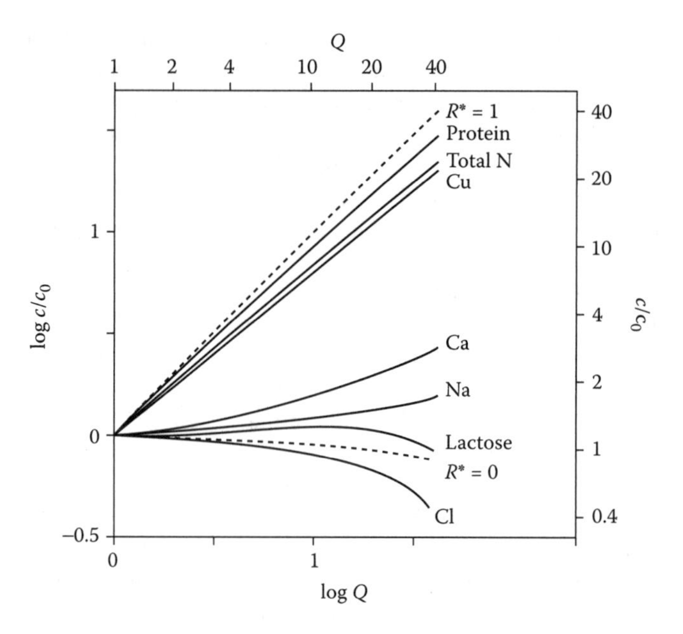

FIGURE 12.6 The ratio of the concentration (c, expressed per m^3 of retentate) of some components in the retentate of ultrafiltered sweet whey to their original concentration (c_0) as a function of the concentration factor Q (= original volume/retentate volume). Approximate examples. (Mostly adapted from J. Hiddink, R. de Boer, and D.J. Romijn, *Neth. Milk Dairy J.*, **32**, 80, 1978.)

higher than that of the original liquid. Figure 12.6 illustrates some of the aspects mentioned. Calculated curves are drawn for $R^* = 1$ and $R^* = 0$. The curve for $R^* = 0$ is not horizontal. This is because at a constant and equal ratio of solute to water (c^* as defined with Equation 12.1) on either side of the membrane, the solute concentration in the total volume (c in Figure 12.6) decreases during concentration, as the water content of the retentate decreases. (In calculating the curve for $R^* = 0$, water not available as a solvent was not taken into account.)

Figure 12.6 shows that there is no total reflection for protein. The reflection for total N is still smaller, because it comprises nonprotein nitrogen (NPN), for which R^* mostly approximates zero. (NPN constitutes almost 25% of the nitrogen in whey!) Obviously, any component that closely associates with protein, such as Cu, is retained in the retentate; to a lesser extent this is true of the counterions of the negatively charged protein, in this case, cations. The figure also shows that Ca^{2+} is present in a relatively higher concentration as a counterion in the diffuse double layer than Na^+, which is in line with theory. Mutatis mutandis co-ions, especially Cl^-, are more than proportionally removed with the permeate, resulting in $R^* < 0$. This is because they have a decreased concentration in the diffuse double layer around the protein molecules.

TABLE 12.2
Composition of the Dry Matter of Retentate Obtained by Ultrafiltration of Whey

Q	1	5	10	20	35	35[a]	20
Dry matter (%)	6.6	10	14	20	25	22	17
pH during ultrafiltration	—	6.6	6.6	6.6	6.6	6.6	3.2
Protein/dry matter (%)	12	34	45	58	70	82	59
Lactose/dry matter (%)	74	51	39	27	17	7	27
Ash/dry matter (%)	8	6	5	4	3.5	2.5	2.7
Citrate/dry matter (%)	2.5	1.8	1.7	1.6	1.4		
Fat/dry matter (%)	1	2	3	4	5	6	4

Note: Approximate examples; Q = concentration factor = volume reduction factor.

[a] Followed by diafiltration, i.e., water is added to increase the volume by a factor of 3, and the mixture is ultrafiltered again.

So far we have discussed reflection. As mentioned in relation to Equation 12.2, retention is the variable of practical interest, and it gradually changes when the concentration factor Q increases. The change in retention is illustrated by the changes in composition of the retentate resulting during ultrafiltering of whey, as given in Table 12.2. One of the columns in the table shows the effect of diafiltration, and it is seen that it considerably enhances the protein content of the dry matter in the retentate. Further 'purification' of the retentate can be achieved by prior RO followed by lactose crystallization and removal of the crystals, and by NF of the retentate to remove part of the salts (see Section 12.4).

12.2.2 Permeate Flux

The volume flux ($m \cdot s^{-1}$) is the quantity q of liquid that passes the membrane per unit time and surface area. Applying the equation of Darcy to a membrane yields

$$q = \left(\frac{B}{h}\right)\frac{\Delta p}{\eta} \tag{12.3}$$

where B is the permeability coefficient of the membrane, h is effective thickness of the membrane, Δp is the transmembrane pressure, and η is viscosity of the permeating liquid. B is approximately proportional to the square of the pore width and to the surface fraction occupied by pores. Consequently, q increases when the pore width is enlarged, but that is at the expense of the selectivity of the membrane. The flux mostly is not precisely proportional to Δp, because at higher pressures B somewhat decreases due to compression of the membrane. Often, (B/h)

is on the order of 10^{-12} m, which implies that the flux of water through an ultrafiltration membrane approximates 400 $kg \cdot m^{-2} \cdot h^{-1}$ at Δp = 100 kPa (1 bar).

The flux achieved during ultrafiltration of whey or skim milk is many times smaller than that during ultrafiltration of water. The following are possible causes:

1. The viscosity of the permeating liquid is higher than that of water, for example, by 20%.
2. Protein molecules immediately adsorb onto the membrane (cf. Figure 14.1), also in the pores, and thereby reduce the effective pore width. The effect can be considerable. Obviously, the narrowing of the pores increases the selectivity.
3. Part of the solutes is retained. Any retention of solute causes a difference in osmotic pressure $\Delta \Pi$ on either side of the membrane, resulting in a somewhat smaller effective pressure difference $\Delta p - \Delta \Pi$. The decrease is small; say, 10%.
4. A concentration gradient is formed because liquid passes the membrane and part of the material in that liquid cannot pass. This was discussed in relation to Figure 12.3. The phenomenon intensifies the effect mentioned under 3, which, however, remains small.
5. A gel layer of protein can be formed at high Q-values (see Figure 12.3), and this causes a further reduction of the flux; the more so, the thicker the layer and the higher the pressure because a higher pressure compresses the gel layer, thereby narrowing its pores. Usually, a gel layer also improves the reflection of many constituents; in other words, it enhances the membrane selectivity.

The preceding facts may explain the effects of some process and product variables on the ultrafiltration flux, at least qualitatively. The composition of the liquid at the pressurized side is of paramount importance because that determines whether a gel layer can be formed or not. The pH strongly affects the solubility of the protein.

Accordingly, the flux will be at a minimum near the isoelectric pH of the protein (Figure 12.7). Moreover, the solubility of calcium phosphates is essential because phosphates are important constituents of the gel layer. Because of this, the following steps increase the flux: removal of calcium (e.g., by electodialysis), increasing the pH to 7.5 (some calcium salts precipitate before the ultrafiltration and hardly enter the gel layer), and preheating (e.g., for 30 min to 55°C, which has a comparable effect). However, the extent of concentration, expressed, for instance, as the protein content of the retentate, has an overriding effect on the flux. Once that concentration is high, a substantial gel layer always forms. This strongly slows down ultrafiltration, despite any measures taken. Incidentally, even a small release of permeate from a highly concentrated retentate causes appreciable further concentration (cf. Figure 24.13).

Formation and thickness of the gel layer depend on the hydrodynamic conditions. This is because a concentration gradient can only form in the laminar boundary layer (Figure 12.3). The intensity of the turbulence determines the

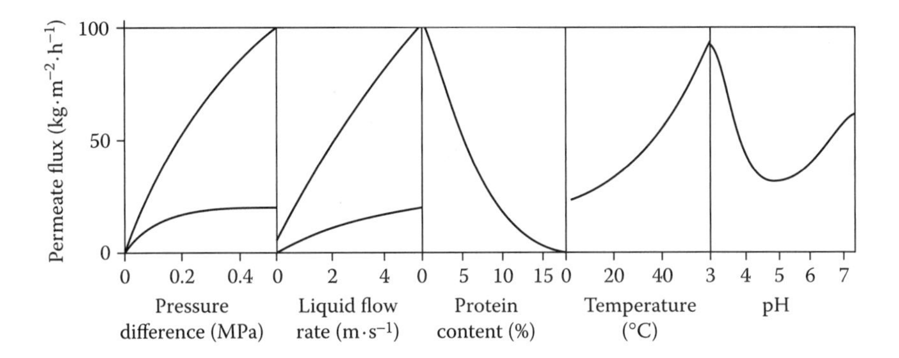

FIGURE 12.7 The influence of some process and product variables on ultrafiltration rate. Approximate examples for whey. Two curves drawn refer to different protein contents, where the upper one refers to nonconcentrated whey.

thickness of the boundary layer. The flow rate of the retentate thus has a considerable effect, as has the geometry of the space containing the retentate. The increase in viscosity due to concentration eventually also causes reduced turbulence; hence, a greater tendency to form a gel layer. Even at constant protein content, i.e., constant Q, the permeate flux eventually decreases, partly due to increasing gel layer density.

Increasing the transmembrane pressure naturally increases the permeate flux, but not proportionally. This is because a higher flux causes a greater concentration gradient, therefore, there is more possibility of formation of a gel layer, which has a counteracting effect on the flux. A higher operation temperature implies a lower liquid viscosity, resulting in a larger flux.

However, we need to also consider the operation time. Fast formation of a gel layer means that the flux rapidly diminishes and that operation must be stopped to clean the installation. A gel layer is formed faster at a higher temperature, because (1) the solubility of calcium phosphate is smaller and (2) at temperatures above 55°C, denaturation of serum proteins will decrease their solubility. Currently, there is a trend to increase the operation time by working at a low temperature. Because this means a smaller flux, the membrane surface must then be larger. The availability of relatively cheap membranes has made it profitable to do this.

12.3 REVERSE OSMOSIS

Reverse osmosis differs from ultrafiltration in the application of much higher pressures (2 to 5 MPa). It removes water against an osmotic pressure. The osmotic pressure Π is considerable. For nonconcentrated milk or whey, $\Pi \approx 0.7$ MPa (7 bar), and it increases during concentration by removal of water according to $\Pi \approx 0.7\ Q^*$, where Q^* = relative increase of the dry-matter content in proportion to water (Subsection 10.1.1). Clearly, the membrane is semipermeable. It does not act as a filter with narrow pores but rather as a layer of material in which water can dissolve and through which it can pass, whereas most of the other components cannot do so or can barely do so. Transport of a component occurs by diffusion through the membrane or, essentially, through the thin layer of the membrane that is semipermeable. Consequently, Equation 12.3 does not apply.

The flux q (in $kg \cdot m^{-2} \cdot s^{-1}$) of a substance x through the membrane is, in principle, given by

$$q = \frac{D^* K \Delta c}{\delta} \tag{12.4}$$

D^* is the effective diffusion coefficient ($m^2 \cdot s^{-1}$) of x in the membrane; its magnitude sharply decreases with increasing molar mass of x. K is the partition coefficient of x, i.e., the ratio of its solubility in the membrane to that in water; it is very much smaller than unity, and it is negligible for large hydrophobic molecules. Δc is the concentration of x ($kg \cdot m^{-3}$) in the retentate minus that in the permeate. δ is the thickness of the membrane (m); it is generally very small because it only concerns the thin active layer of the membrane.

Obviously, the retentate obtained by reverse osmosis differs somewhat from the concentrate after evaporation, and the permeate is by no means pure water. The reflection coefficient R^* (Equation 12.1) of small molecules is 0.75 to 0.99, greatly varying with the composition of the membrane. Urea can pass to some extent, and even lactose, some salts, and low-molar-mass peptides may do so. Accordingly, bacteria can grow in the permeate. Volatile flavor substances are satisfactorily retained.

R^* of most components increases with the transmembrane pressure Δp and decreases with increasing difference in osmotic pressure $\Delta \Pi$ on either side of the membrane (hence, with increasing retentate concentration). For an aqueous solution of solute x, the water flux will be proportional to $\Delta p - \Delta \Pi$ and the flux of x to $\Delta p + \Delta \Pi$. In other words, Equation 12.1 should be adjusted as follows:

$$R^* = 1 - \frac{(q_{x,0}/c^*)(\Delta p + \Delta \Pi)}{q_{w,0}(\Delta p - \Delta \Pi)} \tag{12.5}$$

where $q_{x,0}$ and $q_{w,0}$ are the solute and water fluxes, respectively, for the hypothetical case that $\Delta \Pi = 0$. For example, $\Delta \Pi / \Delta p$ initially is about 0.2, and it becomes 0.8 toward the end of the process. Suppose now that $R^* = 0.9$ for $\Delta \Pi = 0$. Then, according to Equation 12.5, during the process R^* decreases from 0.85 to 0.1. R^* of some compounds can even become negative if the liquid is highly concentrated. Fortunately, R^* of most solutes is much higher, and it decreases, for instance, from 0.985 to 0.91 when $\Delta \Pi / \Delta p$ changes from 0.2 to 0.8. Moreover, the total amount of permeate released toward the end of the process is only small. On the other hand, a concentration gradient developing near the membrane (Figure 12.3) raises the effective $\Delta\Pi$ above the average value. Furthermore, increased concentration reduces the retention R at constant reflection R^*. All in all, the last bit of permeate released during reverse osmosis may be more like ultrafiltration permeate than like water. This, of course, depends on the type of membrane involved.

Applying reverse osmosis to whey or skim milk results in a smaller flux than when pure water is used. This is comparable to conditions during ultrafiltration; however, among the causes listed in Subsection 12.2.2, the first two will hardly have an effect, because the viscosity of the permeate is and remains similar to that of water, and there are no pores that can be fouled. The increase of the osmotic pressure, on the other hand, is of paramount importance. The water flux is proportional to $\Delta p - \Delta\Pi$ (see the preceding text). Suppose that skim milk is concentrated to 30% dry matter (a higher concentration is usually not reached) and that a pressure difference $\Delta p = 4$ MPa is applied. Initially, $\Delta p - \Delta\Pi = 4 - 0.7 = 3.3$ MPa. For 30% dry matter, $Q = 30/9.3 = 3.23$ and $Q^* = 4.18$. This will lead to $\Delta\Pi \approx 2.9$ MPa; hence, $\Delta p - \Delta\Pi = 1.1$ MPa. In other words, the effective transmembrane pressure is reduced to one third, or even less because the concentration gradient near the membrane can increase considerably. Finally, a gel layer may form, especially at high Δp; a high Δp initially causes a large flux, and therefore, a strong concentration gradient. This layer reduces the flux still further. Figure 12.8A illustrates all of these aspects.

Figure 12.8B gives examples of the flux at increasing concentration, as obtained during reverse osmosis of some liquids. The differences involved greatly depend on differences in formation of a gel layer. Micellar casein readily forms such a layer. The proteins in whey do not form a gel layer at pH ≥ 6, but they do at pH = 4.6, at which point they are far less soluble. Calcium phosphate is saturated in

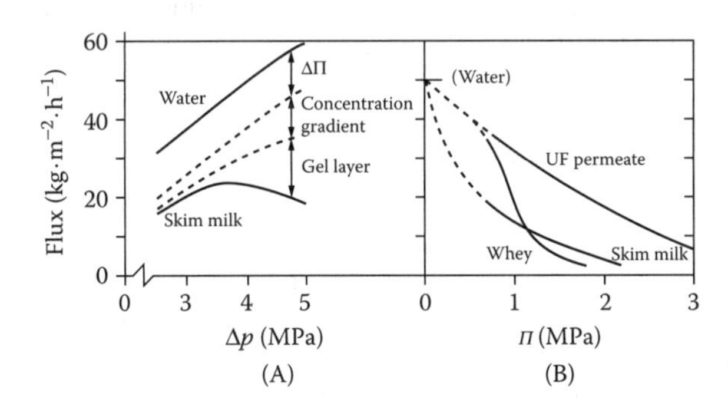

FIGURE 12.8 The effect of some process and product variables on the permeate flux during reverse osmosis. Approximate examples for pH ≈ 6.7 and 30°C. (A) Flux of pure water and slightly concentrated skim milk ($Q^* \approx 1.5$) as a function of the transmembrane pressure Δp. The probable causes of the flux difference are shown schematically. (B) Flux of some liquids as a function of the concentration, expressed in the osmotic pressure Π of the retentate. (Data from J. Hiddink, R. de Boer, and P.F.C. Nooy, *J. Dairy Sci.*, **63**, 204, 1980.)

milk and whey; accordingly, it plays a part in the formation of a gel layer, especially in whey at neutral pH. This is because casein can accommodate insoluble calcium phosphate far better than serum proteins can. At pH = 6, calcium phosphate is less readily supersaturated. Therefore, in reverse osmosis of whey the following results for the water flux (in $kg \cdot m^{-2} \cdot h^{-1}$) may be found:

pH of whey	6.6	6.0	4.6
Initial flux	37	37	27
Flux when $Q \approx 3$	5	20	15

Figure 12.8B also shows that the flux remains high during reverse osmosis of an ultrafiltrate of skim milk or whey. This is because little, if any, gel layer is formed. Accordingly, a much higher Δp can be applied without the flux decreasing again (as in Figure 12.8A).

Reverse osmosis is never applied to (whole) milk and rarely to skim milk: the flux is small and the maximum concentration attainable is low. The process is extensively applied to whey and to various ultrafiltrates, especially as a first step in the manufacture of powder. Concentration of cheese whey to a dry-matter content of 25 to 28% is possible, and the retentate is generally further concentrated by evaporation before spray drying (however, it is currently more economical to concentrate by NF to about 18% dry matter before evaporation). The permeate flux is virtually proportional to the viscosity of water (see Table A.10), and therefore, it strongly increases with temperature. If RO is done at low temperature, crystalli-zation of lactose can readily occur (see Figure 10.11), which should be avoided. At high temperature, say, 55°C, lactose crystallization does not pose a problem.

The technical operation is briefly discussed in Subsection 12.1.4. The very high pressures needed (2 to 4 MPa) necessitate several adaptations, as compared to, for instance, ultrafiltration.

12.4 DESALTING

Desalting is for the most part applied to whey. It can be achieved in various ways. In all of these, ions are removed and not undissociated salts (see Subsection 2.2.2). This implies, for instance, that at a low pH organic acids are hardly removed because the carboxyl group is incompletely dissociated. The presence of proteins, which generally have a negative charge, impedes the removal of counterions (especially Ca^{2+}) and enhances that of co-ions (for the most part Cl^-).

Electrodialysis (ED). Here the ions are removed from the whey due to an electric potential gradient and the presence of two kinds of membranes, one kind being permeable to anions and the other to cations. Transmembrane pressure is zero. Generally, the whey is preconcentrated by RO or evaporation to about 20% dry matter before ED; this increases the ion concentrations and decreases the costs. Besides desalted whey, two salt solutions are produced that must be regenerated.

The rate of removal varies with the kind of ions. Examples of the percentages removed when desalting whey are:

Removal by	**ED 40%**	**ED 60%**	**NF 45 %**
K + Na	42	64	45
Ca	24	35	6
Cl	71	89	54

where a result for nanofiltration is included for comparison; the percentage figures of ED and NF refer to the overall proportion of salts removed. ED can achieve 80% removal but one generally stops at 60% for economic reasons. To achieve further desalting, ED can be followed by ion exchange.

Ion Exchange. Whey can be passed over ion exchange columns for desalting. The removal rate of the various ions depends on the properties of the resins applied, but very low salt contents can be reached. A disadvantage is that the resin must be frequently regenerated, which requires large quantities of chemicals and water.

Nanofiltration. Although this method is less efficient for salt removal than ED, it has the advantage that the liquid is also concentrated, which often is desired anyway. Consequently, it is generally more cost-effective than ED, and it has become the method of choice for partial desalting of whey.

Nanofiltration evolved because membrane materials were developed that have a high solubility — therefore, a high permeability — for monovalent ions and water but not for other components. Generally, the permeability to other salts is low, whereas proteins, lactose, urea, and other small nonionic molecules are almost totally reflected. Examples of the percentage removed for some ions as a function of pH are given in the following table:

pH of whey	4.6	5.8	6.6
Ca + Mg	4	2	2
Na + K	62	46	35
Anions	40	37	29

The anions removed consist for the most part of Cl^-. It is seen that desalting works better at a lower pH.

In most cases, the whey is concentrated to 20% dry matter and an overall salt removal of 40 to 50%. Considerably greater removal of monovalent ions can be reached by diafiltration; most other ions are still retained. Also, the whey obtained from Cheddar curd after salting, which contains a very high NaCl content, can be effectively desalted.

The membranes are comparable to those used for RO in structure and composition. However, the difference in osmotic pressure between retentate and permeate is not very high. This is because the concentration factor reached is not very high, and the monovalent ions passing the membrane make up a significant part of the osmotic pressure of whey. Hence, the transmembrane pressure need not be as high as for RO; it is generally 0.2 to 0.4 MPa. This means that the technical operation is similar to that applied in ultrafiltration. Nanofiltration is generally done at a low temperature, e.g., 25°C, because the selectivity of the separation decreases with increasing temperature.

Suggested Literature

A clear and comprehensive textbook, which emphasizes basic aspects of all membrane processes, but does not give much on applications: M. Mulder, *Basic Principles of Membrane Technology,* Kluwer Academic, Dordrecht, 1991, 2000.

A comprehensive handbook of both theoretical and practical aspects of all kinds of membrane separation: K. Scott, *Handbook of Industrial Membranes,* Elsevier, Oxford, 1995.

Enlightening studies of ultrafiltration and reverse osmosis of dairy liquids: J. Hiddink, R. de Boer, and D.J. Romijn, *Neth. Milk Dairy J.,* **32**, 80, 1978; J. Hiddink, R. de Boer, and P.C.F. Nooy, *J. Dairy Sci.,* **63**, 204. 1980.

Newer aspects discussed: New applications of membrane processing, IDF Special Issue 9201, 1992; Advances in membrane technology for better dairy products, IDF Bulletin 311, 1996.

13 Lactic Fermentations

If raw milk is stored, it spoils by microbial action. At moderate temperatures, lactic acid bacteria are generally predominant, and the milk spontaneously becomes sour. Nearly all types of fermented milk products are based on such souring activity of lactic acid bacteria. In this chapter, some of the basic microbial aspects are discussed, as well as some principles of processing, especially starter manufacture and handling. A starter is a culture of specific microorganisms added to milk or another liquid to induce the desired fermentation. Manufacture of fermented milks, with some emphasis on yogurt, is treated in Chapter 22. Lactic acid fermentation is also of importance for cultured butter (Chapter 18) and is essential in the manufacture and ripening of cheese (Part IV of this book).

13.1 LACTIC ACID BACTERIA

Lactic acid bacteria are the prime agents in producing soured (fermented) milk and dairy products. Although they are genetically diverse, common characteristics of this group of bacteria include being Gram-positive, nonmotile, and non-spore-forming. Lactic acid bacteria are unable to produce iron-containing porphyrin compounds, such as catalase and cytochrome. Thus, they grow anaerobically but are aerotolerant. They obligatorily ferment sugars with lactic acid as the major end product. They tend to be nutritionally fastidious, often requiring specific amino acids and B vitamins as growth factors.

13.1.1 Taxonomy

There are currently 12 genera of lactic acid bacteria, of which 4 contain organisms used in dairy fermentations: *Lactococcus, Leuconostoc, Streptococcus*, and *Lactobacillus*. A fifth genus, *Enterococcus*, is occasionally found in undefined starter cultures. Important phenotypic taxonomic criteria to discriminate between the genera and species are: morphological appearance (coccus or rod), fermentation end products (homofermentative or heterofermentative), carbohydrate metabolism, growth temperature range, optical configuration of the lactic acid produced, and salt tolerance. These major classification properties of the five genera and some of the occurring species are summarized in Table 13.1.

Modern genetic methods allow certain specific DNA sequences in bacteria to be selected and determined. Among them are sequencing of 16S rDNA, often in combination with randomly amplified polymorphic DNA (RAPD) techniques. The 16S rDNA gene is relatively large (15,000 nucleotides), and some of its sequences are highly conserved, whereas others are variable. The conserved

TABLE 13.1
Classification Characteristics of Some Lactic Acid Bacteria Involved in Fermentation of Milk and Dairy Products

Genus	Species	Morphology	Growth at 10°C	Growth at 45°C	Temperature Dependence	Fermentation Lactose[a]	Lactic Acid isomer[b]	Citrate Metabolism	Diacetyl Production	EPS[c] Formation	Tolerance Against 6.5% Salt
Lactococcus	*L. lactis* ssp. *lactis*	Coccus	+	–	Mesophilic	Homof	L	–	–	+/–	–
	L. lactis ssp. *lactis* biovar. *diacetylactis*	Coccus	+	–	Mesophilic	Homof	L	+	+	–	–
	L. lactis ssp. *cremoris*	Coccus	+	–	Mesophilic	Homof	L	–	–	+/–	–
Leuconostoc	*Leu. mesenteroides* ssp. *cremoris*	Coccus	+	–	Mesophilic	Heterof	D	+	+	+/–	–
	Leu. lactis	Coccus	+	–	Mesophilic	Heterof	D	+	+	+/–	–
Enterococcus	*E. faecalis*	Coccus	+	–	Mesophilic	Homof	L	+/–	–		+
	E. faecium	Coccus	+	–	Mesophilic	Homof	L	+/–	–		+
Streptococcus	*S. thermophilus*	Coccus	–	+	Thermophilic	Homof	L	–	–	+	–
Lactobacillus	*Lb. helveticus*	Rod	–	+	Thermophilic	Homof	D L	–	–		–
	Lb. delbrueckii ssp. *bulgaricus*	Rod	–	+	Thermophilic	Homof	D	–	–	+	–

Lb. delbrueckii ssp. *lactis*	Rod	–	+	Thermophilic	Homof	D	–	–	–
Lb. acidophilus	Rod	–	+	Thermophilic	Homof	D L	–	–	–
Lb. casei	Rod	+	–	Mesophilic	Homof[d]	L	–	–	+/–
Lb. plantarum	Rod	+	–	Mesophilic	Homof[d]	D L	–	–	+/–
Lb. fermentum	Rod	–	+	Thermophilic	Heterof[e]	D L	–	–	

[a] Homof: homofermentative; heterof: heterofermentative; [d]some strains are facultatively heterofermentative; [e]obligatively heterofermentative (see Subsection 13.1.2.1).

[b] L: dextrorotatory lactic acid; D: levorotatory lactic acid (see Subsection 13.1.2.2).

[c] EPS: exocellular polysaccharide.

+/–: some strains do, others do not.

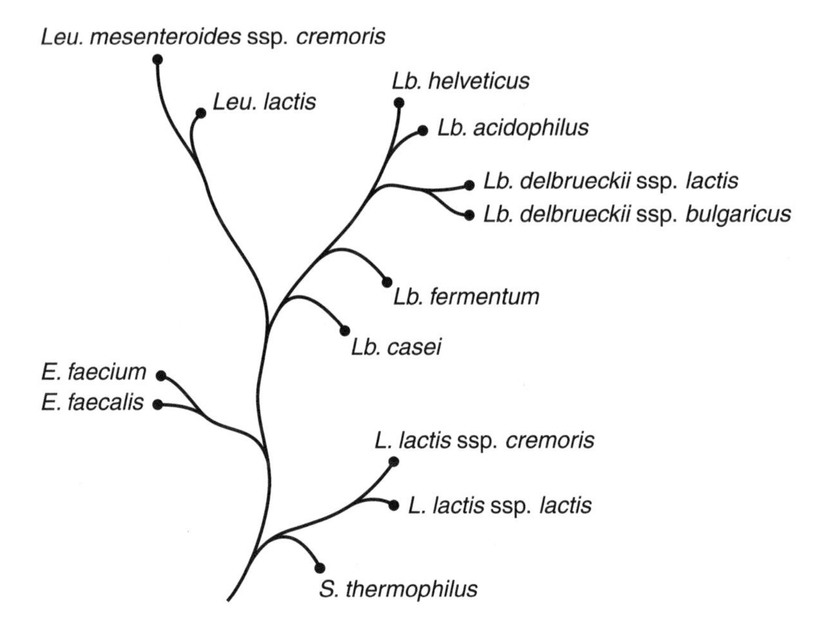

FIGURE 13.1 Phylogenetic tree showing the relationships among some lactic acid bacteria. *E.: Enterococcus; L.: Lactococcus; Lb.: Lactobacillus; Leu.: Leuconostoc;* and *S.: Streptococcus*. (Adapted from P.F. Fox et al., *Fundamentals of Cheese Science,* Aspen, Gaithersburg, MD, 2000.)

region contains genus-specific sequences, whereas the variable regions contain species- or subspecies-specific sequences. The RAPD technique is often used to distinguish strains of a bacterial species, because it forms strain-specific patterns of DNA sequences by the amplification with arbitrarily chosen primers of approximately ten nucleotides. This technique has been characterized as a chromosomal fingerprinting method. Both methods are used to accurately determine phylogenetic relationship among bacteria. These relationships can be expressed in relative distances between genera and species, which results in a phylogenetic tree as visualized in Figure 13.1.

13.1.2 Metabolism

All lactic acid bacteria require organic carbon as a source of carbon and energy. They cannot derive energy via respiratory activity, and therefore, they rely primarily on reactions that occur during glycolysis to obtain energy in the form of adenosine triphosphate (ATP). There are other means by which these organisms can conserve energy and save ATP, however, and this energy would ordinarily be used to perform functions like nutrient transport.

During the growth of bacteria in milk, the major energy source is lactose, which is converted into lactic acid and other products via two different metabolic routes. Another metabolic feature of some lactic acid bacteria is the ability to metabolize citric acid, which is a minor component in milk. It may be converted into diacetyl, which is an important flavor compound in fermented dairy products. Dairy lactic acid bacteria are also well adapted to use casein as a source of nitrogen. They are able to degrade this protein and the derived peptides to satisfy their amino acid requirement. All these metabolic activities will be described in detail in the following subsections.

13.1.2.1 Metabolism of Lactose

Uptake and hydrolysis. The first step in the metabolism of lactose is its transport into the cell. There are two main systems used by lactic acid bacteria to transport lactose across the cell membrane. One is the group translocation mechanism, in which lactose is phosphorylated during passage across the cytoplasmic membrane. The initial source of phosphate is the energy-rich glycolysis intermediate phosphoenol pyruvate (PEP), and the mechanism is called the phosphoenol pyruvate phosphotransferase system (PEP-PTS). Four proteins are involved in this system (Figure 13.2). One, enzyme II (EII), is a cytoplasmic membrane protein that translocates a specific sugar from the outside to the inside of the cytoplasmic membrane, where it is phosphorylated by a second sugar-specific protein, enzyme III (EIII), which resides in the cytoplasm. The two other proteins, enzyme I (EI) and HPr (histidine-containing protein), are nonspecific cytoplasmic proteins and are involved in transferring a phosphate residue from PEP to EIII. As a consequence of these reactions, lactose-phosphate accumulates intracellularly at a net cost of energy-rich phosphate. The lactose-phosphate is hydrolyzed to glucose and galactose-6-phosphate by a phospho-β-galactosidase (P-β-gal). Glucose is converted to glucose-6-phosphate, and both sugar phosphates are metabolized further. The PEP-PTS system is characteristic for lactococci.

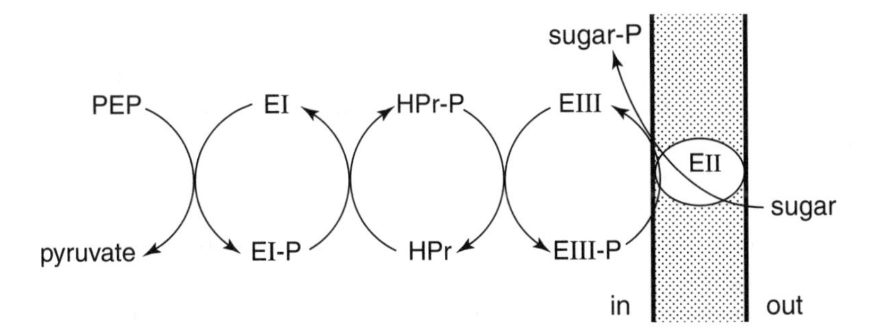

FIGURE 13.2 Schematic representation of the phosphotransferase (PTS) system of sugar transport over the membrane. (Adapted from V. Monnet, S. Condon, T.M. Cogan, and J.-C. Gripon. In: T.M. Cogan and J.-P. Accolas, Eds., *Dairy Starter Cultures,* VCH Publishers, Cambridge, U.K., 1996.)

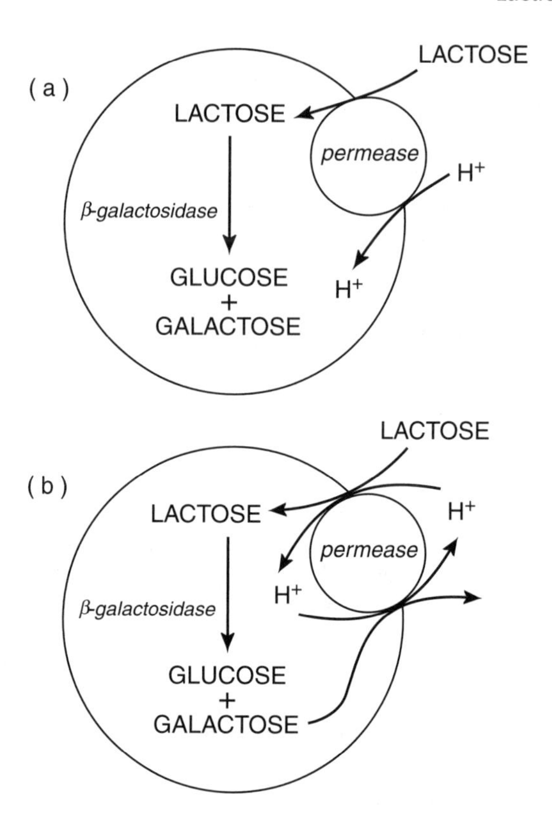

FIGURE 13.3 Models for uptake of lactose via lactose permease. (a) Lactose/H^+ symport; (b) lactose/galactose antiport. (Adapted from V. Monnet, S. Condon, T.M. Cogan, and J.-C. Gripon. In: T.M. Cogan and J.-P. Accolas, Eds., *Dairy Starter Cultures,* VCH Publishers, Cambridge, U.K., 1996.)

The second system for transport of lactose into cells of lactic acid bacteria involves cytoplasmic proteins (permeases) that translocate the sugar into the cell without chemical modification. Situated in the cytoplasmic membrane, the lactose permease translocates lactose together with protons into the cytoplasm (Figure 13.3a). The lactose permease is an active transport system, and the energy is provided in the form of a proton motive force developed by the transmembrane ATPase at the expense of ATP hydrolysis. Some thermophilic lactic acid bacteria, when grown on lactose, excrete the galactose portion of the molecule in proportion to the amount of lactose taken up (Figure 13.3b). Most of these strains are not able to metabolize galactose further. In these bacteria, a single transmembrane permease simultaneously translocates lactose molecules into the cytoplasm and galactose molecules out of the cell. The energy generated through galactose efflux supports lactose intake. The hydrolysis of lactose into glucose and galactose is accomplished by β-galactosidase (β-gal). This permease uptake system is characteristic of dairy starter bacteria other than *Lactococcus* species.

Further metabolism. In lactococci, the subsequent metabolizing of glucose-6-phosphate is via the glycolytic or Embden–Meyerhof (EM) pathway, and galactose-6-phosphate enters the tagatose pathway (Figure 13.4). Characteristic of these pathways is the presence of aldolases, which are necessary for the hydrolysis of hexose diphosphates to glyceraldehyde-3-P. The only fermentation product of these pathways is lactic acid; hence, this metabolism is named the *homofermentative lactic acid fermentation.* One mole of lactose is transformed to 4 moles of lactic acid. Metabolism of glucose and galactose by thermophilic lactic acid bacteria also occurs via glycolysis, but some of these bacteria ferment only the glucose moiety of lactose, and galactose is secreted as an exchange molecule for the transport of lactose into the cell (see preceding text). In this case the metabolism of lactose yields only 2 mol of lactic acid.

Leuconostoc species and some lactobacilli ferment glucose via the phosphoketolase pathway, and galactose is transformed to glucose-1-phosphate via the Leloir route (Figure 13.4). The presence of glucose-6-P dehydrogenase and phosphoketolase results in the conversion of 6-P-gluconate to CO_2 and a pentose-5-P, which in turn is converted to glyceraldehyde-3-P and acetyl-P. The conversion of glyceraldehyde-3-P to lactic acid is by the glycolytic pathway and acetyl-P is converted to ethanol. Thus, besides lactic acid, other products are formed, and therefore this metabolism is named the *heterofermentative lactic acid fermentation.* The heterofermentative lactic acid bacteria lack the key enzyme of the homofermentative pathway, fructose-1,6-diphosphate aldolase. Some lactobacilli, containing this enzyme and fermenting lactose under normal conditions almost entirely to lactic acid, ferment lactose under carbohydrate-limiting conditions to lactic acid, acetic acid, ethanol, and formic acid. They appear to express, then, the activity of several enzymes for the conversion of pyruvate to products other than lactic acid (see Subsection 13.1.2.2). These lactobacilli are often called *facultatively heterofermentative* to distinguish them from the *obligatively heterofermentative* lactobacilli.

Some homofermentative as well as some heterofermentative lactic acid bacteria use the permease system to transport lactose into the cell. These homofermentative bacteria produce fructose-6-phosphate from glucose-6-phosphate, whereas these heterofermentative bacteria produce 6-phosphogluconate, as visualized in Figure 13.4.

Energy yield. Every pathway involves consecutive reaction steps catalyzed by several enzymes. The free-energy content of a metabolite is lower than that of the preceding product (a thermodynamic prerequisite), but not every reaction step provides useful energy. The sequence of the reactions is such that most of the free-energy transfer occurs in one step, leading to formation of an energy-rich phosphate, i.e., ATP, a coenzyme responsible for the transport of phosphate and energy. Formation of ATP from adenosine diphosphate (ADP) occurs at the expense of a reduction of nicotinamide adenine dinucleotide (NAD^+), a coenzyme essential for transport of electrons and hydrogen, to NADH. NADH has to be oxidized again for the fermentation to proceed. This is because the amount

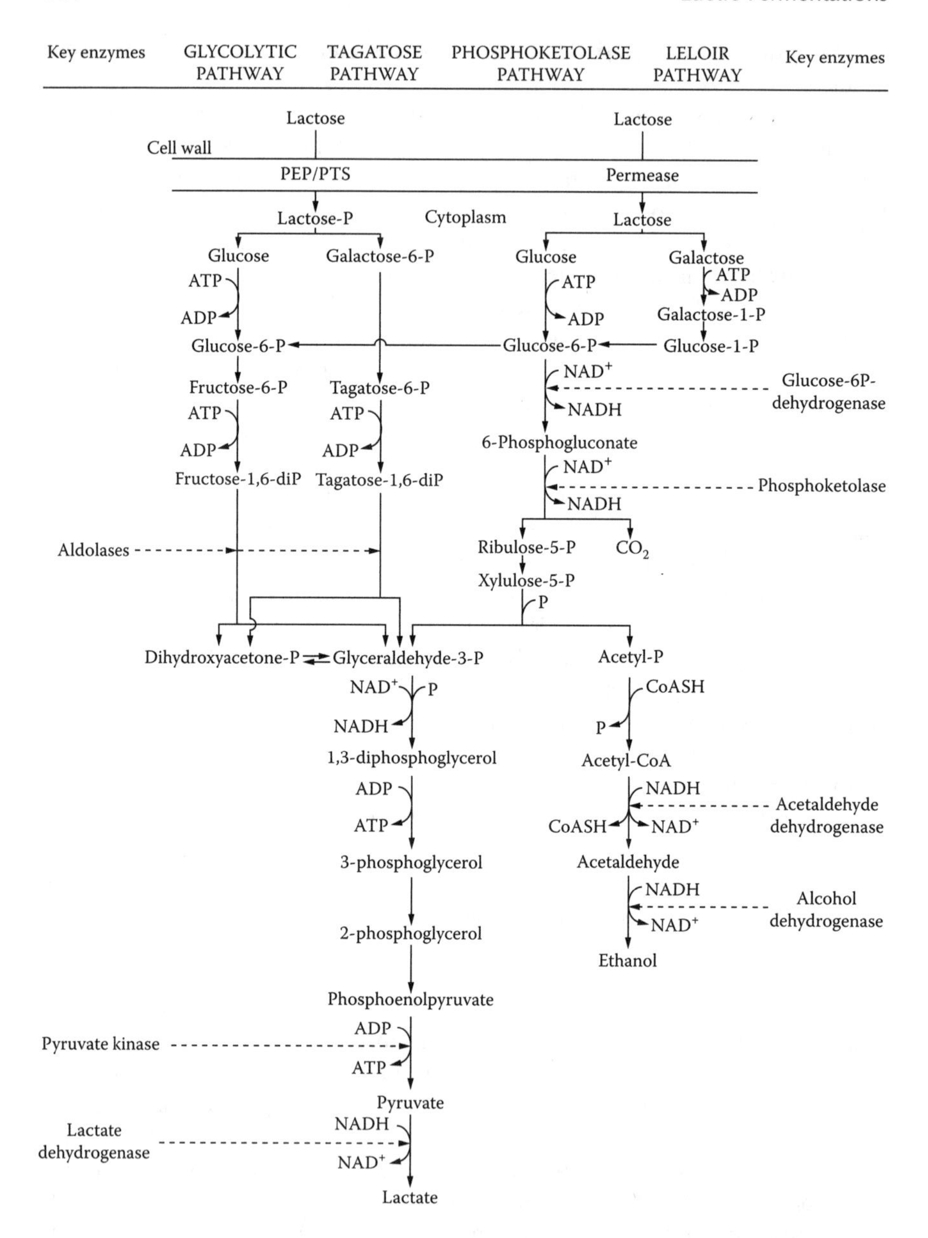

FIGURE 13.4 Metabolism of lactose in lactic acid bacteria. (Adapted from T.M. Cogan and C. Hill. In: P.F. Fox, Ed., *Cheese: Chemistry, Physics and Microbiology,* Vol. 1, *General Aspects,* 2nd ed., 1993.)

of NAD^+ in the bacterial cell is very small. During optimal homofermentative fermentation of 1 mole of lactose to lactic acid, 4 moles of ATP are formed:

$$\text{lactose} + 4\ H_3PO_4 + 4\ \text{ADP} \longrightarrow 4\ \text{lactic acid} + 4\ \text{ATP} + 3\ H_2O$$

Up to 95% of the sugar is converted to lactic acid, and the homofermentative lactic acid bacteria behave as 'lactic acid pumps.'

During the heterofermentative formation of lactic acid, the yield of ATP per mole of lactose is only 2:

$$\text{lactose} + 2\ H_3PO_4 + 2\ \text{ADP} \longrightarrow 2\ \text{lactic acid} + 2\ \text{ethanol} + 2\ CO_2 + 2\ \text{ATP} + H_2O$$

The formation of ethanol requires much energy. Accordingly, the presence of another hydrogen acceptor causes preferential conversion of acetyl-P to acetic acid by which one ATP is formed (see the following subsection).

13.1.2.2 Pyruvate Metabolism

Type of lactic acid formed. Lactic acid bacteria mainly accomplish the regeneration of NAD^+ by reducing pyruvate to lactate by lactate dehydrogenase. Lactate can exist as the D or L stereoisomer, and lactate dehydrogenases specific for each isomer are found in these bacteria. Certain lactic acid bacteria contain both enzymes, which often differ in activity so that the ratio between the amounts of both isomers produced may vary widely. If the ratio is 1, the product obtained is called a racemic mixture (DL). Some bacterial strains can form such a mixture because they contain lactate racemase, in addition to lactate dehydrogenase. The racemase can transform one isomer into the other.

Other products from pyruvate. Lactate is by far the major product formed from pyruvate by lactic acid bacteria in fermentations involving high concentrations of lactose or glucose. However, at very low concentrations of these sugars, other end products, such as formate, acetate, ethanol, and acetoin, may be formed in significant proportions. The first step in this alternative route of pyruvate metabolism is the conversion of pyruvate to acetyl CoA. This is accomplished by an oxygen-sensitive enzyme, pyruvate formate lyase, which produces formate and acetyl CoA. Alternatively, pyruvate may be reduced by the pyruvate dehydrogenase pathway, during which acetyl CoA and CO_2 are formed, and NAD^+ is reduced to NADH. Acetyl CoA can be converted either to acetate via acetyl phosphate to generate additional ATP or to ethanol via acetaldehyde to redress the NAD^+/NADH imbalance caused by the dehydrogenase step (Figure 13.5). In certain thermophilic streptococci the pyruvate formate lyase enzyme is in the active form even when there is an excessive amount of sugar present. This enzyme accounts for the formate production by *Streptococcus thermophilus*, which is of importance during yogurt manufacture (Section 22.4).

Reduction of oxygen. Fermentation pathways are generally regarded as anaerobic pathways, because ATP can be generated in the complete absence of oxygen.

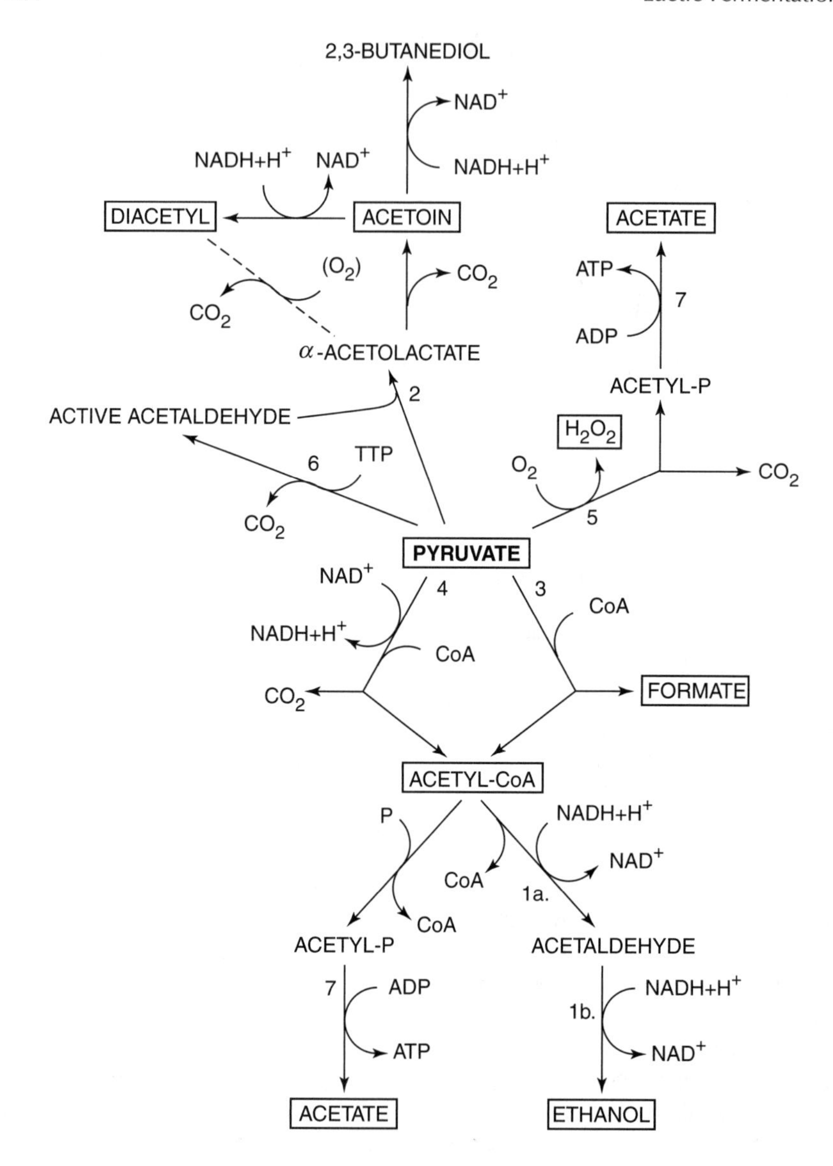

FIGURE 13.5 Pathways for the alternative fates of pyruvate. Dashed arrow denotes a nonenzymatic reaction. Important metabolites and end products are framed. Selected enzymatic reactions are numbered: (1a) acetaldehyde dehydrogenase, (1b) alcohol dehydrogenase, (2) acetolactate synthase, (3) pyruvate formate lyase, (4) pyruvate dehydrogenase, (5) pyruvate oxidase, (6) pyruvate decarboxylase, and (7) acetate kinase. (Adapted from L. Axelsson. In: S. Salminen and A. von Wright, *Lactic Acid Bacteria,* 2nd ed., Dekker, New York, 1998.)

Lactic acid bacteria are aerotolerant, i.e., they can generally cope with toxicity problems caused by oxygen or oxygen metabolites such as superoxide and H_2O_2. Oxygen is usually reduced to water in these bacteria by either of two mechanisms involving flavoproteins:

$$(1) \quad 2\ NADH + 2\ H^+ + O_2 \xrightarrow{NADH(H_2O)oxidase} 2\ NAD^+ + 2\ H_2O$$

$$(2) \quad NADH + H^+ + O_2 \xrightarrow{NADH(H_2O_2)oxidase} NAD^+ + H_2O_2$$

$$NADH + H^+ + H_2O_2 \xrightarrow{NADH\ peroxidase} NAD^+ + 2\ H_2O$$

These reactions cause the redox potential to decrease substantially. Moreover, oxidation of NADH to NAD^+ enables the bacteria to perform alternative dissimilation processes that yield energy. As a result, heterofermentative lactic acid bacteria can produce acetic acid rather than ethanol from acetyl-P, thereby doubling the ATP yield. An NADH oxidase combined with a pyruvate oxidase has been found in some of the lactic acid bacteria but not in most lactococci. Pyruvate oxidase catalyzes the conversion of pyruvate to acetyl-P, which is hydrolyzed into acetic acid and ATP (reaction 5 in Figure 13.5).

The structural formulas of important metabolites are given in Table 13.2.

An important alternative metabolic pathway through pyruvate is the citrate metabolism, which will be described in the next section.

13.1.2.3 Citrate Metabolism

The homofermentative *Lactococcus lactis* ssp. *lactis* biovar. *diacetylactis* and the heterofermentative leuconostocs, especially the strains of *Leuconostoc mesenteroides* ssp. *cremoris*, metabolize citrate. It is not used as an energy source but is metabolized only in the presence of a fermentable sugar such as lactose. Additional pyruvate is formed during citrate metabolism, so that more of it becomes

TABLE 13.2
Chemical Structure of Some Products in the Metabolism of Pyruvate

Compound	Structure
Pyruvic acid	CH_3–CO–COOH
Lactic acid	CH_3–CHOH–COOH
Acetaldehyde	CH_3–CHO
α-Acetolactic acid	CH_3–CO–C(COOH)OH–CH_3
Diacetyl	CH_3–CO–CO–CH_3
Acetoin	CH_3–CO–CHOH–CH_3
2,3-Butanediol	CH_3–CHOH–CHOH–CH_3

available than is required for oxidation of NADH (released during the sugar fermentation). Accordingly, specific end products can be formed; these include acetic acid, CO_2, and 'C_4 products,' including diacetyl, which is an important flavoring compound in some milk products. The organisms involved are sometimes called aroma-forming bacteria.

Citrate is transported into the cell by citrate permease. At first, citrate is hydrolyzed into acetate, CO_2, and pyruvate by citrate lyase according to

$$COOH–CH_2–C(OH)COOH–CH_2–COOH \rightarrow$$
$$CH_3–COOH + CO_2 + CH_3–CO–COOH$$

Formation of diacetyl from pyruvate can then occur via α-acetolactate formed by condensation of active acetaldehyde and another molecule of pyruvate. α-Acetolactate is decarboxylated to acetoin (no flavor compound); the formation of diacetyl from α-acetolactate depends on the redox equilibria in the system. α-Acetolactate is an unstable molecule and, at low pH, may be nonenzymatically decarboxylated to acetoin or, in the presence of oxygen, oxidatively to diacetyl. In the latter case, α-acetolactate should be extracellular; production of diacetyl via the α-acetolactate pathway appears to be most common (see Figure 13.5).

During the growth of *diacetylactis* strains in milk, the contents of diacetyl and acetoin keep increasing as long as citrate is present. Citrate suppresses the synthesis of both diacetyl and acetoin reductases. Accordingly, once citrate has become exhausted, reduction in the levels of both diacetyl and acetoin occurs by formation of acetoin and 2,3-butanediol, respectively. However, the pH has become low in the meantime and so have the activities of the reductases mentioned.

During growth of the leuconostocs in milk, the metabolism of pyruvate depends on the pH. Production of diacetyl and acetoin only occurs below pH $\approx$ 5.5. The explanation is uncertain, but various intermediate metabolites of the heterofermentative sugar fermentation have been shown to inhibit the formation of acetolactate synthase; during the initial growth stage, the production of α-acetolactate is thereby suppressed. There is no NAD^+/NADH involved in the conversion of pyruvate to diacetyl and acetoin via α-acetolactate. Reduction of diacetyl and acetoin requires NADH. It enables the leuconostocs to oxidize NADH and to produce acetic acid (with a yield of ATP) rather than ethanol during sugar metabolism. It follows that flavoring compounds are lost, to an extent that depends on the bacterial strain involved. All in all, leuconostocs have a much stronger reducing capacity than the *diacetylactis* strains.

Strains of *Streptococcus thermophilus* and *Lactobacillus delbrueckii* ssp. *bulgaricus* cannot metabolize citrate. Therefore any diacetyl and acetoin must be formed from pyruvate produced during sugar metabolism.

13.1.2.4 Regulation of Metabolism

Sugar fermentation pathways are controlled at two levels. Control may be exerted on the level of gene expression, which determines the intracellular concentration of fermentation pathway enzymes. This coarse control is usually fine-tuned by key steps in pathways catalyzed by enzymes whose activity is modulated by fermentation metabolites. It is generally believed that most of the genes for the major pathway enzymes, such as the EM pathway, are permanently expressed. Regulation at the level of gene expression and transcription is usually observed only in genes involved in the transport of sugars or hydrolysis of disaccharides. For the fine regulation at the level of enzyme activity, a number of key steps are involved, including the PEP-PTS transport enzymes, phosphofructokinase, pyruvate kinase, and lactate dehydrogenase (see Figures 13.2 and 13.4).

The PEP-dependent lactose PTS relies on the availability of protein phosphates, which are phosphorylated during the conversion of PEP into pyruvate by pyruvate kinase (Figure 13.2). If glycolysis is minimal, especially if there is no sugar left, the amounts of hexose diphosphates also decrease. Moreover, accumulation of inorganic phosphate occurs in the cell. Because inorganic phosphate slows down any pyruvate kinase activity present, PEP and its precursors 3-phospho and 2-phosphoglycerol (PEP potential) will pile up. When sugar becomes available again, the inorganic phosphate content decreases, and immediately the PEP potential is fully used for the sugar uptake.

Cells that metabolize sugars via the glycolytic pathway, form fructose-1,6-diphosphate in a reaction catalyzed by phosphofructokinase from fructose-6-phosphate. Fructose-1,6-diphosphate is actually a key regulator of the metabolic flux through this pathway. It activates the conversion of PEP by pyruvate kinase and the subsequent reduction of pyruvate into lactic acid. At a high rate of fermentation, thus when the intracellular level of fructose-1,6-diphosphate is high, pyruvate is almost exclusively converted to lactate. At a slow rate of lactose fermentation, the end product spectrum changes, as is shown in Subsection 13.1.2.2. This is due to a decrease in the activity of pyruvate kinase and also in that of lactate dehydrogenase, attributed to a decrease in fructose-1,6-diphosphate and an increase in inorganic phosphate.

Pyruvate formate lyase, which is able to convert pyruvate to acetyl CoA (Subsection 13.1.2.2), is very sensitive to oxygen, has a lower affinity for pyruvate than lactate dehydrogenase has, and is inhibited by triose phosphates. Thus, under homolactate fermentation conditions, in which lactate dehydrogenase is activated by fructose-1,6-diphosphate, pyruvate formate lyase is inhibited by triose phosphate levels. Conversely, at low rates of fermentation, the concentrations of fructose-1,6-diphosphate and of the inhibitors of pyruvate formate lyase are low, permitting greater diversion of pyruvate to acetyl CoA and formate.

Some aspects of the regulation of glycolysis in lactococci are summarized in Table 13.3.

13.1.2.5 Production of Acetaldehyde

Acetaldehyde is predominantly accumulated by lactic acid bacteria that have no alcohol dehydrogenase enzyme. These bacteria therefore cannot reduce acetaldehyde (formed via the pyruvate formate lyase pathway or the pyruvate dehydrogenase

TABLE 13.3
Stimulation (+) and Inhibition (–) of Some Enzymes by Various Compounds in the Glycolysis Pathway of Lactococci

Enzyme	FDP	Pi	PEP	Triose-P
PEP-PTS	–	–	+	
Phosphofructokinase	+	–	–	
Pyruvate kinase	+	–		
Lactate dehydrogenase	+	–		
Pyruvate formate lyase	–			–

Note: FDP = fructose-1,6-diphosphate; Pi = inorganic phosphate; PEP = phosphoenol pyruvate; triose-P = triose phosphates; PEP-PTS = phosphoenol pyruvate phosphotransferase system.

pathway; see Figure 13.5) to ethanol. Examples of acetaldehyde-accumulating bacteria are found among strains of *Lactococcus lactis* ssp. *lactis* biovar. *diacetylactis* and *Lactobacillus delbrueckii* ssp. *bulgaricus*. The latter bacterium and *Streptococcus thermophilus* also produce acetaldehyde from the free amino acid threonine according to:

$$\text{threonine} \xrightarrow{\textit{threonine aldolase}} \text{acetaldehyde} + \text{glycine}$$

Far more acetaldehyde is produced via this pathway than via the other one.

It is an important feature of yogurt bacteria to produce this characteristic flavoring component, which is essentially a product of the protein degradation (see Subsection 13.1.2.6). The glycine formed is converted to serine by serine hydroxymethyltransferase, both compounds being major sources of one-carbon units necessary for the biosynthesis of folic acid derivatives, which is also characteristic of yogurt bacteria.

13.1.2.6 Protein Metabolism

Growth in milk. Lactic acid bacteria require many nutrients. Milk contains insufficient amounts of immediately available nitrogenous compounds (i.e., low-molar-mass peptides and amino acids) to sustain the growth of the bacteria. A prerequisite for good growth in milk is the presence of a proteolytic enzyme system in the bacterial cell, which consists of enzymes associated with the cell envelope as well as intracellular enzymes. The consecutive enzymes hydrolyze the large protein molecules to assimilatable components (Figure 13.6). Not surprisingly, the presence of this system in lactic acid bacteria on the one hand, and their

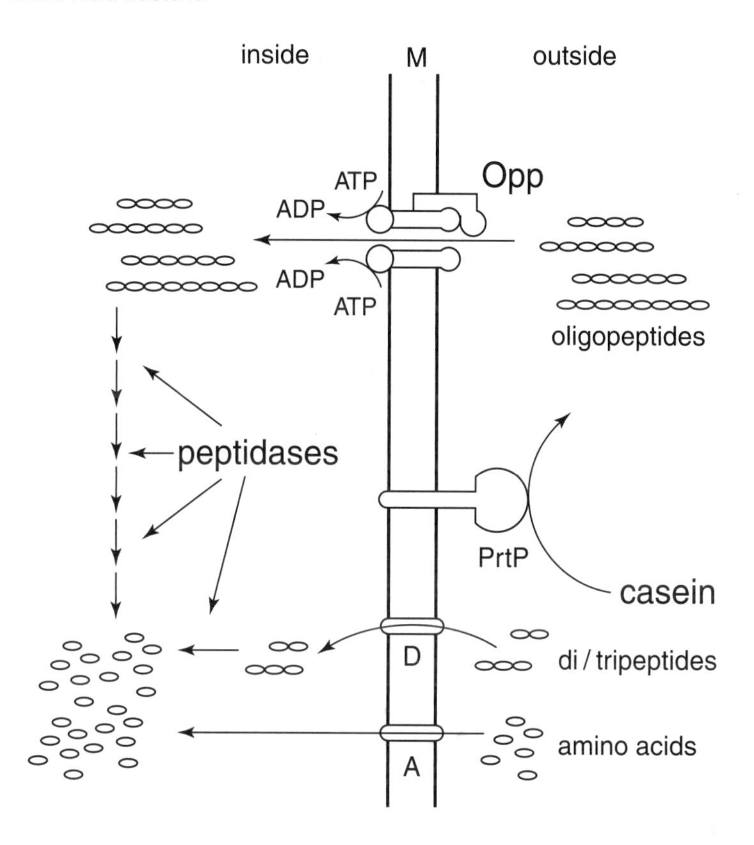

FIGURE 13.6 Model of the proteolytic pathway in *L. lactis*. Included is also the transport of di- and tripeptides and free amino acids. PrtP = membrane-anchored proteinase; Opp = oligopeptide transport system; D = di/tripeptide transport system(s); A = amino acid transport system(s); M = cytoplasmic membrane. (Adapted from L. Axelsson. In: S. Salminen and A. von Wright, Eds., *Lactic Acid Bacteria,* 2nd ed., Dekker, New York, 1998.)

growth and acid production rate on the other hand, are significantly correlated. Strains missing cell wall proteinases (called prt$^-$ strains) hardly grow in milk. In mixed cultures, these strains rely on the nitrogenous compounds produced by prt$^+$ strains. The quantity of cell wall proteinase formed by prt$^+$ strains is greatly reduced if sufficient small peptides are present in the culture medium.

Proteolytic system. The lactococcal proteinase (PrtP in Figure 13.6) is a membrane-anchored serine proteinase that is synthesized as an inactive preproteinase within the cell and subsequently transformed into a mature active proteinase at the outside of the membrane. PrtP exists in at least two variants in lactococci with somewhat different specificities in the degradation of milk casein. The information about the proteinases in other starter bacteria is limited, but available data indicate that they are similar to the lactococcal proteinases. Proteolytic activity is lacking in *Streptococcus thermophilus*, which explains the protocooperative relationship between this organism and the proteinase-positive *Lactobacillus delbrueckii* ssp. *bulgaricus* in yogurt cultures (see the preceding text).

Peptides formed by the proteinase activity and containing up to about eight amino acid residues can be transported across the cytoplasmic membrane into the cell. Various transport systems have been identified, including those for oligopeptides, tripeptides, dipeptides, and amino acids. Inside the cell, the peptides are hydrolyzed by peptidases to the individual amino acids necessary for the synthesis of proteins required for growth.

Amino acid conversion. The formation of amino acids from protein is also important in the ripening of cheese (Chapter 25). During the ripening, the starter bacteria may die and lyse, and their intracellular peptidases may act on any peptides present around the cells. The amino acids produced are considered to be important precursors of the flavor compounds that are characteristic of mature cheese. These flavor compounds are formed by biochemical and chemical conversion of the amino acids (see Subsection 25.5).

Arginine metabolism. Many lactic acid bacteria produce NH_3 from arginine and simultaneously produce 1 mol of ATP per mol of arginine metabolized. Ornithine, also produced in this reaction, is exported from the cell by a transport system, which drives the uptake of arginine. *Lactococcus lactis* ssp. *lactis* produces NH_3 via this pathway, whereas *Lactococcus lactis* ssp. *cremoris* does not, owing to the lack of one of the enzymes for this pathway.

13.1.2.7 Lipolytic Activity

The lipolytic activity of lactic acid bacteria is limited. It mainly concerns the hydrolysis of di- and monoglycerides formed from triglycerides by foreign lipases. Lactic acid bacteria may have esterase activity, which facilitates esterification of fatty acids. Lipolysis by lactic acid bacteria during ripening of cheese is discussed in Chapter 25.

13.1.2.8 Formation of Exopolysaccharides

Many strains of lactic acid bacteria can produce exopolysaccharides (EPS). EPS can be present as a capsule, closely or loosely attached to the bacterial cell, or be excreted in the milk. Although the chemical composition of EPS appears to vary from strain to strain, many of them have been shown to contain galactose, glucose, and rhamnose moieties. There are thermophilic as well as mesophilic EPS-producing lactic acid bacteria. Most strains of *Streptococcus thermophilus* and *Lactobacillus delbrueckii* ssp. *bulgaricus* produce ropy EPS, and this property is widely used to improve the rheological quality of stirred yogurt (Section 22.4). Some strains of *Lactococcus lactis* sspp. *lactis* and *cremoris* also produce EPS (Section 22.2).

13.1.3 Genetics

The use of genetics to analyze the physiological and biochemical properties of lactic acid bacteria has proved very successful in explaining these properties on a molecular level. Significant information has been gained about their control and regulation. This is particularly true for many of the essential functional properties in the metabolism of carbohydrates and proteins (Subsection 13.1.2). The discovery that the genes for several of these properties are located on extra-chromosomal pieces of DNA, which are called plasmids, has greatly facilitated the development of knowledge on the genetics and physiology of lactic acid bacteria. The structure and function of both plasmid-linked and chromosomal genes have been further elucidated. This has resulted in physical and genetic maps of whole *plasmids* and *genomes* of these bacteria.

Plasmids are much smaller than the chromosomes; their size ranges from about 3,000 to 60,000 base pairs. Several properties of lactic acid bacteria, which are important for dairy fermentations, tend to be encoded on plasmids. These properties include proteinase production, enzymes involved in the uptake and metabolism of lactose, transport of citrate, exopolysaccharide production, bacteriocin production (Subsection 13.1.4), and phage resistance (Subsection 13.3.3).

Several plasmids can occur in the cell. The plasmid DNA replicates independently of the chromosomes. During cell division, each daughter cell acquires a replica of the chromosomal DNA and, in most cases, replicas of the plasmids. It may happen that a daughter cell is not provided with the replica of a certain plasmid, and this cell will not show the property encoded on this plasmid. Thus, certain important properties of starter bacteria may be lost upon prolonged cultivation, which should therefore be avoided. Cells that lose the proteinase plasmid become proteinase negative (prt^-) and consequently grow slowly in milk (Subsection 13.1.2.6). Those that lose the lactose plasmid are unable to metabolize lactose and, therefore, cannot grow in milk.

Plasmid DNA can be transferred from one cell to another cell of the same species by conjugation, a process that has been frequently reported for lactococci. Several plasmids are conjugative: they mediate effective cell-to-cell contact and, thereby, their own transfer. This has resulted in transfer of important plasmid-coded traits and in construction of cells with a combination of desired properties. Also, physical techniques have been developed to transfer plasmid DNA propagated in one cell into another cell. The use of these genetic-engineering techniques on plasmids has played a major role in understanding the genetics and metabolism of lactic acid bacteria.

Lactic acid bacteria have been involved considerably in the recent explosion of *genome sequence* determinations. Presently, the genomes of several representatives of these bacteria have been sequenced. Comparison of the genome sequences of multiple lactic acid bacteria species and strains is expected to provide a critical view of microbial evolution, including the genetic events leading to their adaptation to specialized environments such as milk.

Genetic tools in the form of DNA probes will become available to search in genomes for specific genes and gene clusters. These specifically designed probes will facilitate the screening of large collections of strains for the presence of certain desired traits, which are required for the successful manufacture of traditional and new fermented dairy products.

13.1.4 Bacteriocins

Many bacteria produce proteins that inhibit the growth of other bacteria, and these proteins are called *bacteriocins*. Generally, they have a narrow target range and inhibit only closely related bacteria. This narrow host range and their proteinaceous nature distinguish them from antibiotics. Bacteriocin production by lactic acid bacteria is rather common. The best known is nisin, which can be produced by *Lactococcus lactis* ssp. *lactis*. It contains 34 amino acids and has a molar mass of 3353 Da. Nisin contains some unusual amino acids, which are formed by posttranslational modification. One of these modified amino acids is lanthionine, and this is why nisin is called a lantibiotic. Nisin has a relatively broad range of activity, inhibiting *Bacillus*, *Clostridium*, *Staphylococcus*, and *Listeria*. It is particularly active against *Clostridium tyrobutyricum*, which can cause late gas production in (semi)hard cheese types (Section 26.2).

Besides nisin, other lantibiotics are produced by lactic acid bacteria, such as lacticin 481 and lactocin S by *Lactococcus lactis* ssp. *lactis* CNRZ 481 and *Lactobacillus sake*, respectively. The number of characterized small bacteriocins not containing lanthionine is even larger. The lactococcins produced by strains of *Lactococcus lactis* ssp. *cremoris* belong to this category. Their range of activity is usually narrower than that of the lantibiotics.

The bacteriocins are encoded either on plasmids or on the chromosome. Most bacteriocins produced by lactic acid bacteria have a bactericidal mode of action that is due to the destruction of the proton motive force involved in transport. Bacteriocin-producing bacteria also have a gene coding for immunity to their own bacteriocin. The synthesis of some bacteriocins is triggered only if target organisms are present.

Bacteriocins are seen as useful aids to increase the safety of foods; some have been identified that inhibit pathogens and food spoilage organisms. Bacteriocins of lactic acid bacteria are, like their producers, generally considered as safe for human consumption. Other properties, such as their heat and acid stability, their activity at relatively low pH, and their activity over a prolonged period, make them very suitable for application in the food industry.

13.2 ACID PRODUCTION

The prime property of lactic acid bacteria is the production of acids, especially lactic acid. Therefore, evaluation of growth is traditionally done by determining the quantity of acid produced by titration. The titration values are often expressed as percent lactic acid. This is generally not justified, as discussed in Section 4.2.

For most applications, knowledge of the pH is of greater relevance. Moreover, pH can be measured in-line, which may be of considerable advantage for monitoring or controlling acidification processes.

The relations between pH, titratable acidity, and bacterial count show several complications:

1. *Acid dissociation.* If milk is acidified (or titrated) with HCl, the acid is fully dissociated at the pH values of interest, but this is not the case for lactic acid. The latter has a p*K* of about 3.9, implying that below an approximate pH value 5.7, the acid is not fully dissociated. Moreover, lactate ions associate to some extent with Ca^{2+} and Mg^{2+} ions. Equation 2.4 allows calculation of the dissociation. Figure 13.7a, curves 1 and 2, shows the difference between HCl and lactic acid, and it is seen to be considerable at low pH.
2. *Buffering capacity.* The quantity of acid needed to produce a given decrease in pH depends on the buffering capacity of the milk, and hence on its composition. Buffering primarily is due to protein and salts. This is discussed in Section 4.2, and Figure 4.1 gives examples. Figure 13.7a, curves 2 and 3, shows the difference between plain and concentrated skim milk, the latter being typical for traditional yogurt. Because the buffering capacity of milk fat globules is virtually zero, the titratable acidity of a cream of x% fat will be approximately $(100 - x)$% of that of the corresponding skim milk.
3. *Other changes* in the milk due to bacterial growth may affect the relation between pH and acidity. Other acids may be formed, e.g.,

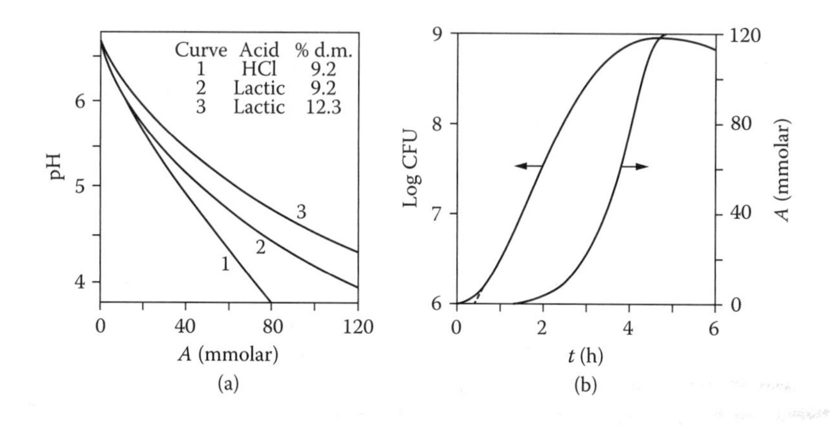

FIGURE 13.7 Acid production by starter bacteria. (a) Titration curves of plain and concentrated skim milk with hydrochloric or lactic acid; d.m. = dry matter. (b) Bacterial count (colony-forming units [CFU], in ml^{-1}) and amount of acid produced (A) as a function of time (t) after a certain quantity of starter (roughly 0.1%) was added to the milk; constant external conditions. Approximate examples.

acetic (pK = 4.7) and carbonic acid (first pK = 6.3). Citric acid may be consumed (Subsection 13.1.2.3), thereby slightly lowering acidity. Proteolysis may somewhat enhance buffering capacity, and excessive lipolysis increases acidity.

4. *Bacterial count vs. acid produced.* Figure 13.7b shows such a relation. The first thing to consider is that bacterial count is commonly given on a log scale, whereas the rate of acid production would, as a first approximation, be proportional to the actual number. For the simplest case, and thus during the exponential-growth phase, the relation between acid produced A (in mol/l) and incubation time t would be

$$A = cN_0\, g(2^{t/g} - 1) \tag{13.1}$$

where c is a proportionality constant (in mol/s), N_0 the initial count per liter, and g the generation time of the bacterium.
5. *Bacterial mass vs. count.* Other things being equal, the acid production rate would be proportional to the bacterial mass, and the number of CFUs per ml may not be proportional to this mass (Subsection 5.1.4). The relation between mass and CFU may vary considerably among species or strains; it may also depend on growth conditions.
6. *Bacterial metabolism* may vary considerably (see Subsection 13.1.2). Some species or strains produce four molecules of lactic acid from one molecule of lactose and others only two; other variations also occur.
7. *Decoupling* of growth and metabolism may occur. Only under fairly ideal and constant conditions, especially in the exponential-growth phase, acid production rate is proportional to bacterial mass. However, when growth slows or even stops, especially due to the accumulation of lactic acid, the enzyme system of the bacteria may still go on converting lactose to lactic acid. This is illustrated in Figure 13.7b (after about 3 h). Decoupling may also occur at other conditions unfavorable for growth, such as low temperature, high salt content, or a combination of these. At more extreme conditions, acid production stops as well.

It may be added that it is the concentration of true lactic acid, i.e., in its undissociated state, that determines whether growth stops. However, the lower the pH, the higher the concentration of undissociated lactic acid at a given total concentration (according to Equation 2.4). The inhibiting effect of other acids, e.g., acetic or sorbic acid, also depends on the concentration undissociated.

The rate of acid production naturally varies greatly with growth conditions such as temperature, pretreatment of the milk (especially heat treatment), oxygen pressure, etc. (see also Subsection 5.1.5). When checking the growth or acidification capacity of various starters in laboratory tests, one should ensure that these growth conditions are exactly the same as during manufacture.

When a starter is added to the milk, the growth conditions for the bacteria are often not quite the same as was the case in the starter, e.g., the heat treatment of the milk and its oxygen pressure may well be different. It generally means that the bacteria have to adapt their enzyme system to some extent before maximum growth rate is attained. This shows up in a lag time — about 0.5 h in the example of Figure 13.7b.

13.3 BACTERIOPHAGES

A *bacteriophage* is defined as a virus that can infect and kill a bacterial cell. Infection occurs if the phage 'fits' the cell; it is then referred to as a homologous phage. Whether or not a phage fits, often depends on the bacterial strain involved. A certain phage strain can usually infect several closely related bacterial strains. The latter strains make up the 'host range' of the phage. One bacterial strain may belong to the host range of various phage strains. Apart from the ability to infect cells, viruses differ from bacterial cells by the absence of a metabolic system (as mentioned in Subsection 5.1.1, viruses cannot be regarded as living particles); hence, their multiplication depends on the biochemical outfit of the host cell. A phage particle in a bacterial cell interferes with the bacterium's metabolism, such that the cell produces phages rather than its own building blocks. Phages are ubiquitous in nature, and they occur in raw milk, though generally in very low numbers. They can proliferate where bacteria are grown in large quantities, as in starters. In such a case, a bacteriophage can be very harmful in the manufacture of fermented products because it can kill the vast majority of the bacteria within the host range of the phage.

13.3.1 Phage Composition and Structure

Phages are smaller than their hosts and can be separated from them by filtration through membranes with a pore size of 0.45 μm. They can only be 'seen' with the electron microscope. Phages consist of a head and a tail (Figure 13.8), both of which contain different proteins. The heads of the phages for lactic acid bacteria are isometric (spherical) or prolate (oblong) in shape with dimensions between 40 and 80 nm. The tail varies in length from 20 to 500 nm and may be contractile or noncontractile. Other features are the presence of a collar between the head and the tail and a distinct base plate at the tail tip, sometimes with additional appendages such as fibers and spikes.

The phage DNA or genome is located in the interior of the phage head, and its size is phage-specific, varying from 2 to $13 \cdot 10^4$ base pairs. All phages for lactic acid bacteria contain a linear molecule of double-stranded DNA, which may become circular if the DNA contains cohesive ends. The genome of the phage contains genes for, among others, structural proteins and a lytic enzyme that causes lysis of the host cell wall.

13.3.2 Phage Multiplication

Phage multiplication may occur in two ways: by the lytic cycle and by the lysogenic cycle. In the lytic cycle, the phages infect and lyse the host cell, whereas

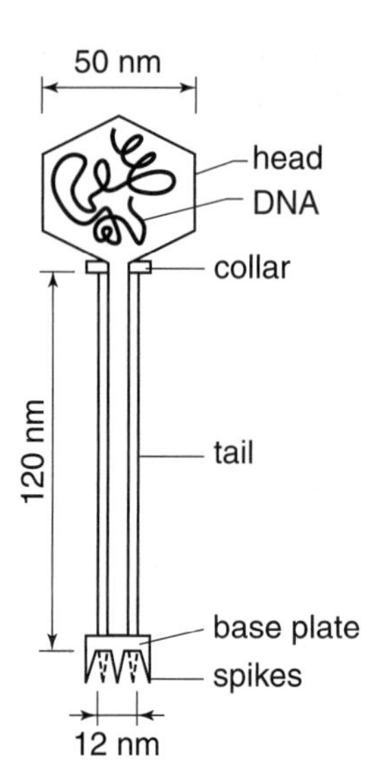

FIGURE 13.8 Schematic drawing of a bacteriophage of *Lactococcus lactis*. The important structural components of the phage are indicated. (Adapted from H. Neve. In: T.M. Cogan and J.-P. Accolas, Eds., *Dairy Starter Cultures,* VCH Publishers, Cambridge, U.K., 1996.)

in the lysogenic cycle, the phages insert their genome into the host chromosome. These phages are called virulent and temperate, respectively.

13.3.2.1 Lytic Cycle

The propagation of a virulent phage in the lytic cycle results in its multiplication and in the release of a new infectious phage progeny. There are several characteristic steps involved (Figure 13.9):

1. *Adsorption of the phage onto the bacterial host.* The first step involves the adsorption of the phage onto a special attachment site on the cell surface of the host. This is a highly specific event, which depends on the presence of appropriate phage receptors. The homologous (see preceding text) phages adsorb to the cell through their tails. The efficiency of this adsorption step depends on the environmental conditions. Many phages require divalent ions, especially Ca^{2+} ions, to become attached to the cell. Accordingly, prevention of bacterial infection by phage can be based on the use of chelating agents, such as EDTA, citrate, or phosphate, that strongly reduce the Ca^{2+} activity.
2. *Injection of phage DNA.* After adsorption of the phage to the cell, the phage DNA is injected from the head through the tail into the bacterial cell. The head remains as an empty ghost at the outside of the cell.

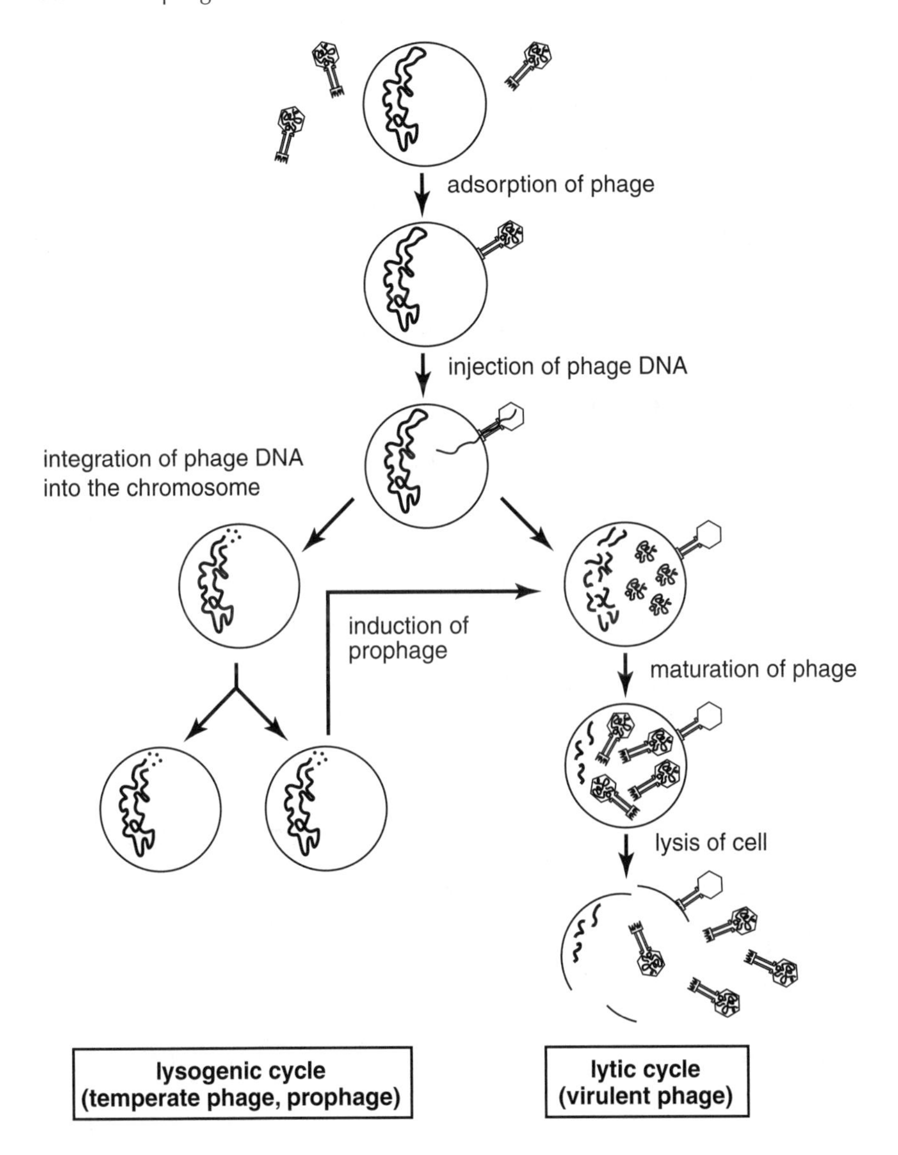

FIGURE 13.9 Schematic drawing of the propagation of virulent (right branch) and temperate (left branch) bacteriophage in a host cell. The cell is shown with its chromosomal DNA. The phage DNA is indicated by dotted lines. (Adapted from H. Neve. In: T.M. Cogan and J.-P. Accolas, Eds., *Dairy Starter Cultures,* VCH Publishers, Cambridge, U.K., 1996.)

3. *Phage maturation.* Once the phage DNA has been injected into the bacterial cell, intricate processes alter the metabolism of the host cell in such a way that DNA and protein are formed only for the phage. Newly synthesized DNA is subsequently packed in a condensed form into the assembled phage heads and, finally, complete phage particles emerge.

4. *Lysis of the host cell.* The lytic cycle is completed when the host cell lyses, releasing the new phage particles into the medium. Lysis is caused by a lytic enzyme called lysin, which is encoded on the phage genome. The phage-enriched medium is often called *phage lysate.*

Moreover, *lysis from without* may occur. Some phages of lactococci are known to mediate intensive lysin production. This excess of lysin ends up in the phage lysate and can destroy the cell walls of other cells, even of cells that do not belong to the host range of the phage. The latter can lyse without being infected by a homologous phage, which is called lysis from without.

13.3.2.2 Growth of Virulent Phages

The lytic growth of a virulent phage on a sensitive host is highly dependent on each phage–host system and is characterized by both a *latent period* and a *burst size*, which are determined in a one-step growth experiment (Figure 13.10). To this end, phages and host cells are mixed in such a ratio that the multiplicity of infection is less than 1, implying that one cell is exposed to one phage at the most. The number of infectious phage particles is monitored periodically during incubation. In the first period, the number of phages remains low and constant because new phages are being synthesized inside the bacterial cell. This is called the latent period and spans the time from initial adsorption to detection of phage

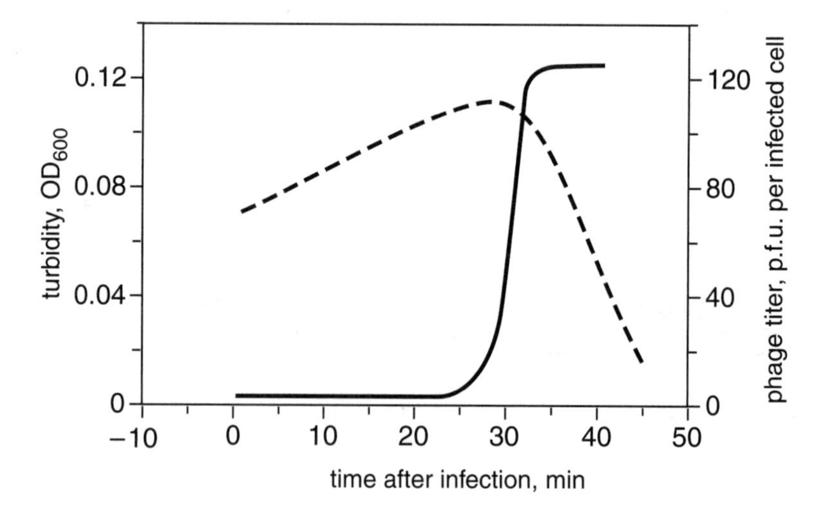

FIGURE 13.10 Results of a one-step growth experiment on *Lactococcus lactis* infected with a lytic phage. The release of progeny phage begins 25 min (latent period) after infection (burst size: 124). The increase in free phage quantity (p.f.u. = plaque-forming units, full line) is accompanied by a decrease in culture turbidity (broken line) due to cell lysis. (Adapted from I.B. Powell and B.E. Davidson, *J. Gen. Virology,* **66**, 2737–2741, 1985.)

progeny after cell lysis. A sudden increase in phage quantity, to a number called burst size, occurs after lysis of host cells by the action of phage lysin.

For lactococcal phages, the latent period varies from 10 to 140 min, and the burst size varies from about 10 to 300 phages. In a growing culture of lactococci, the multiplication of phages is very rapid, much more so than the growth of bacteria. Assuming a latent period of 60 min and a burst size of 150 phages, one phage will result in a progeny of 22,500 phages (150 × 150) after 2 h. If each phage still has one bacterium to its disposal, in 3 h the number of phages will be around $3.4 \cdot 10^6$. Keeping in mind that the generation time of a *Lactococcus* cell is around 60 min, one cell will produce only 8 new cells in these 3 h. Thus, the phages will soon outnumber the bacterial cells. This clearly indicates the serious problems that occur after infection of a lactococcal fermentation with phages.

Phage numbers in a sample can be estimated by carrying out a plaque assay. In this procedure, serial dilutions of the phage suspension are mixed with a high number of phage-sensitive cells in molten sloppy agar (0.7%) at 45°C, and the mixture is poured over prehardened agar in a Petri dish. After incubation at the optimum temperature of the host for up to 8 h, clear halos, called *plaques*, can be seen in a lawn of bacterial growth. This is due to multiplication and spreading of phage particles through the lawn around the infected host cell, causing a visibly distinct area of lysis and clearing in the lawn. Each plaque is considered to have arisen from one phage, and counting the number of plaques and multiplying it by the dilution factor yields the number of phages (plaque-forming units) in the sample.

13.3.2.3 Lysogenic Cycle

The lysogenic cycle is an alternative way of phage propagation (Figure 13.9). Adsorption and injection of DNA occur as in the lytic cycle, but instead of phage multiplication, the phage DNA is inserted into the bacterial chromosome. The recombination event occurs at a specific region of homology between the phage DNA and the host cell DNA. The phage DNA is replicated synchronously with the bacterial DNA, giving rise to a progeny of lysogenic cells. In these circumstances, the phage is called a *temperate phage* or *prophage*. The majority of lactic acid bacteria are lysogenic, and in this state the cell is immune to attack of its own phage or closely related phages. A bacterial cell can be the carrier of more than one prophage and thereby be resistant to various phages.

Excision of the prophage from the host chromosome can induce the temperate phage to become virulent and multiply (Figure 13.9). This can occur either spontaneously or be induced by UV light or by treatment with mutagenic agents such as mitomycin C. The number of free phages in a culture of lysogenic bacteria will generally remain small, because the temperate phages liberated from the host cell cannot proliferate as virulent phage on the immune intact lysogenic cells. The free phages can only be detected by using a susceptible strain, called *indicator strain*.

13.3.2.4 Pseudolysogeny

Many mixed-strain starter cultures (see Section 13.4) are permanently infected with a low number of virulent phages. These are called *own phages* to distinguish them from disturbing phages, and they multiply on a limited number of phage-sensitive cells in the culture. The released phages cannot infect the rapid-acid-producing cells, which are phage-insensitive. The chronic phage-carrying stage of the culture contributes significantly to its phage insensitivity. This phenomenon is called pseudolysogeny. The culture continues growing normally as long as no infection with disturbing phages occurs.

13.3.3 Phage Resistance Mechanisms

Bacteria have developed several phage resistance mechanisms in the course of time. Inhibition of phage adsorption, restriction/modification systems, and abortive infection mechanisms are found in lactic acid bacteria (Figure 13.11). All these

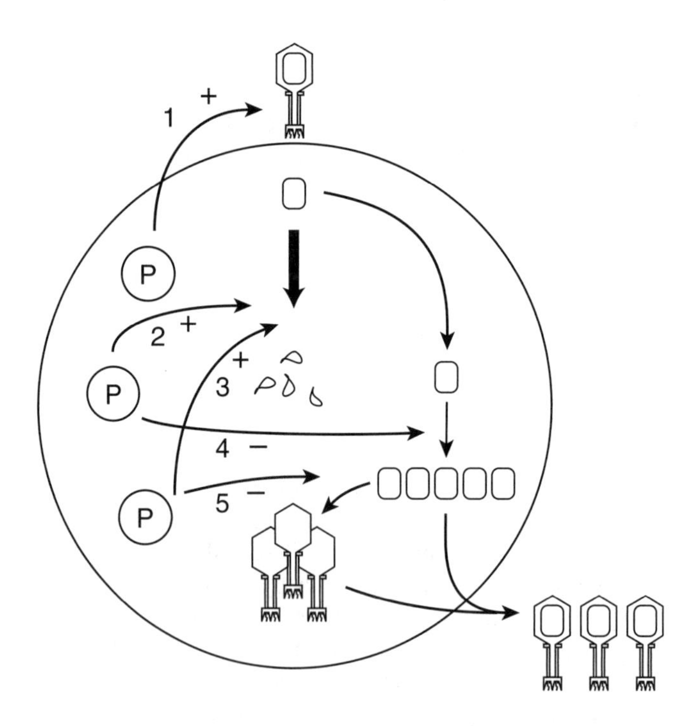

FIGURE 13.11 Plasmid-coded phage resistance mechanisms in *Lactococcus lactis*. (1) Adsorption inhibition; (2) and (3) restriction/modification systems; (4) abortive infection mechanism, which interferes with phage DNA replication; (5) abortive infection mechanism, which affects phage protein synthesis. P = plasmid; + = cells survive; − = cells die. (Adapted from G.F. Fitzgerald and C. Hill. In: T.M. Cogan and J.-P. Accolas, Eds., *Dairy Starter Cultures,* VCH Publishers, Cambridge, U.K., 1996.)

mechanisms are commonly encoded on plasmids. The lysogenic cycle (Subsection 13.3.2.3) also causes a form of phage insensitivity.

Adsorption inhibition. In adsorption inhibition, the receptor sites for the phage at the cell surface are masked, so the phage cannot attach to the cell and, hence, does not initiate its lytic cycle. In some *Lactococcus lactis* ssp. *cremoris* strains, the masking agents have been identified as polymers containing rhamnose and/or galactose, whose biosynthesis is plasmide-coded.

Restriction/modification systems. Restriction/modification involves two enzymes, one of which, the restriction enzyme, hydrolyzes incoming phage DNA. The other, the modification enzyme, modifies the host DNA, so that it becomes immune to the restriction enzyme. Usually, the host DNA is modified by methylation of some of the nucleic acid bases. This defense mechanism is active only after adsorption of the phage and injection of the DNA. In *Lactococcus lactis* species, several plasmide-coded restriction/modification systems have been identified and characterized.

Abortive infection. Abortive infection is the term used for phage resistance mechanisms that are not due to adsorption inhibition and restriction/modification. The presence of the abortive infection mechanism results in the total loss of plaque formation or at least to a reduction in plaque size. This is due to a reduction in burst size and, in many cases, an extension of the latent period. Abortive infection generally results in the death of the infected cells. The occurrence of several different abortive infection mechanisms is known for lactococcal strains. The mechanisms are generally based on inhibition of phage DNA replication or on repression of the synthesis of structural phage protein. The genes for abortive infection are located on plasmids.

Improvement of phage resistance. The majority of phage resistance mechanisms in lactococci are plasmid-coded, and many of these plasmids are conjugative (Subsection 13.1.3). This fact has been used to improve the phage resistance of phage-sensitive starter cultures. Specific cultivation techniques have been developed to select phage-resistant transconjugants, which have picked up the resistance plasmid, in the culture of phage-sensitive starters. This technique has been used successfully in the production of phage-resistant strains for commercial use.

13.3.4 Inactivation

Usual processes to destroy microorganisms (including heat treatment, disinfection, and gamma and ultraviolet irradiation) can also be applied to inactivate phages, i.e., destroy their ability to infect. The phage type and the composition of the culture medium largely determine the heat resistance of phages. Inactivation is fastest in pure water and is considerably slowed down by the presence of proteins or salts, especially those of calcium and magnesium. Several phages of lactic acid bacteria survive low pasteurization of milk, i.e., 15 s at 72°C. Inactivating the most heat-resistant phages requires a heat treatment of 1 min at 95°C.

Also, several disinfectants inactivate phages. This is true of hypochloric acid: concentration of available chlorine, temperature, contact time, and type of phage strains involved determine the effectiveness of the disinfectant. Many bactericides have no effect on the inactivation of phages. These include:

- Metabolic toxins; phages have no metabolism.
- Solvents for fat, such as ether and chloroform. Most of the phages do not contain lipids.
- Antibiotics.

13.4 ECOLOGICAL ASPECTS

Microorganism always show interactions between themselves and with their environment, which comprises their *ecology*. Fermented milk products, and hence the starters for lactic fermentations (see Section 13.5), practically always contain a mixture of lactic acid bacteria and, in some cases, also other microorganisms. These mixtures can be complex and variable in composition due to several types of interactions between the species, subspecies, and strains of lactic acid bacteria present. The cheese starters very often contain a mixture of strains from the genera *Lactococcus* and *Leuconostoc*, whereas for the production of yogurt, a mixture of thermophilic streptococci and lactobacilli is required. Shifts of the population can occur in these mixtures. To guarantee and control the fermentation process and the quality of the fermented product, a starter of constant composition is needed; this means that its potential instability should be taken into account. There are various mutual growth relations between bacteria in general and between lactic acid bacteria in particular:

1. One organism can promote the growth of another without benefiting itself. This is referred to as *commensalism*. An example is the presence of prt^+ lactococci enabling the growth of prt^- variants in milk, as mentioned in Subsection 13.1.2.6. When organisms can grow independently but do so better when together, it is called *mutualism* or *protocooperative growth*. A good example is the mutual growth stimulation of *Streptococcus thermophilus* and *Lactobacillus delbrueckii* ssp. *bulgaricus* in yogurt (Subsection 22.4.1). Another example is the growth of *Leuconostoc mesenteroides* ssp. *cremoris* in a mixed culture. Unlike many strains of *Leuconostoc lactis*, the former leuconostoc (prt^-) hardly grows in milk but shows satisfactory growth in a culture with prt^+ lactococci. Conversely, *Leuconostoc mesenteroides* ssp. *cremoris* enhances the growth of lactococci by production of CO_2.
2. During growth of a mixed population, there is a constant competition for nutrients, especially if their concentration is growth-limiting. In milk, this is the case for free amino acids, B vitamins, and some essential trace elements such as Mn^{2+}. Bacterial strains with the highest affinity for these limiting nutrients may be favored during growth in milk. Thus, they compete successfully against others, which ultimately results in a shift in the composition of the population (Subsection 13.5.3).

3. Lactic acid bacteria may produce inhibitors toward which some strains in the population can be sensitive. Undissociated acids, such as lactic, acetic, and formic acid, formed during metabolism of sugars, belong to these potential inhibitors and may slow down the growth of certain acid-sensitive strains in the population. Another potential inhibitor is hydrogen peroxide, formed during oxidation of NADH by flavoproteins (Subsection 13.1.2.2). Some strains produce bacteriocins (Subsection 13.1.4) with a specific inhibitory effect on certain members of the population, thereby modifying its composition.
4. Generally, a population composed of several types of bacteria is genetically unstable. Mutants with altered properties may arise, e.g., because of a chromosomal mutation or loss of a plasmid. Especially under environmental conditions in which the mutant has a growth advantage over the original strain, this may cause a severe change in the population composition. The presence of antibiotics may favor growth of antibiotic-resistant mutants, and an infection by bacteriophages may give rise to phage-resistant mutants. (Subsection 13.3.3).

Also, the presence of temperate bacteriophages in lysogenic bacteria (Subsection 13.3.2.3) poses a potential genetic instability because these phages can become virulent and, after release, proliferate on sensitive strains, which results in a shift in the composition of the population.

Lactic acid bacteria may be considered the autochthonous microorganisms in the milk ecosystem. Environmental conditions (especially temperature) and mutual interactions between the lactic acid bacteria, however, determine how their population in milk is composed. In Subsection 13.5.3, important factors influencing shift in the flora of specific mixtures of starter bacteria are discussed.

13.5 STARTERS

A starter is a culture of one or more species or strains of lactic acid bacteria that is added to milk to ferment it. Sometimes the starter also contains non-lactic acid bacteria, whereas in some other cases the latter are added separately to the milk.

Traditionally, a starter is obtained via growth of lactic acid bacteria in milk at a suitable temperature. The starter is subsequently maintained by propagating and growing it in a fresh portion of milk. Currently, special growth media other than milk are also utilized to avoid multiplication of bacteriophages during starter manufacture (see Subsection 13.5.5).

13.5.1 Composition

Table 13.4 gives a survey of the composition of starters as used in the manufacture of some fermented dairy products. On the basis of their composition, starters can also be classified as:

TABLE 13.4
Organisms Present in Various Types of Starter and Their Use in the Manufacture of Fermented Milk Products

	Mesophilic				Thermophilic	
	Aromatic			Nonaromatic		
	L	DL	D	O	In Cheese	In Soured Milks
Organism						
Lactococcus lactis ssp. *lactis*	+	+	+	+		
Lactococcus lactis ssp. *cremoris*	+	+	+	+		
Lactococcus lactis ssp. *lactis* biovar. *diacetylactis*		+	+			
Leuconostoc cremoris/ Leuconostoc lactis	+	+				
Streptococcus thermophilus					+	+
Lactobacillus delbrueckii ssp. *bulgaricus*					+	+
Lactobacillus helveticus					+	
Lactobacillus delbrueckii ssp. *lactis*					+	
Lactobacillus acidophilus						+[a]
Applied in						
Butter	+	+				
Cultured buttermilk	+	+				
Sour cream	+	+				
Yogurt						+
Fresh cheese	+	+		+[b]		
Gouda-type cheeses	+	+	+			
Cheddar-type cheeses[c]				+		
Emmentaler[d]					+	

[a] Only for acidophilus milk and for some types of special yogurt (see Section 22.2).
[b] Used in the manufacture of cottage cheese (see Subsection 27.2.2).
[c] Cheese from which eyes are absent (blind cheese).
[d] And other cheese varieties in which high temperatures are applied during manufacture.

1. *Single-strain.* Every starter consists of a pure culture of one strain.
2. *Multiple-strain.* These consist of a defined mixture of pure cultures of a few (say, six) strains of different species of bacteria or of different strains of one species. The preponderance of each strain can change.
3. *Mixed-strain.* These are natural starters, consisting of an undefined mixture of strains of different species of bacteria. The composition of this type of starter is based on a dynamic equilibrium between various starter bacteria, and it can change considerably during use (see Section 13.4).

In mesophilic starters (Table 13.4), a single-strain starter consists of a strain of *Lactococcus lactis* ssp. *cremoris* or, less frequently, a strain of *Lactococcus lactis* ssp. *lactis* or its biovariant *diacetylactis.* Multiple-strain cultures contain strains of *Lactococcus lactis* ssp. *cremoris* and/or *lactis*, often combined with the biovariant *diacetylactis* or with *Leuconostoc cremoris* and *Leuconostoc lactis.* The mixed cultures are categorized as O, L, DL, and D types; L refers to leuconostocs and D to diacetylactis being present; in an O type, neither of these two groups occur; and DL contains both.

Thermophilic starters almost always consist of two organisms, *Streptococcus thermophilus* and either *Lactobacillus helveticus*, *Lb. delbrueckii* ssp. *lactis,* or *Lb. delbrueckii* ssp. *bulgaricus*. For the manufacture of cheese, such as Emmentaler (Table 13.4), these organisms are generally grown individually, but they are grown together for yogurt production to benefit from their mutual growth stimulation (Section 13.4). Similar to mesophilic mixed cultures, thermophilic mixed cultures may contain several strains of each species.

Strains of thermophilic lactic acid bacteria have, in recent years, been included as a starter besides the traditional mesophilic starter for some semihard cheese types to improve the flavor. They are cultivated separately from the latter starter and are referred to as *adjunct starters* (see Subsection 25.7.5). Mesophilic lactobacilli may also be used as an adjunct starter.

Some fermented milks, such as kefir and kumiss (Section 22.2), have a combined lactic acid and alcohol fermentation. Various kinds of lactic acid bacteria and lactose-fermenting yeasts are involved. Besides lactic acid bacteria, other organisms are also used in the manufacture of certain cheese varieties (Chapter 27). Examples are propionic acid bacteria in the manufacture of Emmentaler, brevibacteria and other coryneforms in the manufacture of surface-ripened cheeses, and molds as a surface or internal flora for soft cheeses.

The starter bacteria greatly affect the properties of the final product. Because of this, dairy manufacturers have increasingly focused on (1) selecting bacterial strains with desirable properties, (2) composing starters by applying suitable strains in appropriate proportions and (3) maintaining starter composition. In doing so, new products may be developed, manufacturing processes optimized, and product properties improved.

13.5.2 Properties

The biochemical conversion of milk components by lactic acid bacteria naturally causes changes in the fermented products. These changes closely depend on the properties of the starter bacteria involved (Table 13.4) and on the type of product. One of the main aspects is acid (especially lactic acid) production from lactose, which affects the preservation, texture, and flavor of the product. Also, other compounds are formed during the fermentation of lactose and citric acid, such as diacetyl, CO_2, and exopolysaccharides, which affect the flavor, texture, and consistency of the product, respectively. Finally, the degradation of protein and fat are essential biochemical events effectuated by the starter organisms, which contribute predominantly to the flavor and texture of the product, especially cheese. Basic aspects of the metabolic activities of starter bacteria have been outlined in Subsection 13.1.2, and the consequences of their properties for the characteristics of the fermented product are discussed in Chapters 18, 22, 24, and 25.

Thus, the conversions by lactic acid bacteria strongly determine shelf life, safety, consistency, and development of flavor and texture of fermented dairy products. Moreover, they may affect the nutritional value (see Subsection 22.5.2, and Section 25.8). The properties of the bacterial species and strains present in the starter determine to what extent the starter contributes to any of the product variables mentioned. In other words, the selection of a starter must be based on the properties desired in the product to be made.

13.5.3 Shifts in Flora

The properties of a fermented milk product should generally be constant and stable. To fulfill this condition, starters of constant activity should be used. In other words, the ratios between the numbers of various bacterial species or strains in the starter and the genetic properties of them should not vary. Ensuring constant bacterial properties is not easy, especially in mixed-strain starters (see Section 13.4). Several factors can be responsible for shifts in the bacterial population when the starter is propagated in the traditional way. Important are:

1. Factors concerning the *composition of the medium* (see also Subsection 5.1.6), including:
   a. Presence of compounds slowing down or preventing growth of a bacterial species or strain in milk. We distinguish:
      - Natural inhibitors, especially agglutinins and the lactoperoxidase –CNS–H_2O_2–system (see Subsections 2.4.3 and 2.5.2). Their concentrations strongly depend on the heat treatment of the milk. This is discussed in Chapter 7.
      - Inhibitors resulting from contamination, e.g., antibiotics and disinfectants.
      - Inhibitors formed by some strains, e.g., hydrogen peroxide and compounds such as nisin, and other bacteriocins, which may show antibiotic action against other strains in the starter (Subsection 13.1.4).

   - Free fatty acids. Fairly low concentrations of low-molar-mass fatty acids (C_4–C_{12}) and of oleic acid may slow the growth of some strains.
  - b. Concentration of CO_2 and other growth-promoting substances, e.g., those formed during heat treatment of milk.
  - c. Concentration of trace elements. Of main importance for growth are Fe^{2+}, Mg^{2+}, Se^{2+}, and Mn^{2+}. The manganese content determines the growth of *Leuconostoc cremoris*. The organism grows poorly at a low Mn^{2+} content in milk, which may cause a starter to turn from the DL type to the D type. The situation is reversed at a high Mn^{2+} content, i.e., a DL starter can become an L starter.
  - d. Concentration of O_2. Excessive concentrations are toxic for growth. At lower concentrations, the effect on the growth varies and depends on the bacterial strain involved (see Section 13.1).
2. *Contamination* of the milk by:
  - a. Bacteriophages. The phages can upset the composition of the starter culture severely and even destroy the starter (see Section 13.3 and Subsection 13.5.5).
  - b. Other lactic acid bacteria or microorganisms.
3. *Mutants* with changed properties can emerge, e.g., because of the loss of plasmid-coded properties. Mutation can also lead to strains with an increased acid production rate.
4. *Mutual growth promotion* of starter bacteria can cause great changes in the ratio of the numbers of the various bacterial strains (see Section 13.4 and Subsection 22.4.1.1).
5. *Incubation conditions* of the starter. Important are:
  - a. Incubation temperature. The equilibrium between bacterial strains in a culture is determined by their relative growth rate or by their survival at the incubation temperature.
  - b. Inoculum percentage. This factor can have a considerable effect on the composition of the flora, as is clearly shown for the yogurt bacteria (Subsection 22.4.1.1).
  - c. Stage of growth of the starter at the time of use. During propagation of the starter, the ratios of the numbers of the bacterial strains keep changing. The composition of the starter culture in each case is determined by:
    - The growth rate of any of the strains as determined by genetic properties and growth conditions; see the preceding points.
    - The different susceptibility of the strains to pH and to metabolites — lactic acid in particular. It causes the growth of various strains to be slowed down to different extents. The bacteria die when the starter is incubated for too long (high lactic acid content, low pH, etc.). The composition of the starter is then increasingly determined by the longest-surviving strains.

Several of the factors mentioned are just as important if pure cultures are used to propagate the starters.

A starter of optimum composition does not suffice for a satisfactory process control. A constant day-to-day rate of acid production is important. The rate of acid production, as partly determined by type and genetic properties of the starter bacteria (slow or rapid strains may be involved) and by inoculum percentage, may change during product manufacture. Several of the factors mentioned have an effect, the most important being contamination by bacteriophages. The processing conditions are also essential. For example, the rate of acid production can be decreased by NaCl or by a heat treatment just below the intensity needed for killing the starter bacteria.

13.5.4 Traditional Starter Manufacture

Traditionally, a starter is propagated by adding an amount of grown culture to a small quantity of intensely heated skim milk. The inoculated milk is incubated until a certain low pH (usually below 4.5) is reached and then cooled. This is the mother culture, which is also used as an inoculum in the preparation of the bulk starter. The latter is the inoculum for the manufacture of the product on the next day. Sometimes an intermediate culture is needed, depending on the volume of the bulk starter (Figure 13.12). The mother culture is also used as inoculum for a new mother culture for the next day.

Mesophilic starters are incubated at about 20°C for about 20 h; the inoculum percentage is 0.5 to 1. Thermophilic starters are incubated at 40 to 45°C for a few hours, after adding a small percentage of inoculum. To preserve the properties of the starter, the mother culture and bulk starter should be prepared under identical conditions, even from day to day. This requires constant incubation time and temperature, inoculum percentage and acidity of the ripe starter, and a fairly constant composition of the milk. Special milk powders are available to make starter milk. Even at the best of times, the traditional method for starter preparation does not guarantee constant composition and properties of the starter (see Subsections 13.5.4.1 and 13.5.4.2).

13.5.4.1 Single-Strain and Defined-Strain Mesophilic Starters

Until 1980, single-strain starters or combinations of these, propagated in the traditional way, were often used for making Cheddar cheese. Initially, cheese was made in each plant from day to day by using the same homofermentative *Lactococcus lactis* strain (isolated from a mixed-strain starter), which was aseptically propagated. However, such strains are highly phage-sensitive, and measures taken to avoid contamination by homologous lytic bacteriophages often failed. This became manifest when cheese-making plants became large; the manufacture of several consecutive batches of cheese on one day readily caused bacteriophages to accumulate in the plant. Figure 13.13 illustrates the effect of a homologous phage attacking a single-strain starter, inoculated in milk. Only a small part of

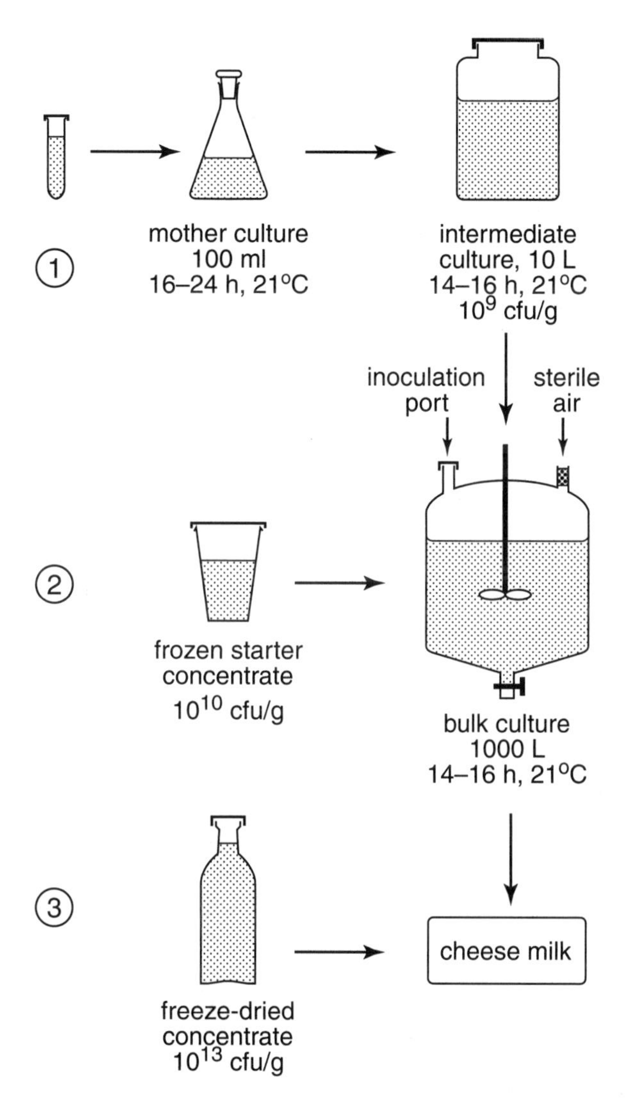

FIGURE 13.12 Schematic presentation of starter manufacture in a dairy for cheese production. 1. Traditional propagation of starter bulk culture. 2. Modern propagation using starter concentrate. 3. Direct vat inoculation.

the bacterial culture is phage-resistant (0.01% in the example). When a normal inoculum percentage is used, resulting in about 10^7 bacteria per ml of inoculated milk, the number of cells drops to 10^3 per ml; thereafter, it increases again due to the growth of resistant variants. Under normal incubation conditions, say 20 h at 20°C, these variants do not grow to 10^9 cells per ml of the starter, which is needed to ascertain a satisfactory rate of acid production. In other words, the starter is unfit for use.

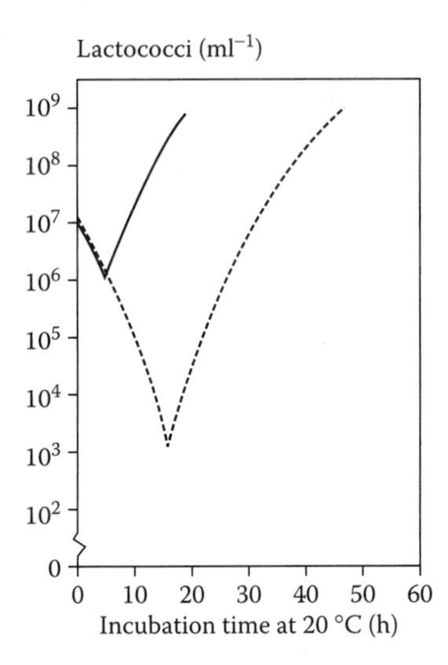

FIGURE 13.13 Behavior of two types of starter after being attacked by a disturbing bacteriophage. Solid line: behavior of a P starter (propagated in practice) containing 10% resistant strains; broken line: behavior of a single-strain starter with 0.01% resistant variants. In both instances, 1% starter was added to milk. (Adapted from J. Stadhouders and G.J.M. Leenders, *Neth. Milk Dairy J.*, **38**, 157, 1984.)

To solve these problems, other single-strain, non-phage-related cultures were introduced, which were applied by rotation. Moreover, use of a starter containing two strains with different properties was found necessary to make Cheddar cheese of satisfactory quality. Also, different combinations of these strains with differing phage sensitivities were applied as defined-strain culture by rotation. The system thus was implicitly based on the idea that the genetic properties of the bacterial strains, with respect to their resistance to phages, and the properties of the phages with respect to their host range, are invariable. It is now clear that this idea was incorrect and that the rotation system can lead to problems (see Subsection 13.5.5).

13.5.4.2 Mixed-Strain Starters

These starters were commonly applied in several regions. At present, they are still being used, though to a decreasing extent. Inoculated and subsequently cultured whey, still being used in certain cheese varieties, can also be regarded as a mixed-strain starter.

The starters are propagated in milk under hygienic conditions, but there is no safeguard against contamination of mother cultures and bulk starters by bacteriophages. Accordingly, the starters contain a great number of phages, about 10^8 per ml, which must be considered remainders of phage attacks. Starters propagated in actual practice (P starters), however, display a marked phage resistance, mainly caused by their phage-carrying character (see Subsection 13.3.2.4). A relatively high number of starter strains, e.g., 10% of all the strains present in the starter, survive an attack of disturbing phages. Usually, such an attack goes unnoticed; the flora shifts to the resistant strains, which, under normal incubation conditions (20 h at 20°C for a mesophilic starter, 1% starter added), can grow to about 10^9 per ml starter (see Figure 13.13). The starter may show a normal rate of acid production.

In actual practice, the rate of acid production may fluctuate, especially in mesophilic starters. Judging by its acid-producing ability, the propagated starter may seem unaffected after a first phage attack, but the resultant flora now contains an increased proportion of strains that are resistant to the first phage. A renewed attack by other phages may damage a larger proportion of the population, causing the rate of acid production to decrease further. In turn, more transfers are needed to attain the original level of acid production rate. Recurrence of the incident intensifies the effect (see Figure 13.14, curve 1).

Furthermore, formation of genetic variants in the starter may contribute appreciably to variations in the rate of acid production. Therefore, though disturbing phages do not emerge, the flora can shift, e.g., to mutants that grow faster. Under practical conditions, the mutants are destroyed soon, again due

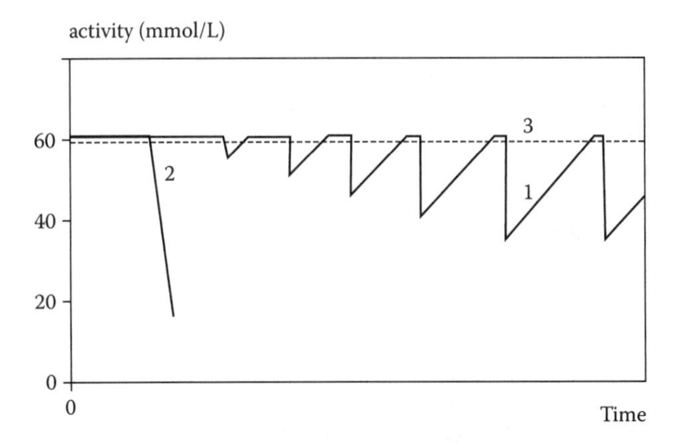

FIGURE 13.14 Fluctuations of the rate of acid production in milk by lab (propagated in laboratory, thus not in practice) and P starters *after* introduction in the plant. 1: P starter, propagated in the traditional way; 2: lab starter, propagated in the traditional way; 3: P starter, introduced as a concentrated mother culture (starter concentrate), while the bulk starter is cultured under complete phage protection. The rate of acid production (activity) is expressed as the titratable acidity of milk inoculated with 1% starter, obtained after 6 h at 30°C. (Adapted from J. Stadhouders and G.J.M. Leenders, *Neth. Milk Dairy J.*, **38**, 157, 1984.)

to contamination by homologous phages (periodic selection). When, however, a P starter is propagated with complete phage protection for a long time in a laboratory, it will eventually be made up of mutants that are highly phage-sensitive; it becomes a 'lab starter.' Consequently, such a starter is rapidly destroyed when it is reintroduced under practical conditions (see Figure 13.14, curve 2).

In actual practice, the propagation of mixed-strain starters can be employed without too many problems. A complete dropout of the starter does not occur because resistant strains are always left behind after a phage attack. However, as stated earlier, the rate of acid production by the starter fluctuates, and it can even drop to below the level desirable for product manufacture; furthermore, the composition of the starter keeps shifting. These properties do not fit the prerequisites for process control in modern cheese manufacture.

13.5.5 Modern Starter Manufacture

In the traditional propagation of starters, mutations, such as those caused by plasmid loss, are responsible for a continuous formation of variants with altered genetic properties, including the phage sensitivity range. This means that a mutant can become a member of the host range of a particular phage or a cluster of phages to which its 'ancestor' was impervious. On top of this, neither the pool of phages in the plant nor their genetic properties are constant. Different kinds of phages are introduced through the raw milk supply; depending on their number, virulence, and the number of host cells in the milk, they can multiply and become a danger to the starter. Incidentally, several phages survive low pasteurization of milk as applied, for instance, in cheese making. Still other phage types are accumulated in the plant by release of temperate phages from lysogenic bacterial cells. Alternatively, throughout multiplication of phages in host cells, their genetic properties can change by mutation, host-induced modification, and recombination of, say, a lytic phage with a prophage.

From this, the risk of contamination by disturbing phages can be inferred to be small if:

1. The quantity of phages in the raw milk is small. This implies that the milk should have a low count of lactic acid bacteria. For cold-stored farm bulk milk, this requirement is usually met.
2. The genetic properties of the starter bacteria are largely fixed. This can be achieved by rapid freezing of the culture and keeping it at a temperature below –35°C.
3. Starter bacteria are selected that are resistant to a great number of phages (see the following text).
4. The number of bacterial strains in the starter is small.
5. During propagation of the starter, contamination by phages is prevented by using clean-room technology and during product manufacture by frequent rigorous cleaning and disinfection.

13.5.5.1 Starter Concentrates

Based on these considerations, a system has been set up (for the first time in New Zealand) that starts from deep-frozen, phage-resistant bacterial strains. Strains isolated from mixed-strain starters (after all, these starters are a satisfactory source of resistant strains) are tested for their resistance against a great number of phages present in the cheese whey of various plants. Subsequently, the fully resistant strains are selected on the basis of their suitability for product manufacture, i.e., rate of acid production, flavor development, resistance to antibiotics, etc. In specialized plants, the suitable strains and their mixtures are cultivated under complete phage protection, then concentrated and distributed while deep-frozen. (To avoid cell damage caused by the freezing procedure, lactose is added to the concentrate as a cryoprotectant; see Section 11.2.) The concentrate can be stored for several months without appreciable loss of properties. Obviously, from a range of selected strains, all kinds of starters can be composed: nonaromatic, aromatic mesophilic, thermophilic, etc.

In the dairy plant, the concentrate represents a mother culture with constant bacterial composition and acid production rate; it serves to prepare the bulk starter culture. Every day the bulk starter milk is inoculated with a new unit of thawed concentrate containing about 10^{10} CFU per ml (Figure 13.12). It is propagated with complete protection from phage development, which requires special provisions for the fermentor, including a phage filter and a device for inoculation of the bulk starter milk. Applying closed machinery (e.g., closed curd-making machines) and maintaining a satisfactory hygienic standard serve to minimize contamination with phages and their accumulation during product manufacture. In practice, the starters can be used for a long time in succession, and a rotation system is not needed. Apart from this, when a disturbing phage emerges, the phage-sensitive strain can be substituted by a resistant one. The number of strains in the starter is restricted to a maximum of six; this is called a multiple-starter system. Sometimes not more than two strains are used (single-pair system).

An older system (developed in the Netherlands) is based on the use of concentrated mother cultures consisting of P starters. Otherwise the two systems do not differ essentially. Mixed-strain starters of a satisfactory acid production rate, fit for making good-quality products, are obtained from various dairy plants. Every starter is immediately deep-frozen to fix the composition of bacteria and phages, along with the properties of the starter, including its natural phage resistance. The starters are rarely transferred. In producing concentrated mother cultures, the proportions of the various strains in the starter should be maintained, especially the ratio of aroma-forming to non-aroma-forming bacteria. A starter can be cultured at a constant pH of, say, 6.0 by continuous titration with alkali, but maintaining the proper ratio of strains is easiest when this is not done. The ripe starter is neutralized with alkali, then concentrated by bactofugation to a concentration of bacteria about 40 times that of a conventional starter; for thermophilic (yogurt) cultures this is about 20 times. After lactose has been added, the product is deep-frozen and transferred to the users.

Again, the latter system ensures a constant bacterial composition of the mother culture as well as an almost complete elimination of fluctuations of the starter activity during product manufacture (see also curve 3 in Figure 13.14). The system is used for cheese as well as for other products.

If these systems are applied, contamination by disturbing phages can only occur during product manufacture. In cheese making, phage multiplication is limited because clotting of the milk hinders diffusion of the phage through the gel.

Thermophilic starters cause far smaller problems than mesophilic starters; the reason is not quite clear. The phages tend to proliferate at a slow rate and to have a restricted host range. For instance, *Lactobacillus delbrueckii* ssp. *bulgaricus* is insensitive to phages that attack *Streptococcus thermophilus*. Furthermore, the nature of the product may play a part. For example, in yogurt manufacture no whey is released, whey being the main pool of phages and the main cause of phage accumulation in cheese plants.

Starter concentrates are nowadays also made from specific strains, which are applied in various dairy fermentations, such as phage-resistant mutants and adjunct starters in cheese manufacture and probiotic strains (Section 22.4) in yogurt manufacture. Also, selected multiple-strain starters, consisting of some three to six *Lactococcus* strains, for use in cheese manufacture are available in deep-frozen concentrates. In most cases, these starters are propagated and grown in milk- or whey-based media, enriched with yeast extract and Ca^{2+}- binding phosphates to promote growth and inhibit phage proliferation, respectively.

13.5.5.2 Direct Vat Inoculation

Direct inoculation of milk with an adequate number of lactic acid bacteria by means of a deep-frozen or freeze-dried concentrated bulk starter culture with up to 10^{13} CFU per gram (Figure 13.12) dispenses with the preparation of a bulk starter. Because there are no propagation steps, the risk of phage attacks is minimized. The direct inoculation method has, however, little advantage over the starter concentrate preparation system, which also guarantees a satisfactory control of the phage problem. A great disadvantage is the high price, so that application should only be considered if there are serious phage problems, which cannot be settled by modernizing the starter preparation, or if the percentage of starter added is small (as in the manufacture of cheese varieties in which very high scalding temperatures are used).

Suggested Literature

A comprehensive book with chapters on taxonomy, genetics, and metabolism of lactic acid bacteria, on bacteriophages, and on starter manufacture: T.M. Cogan and J.-P. Accolas, Eds., *Dairy Starter Cultures,* VCH Publishers, New York, 1996.

For the classification and physiology, see: L. Axelsson, *Lactic acid bacteria: classification and physiology* in S. Salminen and A. von Wright, Eds., *Lactic Acid Bacteria, Microbiology and Functional Aspects,* 2nd ed., Dekker, New York, 1998.

Other books with much information: B.A. Law, Ed., *Microbiology and Biochemistry of Cheese and Fermented Milks,* 2nd ed., Blackie, London, 1997, especially Chapter 2 (Classification and identification of bacteria important in the manufacture of cheese), Chapter 9 (Proteolytic systems of dairy bacteria), and Chapter 10 (Molecular genetics of dairy lactic acid bacteria). Further, see: T.M. Cogan and C. Hill, Chapter 6, in P.F. Fox, Ed., *Cheese: Chemistry, Physics and Microbiology,* Vol. I, *General Aspects,* 2nd ed., Chapman and Hall, London, 1993. See also the new edition of this book: P.F. Fox, P.L.H. McSweeney, T.M. Cogan and T.P. Guinee, Eds., *Cheese: Chemistry, Physics and Microbiology*, Vol. 1, *General Aspects*. 3rd ed., Elsevier Academic Press, Amsterdam, 2004, especially chapters on starter cultures by E. Parente and T.M. Cogan (general aspects), M.J. Callanan and R.P. Ross (genetic aspects), S. McGrath, G.F. Fitzgerald and D. van Sinderen (bacteriophage), and J.-F. Chamba and F. Irlinger (secondary and adjunct cultures).

An extended book on all aspects of cheese, with a clear chapter on starter cultures: P.F. Fox, *Fundamentals of Cheese Science,* Aspen, Gaithersburg, MD, 2000.

A comprehensive review on phages in starter bacteria: G.E. Allison and T.R. Klaenhammer, *International Dairy Journal,* 8, 207–226, 1998.

14 Fouling and Sanitizing

In the dairy industry, cleaning and disinfection are essential operations. Fouling occurs because milk residues remain on the surfaces of the equipment. Residues of milk that have dried up are difficult to remove. Excessive fouling is costly because milk is lost, increased concentrations of detergents are required, and consequently more wastewater is produced.

Fouling especially occurs during heating of milk, which results in the formation of a deposit on metal surfaces that is difficult to remove. Deposit formation reduces the rate of heat transfer and the flow rate of milk in the equipment. Eventually, the equipment will stop operating. In a multiple-effect evaporator with, say, six effects, the costs due to fouling (milk losses and cleaning) can account for more than half of the total running costs (including machinery, energy, etc.). Cleaning of equipment is necessary to reduce all of these problems and to prevent the growth of microorganisms in milk residues, which is highly undesirable. Several microbes can readily grow on surfaces containing a thin film of milk deposit.

A special case is fouling of membranes, which is discussed in Chapter 12.

14.1 DEPOSIT FORMATION

Fouling of a surface always starts with adsorption, for the most part of proteins. From milk and its derivatives, first of all, serum proteins will adsorb onto a metal surface, irrespective of whether that surface is hydrophilic or hydrophobic. Adsorption occurs almost instantaneously (in about 1 ms) and is not fouling: it concerns a monomolecular layer. However, onto this adsorbed layer more serum protein, as well as other materials, including calcium phosphates, casein micelles, fat globules, and bacteria (often by means of their glycocalix), are deposited. Figure 14.1 illustrates the difference between the initial fast adsorption of protein and the subsequent much slower deposition.

It may be useful to give some ideas about the quantities involved. Adsorbed, i.e., monomolecular, protein layers have surface loads for the most part varying from 1 to 3 $mg \cdot m^{-2}$. A deposit layer of 1 mm thickness consisting of 20% protein would correspond to a protein load of about 200 $g \cdot m^{-2}$, i.e., 10^5 times that of an adsorbed layer. A reasonable value for the rate of deposit formation in a heat exchanger when pasteurizing milk would be 30 $g \cdot m^{-2} \cdot h^{-1}$. It would then take about 7 h for a 1-mm layer to form. In some situations, faster deposition rates are observed (see the following text).

Two *types of deposited material* are often distinguished, arbitrarily called A and B. Type A is typically formed at moderate temperatures, say 80°C. It consists, for instance, of 35% 'ash' and 50% protein in the dry matter, and it is yellowish,

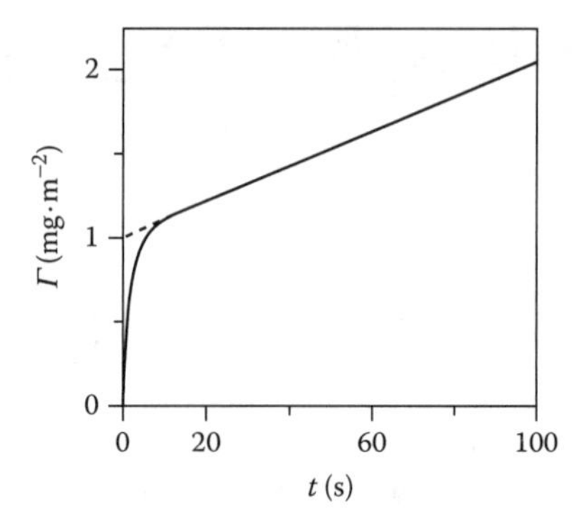

FIGURE 14.1 Deposition of protein from a very dilute solution of serum proteins onto a chromium oxide surface at 85°C. Γ is the protein load and t is the time. Approximate results. (Adapted from experiments by T. Jeurnink et al., *Colloids and Surfaces B: Biointerfaces,* **6**, 291–307, 1996.)

voluminous, and curd-like. Table 14.1 gives an example of its composition compared with that of skim milk. It is seen that about 10% of the deposited dry matter is not accounted for; it may be that this concerns Maillard products because more is formed at a higher temperature. Type-B deposit is typically formed at high temperatures, usually, greater than 100°C. It contains more than 70% 'ash' (for the most part, calcium phosphate) and some protein. It looks compact, gritty, and grayish, and it is also called *milkstone* or *scale.*

It is obvious that two milk components are preferentially deposited, i.e., serum protein and calcium phosphate. These are indeed the substances that become

TABLE 14.1
Composition of the Dry Matter of Skim Milk and of the Deposit Formed in a Heat Exchanger when the Skim Milk Is Heated to 85°C

	Percentage of Dry Matter in	
Component	**Skim Milk**	**Deposit**
Serum protein	6	34
Casein	31	11
Total salts	8	45
Ca	1.3	16
PO_4, inorganic	2.0	23
Sugars	53	0.0
Other	2	10

Approximate example. *Source:* Adapted from results by T.J.M. Jeurnink et al., *Neth. Milk Dairy J,* **50**, 407–426, 1996.

insoluble and supersaturated, respectively, at high temperatures. For serum proteins this follows from Figure 7.9E, and for inorganic phosphate from Figure 2.8, which show that below 70°C, the changes occurring with these substances are very slow; this agrees with the observation that below 70°C, little fouling occurs. It is thus clear as to which substances are primarily deposited and why, but the mechanism needs elaboration. Three mechanisms can be envisaged:

1. *Temperature gradient near the wall:* When a vessel containing a liquid is heated from the outside, its inner surface will have a higher temperature than the liquid. If a reaction leading to insolubility proceeds faster at a higher temperature, some of the material may be deposited on the wall, and its concentration close to the wall is thereby lowered. This implies that a concentration gradient is formed, inducing further transport of material to the wall where part of it gets deposited. This occurs in the heating section of a heat exchanger, where fresh liquid continuously flows past the wall. A temperature gradient, and thereby a concentration gradient, will only exist in a laminar boundary layer near the wall, because any turbulence farther away causes intensive mixing (see Figure 14.2). Nevertheless, this situation implies a continuous deposition, the more so (1) as the temperature gradient is greater (larger ΔT between wall and liquid) and (2) as the turbulence in the liquid is less intensive, because that makes a thicker laminar boundary layer.

   The importance of this mechanism in milk heat exchangers can be evaluated from the effect of ΔT on fouling. It turns out that for the same milk temperature, fouling is stronger in the heating section ($\Delta T > 0$)

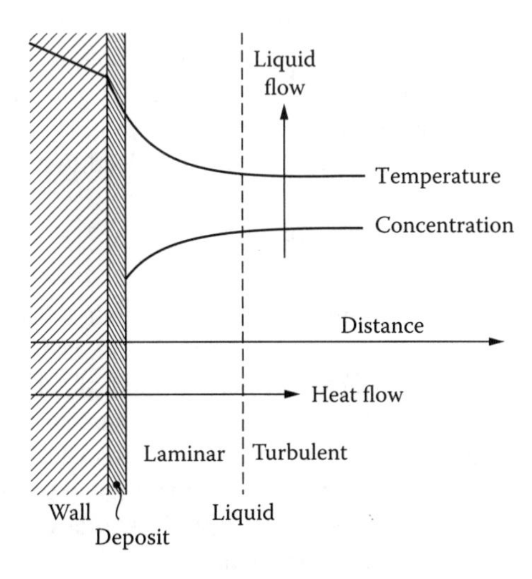

FIGURE 14.2 Temperature gradient near a heated wall in a flow-through heat exchanger and its effect on the concentration gradient of a material that becomes insoluble at high temperature and is thereby deposited onto the wall (schematic).

than in the holding ($\Delta T = 0$) or cooling sections ($\Delta T < 0$), but not greatly so. Consequently, a temperature gradient is not a necessary condition for fouling to occur. On the other hand, a very large ΔT does enhance fouling, especially for concentrated milks.

2. *Competition between surfaces:* Insoluble material may also be deposited on the surface of any particles present in the liquid. In milk, casein micelles are the obvious candidates, and it is, indeed, known that the insoluble serum proteins formed upon heating, as well as calcium phosphate becoming supersaturated at high temperatures, become associated with the micelles. Because casein micelles are very close to each other in milk (on the order of 0.1 μm), and hence to the wall of the vessel, it is only the material that is very close to the wall that can become deposited onto its surface. However, in a heat exchanger, there is a continuous flow of liquid past the wall. The specific surface area A of the casein micelles can be calculated from the values given in Table 3.5. This yields:

$$A = 6\ \varphi/d_{vs} \approx 6 \times 0.06/0.1 = 3.6\ \mu m^{-1} = 3600\ m^2/l$$

In a plate heat exchanger, 1 liter of milk would be, at any point of time, in contact with about 0.5 m^2 of heating surface. In the first approximation, it follows that 0.5/3600, corresponding to 0.014% of the material involved, would be deposited onto the metal surfaces. The proportion has been determined for β-lactoglobulin at 0.14%. This would imply that the metal surface is about 10 times as 'attractive' as the surface of casein micelles for protein to deposit on. Although the proportion deposited is quite small, the continuous supply of fresh milk means that eventually a substantial layer of deposit can be formed. In a heat exchanger, about 1000 l of milk may pass any surface per hour; for the example just given, this then would lead to a deposition of

$$3.2 \times 0.0014 \times 1000/0.5 = 9\ g \cdot m^{-2} \cdot h^{-1}$$

of β-lactoglobulin (assuming the milk to contain 0.32% of it).

Even if no particles are present in the liquid, as in whey, a similar competition may occur. In whey, denaturation of serum protein leads to the formation of aggregates. These then compete with the walls of the heat exchanger for further deposition of material that becomes insoluble. However, the deposition on the aggregates is less efficient than that on casein micelles, and whey gives a faster fouling rate, for proteins as well as calcium phosphate, than milk under the same conditions.

3. *Air or vapor bubbles:* At a hot surface in contact with a liquid, air bubbles of about 1 mm size may readily form if the liquid contains

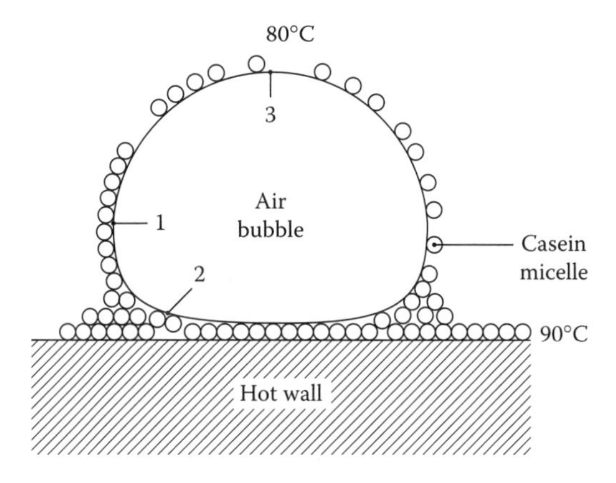

FIGURE 14.3 Air bubble formed at a heated surface and processes occurring near it (schematic and not to scale).

sufficient air for it to become supersaturated at the high temperature. The air in the bubbles is, of course, saturated with water vapor. If a bubble remains at the surface, it can considerably enhance fouling. This is illustrated in Figure 14.3 for casein micelles, but it applies equally well to other substances depositing, say, serum protein. Protein adsorbs onto the bubble (1). The bubble acts as an insulator, whereby the temperature at (2) is higher than at (3), say 90°C and 80°C, respectively. This causes water to evaporate near (2) into the bubble, and to condense near (3), whereby water and heat are transferred. Figure 10.7, shows that the relative water vapor pressures would be 0.72 and 0.48, respectively, at the two temperatures, and a substantial driving force for water transport thus exists. The liquid near (2) now becomes concentrated, leading to (greatly) enhanced deposition of protein. To overcome this problem, the milk should be evacuated before or during heat treatment, or the milk should be kept at a sufficiently high pressure during heating.

It appears that β-lactoglobulin plays a key role in most fouling of dairy liquids. As discussed in Subsection 7.2.2, it is transformed into another form at high temperature, where the –SH group buried in the interior becomes reactive, whereby it can induce formation of –S–S– cross-links with other protein molecules. This reaction, as well as the formation of protein dimers and trimers, probably occurs in the bulk of the liquid and not at a surface. If, however, this aggregation reaction has been allowed to proceed almost completely, which means that termination of the polymerization reaction has occurred in whey or that the serum proteins have become almost fully associated with the casein micelles in milk, further deposition would not occur. It is indeed observed that milk or whey that had been heated to such an extent that nearly all β-lactoglobulin has become insoluble (see Figure 7.9E) hardly shows any protein fouling.

Deposition of calcium phosphate just continues, and the higher the temperature, the faster it proceeds. This explains the formation of milkstone at high temperature, where the serum proteins have already been denatured. When milk is UHT heated in a heat exchanger, the part up to 70°C may show very little fouling; between 70°C and 100°C, a type-A deposit forms; and at higher temperatures, including the first part of the cooling section, type-B deposits form.

Casein micelles generally show little tendency for depositing, because of the repulsion caused by their hairy layers (Subsection 3.3.3). They become, to some extent, entrapped in the deposit via the association with denatured serum protein. Some proteinases, especially those produced by psychrotrophic bacteria, can cause splitting of κ-casein, thereby diminishing the repulsion caused by the hairy layer. This leads to enhanced deposition of micellar casein during heat treatment.

Some fat globules also become enclosed in the deposit, presumably because serum protein becomes associated with the fat globule membrane upon heat denaturation. This provides possibilities for deposition of the globules at a denatured serum protein layer; the fat content of a deposit from fresh milk is, say, 5% of the dry matter. The globules in homogenized milk have completely different surface layers (Section 9.5), consisting mainly of micellar casein and some native or denatured serum proteins. These globules are readily incorporated in the deposit, which then can have a much higher fat content up to 40% of the dry matter.

Variables. In practice, the following are important factors affecting fouling in heat exchangers:

1. *Temperature:* This has been extensively discussed. Below 60°C, hardly any deposit is formed.
2. *Preheating:* The formation of type-A deposit is largely governed by the denaturation of β-lactoglobulin. Preheating to such an extent that by far most of the β-lactoglobulin is denatured will greatly diminish fouling. This is especially important for the concentrating of milk in an evaporator. Also, for heat treatment of plain milk, it can be useful. When heating milk quickly to a temperature at which β-lactoglobulin denatures at a reasonable rate and then keeping the milk at that temperature for a while in a vessel with a small ratio of surface area to volume, the denaturation causes far less fouling.
3. *Temperature difference between heating medium and liquid:* Although this does not have an overriding effect in most cases, an excessive difference should certainly be avoided.
4. *Formation of gas bubbles:* The formation of gas bubbles at the heated surface markedly accelerates deposit formation, as discussed earlier. Degassing of the milk prior to heat treatment can therefore diminish fouling. Rapid streaming or agitation can dislodge bubbles from the surface.

5. *Acidity:* The lower the pH of the milk, the quicker a protein deposit forms. Milk or whey that has turned sour by bacterial action can readily cause a heat exchanger to become blocked. This is presumably caused by the lower solubility of denatured serum protein at a lower pH (see Figure 7.2.A). Mineral deposition is less at a lower pH. This is due to the greatly enhanced solubility of calcium phosphates at a lower pH (see Figure 2.7).
6. *Concentrated milk:* This shows much quicker deposit formation than plain milk. (In the pipes of a falling film evaporator, deposit layers may reach a thickness of 1 cm.) In addition to a higher concentration of reactive components, the higher viscosity (hence, higher wall temperature) and the lower pH play a role.
7. *Homogenization:* Homogenization prior to heat treatment enhances deposition of globular fat, as discussed earlier.
8. *Other milk properties:* There is a large variation among different lots of milk, but the explanation is not always clear. Colostrum causes serious fouling during pasteurization due to its high serum protein content. Protein degradation caused by proteolytic enzymes from psychrotrophs can markedly enhance deposit formation, as mentioned.
9. *Cold storage:* Cold storage of milk reduces deposit formation during heat treatment (unless excessive growth of psychrotrophs occurs). The explanation for this is not clear.

A completely different type of deposit, called *biofilms*, can occasionally be formed by bacteria. *Streptococcus thermophilus*, for instance, may grow on the surface of a heat exchanger for milk at temperatures between 40 and 50°C. Only a very small fraction of the bacteria can attach themselves to the wall, but as soon as one cell is present at the wall at a suitable temperature it can grow out, forming a biofilm; this is a mixed layer of bacterial cells and milk components. Several bacteria can form similar biofilms under specific conditions.

14.2 CLEANING

Cleaning is primarily aimed at (1) the removal of material that interferes with the proper functioning of processing equipment and (2) the prevention of contamination of the products to be made. Microorganisms form the most important contaminants; they can grow in nearly all deposits. Disinfection without prior and thorough removal of all deposits will generally fail. A surface that looks clean may nevertheless have a thin layer (up to at least 10 μm) of milk components that can support significant growth of microbes.

Furthermore, cleaning operations should not be harmful to the staff or lead to pollution of the environment, they should not damage equipment nor corrode materials, and the costs should be reasonable.

Variables. Cleaning requires detergents (cleaning agents) to dissolve or finely disperse the deposit, and moreover, the fouling material has to be removed. Important factors affecting the results are the following:

1. *Nature of the deposit:* This primarily determines the kind of cleaning agents needed. Deposits formed during heat treatment are discussed in the previous section and those formed in membrane processes in Chapter 12. Other fouling may occur, e.g., a thin layer of butter in a churn.
2. *Cleaning agents:* Alkaline solutions (e.g., 1% NaOH, i.e., 0.25 M) can, in principle, dissolve proteinaceous materials and hence type-A deposit and most membrane fouling. Too weak a solution causes insufficient dissolution, but too high a concentration of NaOH tends to change the deposit into a rubbery layer that is hard to remove. Acid solutions (generally phosphoric or nitric acid of about 0.2 molar) serve to dissolve type-B deposit (scale). Alkali tends to enhance scale deposition. If the water used has considerable hardness, a water conditioner, e.g., a polyphosphate that keeps calcium in solution, can be added to minimize such deposition.

   Usually, alkali treatment is followed by a separate acid rinse. If fouling is not excessive and involves little scale, a combined detergent often is used, containing an alkaline substance and a calcium chelating agent, such as ethylene diamine tetra-acetic acid (EDTA).

   Surfactants, i.e., soap-like materials, are added in some cases. A soap solution can loosen a fatty deposit from a surface and keep oil molecules dispersed inside the soap micelles. Hence, a surfactant is certainly needed when a fatty layer is present. It is not necessary to remove fat globules, which are readily dispersed in water (because of their surface layers).

   In the cleaning of ultrafiltration membranes, etc., proteolytic enzymes are occasionally used.
3. *Duration:* Dissolution of the deposit may take a long time if the detergent has to diffuse into the layer, as is often the case. Timescales are several minutes to half an hour. For very thick layers, the diffusion time may be hours and then other mechanisms come into play; see the following item.
4. *Agitation:* This is essential because the dispersed deposit has to be removed, which is generally achieved by liquid flow through the equipment. The flow should be turbulent, and the higher the Reynolds number, the thinner the laminar boundary layer (through which all transport is by diffusion); therefore, the quicker the overall process. Turbulence also causes local pressure fluctuations, and these are essential in the removal of thick deposit layers. The fluctuations can cause cracks to form in the deposit, with the result that lumps of it become dislodged, which are then transported and further broken down in the turbulent flow.

5. *Temperature:* At a higher temperature, nearly all the processes involved proceed faster. The liquid viscosity is lower, so the Reynolds number and diffusion coefficients are larger. Chemical reactions involved in dissolution of the deposit may also proceed faster. If a fatty layer is present, melting of all the crystalline fat is needed for the fat to be removed. On the other hand, at very high temperature, say 100°C, additional calcium phosphate may be deposited, especially in an alkaline solution.
6. *Nature of the surface:* Some materials can show stronger fouling than others in specific cases, but the literature results are conflicting. For efficient cleaning, the surface should be very smooth. The cleaning agents should not corrode the surface. Most stainless steel is very resistant to corrosion, except against strong acids for a prolonged time. Also, borosilicate glass is very resistant. Most problems arise with polymeric materials, especially as used in membranes for ultrafiltration or reverse osmosis.
7. *Internal geometry of the equipment:* From what has been said about the role of agitation, it may be clear that the Reynolds number should be high in all parts of the equipment. This means absence of dead ends and of sharp corners. These prerequisites should be taken into account when designing and manufacturing the equipment. The worst problem is the presence of sharp crevices in the material or the construction. In such a crevice, deposit forms and cannot be removed by the common cleaning methods.

Procedures. Cleaning operations can be performed in various ways, but 'cleaning in place' (CIP) is generally preferred because of its convenience (the equipment need not be dismantled and control is relatively simple), efficiency, and low cost. Two variants of CIP may be distinguished. For heat exchangers and similar equipment of relatively small internal volume, the apparatus is more or less used as when making a product, but now with cleaning liquid; it is thus completely filled. Often, the cleaning liquid is recirculated for a while. It may take a long time before the desired temperature is reached, and a separate heat exchanger for bringing the liquid to the desired temperature, say, 80°C, is sometimes used. For equipment containing relatively large vessels such as tanks, spraying of cleaning liquids is applied. Care must be taken that every spot at the inner surface is reached and with sufficient intensity. Similar conditions hold for apparatuses such as bottle washers.

CIP often occurs in the following steps.

1. *Prerinsing:* Vigorous prerinsing with water can remove some 80% to 90% of the residual (i.e., not-deposited) material in the equipment. Especially for viscous products (stirred yogurt and evaporated milk), it may take a long time before all of the product is washed away. To limit loss of milk and production of wastewater, most milk residues should be removed before prerinsing.

2. *Cleaning steps:* A much-applied method is first cleaning with an alkali and then with an acid solution. The alkali removes most of the deposit, leaving a certain amount of scale. After rinsing with water to remove most of the alkali, nitric or phosphoric acid is introduced to remove the scale. Instead of this two-step cleaning, one-step cleaning by using compound detergents can be applied. Cleaning with alkali and acid is always done after serious fouling of equipment (e.g., heat exchangers for sterilizing milk, and evaporators), whereas compound detergents are generally used for less tenacious fouling, as occurs for low pasteurization of plain milk.
3. *Final rinsing:* Rinsing with water is meant to remove cleaning agents. After acid cleaning, the acid should exhaustively be washed away, particularly when disinfection with sodium hypochlorite is subsequently applied.

Achieving a satisfactory separation of the consecutively used liquids is needed to restrict consumption of water, loss of chemicals, and cost of wastewater treatment. Efficient separation is facilitated by in-line conductivity measurements because the electrical conductivity of the various liquids often differs substantially; determination of the pH, temperature, or turbidity can also be applied.

14.3 DISINFECTION

The aim of disinfection is to kill the microorganisms present on surfaces, and thereby prevent contamination of the product during manufacture and packaging. A satisfactory disinfection does not necessarily kill all microorganisms present but reduces their number to a level at which any quality and health risk can reasonably be assumed to be absent. Disinfectants can attack only if they can reach the microorganisms. Moreover, the action of a disinfectant is often restricted because it becomes inactivated by organic compounds present, although this varies among disinfectants (see Table 14.2). Because of these complications, microorganisms embedded in product remnants or deposits, have an increased survival probability even if a strong disinfectant is used. Surviving organisms, then, may proliferate to considerable numbers during the period between disinfection and the next processing run. Clearly, good cleaning should precede any disinfection. Combined cleaning and disinfection can only be used if merely a loosely connected deposit is present and prerinsing removes most of it.

Most microorganisms are removed during cleaning. Moreover, some cleaning agents such as strong alkali and nitric acid solutions have a disinfecting action. Accordingly, a separate disinfection step is required only if too many microorganisms have been left after the cleaning. Aseptic filling machines should be disinfected just before manufacture starts rather than after the cleaning. This will not be necessary in other cases, if the cleaning is satisfactory.

TABLE 14.2
Properties of Some Frequently Used Disinfectants[a]

Property	Na-Hypochlorite	Iodophors	Quaternary Ammonium Compounds	Peroxy-Acetic Acid
Active Against				
Gram-positive bacteria	+ +	+ +	+ +	+ +
Gram-negative bacteria	+ +	+ +	+	+ +
Bacterial spores	+ +	+	+/–	+ +
Yeasts	+	+ +	+	+
Molds	+ +	+	+/–	+ +
Bacteriophages	+ +	+	+/–	+ +
Activity				
At high soil load	–	+	+ +	+
At low temperature	+ +	–	–	+ +
At low pH	+	+ +	+/–	+ +
At high pH	+	–	+	+
In hard water	+ +	+/–	–	+
Other Aspects				
Toxic	+/–	–	–	–
Corrosive	+	+/–	–	+/–
Foaming	–	+	+ +	–
Price	Low	High	Moderate	Moderate

Note: ++ = strong; + = reasonably; +/– = depends on conditions; – = little or not.

[a] For the most part from data by B.R. Cords et al.; see suggested literature.

Heat or chemical agents can be used for disinfection. In the former method, hot water or steam are used. Maintaining the required minimum temperature over the entire surface for a sufficient time is essential. High temperatures cause denaturation of proteins left, and these then can precipitate upon the equipment. Prior to heat disinfection, rigorous cleaning therefore remains necessary. Disinfection by heat, especially with steam, has the additional benefit of enhancing the subsequent drainage and drying of the machinery, thereby diminishing the risk of bacterial growth. Moreover, after heat disinfection, no disinfectant residues are left.

In the dairy industry, an aqueous solution of sodium hypochlorite, i.e., NaOCl, is the most important chemical used for disinfection. Sodium hypochlorite is prepared by injecting chlorine into a NaOH solution. The following reaction occurs:

$$2\ NaOH + Cl_2 \rightarrow NaOCl + NaCl + H_2O \qquad (14.1)$$

and in aqueous solutions:

$$NaOCl + H_2O \rightarrow NaOH + HOCl \tag{14.2}$$

$$HOCl \rightleftarrows H^+ + OCl^- \tag{14.3}$$

The undissociated HOCl is bactericidal. Its degree of dissociation is smaller for a lower pH, and the bactericidal effect is maximal at about pH 5. However, at pH 5, hypochloric acid is unstable and very corrosive. Because of this, the concentrated Na-hypochlorite solution is stabilized by the addition of a small excess of NaOH to achieve a pH of 8 to 9. During application, the dilution with water lowers the pH to a value at which the bactericidal agent is active. It is advisable that any acid left after the preceding cleaning step is removed because chlorine gas is formed in an acid atmosphere. Chlorine gas is poisonous if inhaled and is also corrosive.

Other disinfectants can also be used. This concerns, for instance, iodophors, i.e., products of a reaction between iodine and a suitable surfactant; quaternary ammonium compounds, i.e., a base of four alkyl groups bound to N (as in NH_4^+), generally with Cl as the counterion; and peroxy-acetic acid CH_3–C (=O) –O–OH (also called peracetic acid). These agents vary with the organisms against which they are most active, the conditions under which they are active, toxicity, corrosiveness, and tendency to foam. The latter is a nuisance when the agent is used in a CIP operation involving spraying. Table 14.2 gives specific information. Hydrogen peroxide (H_2O_2) is especially used to disinfect packaging material in aseptic filling operations.

Residues of cleaning agents and disinfectants should not contaminate the finished products, and a final rinsing step is therefore essential. Some agents such as quaternary ammonium compounds can adsorb onto surfaces of equipment and such residues can eventually be taken up into the product.

Suggested Literature

An interpretive review on fouling and cleaning: T.J.M. Jeurnink, P. Walstra, and C.G. de Kruif, Mechanisms of fouling in dairy processing, *Neth. Milk Dairy J.*, **50**, 407–426, 1996.

Much practical information, especially concerning the processes applied in the U.S.: B.R. Cords, G.R. Dychdala, and F.L. Richter, Cleaning and sanitizing in milk production and processing in: E.H. Marth and J.L. Steele, Eds., *Applied Dairy Microbiology,* 2nd ed., Chapter 14, Dekker, New York, pp. 547–585, 2001.

15 Packaging

Packaging is an essential process step in the manufacture of most foods. The following are the main *objectives*:

1. Containment, i.e., separating the food from the environment: It involves partitioning of the product into units that can be handled during distribution, storage, transport, and final use. It prevents contamination of the environment with the food material, which would cause hygienic problems. It generally guarantees the integrity and the quantity of the contents.
2. Protection of the product from outside influences: This implies prevention of contamination with microorganisms and chemical compounds (for example, oxygen and flavor compounds) or dirt particles, and exclusion of radiation, especially light. Packaging often is an essential part of food preservation. Protection also implies preventing (or minimizing) loss of components, such as water and flavor substances, to the environment.
3. Convenience for the consumer: An obvious point is that it should be easy to open the package and close it again. Minimizing contamination after a portion has been taken out of the package is also of importance, as is easy stacking of the containers (for example, a package containing milk that can be laid down after it has been opened). Packaging in a range of portion sizes is convenient for those who need either a little or a lot of food per day. Modern packaging systems often enable the distribution of foods that are ready to eat (for example, dairy desserts), or that need only little preparation (for example, heating in the package).
4. Providing information: This may be factual information regarding the quality of the product, its composition, nutritive value, keeping quality (day before which it should be consumed), manner of storage, how to handle the product, and so forth. Moreover, several marketing messages may be printed on the label, from the brand name, to possible applications of the product (recipes) and potential benefits of its consumption.

Some aspects mainly related to the first two objectives will be discussed, with some emphasis on packaging of liquid milk products.

15.1 DISTRIBUTION SYSTEMS

These vary widely, according to product type and local practices and facilities.

Milk and milk products may be sold unpackaged. The product is kept in a vat, and the desired amount is poured or ladled from the vat into a smaller vat belonging to the consumer. The method is cheap with respect to processing and packaging material, but it is labor intensive. An important drawback is that contamination by microorganisms is inevitable. The contaminated milk will rapidly spoil and may contain pathogens, and it is highly advisable that the user boil the milk and clean the vat.

Packaging in glass bottles (currently also in PET or polycarbonate bottles) has the advantage that the bottles can be used many times, but the drawback is that their return, and especially their cleaning and subsequent inspection, is laborious and expensive. The disadvantage of the great weight of glass bottles may be acceptable in the case of home delivery.

Most milk is distributed in single-service containers. Containers for durable milk products are often made of tinplate or various synthetic materials. For less durable products, plastics or laminates of cardboard and plastic are often shaped into cartons, sachets, or small cups. The contents may range from about 10 ml (coffee cream) to 3785 ml (beverage milk in some countries).

Another important variable is whether the packaged product is sufficiently stabilized or still has to be processed (for example, cooling, sterilization, and shaking) or transformed (for example, lactic acid fermentation, often with CO_2 formation). In-bottle or in-can sterilization implies heating under pressure in a moist atmosphere, and is predominantly applied to products packaged in glass or plastic bottles, or in cans; close control of the heating intensity and of the closure of the packages is required.

Still another variable is the stage at which the package is made. Compare the use of a previously prepared package that needs only to be closed after filling (glass bottle, can, and some cartons and plastic bottles) to that of a package that is made and filled simultaneously (formation of cartons, blowing of bottles from extruded plastic, and pressing of plastic cups from a foil). In one system, a vertical cylinder is formed from laminated packaging material (cardboard and plastic). It is supplied with milk while it is rapidly pulled down. The filled, moving tube is sealed and cut at regular distances so that tetrahedral or brick-shaped packages are formed. During filling, particular measures may be taken to prevent microbial contamination (aseptic packaging).

The manufacturer's selection of a particular packaging system depends on the specific requirements for the package, the extent to which the process can be fitted into the whole operation, the reliability, and the costs involved. Among other important aspects are environmental pollution and restrictions of the use of nonreturnable packages.

15.2 PACKAGING MATERIALS

Several widely varying materials are in use. The extent to which they meet various requirements and preferences will be briefly discussed. A number of characteristics are listed in Table 15.1. The data involved are highly approximate because they can vary widely according to the precise composition and method of manufacture. The following packaging material criteria are far from exhaustive.

TABLE 15.1
Properties of Some Packaging Materials

							Permeability to			
							H_2O	O_2	CO_2	
Material	Strength	Flexibility	Sealability	Resists sterilization	Resists freezing	Transparency	10^{-12} kg.$m^{-1}.s^{-1}$	10^{-18} kg.$m^{-1}.s^{-1}.Pa^{-1}$		Fat
Glass	Brittle	0	—	Yes	No	Clear	0	0	0	0
Tinplate	Great	Small	—	Yes	Yes	0	0	0	0	0
Aluminum foil	+ + +	+ +	Not	Yes	Yes	0	< 0.1	0.002	0.003	0
Paper or cardboard	+ +	+ +	Not	No	Yes/no	+	great	great	great	great
Cellophane	+ +	+ + +	Good	No	Yes	Clear	100	1	10	tr
Coated cellophane	+ +	+ + +	Good	No	Yes	Clear	1	0.1	0.1	tr
Polyethylene, L.D.[a]	+	+ + +	Good	No	Yes	+ + +	4	40	200	+ + +
Polyethylene, H.D.[a]	+ +	+ +	Good	No	Yes	+ + +	1	10	50	+ +
Polyvinyl chloride	+ +	+ +	Fair	No	No	Clear	10	1	10	+ +
Polyamide (nylon)	+ +	+ +	Poor	Yes	Yes	Clear	40	0.3	1	tr
CPET[b]	+ + +	+ +	Good	Yes	Yes	Clear	5	0.3	2	tr
Polycarbonate	+ + +	+	Not	Yes	Yes	Clear	500	10	50	
Polypropylene	+ + +	Depends	Depends	Yes	No	Clear	3	20	100	+ +
Polystyrene	+ + +	+	Not	No	Yes	+ + +	30	20	100	+ +

Note: 10^{-12} kg·m^{-1}·s^{-1} corresponds to 3.5 g·m^{-2}·d^{-1} at a layer thickness of 25 μm; 10^{-18} kg·m^{-1}·s^{-1}·Pa^{-1} corresponds to 0.35 g·m^{-2}·d^{-1} at a pressure difference of 1 bar and a layer thickness of 25 μm.

[a] L.D. = low density; H.D. = high density.

[b] CPET = crystalline poly(ethylene-terephthalate).

Processability. Is the material brittle, pliable, or moldable? Is it available in the desired thickness (e.g., cellophane can only be made thin-walled)? Is it suitable for being sealed (especially by heat sealing) or is it suitable for lamination (adhesiveness)? Can it readily be cleaned and sterilized? Is it resistant to high temperatures, for example, during in-bottle sterilization?

Resistance. Does the material resist damage? In other words, is it strong enough (this depends very much on its thickness) and wear resistant? Can it withstand fluctuations in pressure and temperature, for example, during sterilization, freezing (some plastics become brittle at low temperature), or gas formation? Is it resistant to a moist atmosphere, that is, does it not soften? Does it show rapid aging? Some plastics rapidly become weak or brittle when exposed to light.

Permeability. Bacteria are generally not let through, provided that the closure of the package is perfect. Passage of a substance through the packaging material may be by diffusion and, consequently, greatly depends on the solubility of the substance in the material. The amount of substance permeating generally is proportional to contact area, time, and concentration difference (for gases often expressed as pressure difference), and inversely proportional to the thickness of the material (see Equation 12.4). Consequently, the permeability can be expressed in, e.g., $kg \cdot m^{-1} \cdot s^{-1} \cdot Pa^{-1}$. Examples are given in Table 15.1. Considering transport of water, the loss of water into air of a certain relative humidity (often 85%) is usually taken as a basis. The permeabilities can greatly depend on the precise composition of the material. Compare, for instance, polyethylene of low and high densities. The latter is more compact due to its large proportion of unbranched chains. Plasticizers (softeners) mostly increase the diffusion coefficients considerably, and the plasticizer content can vary widely. Most plastics are hydrophobic, so the permeability to hydrophobic components (e.g., fat) is fairly large. Compare also the permeabilities of CO_2 and O_2 in Table 15.1.

The permeability also depends on temperature (T). In most materials, the permeability increases as T increases, because the diffusion coefficient tends to increase. On the other hand, the permeability for a given chemical compound is also proportional to its solubility in the packaging material, and the solubility may either increase or — for most gases — decrease as T increases. Hence, it is difficult to predict the temperature dependence of the permeabilities.

The preceding relations often do not apply if the layer becomes very thin (for example, 25 μm or less), because such a thin film can contain perforations. Aluminum foil is a good example because the permeability of aluminum to almost all substances is effectively zero, but any perforations cause trouble. Their number increases considerably with decreasing thickness of the foil and depends, moreover, on the production process and further handling in the dairy, during distribution, and so forth.

The permeability of the packaging material naturally depends on its thickness. Often, containers composed of layers of different materials, so-called laminated foils, are applied. If the permeability to a certain component in a packaging material of a given thickness is designated as b (expressed in, for example, $kg \cdot m^{-2} \cdot s^{-1}$), the total permeability of a laminate can be calculated from $1/b_{total} = \Sigma(1/b_i)$.

Release of components of the material into the food depends on the type of food (pH, presence of fat, and so forth) and on the temperature. Plastics may release plasticizers, if still present, especially to high-fat products. Cans can release iron, tin, etc., and because of this, tinplate is always coated, that is, supplied with a plastic layer. Uncovered cardboard may release several substances into the milk. Generally, legal requirements apply to the release of several components.

Heat Insulation. Often a well-insulating package is not desirable, because after packaging heating and/or cooling are to be applied. Although most plastics have poor heat conductivity, the layer often is too thin for satisfactory insulation. If insulation is needed, expanded polystyrene (polystyrene foam) can be applied.

Light Transmission. For many foods a transparent package is desirable so that the user can see the contents. The drawback for milk products is that light-induced flavors (cardboard or sunlight flavor, and oxidized or tallowy flavor) may develop. Cardboard is not transparent but is certainly not impermeable to light. Glass can be browned (it is the short-wavelength light that is most harmful), but brown glass is often considered unattractive. Most plastics are quite transparent. Fillers can be applied to give color, and TiO_2 is often used for a white color.

Printability of the material often is important for the trade.

Laminates. It will be clear that in many instances no single packaging material meets all requirements. Because of this, laminates are applied. In a milk carton for durable, aseptically packaged products, we may find (going from outside to inside):

Polyethylene: for water repellance.
Paper: for printing.
Cardboard: for firmness.
Polyethylene: for making cardboard adhere to aluminum.
Aluminum: against passage of light and all substances.
Polyethylene: for good sealability; sealing here means closing the filled package by pressing while heating.

All the layers are very thin (e.g., 20 μm, with aluminum foil being even thinner), except for the cardboard; a 1-liter package weighs about 25 g and a glass bottle from 400 to 600 g.

15.3 FILLING OPERATION

There are various methods to fill a package with a certain amount of liquid. Weighing is rarely applied. Bottles are usually filled to a certain level, but for highly viscous products a measuring pump should be used; one or a few turns of a plunger determines the amount of product delivered, nearly independently of the product viscosity. Sometimes the filling step itself can cause problems because the high strain rates applied may change the consistency of the product, which then becomes too thin. Accordingly, high-speed filling machines may be unsuitable for products such as yogurt and custard.

The extent of *contamination by bacteria* during packaging is essential for the keeping quality of the milk product. Relatively simple measures may yield substantial results, but strictly aseptic packaging is far more difficult to achieve. For less durable products, contamination should be rigorously avoided if the product is heated before being packaged. Accordingly, the packaging material should be devoid of pathogenic microorganisms and contain few if any bacteria that can grow during its storage. Satisfactory standards of hygiene during manufacture, transport, and storage of the packaging material will prevent many problems, because the materials involved are very poor substrates for microorganisms. Moreover, high temperatures and little water are used during manufacture of packaging materials.

Packages intended for repeated use (bottles) should be thoroughly cleaned before filling, and after cleaning they should be examined to remove dirty and damaged bottles. It is a known fact that consumers may put a milk bottle to other uses or insert into it objects that may barely be removable. This involves a certain danger but produces no appreciable health hazard. After cleaning, the package is disinfected, for example, with a sodium hypochlorite solution of 10 ppm activated chlorine if the milk product is not reheated. The bacterial count should not exceed 50 per bottle. If the product is heated after packaging, the packaging material causes few or no bacteriological problems. A major point is that leakage of the closure due to pressure differences occurring during cooling must be prevented.

In *aseptic packaging* of durable products, spoilage of fewer than 1 in 10^5 packages — and preferably less — may be considered acceptable. Pipes, storage tank, and surfaces of the packaging machine come into contact with the sterilized product and have to be sterilized. The same holds true for the packaging material. Laminated paper has been shown to contain, say, 10 organisms per 100 cm^2, among which about 3% are spores. The inner surface of a 1-liter carton is about 800 cm^2 and will thus on average be contaminated by about 2.5 spores. These spores are the most heat resistant, and hence their number must be reduced to less than 10^{-5} per package. Furthermore, the packages should be aseptically closed; an atmosphere with overpressure and sterile air is usually applied.

Sterilization of the packaging material should not impair that material. Consequently, steam or hot water heating often is not possible. In most cases, sterilization with a hot (60 to 80°C) and concentrated (20 to 35%) solution of H_2O_2 is applied. Hot air (>100°C) can readily remove residues of H_2O_2, and it provides an additional sterilizing effect. H_2O_2 has an advantage over other liquid disinfectants in that it causes no serious problems with respect to residues left in the milk. Gaseous disinfectants such as ethylene oxide have a slow spore-killing action and can only be applied if a long reaction time (several hours) is feasible. Because suitable light sources have been developed, sterilization by UV irradiation is becoming increasingly prevalent, especially for packaging materials and machines that are less readily sterilized by H_2O_2. UV light of 200 to 280 nm accounts for the sterilizing effect. If dust particles have become attached to the packaging material, H_2O_2 will produce better effects due to its rinsing effect, whereas UV irradiation will be less effective due to particle shade. Clean-room techniques combined with irradiation are sometimes applied.

Aseptic packaging has to be meticulously checked. Not only must the packaged product be examined, but so must all preceding steps, as well as the operators, which are potential carriers of pathogens. If just one bacterium reaches the product, and that bacterium is pathogenic and can proliferate (for example, *Staphylococcus aureus*), the result could be disastrous. In addition to regular sampling during production, further samples should be taken at the times or in situations known to be associated with an increased risk of contamination. It is advisable to incubate these samples long enough, in most cases from 5 to 7 days at 30°C to allow sublethally damaged bacteria also to grow to detectable counts. The products should only be delivered if the result of the shelf-life test is satisfactory.

Suggested Literature

A general overview of the packaging of foods: G.L. Robertson, *Food Packaging: Principles and Practice,* Dekker, New York, 1993, which also has a chapter on milk products.

The packaging of dairy products: *Technical Guide for the Packaging of Milk and Milk Products,* IDF Document 143, 1982.

Part III

Products

16 Milk for Liquid Consumption

Liquid milk can be delivered to the consumer after various heat treatments: none (raw milk), pasteurized or sterilized, and either packaged or not (although sterilized milk is, of course, always packaged). The properties of liquid milk that require the most attention are safety to the consumer, shelf life, and flavor. Safety is, of course, essential and consumption of raw milk cannot be considered safe. Consequently, the delivery of raw milk is prohibited or severely curtailed in many countries. Likewise, delivering milk that is not packaged may involve health hazards.

The relative importance of other quality marks depends on usage. Milk can be consumed as a beverage, in which case flavor is of paramount importance. Most consumers tend to dislike a cooked flavor and, therefore, low-intensity pasteurization is generally preferred. Others use milk primarily in coffee or tea, in cooking, in baking, etc., where the absence of a cooked flavor is mostly not essential (if not too intense) and shelf life may be the most important quality mark. Consequently, sterilized milk is often favored. One may even use milk preserves like evaporated milk, dried milk, or — for some uses — sweetened condensed milk.

Liquid milk may vary in composition. Often fat content is standardized to a value near that of average raw milk, but low-fat (semiskim) and skim milks are also sold. Fortification with solids-not-fat or with protein is occasionally applied. Standardization to a specified protein content by means of ultrafiltration is another possibility, but it may not be allowed. Most countries have legal requirements for a minimum solids-not-fat or protein content.

Basic aspects of the processes involved have been discussed. See especially Chapter 7.

16.1 PASTEURIZED MILK

Pasteurized beverage milk must be safe for the consumer and have a shelf life of a week or longer when kept refrigerated. Flavor, nutritive value, and other properties should deviate only slightly from those of fresh raw milk.

The following contaminants can in principle be harmful to the consumer:

- Pathogenic microorganisms, which may already be in the milk while in the udder, or be incorporated during or after milking. Most of these do not survive pasteurization, but they may also enter the product by recontamination.
- Toxicants taken up by the cow (e.g., with the feed) and entering the milk during its synthesis.

- Antibiotics, used to treat (the udder of) the cow.
- Disinfectants used on the farm or in the plant.
- Bacterial toxins formed during keeping of the milk.
- Other toxicants entering the milk by contamination during and after milking.
- Radionuclides.

Pathogenic microorganisms can be killed by heat treatment. Most (chemical) contaminants cannot be eliminated in this way. Obviously, proper cattle management and adequate methods of collecting and handling the milk are necessary to prevent health hazards. Regular checks for the absence of contaminants are needed.

With reference to the shelf life and safety of the milk, most countries have legal requirements for the maximum number of microorganisms (colony count) and coliforms, and for the absence of the enzyme alkaline phosphatase. To meet these requirements, the original milk should not contain too many heat-resistant bacteria (Chapter 5), the pasteurization step should be checked (recording thermometer and flow diversion valve), and contamination of pasteurized milk with microorganisms (or with raw milk) should be minimized.

A cream layer is less desirable, especially when nontransparent packaging material is used. It can be prevented by homogenization, which implies that the pasteurizing intensity should be adapted to avoid lipolysis. The more intense the heat treatment, the more the flavor of the milk will differ from that of raw milk.

16.1.1 Manufacture

Figure 16.1 gives an example of the manufacture of pasteurized milk for liquid consumption. See also Figure 7.17.

The importance of *thermalization* to prevent fat and protein breakdown by heat-resistant enzymes of psychrotrophic bacteria is discussed in Chapter 7 (see also Subsection 6.4). But as a rule, the keeping time of pasteurized milk is too short to cause noticeable decompositions by these enzymes, unless the original milk had a high count of psychrotrophs. Furthermore, thermalization at a rather high temperature (say 20 s at 67.5°C) causes a considerable inactivation of milk lipase (about 50%) and permits a somewhat lower pasteurization temperature in the manufacture of homogenized milk. Despite these obvious advantages of thermalization, dairy plants often only cool the milk (mainly to save on costs), taking the risk of some growth of psychrotrophs.

Separation is needed to adjust to the desired fat content. If homogenization is omitted, only a part of the milk will be skimmed, while the skim milk volume obtained should suffice to standardize the milk.

Homogenization serves to prevent the formation of a cream layer in the package during storage. Many users dislike such a layer. In low-pasteurized milk (alkaline phosphatase just inactivated), a loose cream layer of agglutinated fat globules forms that can be easily redispersed throughout the milk. In high-pasteurized milk, the cold agglutinin has been inactivated and a cream layer forms far more slowly, but

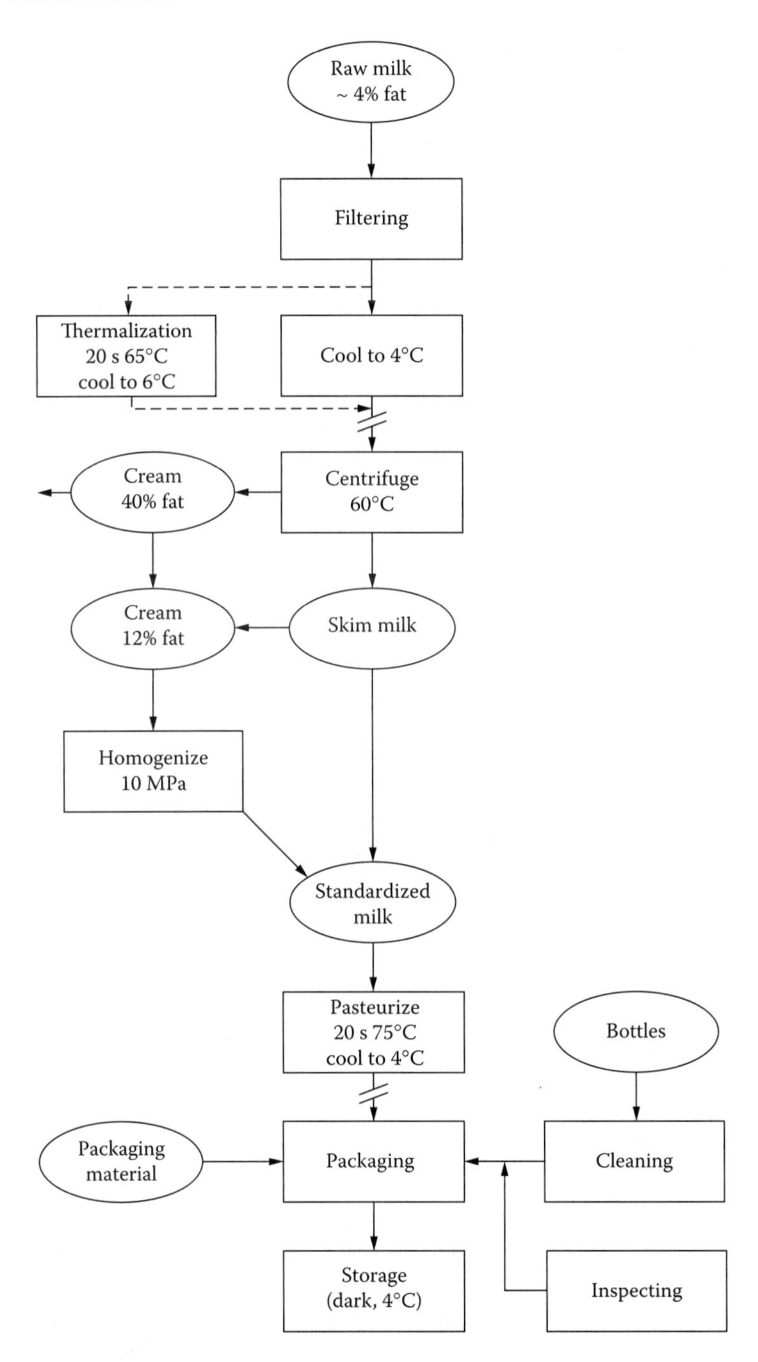

FIGURE 16.1 Example of the manufacture of homogenized, pasteurized beverage milk.

then it is a compact, hardly dispersible layer; a solid cream plug may even result from partial coalescence of the fat globules. Therefore, this milk is usually homogenized. As a rule, not all of the milk is homogenized but only its cream fraction (partial homogenization) to reduce cost. Obviously, all milk should then be separated. Homogenization clusters should be absent after the homogenization; therefore, the fat content of the cream should be rather low (10% to 12%) and the homogenizing temperature not too low (≥55°C); moreover, two-stage homogenization should be applied (see Chapter 9). Usually the homogenization precedes the pasteurization to minimize the risk of recontamination. Because milk lipase then is still present, the milk should immediately be pasteurized.

After partial homogenization the milk may still cream due to cold agglutination. This results from the agglutinin in the skim milk after warm separation being not fully inactivated by the subsequent pasteurization (see Figure 16.2). In spite

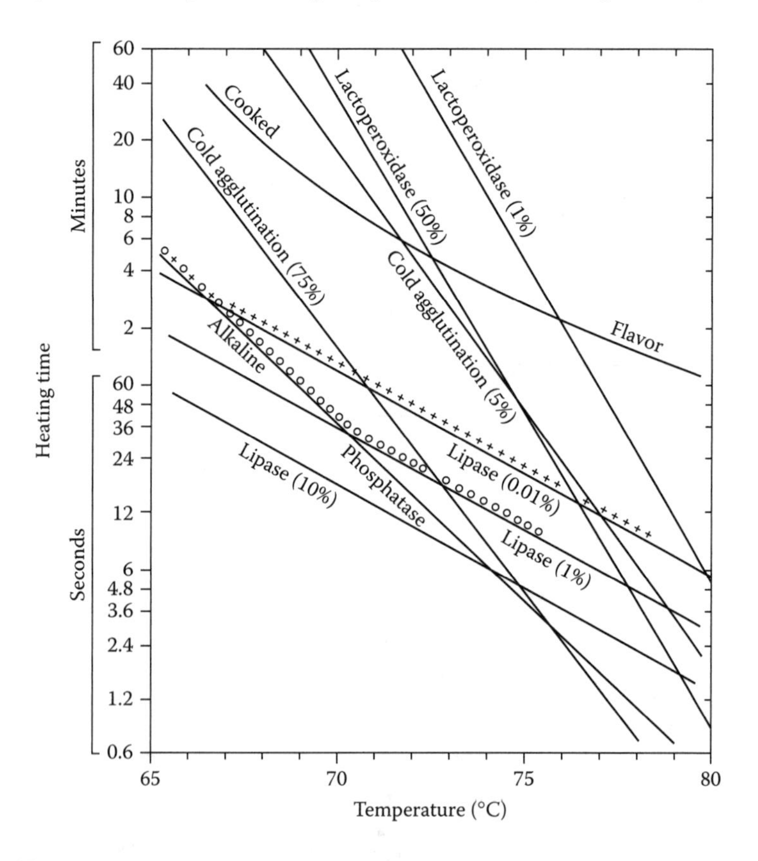

FIGURE 16.2 The heating time of milk needed to obtain certain effects as a function of temperature: inactivation of alkaline phosphatase to become 'nondetectable'; inactivation of lipoprotein lipase, cold agglutination, and lactoperoxidase; and generation of a noticeable cooked flavor. The figures on the curves denote the approximate proportion of the activity left. Lower limits for low pasteurization of nonhomogenized (○○○○○) and of homogenized (+++++) milk are indicated.

of a low ratio of agglutinin to fat surface area, the fat globules can agglutinate if the raw milk contained much agglutinin.

Homogenized milk has an increased tendency to foam, especially at low temperature.

Standardization with respect to fat content is described in Section 6.5. It can be done by adding skim milk (or cream) to the milk in the storage tank or by continuous standardization.

Pasteurization ensures the safety and greatly enhances the shelf life of the product. See Subsection 7.3.4 for the kinetics of killing bacteria. A mild heat treatment, e.g., 15 sec at 72°C, kills all pathogens that may be present (especially *Mycobacterium tuberculosis*, *Salmonella* spp., enteropathogenic *E. coli*, *Campylobacter jejuni*, and *Listeria monocytogenes*) to such an extent that no health hazard is left. Some cells of some strains of *Staphylococcus aureus* can survive the heat treatment, but they do not grow to the extent as to form hazardous amounts of toxins. Such pasteurization inactivates alkaline phosphatase to the extent as to be no longer detectable (the enzyme may, however, regenerate slightly after keeping the product for some days, but this is especially true of pasteurized cream). Most of the spoilage microorganisms in raw milk, such as coliforms, mesophilic lactic acid bacteria, and psychrotrophs, are also killed by low pasteurization. Among those not killed are heat-resistant micrococci (*Microbacterium* spp.), some thermophilic streptococci, and bacterial spores. But these microorganisms do not grow too quickly in milk, except *Bacillus cereus*. The latter organism is pathogenic if present in large numbers, but prior to this the milk has become undrinkable because of its off-flavor.

Among the undesirable enzymic decompositions lipolysis (as caused by the natural milk lipoprotein lipase) is of special importance. Figure 16.2 shows the time–temperature relations that reduce the activity of the enzyme to 10^{-1}, 10^{-2}, and 10^{-4}, respectively. Homogenized milk is highly susceptible to lipolysis because of its readily accessible substrate; therefore, it should be rather intensely heated (e.g., 20 s at 75°C) to reduce its lipase activity to 10^{-3} or 10^{-4}. A decrease to 10^{-2} suffices for nonhomogenized milk, which implies a heating of, say, 15 s at 72.5°C. Plasmin is not inactivated by pasteurization (see Figure 7.9B); but the keeping time of pasteurized beverage milk generally is too short to cause problems.

After low pasteurization of the milk (15 s at 72°C), sufficient natural substances inhibiting bacterial growth remain intact, but a somewhat higher pasteurization temperature, as is needed for homogenized milk, clearly decreases their effect (see Figure 16.2). It mainly concerns the immunoglobulins; their inactivation runs parallel to that of the agglutinins that determine the creaming properties. The agglutinins against bacteria (e.g., inhibitors of *B. cereus*) are also inactivated by homogenization and thus are absent in homogenized milk; they may remain active upon partial homogenization. The lactoperoxidase–thiocyanate–H_2O_2 system is less heat sensitive; its inactivation becomes perceptible at temperatures greater than 76°C, when heating for 15 s. The effect of the presence or absence of inhibitors does, however, depend on the bacterial flora present. In high-pasteurized

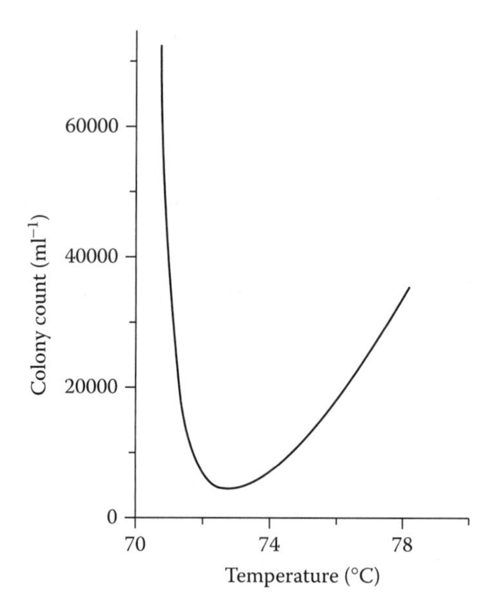

FIGURE 16.3 Example of the influence of the pasteurization temperature (pasteurization for about 20 s) on the bacterial count of unhomogenized milk after keeping it for 7 d at 7°C. Data courtesy of M.P. Kimenai.

milk (e.g., 15 s 85°C) the bacterial growth inhibitors are eliminated and, despite its lower initial bacterial count, the milk may have a shorter shelf life than has low-pasteurized milk. These effects are illustrated in Figure 16.3. High-pasteurized milk is often heated in the bottle, and this improves its keeping quality because recontamination cannot occur; however, it also causes a distinct cooked flavor.

In the manufacture of low-pasteurized beverage milk, flow-through heating is commonly applied, as a rule in a plate heat exchanger. The time–temperature combination selected is a compromise between sufficient inactivation of milk lipase and conservation of its ability to inhibit bacterial growth. Usually the temperature is adjusted, but as is seen in Figure 16.2, adjusting the length of time at a constant temperature may give better results (note that the slopes of the curves differ). On pasteurizing homogenized milk, the agglutinins should be inactivated to such an extent as to prevent creaming of the milk. A cooked flavor may sometimes be observed.

High-pasteurized milk has a somewhat whiter color (as has ultra-high-temperature [UHT] short-time heated milk; see Section 16.2), for the most part due to its homogenization. A more intense heating causes browning due to Maillard reactions. Occasionally, heating to over 100°C is applied to kill spores of *B. cereus*, thereby enhancing shelf life.

Packaging of low-pasteurized beverage milk is generally done in single-service containers such as cartons. A certain quantity of milk is still filled in glass or plastic bottles (see Chapter 15). Great care should be taken to ensure hygiene during packaging in terms of the safety of the product, but especially because of the effect of recontamination on the shelf life of the product; aseptic packaging would be desirable. The temperature of the milk may increase by about 1 K during packaging due to the transportation in pipelines and on conveyor belts, and due to the use of sealing machinery. Because recooling of packaged products is slow, especially if piled up closely, such temperature increase should be anticipated by deeper cooling after pasteurization.

16.1.2 Shelf Life

Shelf life is the time during which the pasteurized product can be kept under certain conditions (e.g., at a given temperature) without apparent undesirable changes. Changes in beverage milk during storage can be distinguished in:

- Decomposition by bacteria growing in the milk, such as acid production, protein breakdown, and fat hydrolysis
- Decomposition by milk enzymes or by extracellular bacterial enzymes, like fat and protein breakdown
- Chemical reactions causing oxidized or sunlight flavor
- Physicochemical changes such as creaming, flocculation, and gel formation, which may, in turn, be caused by the above-mentioned changes

Changes caused by bacteria growing in the milk mostly do not become noticeable before their count amounts to 5×10^6 to 20×10^6 ml^{-1}, depending slightly on the bacterium species involved. If *B. cereus* is the spoilage organism, the limit taken is 10^6 ml^{-1}. Such counts should, however, not yet have been attained at the moment of purchase by the user. Pasteurized beverage milk should keep for, say, a week after purchase, provided it is kept refrigerated (below 7°C). Sometimes, a 'day of ultimate sale' is given with the product; in other cases an 'ultimate day of consumption' (or minimum guaranteed shelf life). Criteria can be formulated for the bacteriological quality of the milk on the dates mentioned.

Enzymatic changes were mentioned in Subsection 16.1.1. Chemical changes especially concern the high susceptibility of low-pasteurized milk to light-induced off-flavors, especially if packaged in transparent bottles. See further Section 4.4.

Deterioration of pasteurized milk is especially caused by growth of microorganisms. It is determined by:

Storage temperature
Extent of recontamination
Growth rate (generation time, *g*) of the bacteria involved
Number of spores of *B. cereus* in the original milk
Activity of substances inhibiting bacterial growth

TABLE 16.1
Generation Time (Hours) of Some Bacterial Strains in Low-Pasteurized Milk at Various Temperatures

Temperature (°C)	4	7	10	20
Bacillus cereus	∞	10	4	1
Bacillus circulans	20	12	10	3
Enterobacter cloacae	8	5	3	1
Pseudomonas putida	6	4	3	1
Listeria monocytogenes		20		
L. monocytogenes, in high-pasteurized milk	30	11	9	2

The storage temperature of the milk is important because the generation time of the microorganisms is highly temperature dependent, as is shown in Table 16.1. There is no real point in lowering the temperature to below 4 to 5°C, because during transit and storage in the distribution network higher temperatures normally prevail, say 7°C. The effect of the temperature on the length of time that pasteurized milk can be kept is shown in Table 16.2.

The growth rate of bacteria depends on the temperature and the bacterium species involved. Starting from a count in the milk of 10 per liter, and with g amounting to 4, 7, and 10 h, a shelf life of 5, 8, and 13 days, respectively, is calculated. Such figures are quite normal. The shelf life of the milk at various temperatures may be predicted if the species of the bacteria involved as well as their initial count and generation time are known. Obviously, the shelf life of the milk depends on the

TABLE 16.2
Examples of the Average Numbers of Days That Low-Pasteurized Milk Can Be Stored at Various Temperatures before It Surpasses the Criteria for Guaranteed Day of Ultimate Sale (A) and Guaranteed Shelf Life (B)

	Average Number of Days to Obtain a Count of					
	5.10^4 ml^{-1} (A)			5.10^6 ml^{-1} (B)		
Milk Samples Taken	4°C	7°C	10°C	4°C	7°C	10°C
Just after the pasteurizer	>14	9.6	5.8	>14	13.6	9.8
From glass bottles	12.8	6.0	4.7	13.5	8.7	7.3
From cartons	>14	7.8	5.2	>14	10.9	7.0

Note: Data kindly provided by M.P. Kimenai.

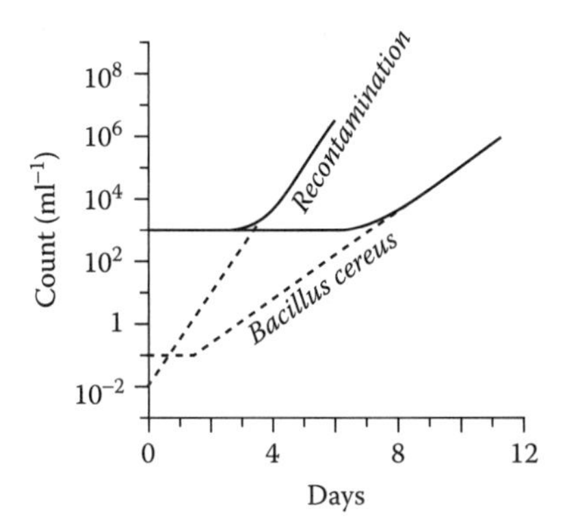

FIGURE 16.4 The bacterial count in low-pasteurized milk during storage at 7°C and the effect of recontamination. Approximate example. Solid line: total count; broken lines: specific flora.

growth possibilities of the bacteria present, whereas the total count just after pasteurization does not give sufficient information, as is illustrated in Figure 16.4.

After pasteurization of the milk its count usually amounts to between 500 and 1000 ml^{-1}, unless many heat-resistant bacteria are present in the original milk. As a rule, the milk is spoiled by 'sweet curdling' caused by *B. cereus* ($g \geq 10$ h at 7°C), unless it is recontaminated (spoilage of recontaminated milk; see the following paragraph). *B. cereus*, forming lecithinase, is also responsible for the 'bitty cream' defect in nonhomogenized milk, i.e., the enzyme coagulates the fat globules in the cream layer that are in the vicinity of a 'colony' of these bacteria. At a storage temperature below 6°C, *B. cereus* cannot grow; deterioration may then be caused by *B. circulans*. High-pasteurized milk, made by heating at about 100°C, is mainly spoiled by *B. licheniformis* or by *B. subtilis*, if the keeping temperature is relatively high. Milk contains, say, 10 spores of *B. cereus* per 100 ml; its shelf life for normal storage conditions amounts to 12 to 14 days if it is not recontaminated.

If the pasteurized milk is *recontaminated*, deterioration is generally faster and of a different nature. This is illustrated in Table 16.2, in which the milk leaving the pasteurizer has not yet been recontaminated, but it commonly becomes so during packaging. The presence of coliforms, detectable after keeping the milk at 20°C, is an indication of recontamination having occurred. The (recontaminated) milk, stored uncooled, turns sour by the growth of, e.g., mesophilic lactic acid bacteria; high-pasteurized milk deteriorates quickest. Below 10°C the milk deteriorates by the growth of psychrotrophs (g = 4 to 5 h at 7°C). The flavor becomes putrid and rancid due to protein degradation and hydrolysis of fat, respectively. Since these psychrotrophs are hardly affected by substances inhibiting bacterial growth, the deterioration rate below 7°C is similar for both high- and low-pasteurized (recontaminated) milk.

The rule is that the more *B. cereus* spores in nonrecontaminated milk, or the heavier the recontamination, the faster the deterioration. Thorough cleaning and disinfection of the filling and lidding machine is needed to avoid recontamination (as far as possible) after flow-through pasteurization. In determining the day of ultimate sale, one usually assumes that some recontamination of the milk occurs.

Frequent and thorough inspections are needed during processing to limit recontamination and to meet the requirements at the day of ultimate sale. To that end, samples may be kept at various temperatures and tested at intervals. The drawback is that the user already received the milk before the result of the shelf-life test is known. Tests have therefore been developed that allow a fairly rapid detection of recontamination by Gram-negative, non-spore-forming bacteria.

16.1.3 Extended-Shelf-Life Milk

Some consumers desire a beverage milk that tastes like low-pasteurized milk, but that can be kept much longer without perceptible quality loss. There are two principles by which such ESL-milk can be produced.

The first involves *UHT-heat treatment*, followed by aseptic packaging. This actually results in sterilized milk. However, a heat treatment of 2 s 140°C or 3 s 135°C will suffice to kill all bacteria (see Figure 16.8), while it can leave the flavor virtually unaltered, provided that *direct* heating is applied. See further Subsections 7.3.1 and 7.3.2. The milk must be free of enzymes produced by psychrotrophs, as these are not inactivated. Plasmin also remains active, causing a bitter flavor. However, when kept refrigerated, this may only become noticeable after about 1 month.

The other principle is physical removal of bacteria and their spores. This can be done by bactofugation, as discussed in Section 8.2. However, this is an expensive method, especially if complete removal of organisms is desired.

Another possibility is removal of microbes by *microfiltration* (see Chapter 12 for membrane processes), which has met with some success. The transmembrane pressures applied are below 1 bar. A high flux and long operating periods can be achieved. The fat globules are also retained, considering that the membrane has a pore size of about 1 μm; therefore, the milk should first be separated. Figure 16.5 gives an outline of a manufacturing process. Some 0.1% to 1% of the total number of bacterial cells passes to the permeate, of *B. cereus* $< 0.05\%$. Stronger reductions, even up to sterility, can be obtained by using membranes with smaller pore size, but that is at the expense of the flux and of the maximum operating time. The amount of retentate is only a small percentage of the initial volume; the protein content is slightly increased, by about 0.5 percentage unit. The retentate is UHT-sterilized along with the cream. The product should be packaged aseptically.

The shelf life of the product is greatly enhanced. On the other hand, part of the product (about 12%) is sterilized; the ensuing cooked flavor is restricted by applying a brief UHT treatment with direct heating. It may be noted, however, that the fat globules (which generate the greater part of the sulfhydryl compounds on intense heat treatment) are in the most intensely heated fraction.

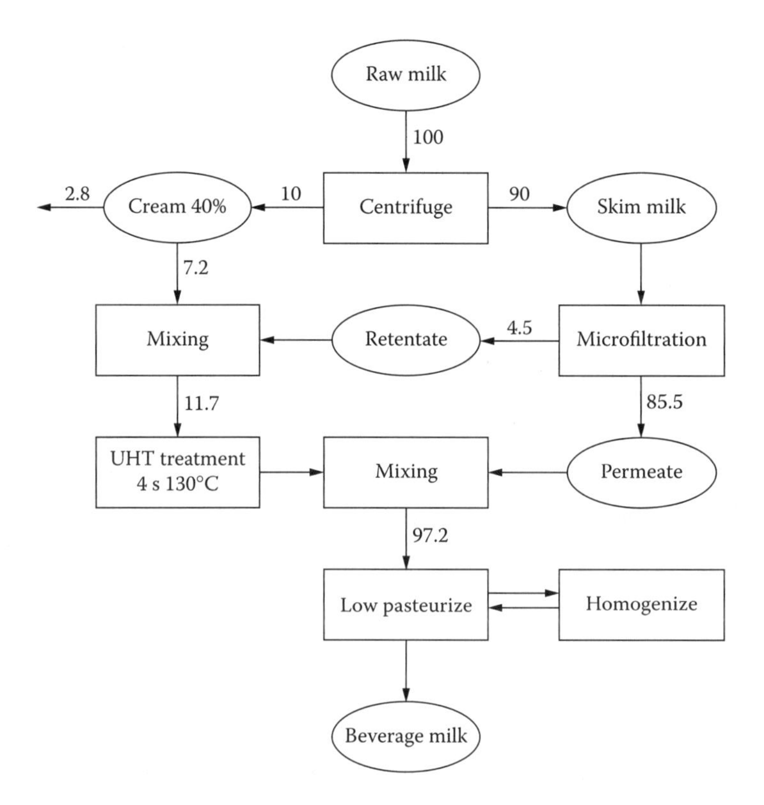

FIGURE 16.5 A manufacturing process for pasteurized beverage milk by using microfiltration. (Adapted from P.J. Pedersen, IDF Special Issue 9201, 1992).

16.2 STERILIZED MILK

16.2.1 Description

Sterilization of milk is aimed at killing all microorganisms present, including bacterial spores, so that the packaged product can be stored for a long period at ambient temperature, without spoilage by microorganisms. Since molds and yeasts are readily killed, we are only concerned about bacteria. The undesirable secondary effects of in-bottle sterilization like browning, sterilization flavor, and losses of vitamins can be diminished by UHT sterilization. During packaging of UHT-sterilized milk, contamination by bacteria has to be rigorously prevented. After UHT sterilization, certain enzymatic reactions and physicochemical changes still may occur.

To achieve the objectives it is necessary that:

- The count of microorganisms, including spores, is reduced to less than 10^{-5} per liter.
- The original milk does not contain enzymes of bacterial origin that cannot be fully inactivated by the heat treatment.

- Enzymes naturally present in milk are sufficiently inactivated.
- Chemical reactions during storage are minimal.
- Physical properties of the milk change as little as possible during treatment and storage.
- The flavor of the milk remains acceptable.
- The nutritive value of the milk decreases only slightly.

These objectives and requirements are hard to reconcile. The most important ones determine what heating process will be selected. Furthermore, factors like processing costs, complexity of the machinery and processing, and, above all, the consumer's wishes must be taken into account.

The inactivation of enzymes and killing of microorganisms are discussed in Subsections 7.3.3 and 7.3.4.

Oxidation causes off-flavors and decomposition of vitamins. Occurrence of these reactions during storage is limited by: intensive heating, causing antioxidants to be formed; deaeration of the milk and excluding air from the package; and using a package that is impermeable to light and oxygen. Furthermore, Maillard reactions can occur, both during the heat treatment (in-bottle sterilization) and during storage (UHT milk) (see Subsection 7.2.3). The latter reactions are responsible for browning, off-flavor, and decreased nutritive value.

Sterilized milk is kept for a long time so that it will show extensive gravity creaming if unhomogenized. Creaming as such is undesirable. Besides, partial coalescence of the closely packed fat globules will lead to formation of a cream plug, which is hard to mix throughout the remaining milk; oiling off may even occur at somewhat elevated temperatures. Therefore, sterilized liquid milk is always homogenized.

If the milk is only in-bottle sterilized, little variation in process conditions is possible; the product obtained can be clearly recognized by the user because of its inevitable sterilized flavor. If the milk is UHT-heated, a sufficient sterilizing effect can readily be achieved, which implies that the appropriate process conditions can be selected on the basis of additional considerations. The flavor can vary from a mild (at, say, 1 s at 145°C, direct heating) to a marked cooked flavor (heating of, e.g., 16 s at 142°C in a heat exchanger with a warming and cooling profile as shown in Figure 7.20, right-hand curve) that can be scarcely distinguished from the flavor of in-bottle sterilized milk. This makes it difficult to correctly characterize UHT-sterilized beverage milk and to clearly inform the consumer. Classification on the basis of the processing equipment involved is insufficient. Therefore, one has tried to characterize UHT milk by means of a chemical change, for which the formation of lactulose is generally used. A standard for UHT milk then would be that it contains less than 600 mg lactulose per liter (see Figure 16.9).

16.2.2 Methods of Manufacture

Figure 16.6 and Figure 16.7 give examples of the manufacture of sterilized milk. Thermalization, separation, and standardization are described in Section 16.1. The proteinases and lipases of psychrotrophs, especially of the genus *Pseudomonas,*

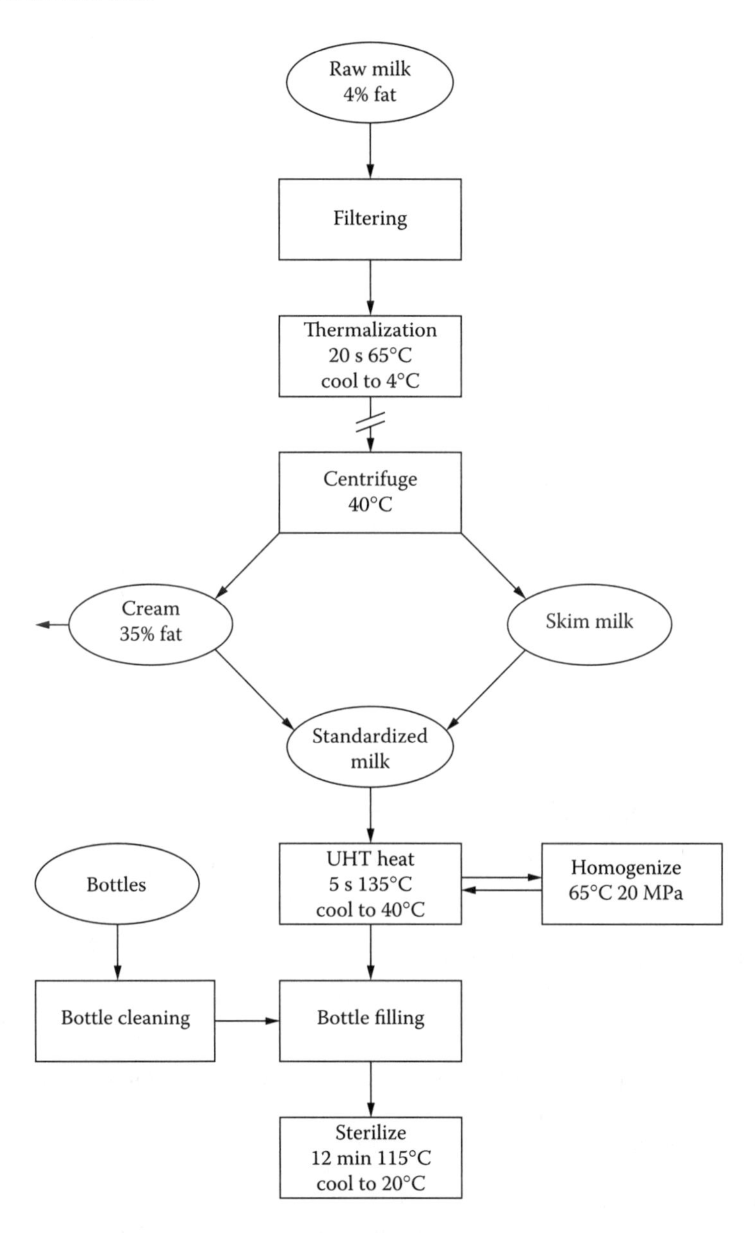

FIGURE 16.6 Example of the manufacture of in-bottle sterilized milk for consumption.

can be very heat resistant and even in-bottle sterilization does not suffice to fully inactivate these enzymes. Therefore, the enzymes should be absent in the raw milk. In particular, the addition of some milk left over for some time should be carefully avoided because in this milk psychrotrophs may have grown extensively.

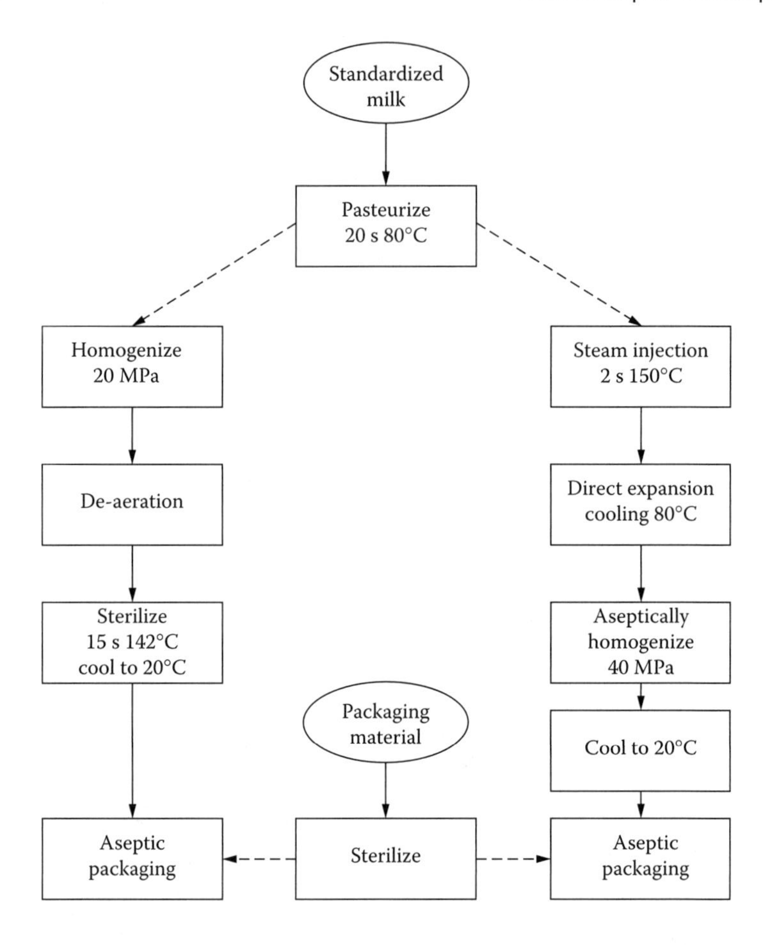

FIGURE 16.7 Examples of the manufacture of UHT-sterilized milk (indirect or direct heating) with aseptic packaging.

These bacteria especially produce heat-resistant enzymes in an (almost) full-grown culture (stationary phase).

The various types of heating processes are:

- In-bottle sterilization
- Flow-through preheating and a mild in-bottle sterilization
- Flow-through sterilization and aseptic packaging

The benefits and disadvantages of these types of heating processes, and the machinery involved, are discussed in Section 7.4. The sterilizing effect required determines the lower limit of the time–temperature relation to be selected. The sterilization intensity also has an upper limit, which is reached when the milk protein starts to coagulate. Nearly all good-quality raw milk is stable enough to

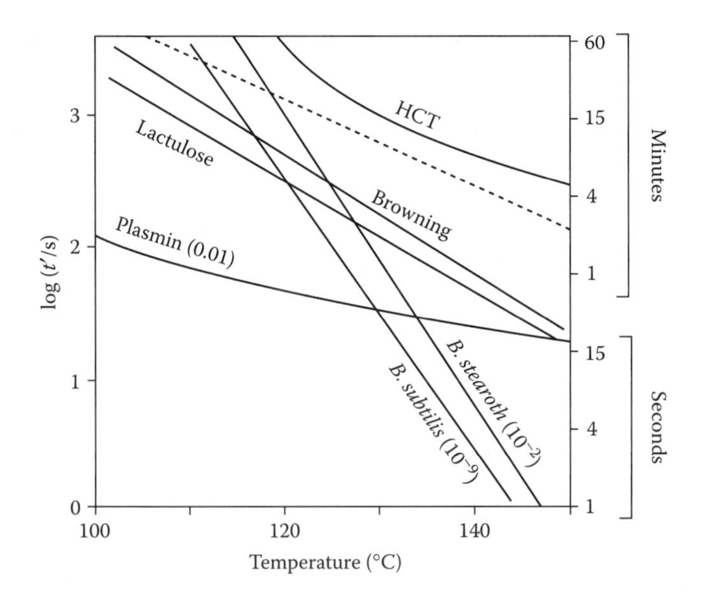

FIGURE 16.8 Changes in milk during sterilization: killing of bacterial spores, inactivation of enzymes, and some undesirable changes such as a significant browning; HCT = approximate heat coagulation time. The broken line very roughly indicates a reduction of the activity of bacterial lipases and proteinases to 0.1. Lactulose corresponds to 600 mg/l.

withstand sterilization (see also Subsection 7.2.4). The heating step in the UHT process with direct heating causes formation of aggregates of casein micelles, which may lead to an astringent mouth-feel and to some sediment on storage of the milk. Heat coagulation is responsible for the aggregates. High-pressure homogenization (often 40 MP is needed) disrupts them; because homogenization must be done aseptically, this needs a specifically designed homogenizer.

The temperature–time regime suitable for sterilization is depicted in Figure 16.8. Generally, a heat treatment above the line given for *Bacillus stearothermophilus* should be selected. Browning of in-bottle sterilized milk is inevitable because at the usual temperature of 115 to 118°C, the curves for sufficient sterilizing effect and significant browning intersect.

UHT sterilization is mostly performed at temperatures above 140°C. Accordingly, the sterilizing effect required is readily attained. But a sufficiently long shelf life at ambient temperature is only obtained if the residual activity of plasmin is at most 1%. Often, the curve for 600 mg lactulose represents the upper limit of UHT sterilization, but at that limit a significant cooked flavor results. Obviously, the suitable heating regime is restricted. For short sterilization times, however, both the selected time–temperature combination and the full thermal load of the product, including heating and recooling, are important (see Subsection 7.3.2).

When indirect UHT heating is applied, oxygen should first be removed from the product by means of deaeration, preferably up to less than 1 mg/kg milk.

In direct heating this is achieved during the evaporative cooling of the product. If a little O_2 is present, it can lead to removal of a slight cooked flavor within a few days, but high O_2 contents cause development of an oxidized flavor and partial loss of some vitamins during keeping. Because the intense heat treatment during in-bottle sterilization forms sufficient antioxidants, deaeration is not necessary in that case (bottles with a crown cork become deaerated during sterilization).

The package for sterilized milk should be impermeable to O_2; on aseptic packaging complete filling should be aimed at (no head space). UHT milk is, moreover, highly susceptible to off-flavors caused by light, so that a package impervious to light is to be preferred (see Chapter 15).

16.2.3 Shelf Life

Spoilage of in-bottle sterilized milk can be caused by insufficient heat treatment, due to which spores of, for instance, *Bacillus subtilis*, *B. circulans*, *B. coagulans*, or *B. stearothermophilus* have survived sterilization. *B. subtilis* has relatively heat-resistant spores, and this bacterium may cause deterioration of in-bottle sterilized milk. If the milk is stored under tropical conditions, it may spoil due to *B. stearothermophilus*, which has very heat-resistant spores. Both a low count of these spores in the original milk and a UHT preheating step can help. *B. stearothermophilus* does not grow below about 35°C. A mild in-bottle sterilization after a UHT presterilization is only possible if during filling not more than a very slight contamination by bacterial spores occurs. If the package is not completely tight (for example, due to an ill-fitting crown cork), then the milk can be recontaminated and so becomes spoiled. Enzymic or oxidative deterioration occurs hardly, if at all, because of the very intense heat treatment.

Deterioration of UHT milk by bacterial growth is usually caused by recontamination. Obviously, the type of deterioration is determined by the species of the recontaminating bacteria. Recontamination by pathogens may even occur, possibly without marked deterioration. Up to now some (rare) cases of food poisoning due to UHT milk contaminated by staphylococci have been reported.

Enzymatic deterioration of UHT milk due to the presence of heat-resistant bacterial enzymes, such as gelation or development of bitter, rancid, or putrid flavors, can only be prevented by a good-quality raw material. Deterioration by plasmin, causing a bitter flavor, will mainly occur in those cases where it is desirable to store UHT milk for a long time (e.g., up to 6 months) and at higher temperature, as in tropical countries. A more intense heat treatment can partially prevent this. Nonenzymatic deterioration of UHT milk during storage may concern: oxidation, influence of light, and Maillard reactions.

The keeping quality of in-bottle sterilized milk is checked by incubation of samples at various temperatures, mostly 30°C and 55°C. After a few days, one can, for instance, determine smell, flavor, appearance, acidity, colony count, or oxygen pressure. The sterility of UHT milk can, in principle, be verified in much the same way. From a statistical viewpoint, however, a check of sterility of a large number of samples of any production run is needed. Measurement of the

O_2 pressure can be done rapidly, but it is only suitable if the product, just after packaging, still contains some oxygen; reduction of O_2 pressure then points to microbial growth. Measurement of the increase in bacterial ATP via bioluminescence is also possible. The sterilized milk should preferably be sold only after the result of the shelf-life test has become known and is satisfactory.

16.3 RECONSTITUTED MILKS

In several regions, there is a shortage of fresh (cows') milk. As an alternative, milk powder can be used to make a variety of liquid milk products. Some common types are the following:

Reconstituted milk: It is simply made by dissolving whole milk powder in water to obtain a liquid that is similar in composition to whole milk. Likewise, reconstituted skim milk can be made.

Recombined milk: It is made by dissolving skim milk powder in water, generally at 40 to 50°C, then adding liquid milk fat (preferably anhydrous milk fat of good quality; Subsection 18.5.1), making a coarse emulsion by vigorous stirring or with a static mixer, and then homogenizing the liquid. This product is similar to homogenized whole milk, except that it lacks most of the material of the natural fat globule membrane, such as phospholipids; see also Chapter 9, Section 9.9.

Other recombined milk products are made. Figure 19.2 gives a manufacturing scheme for recombined evaporated milk.

Filled milk: It is like recombined milk, except that instead of milk fat, a vegetable oil is used to provide the desired fat content. See Subsection 16.5.1.

Toned milk: It is a mixture of buffaloes' milk and reconstituted skim milk. The high fat content of buffaloes' milk (e.g., 7.5%) is thereby toned down.

16.4 FLAVOR

Good flavor is, of course, an essential quality mark of beverage milk. Fresh milk has a fairly bland flavor, where full-cream milk has a richer taste than (partly) skimmed milk. The main aspect is, however, the absence of off-flavors. The fresh milk may already have off-flavors (see Subsection 4.4). These can mostly not be removed, although flavor compounds formed by heating may to some extent mask off-flavors; the first effect of heat treatment mostly is that the typical 'cowy' flavor of fresh milk is reduced (or masked), so that the flavor becomes even more bland. Flash boiling of milk, as occurs in the cooling section of a direct UHT heater or in a vacreator, may reduce some off-flavors.

Microbial growth, either before or after processing, may cause various off-flavors. Unclean or even putrid flavors are mainly caused by some psychrotrophs; some, e.g., *Pseudomonas fragii*, cause fruity flavors. Lactic acid bacteria eventually cause the milk to turn sour, but other defects, such as a malty flavor (caused by *Lactococcus lactis* ssp. *maltigenes*) may also occur. *Bacillus circulans* occasionally causes a phenolic flavor in in-bottle sterilized milk. Growth of *B. cereus* in pasteurized milk readily leads to a very unclean flavor; this is fortunate because it prevents the consumer from drinking such milk, which might contain sufficient toxin to be hazardous. Lactic acid bacteria cannot grow at refrigerator temperatures, but some strains of *B. cereus* can grow at 7°C.

Milk enzymes may cause a bitter flavor due to proteolysis by plasmin, as may occur in UHT milk, and a soapy-rancid flavor, due to lipolysis by lipoprotein lipase in low-pasteurized milk. However, a soapy-rancid flavor is mostly due to lipolysis occurring prior to pasteurization or to action of heat-resistant microbial lipase originating from psychrotrophs. Lipolysis is discussed at length in Subsection 3.2.5.

Lipid oxidation (see Subsection 2.3.4) may occur due to contamination of the milk with Cu or due to exposure to light. The resulting off-flavor is often called 'tallowy' or just oxidized flavor, but in some cases a cardboard-like flavor develops. The latter may be caused by oxidation of phospholipids, and it can also develop in skim milk. The susceptibility of milk to obtain oxidized flavor is greatly variable, but it appears that rigorous exclusion of Cu (which implies a contamination of less than, say, 3 μg Cu per kg) and of light is effective in preventing the defect to occur.

Exposure to light can be highly detrimental to milk flavor, and 10 min of direct sunlight on pasteurized milk in a glass bottle or 10 h of fluorescent light on milk in a carton (that is not provided with an aluminum foil layer) may be sufficient to produce defects. The off-flavor is formed not immediately but during several hours after illumination. It may concern oxidized flavors, but also a quite different 'sunlight' flavor. The latter is mainly due to oxidation of free methionine to methional (CH_3–S–CH_2–CH_2–CHO) and to free thiols formed from sulfur-containing amino acid residues; the presence of riboflavin is needed for the sunlight flavor to develop.

Heat treatment leads to a change in flavor, the appreciation or dislike of which varies greatly among consumers. Every type of heat treatment causes its own flavor profile depending on the total thermal load of the process. The main flavor profile elements are cooked flavor, UHT ketone flavor, and sterilized-caramelized flavor. Cooked flavor is mainly caused by the presence of H_2S liberated after denaturation of protein (mainly from the fat globule membrane) during high pasteurization and boiling. UHT milk has also a cooked flavor but, in addition, it has a ketone flavor that predominantly originates in the lipid fraction and is due to methyl ketones and, to a lesser extent, to lactones and sulfur compounds. The cooked flavor as well as the ketone flavor greatly depend on the type of UHT process used. Often, the mild cooked flavor of UHT milk disappears partly during the first week after manufacture due to oxidation of the reducing sulfur compounds. Figure 16.9 gives an example of the flavor score of UHT milk, heated at various intensities.

When applying a more intense heat treatment, a caramel-like flavor develops, caused by certain Maillard and caramelization products; it is also called 'sterilized-milk flavor.' It can be perceptible in milk sterilized by indirect UHT-heat treatment. It is the dominant flavor of in-bottle sterilized milk, where it masks cooked and ketone flavors.

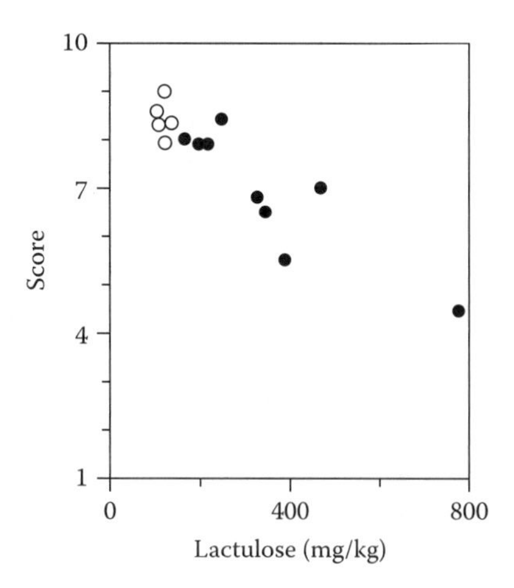

FIGURE 16.9 Average flavor score of UHT milk by a taste panel. Heating at various intensities is expressed by lactulose content. Direct (○) or indirect (●) heating. (Adapted from P. Eberhard and P.U. Gallmann, Federal Dairy Research Institute, Liebefeld-Bern, Switzerland.)

16.5 NUTRITIVE VALUE

For nutritive aspects of the various milk components see Subsections 2.1.2, 2.2.4, 2.3.3, and 2.4.5, and Table 2.18. In this section we will consider changes in nutritive value due to deliberate changes in composition or due to processing and storage.

16.5.1 Modification of Composition

Milk is a good source of several nutrients. Nevertheless, milk of modified composition is produced, often because of its (presumed) nutritional benefits.

Standardization of *fat content* is quite general, for instance in the form of whole milk (e.g., 3.5% fat), low-fat milk (e.g., 1.5%), and skim milk (< 0.1%). With the fat, the fat-soluble *vitamins* are removed. Vitamins A and D then are often added. In some countries, whole milk is also fortified with vitamin D (e.g., 20 μg calciferol per liter). Other vitamins are rarely added.

Another change may be the replacement of milk fat by fats of other sources, almost exclusively of vegetable origin. This is a form of 'filled milk.' Originally, filled milk often contained coconut oil, but that is very low in vitamins A and D, which therefore need be added. Currently, vegetable oils with a high content of polyunsaturated fatty acid residues are generally used; these tend to be rich in vitamins D and E. The latter is also needed as an antioxidant because the oils are quite sensitive to autoxidation; in some cases, other antioxidants are added.

Although milk has a relatively high content of *calcium*, which is, moreover, well absorbed, calcium-fortified milk is also produced. One may add Ca in the form of calcium lactate or bicarbonate. Another possibility is adding a milk salt mixture obtained from an ultrafiltration permeate of acid whey.

For people suffering from *lactose intolerance*, lactose-free milk has been produced. The milk is UHT-heated, then a small quantity of an aseptic preparation of the enzyme lactase is added, and the milk is aseptically packaged. After a few days of storage, the lactose is almost fully hydrolyzed into glucose and galactose. This has not become a success. The product is relatively expensive, and most consumers consider the taste too sweet. Lactose intolerant people are better off consuming fermented milks (see Subsection 22.5.2).

Several special products are produced for specific groups of consumers, e.g., for people suffering of a certain disease or allergy, and for some age groups. See Section 16.6 for infant formulas. Moreover, there is a trend to develop "functional foods" based on milk, which are alleged to give specific health benefits.

16.5.2 Loss of Nutrients

The nutritive value of pasteurized and UHT-sterilized milk changes little by the heat treatment and during storage. In-bottle sterilized milk shows a somewhat greater loss of nutritive value. Of special importance are the decrease of available lysine and the total or partial loss of some vitamins. Some data are given in Table 16.3.

Maillard reactions are responsible for the partial loss of lysine. They occur to some extent in UHT-sterilized milk during storage and in in-bottle sterilized milk during heating. The loss of lysine is not serious in itself because in milk protein, lysine is in excess.

TABLE 16.3
Approximate Loss (in %) of Some Nutrients in Milk during Heating and Storage

Treatment	Available Lysine	Vitamin B_1 (Thiamin)	Vitamin B_6 (Pyridoxal)	Vitamin B_9 (Folic Acid)	Vitamin B_{12}	Vitamin C
Pasteurization	0	5–10	0–5	3–5	3–10	5–20
UHT sterilization, directly	0	5–15	5–10	10–20	10–20	10–20
UHT sterilization, after 3 months storage[c]	2	10–20[a,b]	20–50[a]	30–100[b]	20–50[b]	30–100[b]
In-bottle sterilization	5–10	20–40	10–20	30–50	30–60	30–60

[a] Dependent on exposure to light.
[b] Dependent on O_2 concentration.
[c] At about 25°C

The losses of vitamins mainly concern vitamin C and some five vitamins of the B group. Vitamins A and E are sensitive to light and/or oxidation, but mostly their concentrations do not decrease in sterilized milk. Losses of vitamins in milk should be evaluated relative to the contribution of beverage milk to the supply of these vitamins in the total diet. Especially the loss of vitamins B_1, B_2, and B_6 are considered undesirable. The loss of vitamin C is generally of minor importance as such (milk is often not an important vitamin C source), but it may affect the nutritive value in other ways. The breakdown of vitamin C is connected with that of vitamin B_{12}; moreover, vitamin C protects folic acid from oxidation.

Loss of vitamins during storage can largely be avoided if O_2 is excluded (see Table 16.3). Vitamins C and B_9 may completely disappear within a few days if much O_2 is present. The loss is accelerated by exposure to light, with riboflavin (vitamin B_2) being a catalyst. Most of the riboflavin disappears on long-term exposure to light. The influence of the package on the permeability to oxygen and light is discussed in Section 15.2.

16.6 INFANT FORMULAS

Breast feeding of young babies is undoubtedly preferable to ensure the development of a healthy child. However, breast feeding is not always possible, and then the baby should be offered a surrogate. Unmodified cows' milk is definitely unsuitable, a reason why specific infant formulas have been developed. These are for the most part based on fractions of cows' milk.

A disadvantage of using infant formulas is that contamination with hazardous microorganisms can happen more readily than with breast feeding. Consequently, strict hygiene should be ensured during use (dissolving, diluting, warming to blood temperature, etc.). Moreover, liquid formulas, whether obtained as such or made by dissolving powder, should be kept in a refrigerator.

16.6.1 HUMAN MILK

The composition of human milk strongly differs from that of cows' milk, as is illustrated in Table 16.4. It should be noticed that the composition of human milk greatly varies, especially among individual mothers. The milk also changes strongly throughout the lactation period, as is indicated in the table by comparing colostrum (milk of the first few days after birth) and 'mature' milk (after two weeks). The table is incomplete: the components given are important for nutrition or differ substantially between human and bovine milk. The table also shows minimum amounts of nutrients recommended for young babies.

Lipids: The total lipid content is as in cows' milk, but the fatty acid residues in the triglycerides show a quite different pattern. Short-chain acids (fewer than 12 C atoms) are hardly present, and the fat contains great amounts of polyunsaturated fatty acids: 18–22 C–atoms, 2–5 double bonds. The content of 'essential' fatty acids is much higher than in cows' milk.

TABLE 16.4
Composition[a] of Human and Cows' Milk and Minimum Requirements for Infant Formulas[b]

Component	Unit	Human Colostrum[c]	Human Milk[c]	Cows' Milk[c]	Required per 300 kJ
Energy	kJ	240	300	290	
Fat	g	2.5	4.2	4.0	3
Linoleic acid	mg		400	70	200
α-Linolenic acid	mg		40	15	35
Cholesterol	mg	25	20	13[d]	
Lactose	g	5.0	6.3	4.6 }	5 (Lactose + other saccharides)
Other saccharides	g	1.8	1.3	0.1 }	
Protein	g	1.6	0.8	3.3	1.3
NPN compounds	g	0.5	0.5	0.1	
Calcium	mg	30	35	115	35
Magnesium	mg	3.5	3	11	4
Zinc	mg	1	0.3	0.4	0.35
Iron	μg	75	80	20	140
Copper	μg	60	40	2	30
Phosphorus	mg	14	14	95	18
Iodine	μg		7	5	3.5
Vitamin A	RE[e]	200	80	45	140
Vitamin D	μg		0.01	0.06	0.7
Vitamin E	mg	1	0.4	0.1	0.5
Thiamin, B_1	μg	2	17	45	30
Riboflavin, B_2	μg	30	30	180	40
Niacin	mg	0.06	0.2	0.5	0.2–0.6
Vitamin B_6	μg		6	65	25
Folic acid	μg		5	5	3
Vitamin B_{12}	μg		0.01	0.4	0.1
Ascorbic acid	mg	4	4	2.2	6

[a] Not complete, approximate.
[b] For babies below 6 months old.
[c] Amounts per 100 g milk.
[d] Cows' skim milk: 2 mg.
[e] Retinol equivalent = μg retinol + μg carotene/6.

Most palmitic acid is esterified in the 2-position, which means that upon lipolysis in the gut a glycerol monopalmitate results, which is readily absorbed by babies, unlike the stearic and palmitic acids resulting from the digestion of bovine milk fat. Hence, the fat used for infant formulas generally consists for the most part of suitable vegetable oils. The latter implies that infant formulas are very poor in cholesterol (about 1 mg per 100 ml), as compared to human milk (see Table 16.4).

TABLE 16.5
Proteins in Human and Cows' Milk

Protein	Human Colostrum	Human Milk, Nature	Cows' Milk
Casein	5[a]	2.5[a]	26
α-Lactalbumin	3	2	1.2
β-Lactoglobulin	0.0	0.0	3.2
Serum albumin	0.4	0.3	0.4
Immunoglobulins	2.5	0.8	0.7
Lactoferrin	3.5	1.5	<0.1
Lysozyme	0.5	0.5	10^{-4}

Note: Approximate averages in grams per kg; incomplete.

[a] Predominantly β- and κ-casein.

Carbohydrates: Human milk contains, besides a relatively high amount of lactose, a substantial quantity of oligosaccharides. These have between 3 and 14 saccharide units, and most of these have a lactose residue and some N-acetyl groups. Their function is not yet fully clear, but it is assumed that some oligosaccharides promote growth of certain bifidobacteria in the large intestine, as these compounds cannot be hydrolyzed by the native enzymes in the gut. Oligosaccharides from various sources are added to some infant formulas; in other cases, lactulose is added, which also stimulates growth of bifidobacteria.

Proteins: See Table 16.5 for protein composition. When feeding a baby with cows' milk, its kidneys have difficulty in processing the large amounts of degradation products of the protein metabolism, especially in combination with the large amount of minerals going along with the casein. Moreover, the casein of cows' milk gives a firm clot in the stomach, and it takes a long time before the clot is sufficiently digested for the resulting peptides to proceed to the small intestine. These problems do not arise with human milk, due to its far lower protein content and small proportion of casein (about 30% of the protein as compared to 80% in cows' milk). When making infant formulas from cows' milk, the protein composition thus needs considerable adjustment.

The composition of the serum protein differs also. Striking are the absence of β-lactoglobulin from human milk, and the presence of a large proportion of antimicrobial proteins, notably immunoglobulin A, lysozyme, and lactoferrin. The amino acid composition of human and bovine milk proteins is not significantly different. Finally, human milk contains a large amount of nonprotein nitrogen. The functions of these compounds are largely unclear; much of the material is indigestible.

Minerals: The content of minerals (inorganic salts) in human milk (about 0.2%) is much lower than in cows' milk (0.6 to 0.7%). This is in accordance with the low protein and the high lactose content (see Subsection 2.7.2.1). The contents of some trace elements, notably iron and copper, are relatively high.

The low content of protein and calcium phosphate in human as compared to bovine milk is clearly related to the relative growth rate of a baby being much slower than that of a calf.

Vitamins: For the most part, the differences between human and cows' milk are fairly small, but the contents of some vitamins are much higher in cows' milk. This seems to pose no problems.

16.6.2 Formula Composition and Manufacture

The composition of the formula should comply with recommended quantities, partly given in Table 16.4 (the recommendations do not greatly differ between countries). Generally, skim milk and sweet whey are used, ratio, e.g., 1 to 5. The whey should not contain more than traces of lactic acid and no added nitrate. Part of the whey should be desalted. Carbohydrate content can be boosted by addition of lactose or a UF permeate. Sometimes oligosaccharides are added (or lactulose). The fat is generally a mixture of vegetable oils, containing sufficient essential fatty acids and oil-soluble vitamins. Vitamin C is usually added, and if need be, vitamins A, D and E. Fortification with Fe and Cu is common practice.

This all concerns a formula for healthy babies of up to 6 months old. For babies over 6 months old, the composition of the mix is different, generally involving a much greater proportion of (skim) milk. Other products are needed for children born preterm, or for children suffering from allergy or some metabolic disease.

The manufacture generally involves wet mixing of the ingredients and pre-emulsification, followed by pasteurization and homogenization. Occasionally, an emulsifier is added, but this is not necessary. After pasteurization either a liquid or a dried product can be made. For the former, the milk is UHT-sterilized, followed by aseptic packaging in cartons. Concentrated products that are sterilized in bottles or cans are also produced. When making powdered formulas, the milk is concentrated by evaporation, followed by spray drying; the pasteurization should be sufficiently intense to kill virtually all pathogens. See Chapter 20 for particulars about powder manufacture, packaging, and storage.

Suggested Literature

General information on beverage milk: *Factors affecting the keeping quality of heat treated milk,* IDF Bulletin 130, Brussels, 1981; *Monograph on pasteurized milk,* IDF Bulletin 200, Brussels, 1986.

Valuable information on UHT heating and aseptic packaging: *New monograph on UHT milk,* IDF Bulletin 133, Brussels, 1981.

Practical information on recombined milk products: *Recombination of milk and milk products,* IDF Bulletin 142, Brussels, 1983.

Much about flavor and its sensory evaluation: F.W. Bodyfelt, J. Tobias, and G.M. Trout, *The Sensory Evaluation of Dairy Products,* AVI, New York, 1988.

Nutritive value of milk: E. Renner, *Milk and Dairy Products in Human Nutrition,* Volkswirtschaftlicher Verlag, München, 1983.

Information on "Human Milk" and on "Infant Formulae": H. Roginski et al., Eds., *Encyclopedia of Dairy Sciences,* Academic Press, 2003, by A. Darragh and by D.M. O'Callaghan and J.C. Wallingford, respectively.

17 Cream Products

Cream is sold in many varieties. The fat content may range from 10% (half-and-half) to 48% (double cream). Although used for several purposes, it is primarily something of a luxury and, therefore, an excellent flavor is of paramount importance. Because of the high fat content, any off-flavor of the fat becomes concentrated. For instance, milk with a fat acidity of 1 mmol per 100 g fat will not be perceived to have a soapy, rancid flavor by most people, but a whipping cream made from it will definitely taste rancid. Therefore, the milk should be impeccable with regard to lipolysis and fat oxidation.

Sometimes anhydrous milk fat is used in cream products and recombination is applied. Such a fat may have an oxidation flavor, and even if impeccable in this respect, the taste of the product may be somewhat less rich because of the absence of components from the milk-fat-globule membrane.

Besides plain cream, some derived products are made, such as sour cream (see Subsection 22.2.1.2) and ice cream. Here we will cover three products, chosen to illustrate most of the important technological and quality aspects.

17.1 STERILIZED CREAM

This cream has about 20% fat (light cream). A good keeping quality is essential because many consumers use it a little at a time or want to have it stored for special occasions. Accordingly, the cream is usually sterilized to guarantee microbial stability. Chemical stability is generally not a problem, although ongoing Maillard reactions can occur during long-term storage. Because of the intense heat treatment, oxidative deterioration scarcely occurs; neither does lipolysis. Physical deterioration may be considerable: gravity creaming and fat clumping or oiling off. Therefore, the cream must be homogenized. If stored for a long time it may thicken with age, form a gel, or become lumpy.

Most of the cream is used in coffee; hence, the name *coffee cream*. Thus, it is important that the cream does not feather in coffee and that it causes sufficient whiteness (i.e., turbidity) after dilution with coffee. Likewise, no oil droplets should appear on the coffee. The sterilization flavor is usually not too objectionable as this is largely masked in the coffee.

Dessert cream is used, for instance, on fruit. A pure flavor is then paramount, as are a white color and a relatively high viscosity. Sometimes a very thick, almost pudding-like cream is made.

17.1.1 Manufacture

Figure 17.1 gives an outline of the traditional manufacturing process for in-bottle sterilized coffee cream. Alternatively, raw or thermalized milk may be skimmed, and the cream obtained may be standardized, pasteurized, and homogenized at the pasteurization temperature. A sterilizing effect of about 9 for *Bacillus subtilis* is usually the norm.

Figure 17.1 also shows the manufacture of ultra-high-temperature (UHT) short-time heated cream. In this case the cream should be homogenized after sterilizing; otherwise, UHT heating will cause coagulation of protein and fat globules and, possibly, coalescence of fat globules.

If a highly viscous cream is desired, the cream will be homogenized at a lower temperature and in one stage in order to produce a maximum of homogenization clusters.

17.1.2 Heat Stability

General aspects of heat stability are discussed in Subsection 7.2.4.

In making sterilized cream it is hard to avoid coagulation during sterilization while at the same time the product is sufficiently homogenized to prevent rapid creaming and (partial) coalescence of fat globules. Homogenization is largely responsible for the poor stability to heat coagulation (see also Section 9.6 and Figure 9.11). Although the heat stability of cream (like that of evaporated milk) can be improved by adjusting the pH and by adding stabilizing salts (e.g., citrate), the main variables are the conditions during the homogenization (see Figure 17.2). It appears that as the surface area of fat globules covered with casein increases, the cream becomes less stable. Because of this, preheating at a high temperature does not help; it causes the serum proteins to precipitate so that a larger part of the oil–water interface is covered by casein. Furthermore, the presence of homogenization clusters will shorten the heat coagulation time (Section 9.7 and Subsection 17.1.4).

The higher the homogenization pressure, the lower the heat stability. However, creaming and (partial) coalescence will cause problems at lower homogenization pressures. Therefore, one must look for a compromise. It is advantageous to make the fat-globule size distribution as narrow as possible (see Section 9.3).

17.1.3 Stability in Coffee

Feathering of the cream in coffee is due to coagulation of the fat globules and runs largely parallel to the heat stability. Consequently, UHT cream is rather susceptible to feathering. In its manufacture (Figure 17.1) no problems arise with heat coagulation, but feathering occurs readily if the homogenization pressure is too high. Moreover, UHT cream is liable to thicken with age (cf. Subsection 19.1.5) or to show aggregation during storage. The latter phenomenon starts with the aggregation of fat globules. Soon this also leads to feathering in coffee. Feathering obviously depends on temperature, pH, and Ca^{2+} activity of the coffee, too. Stability in the

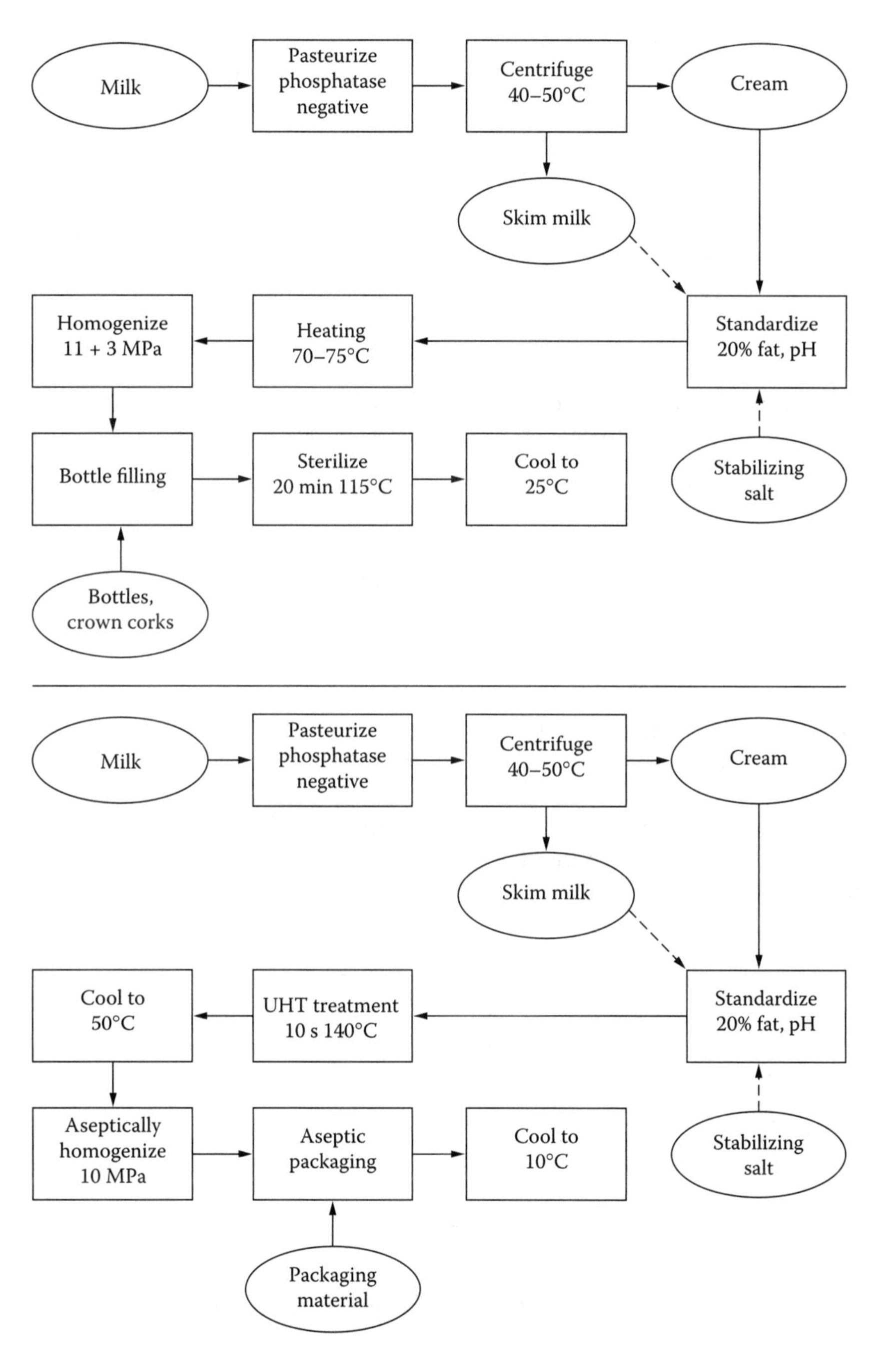

FIGURE 17.1 Examples of the manufacture of coffee cream (top) and dessert cream (bottom). Stabilizing salt added is 0.15% $Na_3C_6H_5O_7 \cdot 5H_2O$.

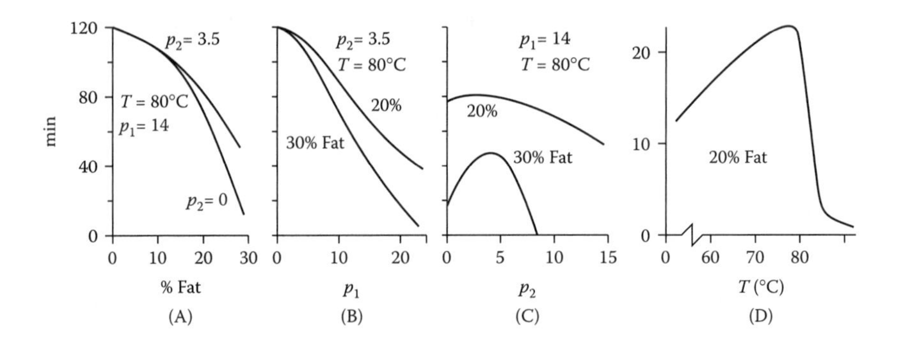

FIGURE 17.2 Heat stability (coagulation time at 120°C) of cream as related to the conditions during homogenization. p_1 is pressure before the first stage, p_2 before the second (MPa); T is homogenization temperature. A, B, C, tests in stationary cans; D, in rotating tubes. (After H. Mulder and P. Walstra, *The Milk Fat Globule,* Pudoc, Wageningen, 1974.)

coffee may be improved by increasing the solids-not-fat content of the cream, presumably as it buffers for H^+ and Ca^{2+} in the coffee.

17.1.4 Clustering

Dessert cream should be somewhat viscous. An obvious way to achieve this is by the formation of homogenization clusters, though thickening agents (carrageenan, alginate) can be added successfully. Parameters affecting clustering are discussed in Section 9.7. Subsection 4.7.1, gives basic aspects of viscosity.

The main factors affecting the viscosity are summarized in Figure 17.3. At a given fat content, the degree of clustering is responsible for the differences

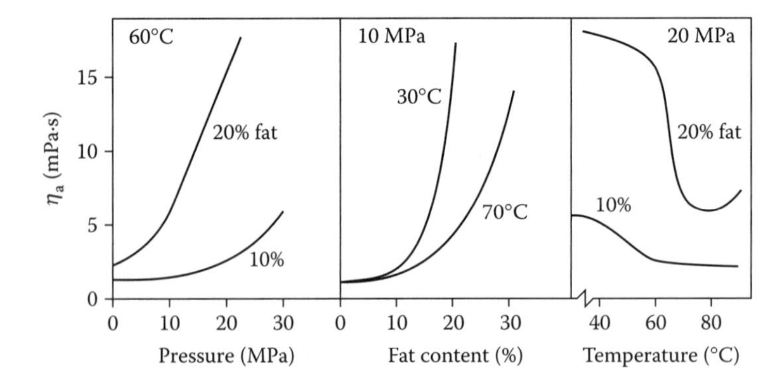

FIGURE 17.3 Influence of some process and product variables on apparent viscosity (η_a) of homogenized cream. Approximate examples. Temperatures and pressures refer to conditions during homogenization. Viscosity measured at room temperature. (Adapted from H. Mulder and P. Walstra, *The Milk Fat Globule,* Pudoc, Wageningen, 1974.)

observed. Clustering increases the viscosity because the effective volume fraction of the fat globules increases: first, because of the plasma entrapped between the fat globules (this part of the plasma is essentially immobilized) and, second, because of the irregular shape of the clusters (causing them to occupy an effectively larger volume when rotating due to the shear). As the fat content of the cream is higher, the increase in viscosity due to a given extent of clustering of the fat globules is stronger. Moreover, the clustering itself is more extensive for a higher fat content.

The viscosity can be reduced considerably by a second homogenization at a much lower pressure: the homogenization clusters are then partly disrupted again (and therefore reduced in size); moreover, the remaining clusters are more rounded. The same can be achieved by exposing the clustered cream to shear, for instance, in a rotating viscometer, and Figure 17.4 shows the ensuing decrease of the (apparent) viscosity with increasing shear rate; the greater the rate, the further the clusters are disrupted. The latter do not reform on release of the shear, as the hysteresis loop shows.

Figure 17.4 shows that the viscosity decreases with increasing shear rate. In other words, the product is 'shear-rate thinning' and has an apparent viscosity. The figure also shows that, if the cream had already been exposed to a high shear rate, the viscosity would remain low even at a subsequent low shear rate. High shear rates should therefore be avoided during pumping and packaging if homogenized cream is to retain its high viscosity. The consumer will usually apply low shear rates, say 20 s^{-1}, when dessert cream is poured, for example.

If cream of a pudding-like consistency should be made it can be homogenized at such a low temperature that a small part of the fat is solid and true clumps of fat globules are formed. The product obtained is, however, very temperature sensitive, so that warming to, say, 35°C causes loss of consistency and oiling off.

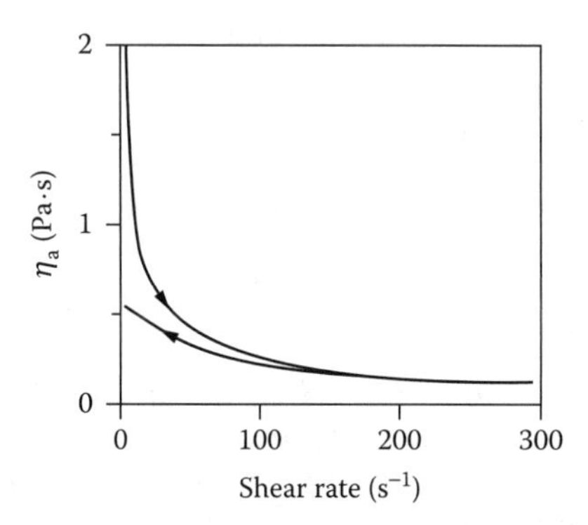

FIGURE 17.4 Example of the influence of the shear rate on apparent viscosity (η_a) of homogenized cream (17% fat, homogenization at 40°C and 21 MPa); measurements at increasing shear rates were followed by those at decreasing rates. (Adapted from H. Mulder and P. Walstra, *The Milk Fat Globule,* Pudoc, Wageningen, 1974.)

Clustered cream hardly shows creaming after homogenization; this is because the whole content of the bottle is like one big cluster. This may be compressed by gravity, causing a separated layer of milk plasma to appear at the bottom of the bottle.

17.2 WHIPPING CREAM

This involves cream of, say, 35% fat. It is primarily designed to be beaten into a foam, often with sugar added. It is mostly available as a pasteurized product in small bottles, plastic cups, or large cans. It is also sold as in-can sterilized cream.

17.2.1 Desirable Properties

The most important specific requirements are:

1. *Flavor:* The product is eaten for its flavor, which obviously must be perfect. Rancid and tallowy flavors in the original milk should be rigorously avoided; this requirement is even more essential than for coffee cream. Not everybody appreciates a sterilization flavor or even a pronounced cooked flavor, and partly because of this, the cream usually is pasteurized.
2. *Keeping quality:* Many kinds of spoilage can occur, but it is often desirable to store the cream for a prolonged time. The original milk should contain not more than a few heat-resistant bacteria; above all, *Bacillus cereus* is a disastrous microorganism in whipping cream (it causes the fat emulsion to become unstable). Nor should growth of psychrotrophs occur in the original milk because they form heat-resistant lipases. To allow for a fairly long shelf life, the pasteurized cream should be packaged under strictly hygienic or even aseptic conditions. Recontamination by bacteria often raises complaints. Therefore, whipping cream is sometimes heated by in-can or in-bottle pasteurization.

   Contamination by even minute amounts of copper causes autoxidation and hence off-flavor. Some coalescence of the fat globules during processing can readily lead to cream-plug formation during storage. The presence of a cream plug implies that the product can hardly be removed from the bottle; moreover, the cream will readily churn rather than whip during the beating in of air.
3. *Whippability:* In a few minutes the cream should easily whip up to form a firm and homogeneous product, containing 50 to 60% (v/v) of air, corresponding to 100 to 150% overrun (the overrun is the percentage increase in volume due to gas inclusion).
4. *Stability after whipping:* The whipped cream should be firm enough to retain its shape, remain stable during deformation (as in cake decoration), not exhibit coarsening of air cells, and show negligible leakage of liquid.

17.2.2 MANUFACTURE

The classical manufacture of whipping cream is fairly simple; an example is shown in Figure 17.5. The pasteurization of the cream should at least be sufficient to fully inactivate milk lipase. Usually, the heat treatment is far more intense in order to improve the bacterial keeping quality, and to form antioxidants (H_2S). The method of heating, as well as the heating intensity, varies widely; holder pasteurization (say, 30 min at 85°C), heating in a heat exchanger (possibly over 100°C), and in-can (bottle) heating (say, 20 min at 103°C) are used. Likewise the manufacturing sequence, separation temperature, and so forth vary widely.

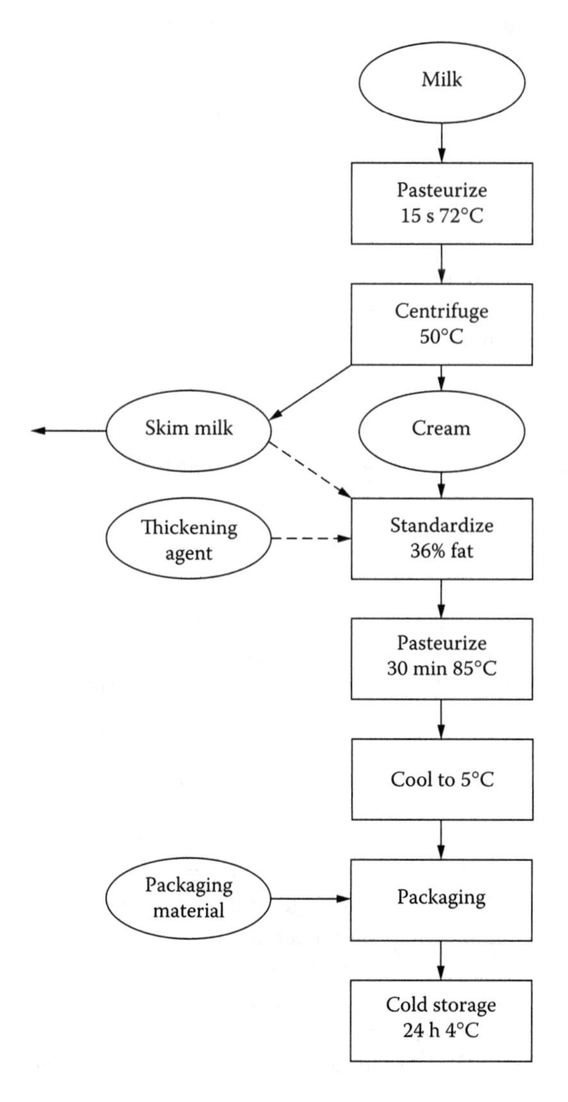

FIGURE 17.5 Example of the manufacture of whipping cream.

Damage, especially (partial) coalescence, of the fat globules should be avoided. The milk and, especially, the cream should be handled gently. The cream should not be processed or pumped unless the fat is completely liquid or largely solid, i.e., only at temperatures below 5°C or above 40°C. Hence, bottle filling of hot cream followed by cooling would be preferable, but it is rather uneconomical.

To be readily whippable on delivery, the cream needs first to be kept refrigerated for a day in order to ensure that all fat globules contain some solid fat. To prevent creaming during storage, a thickening agent is generally added (say, 0.01% κ-carrageenan). This causes a small yield stress in the cream, for instance, of 10 mPa, which is sufficient to arrest any motion of the fat globules.

Modified Products. The manufacturing process can be modified in various ways. The cream can be sterilized, usually by UHT heating followed by aseptic packaging. To keep the cream stable during processing and storage, it generally has to be homogenized. This tends to impair whippability and additional measures have to be taken (see Subsection 17.2.3). Even then, temperature fluctuations during storage of the cream may cause 'rebodying,' a form of partial coalescence (see Subsection 3.2.2.2, item 7), which makes the product unsuitable for whipping.

Another modification is (partial) replacement of milk fat by vegetable fat; this involves recombination. A quite different product is *instant whipping cream.* Cream is packaged in an aerosol can in an N_2O atmosphere under a pressure of, say, 8 bar. Upon pressure release, the cream leaves the can through a nozzle that causes it to be instantly converted into a foam. Ready-made whipped cream is also produced, usually in a frozen form. All these products necessitate adjustment of composition, especially of the surface layers of the fat globules.

17.2.3 The Whipping Process

See Subsection 3.2.2.2 for partial coalescence and Subsection 3.2.3 for interactions between fat globules and air bubbles.

Classical whipping of cream in a beater will primarily be considered. The following *processes* occur:

1. Largish air bubbles are beaten into the cream.
2. Air bubbles are broken up into smaller ones, in a manner comparable to the breakup of fat globules in a homogenizer.
3. Air bubbles collide with each other and may coalesce.
4. Protein adsorbs onto the air–water interface, whereby the coalescence rate of the bubbles is greatly decreased.
5. Air bubbles may coalesce with the air above the cream, hence disappear. The rate of processes 3, 4 and 5 is higher for a larger volume fraction of air (φ_a), larger bubbles, and a lower viscosity of the system.
6. Fat globules collide with air bubbles and become attached to them (cf. Subsection 3.1.1.5).

7. Some liquid fat from the fat globules spreads over the air–water interface.
8. Partial coalescence (clumping) of fat globules occurs. This may happen in the plasma phase, owing to the high velocity gradient, and also at air bubble surfaces, because of coalescence of air bubbles. The resulting decrease in bubble surface area pushes the adsorbed globules closer to each other, and the liquid fat on the air–water interface can act as a sticking agent. Eventually, fairly large clumps are formed.

These processes occur simultaneously, although the rate of (1) soon decreases, because the system becomes quite viscous, and (8) begins sluggishly.

The whipping process should lead to a specific *structure*, in which (1) the air comprises 50 to 60% of the volume, (2) the air bubbles are 10 to 100 μm in diameter, (3) the bubbles are fully covered by fat globules and fat globule clumps, and (4) the clumped fat globules make a space-filling network throughout the plasma phase. This network also makes contact with the bubbles. In this way a firm, smooth, and relatively stable product results.

17.2.3.1 Rates of Change

Whether this result will be obtained depends on the relative rates at which the processes mentioned occur. Assuming that small air bubbles are formed fast enough, two rates are presumably essential. The first is the rate of attachment of fat globules to the bubbles. An almost full coverage with globules or small clumps is needed to prevent coalescence of bubbles. The second is the rate of partial coalescence. If it is too slow, a solid network will not be formed in a reasonable time. If it is too fast, visible butter granules will form (see Subsection 18.2.2); a satisfactory network is not formed either. To put it with some oversimplification, a balance between whipping and churning is needed. Figure 17.6 illustrates the rates of changes occurring during whipping. It is seen that churning eventually predominates; hence, beating should be stopped before the clumps become very large.

Several factors affect the various rates. Foremost is beating speed. The wires of the beaters should move through the liquid at a rate of at least 1 m/s to achieve whipping within a reasonable time. When increasing the beating rate, for example from 1 to 3 m/s, the whipping time markedly decreases, say, from 10 to 1 min. However, the time also depends on bowl size and configuration of the beating apparatus. With increasing beating rate, the rates of most of the changes mentioned increase, especially of processes 1, 2, 6, and 8. Fat content also has a large effect, as shown in Figure 17.7, but the influence depends on the beating intensity. The faster the beating, the lower the fat content at which a stable foam can be formed and the higher the overrun. At quite low fat content, say < 20%, not enough fat globules are present to stabilize the air bubbles; the overrun then decreases with decreasing fat content.

As mentioned, the rate of *partial coalescence* is of paramount importance. It naturally increases with increasing beating rate and fat content. Moreover, the rate increases as the size of the clumps increases (see Figure 17.6). Another

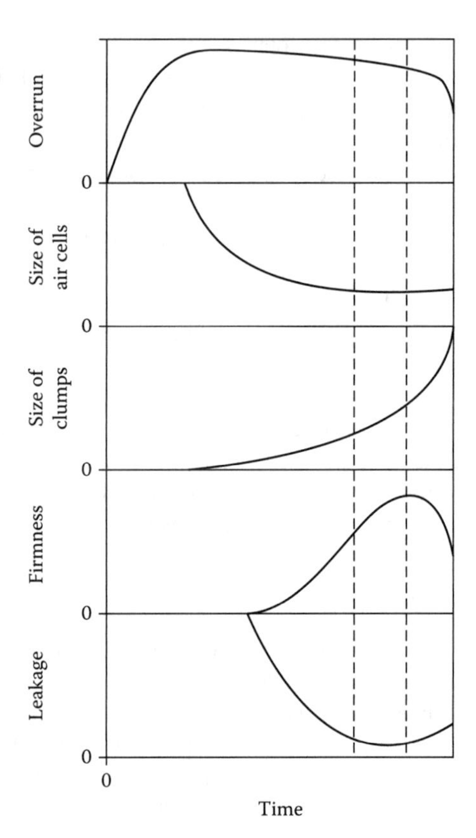

FIGURE 17.6 Processes occurring during whipping of cream. The parameter of firmness may be the time needed to lower a weight into the product; leakage means the amount of liquid drained from a certain volume in a certain time. Between the broken lines the product is acceptable. Approximate results after various sources.

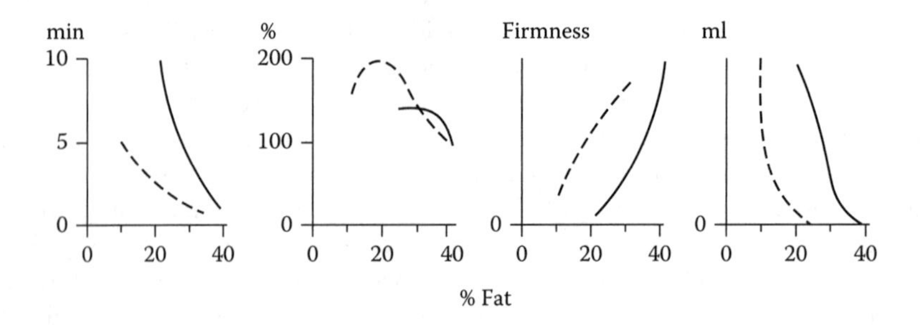

FIGURE 17.7 Properties of whipping cream. Whipping time (min), overrun (%), firmness (approximately a yield stress), and leakage of liquid (ml) as a function of the fat content, for conventional whipping cream (—) and for a product with emulsifier added (---). Approximate examples. (After H. Mulder and P. Walstra, *The Milk Fat Globule,* Pudoc, Wageningen, 1974.)

important variable is the solid fat content (see Figure 3.12c). In practice, this means that the beating temperature is an essential variable. At 5°C, partial coalescence is relatively slow, and it needs the presence of air bubbles; if the fat has a very high solid fat content at a low temperature, the cream can hardly be whipped. At higher temperatures, the clumping proceeds faster, it will also occur in the plasma, and it may become too fast to allow sufficient air inclusion. If the fat is almost fully liquid, whipping is impossible.

In *homogenized cream*, partial coalescence is too slow, because the fat globules are too small and their proteinaceous surface layers provide good stability. However, homogenization at low pressure (1 to 4 MPa), preferably, in two stages (e.g., 2 and 0.7 MPa at 35°C) gives rise to small homogenization clusters of, say, 15 μm, and such a cream can be whipped. Another measure to enhance whippability is the addition of a suitable small-molecule surfactant (usually called emulsifier) that displaces (part of) the protein from the globule surface. This enhances the susceptibility of the globules to partial coalescence and markedly affects whipping properties (see Figure 17.7). The results depend greatly on the type of surfactant in a manner not fully understood. Glycerol monooleate and Tween 20 are among the types used.

17.2.3.2 Stability

Whipped cream is not completely stable but subject to physical changes. The main instabilities arise from:

1. *Leakage* of plasma from the product: If the structure of the whipped cream is as described earlier, leakage can be insignificant. Figure 17.6 and Figure 17.7 show the effect of some variables. Leakage can also be impeded by adding a thickening agent, but fairly high concentrations are needed.
2. *Ostwald ripening* (see Subsection 3.1.1.4): This occurs in almost every foam, because the bubbles differ substantially in size and the solubility of gases in water is relatively high. In a traditional whipped cream, the rate is slow. This is due to the air bubbles being for the most part covered by a layer of fat globules and clumps; this implies that the bubbles can hardly shrink. In a whipped cream of low fat content and high overrun, Ostwald ripening can be appreciable.
3. *Collapse* of the foam: If Ostwald ripening is substantial and coalescence of air bubbles also occurs, the volume of the product decreases on storage. In most whipped creams this is a slow process.
4. *Sagging*: Even if no collapse occurs, the whipped cream may sag under its own weight, if the product is insufficiently firm. A yield stress of about 300 Pa should suffice to assure 'shape retention' in most situations.

Traditional whipped cream is stable for a few hours, but this is not true of all modified creams. Especially, the whipped cream emerging from an aerosol can is quite unstable. During the very fast process of bubble formation and expansion, there is hardly any possibility of fat globules to become attached to the bubbles or to clump. To give the product firmness, it should be a closely packed bubble system, which needs a bubble volume fraction of at least about 0.8 (overrun 400%). Moreover, N_2O is highly soluble in water, causing rapid Ostwald ripening. This results in extensive collapse and in, say, 30 min almost no foam is left.

17.3 ICE CREAM

There are numerous types of edible ice, essentially mixtures of water, sugar, flavor substances, and other components, which are partly frozen and beaten to form a rigid foam. In most types, milk or cream is an important ingredient. Some examples of the composition are given in Table 17.1. Nowadays, a part of the milk solids-not-fat is often substituted by whey constituents to lower ingredient costs. In some countries, the milk fat is often substituted by vegetable fat, for instance, partly hydrogenated palm kernel oil. Dairy ice cream is the product discussed here.

Furthermore, soft serve, ordinary, and hardened ice cream are distinguished. Soft ice is eaten while fresh. It is made on the spot, its temperature is usually –3 to –5°C, and, therefore, it still contains a fairly large amount of nonfrozen water (see Section 11.2); generally, its fat content and overrun are rather low. Hardened ice cream, usually packaged in small portions and sometimes supplied with an external chocolate coating, is much lower in temperature (say, –25°C). The solution remaining is in a glassy state, and it has a shelf life of several months. Ordinary ice cream has a lower temperature than that of soft ice cream (–10 to –15°C), but is not so cold as to be entirely solid; it is stored for a few weeks at the most in cans, from which portions can be ladled out.

Milk or cream of impeccable flavor is needed, especially with respect to rancidity and autoxidation. The latter defect may occur in hardened ice cream because it is stored for long periods, and its water activity is rather low; it contains a great deal of oxygen. Therefore, contamination by copper has to be rigorously avoided.

TABLE 17.1
Approximate Composition (Percentage by Weight) of Some Types of Ice Cream

Constituent	Dairy Ice Cream	Ice Milk	Sherbet	Ice Lolly
Milk fat	10	4	2	0
Nonfat milk solids	11	12	4	0
Added sugar	14	13	22	22
Additives	0.4	0.6	0.4	0.2
% Overrun[a]	100	85	50	~0
Edible energy, kJ/100 ml	390	300	340	370

[a] % overrun means the relative increase in volume by air beaten in.

Soft ice cream often causes microbiological problems, though it is kept cold and its high sugar content may, to some extent, act as a preservative. Pathogenic organisms will not grow, but they are not killed. Bacteria are enabled to grow if the temperature becomes too high locally or temporarily, as can easily happen with the practices at vending places. Abundant growth can occur in poorly cleaned processing equipment and in the mix, if stored for too long. Hence, strict hygienic measures have to be taken. Large numbers of enterobacteria (*E. coli*, *Salmonella* spp.) are frequently found.

17.3.1 Manufacture

Figure 17.8 gives a flowchart in which cream is the starting material. As follows from Table 17.1, additional milk or whey solids are generally added (otherwise the content of milk solids-not-fat would be about 7%). Often one starts from skim milk powder and sweet-cream butter or anhydrous milk fat. Other ingredients are whey powder or demineralized whey. Thus, recombination is needed in these cases.

The first stages of the manufacture need little elaboration. Composing the *mix* is relatively simple. The additives are 'emulsifier,' stabilizer (a thickening agent, usually a mixture of polysaccharides), and flavor and color substances. The role of the additives is discussed in Subsection 17.3.3. Clearly, ingredients such as fruit pulp and ground nuts should be added after the homogenization.

Pasteurization of the mix primarily serves to kill pathogenic and spoilage microorganisms. Additives added after homogenization should usually be pasteurized separately. The second important objective is to inactivate lipase because it is still a little active even at a very low temperature. Bacterial lipases should thus be prevented from occurring. Finally, quite intense heating of the mix is desirable (especially for hardened ice cream) to decrease its susceptibility to autoxidation; a cooked flavor may be undesirable, according to the added flavor substances.

Homogenization is specifically meant to give the ice cream a sufficiently fine, smooth texture (see Subsection 17.3.2). Excessive formation of homogenization clusters should be avoided as it causes the mix to become highly viscous and the desirable fine texture not to be achieved; consequently, the homogenization pressure should be adapted to the fat content, to the pasteurization intensity, and, if need be, to the further composition of the mix (see Section 9.7).

Cooling and *ripening* (keeping cold for some time) are desirable for two reasons. The fat in most of the fat globules should largely be crystallized before the ice cream mix enters the freezer; it is important to note that considerable undercooling may occur because the fat globules are very small (Subsection 2.3.5.2). Certain stabilizers such as gelatin and locust bean gum need considerable time to swell after being dispersed. Some added emulsifiers need considerable time at low temperature to displace protein from the fat globules (Subsection 17.3.2).

Freezing implies rapid cooling of the mix to a few degrees below zero; in this way, ice is formed while air is beaten in. This must run simultaneously: after the bulk of the water is frozen, any beating in of air becomes impossible, and

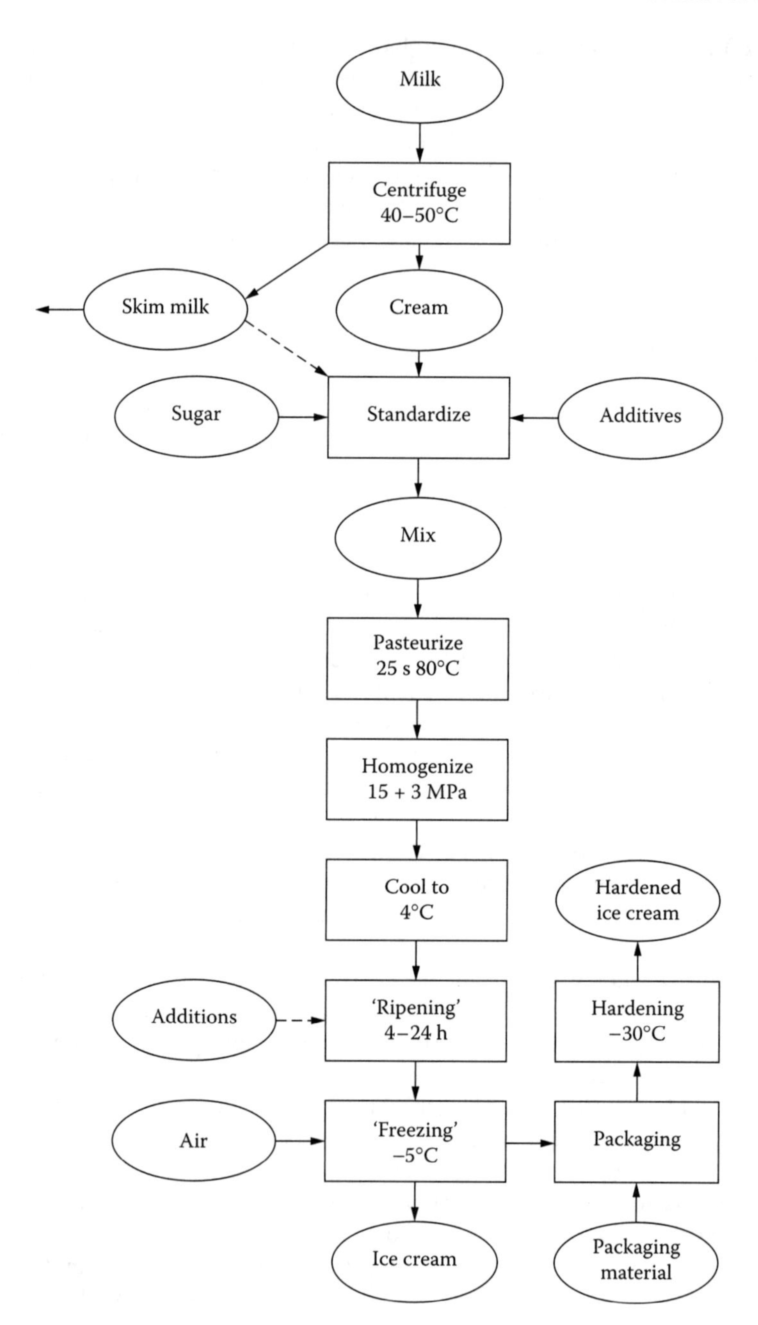

FIGURE 17.8 Example of the manufacture of ice cream.

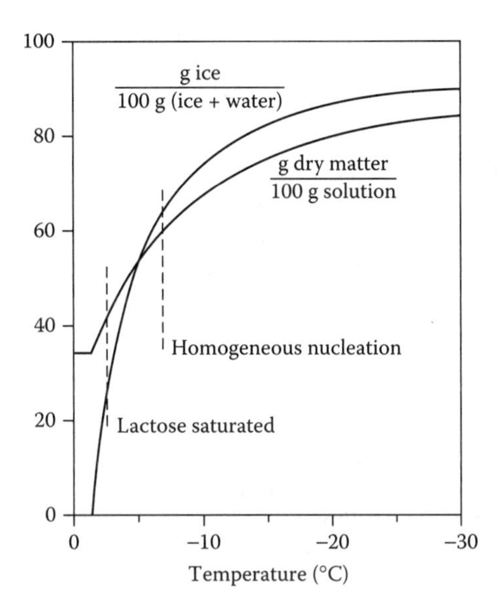

FIGURE 17.9 Freezing of ice cream mix. Approximate quantity of frozen water and concentration of the remaining solution, assuming that the ice is in equilibrium with the liquid and that no other constituents crystallize. The estimated temperatures for saturation of lactose and for its homogeneous nucleation are also indicated.

freezing after air is beaten in leads to insufficient churning of the fat globules (see next subsection) and can damage the foam structure. Moreover, the vigorous beating enables rapid cooling, because of which small ice crystals can be formed. Figure 17.9 gives the approximate amount of frozen water as a function of temperature; a different composition causes a somewhat different curve. Usually, freezing is done in a scraped-surface heat exchanger — essentially a horizontal cylinder that is cooled externally by means of direct evaporation (−20 to −30°C) and equipped with a rotating stirrer (150 to 200 r.p.m.) that scrapes the wall. A layer of ice is formed on the wall. Pieces of ice are broken from the layer by the scraper and are distributed throughout the mass. A layer of ice about 50 μm thick is left. In its simplest design the cylinder is partly filled and the stirrer beats air cells into the mix. In continuously working machinery, air and mix enter the equipment in predetermined volume quantities (allowing the overrun to be exactly adjusted) while the stirrer reduces the air cells in size. The process of manufacture takes a few minutes. The mix leaves the freezer at −3.5 to −7°C. A second heat exchanger may be applied, in which the mix is cooled further, while stirred, to about −10°C without additional beating in of air. Deeper cooling cannot be achieved in a flow-type exchanger because the product becomes too firm.

Packaging of ice cream often is a complicated operation, especially if mixtures or exceptional shapes are wanted. In the latter case the packaging step may be associated with the start of the hardening in order to give the portions appropriate shape retention.

The *hardening* process serves to rapidly adjust the temperature of the ice cream to such a level as to retain its shape and to give it a sufficient shelf life with respect to chemical and enzymatic reactions, as well as to the physical structure. The packaged ice cream can be passed through a so-called hardening tunnel, in which very cold air (say, −40°C) is blown past the small packages for some 20 min. Likewise, packaged ice cream can be passed through a brine bath of low temperature.

17.3.2 Physical Structure: Formation and Stability

The chemical composition of an ice cream mix with air on top is exactly equal to that of the corresponding ice cream. All the same, the differences in appearance, consistency (mouthfeel), and flavor are huge; these are caused by the difference in physical structure. This is illustrated in Figure 17.10. When half of the water is frozen (about −5°C) the following structural elements can be distinguished (d = diameter, φ = volume fraction):

Ice crystals: d = 7–170 μm, on average about 50 μm, $\varphi \approx 0.3$
Lactose crystals: length ≈ 20 μm, $\varphi \approx 0.005$; not always present
Air cells: d = 60–150 μm, $\varphi \approx 0.5$
Thickness of foam lamellae: 10–20 μm
Fat globules: d < 2 μm, $\varphi \approx 0.06$ (including globules in clumps)
Fat globule clumps: up to 10 μm in size

The size of the ice crystals depends on the stirring intensity and on the cooling rate during freezing; the quicker the freezing, the smaller the crystals. Hardening

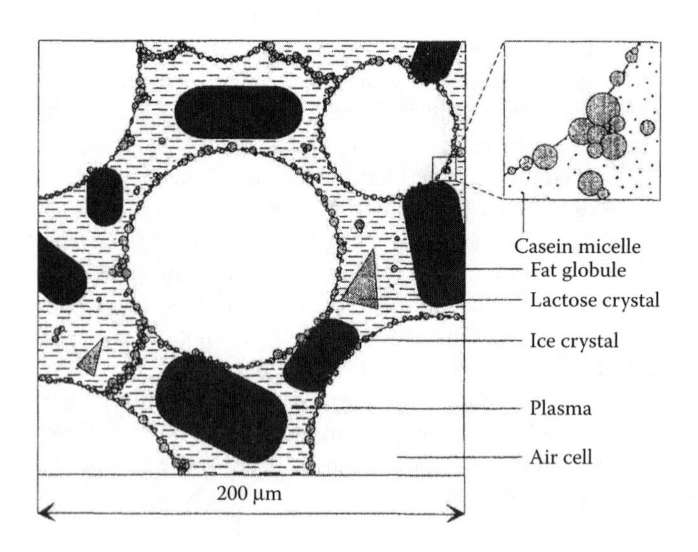

FIGURE 17.10 Schematic presentation of the structure of ice cream at about −5°C.

causes the ice crystals to increase in size, roughly, by a factor of two. Immediately after freezing, no lactose crystals are present. To be sure, the temperature is below that for saturation of lactose, as is seen in Figure 17.9, but it is still above that for its homogeneous nucleation. Only after deep cooling can lactose crystals form.

The preceding summing up with respect to the structure does not yet complete the picture. Microscopically, many air cells can be observed to be somewhat deformed by the ice crystals. This is not surprising when one considers that the system is more or less 'fully packed,' i.e., the combined volume fraction of the structural elements is about 0.8. Furthermore, the air cells are almost entirely covered with fat globules and their clumps. The clumped fat globules, together with the air cells to which they are attached, form a continuous network throughout the liquid (see Figure 17.10). This has the following important effects:

1. The air cells become stabilized by the fat globules (see below).
2. After the ice crystals have melted (in the mouth, for instance) the mass retains some firmness ('stand-up'); this is illustrated in Figure 17.11, in which the extent of fat clumping is expressed as a *churned-fat index* (which may be determined by examining what proportion of the fat creams rapidly after complete melting of the ice crystals). The stand-up is a valued organoleptic property.
3. The clumping (partial coalescence) of the fat globules changes the texture, i.e., the ice cream looks less glossy and therefore appears more attractive to most people. This property is called *dryness*, and it correlates very well with the experimentally obtained churned-fat index.
4. Ice cream having insufficient dryness sticks to the processing equipment, which may interfere with the packaging operation, etc.

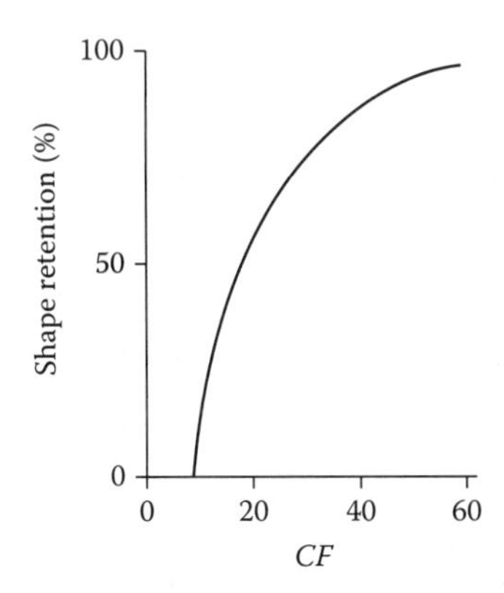

FIGURE 17.11 The influence of the extent of clumping of the fat globules, expressed as churned-fat index (*CF*), on the shape retention of ice cream. The retention is the height of a cube of ice cream after keeping it for some time at room temperature, expressed as a percentage of its initial height. (After H. Mulder and P. Walstra, *The Milk Fat Globule,* Pudoc, Wageningen, 1974.)

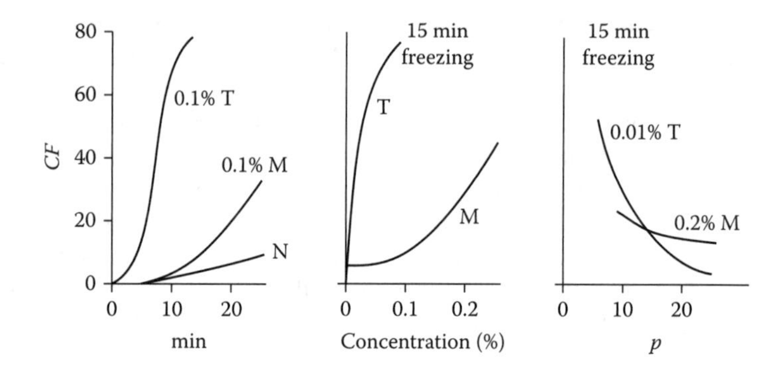

FIGURE 17.12 Influence of freezing time (min), concentration, and nature of the emulsifier and homogenizing pressure (p, in MPa) on the clumping of fat globules, expressed as churned-fat index (CF); M = glycerol monostearate; T = Tween 80; and N = no emulsifier. Approximate examples. (From H. Mulder and P. Walstra, *The Milk Fat Globule,* Pudoc, Wageningen, 1974.)

The mentioned network of clumped fat globules is formed during freezing (see Figure 17.12). Although the air bubbles become almost completely covered with fat globules, flotation churning presumably does not occur, because too little liquid fat is available to spread over the air bubbles. In all likelihood, clumping is predominantly caused by mechanical forces, i.e., the fat globules are pushed together during beating because of the presence of ice crystals and are damaged by them. The lower the temperature, the more the ice and the faster the clumping.

If unhomogenized cream would be taken, all the fat globules together would just suffice to cover air cells of 100 μm diameter ($\varphi = 0.5$) throughout. But natural milk fat globules clump rapidly, and the aggregates formed are not nearly sufficient to fully 'encapsulate' such air cells. This implies that the cells would be unstable during and after freezing (coalescence and Ostwald ripening may occur) and large air bubbles develop, causing a coarse texture. The fat globules become much smaller in size by homogenization of the cream, and then they can cover a much larger air cell surface even after clumping (unless extensive homogenization clusters have been formed). The homogenized fat globules will, however, hardly clump, so the desired network of clumped fat globules does not form. But these globules increasingly tend to clump if a suitable, small-molecule surfactant (usually called emulsifier) is added (Figure 17.12). This is due to the emulsifier displacing part of the protein from the surface layers, which decreases the stability of the fat globules to partial coalescence; the globules also become more readily attached to the air bubbles.

Ideally, the structure of freshly made ice cream would remain unchanged, but this is not the case. The air bubbles in soft ice rapidly increase in size, say, from 25 to 50 μm in half an hour. Consequently, soft ice has a short shelf life. Presumably, both Ostwald ripening and coalescence contribute to the coarsening. The process is slower at a higher churned-fat index. The explanation must be that air bubbles that are better covered with fat globules are more stable.

The coarsening is enhanced during the early stages of hardening, when growing ice crystals press bubbles close to each other, thereby inducing coalescence. This process can also lead to *channelling*, i.e., the establishment of air connections between air bubbles and with the outside air. This results in loss of air and, therefore, in shrinkage of the ice cream when its temperature is increased. At a constant lower temperature, the coarsening is slower, presumably because of increased viscosity of the liquid phase. Below about –28°C, the liquid changes into a glassy state (see Section 11.2) and, virtually, no further physical changes occur. However, autoxidation can go on, albeit slowly. Fluctuations in temperature strongly enhance coarsening and shrinkage.

Ice crystals and, if present, lactose crystals are also subject to Ostwald ripening, except when the temperature is below the glass transition point. Temperature fluctuations also worsen these changes, often leading to crystals that are large enough to be perceived in the mouth.

17.3.3 Role of the Various Components

Fat is of special importance for the flavor and for a solid structure to be formed during freezing and therefore for consistency, appearance, and melting resistance. A high fat content leads to a dry, almost grainy texture, a low fat content to a smooth, homogeneous, somewhat slimy texture.

Milk solids-not-fat contribute to the flavor. They are also responsible for part of the freezing-point depression and for an increased viscosity. The protein partly serves to stabilize the foam lamellae during air incorporation; it is essential for the formation of fat-globule membranes during homogenization. Lactose can crystallize at low temperature. The crystals formed should be small in order to prevent sandiness. To that end, cooling should be quick during freezing, and afterward temperature fluctuations should be avoided.

Sugar, often sucrose, is essential for the taste and for the freezing-point depression. Too little sugar may cause too much ice to be formed; too much sugar often makes the ice cream overly sweet. To overcome this, part of the sucrose may be replaced by a substitute such as glucose syrup, which is less sweet and leads to a greater freezing-point depression per kg sugar. The sugar also causes a higher viscosity, especially when most of the water has been frozen. However, the most important role of the sugar is that it causes far less water to freeze than otherwise would be the case. As a result, the consistency of the ice cream is softer and its mouthfeel less cold.

The role of the *stabilizer* or, more properly speaking, of the thickening agent is not quite clear. Among those used are gelatin, alginate, carrageenan, pectin, locust bean gum, guar gum, xanthan, carboxymethylcellulose, and mixtures. Of course, these substances affect the consistency and, consequently, also the heat transfer during the freezing. If little clumping of fat globules occurs as, for instance, in low-fat ices, the desired firmness and prevention of excessive Ostwald ripening of air bubbles must be achieved by means of thickening agents. However, these agents may cause the consistency of the product to become somewhat slimy in the mouth. Furthermore, the thickening agents are often assumed to counteract the Ostwald ripening of ice and lactose crystals, and even to prevent crystallization of lactose. Many thickening agents at high concentrations (as is the case in ice cream at low temperature) do, indeed, lower the crystallization rate and thereby slow down Ostwald ripening, but it is very unlikely that they can inhibit crystallization.

Emulsifier is not needed in the proper sense of the word (more than sufficient protein is present during homogenization) and it does not play a significant role in foam formation either. It serves to stimulate the fat globules to clump and to become attached to the air bubbles. The emulsifiers used include egg yolk, monoglycerides, poly(oxyethylene) sorbitan esters (Tweens), and esters from citric acid and monoglycerides.

Flavoring agents are self-evident. Sometimes an antioxidant is added.

Naturally, *ice crystals* are essential for the consistency and for the coolness in the mouth. Moreover, the low temperature causes the sweetness to be less intense. The crystals should not be too large; hence, freezing should be fast and the storage temperature should not fluctuate.

Air cells play a threefold part. They make the ice cream light; otherwise it would be too rich. They soften its consistency and thereby make it deformable in the mouth. They moderate the coldness by lowering the rate of heat transfer; otherwise the ice cream would be far too cold in the mouth. The amount of air may be bound to a maximum because, according to statutory requirements, the density of the ready-made ice cream may not be below a given value, generally 500 kg/m^3.

Suggested Literature

Aspects of coffee cream and whipping cream: H. Mulder and P. Walstra, *The Milk Fat Globule,* Pudoc, Wageningen, 1974.

A practical book about ice cream: R.T. Marshall and W.S. Arbuckle, *Ice Cream,* 5th ed., Chapman and Hall, New York, 1996.

More fundamental information on ice cream: K.G. Berger, Chapter 9, in K. Larsson and S.E. Friberg, Eds., *Food Emulsions,* 3rd ed., Dekker, New York, 1997.

See also two symposium reports: W. Buchheim, Ed., *Ice Cream,* International Dairy Federation, Brussels, 1998; H.D. Goff and B.W. Tharp, Eds., *Ice Cream II,* International Dairy Federation, Brussels, 2004.

18 Butter

18.1 DESCRIPTION

Butter is generally made from cream by churning and working. It contains a good 80% fat, which is partly crystallized. The churning proceeds most easily at a temperature of around 15 to 20°C. Therefore, butter typically is a product originating from regions having a temperate climate. In addition to accumulated practical experience, a good deal of science has now been incorporated in butter making, enhancing the shelf life and quality of the product and the economy of manufacture.

Some *variants* occur: butter from cultured (soured) or from sweet cream and butter with or without added salt. Formerly, the salt was added as a preservative, but nowadays it is mainly added for the flavor; moreover, souring of the cream inevitably occurred (due to the duration of the gravity creaming), and now it is practiced intentionally. It enhances the keeping quality (although this hardly makes a difference when applying modern technology), and it greatly influences the flavor.

The following are the most important specific requirements for the product and its manufacture:

1. *Flavor:* Off-flavors of the fat are to be avoided, especially those caused by lipolysis, but also those due to volatile contaminants. The latter mostly dissolve readily in fat; examples are off-flavors caused by feeds such as silage and *Allium* (onion) species. If the cream is heated too intensely, the butter gets a cooked flavor. Moreover, careful attention has to be paid to the souring (see the following text).
2. *Shelf life:* Spoilage by microorganisms may cause several off-flavors (putrid, volatile acid, yeasty, cheesy, and rancid). In cultured-cream butter, spoilage usually involves molds and yeasts, the pH of the moisture being too low (~4.6) for bacterial growth. Lipolysis causes a soapy-rancid flavor; no lipases formed by psychrotrophs should be present in the milk. Furthermore, autoxidation of the fat can occur, especially at prolonged storage, even at a low temperature (–20°C), leading to a fatty or even a fishy flavor.
3. *Consistency:* Butter derives its firmness largely from fat crystals that are aggregated into a network. Butter should be sufficiently firm to retain its shape; likewise, oiling off (that is, separation of liquid fat) should not occur. On the other hand, the butter should be sufficiently soft so as to be easily spreadable on bread. The consistency can cause problems, because the firmness and the spreadability depend strongly on the composition of the fat and on the temperature.

4. *Color and homogeneity:* These rarely pose problems.
5. *Yield:* Some fat is lost in the skim milk and in the buttermilk. If the water content is below the legal limit (for example, 16%), this also means a loss of yield.
6. *By-products:* Buttermilk is sometimes desirable, but often it is not, owing to insufficient demand. Sour-cream buttermilk is only applicable as a beverage (or as animal feed), but it keeps poorly due to rapid development of an oxidized flavor (Section 22.3). Sweet-cream buttermilk can more readily be incorporated in certain products.

18.2 MANUFACTURE

18.2.1 PROCESSING SCHEME

Figure 18.1 gives a schematic example of more or less traditional butter making from sour cream. Figure 18.2 shows a schematic representation of the physical changes involved.

The *skimming* is mainly done for economical reasons: (1) reduction of fat loss (e.g., the fat content of buttermilk is 0.4% and that of skim milk is 0.05%; this means that removal of 1 kg of skim milk from the liquid to be churned will result in an additional yield of about 4 g of butter), (2) reduction of the size of the machinery (especially the churn), and (3) reduction of the volume of buttermilk. Hence, a high fat content of the cream (e.g., 40%) has advantage, also because it counteracts development of off-flavors (see Subsection 18.3.3). If a continuous butter-making machine is used, the fat content of the cream is often taken even higher, for example, up to 50%.

Pasteurization serves to kill microorganisms, inactivate enzymes, make the cream a better substrate for the starter bacteria, and render the butter more resistant to oxidative deterioration (see Subsection 18.3.3). Overly intense heating causes a cooked or gassy flavor. Sometimes the cream is pasteurized in a vacreator, which involves the hot cream being put under vacuum to cool, due to which some compounds causing off-flavors are removed.

The *starter* (see Section 13.5) should produce lactic acid and 'aroma' (i.e., primarily diacetyl). Moreover, the starter should not be strongly reducing, because this causes loss of diacetyl (by reduction to acetoin and 2,3–butanediol). Aroma formation is extensively discussed in Subsections 13.1.2.2 to 13.1.2.4.

Use is made of a mesophilic mixed-strain starter culture, containing the acid producers *Lactococcus lactis* sspp. *lactis* and *cremoris*; and the aroma producers *Leuconostoc mesenteroides* ssp. *cremoris* and *Lactococcus lactis* ssp. *lactis* biovar. *diacetylactis*. The ratio between the numbers of acid and aroma producers is critical. If too few aroma producers are present, they will be overgrown by the others and little diacetyl would be formed. If the aroma bacteria predominate, too little lactic acid would be formed, and the pH would remain too high to produce sufficient diacetyl. Moreover, a careful selection of strains is needed, with respect to the reducing capacity of the starter, and the formation of acetaldehyde by the

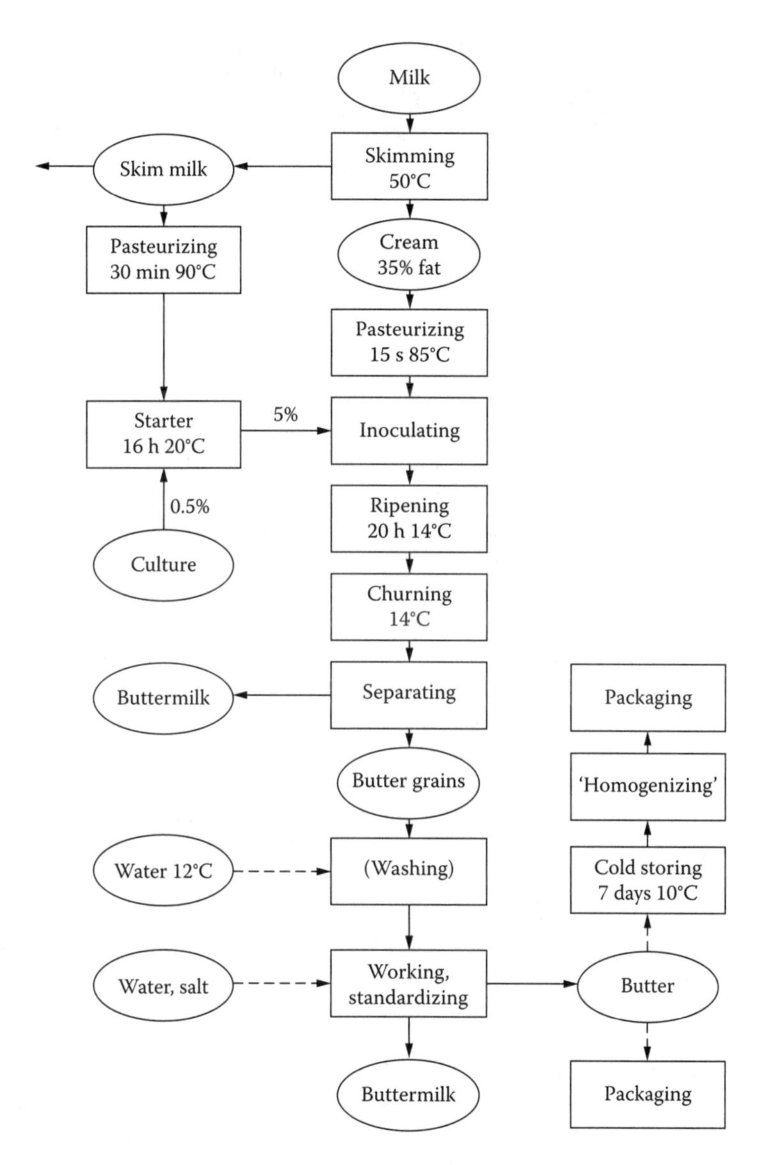

FIGURE 18.1 Example of butter making from ripened cream.

diacetylactis strains, because too much of it causes a yogurt-like flavor. The starter is usually cultured from a frozen concentrate. The inoculum percentage is high because the bacteria have to grow at a suboptimal temperature.

The purpose of *ripening* is to sour the cream and to crystallize fat. Without solid fat, churning is impossible, and too little solid fat goes along with excessive fat loss in the buttermilk. The method of cooling (temperature sequence) affects the butter consistency (see Subsection 18.3.2).

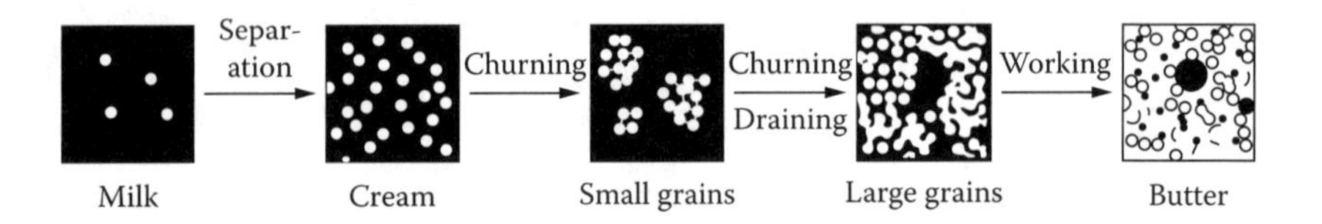

FIGURE 18.2 Stages in the formation of butter. Greatly simplified, not to scale. Black represents the aqueous phase; white represents fat. (After H. Mulder and P. Walstra, *The Milk Fat Globule,* Pudoc, Wageningen, 1974.)

The *churning* is in most cases achieved by beating in of air (see Subsection 18.2.2). It can be done in a churn, mostly consisting of a large vessel (tub, cylinder, cube, or double-ended cone) with so-called dashboards, which is partly (at most half) filled with cream, and which is rotated at several revolutions per minute (r.p.m.). The churning then takes, say, 20 min. There are also churns with a rotary agitator (for example, 20 r.p.m.). The latter principle is also applied in the frequently used continuous butter-making machine according to Fritz (see Figure 18.3). Here the paddle turns very quickly (500 to 3000 r.p.m.) and the cream stays in it for less than 1 min. To achieve this, high-fat cream (about 50% fat) has to be used. These machines can have very large capacities.

The churning should proceed rapidly and completely (low fat content in the buttermilk), and the formed butter grains should have the correct firmness to allow for efficient working. The size of the butter grains can be varied by continuing the churning for various lengths of time after grains have formed. Very fine butter grains (on the order of 1 mm) are hard to separate from the buttermilk, especially in continuous machines.

If the butter grains are not too large, their firmness can to some extent be affected by *washing*, that is, via the wash-water temperature. The washing consists of mixing the butter grains with cold water, after which the water again is drained off. This reduces the dry-matter content of the butter moisture. Formerly, washing was done to improve the keeping quality of the butter, but nowadays it is only done to control the temperature, if needed.

The *working* (kneading) is done (1) to transform the butter grains into a continuous mass; (2) to finely disperse the moisture in the butter; (3) to regulate the water content; and (4), if desired, to incorporate salt (see Subsection 18.2.3). Working consists of deforming the butter. This can, for instance, be achieved by squeezing the butter through rollers, by allowing it to fall from a height, or by squeezing the butter through perforated plates (in the continuous machines). During the working, the water content is regularly checked and, if need be, additional water is added to arrive at the accepted standard value.

The butter can now be immediately *packaged*, for example, in a retail package. Often one wants the butter after the working to be soft enough to be pumped from the churn-and-worker by a suitable positive pump. Sometimes, the butter is allowed to set (see Subsection 18.3.2), or it is for some other reason kept for

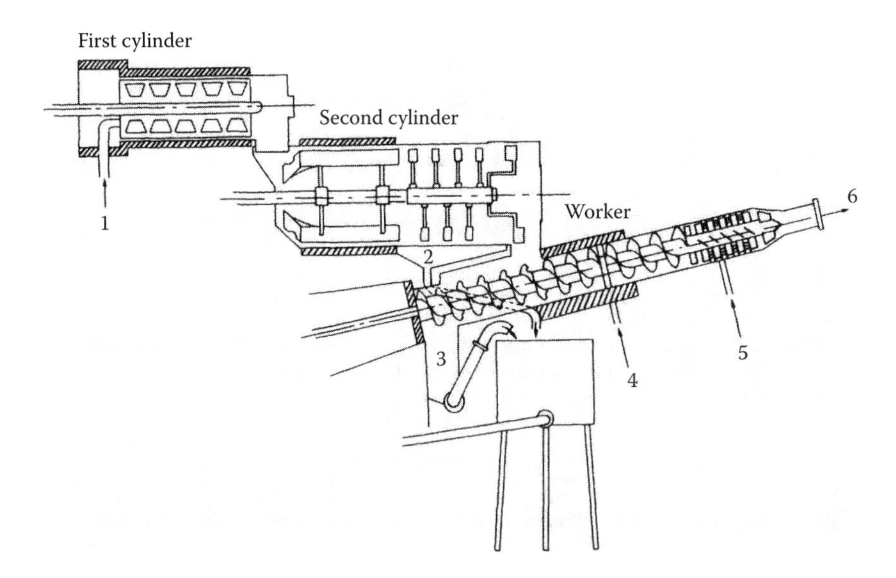

FIGURE 18.3 Example of a continuous butter-making machine according to Fritz; highly simplified diagram. The cream enters at (1) and is very intensively churned in the first cylinder (turning speed of beater, for example, 2000 r.p.m.), yielding very fine butter grains. In the second cylinder (say, 30 r.p.m.), the grains are churned into larger ones, allowing the buttermilk to drain off via a sieve (2). The grains fall in the worker, where they first are kneaded together by the worm, with the residual buttermilk being drained off (3). The mass may be chilled with water (4). The butter is now squeezed through a series of perforated plates and leaves the machine as a strand (6). To adjust the butter composition, additional water, brine, and so forth, can be incorporated during working (5). Some machines are equipped with two worker sections in series.

some time before packaging. It is then too firm to pass through the packaging machine, and it must be passed through a butter 'homogenizer' to soften it; this may also prevent the moisture dispersion from becoming too coarse during packaging.

18.2.2 The Churning Process

The instability of fat globules and their interactions with air bubbles are discussed in Subsections 3.2.2 and 3.2.3 (see especially Figure 3.13).

During churning, air is beaten into the cream and disrupted into small bubbles. The fat globules touch these bubbles, often spread part of their membrane substances and some of their liquid fat over the air–water interface, and become attached to the bubbles; one bubble catches several globules. This resembles flotation, although in true flotation the foam is collected. In the churning process, however, the air bubbles keep moving through the liquid and collide with each other. They thus coalesce, and in this way their surface area diminishes. As a consequence, the adhering fat globules are driven toward one another.

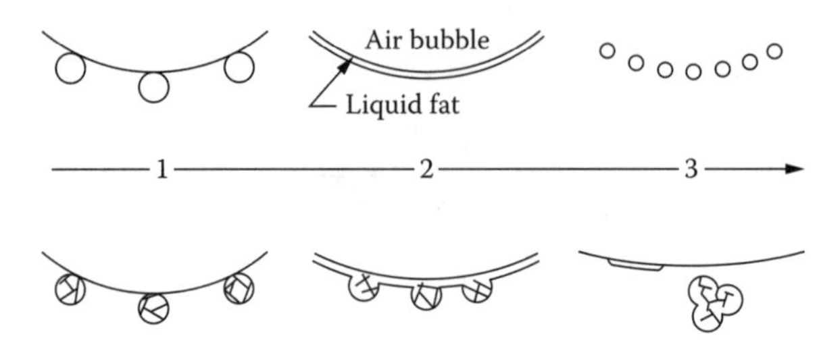

FIGURE 18.4 Schematic representation of the interactions between fat globules and air bubbles during churning. If the fat is liquid, the fat globules are disrupted by beating in of air (top). If the fat globules contain solid fat, clumps are formed (bottom).

Now the liquid fat acts as a sticking agent, and the fat globules are clumped together. In this way small fat clumps are formed. All these changes are illustrated in Figure 18.4.

The clumps, in turn, participate in the churning process, resulting in still larger clumps. When the clumps become larger, direct collision between them increasingly occurs, and the clumps now grow without the air bubbles any longer playing an important part. Flotation thus predominates initially and mechanical clumping (partial coalescence) later on. In addition, more and more liquid fat and membrane material is released (it first spreads over the air bubbles and partly desorbs when the bubbles coalesce); this is called *colloidal fat*, and it consists of tiny liquid fat droplets and membrane remnants. Toward the end of the churning little foam has been left, presumably because too few fat globules remain to cover the air bubbles and thereby stabilize these bubbles. These changes are illustrated in Figure 18.5.

Several factors affect both the rate and the efficiency of the churning process, and Figure 18.6 gives examples of the relations. The type and filling level of the churn, as well as the turning speed of the churn, naturally influence the churning process. The churning time decreases with increasing fat content, but less sharply than would be expected on account of the increasing probability of collision between fat globules; the churning time then would vary inversely with the square of the fat content (see Subsection 3.1.3.1). Obviously, the flotation churning is a very efficient process; even milk can be churned readily, and only if cream of very high fat content is used does the fat content of the buttermilk increase. The influence of the size of the fat globules is as expected. Homogenized milk cannot be churned.

The proportion of solid fat is crucial (see Figure 3.12c). If the fat is fully liquid, a kind of homogenizing rather than churning occurs (Figure 18.4, upper row). Also, if the globules contain very little solid fat, then the cream does not churn; the clumps formed are soon pulled to pieces. But for the rest, the higher the proportion of solid fat, the slower the churning and the lower the fat content in the buttermilk. If the fat globules contain relatively little liquid fat, they can

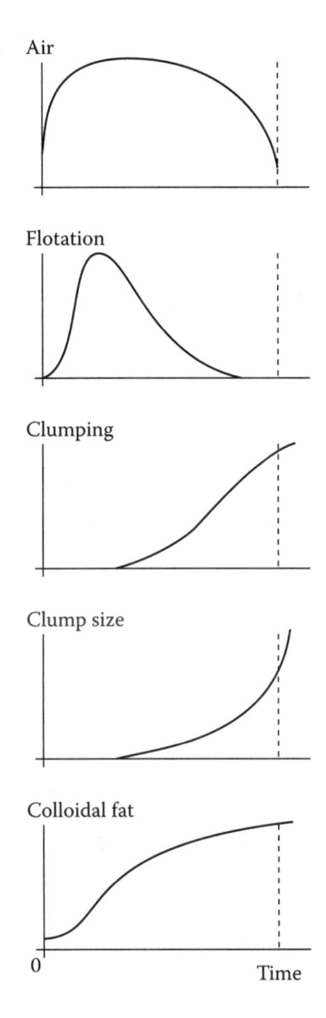

FIGURE 18.5 Events taking place during traditional churning. Shown are the amount of air entrapped in the cream; rate of flotation (extent to which fat globules attach to air bubbles); rate of clumping; size of the clumps or butter grains; and amount of fat in a colloidal state (i.e., not recoverable by centrifugation). The broken line indicates the point of "breaking" (formation of clearly visible butter grains). Approximate examples. (Adapted from H. Mulder and P. Walstra, *The Milk Fat Globule,* Pudoc, Wageningen, 1974.)

still be attached to the air bubbles, and the first stages of the churning, in which flotation predominates, do occur; but mechanical clumping hardly takes place, and the temperature should be raised to allow the formation of butter grains.

The temperature therefore has a considerable effect on the churning, but so does the temperature history (see Subsection 2.3.5.2). If the precooling is not sufficiently deep, undercooled (that is, liquid) fat globules will still be present, and

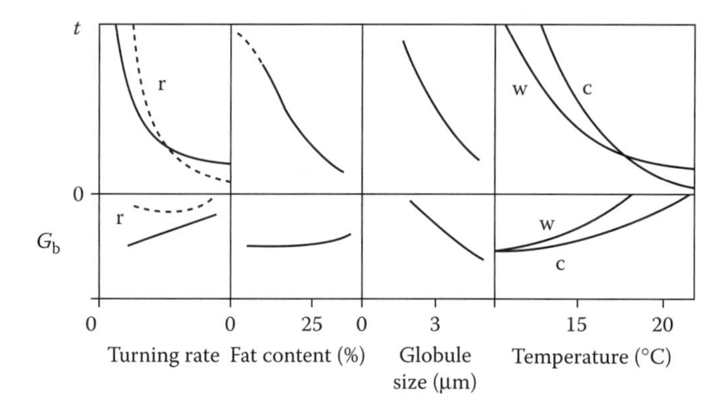

FIGURE 18.6 The effect of some variables on the churning time (t) and the efficiency (as percentage of fat in buttermilk, G_b) in a traditional churn. Variables are the turning rate of churn or agitator; fat content of the cream; average fat globule size; and churning temperature. r = churn with rotary agitator; c = cream kept cool before the churning; w = cream kept warm before bringing it to churning temperature. Approximate results. (Adapted from H. Mulder and P. Walstra, *The Milk Fat Globule,* Pudoc, Wageningen, 1974.)

the fat content of the buttermilk will substantially increase. Whether or not the cream is soured does affect the churning process, but the explanation is not clear.

Note in Figure 18.6 that rapid churning mostly coincides with a high fat content in the buttermilk (except with regard to the influence of the fat globule size). If much liquid fat can spread over the air bubbles, disruption of fat globules occurs. Furthermore, the smallest fat globules may easily escape the churning process if it proceeds very rapidly. Sweet-cream buttermilk can be separated centrifugally (sour buttermilk cannot), but its fat content remains rather high (for example, 0.2%) due to the colloidal fat.

The churning proceeds very fast in a continuous butter-making machine, as the beater turns at a very high rate. Accordingly, the fat content of the buttermilk tends to be high, unless the cream is deeply precooled (4°C) and is churned at a low temperature (8 to 12°C). A high-fat cream then is required for sufficiently rapid churning. Presumably, flotation churning is less important here than in a traditional churn. A high-fat cream can also be churned by means of a rapidly rotating paddle without the beating in of air; however, the churning time is usually somewhat longer, and the fat content in the buttermilk slightly higher.

18.2.3 Working

A partial phase inversion does occur during the churning; a continuous fat phase has developed in the butter granules (see Figure 18.2). But in the whole mass of butter grains, the aqueous phase is still continuous. The working accomplishes a further phase inversion. In this stage excessive moisture is squeezed out and the remaining moisture droplets are disrupted into smaller ones. This does not concern the very small moisture droplets that are left between individual clumped fat globules; these are too small (on average about 2 μm) to be disrupted by the working.

During the working the butter is deformed, and therefore velocity gradients (Ψ) occur. The deformation (flow) causes a shear stress $\Psi \times \eta$, where η is the viscosity. The flowing mass, roughly a mixture of liquid fat and aggregated crystals, does not have true viscosity, but due to the presence of the crystals the effective viscosity (η_{eff}) is high. The flow also exerts a stress on the moisture droplets, thereby deforming them. If the stress involved exceeds the Laplace pressure of the droplet ($4\gamma/d$, where d is the droplet diameter and γ the interfacial tension oil-plasma, amounting to about 15 $mN \cdot m^{-1}$), the droplet is disrupted. This is illustrated in Figure 18.7A and Figure 18.7B. Disruption is more effective in

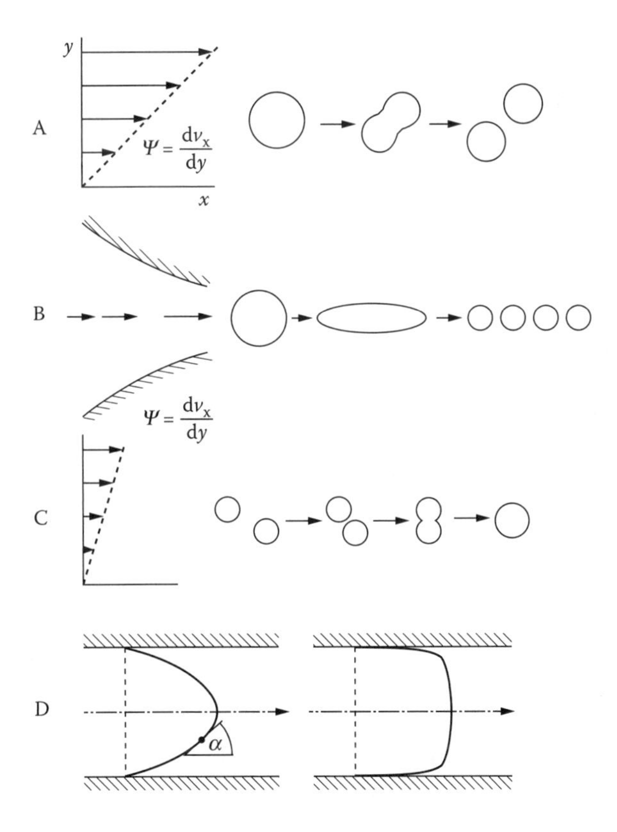

FIGURE 18.7 Diagrammatic representation of the disruption and coalescence of moisture droplets during the working of butter (or margarine). (A) Disruption in a flow with simple shear, i.e., velocity gradient normal to the direction of the flow. (B) Disruption in a converging flow (= extensional flow), i.e., velocity gradient in the direction of the flow. (C) Encounter and coalescence of (small) droplets in flow; shear stress too small to disrupt the droplets. (D) Velocity patterns in Poiseuille flow (left) and in partial plug flow (right).

extensional flow than in simple shear flow. Extensional flow always occurs during working. Obviously, at a higher velocity gradient (more intensive working) or at a higher effective viscosity of the butter (greater amount of solid fat, e.g., at lower temperature), smaller droplets can result. (*Note*: In some, but not all, practical situations essentially a given stress is applied. A higher viscosity then does not result in a higher shear stress, because a proportional decrease in velocity gradient occurs.)

During the working of butter, the velocity gradient varies widely from place to place and from one moment to the other. Figure 18.7D (left) illustrates the velocity pattern in the case of true Poiseuille flow. It shows that the velocity gradient, proportional to $\cot \alpha$, depends strongly on position. The differences may even be much greater due to the fact that the mass of oil and crystals shows a kind of plug flow (Figure 18.7D, right). This is because the mass has a yield stress; when the butter flows along a wall (for example, through an opening), the shear stress in the mass may be largest near that wall, so that the mass yields there and thereby locally decreases the effective viscosity. In other words, a small proportion of the butter deforms with a strong velocity gradient; a larger proportion with a very weak gradient.

Because of the velocity gradient, the droplets collide with each other and can coalesce, assuming that the shear stress is small enough to prevent redisruption. This is shown in Figure 18.7C. The collision frequency of a droplet (diameter d_1) with other droplets (number per unit of volume N_2, diameter d_2) is about equal to $(d_1 + d_2)^3 N_2 \Psi/6$ (see Subsection 3.1.3.1). The probability for coalescence will therefore be greater for larger droplets, but such droplets will be easier disrupted as well. A kind of steady state of disruption and coalescence now develops (assuming that Ψ and η_{eff} remain constant), and this is reflected in a certain droplet size. But a wide droplet size distribution will result because, as mentioned, Ψ varies widely, so that during working disruption predominates in some places, and coalescence in others. An example is given in Figure 18.8.

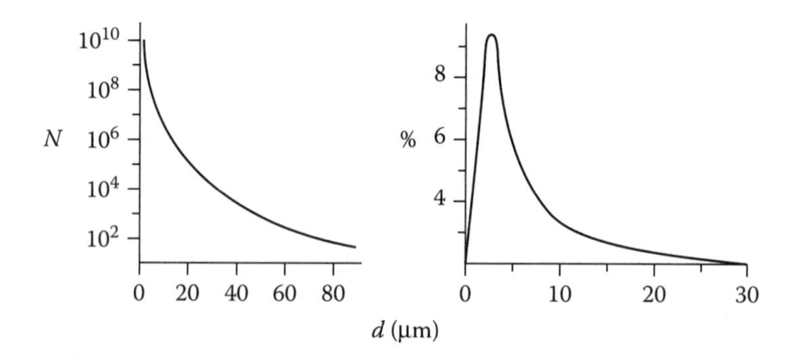

FIGURE 18.8 Example of the size frequency distribution of moisture droplets in well-worked butter. N is the number frequency (per ml per μm class width); % refers to the volume of the moisture per μm class width; and d is the droplet diameter. (After H. Mulder and P. Walstra, *The Milk Fat Globule,* Pudoc, Wageningen, 1974.)

The mentioned considerations can help in finding out the actual processing conditions needed to obtain a product with desired properties. One additional point should be mentioned. The (partial) plug flow occurs especially at low temperature (much solid fat), so that a very wide size distribution with numerous large droplets is obtained, despite the high effective viscosity. Increasing the working speed causes the droplets to become smaller; the butter becomes 'dry'. Likewise, moisture can be incorporated into the butter. During working at a very slow speed, larger droplets emerge again, especially at low temperature; the butter becomes 'wet,' i.e., visible droplets appear. In that way, excessive moisture can be worked out of the butter. The butter can also readily become wet during repackaging in retail packages, where weak velocity gradients prevail.

The moisture dispersion, that is, the fineness of the droplets, is of great importance for the keeping quality. Contamination by microorganisms cannot be fully prevented, so butter can spoil. But if there are, say, 10^3 ml^{-1} microorganisms and 10^{10} ml^{-1} moisture droplets (see Table 18.1), then only a negligible part of the moisture would be contaminated, and because the microorganisms cannot pass from droplet to droplet, the spoilage will be negligible. Naturally, it is especially the large droplets that become contaminated. The fraction of the moisture being contaminated is proportional to the colony count and approximately proportional to the volume-average volume $(\pi/6)\sum N_i d_i^6 / \sum N_i d_i^3$ of the droplets. If some large droplets ('free moisture') are left in the butter, growth of microorganisms readily occurs. This is illustrated in Figure 18.9.

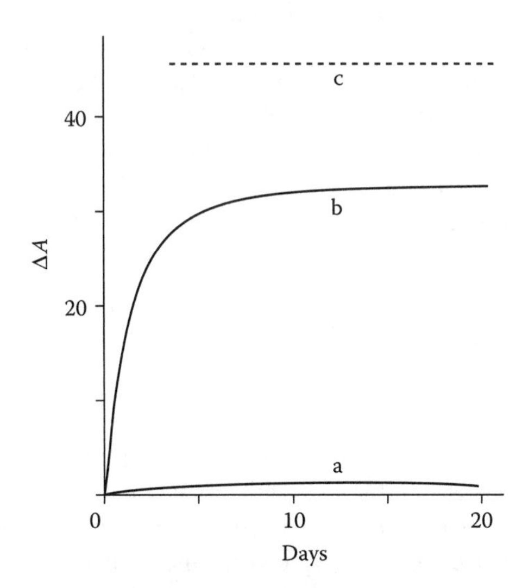

FIGURE 18.9 Increase in acidity (ΔA, mM) of the serum of butter kept at 17°C. Just before churning, 10% starter had been added to the sweet cream. (a) Butter worked until dry. (b) Butter worked less well. (c) Final ΔA expected if all droplets were to acidify. (Data provided by Mulder and Zegger, unpublished.)

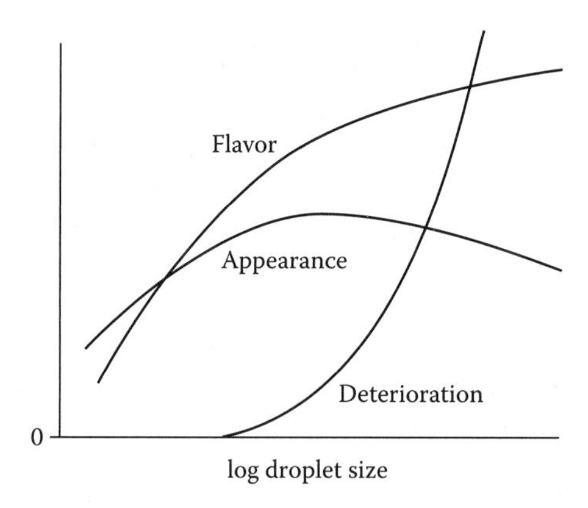

FIGURE 18.10 Intensity of perceived flavor (cultured cream butter), appearance score (color and gloss), and rate of deterioration by microorganisms, as a function of the average aqueous droplet size of butter. Meant to illustrate trends.

The smaller the droplets, the paler the butter due to the stronger light scattering. Apart from that, the color is mainly determined by the β-carotene content. Sometimes coloring matter is added.

The smaller the droplets, the flatter the taste of the butter: salt as well as aroma are much better perceived if the butter is slightly wet. Hence, working dry should not be overdone. Especially in the continuous machines, the working (squeezing through a set of perforated plates) is very intense. By this type of working, many fat globules also are fragmented and the butter becomes somewhat 'greasy' in consistency. All this does not alter the fact that in many cases the problem is to obtain sufficiently fine droplets rather than the converse.

Figure 18.10 serves to illustrate the approximate influence of average droplet size on the quality aspects mentioned earlier. It follows that there is an optimum droplet size.

18.3 PROPERTIES

18.3.1 MICROSTRUCTURE

Figure 18.11 and Table 18.1 illustrate the microstructure of butter.

A striking and important difference with, for example, margarine is the presence of several partly intact fat globules. Their number depends on the method of manufacture, and it decreases strongly with intensive working. Note that most crystals in the fat globules are tangentially arranged.

The continuous phase is liquid fat. Sometimes a continuous aqueous phase persists, especially in insufficiently worked butter. This aqueous phase partly passes through the surface layers of the fat globules. The fact that displacement of water through butter can occur generally has another cause: approximately

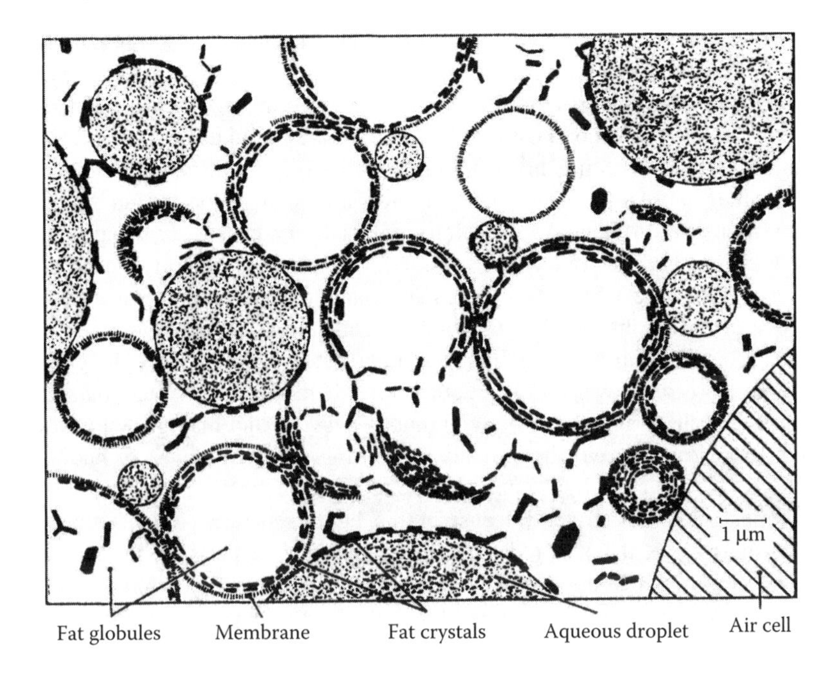

FIGURE 18.11 Butter microstructure at room temperature. Liquid fat is white. Membrane thickness is greatly (about ten times) exaggerated. (Adapted from H. Mulder and P. Walstra, *The Milk Fat Globule,* Pudoc, Wageningen, 1974.)

TABLE 18.1
Structural Elements of (Conventional) Butter

Structural Element	Approximate Number Concentration (ml^{-1})	Proportion of Butter (%, v/v)	Dimension (μm)
Fat globule[a]	10^{10}	5–30[c]	1–5
Fat crystal[b]	10^{13}	10–40[d]	0.01–2
Moisture droplet	10^{10}	15	1–25[c]
Air cell	10^{6}	~2	>20

[a] With (for the greater part) a complete membrane.
[b] At higher temperatures mainly inside the fat globules, at low temperatures forming solid networks.
[c] Strongly depends on the intensity of working.
[d] Strongly depends on the temperature.

0.2% (v/v) water can dissolve in liquid fat, which implies that water can diffuse through the continuous oil phase.

The moisture droplets consist, in principle, of (sour) buttermilk, but they are not always identical in composition. Differences are found due to water addition, washing, and the working in of starter, salt, or brine. Differences in osmotic pressure then cause a slow water transport toward the most concentrated droplets. Hence, moisture droplets in the vicinity of a salt crystal mostly disappear, and the salt crystal changes into a large droplet; accordingly, the butter turns 'wet.'

The number and size of the fat crystals greatly depend on the temperature and on the temperature history. A significant part of the crystalline fat may be inside the fat globules because during churning liquid fat is extruded from the globules, mainly by spreading over the air bubbles. But there are also crystals outside the globules, and these aggregate to a continuous network and may grow together to form a solid structure, which is mainly responsible for the firmness of butter. The crystals inside the globules do not participate in this network and, therefore, they hardly make the butter firmer. Because of this, butter generally contains more solid fat than margarine, if both products are equally firm; this results, in turn, in the butter feeling cooler in the mouth (due to the greater heat of melting).

The crystals outside the fat globules thus make up a continuous network, in which part of the water droplets (often with crystals attached to their surface) and damaged fat globules may participate. This network retains the liquid fat as a sponge. When the temperature increases, many crystals melt and the network becomes less dense and coarser. Because of this, destabilization can eventually occur, that is, the butter separates oil. Oiling off occurs more readily (at equal solid fat content) if the crystals are coarser.

Air cells always occur in butter, unless the working is done in vacuum (which is possible in some continuous machines). Moreover, butter contains up to about 4% (v/v) of dissolved air.

18.3.2 Consistency

Butter is a plastic (or ductile) material: it can be permanently deformed without losing its coherence and solid properties. The consistency of such a material is defined as its resistance to permanent deformation. In the case of butter, several factors are of importance. Butter should be firm against sagging under its own weight; sagging may also go along with oiling off, which is quite undesirable. Butter should be readily spreadable, without being too 'short' or crumbly. It should be easily deformed in the mouth, without feeling greasy; the latter implies that the fat should completely melt at 35°C.

The consistency of butter is primarily determined by the properties of the fat crystal network. Fat crystallization is treated in Subsection 2.3.5; Subsection 2.3.5.7. discusses network formation and structure. It follows that a plastic fat consists of a moderately homogenous network of crystals, filled with oil. The crystals are small and platelet-shaped. Initially, the crystals are kept together by van der Waals forces, but they soon become much more strongly bonded to each other due to sintering.

When a small stress (σ) is applied to a sample of a plastic fat, it shows *elastic* (reversible) deformation: when the stress is released, the original shape of the sample is recovered. The ratio of stress to strain (ε), that is, the relative deformation, is called the elastic modulus. Many rheologists determine the modulus because it is considered an important parameter to characterize a solid material. In a plastic fat, however, above a very small value of the strain the magnitude of σ/ε starts to decrease and the deformation becomes (partly) permanent. For butter, the critical strain at which this occur is about 1%; for margarine, it is even smaller, say, 0.2%. This difference is perceptible when spreading the product: butter exhibits a more elastic behavior. The explanation is presumably that the butterfat crystals are relatively thinner, and can therefore more readily be bent, than those of margarine fats.

When a slowly increasing stress is applied to a plastic fat, an increasing number of bonds in the crystal network are broken, which implies an irreversible change in structure. If the stress becomes large enough, the material will show *yielding*, meaning that it starts to flow; the stress at which this occurs is called the *yield stress* (σ_y). This is illustrated in Figure 18.12A. (Because the value of σ_y depends on the sensitivity of the instrument used, some workers prefer to use the extrapolated Bingham yield stress σ_B.) The strain at yielding is on the order of 25%. Nearly all important consistency aspects of a plastic fat depend on the yield stress rather than on the modulus. Most methods used in practice to determine butter 'firmness' correlate well with the yield stress. The methods involve measurement of the penetration distance of a cone in the sample; or of the force needed for wire cutting, or pushing a probe into, the sample.

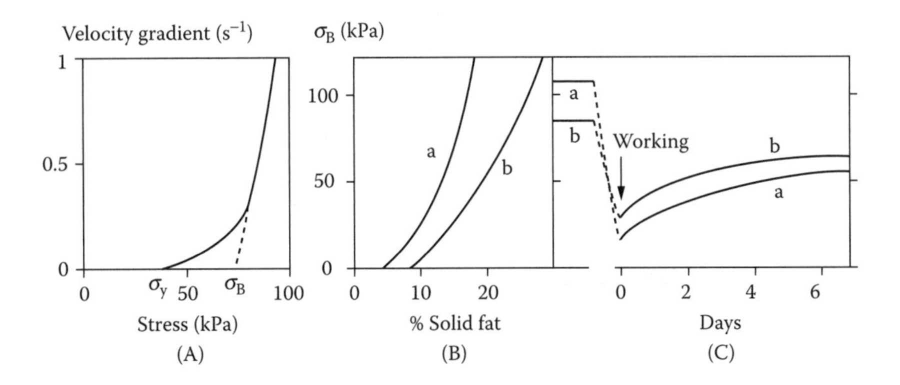

FIGURE 18.12 (A) Deformation rate of butter as a function of the stress applied; σ_y = yield stress. (B) Firmness as Bingham yield stress (σ_B) as a function of the percentage of solid fat. (C) Effect of working and of the length of time after working on σ_B. a = plastic fat or margarine; b = butter. Approximate examples. (Adapted from P. Walstra and R. Jenness, *Dairy Chemistry and Physics,* Wiley, New York, 1984.)

The value of the yield stress strongly increases with increasing solid fat content, as illustrated in Figure 18.12B. The extent to which sintering has occurred also has a strong effect. For the same solid fat content, the yield stress tends to be larger if the crystals forming the network are smaller. Working the fat, which means strongly deforming it, considerably decreases its firmness. This is called *work softening*. After working, the value of σ_y increases again, due to aggregation of network fragments into a space-filling structure, and due to sintering. The changes are illustrated in Figure 18.12C.

As compared to a plastic fat, butter has additional structural elements. Part of the fat is in fat globules, and crystals in these globules do not participate in the fat crystal network. If this is a substantial portion of the solid fat, the firmness of the butter will be decreased. This may be one of the reasons that butter tends to have a lower firmness than margarine for the same solid fat content (see Figure 18.12B). Another reason for this may be that the fat globules interrupt the crystal network, making it less homogeneous; the latter tends to decrease the yield stress of a material. The aqueous droplets also interrupt the crystal network. However, clear correlations between firmness and moisture content or droplet size have not been established; presumably the variation in these parameters is too small.

The question now is: What measures can be taken, in practice, to influence butter firmness? Important variables are:

1. The *temperature*, which has a very strong effect, as shown in Figure 18.13, for the most part caused by its influence on solid fat content. Although

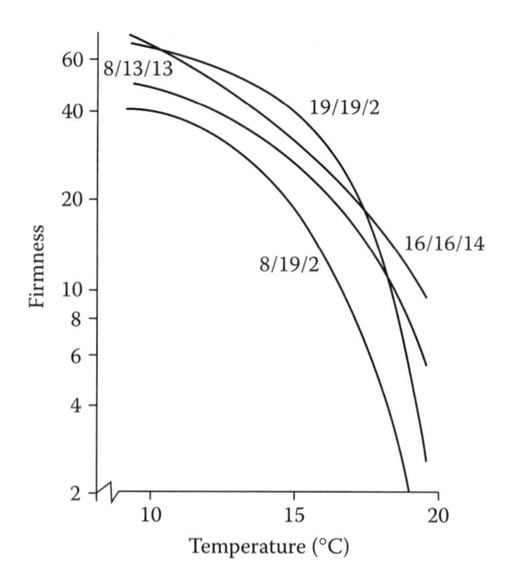

FIGURE 18.13 Firmness (as measured by a penetrometer; arbitrary units) of butter as a function of the temperature of measurement. Butters from the same cream, but with different cream temperature treatment (indicated on the curves in °C). (Courtesy of H. Mulder.)

at one temperature the values may vary by a factor up to three, all the curves show a decrease in firmness by a factor of about 30 when going from 10 to 20°C. This poses important problems: at refrigerator temperature, the butter may be virtually unspreadable, and at room temperature, it may exhibit oiling off.

2. The *fat composition* has a considerable effect, and the milk to be used for butter making may, therefore, be selected, for example, according to region of origin. Alternatively, cream can be frozen and, in another season, be churned together with fresh cream. Similarly, firm and soft butter can be worked together. The fat composition can be affected via the feed of the cow, but this has not yet come into practice.
3. The *method of manufacture* has a clear effect, especially the temperature treatment of the cream (and, if need be, the butter grains). Figure 18.13 gives some trends; the dependence of the firmness on the temperature can also be somewhat affected. (*Note*: The temperature treatment of the cream is usually indicated as a/b/c, where a, b, and c are the successive temperatures.)

   The temperature treatment of the cream is constrained by certain conditions:

   a. The souring should be adequate (sufficiently high temperature for enough time).
   b. The cooling of acidified high-fat cream proceeds very slowly because of the low heat-transfer coefficient.
   c. The churning must proceed satisfactory, and this can only be achieved within a relatively narrow temperature range.
   d. The fat content of the buttermilk should not become too high, implying that the liquid fat content should not be too high during churning.

   An example is constant temperature, e.g., 13/13/13°C, especially applied to obtain firm butter. As a result, the fat content in the buttermilk is often high. To make soft butter, cooling in steps is often used; a short deep precooling is necessary to obtain sufficient nucleation, for example, 8/20/14°C. (This is often called the Alnarp method.) Due to the cooling in steps, somewhat less solid fat will result according to the compound crystal theory, but the differences are small. On average, the crystals will also be larger and, possibly, a greater proportion of the solid fat will be inside fat globules. A quantitative explanation is lacking. In the continuous butter making from sweet cream (see also Section 18.4), 4/4/12°C treatment often is applied: after separation the cream is cooled, kept cool in large tanks, raised to the churning temperature by means of a heat exchanger, and churned.

   As mentioned, the higher the proportion of fat in fat globules, the lesser may be the firmness. Very intensive or prolonged working can reduce the quantity of globular fat.

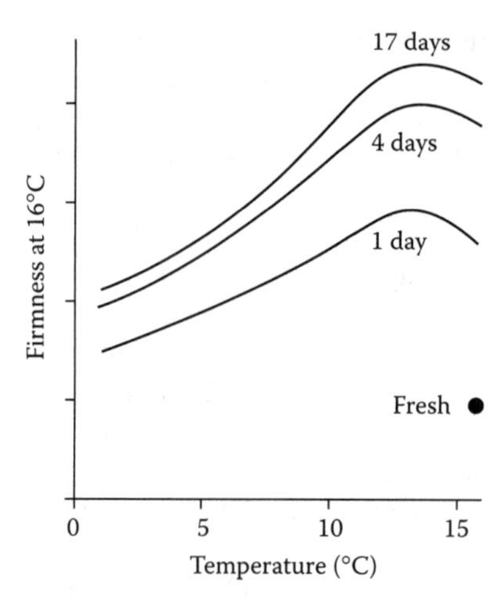

FIGURE 18.14 Effect of temperature and time of storage on the firmness (measured with a penetrometer; arbitrary units) of butter, determined at 16°C. The point indicates firmness of the fresh butter. (Courtesy of H. Mulder, *Zuivelonderzoek*, Vol. 2, The Hague, Algemeene Nederlandsche Zuivelbond FNZ, 1947.)

4. The *storage conditions* have a considerable effect on the consistency. To start with, the butter will always set, which occurs faster at a higher temperature (see Figure 18.14). The cause is additional sintering due to changes in the crystallization, such as the rearrangement of compound crystals and polymorphic changes; these changes proceed more slowly if less liquid fat is available (lower temperature). The setting may continue for a very long time, be it at an ever-decreasing rate. It will be accelerated if the butter temperature is temporarily raised. Then solid fat melts, slowly solidifying again on subsequent cooling, so that a more solid structure can be formed. In that case, the firmness may increase by as much as 70%. In particular, butter made according to the Alnarp method is sensitive to temperature fluctuations. Because these readily occur during purchase and use of the butter, the favorable effect of a certain temperature treatment of the cream on the spreadability of the butter may be disappointing in practice.
5. *Working* of the butter causes a strong reduction of its firmness because solid structures are broken (Figure 18.12C). It is true that the butter sets again, but not up to the original firmness. Hence, to obtain firm butter, packaging should be done immediately after manufacture while the butter is still very soft (the packaging itself involves intensive working); the packaged butter then can set fully, especially if it is not stored too cold. If it is desirable to make soft (spreadable) butter, the best policy is to first let the butter set for a considerable time after

manufacture and to package it afterward. To allow packaging, the butter must at first be worked soft in a butter homogenizer.

18.3.3 Cold Storage Defects

To keep butter for a long time, it should be stored at a temperature of, say, –20°C. If the butter has been well made, and if the original milk did not contain too many bacteria with thermoresistant lipases, it can keep for a very long time in cold storage. It will then deteriorate by autoxidation of the fat, leading to flavor defects after a period of 1 month to 2 yr. The factors affecting autoxidation are discussed in Subsection 2.3.4.

The keeping quality in cold storage greatly depends on the method of manufacture. Processing variables having an effect are described in the following text:

1. Contamination with even minute quantities of copper should be strictly prevented.
2. By cooling the milk for some time before its use (e.g., for at least 2 h at 5°C), a part of the copper on the fat globules moves to the plasma; this may restrict the autoxidation. Moreover, the cooling causes a migration of protein to the plasma, and this is precisely the protein that liberates H_2S during heat treatment. In this way, a cooked flavor after heat treatment can be limited. In many regions most of the milk is already kept cool for a time in the bulk tank on the farm.
3. Heating of milk or cream causes migration of copper from the plasma to the fat globules (see Figure 18.15A). Pasteurization of the milk should therefore be avoided because more copper is available to migrate in milk than in cream. Even during thermalization the temperature should not be too high.
4. Due to souring of the cream (or the milk), a considerable part (30 to 40%) of the 'added' copper (i.e., copper that entered by contamination) moves to the fat globules. Because of this, butter from sour cream is much more affected by autoxidation than that from sweet cream.
5. With reference to the migrations mentioned in the preceding two items, it is important to adjust the fat content of the cream to a high level because this causes a lower copper content in the butter.
6. Heating of the cream largely prevents the migration during souring mentioned in item 4 (see Figure 18.15B). In all probability, copper becomes bound to low-molar-mass sulfides, especially H_2S, formed by the heat treatment. This causes a strong reduction of the autoxidation in sour-cream butter. Therefore, it would be desirable to pasteurize the cream very intensively, but then much H_2S is formed, yielding butter with a gassy or cooked flavor. Although this flavor defect decreases slightly during storage, it is objectionable. The pasteurization conditions should therefore be optimized (not too high and not too low); of course, the smaller the spread in holding time, the better.

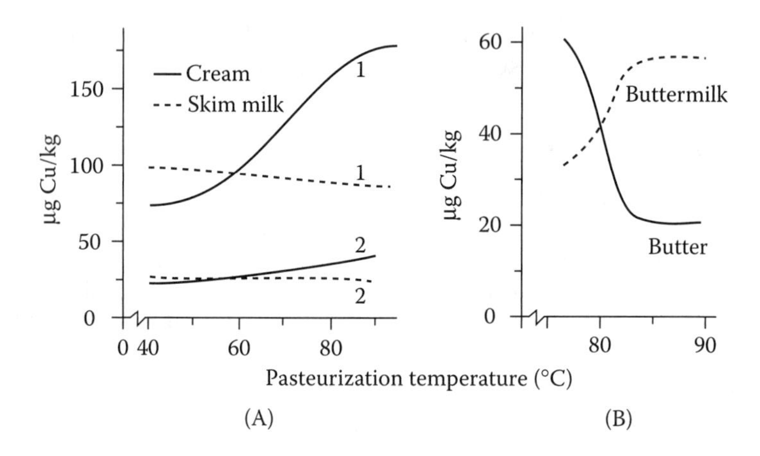

FIGURE 18.15 The partition of copper over the products as a function of the pasteurization temperature (pasteurizing for 15 s). A. Heating of the *milk* before skimming; examples of milk with a high (1) and with a low (2) copper content. B. Heating of the *cream* and estimating the quantities *after souring* and churning. (After H. Mulder and P. Walstra, *The Milk Fat Globule,* Pudoc, Wageningen, 1974.)

7. Adding salt to sour-cream butter considerably accelerates the autoxidation. In sweet-cream butter a high salt content has an oxidation-diminishing effect.
8. The lower the storage temperature, the longer the keeping quality.

18.4 CULTURED BUTTER FROM SWEET CREAM

It is often a problem to dispose of sour-cream buttermilk because it has a very short shelf life and the demand for it as a beverage often is small. Moreover, it cannot be pasteurized. Sweet-cream buttermilk can be processed much more easily. On the other hand, several markets prefer aromatic butter, which has to contain acid (lactic acid) and aroma substances (mainly diacetyl). Attempts have been made to churn sweet cream and to add starter to the butter granules afterward and working it into the butter, but with disappointing results: the flavor remains almost the same as that of sweet butter. This is not surprising, because the souring and the diacetyl production in the butter can hardly occur if the moisture has been well dispersed (see Figure 18.9); the pH also remains high.

NIZO (Ede, the Netherlands) has developed an alternative manufacturing process. Sweet-cream butter grains are worked together with a very aromatic starter and a concentrated starter permeate, essentially a lactic acid solution. A flow chart of the butter making is shown in Figure 18.16; churning and working can proceed in a churn-and-worker or in a continuous butter-making machine. The initial water content (after the first working) should be low so as not to exceed the 16% limit afterward. In preparing the starter permeate, a partly delactosed whey is soured by *Lactobacillus helveticus*; then the liquid is purified by

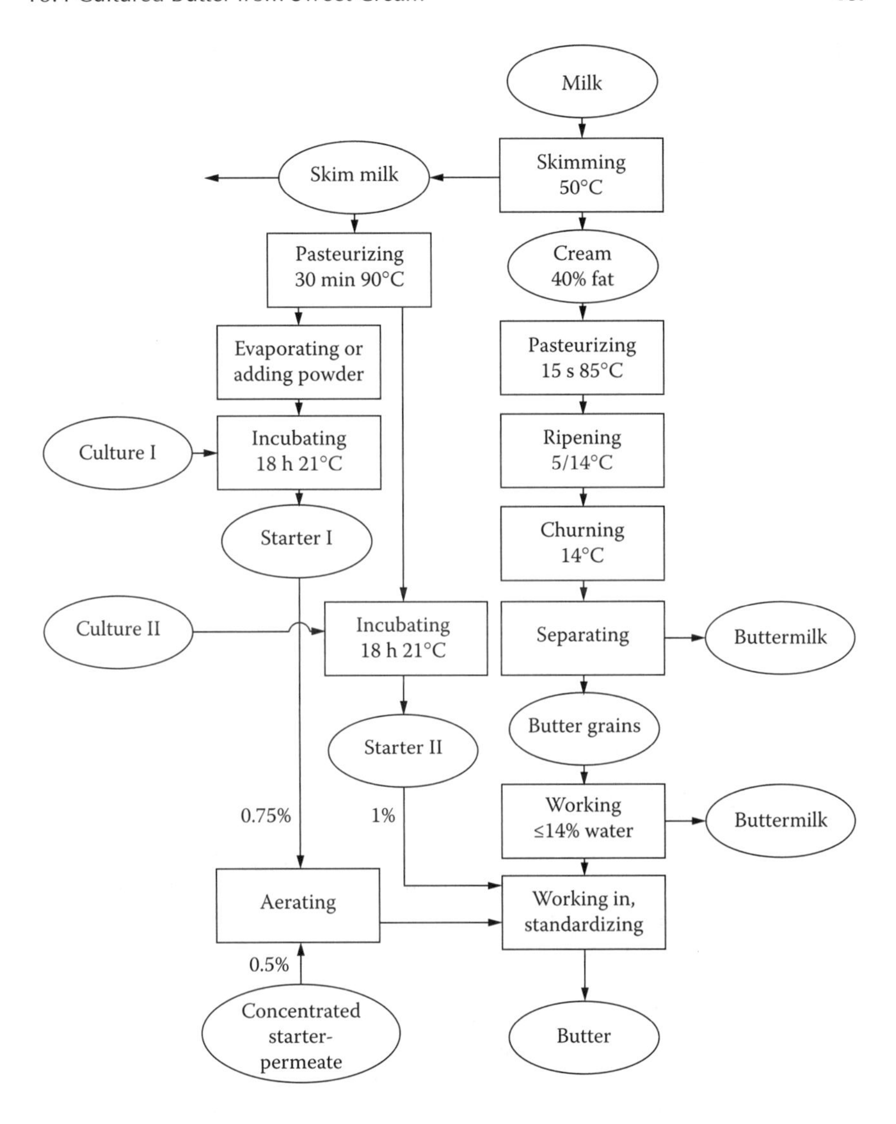

FIGURE 18.16 Example of the manufacture of aromatic butter according to the NIZO method.

ultrafiltration, and the permeate concentrated by evaporation. The lactic acid content of the permeate then is about 16% (acidity 1.8 M).

At first a normal aromatic starter was worked in, apart from the starter permeate. This had the disadvantage that the aroma must develop after the butter making and that the aroma formation depends too much on the conditions; hardly any aroma is produced if the butter is immediately transferred to cold storage, whereas at room temperature the aroma production can be excessive if the butter has a not-very-fine moisture dispersion. Therefore, a very aromatic starter (containing a selected *diacetylactis* strain) is grown in evaporated skim milk (starter I in Figure 18.16); it is mixed with the starter permeate and subsequently aerated. In this way sufficient diacetyl is formed to give the butter a quantity of >1 mg/kg from the very beginning; the bacteria are killed by the high lactic acid content, so that no reduction of diacetyl can occur. Starter II contributes some other flavor compounds and can cause a continued formation of diacetyl; the concentration of diacetyl in the butter eventually is between 1.5 and 2.5 mg/kg.

It is obvious that moisture droplets will occur in the butter that differ in composition: sweet buttermilk, starter, and starter with starter permeate. Some compounds can migrate, e.g., water and lactic acid (both slowly), and especially diacetyl, which is fairly fat soluble (the water/oil partition coefficient approximately equals unity).

The processing is rather more complicated than for traditional butter making, but that need not be a problem in large centralized plants. Moreover, the following may be important differences compared to traditional butter making:

1. Sweet-cream buttermilk is obtained (originally the main purpose of the process).
2. Less starter is needed.
3. The desired starters are very phage sensitive (especially starter I). Contamination by bacteriophages should therefore be prevented.
4. The solids-not-fat content of the butter increases slightly, and hence the yield.
5. The aromatic flavor of the cultured butter is more pronounced.
6. The copper content of the butter can be much lower (see Subsection 18.3.3, item 4). Not surprisingly, the butter is very stable to autoxidation.
7. The quantity of free fatty acids in the butter can be lower. In the cream a partition equilibrium exists between free fatty acids in the fat and in the plasma. The lower the pH, the less the dissociation of the fatty acids (p$K_a \approx 4.8$); hence, the higher their solubility in the fat and the lower in water. Moreover, it appears that the acidity of the fat is somewhat increased by churning at low pH (explanation unknown). Hence, butter from sour cream will contain a larger amount of free fatty acids and will more readily attain a soapy-rancid flavor.
8. Because during cream ripening the souring need not be considered, one is more free to select the best temperature treatment to control butter consistency. In general, there are more degrees of freedom in the method of processing.
9. The sweet cream can far more easily be pumped and passed through a heat exchanger.

Apart from item 3, these are all advantageous to the manufacturer.

18.5 HIGH-FAT PRODUCTS

Besides butter, there are other high-fat products in which fat is the continuous phase. Such products are made for various reasons.

Traditionally, butter was melted down to increase its keeping quality, i.e., after heating the butter, the formed butter oil was separated. The rendered butter so obtained had a long shelf life. Nowadays, this product is designated anhydrous milk fat; it may be used as such in the kitchen because, contrary to butter, it allows heating at a high temperature. In a country such as India where the temperature may be too high for butter making, *ghee* is made from buffaloes' milk.

Separated fat can be modified and/or fractionated in various ways. The main aim may be to change the crystallization behavior of the fat. The modified fat can be used in recombined butter, in chocolate (as a partial substitute for cocoa butter), in bakery products (such processes as kneading of dough and paste, and beating in of air require a constant melting behavior and often a high final melting point), in recombined cheese (restricting oiling off), or in instant milk powder (liquid fat yields better instant properties).

From milk fat and skim milk, *recombined butter* can be made. The purpose may be to enhance the value of poor-quality cream or to obtain a product with other properties, such as different firmness (spreadability), a higher water content, or a higher content of polyunsaturated fatty acids. For example, several kinds of spread are made, especially low-fat spreads. To achieve this, blends of milk fat and vegetable oil can be used. Alternatively, a vegetable oil (for example, 25 to 30% refined soybean oil) can be worked into butter to obtain a spreadable mixture, such as the Swedish Bregott.

Some processes and products are now briefly discussed.

18.5.1 Anhydrous Milk Fat

The most important general requirement is that the fat be very pure, and stable to autoxidation. To secure this, good-quality fresh milk should be used. Contamination by traces of copper is highly detrimental to quality. The water content should not exceed 0.1%, because otherwise moisture droplets may form at low temperature. If the water content is higher (up to 0.4%), the product is usually designated 'butter oil.'

There are numerous manufacturing processes. As a rule, one starts from butter. Alternatively, one can make high-fat cream (by centrifuging twice) and accomplish a phase inversion in it: If a very concentrated oil-in-water (o/w) emulsion is destabilized, a water-in-oil (w/o) emulsion is usually formed. To achieve this, high-fat cream can be passed through an agitator, a special pump, or even a homogenizer; often, the phase inversion is facilitated by first subjecting the cream to washing, i.e., dilution with water and reseparation. If cream with 82% fat is passed through a scraped-surface heat exchanger while being cooled sufficiently for fat crystallization to occur, then butter is formed. When made from sweet cream, the butter can be very durable.

In the manufacture of anhydrous milk fat, the cream destabilization usually occurs at high temperature, and a butter oil containing plasma droplets is obtained; often, a more or less complete separation into two layers immediately sets in. This can also be achieved by melting of butter. Subsequently, a separation has to take place by decanting or, in common practice, by centrifuging; for this purpose a special separator is needed. (Generally the plasma is once again passed through a normal cream separator, yielding a little cream.) The fat obtained in this way is very pure; if the temperature during the melting and separating was not too high, it will be almost free from polar lipids.

Another method of working is based on evaporation of water by heat treatment from butter or fat cream, or from an intermediate product, for example, butter grains; alternatively, washed cream can be used. When the evaporation is done in an open vat, that is, at atmospheric pressure, the temperature becomes high, up to 120°C, and the product so obtained is similar to ghee. Furthermore, cream or butter can be dried in a vacuum evaporator or spray drier. In all cases, the nonfat solids are left dispersed in the fat. These can be removed by decanting, filtering, or centrifugation.

In some tropical regions, where the temperature is too high to make butter without refrigeration, *ghee* was and is produced, especially in India. Traditionally, it was made from cream of buffaloes' milk. This milk has fairly large fat globules, which do not exhibit cold agglutination, but cream sufficiently fast at high temperature. During creaming, the product turns sour. Traditionally, the cream was heated over an open fire until all the water was boiled off; precipitated solids were removed by decantation. In this way a product with a typical flavor and a soft and somewhat grainy texture was obtained, with a reasonable shelf life. Currently, ghee is produced by a variety of methods, from cows' or buffaloes' milk. Cream can be obtained by centrifugal separation, possibly followed by starter addition and fermentation; butter is also used for ghee manufacture. Heating to remove water can be done in various apparatuses, and solids are often removed by filtration.

Milk fat made by one of the mentioned processes usually contains about 0.4% water. On cooling, droplets are formed (the solubility of water in milk fat is 0.1, 0.2, and 0.4% at 10, 40, and 90°C, respectively), and the product can spoil rapidly. Therefore, vacuum drying is generally applied, for example, at 40°C and 2 kPa (= 0.02 bar). It causes a decrease of the water content to below 0.1%. Also, the oxygen content decreases significantly. A product made in this way may keep for some years if made from milk without any incipient autoxidation, if stored in isolation from air and light, and if copper contamination has been rigorously prevented.

18.5.2 Modification of Milk Fat

The most widely used modification is *fractionation* by means of crystallization. After solidification of milk fat at a certain temperature in such a way as to form fairly large crystals, the fat can mechanically be separated into a solid and a liquid

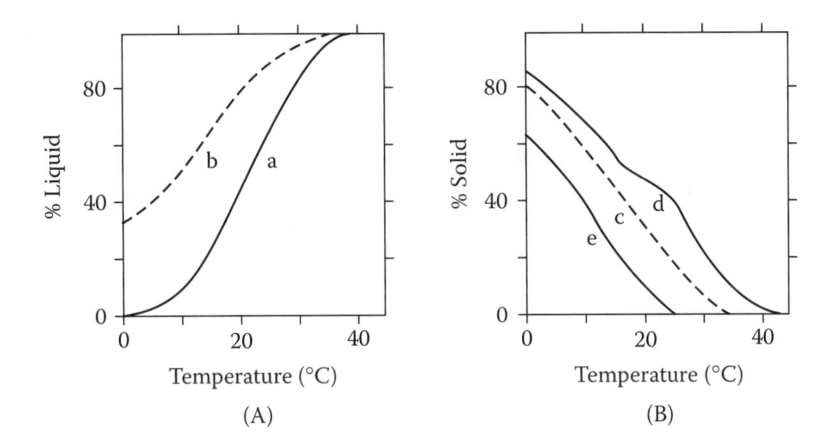

FIGURE 18.17 Examples of fractionation of milk fat. (A) Amount of liquid fat that was separated as a function of the solidification temperature applied (curve a), in comparison with the amount of liquid fat actually present in the fat (curve b). (B) The percentage of solid fat as a function of the melting temperature applied. The solid contents of the 'solid' (curve d) and the 'liquid' (curve e) fractions, obtained by fractionation at 25°C, are compared with the original fat (curve c).

portion. The purpose is to obtain fractions with differing melting behavior. The composition is also altered in another respect because fat-soluble components, such as carotenoids, vitamins, and flavor compounds, become concentrated in the liquid fraction. Accordingly, it remains to be seen as to whether a solid fraction may still be considered milk fat.

The success of a single fractionation is less than expected. The separation is incomplete (see Figure 18.17A) because the network of fairly small crystals readily retains liquid fat. To be sure, one can attempt to form large crystals by cooling the fat very slowly. This may cause formation of spherulites. Spherulites are sphere-shaped crystals, but they are made up of a great number of ramified radial needles, between which, again, liquid fat is held. A far better fractionation can be achieved by crystallization of the fat from acetone, but this is an expensive method; moreover, use of the product obtained in foods may not be allowed.

Furthermore, the difference in melting curves of the various fractions is disappointing (Figure 18.17B). This must be ascribed at least partly to the strong tendency of milk fat to form compound crystals. Fractionation in steps, combined with optimization of the separation method, can yield much better results, although it takes a long time (a few days). Milk fat fractions are frequently used in practice, especially to make butter more spreadable.

Milk fat can also be modified chemically, but the products can no longer be called milk fat. *Hydrogenation* (by using H_2 and a catalyst at high temperature) decreases the number of double bonds and thereby increases the high-melting proportion of a fat; this is why the process is often called hardening.

It also causes several other changes, such as displacement of remaining double bonds and cis-trans isomerization. Hydrogenation can be applied to make cocoa butter replacers from milk fat. *Interesterification* causes the distribution of the fatty acid residues over the positions of the triglyceride molecules to become increasingly random. It can be achieved by heating the fat in the presence of a catalyst, such as sodium methoxide. (At very high temperatures, say, 150°C, it also occurs without a catalyst being present.) After interesterification, the melting range of milk fat is shifted to higher temperatures. Similar effects occur on cis-trans isomerization. None of these chemical modifications is applied to milk fat on a commercial scale.

18.5.3 Recombined Butter

Milk fat and skim milk can be recombined to yield a butter-like product. The manufacturing process used for and physical structure of the product are then practically identical to those of margarine. The difference lies is in the composition.

A disadvantage is that recombined butter is firmer than natural butter of the same fat composition; the texture is also slightly different. This is largely due to the product not containing fat globules, so that all fat crystals can participate in networks and solid structures. Accordingly, the firmness has to be adjusted by a different method, generally via the fat composition. The final melting point of the fat should be below body temperature because the butter should fully melt in the mouth. If temperature fluctuates widely, a coarse texture readily develops, due to the fat crystals becoming very large.

A simplified outline of the manufacturing process is given in Figure 18.18. It is largely self-explanatory. In the production of margarine, the fat to be used is subjected to various treatments, including degumming, alkali refining, bleaching, deodorizing, and partial hydrogenation.

The treatment of the aqueous phase mainly serves to impart a good flavor and to improve the keeping quality (preventing spoilage by microorganisms, lipase, etc.). Usually, in the production of recombined butter, hardly any additives are used, whereas margarine may contain an antioxidant with or without synergist, coloring matter, added vitamins A and D, fat-soluble flavoring agents, and an emulsifier.

Emulsification serves to obtain a fairly homogeneous mixture of constant composition. The w/o emulsion is not stable, and the liquid must be agitated at all times until fat crystals have been formed. The crystals stabilize the moisture droplets by adsorbing onto the interface (Pickering stabilization). The emulsifiers (in margarine, e.g., monoglycerides and soy bean lecithin) serve other purposes. Monoglycerides seem to prevent the fat crystals from flocculating so strongly into a network that they can insufficiently adsorb onto the moisture droplets. Moreover, the emulsifier plays a twofold role during the heating of the product in a frying pan. Due to melting of the fat, the moisture droplets become unstable, and if they now quickly flow together into large droplets, these can start to splash

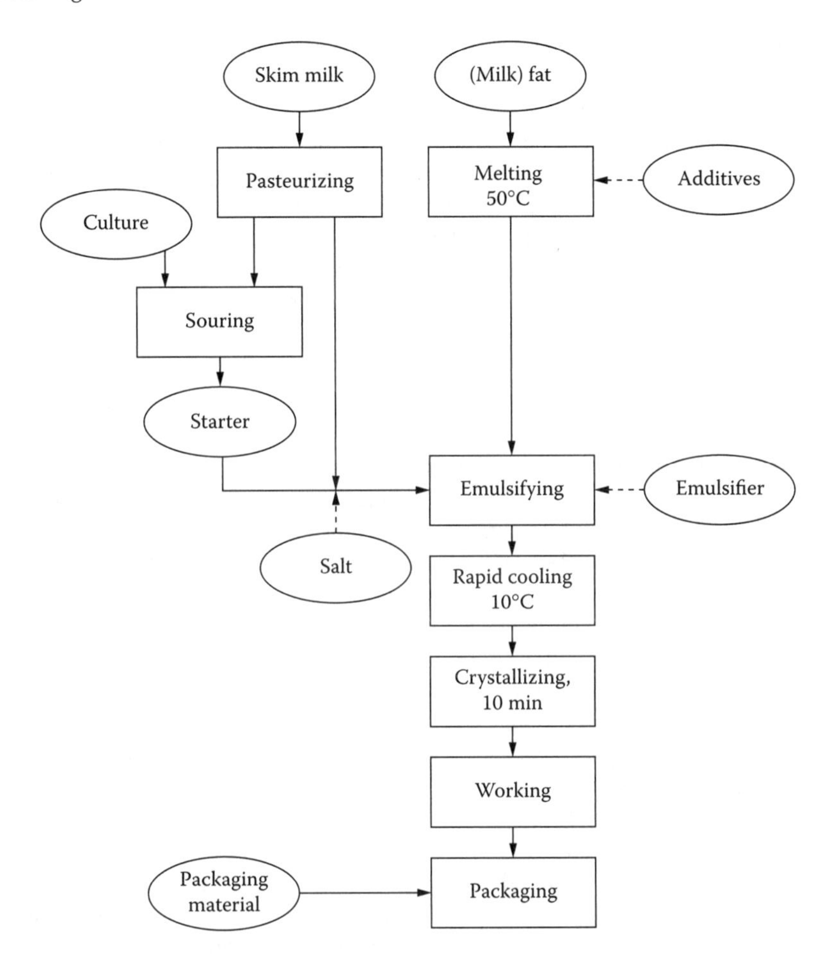

FIGURE 18.18 Example of the manufacture of recombined butter (or margarine).

very annoyingly when reaching the boiling point. The emulsifier slows down such coalescence. Furthermore, lecithin contributes to the typical aroma during frying. Therefore, in the case of recombined butter, part of the skim milk is often replaced by sweet-cream buttermilk; in this way the composition also becomes closer to that of natural butter.

The cooling is usually performed in a scraped-surface heat exchanger; otherwise the heat transfer would proceed far too slowly, causing the formation of overly large crystals. Due to the scraping and stirring, the moisture droplets, which during emulsification may still be up to 1 mm in size, are disrupted into much smaller ones. A further reduction occurs by working, which consists of forcing the mass through a small opening (valve) or a perforated plate (see Subsection 18.2.3).

The working also serves to destroy solid structures of fat crystals. These structures should by then already have been formed, which means that by far most of the crystallization should already have taken place. This is achieved by passage through a crystallization tube (B unit) before the working. However, milk fat crystallizes very slowly and, accordingly, it is generally impossible to achieve a fairly complete crystallization in one processing step (as is common in margarine manufacture). Consequently, the butter can still set considerably after manufacture. During crystallization, cooling often has to be done again. If, for instance, 20% of the fat crystallizes, then the heat of crystallization suffices to increase the blend temperature by 8°C.

18.5.4 Low-Fat Butter Products

Low-fat butter products have long been in existence (from the early 1940s), but they have received renewed interest following the success of low-fat margarine. These 'spreads' have a water content of, say, 40%. There are also spreads that consist partly of a vegetable fat and that are held to be better for health than butter by some people.

Manufacture of low-fat butter cannot be achieved by churning, because the formation of a product with a discontinuous aqueous phase fails. Most surface-active substances in cream are water soluble, and this largely prevents the formation of a w/o emulsion. (Only at a very high fat content and in the presence of fat crystals is a kind of phase inversion from cream to butter possible. Melting away of the crystals destabilizes any formed w/o emulsion and results in a continuous aqueous layer.)

In the production of the butter, a manufacturing process comparable to that depicted in Figure 18.18 should be applied, and a suitable (that is, oil-soluble) emulsifier should be added. It may be a problem to make the moisture droplets sufficiently small and to prevent them from coalescing in the product, especially at a somewhat higher temperature. Therefore, a gelling agent (such as gelatin) may be added. It causes the droplets to become solid-like, unable to coalesce. At the least, a thickening agent should be added to the aqueous phase. This may be a protein mixture, such as serum protein that has been coagulated to yield aggregates of about 1 μm in size. The agent causes the flavor and mouthfeel of the product to be more similar to butter. Flavor substances as well as aroma-forming starter bacteria may be added. All the same, there is no common appreciation of the flavor of such a 'butter.' One of the aspects involved may be that the product contains less crystalline fat than butter and therefore feels less cool in the mouth.

Spreadability is generally not a problem. However, if the butter product has a relatively high content of liquid fat, it may become too soft, and at room temperature it may not retain its shape, exhibiting oiling off. A product with a small number of fat crystals may well have large moisture droplets, which can readily cause microbial spoilage (Subsection 18.2.3). A preservative can then be added.

An alternative technique for the manufacture of low-fat butter is to incorporate a fairly viscous aqueous compound (such as, a pasteurized caseinate solution) into natural butter. Some authors say that in this way a bicontinuous system is formed. In any case, a preservative is needed, and this is certainly true for spreads of the o/w type, essentially fat cream that has been transformed into a spreadable product by the use of thickening and gelling agents. Naturally, the latter products behave differently from natural butter. For example, the included moisture will migrate into the bread onto which it is spread. Obviously, making a good-quality low-fat butter is far from easy.

Suggested Literature

There is very little recent literature about the manufacture and properties of butter. General and comprehensive information: H. Pointurier and J. Adda, *Beurrerie Industrielle: Science et Technique de la Fabrication du Beurre,* La Maison Rustique, Paris, 1969.

Physicochemical aspects of churning, butter, and anhydrous milk fat: H. Mulder and P. Walstra, *The Milk Fat Globule,* Pudoc, Wageningen, 1974.

Additional information: Chapter 4 (Physical chemistry of milk fat globules) and Chapter 5 (Crystallization and rheological properties of milk fat) of P.F. Fox, Ed., *Advanced Dairy Chemistry,* Vol. 2, *Lipids,* 2nd ed., Chapman and Hall, London, 1995.

19 Concentrated Milks

Concentrated milks are liquid milk preserves with a considerably reduced water content. Water is removed by evaporation. Preservation is achieved either by sterilization, leading to a product called *evaporated milk*, or by creating conditions that do not allow growth of microorganisms. The latter is generally realized by addition of a large quantity of sucrose and exclusion of oxygen. The resulting product is called *(sweetened) condensed milk.*

These products were initially meant for use in (usually tropical) regions where milk was hardly or not available. The milks were packaged in small cans. The contents were often diluted with water before consumption to resemble plain milk. Currently, alternative products are used more often, such as whole milk powder or recombined milk. For the concentrated milks, some alternative forms of use developed, and processing and packaging have been modified. The consumption of sweetened condensed milk has greatly declined.

19.1 EVAPORATED MILK

Evaporated milk is sterilized, concentrated, homogenized milk. The product can be kept without refrigeration and has a long shelf life; it is completely safe for the user. After dilution, flavor and nutritive value of the product are not greatly different from that of fresh milk. A major problem with sterilization is the heat stability; the higher the concentration of the milk, the lower its stability. That is why concentrating cannot be by more than about 2.6 times, which corresponds to a level of about 22% solids-not-fat in the evaporated milk.

Currently, bottled evaporated milk is often used in coffee in certain countries. It can be added while cold because a fairly small amount is involved as compared to nonevaporated milk. After the bottle has been opened, the milk can be kept in a refrigerator for up to 10 days because it initially contains no bacteria at all and because contaminating bacteria grow somewhat more slowly owing to the reduced water activity, which is about 0.98.

Table 19.1 gives the composition of some kinds of evaporated milk and skim milk.

19.1.1 Manufacture

Figure 19.1 outlines manufacturing processes of in-bottle- and UHT-sterilized evaporated whole milk. Several variations are possible. Some process steps are discussed in more detail in the following text.

TABLE 19.1
Approximate Composition of Some Kinds of Evaporated Milk

Type	Fat (%)	Solids-Not-Fat (%)	Concentration Factor
Evaporated milk, American standard	7.8	18.1	2.1
Evaporated milk, British standard	9	22	2.6
Low-fat evaporated milk	4	20	2.25
Evaporated skim milk	0.1	22	2.35

Preheating serves to enhance the heat stability of the evaporated milk, inactivate enzymes, and kill microorganisms, including a significant proportion of the bacterial spores present. The heating temperature–time relationship is usually selected on the basis of heat stability. Formerly, a long heat treatment (e.g., 20 min) at a temperature below 100°C was often applied. Currently, UHT treatment is generally preferred. It reduces the number of spores in the milk considerably, and therefore a less intensive sterilization suffices.

Concentrating. The milk is usually concentrated by evaporation (see Section 10.2). Standardization to a desired dry-matter content is of much concern. A higher concentration causes a lower yield and a poorer heat stability. Continuous standardization is usually applied via determination of the mass density. Based on that parameter, either the raw milk supply or the steam supply is adjusted; it is obvious that density and dry-matter content of the raw milk must be known. Alternatively, standardization can be based on refractive index determination. The milk can also be concentrated by reverse osmosis, but this is rarely done.

After concentrating, the manufacturing processes for in-bottle-sterilized and UHT-sterilized evaporated milk differ. The former process is discussed first.

Homogenization serves to prevent creaming and coalescence. It should not be too intensive because the heat stability becomes too low.

Stabilization. To ensure that the evaporated, homogenized milk does not coagulate during sterilization and at the same time does acquire a desirable viscosity, a series of sterilization tests is often done on small quantities of the evaporated milk to which varying amounts of a stabilizing salt (for the most part, Na_2HPO_4) are added. The tests are needed because variation occurs among batches of milk. Essentially, the addition of the salt means adjusting the pH (see Subsection 19.1.3). Because further processing must be postponed until the test results are available, this necessitates cooling the evaporated milk after its homogenization and storing it for a while. However, long-term storage should be avoided to prevent bacterial growth; moreover, cold storage of the milk increases the tendency of age thickening (see Subsection 19.1.5). The stabilizing salt is added as an aqueous solution, which dilutes the evaporated milk slightly. Therefore, the milk is often concentrated somewhat too far and is restandardized to the correct dry-matter content during stabilization.

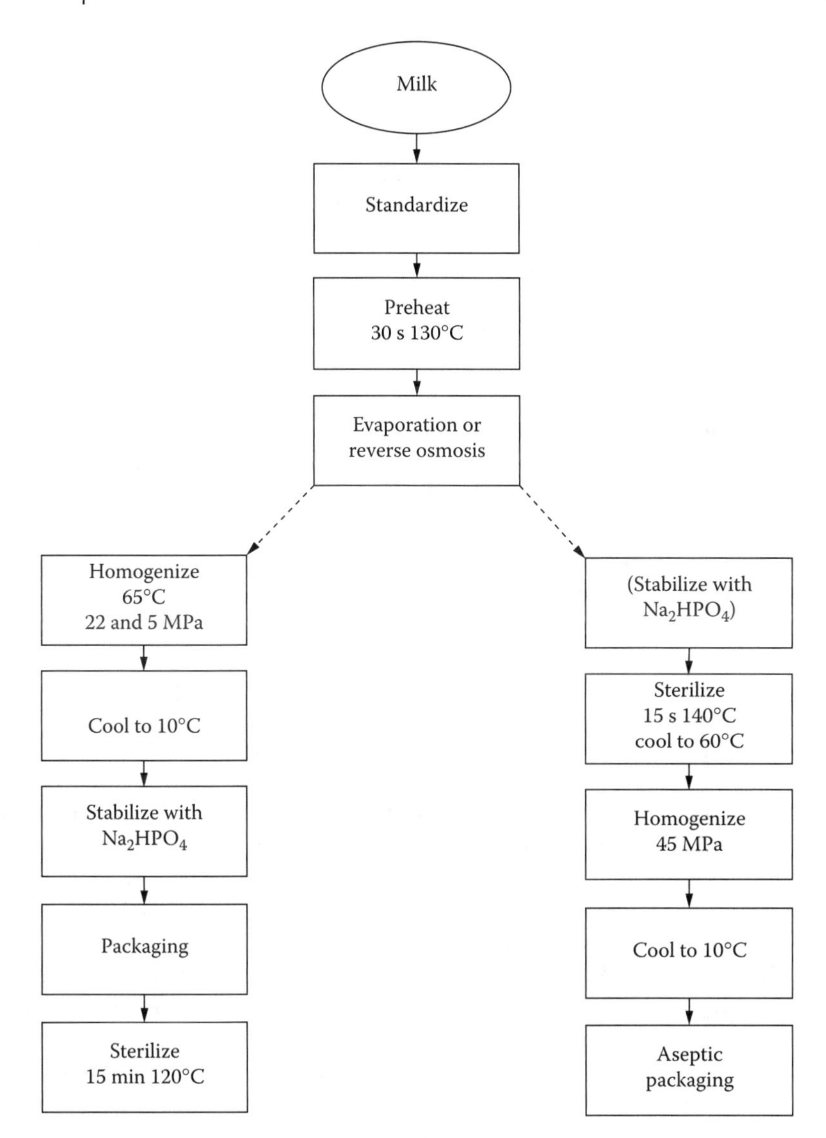

FIGURE 19.1 Examples of the manufacture of in-bottle-sterilized (left) and UHT-sterilized (right) evaporated whole milk.

Packaging in cans is common. The tin plate of the cans is coated (provided with a protective layer of a suitable polymer) to prevent iron and tin from dissolving in the product. After filling, the cans may be closed by soldering, but mechanical sealing is currently preferred. Evaporated milk intended for use in coffee is usually packaged in bottles that are closed with a crown cork or a screw cap.

Sterilization. In-bottle or in-can sterilization can be applied batchwise (in an autoclave) or continuously. Machines that have rotary air locks (to maintain the pressure) may be applied for cans and hydrostatic sterilizers for bottles (see Figure 7.21).

The sterilization is primarily aimed at killing all bacterial spores — reduction to, say, 10^{-8} spores/ml — and inactivating plasmin. Lipases and proteinases from psychrotrophs should be absent from the raw milk because these enzymes would be insufficiently inactivated. The most heat-resistant spores are those from *Bacillus stearothermophilus*. This bacterium does not grow at moderate temperatures but may do so in the tropics. At 121°C, the D value of the spores is some 4 to 7 min. The preheating as given in Figure 19.1 suffices for a sterilizing effect S almost equal to 1, whereas the sterilization gives S almost equal to 3 at most and, hence, added together giving S less than or equal to 4. Contamination by these spores should therefore be slight, and growth of the organism occurring in the evaporator, possibly followed by sporulation (e.g., during intermediate cold storage), should rigorously be avoided (see also Section 20.3). If the sterilizing effect is adequate for *B. stearothermophilus*, then *B. subtilis*, *Clostridium botulinum*, and *C. perfringens* are also absent (see Table 7.4).

UHT sterilization kills bacterial spores more effectively than in-bottle sterilization. The combination of preheating and UHT treatment of the concentrate as shown in Figure 19.1 suffices to inactivate plasmin. Preheating is also required to prevent excessive heat coagulation in and fouling of the UHT sterilizer. Some heat coagulation nearly always occurs, and the subsequent homogenization also serves to reduce the size of the protein aggregates formed. Aseptic homogenization must be applied. Indirect UHT sterilization in a tubular heat exchanger allows the pump of the homogenizer to be fitted before the heater and the homogenizing valve behind it. Thereby, the risk of recontamination with bacteria is diminished. The addition of stabilizing salt can often be omitted if UHT sterilization is applied, or the amount to be added is not so critical that sterilization tests must be carried out. It implies that the whole process from preheating up to and including aseptic packaging can proceed without interruption. Aseptic packaging and suitable packaging materials are discussed in Chapter 15.

Recombination. Manufacture of evaporated milk by means of recombination is briefly outlined in Figure 19.2. The skim milk powder used has to comply with strict requirements. The powder must have been made from skim milk that is heated so intensely (e.g., for 1 min at 130°C) that the recombined concentrated milk after its homogenization is sufficiently heat stable. The count of *B. stearothermophilus* spores should be so low that a moderate sterilization of the evaporated milk suffices. Sometimes, up to 10% of the skim milk powder is replaced by sweet-cream buttermilk powder to improve the flavor of the product. The copper and peroxide contents of the anhydrous milk fat should be low to avoid flavor deterioration. A high level of calcium in the water used can cause problems with heat stability. 'Filled evaporated milk' is also made. A fat different from milk fat is used.

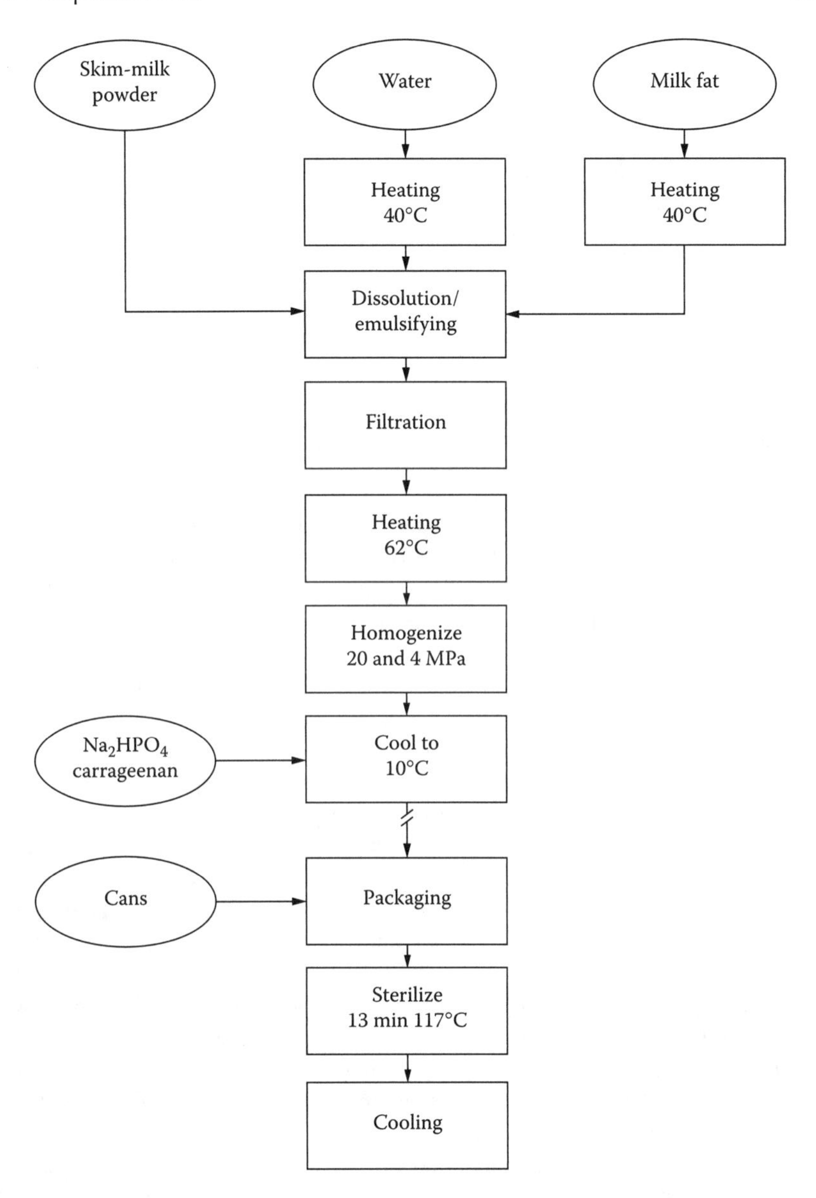

FIGURE 19.2 Example of the manufacture of recombined evaporated milk.

19.1.2 Product Properties

Maillard reactions are of paramount importance for the *flavor* and *color* of evaporated milk. Obviously, temperature and duration of the heat treatment during manufacture determine the initial concentration of the reaction products, but

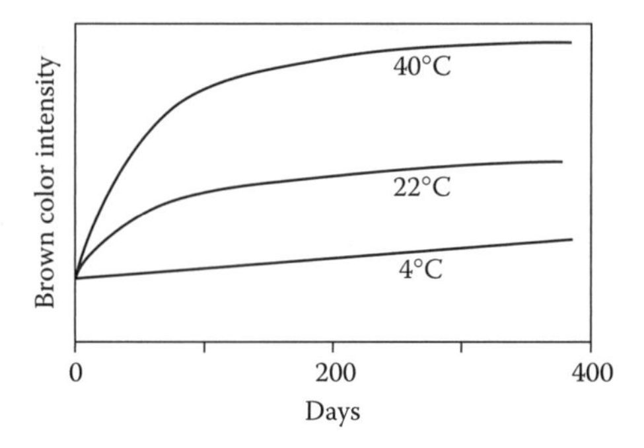

FIGURE 19.3 Browning (arbitrary units) during storage of evaporated milk at various temperatures. (Adapted from S. Patton, *J. Dairy Sci.*, **35**, 1053, 1952.)

ongoing Maillard reactions occur during storage, especially at a high temperature (see Figure 19.3). The milk eventually develops a stale flavor also due to Maillard reactions. The flavor after a long storage time differs considerably from that directly after intense heating. This is because the complicated set of reactions involved leads to different reaction products at different temperatures. A sterilized-milk flavor may be appreciated by some people when the milk is used in coffee. Off-flavors due to autoxidation need not occur.

When the milk is used in coffee, the brown color is often desirable to prevent the coffee from acquiring a grayish hue. The brown color depends greatly on the Maillard reactions, although the color of the fat plays a part.

The *viscosity* of evaporated milk is often considered an important quality mark. Many consumers prefer the milk to be viscous. This can be achieved by sterilization in such a way that visible heat coagulation is barely prevented. UHT-evaporated milk is always less viscous and, therefore, κ-carrageenan is often added (see Subsection 19.1.4).

If the original milk contains bacterial lipases and proteinases due to growth of psychrotrophs, these *enzymes* may remain active in the evaporated milk and lead to strong deterioration, i.e., soapy-rancid and bitter flavors, and to age thinning. Evaporated skim milk may even become more or less transparent due to proteinase activity.

The *nutritional value* of evaporated milk can be significantly decreased as compared to that of plain milk. In-container sterilization can destroy up to 10% of the available lysine, about half of the vitamins B_1, B_{12}, and C, and smaller propertions of vitamin B_6 and folic acid. All of these changes are far smaller when UHT heating is applied.

19.1.3 Heat Stability

The mechanisms of heat coagulation of milk and the factors affecting the heat coagulation time (HCT) are discussed in Subsection 7.2.4. As mentioned, concentrated

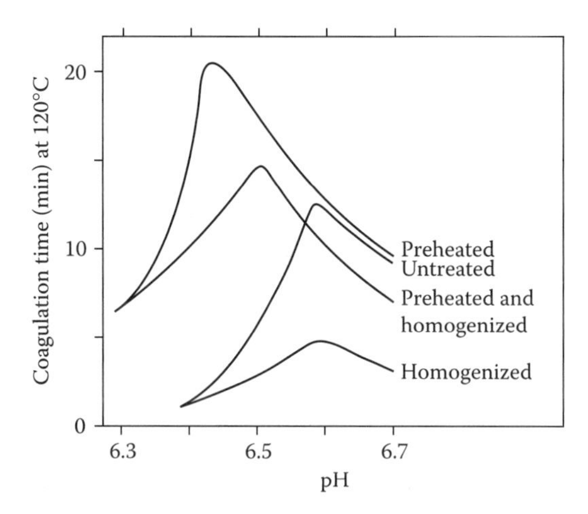

FIGURE 19.4 Influence of preheating and homogenization on the heat stability of evaporated milk (British standard) as a function of its pH (measured at room temperature before sterilization). Approximate examples.

milk is far less stable during sterilization than nonevaporated milk, and the fairly intensive homogenization applied decreases the heat stability further. Moreover, evaporated milk should increase in viscosity during sterilization. Essentially, the viscosity increases by incipient coagulation. A subtle process optimalization is needed to meet these requirements.

In any case, the milk must be preheated before evaporation in such a way that most serum proteins are denatured (see Figure 7.9E). Otherwise, the evaporated milk forms a gel during sterilization due to its high concentration of serum proteins. Preheating is, for example, for 3 min at 120°C. Figure 19.4 shows the significant effect of preheating.

The pH should always be adjusted. Preheating and evaporation have lowered the pH to about 6.2 (American standard) or 6.1 (British standard), and that is far below the optimum pH. In practice, $Na_2HPO_4 \cdot 12H_2O$ is usually added, but NaOH can also be used.

The effect of homogenization is also illustrated in Figure 19.4 (see also Section 9.6). The influence of some variables such as the homogenization temperature differs from that in sterilized cream (Subsection 17.1.2). Homogenization of evaporated milk does not lead to formation of homogenization clusters. It is often observed that a slight homogenization increases HCT (Figure 19.5), which cannot be easily explained.

Figure 19.5 summarizes the approximate effect of some variables on the heat stability. Clearly, UHT heating of evaporated milk after homogenization is not possible. Even traditional sterilization is difficult if the milk is highly concentrated or if the evaporated milk is intensely homogenized. There are some other factors affecting heat stability. It can be improved by lowering the calcium content of

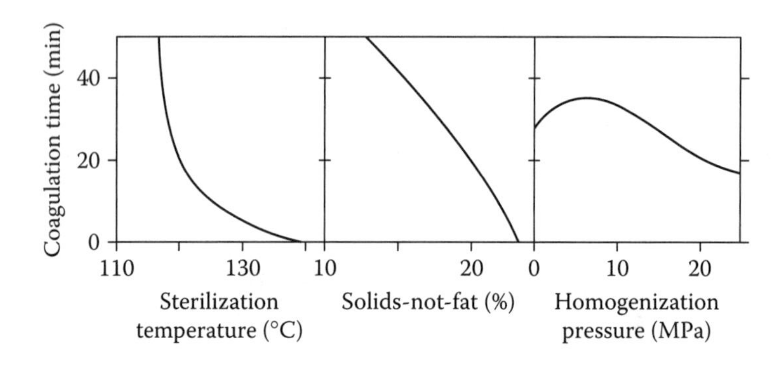

FIGURE 19.5 Heat stability measured at 120°C (unless otherwise stated) and at the optimum pH of preheated homogenized evaporated milk. Approximate examples of the influence of some variables.

the milk before evaporation by means of ion exchange. Addition of 15 mmol H_2O_2 (0.05%) or of about 15 μmol Cu^{2+} (from 0.5 to 1 mg·kg^{-1}) after preheating but before evaporation tends to increase the heat stability.

19.1.4 Creaming

Creaming of evaporated milk eventually leads to formation of a solid cream plug that cannot be redispersed. This may be due to bridging of adjacent fat globules because of 'fusion' of the fragments of casein micelles in their surface layers. Accordingly, intensive homogenization is necessary.

The newly formed surface layers during homogenization can be fairly thick. The preheating leaves hardly any dissolved serum proteins, and the evaporation and sterilization steps increase the average diameter of the casein micelles. Especially after homogenization at high pressure and low temperature, the layers may be thick enough for the smaller globules to have a higher density than the plasma; consequently, they settle rather than cream. As a result, the fat content of evaporated milk in both the top and bottom layers of a can that has been stored undisturbed for several months is often found to be higher than that in the middle.

A higher viscosity of the evaporated milk often involves a slower creaming, but the relations are not straightforward. To begin with, it is primarily the viscosity of the plasma phase, not that of the product, that determines creaming rate. Generally, a high viscosity is due to an approaching heat coagulation. The homogenized fat globules tend to participate in this coagulation and hence to form clusters, which will cream rapidly. In fact, measures that counteract the heat coagulation, such as the addition of traces of copper (Subsection 19.1.3), usually lead to a decreased creaming, despite the lower viscosity resulting from such additions. In any case, the creaming in evaporated milk has been insufficiently investigated. The viscosity of the plasma can depend strongly on shear rate, as is illustrated by curve 2 in Figure 4.6 (albeit of an evaporated milk, not its plasma, that had shown age thickening). The values for plasma at the very low shear rates relevant for creaming, for the most part between 0.01 and 0.1 Pa·s, are generally not known. Creaming rates as predicted in Table 9.3 can agree approximately with observed values. In UHT-sterilized evaporated milk, the plasma viscosity is relatively low as long as age thickening is negligible, and creaming tends to be much faster. Hence, κ-carrageenan is often added to decrease creaming rate.

As already stated, homogenization has an adverse effect on the heat stability and, consequently, the homogenization pressure cannot be high. It is especially the largest fat globules that exhibit creaming, and it is therefore advisable to aim at having the relative width (c_s) of the globule size distribution as small as possible. The width is greatly affected by the type of homogenizer used. Two-stage homogenization is often used, but its effect on c_s is negligible (neither is this type of homogenization required to break up homogenization clusters because these are not formed). Homogenizing twice does lead to a lower c_s. Alternatively, if need be, the lightly homogenized evaporated milk may be separated by centrifugation and depleted of the largest fat globules. The cream obtained can be added to the unhomogenized concentrate.

UHT-evaporated milk can be homogenized far more intensely because the sterilization precedes the homogenization. Intense homogenization is also required to prevent excessive creaming because the viscosity of the plasma phase is much lower than in conventional evaporated milk.

19.1.5 Age Thickening and Gelation

When evaporated milk is kept, its viscosity may initially decrease slightly (Figure 19.6). This may be explained in terms of the casein micelle aggregates (formed during sterilization) changing from an irregular to a spherical shape, as a

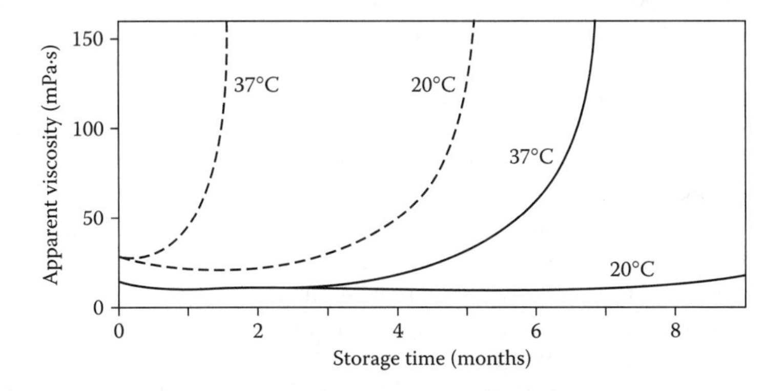

FIGURE 19.6 Age thickening of evaporated milk at two temperatures. Approximate examples for UHT-evaporated milk. Polyphosphate added (—) or not added (----). Viscosities obtained at high shear rate (>10 s^{-1}). Approximate results after various sources.

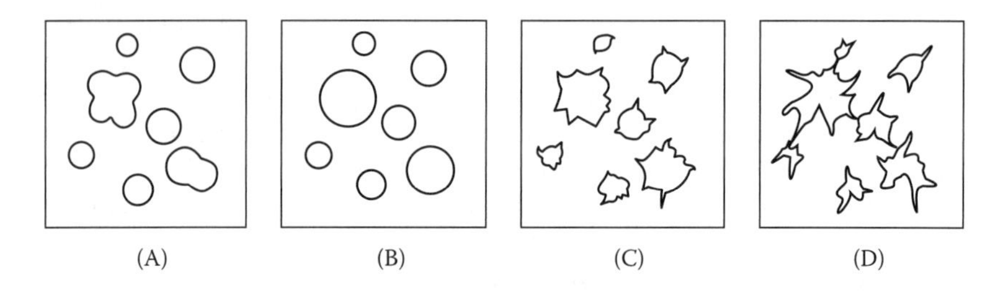

FIGURE 19.7 Schematic picture of the change observed in the casein micelles of evaporated milk during storage. The apparent viscosity is at a minimum in stage B.

result of which the effective volume fraction decreases; see in Figure 19.7 the change from stage A to stage B. Subsequently, the viscosity tends to increase, and it becomes strongly dependent on the shear rate, as illustrated in Chapter 4, Figure 4.6, curve 2. Soon the milk displays a yield stress, and a gel is formed that firms rapidly. The mechanism involved is not quite clear. In most cases, gelation is not caused by proteolytic enzymes, nor are Maillard reactions responsible, although the latter parallel gelation. Moreover, gelation is not related to heat coagulation. For instance, it does not depend significantly on the pH, and its rate increases rather than decreases after lowering of the calcium content. Electron microscopy reveals that thread-like protrusions appear on the casein micelles, which eventually form a network. This is illustrated in Figure 19.7, stages C and D. It is likely that a slow change in the micellar calcium phosphate is at least partly responsible for the changes observed, but a definitive explanation is still lacking.

Age thickening and gelation tend to occur fast in UHT-evaporated milk. It may then be due to proteolysis caused by enzymes released by psychrotrophs, but also if such enzymes are absent, fast age gelation occurs. A more intense sterilization after evaporation delays gelation. Gelation is faster in a more concentrated milk (see Figure 10.10C) and at a higher storage temperature (Figure 19.6). Addition of sodium polyphosphate (about 0.4% in the dry matter) delays gelation considerably; the higher the molar mass of the phosphate, the more effective it is. Addition of citrate or orthophosphate often accelerates gelation, presumably because of binding of calcium. Polyphosphates may be hydrolyzed to yield orthophosphate, especially during heating. Consequently, addition of polyphosphate does not counteract gelation of in-bottle-sterilized evaporated milk, to the contrary.

Conventional evaporated milk only gels if kept for a long time at a high temperature (as in the tropics). Extensive Maillard reactions also occur. Rapid gelation can occur, however, if the evaporated milk before its sterilization is kept refrigerated at, say, 4°C for a few days.

All in all, adequate measures can be taken to delay gelation of the evaporated milk for a considerable time. Gelation can be examined by suspending a can of evaporated milk on a torsion wire and checking whether the milk has elastic properties. If so, the can will remain oscillating for a while when it is given a turn and then released.

19.2 SWEETENED CONDENSED MILK

Sweetened condensed milk is milk that is concentrated by evaporation, to which sucrose is added to form an almost saturated sugar solution, after which it is canned. The high sugar concentration is primarily responsible for the keeping quality of the product and for its fairly long shelf life, even after the can has been opened, although it then will eventually become moldy.

Table 19.2 gives compositions of two kinds of sweetened condensed milk. Sweetened condensed skim milk is also made. The milk is highly concentrated: the mass concentration ratio, Q, ranges from 4.6 to 5, and the increase in concentration relative to water, Q^*, is 7.3 to 8.5. Because of this and the high sugar content, the product is highly viscous, i.e., η_a is approximately 2 Pa·sec, about 1000 times the viscosity of milk. The product is somewhat glassy in appearance because the fat globules show little light scattering as the refractive index of the continuous phase is almost equal to that of fat. The turbidity of the product is largely due to lactose crystals. Most of the lactose crystallizes because of its supersaturation (see Subsection 19.2.2).

19.2.1 MANUFACTURE

Figure 19.8 is a flow diagram of a typical manufacturing process for sweetened condensed milk. The process steps are briefly discussed in the following text:

Heating: Pathogens and potential spoilage organisms must be killed. Among the enzymes, milk lipase should primarily be inactivated; bacterial lipases are not inactivated and, if present, can cause severe rancidity. Deterioration caused by proteinases has not been reported. The heating intensity considerably affects viscosity and also age thickening and gelation of the product, so the actual heat treatment must be adjusted to these properties. UHT heating at about 130 to 140°C is commonly applied.

TABLE 19.2
Approximate Compositions of Two Kinds of Sweetened Condensed Milk

	American Standard	British Standard
Fat content (%)	8	9
Milk solids-not-fat (%)	20	22
Lactose (%)	10.3	11.4
Sucrose (%)	45	43.5
Water (%)	27	25.5
Lactose/100 g water (g)	38.3	44.6
Sucrose/100 g water (g)	167	171
Concentration factor Q	4.60	5.00

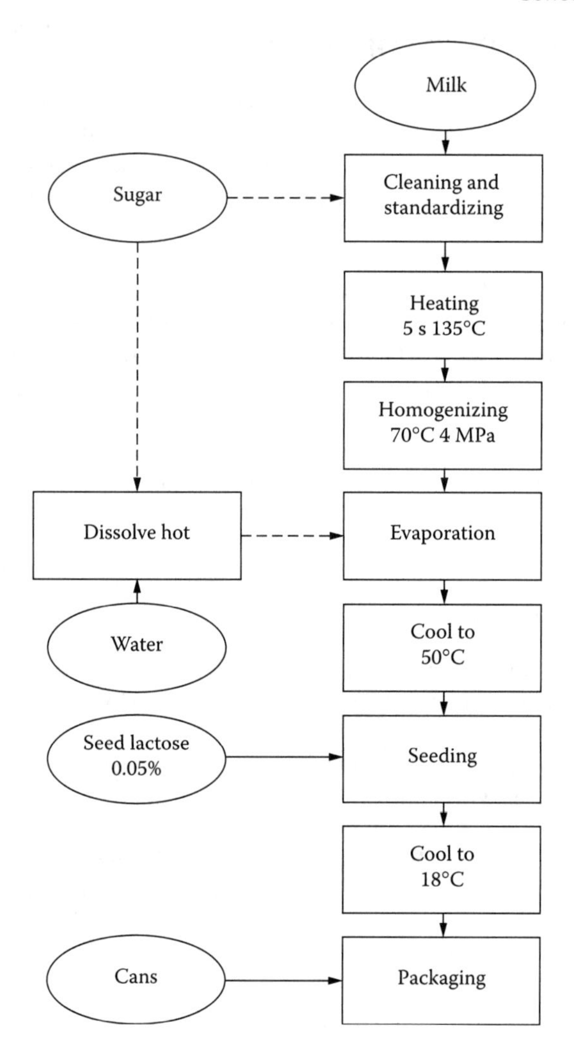

FIGURE 19.8 Example of the manufacture of sweetened condensed milk.

Homogenization: Creaming is often not a major problem, and therefore homogenization is not always done. Currently, however, sweetened condensed milk is made less viscous (and exhibits less thickening) than previously. The density difference between fat globules and continuous phase is large, over 400 $kg \cdot m^{-3}$; for a viscosity of the continuous phase of 1 Pa·s, the creaming rate would be about 1% of the fat per day. This is too high, so that homogenization is often done, although at low pressure, i.e., 2 to 6 MPa.

Sugar: This can simply be added to the original milk. The amount added can be adjusted readily, and the sugar is pasteurized along with the milk.

However, this procedure causes fairly extensive Maillard reactions during heating and evaporation, and above all, a faster age thickening. Alternatively, a concentrated sugar solution, which should be sufficiently heat-treated to kill any osmophilic yeasts, is added at the end of the evaporation step. The sugar should be refined and be devoid of invert sugar to prevent excessive Maillard reactions.

Concentration: This is usually done by evaporation. A falling-film evaporator is generally used to remove the bulk of the water and a circulation evaporator to remove the remainder. Relatively high temperatures (up to 80°C) are often applied, which implies a lower viscosity in the evaporator but a higher initial viscosity of the final cooled product. The low water content of the sweetened condensed milk implies a high viscosity and boiling point. Evaporation in continuously operating equipment with many effects is, therefore, not easy. Fouling, thus, readily occurs. It is difficult to accurately adjust the desired water content, which is mostly monitored by means of refractive index.

Cooling and seeding: In these steps, formation of large lactose crystals must be avoided. Consequently, seed lactose is added. Before that, the condensed milk must be cooled to a temperature at which lactose is supersaturated so that the seed lactose does not dissolve. However, the temperature must not be so low that spontaneous nucleation can occur before the seed crystals are mixed in. After seeding, cooling should be continued to crystallize the lactose.

Packaging: Packaging in cans is common. The cans are then covered with a lid and the seams are sealed. Cans and lids are first sterilized, e.g., by flaming. The packaging room is supplied with air purified through bacterial filters.

19.2.2 Keeping Quality

19.2.2.1 Microbial Spoilage

Sweetened condensed milk is not sterile. It contains living microbes and spores. The low water activity (about 0.83) or, rather, the high sugar content prohibits growth of most but not all microorganisms.

Deterioration usually occurs by osmophilic yeasts, most of which belong to the genus *Torulopsis*. The yeasts often cause gas formation (bulging cans), a fruity flavor, and coagulation of protein. Coagulation may result from ethanol production. As a result, the product becomes unacceptable. The yeasts do not start easily, especially if the sugar concentration is high. It may thus take several weeks for incipient growth to be perceptible.

Some micrococci may grow in sweetened condensed milk, although slowly, especially if water activity and temperature are high. Presumably, the presence of oxygen is required. It may happen that they grow to reach a colony count of, say, 10^5 ml^{-1} and then stop growing, without causing noticeable defects. If they keep growing, coagulates eventually form and several off-flavors develop.

Some molds, especially strains of *Aspergillus repens* and *A. glaucus*, can grow as long as oxygen is present. If so, fairly firm colored lumps are formed and an off-flavor develops. One spore in one air bubble can cause such a lump.

Obvious remedies for microbial spoilage include the killing of all saprophytes and mold spores in the milk and in the sugar. Bacterial spores cannot germinate in sweetened condensed milk. Growth of harmful microorganisms in the dairy plant should be rigorously avoided. No sugar and residues of the milk should be left about. Satisfactory hygienic standards must therefore be maintained, especially in the packaging room. Harmful microorganisms cannot grow during concentrating, but the machinery must be thoroughly cleaned, immediately after evaporation. Mold spores can be removed by air filtration.

The packaging machine should fill the cans very accurately with a safety margin of 1 g. Too little condensed milk in the cans means that more air is left, which increases the chance of growth of molds and micrococci. If the cans are overfilled, the milk may spill over the side and encourage growth of osmophilic yeasts.

19.2.2.2 Chemical Changes

The main change in sweetened condensed milk during storage is presumably *age thickening* and, finally, gelation (see also Subsection 19.1.5). Sweetened condensed milk is far more concentrated than evaporated milk. Nevertheless, it does not thicken markedly faster with age. It is usually assumed that added sucrose inhibits age thickening; other sugars or hexitols have a similar effect. Sucrose increases the Ca^{2+} activity. A difference with evaporated milk is that an initial decrease in viscosity before age thickening is not observed. The viscosity increases almost linearly with time.

The following are the main factors affecting age thickening:

1. *Type of milk:* Variation — often seasonal — occurs among batches of milk.
2. *Preheating of milk:* The more intense the heat treatment, the higher the initial viscosity, and the sooner a gel can form. Hence, UHT heating is now generally applied.
3. *Stage at which sugar is added:* The later in the evaporating process, the less the age thickening.
4. *Concentration factor:* The higher the concentration factor, the more the age thickening. That explains why sweetened condensed milk of the British standard thickens faster with age than that of the American standard.
5. *Stabilizing salts:* The influence of added salts varies widely and depends on, for example, the stage at which it is added. Salts are added up to, say, 0.2%. Adding a small amount of sodium tetrapolyphosphate (e.g., 0.03%) mostly delays are thickening considerably, whereas adding more may have the opposite effect.

6. *Storage temperature:* Age thickening considerably increases with storage temperature, $Q_{10} \approx 3.4$. At tropical temperatures, gelation inevitably occurs within about a year.

Ongoing *Maillard reactions* are likewise inevitable. Brown discoloration is stronger as the storage temperature is higher, the milk is evaporated to a higher concentration, and more intense heating is applied. Additional Maillard reactions occur if the added sucrose contains invert sugar.

Autoxidation of fat can occur because the packaged product contains a little oxygen and may not have been heated sufficiently for antioxidants to be formed. Obviously, any copper contamination should be rigorously avoided.

19.2.2.3 Lactose Crystals

Sweetened condensed milk contains around 38 to 45 g lactose per 100 g water, as shown in Table 19.2. Figure 2.3 shows the solubility of lactose at room temperature to be about 20 g per 100 g water, but in sweetened condensed milk the solubility is about half as much due to the presence of sucrose (see Figure 19.9). It implies that 75% of the lactose tends to crystallize, meaning about 8 g per 100 g sweetened condensed milk. Due to the high viscosity, nucleation will be slow and only a few nuclei would be formed per unit volume of milk, leading to large crystals.

Without special measures, the product will obtain a relatively high quantity of large crystals. These crystals settle and are responsible for a sandy mouthfeel. Although the crystals may not be so large as to be felt singly in the mouth, they can be large enough to cause a nonsmooth impression. To avoid this, they should be smaller than about 8 μm in length.

Preventing crystallization is not possible and, accordingly, a large number of crystals should be obtained. Satisfactory results can be reached by using

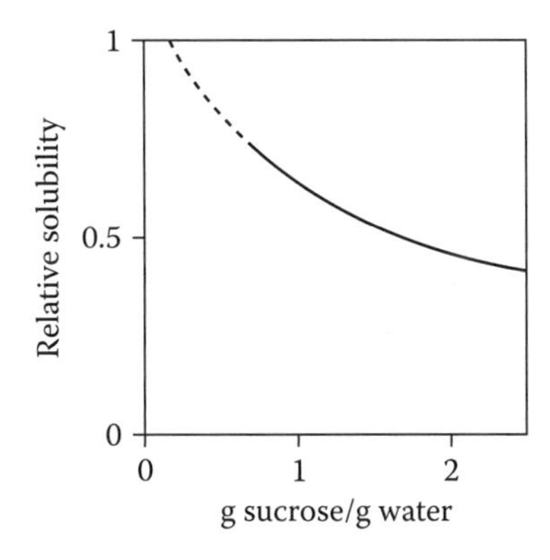

FIGURE 19.9 Influence of sucrose on the relative solubility of lactose (per unit mass of water). Approximate results at about 40°C.

seed lactose. Adding 0.03% seed lactose represents 0.004 times the amount of lactose to be crystallized. The final size of the crystals in the product should not exceed 8 μm. Consequently, the seed lactose would contain enough seed crystals (one per crystal to be formed) if its crystal size does not exceed about $(0.004 \times 8^3)^{1/3}$, i.e., 1.25 μm. Such tiny crystals can be made by intensive grinding of α-lactose hydrate.

Suggested Literature

A general overview, on an introductory level: M. Carić, *Concentrated and Dried Dairy Products,* VCH, New York, 1994.

Aspects of heat coagulation and age gelation: P.F. Fox and P. McSweeney, Eds., *Advanced Dairy Chemistry,* Vol. 1, *Proteins,* 3rd ed., Kluwer Academic, New York, 2003, Chapter 19 (J.E. O'Conell and P.F. Fox, Heat-induced coagulation) and Chapter 20 (J.A. Nieuwenhuijse and M.A.J.S. van Boekel, Changes in sterilized milk products).

20 Milk Powder

This chapter discusses spray-dried powders made from whole milk, skim milk, and, to a lesser extent, whey. Basic principles of evaporation and drying are described in Chapter 10.

20.1 OBJECTIVES

We may distinguish the following objectives:

1. The main purpose of the manufacture of milk powder is to convert the liquid perishable raw material to a product that can be stored without substantial loss of quality, preferably for some years. Decrease in quality mainly concerns formation of gluey and tallowy flavors (due to Maillard reactions and autoxidation, respectively) and decreasing nutritive value (especially decrease in available lysine). If the water content becomes very high and the storage temperature is high, caking (due to lactose crystallization) and enzymic and even microbial deterioration can occur; however, such problems are avoidable.
2. The powder should be easy to handle. It should not dust too much or be overly voluminous. It should be free-flowing, i.e., flow readily from an opening, and not stick to the walls of vessels and machinery. The latter requirement is especially important for powder used in coffee machines, etc.
3. After adding water the powder should be reconstituted completely and readily to a homogeneous mixture, similar in composition to the original product. Complete reconstitution means that no undissolved pieces or flakes are left and that neither butter grains nor oil droplets appear on top of the solution. 'Readily reconstituted' means that during mixing of powder and water no lumps are formed, because these are hard to dissolve. In the ideal situation the powder will disperse rapidly when scattered on cold water; this is called *instant powder*. Special processing steps are needed to achieve this property. The importance of instant properties closely depends on the kind of application.
4. According to its intended use the reconstituted product should meet specific requirements. If the use is beverage milk, the absence of a cooked flavor is of importance. If the powder is to be used for cheese making, the milk should have good clotting properties. If used to make recombined evaporated milk, satisfactory heat stability is necessary. So there

TABLE 20.1
Approximate Composition (% w/w) of Some Types of Powder

	Powder From			
Constituent	**Whole Milk**	**Skim Milk**	**Whey**	**Sweet-Cream Buttermilk**
Fat	26	1	1	5
Lactose	38	51	72–74	46
Casein	19.5	27	0.6	26
Other proteins	5.3	6.6	8.5	8
'Ash'	6.3	8.5	8	8
Lactic acid	—	—	0.2–2	—
Water	2.5	3	3	3

are several widely divergent requirements that cannot be reconciled in one powder. For instance, it is not possible to make whole milk powder that has no cooked flavor and at the same time develops no oxidized flavor during storage. With respect to the intensity of the heat treatment, milk powders are classified as low-, medium-, or high-heat (Subsection 20.4.6).

5. The product must be free of health hazards, be it toxic substances or pathogenic organisms. Besides general hygienic measures and checks prevailing in the dairy industry, there are some specific considerations.

The approximate *composition* of some types of powder is given in Table 20.1. There are other kinds of powder, e.g., cream powder, ice cream mix powder, infant formulas of various types, calf milk replacers, etc. All of these products have specific requirements.

Because the composition of the raw material varies, the composition of powders also varies. Accordingly, one has to tolerate a certain margin. This offers the possibility for (limited) adulteration; for instance, buttermilk powder or whey powder can be added to (skim) milk powder. The presence of a foreign powder can mostly be detected microscopically, but admixture of a small percentage of another liquid before the drying generally cannot so easily be established. Because whey is cheaper than skim milk, this kind of 'adulteration' sometimes occurs.

20.2 MANUFACTURE

Figure 20.1 gives a flow sheet for the manufacture of whole milk powder. Many steps involved will be self-evident. Intense pasteurization is needed to obtain resistance to autoxidation. The concentrate is not always homogenized, especially if atomization is done by means of a nozzle, because the fat globules are effectively disrupted in the nozzle (Subsection 10.4.2). Homogenization of highly

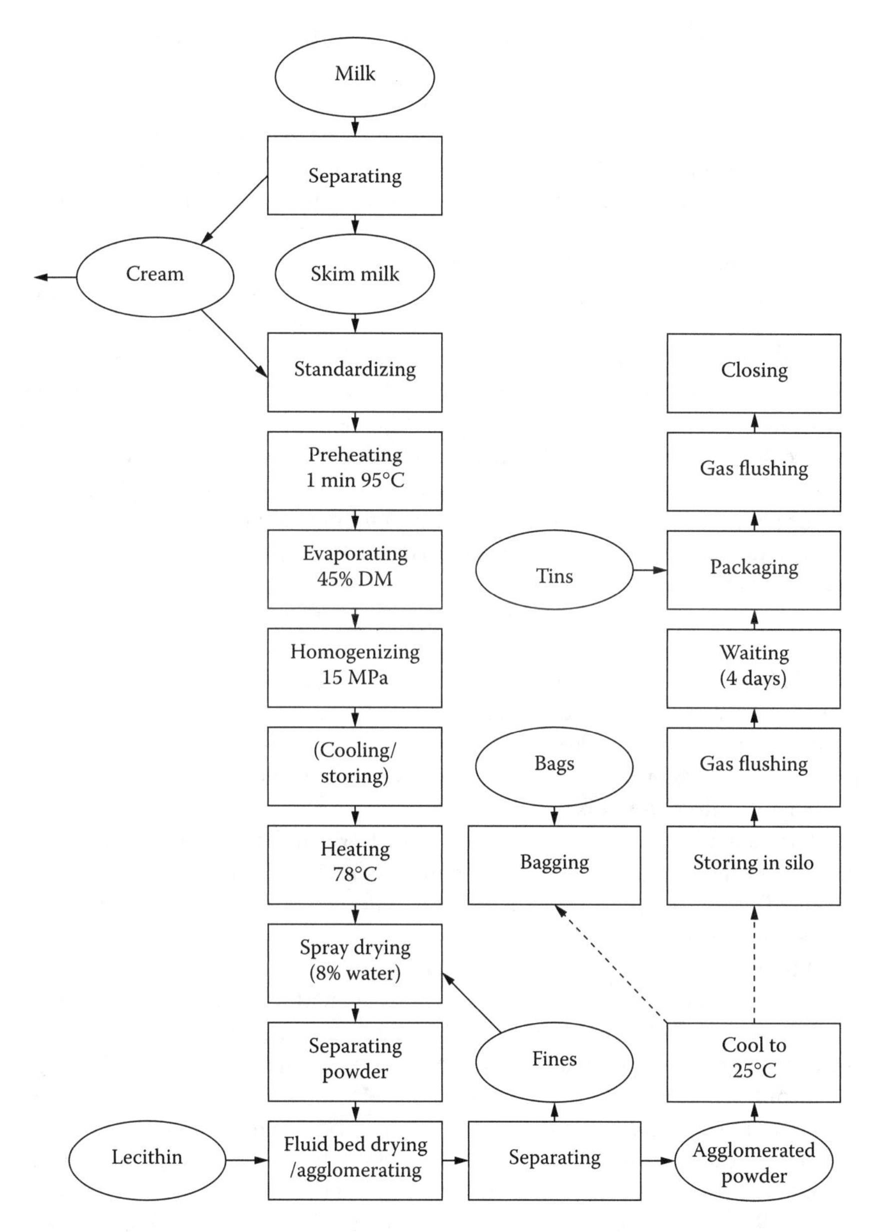

FIGURE 20.1 Example of the manufacture of whole milk powder.

concentrated milk considerably increases its viscosity (because the transfer of large casein micelles to the fat globules gives the latter such an irregular shape as to increase the effective volume fraction of fat globules plus casein micelles). This increase leads to coarser droplets during atomization, with all drawbacks involved (Subsection 10.4.2). Consequently, if the concentrated milk is not homogenized, evaporation can continue up to a higher dry-matter content. Storage (buffering) of the concentrate before atomization is not always applied; it is done mainly to overcome differences in capacity between evaporator and drier. However, the concentrate should not be kept warm for more than a short time to prevent the growth of microorganisms. A refrigerated concentrate generally is too viscous to be atomized readily, and it is therefore heated. The latter must be done just prior to atomization because otherwise the viscosity increases again (age thickening) (see, for example, Figure 10.10C). The heating can at the same time serve to kill bacteria that may have recontaminated the concentrate.

Lecithinizing during the drying in the fluid bed is not always applied; it is meant to obtain instant properties (Subsection 20.4.5.1). The so-called gas flushing, essentially displacing air by N_2 or a mixture of N_2 and CO_2, is to remove a considerable part of the oxygen and thereby to improve the stability toward autoxidation (Section 20.5); it can be done once or twice. If it is not done, the powder may be packaged into multilayer paper bags with a polyethylene inner layer. Whole milk powder, however, is often packaged in tins or in plastic containers to minimize oxygen uptake.

For the manufacture of *skim milk powder* the pasteurization can be less intense (at least phosphatase negative), according to the intended application (Subsection 20.4.6). Homogenization is omitted, and the milk can be concentrated to a somewhat higher solids content. Nor are lecithinizing and gas flushing carried out. Sometimes vitamin preparations are added, especially vitamin A. This can be achieved by dry mixing afterward, or by emulsifying a concentrated solution of vitamin A in oil into a part of the skim milk.

The manufacture of *whey powder* is largely similar to that of skim milk powder. At first, curd fines should be removed from the whey by filtration or by means of a hydrocyclone, and the whey should be separated. A problem is the processing of sour whey, which causes rapid fouling of the machinery. Sour whey (or skim milk) can be neutralized with alkali.

Whey can be evaporated to at least 60% dry matter, but then crystallization of lactose readily occurs (see Figure 10.11). An alternative operation is to allow the lactose in the evaporated whey to crystallize as completely as possible, e.g., by keeping it for 3 h at 25°C while stirring. If the dry-matter content is over 60%, seeding with lactose crystals is not necessary. Atomization should be with a disk, as a spray nozzle would become blocked. The precrystallized whey powder then obtained has some attractive properties, especially in regard to caking (Section 20.5). An additional advantage for the manufacturer is the higher yield: The conventional methods for determining the water content of powders do not remove the bulk of the water of crystallization of α-lactose monohydrate; crystallization of 80% of the lactose yields up to a maximum of 3% more whey powder. In this way precrystallized skim milk powder also can be made, but then a longer crystallization time and seeding with lactose crystals are needed.

20.3 HYGIENIC ASPECTS

The requirements for the bacteriological quality of milk powder partly depend on its intended use and, in connection with this, also on the manufacturing process. For example, whether the powder is meant for direct consumption or whether it is subjected to a heat treatment after reconstitution (e.g., for recombined milk) is important. The heat treatment during the manufacture of (skim) milk powder, classified as 'low heat' (Subsection 20.4.6), usually is not more intense than the heat treatment during low pasteurization (say, 72°C for 20 s); consequently, many bacteria may survive the manufacturing process.

The causes for milk powder to be bacteriologically unacceptable or even unsafe can be of three kinds:

1. In the fresh milk, bacteria are present that are not killed by the heat treatments to which the milk is subjected before and during drying.
2. Conditions during the various process steps until the product is dry do allow growth of some bacterial species.
3. During manufacture, incidental contamination may occur. The level of contamination is generally low and remains low if the bacteria involved cannot grow.

In establishing the bacteriological quality of the powder, the species of bacteria involved should be considered. Table 20.2 gives an overview that applies for one example of the manufacturing process. Then the cause of the contamination may be deduced, as may the measures that must be taken to improve the quality. The table also indicates some attention points for checking hygiene.

20.3.1 Bacteria in the Original Milk

See Chapter 5 for general aspects.

In deep-cooled milk, psychrotrophic Gram-negative rods can develop during prolonged storage (e.g., *Pseudomonas* spp.). As is well known, these bacteria do not survive even a mild heat treatment. Proteinases and lipases formed by these rods may survive and become incorporated into the powder. Prevention of the growth of these bacteria (refrigeration, limiting storage time, and thermalization) is discussed in Section 6.4. Contamination and growth during storage of the thermalized milk should be avoided.

Of greater importance are heat-resistant bacteria and bacterial spores. They survive low pasteurization (72°C for 15 s), and most are not killed during evaporation and drying. Due to concentrating, the powder contains about ten times as many bacteria per gram as the milk immediately after preheating. A more intense pasteurization will kill the heat-resistant cocci (e.g., *Enterococcus faecalis*, *S. thermophilus*) and in a high-quality medium-heat or high-heat milk powder only bacterial spores and *Microbacterium lacticum* can originate from the original milk.

TABLE 20.2
Example, for a Given Process Scheme, of the Events Occurring with Important Bacteria and Bacterial Enzymes during Manufacture of Dried (Skim) Milk

Process Step	Raw Milk	Thermalizing	Storage	Pasteurization[a] Regen.[b]	Pasteurization[a] Low	Pasteurization[a] High	Evaporation Eff. 1–3	Evaporation Eff. 4–6	Balance Tank	Drying	Packaging	Storage
Temperature (°C)	5	65	6	6–65	72	95	70–58	58–42	40	40–100	25	Ambient
Duration (min or days)	3 d	1/4 m	2 d	1 m	1/4 m	1/4 m	6 m	8 m	2 m	4 m	3 d	
Dry-matter content (%)	9	9	9	9	9	9	9–30	30–50	50	50–96	96	96
Psychrotrophs	G*	K	C, (G)*	(S)	K	K						
Heat-resistant enzymes	F	S	(A, F)	S	S	S······	······	······	······	······	······S	P
Staphylococcus spp.	C	(S)	S	(S)	K	K	(C)	(G)	C*, G	(S)	(C)*	P
Enterobacteria	C	(S)	S	(S)	K	K			C*	(S)	C*	P
Salmonella spp.	C	(S)	S	(S)	K	K			C*	(S)	C*	P
Streptococcus thermophilus	(C)	G*	S	G*	S	K	S······	······	······	······	······S	P
Enterococci	C	S······	······	······	······S	K	S	G	G*	(S, C)	S	P
Bacillus cereus	C	S, (C)	C	C*	S	(S)	(C)	(C)	(C)*	S	S	P
B. stearothermophilus	(C)	C	S	C	S	S	G	G	G*	S	S	P
Clostridium spp.	C*	S······	······	······	······S	S······	······	······	······	······	······S	P

Notes: A = active; C = contamination possible; Eff. = effects; F = formed; G = growth; K = killed; P = can be present; S or S-----S = survive; ( ): = partly, occasionally; and * = attention point.

[a] Two alternative pasteurization intensities are given;
[b] Regen. = heat regeneration section.

Among the aerobic and anaerobic spore-forming bacteria, *Bacillus cereus* and *Clostridium perfringens* are especially important to the powder quality. If the reconstituted milk is to be used for cheese making, a very low count of gas-forming anaerobic spore formers (*C. tyrobutyricum* and *C. butyricum*) may be essential. All of these bacteria are likely to originate mainly from contamination during milking (dung, soil, and dust). A low count of anaerobic spore formers points to a good-quality silage (silage being the source of most of the clostridia: feed → cow → dung → milk). But the pathogenic *C. perfringens* usually does not originate from the silage, though it may from the dung. Hence, a low count of anaerobic spore formers need not be an indication of the absence of *C. perfringens*. Likewise, the total count of aerobic spore formers is not always an indication of the spore number of *B. cereus*. Usually, the total count is higher during winter, but the count of *B. cereus* may be highest in summer and autumn. This probably is because (1) contamination with *B. cereus* is heavier on pasture, and (2) at higher environmental temperatures *B. cereus* can develop and sporulate in imperfectly cleaned and disinfected equipment outside the operating periods. To kill bacterial spores, heat treatment at 90 to 110°C for 10 to 20 s is insufficient; UHT treatment should be used. The *D* value is about 4 s at 125°C for *B. cereus* as well as for *C. perfringens*, so that heating for 15 to 20 s at that temperature would cause a sufficient reduction of these spores.

20.3.2 Growth during Manufacture

Temperature and water activity during successive steps in manufacture are such that some thermophilic bacteria can readily grow and they are not or insufficiently killed during drying. The type of bacterium often is characteristic of the cause of the contamination. Some examples will be discussed.

In the regeneration section of pasteurizers and thermalizers (and possibly in the part of the evaporator plant where the milk is heated in counterflow), *S. thermophilus,* in particular, can develop. The bacterium grows fastest at 45°C but scarcely multiplies at temperatures over 50°C. It generally does not grow in the evaporator, because the temperature is too high in the first effects, whereas in the later effects a_w will be too low. Because *S. thermophilus* is moderately heat resistant, relatively high counts may result, especially in low-heat milk powder and in whey powder. In medium and high-heat powder the bacterium is killed during manufacture. Determination of the count of *S. thermophilus* in the milk just before the preheater may give a good indication of the fouling of the heating section of plant and of the moment at which it should be cleaned. In some drying plants that employ a wet washer to recover powder fines, the outlet air is brought into contact with a film of milk rather than water, thereby preheating the milk and saving energy; this implies that the preheated milk acquires the wet bulb temperature (about 45°C), which leads to ideal growth conditions for *S. thermophilus*.

The conditions in the second half of the evaporator and in the balance tank are not optimal for *S. thermophilus*. Enterococci (*E. faecium* being an important representative), in particular, will start to grow. If the milk is properly preheated and the plant is satisfactorily cleaned and sterilized, it will take a rather long time, however, before substantial counts of *E. faecium* have developed. In actual practice, these prerequisites are not always met and in milk powder with a relatively high count, *E. faecium* often is the predominant species.

Likewise, the conditions in the second part of the evaporator and in the balance tank are favorable for growth of *Staphylococcus aureus*. The bacterium generally is killed by pasteurization, and strains of *S. aureus* in milk powder have been shown to have phage characteristics different from the strains in raw milk. Presumably, they originate from direct or indirect human contamination. Heat-stable enterotoxins can be formed and the amounts formed at counts of 10^7 to 10^8 per ml may cause the powder to be a health hazard. Although *S. aureus* is not heat resistant, the conditions during drying appear to be such that complete killing is not achieved. Roughly 10^{-5} to 10^{-1} of the initial count of these bacteria have been found to survive under various practical conditions. This means that *S. aureus* can at least be found in 1 g of fresh powder if its count before the drying was so high that production of enterotoxins could have occurred.

Bacillus stearothermophilus (ssp. *calidolactis* in particular) can readily grow at higher temperatures. Its growth range is from 45°C to 70°C, with an optimum near 60°C. It can also grow in concentrated milk and therefore throughout the equipment between preheater and drier. Moreover, the bacterium can form spores under these conditions, which further limits its killing during drying. Obviously, some growth of *B. stearothermophilus* will always occur during manufacture, even in a well-cleaned and disinfected plant. However, under most conditions this will not cause problems.

With regard to the growth of bacteria during the manufacturing process, a number of measures will have to be taken to prevent the equipment from becoming a kind of fermentor for bacteria. Special attention should be paid to the temperature, to the time for which the product stays in the equipment, and to matching of the capacities of the various parts of the plant. Multiple-effect falling-film evaporators are more satisfactory in this respect than are flash evaporators. (If it is desirable to achieve as high a concentration factor as possible, it should be considered that the viscosity of the product in the last effect will be quite high — hence, also, its holdup time.) The balance tanks need special attention; it is recommended that the volume of the concentrate balance tank be kept as small as possible. Mostly, there are two tanks, making possible a change every 2 h or so, and the cleaning of one while the other is in use. Contamination of the product flow from outside should be prevented, with particular reference to *S. aureus*.

The concentrated milk may be pasteurized just before it enters the drier, and this is increasingly being done. It is especially successful with regard to the killing of *E. faecalis* and *E. faecium*, provided that the temperature applied is high enough. Heating for 45 s at 72°C has little effect; 45 s at 78°C causes a considerable reduction. Due to the decrease in water content, the D and Z values have slightly increased (see Table 7.5).

20.3.3 Incidental Contamination

A distinction can be made between contamination before drying (wet part) and during or after drying (dry part). Bacteria involved in these types of contamination generally do not grow during the process and contribute little to the count of the powder.

Contamination after preheating and before drying can readily occur if the equipment has been insufficiently cleaned. This is only important if the bacteria involved can survive the drying (and the pasteurization before drying, if carried out). In spite of the high inlet and outlet temperatures of the air, the concentrate droplets usually will not attain a high temperature (Subsection 10.4.4); moreover, the heat resistance of the bacteria increases markedly with the dry-matter content. About 70% of *E. faecalis* and *E. faecium* survive during drying, whereas survival of *S. aureus* varies widely. About 10^{-4} to 10^{-5} of the initial count of *Salmonella* spp. and *E. coli* will survive. Based on these facts, and due to the relatively low level of contamination, the powder may be expected to contain no appreciable counts of enterobacteria immediately after leaving the drier. This is confirmed in actual practice.

Contamination of the powder can occur at many places — in the spray drier, during fluid bed drying, and during packaging. The species of contaminating bacteria can vary widely, but it usually concerns species that can grow in wet remnants of milk powder in the drier or in the surroundings of the manufacturing line. Contamination via (in)direct human contacts should also be considered (e.g., *S. aureus*). Bacteria can easily survive in dry powder, and undesirable bacteria can start to grow if the water content increases to over 20%. The supply of cooling air into the spray drier and into the fluid bed drier can be a source of direct contamination. It may also be responsible for indirect contamination because it gives, at certain sites, better conditions for survival and growth of bacteria in remnants of not fully dried powder. Special precautions are needed if the drier and its accessories have been wet-cleaned. To restrict such incidental contamination, the plant and its surroundings should be rigorously freed of remnants of (wet) powder.

Among the kinds of bacteria found in this type of contamination, enterobacteria are of special interest. It generally concerns coliforms; this is possibly caused by lactose being the only carbon source in this environment. Nevertheless, the use of coliforms as an indicator organism is of restricted value for milk powder: If coliforms are absent, salmonellae or other pathogenic bacteria may still be present.

20.3.4 Sampling and Checking

Bacteria originating from contamination or growth prior to drying will usually be homogeneously distributed throughout the powder and cause no problems with sampling. This is different for bacteria originating from incidental contamination of the powder, which may be distributed quite inhomogeneously. It is not possible to devise sampling schemes that guarantee detection of incidental contamination. To ensure that a product is bacteriologically safe, not only should the powder be sampled but samples must also be taken at sites that are potential sources of contamination. Table 20.2 indicates examples of such attention points.

20.4 POWDER CHARACTERISTICS

Several properties of powdered products, for the most part of a physical nature, affect the quality and the suitability of the powder in specific applications. These properties, as well as the — often considerable — influence of process variables on them, are discussed in this section.

20.4.1 The Particle

A milk powder particle generally consists of a continuous mass of amorphous lactose and other low-molar-mass components in which fat globules, casein micelles, and serum protein molecules are embedded. The lactose generally is in a glassy state (Subsection 10.1.4), the time available for its crystallization being too short. If, however, precrystallization has taken place, large lactose crystals may be present (some tens of micrometers in size). When precrystallized whey powder is examined microscopically, most of the particles look more like lactose crystals of the tomahawk shape to which other material adheres. If lactose has been allowed to crystallize afterward due to water absorption (Section 20.5), its crystals are generally small (about 1 μm).

Most fat globules are less than 2 μm, but a small proportion of the fat (e.g., 2% of it) is to be found as a thin layer on parts of the surface of the powder particles (see the following text). The formation of vacuoles in the powder particles is discussed in Subsection 10.4.2.2. The vacuole volume v of most powders varies from 50 to 400 ml·kg^{-1}. An impression of the shape of the powder particles is in Figure 10.17.

Examples of *particle size distributions* of milk powder are in Figure 10.15. Usually, d_{vs} is between 20 and 60 μm, and the distribution is relatively wide: c_s ranges from 0.4 to 0.7. The particles in agglomerated powder are much larger, up to 1 mm, and are irregular in shape. Such a powder usually contains very few separate particles smaller than, say, 10 μm. Within one sample of nonagglomerated powder the larger-sized particles have on average a higher vacuole volume, partly because a drying droplet shrinks more strongly if it contains no vacuoles. The factors affecting particle size distribution are discussed in Subsection 10.4.2.1.

20.4.2 Extractable Fat

The term 'free fat' is often used, but it is literally incorrect as it suggests that some fat is entrapped in the powder particles without a surrounding membrane. So-called free fat is determined by extraction with an organic solvent. In fact the fat is then extracted from all fat globules that are in contact with the surface of the particle, or with a vacuole, with pores or cracks, etc. This is illustrated in Figure 20.2. Cracks can readily be formed if the lactose is in a glassy state. Obviously, the amount of extractable fat will be higher as the powder particles are smaller, a greater number

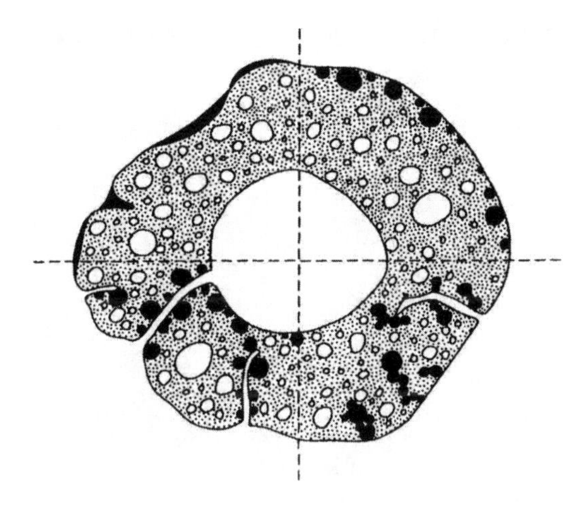

FIGURE 20.2 Diagrammatic model of a milk powder particle, indicating which part of the fat (dark) can be extracted. In each quadrant another type of extractable fat is indicated. (From T.J. Buma, *Neth. Milk Dairy J.*, **25**, 159–174, 1971.)

of vacuoles have been entrapped, and the drying has been performed at a higher temperature (giving more cracks; see also Figure 10.16B and Figure 10.24). Increasing the water content of the powder, as can occur due to uptake of water from the air, decreases the amount of extractable fat, as cracks are closed due to swelling. (The latter implies that the lactose is losing its glassy state.) The methods of extraction are, moreover, quite empirical and the resulting values greatly depend on small variations in the analytical procedure.

The content of extractable fat is often used as a quality mark, but this makes little sense, as it hardly correlates with other product properties (except sometimes with the rate at which vacuoles in the powder particles become filled with air). The upper-left quadrant in Figure 20.2 indicates that part of the fat may be located on the surface of the powder particles. Such fat may be estimated by means of a very brief (say, 3 seconds) extraction at low temperature. It correlates more or less with the contact angle for wetting of the powder and consequently with the dispersibility (more surface-fat → poorer wettability) (see Subsection 20.4.5.1). It would be far better to estimate the fraction of the particles' surface that is covered with fat: an impression can be obtained by light microscopy of powder stained with an oil-soluble dye.

The amount of extractable fat can be reduced by intensive homogenization, but only in milk powder with less than 20% fat does this affect the dispersibility — and then only slightly.

20.4.3 Free-Flowingness

Free-flowingness refers to the ability of a powder to be poured. The property may be estimated by pouring out a heap of powder under standardized conditions and

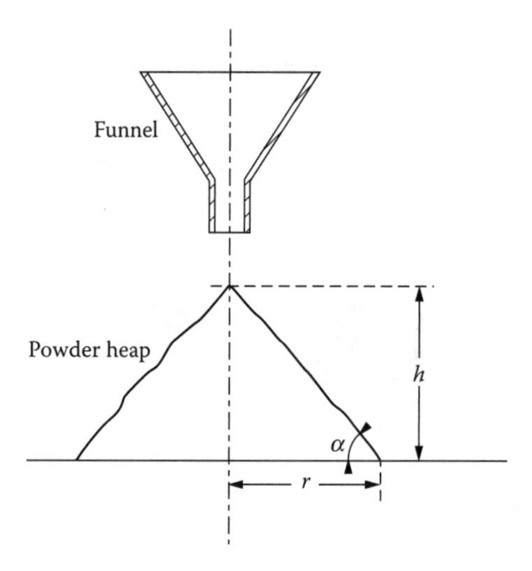

FIGURE 20.3 Determination of the angle of repose α of a powder heap, formed after pouring the powder through a funnel stem. Free-flowingness can be expressed as cot $\alpha = r/h$.

measuring its angle of repose α. This is illustrated in Figure 20.3. The more readily the powder flows, the smaller the angle is. The cot α, therefore, is an appropriate measure of free-flowingness. If the powder is not free-flowing it may be called *sticky*: the particles do not readily tumble over each other. Consequently, the porosity ε — i.e., the volume fraction of voids between the particles — will be high.

Examples are in Table 20.3. It shows that skim milk powder is more free-flowing than whole milk powder, and that agglomeration decreases the free-flowingness, at least for skim milk powder. With increasing water content of the powder, the free-flowingness may at first increase slightly, but it considerably

TABLE 20.3
Examples of the Cotangent of the Angle of Repose α (as a Measure of Free-Flowingness) and the Porosity ε (Lightly Bulked) of Some Types of Powder

Powder	Additive	cot α	ε
Whole milk	—	0.45	0.74
Skim milk	—	0.97	0.57
Instant skim milk	—	0.75	0.73
Whole milk	2% $Ca_3(PO_4)_2$	1.19	0.56
Skim milk	Same	1.28	0.54
Instant skim milk	Same	0.93	0.63
Whole milk	0.5% SiO_2	1.23	0.51

decreases at still higher water contents (e.g., > 5%). Usually, the free-flowingness is slightly better at a lower temperature. It can considerably be improved by adding some inert very fine powders such as SiO_2, Na-Al-silicate, or $Ca_3(PO_4)_2$ (see Table 20.3). Such additions are applied, for instance, in milk powder for coffee machines.

20.4.4 Specific Volume

The *density* of a powder may be defined in various ways. The density of the particle material, i.e., excluding the vacuoles, is called the *true density* ρ_t. Approximately we have:

Dried whole milk	$\rho_t = 1300\ \text{kg}\cdot\text{m}^{-3}$
Dried skim milk	$\rho_t = 1480\ \text{kg}\cdot\text{m}^{-3}$
Dried whey	$\rho_t = 1560\ \text{kg}\cdot\text{m}^{-3}$

The *particle density* ρ_p includes the vacuoles and is therefore given by:

$$\rho_p = \frac{\rho_t}{1 + v\rho_t} \tag{20.1}$$

where the vacuole volume v is in $\text{m}^3\cdot\text{kg}^{-1}$. In skim milk powder ρ_p usually is 900 to 1400 $\text{kg}\cdot\text{m}^{-3}$. The *bulk density* of the whole powder, ρ_b, is given by

$$\rho_b = \rho_p(1 - \varepsilon) = \rho_t \frac{1 - \varepsilon}{1 + v\rho_t} \tag{20.2}$$

where ε is the *porosity* or void volume fraction. Generally, ε varies from 0.4 to 0.75, but it depends on the method of handling the powder. It decreases considerably when a lightly bulked powder is set by means of tapping or shaking, e.g., from 0.70 to 0.45 for whole milk powder and from 0.55 to 0.40 for skim milk powder. ε depends also on other factors; the relation with free-flowingness was discussed in the preceding subsection. All in all, ρ_b can vary widely, for instance, from 300 to 800 $\text{kg}\cdot\text{m}^{-3}$.

Specific volume is equal to $1/\rho_b$. A distinction is often made between specific *bulk* volume (if the powder is lightly bulked) and specific *packed* volume (if the powder is allowed to set, for instance, by tapping). For the estimation a tapping apparatus may be used, which repeatedly causes a graduated cylinder with powder to fall on a solid base. Obviously, the packed volume is of great importance in connection with the mass of powder that can be stored in a certain package. Preferably, the volume should not be greatly reduced by tapping as, otherwise, a can full of powder when filling may later appear to be partly empty when opened by the consumer.

Equation 20.2 shows that the variables determining ρ_b are vacuole volume (v) and porosity (ε). The process variables affecting v are summarized in Figure 10.16.

Generally, factors causing a higher viscosity of the concentrate during atomization will lead to a lower v and thus to a higher ρ_b. Other factors affect ε (see, e.g., Table 20.3), and not all of those are easy to explain. In general, ε will be higher if the powder particles are more irregular in shape and differ less in size. Accordingly, agglomeration, as well as removal of small powder particles, clearly increases ε, hence decreases ρ_b. ε may decrease slightly when increasing the water content of the powder; presumably, it causes the particle surface to become smoother. Pre-crystallization of lactose makes the particles more angular and slightly increases ε.

Obviously, ρ_b increases due to shaking, tapping, or vibrating. In general the effect produced is reversible. But in an agglomerated powder the agglomerates can be disrupted; consequently, ρ_b is irreversibly increased, and the instant properties are impaired (Subsection 20.4.5.1).

20.4.5 Dissolution

When dissolving milk powder in water, the following stages can be distinguished:

- The powder is dispersed, implying that the powder particles will become fully wetted, which may take some time.
- Subsequently, the soluble components dissolve, and the colloidal particles (fat globules and casein micelles) become dispersed in the solution. This will take several minutes. Generally, a small part does not dissolve.
- After a solution is obtained, it may take about a day before its composition, especially the salt distribution, has more or less reached equilibrium. This process may be speeded up by heating the reconstituted milk to 50°C to 60°C, and then cooling it again.

In this section, the first process, i.e., the dispersion of powder, and the partial insolubility will be further discussed.

20.4.5.1 Ease of Dispersing; Instant Powder

Dissolving a powder in hot water by using a high-speed agitator causes few problems. However, dispersing it in cold water under household conditions may be far from easy. The dispersibility primarily depends on the rate at which water fully penetrates a mass of powder. If this happens quite fast, the powder is said to have 'instant' properties.

The following phenomena, which occur when an amount of powder contacts water, are of importance:

1. The powder should be wettable by the water. Consider two powder particles close together on a water surface (Figure 20.4). The wetting depends on the *contact angle* (θ, as measured in the aqueous phase)

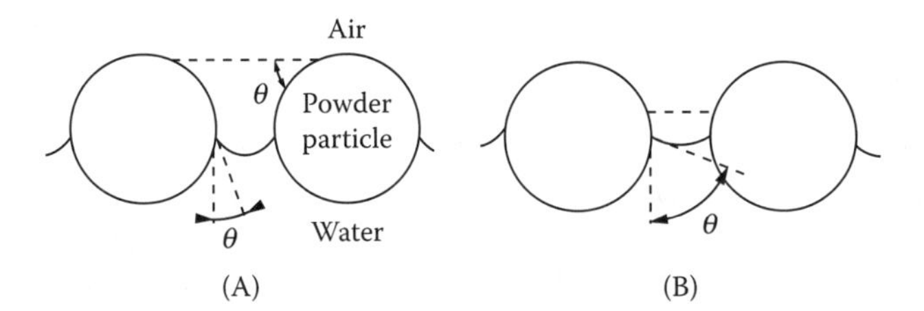

FIGURE 20.4 Capillary rise of water between two powder particles. θ = angle of contact. The broken line denotes the situation in which (for cylinders) the water surface is flat. (A) $\theta = 20°$. (B) $\theta = 70°C$. Highly schematic.

of the system consisting of dried milk, water, and air. This is discussed in Subsection 3.1.1.5. For a hydrophobic solid θ is large; for a hydrophilic solid it is small. If θ is less than 90°, the particles are wetted. For dried skim milk θ is about 20°; for dried milk about 50°. This would imply that water is always sucked into the pores between the particles by capillary forces: if the water surface is curved, the capillary force acts in the direction of the concave side, i.e., upward in the case of Figure 20.4. In skim milk powder this is indeed the case in practice. But the figure shows that the maximum height to which the water can rise between two powder particles — which is until a flat water surface is reached — is greater for a smaller θ. In actual practice, the situation is more complicated (three-, not two-dimensional situation; surface of the powder particles is not smooth), but in a qualitative sense the same relations hold. During wetting of a powder mass there is an effective contact angle θ_{eff}, which is significantly greater than the contact angle on a plane surface. In whole milk powder θ_{eff} may be greater than 90°, especially if the fat is partly solid; the consequence is that water will not penetrate the powder mass, or does so only locally. The remedy is to cover the powder particles with a thin layer of lecithin, thereby considerably decreasing θ_{eff}.

2. The speed of penetration of the liquid into the powder is slower for a larger θ_{eff}. Further, the speed is about proportional to the average diameter of the pores between the particles and inversely proportional to the liquid viscosity. The smaller the powder particles, the wider their spread in size (allowing smaller particles to fill the voids between the larger ones), and the smaller the porosity, the narrower the pores will be. In most skim milk powders, the pores are too narrow to permit rapid penetration of water.
3. Problem 2 above is aggravated by the pores becoming narrower during wetting. This is because the surface tension of an air–water interface between two particles pulls these partly wetted particles closer to each other. Due to this *capillary contraction* the power may decrease by 30 to 50% in volume upon wetting.

4. Penetration of the aqueous solution is also slowed down by dissolution of powder constituents in the water, thereby increasing liquid viscosity. Lactose, the main constituent, is present in an amorphous state, so it dissolves quickly to form a highly viscous solution.
5. Due to phenomena 2 to 4, the penetration of water will soon stop. Lumps of powder are formed, which are dry inside and have a very viscous outer layer of highly concentrated milk. Such lumps dissolve very sluggishly.
6. Other properties of the powder may slightly affect the dispersibility — for instance, the force needed to pull adjoining particles apart — or the particle density, which determines whether the particles will sink. Generally, these aspects are not decisive.

To give the powder *instant properties*, it thus should be agglomerated, preferably in such a way that at most a few fine particles are left. This causes the pores through which the water primarily penetrates to be much wider. Hence, the agglomerates are readily dispersed and they subsequently can dissolve. It is furthermore important that the agglomerates are strong enough to avoid their disintegration when the powder is subjected to external forces, as during packaging and shaking. These properties depend on the conditions in the fluid bed drier (Subsection 10.4.5). In addition, whole milk powder should be lecithinized to ensure a small contact angle.

20.4.5.2 Insolubility

Insolubility can be estimated in various ways. In all tests, powder is dissolved in water under standardized conditions (concentration, temperature, and duration and intensity of stirring), and then the fraction that has not been dissolved is estimated (e.g., volumetrically after centrifugation or via determination of dry matter). The result is called the *insolubility index*. The insoluble fraction, essentially the material that sediments during centrifugation, will predominantly consist of casein. In whole milk powder aggregates of coagulated protein with entrapped milk fat globules (the so-called white flecks) may float to the surface; the quantity involved usually is more than the sediment. Hence, the insolubility found closely depends on the method used. Some examples are in Table 20.4.

The insolubilization of a fraction of the milk powder is related to heat coagulation (Subsection 7.2.4). Consequently, the extent to which it occurs greatly depends on the time during which the drying material is at high temperature and on the degree of concentration during drying. The determining factors are discussed in Subsection 10.4.4.5, and the influence of some product and process variables is summarized in Figure 10.24. Furthermore, preheating has an effect: high preheating → higher viscosity → larger droplets on atomization → more intense heat treatment during drying → increased insolubility. For high-preheated milk, however, evaporation to a given viscosity generally implies a lower degree of

TABLE 20.4
Some Examples of the Insolubility of Whole Milk Powder, Determined by Different Methods

Powder Sample	ADMI	ZKB	CCF	
			x	y
1	0.02	0.04	0.2	0
2	0.02	0.04	2.2	0.1
3	0.04	0.07	2.1	0.1
4	0.08	0.04	1.3	0.1
5	0.10	0.04	2.6	0.3
6	0.16	0.12	1.6	0.2
7	1.1	—	2.8	0.5
8	3.4	—	0.9	0.8

Note: ADMI: ml sediment per 50 ml of reconstituted milk, dissolving at 24°C with intensive stirring, centrifuging; ZKB: as ADMI, but at 50°C and less intensive stirring; CCF: in grams per 30 g of powder, dissolving at 20°C with gentle stirring, centrifuging; x is excess of dry matter in top layer, and y the same in bottom layer.

concentration, and heat coagulation and insolubilization during drying would be less. Finally, homogenization of the concentrate will increase the insolubility (specially the term x in the CCF-test: see Table 20.4), but this may hardly be noticeable, as considerable homogenization occurs anyway during evaporation and atomization. Often, the insolubility of whole milk powder is indeed higher than that of skim milk powder. Homogenization of the concentrate may also lead to greater insolubility if it considerably raises its viscosity, increasing the droplet size and thereby the drying time.

20.4.6 WPN Index

Heat treatment of the original product or the concentrate can cause denaturation of serum proteins; the conditions during spray drying are rarely such as to cause extensive heat denaturation. The extent of denaturation is an important quality mark in relation to the use of milk powder. For instance, if the powder is to be used in cheese making, practically no serum protein should have been denatured in view of the rennetability; in infant formulas, on the other hand, the rennetability should be poor.

The extent of the denaturation of serum protein can be used as a measure of the heating intensity applied. This is true also where denaturation by itself may be of no importance, but other changes associated with intense heat treatment are. An example is the flavor of a powder to be used in beverage milk, which requires a mild heat treatment. An intensive heat treatment is needed for some other uses, for instance, to acquire good stability against heat coagulation in the manufacture of recombined evaporated milk, or a high viscosity of the final product when making yogurt from reconstituted milk. It is also desired if milk powder is used in milk chocolate; presumably, Maillard products contribute to its flavor.

The *whey protein nitrogen index* (WPN index) is generally used to classify milk powders according to the intensity of the heat treatment(s) applied during manufacture. To that end, the amount of denaturable serum protein left in the reconstituted product is estimated, usually by making acid whey and determining the quantity of protein that precipitates on heating the whey. This can be done by Kjeldahl analysis of protein nitrogen or by means of a much easier turbidity test that is calibrated on the Kjeldahl method. The result is expressed as the quantity of undenatured serum protein per gram of skim milk powder. The classification is as follows:

WPN $\geq$ 6 mg N per gram: low-heat
6 > WPN > 1.5 mg N per gram: medium-heat
WPN $\leq$ 1.5 mg N per gram: high-heat

The validity of the test is moderate. Due to (1) the natural variation in the amount of denaturable protein in raw milk (recalculated as WPN index 6.5 to 10) and (2) the limited reproducibility of the test, a substantial portion of the serum protein (up to 40%) may have been denatured in a powder classified as low-heat, and in a high-heat powder up to 23% may be undenatured. Of course, the average validity is far better.

If one wants to make low-heat powder, one should: (1) apply low pasteurization (e.g., 15 s 72°C), (2) begin the evaporating process not above 70°C (and preferably somewhat lower), (3) evaporate the milk not too far, (4) keep the concentrate temperature below 60°C, (5) cool the concentrate if it is to be kept for a fairly long time, (6) maintain the outlet temperature during spray drying at a low level, and (7) mix air and droplets in the drying chamber such that no local overheating of the drying droplets can occur. If one wants to make high-heat powder, the milk should be intensely heated, for example, 5 min 90°C or 1 min 120°C; often still higher temperatures are used. A more intense preheating causes a higher viscosity of the concentrate at the same dry-matter content (Figure 10.10), with all of its consequences (Subsection 10.4.4.5).

20.4.7 Flavor

Dissolved milk powder often has a cooked flavor, which results from the flavor compounds formed during preheating and possibly during evaporation. During drying, conditions are mostly not such that off-flavors are induced. On the contrary, a considerable part of the volatile sulfhydryl compounds (especially H_2S) is removed. A cooked flavor mainly results from methyl ketones and lactones formed by heating of the fat (they are almost absent in skim milk powder) and from Maillard products. Flavor changes during storage are discussed in Section 20.5.

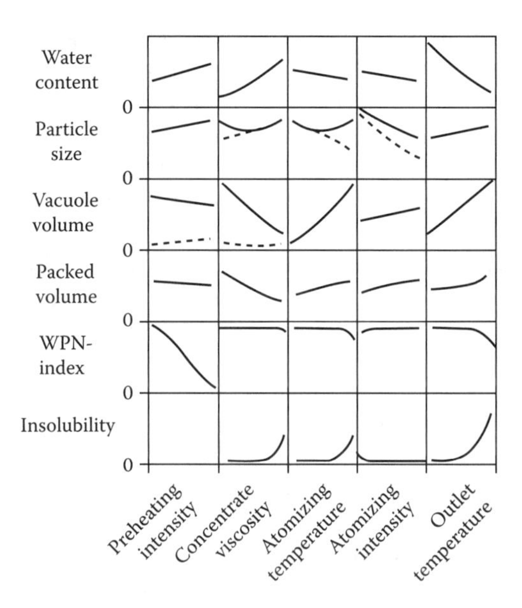

FIGURE 20.5 The influence of the intensity of preheating of the milk, of the concentrate viscosity (hence, extent of evaporation), of the temperature and the intensity (disk speed, pressure) of atomization, and of the outlet temperature of the drying air on some properties of spray-dried milk. Very approximate. (---) denotes conditions at which vacuoles develop hardly or not at all.

20.4.8 Conclusions

Figure 20.5 summarizes the influence of some process variables on six product properties. Such a summary is inevitably an oversimplification, as the relationship between two parameters may depend on other variables. For instance, the curves for the WPN index (except for the extreme left-hand one) have been drawn for a mild preheating; otherwise, they would be at a much lower level. Moreover, it is mostly not possible to vary only one process variable. Except in the upper row, it has been assumed that the conditions (usually the concentrate supply in the drying chamber) have been adjusted in such a way that the water content of the powder remains constant. An exception is the outlet temperature: if this is very low the water content of the powder will inevitably be high, and additional drying should then follow, for instance in a fluid bed. It may be added that such a second-stage drying provides increased possibilities to make powder with various desirable characteristics.

20.5 DETERIORATION

The most important variable determining the rate of undesirable changes in milk powder is the *water content*. When comparing different types of powder, it is probably easiest to consider water activity (a_w), as for instance in Figure 10.5.

Examples of the relation between water activity and water content are shown in Figure 10.3 and Figure 10.18. The relationship depends on the composition of the product; some examples are given in Table 20.5. The higher a_w of whole milk powder as compared to skim milk powder of the same water content is caused by the fat not affecting a_w. Whey powder has an a_w slightly different from that of skim milk powder because in a dry product the soluble constituents (especially sugar and salts) decrease a_w somewhat less than casein. This is only true, however, as long as all lactose is amorphous, which often does not apply to whey powder. The data in Table 20.5 show that a_w is considerably reduced if lactose crystallizes without absorption of water by the powder (compare rows 1 and 5), at least if a_w is less than about 0.5 (see Figure 10.3b). Crystalline lactose binds water very strongly, and that is also why the usual oven-drying methods to estimate the water content do not include the bulk of the water of crystallization. If the water content excluding the water of crystallization is taken as a basis, then a_w is even higher for the powder with crystallized lactose; compare rows 1 and 6 in Table 20.5.

It is thus advisable to make milk powder sufficiently dry and to keep it in that condition. If it is not hermetically sealed from the outside air, it will attract water in most climates. The higher the temperature, the higher the water activity; see Table 20.5 (compare rows 1 and 4) and Figure 10.19. Because several reactions are faster at a higher a_w, this implies that a temperature increase may well cause an extra acceleration of deterioration.

The latter effect may be especially strong if the powder loses its glassy state. Figure 10.4b shows that a lactose–water mixture will be at most ambient temperatures in the glassy state if its water activity is below 0.3. Because lactose is the dominant component of the amorphous material in a powder particle, about the same relation is supposed to hold for the powder, which has been experimentally confirmed. This means that most dairy powders are in the glassy state (i.e., the

TABLE 20.5
Approximate Water Activity of Various Kinds of Spray Powder as a Function of the Water Content and of Some Other Variables

Powder Made of:	Temperature (°C)	State of Lactose	Water Content (% w/w)			
			2	3	4	5
1 Skim milk	20	Amorphous	0.07	0.13	0.19	0.26
2 Whole milk	20	Amorphous	0.11	0.20	0.30	0.41
3 Whey	20	Amorphous	0.09	0.15	0.20	0.26
4 Skim milk	50	Amorphous	0.15	0.24	0.33	0.42
5 Skim milk	20	Crystalline[a]	0.02	0.04	0.06	0.12
6 Skim milk	20	Crystalline[b]	0.09	0.16	0.25	0.38

[a] Water content of powder includes water of crystallization; the lactose is crystallized insofar as sufficient water is present for the crystallization.
[b] Water content of powder does not include water of crystallization.

nonfat part of the material), except if the water content is high and the temperature is also high. A change in conditions leading to a glass–liquid transition will strongly accelerate most reactions and physical changes occurring in a powder.

Microbial and *enzymic* deterioration are rare in milk powder. For microbial deterioration to occur, a_w should increase to over 0.6 (and for the majority of microorganisms much higher); such a high a_w can only be reached if the powder is exposed to fairly moist air. Deterioration then is often caused by molds. Enzymatic hydrolysis of fat has been observed at $a_w \geq 0.1$, although extremely slow. Accordingly, whole milk powder must be free of lipase. Milk lipase will always be inactivated by the intense pasteurization of the milk as applied in the manufacture of whole milk powder. This is by no means ensured, however, for bacterial lipases. Hence, not too many lipase-forming bacteria should occur in the raw milk. Proteolysis in milk powder appears highly improbable and has not been reported.

Of course, enzymic deterioration of liquid products made from the milk powder can occur if enzymes are present before the drying, as drying usually does not cause substantial inactivation of enzymes (see Subsection 10.4.4).

Caking. What may be noticed first when milk powder or whey powder absorb water from the air, is the formation of lumps; eventually, the whole mass of powder turns into a solid mass (cake). Crystallization of lactose is responsible, as it causes the powder particles, largely consisting of lactose, to grow together (to sinter). Because water is needed for crystallization of α-lactose, caking does not occur at low a_w, say, below 0.4. At a higher temperature crystallization can occur far more readily, a_w being higher; moreover, the viscosity of the highly concentrated lactose solution (essentially the continuous phase of the powder particles) is lower, causing nucleation, hence crystallization, to be faster.

The susceptibility to caking, especially high in whey powder, is considerably reduced if most of the lactose is crystallized before the drying (in the concentrate). Such precrystallized powder is usually called 'nonhygroscopic,' which may be a misnomer because the powder concerned does not attract less water (this is determined by its a_w in relation to that of the air), but the consequences are less noticeable.

Maillard reactions increase considerably with water content (see Figure 10.5 and Figure 10.12) and with temperature. They lead to browning and to an off-flavor. The 'gluey' flavor that always develops during storage of dry milk products with a too-high water content is usually ascribed to Maillard reactions; the main component appears to be *o*-aminoacetophenone. If extensive Maillard reactions occur, they are always accompanied by insolubilization of the protein. Accordingly, the insolubility index increases when milk powder is stored for long at a high water content and temperature; at a normal water content the ADMI number (see Table 20.4) may increase to 0.5 in 3 years time.

Autoxidation of the fat and the ensuing tallowy off-flavor pose a difficult problem when storing whole milk powder. The rate of autoxidation strongly increases with decreasing a_w (see Figure 10.5); however, to prevent other types of deterioration (especially Maillard reactions) a_w should be as low as possible. The effective Q_{10} of the autoxidation reaction in milk powder is relatively low (about 1.5) because a higher temperature also causes higher a_w.

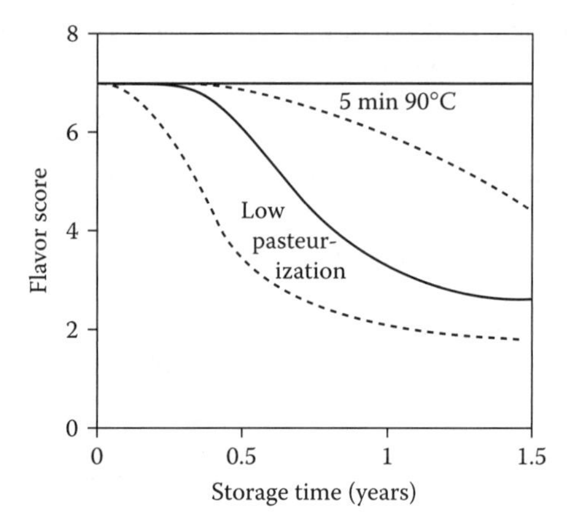

FIGURE 20.6 Influence of preheating and of gas flushing, whether (—) or not (---), of whole milk powder on its flavor score (scale 0 to 8) during storage at room temperature under exclusion of air. (Data from E.A. Vos and J.J. Mol, NIZO-Mededelingen M12, 1979.)

To keep the autoxidation within reasonable limits for a long time, a number of measures should be taken (see also Subsection 2.3.4):

1. The milk should be intensely heated to form antioxidants (see Figure 20.6). The problem is, of course, that the heat treatment also causes a distinct cooked flavor.
2. The water content of the powder should be adjusted as high as possible without causing Maillard reactions to occur too fast; the most suitable water content is generally 2.5 to 3%.
3. Oxygen should be removed as effectively as possible (by gas flushing; see Figure 20.6). A problem is that the vacuoles in the powder particles contain some air, hence, O_2. Either the powder should contain hardly any vacuoles or the gas flushing should be repeated after a few days. Equilibrating the gas inside and outside the vacuoles by diffusion takes several days in whole milk powder (several weeks in skim milk powder). Equilibration is faster if the powder particles have a greater number of cracks (see Subsections 20.4.1 and 20.4.2).
4. The powder should be packaged in such a way that air and light are kept out. Generally, this implies packaging in cans.
5. Rigorous measures should be employed against contamination of the milk with copper.
6. Intensive homogenization of the concentrate should be carried out.

A so-called 'toffee flavor' can develop during storage of whole milk powder with a high water content at high temperature. It can be ascribed to formation of δ-decalactone and related compounds in the fat. The compounds involved are not formed by oxidation.

Loss of nutritive value during storage primarily concerns loss of available lysine due to Maillard reactions. Storage at 20°C at a normal water content does not cause an appreciable loss; at 30°C, a loss of 12% after storing for 3 years has been reported. Extensive Maillard reactions cause a decrease in protein digestibility and formation of weak mutagens.

Extensive autoxidation results in formation of reaction products between hydroperoxides and amino acid residues (this partly gives methionine sulfoxide), and between carbonyl compounds and ε-amino groups; this may cause the biological value of the protein to decrease slightly. Of greater concern is the loss of vitamin A in vitamin-fortified skim milk powder, due to its oxidation. This especially occurs if the vitamin preparation is dissolved in oil and then emulsified into the skim milk before atomization. Usually, dry added preparations are more stable. It is, however, very difficult to homogeneously distribute a minute amount of a powder throughout a bulk mass.

20.6 OTHER TYPES OF MILK POWDER

Roller-dried milk looks completely different from spray powder in the microscope. It consists of fair-sized flakes. Due to the intense heat treatment during the drying it has a brownish color, a strong cooked flavor, and the availability of lysine has been considerably reduced, by 20 to 50%.

Freeze-dried milk consists of coarse, irregularly shaped, and very voluminous powder particles, which dissolve readily and completely. However, the fat globules show considerable coalescence, unless intense homogenization has been applied. In most cases, damage due to heat treatment is minimal.

Suggested Literature (See also Chapter 10)

A general overview on an introductory level: M. Carić, *Concentrated and Dried Dairy Products,* VCH, New York, 1994.

Properties of dried milk products: S.T. Coulter and R. Jenness, in: W.B. van Arsdell et al., Eds., *Food Dehydration*, 2nd ed., Vol. 2, AVI, New York, pp. 290–346, 1973.

A survey of the bacteriology of spray-dried milk powders: J. Stadhouders, G. Hup, and F. Hassing, *Neth. Milk Dairy J.*, **36**, 231–260, 1982.

21 Protein Preparations

A wide variety of dried, milk-protein-rich products are produced. Traditionally, isolated casein was used for the manufacture of synthetic wool and buttons, for sizing paper (i.e., to make it smooth and easy to write on), etc. These uses have much diminished. Currently, the main use of various types of casein and of whey-protein preparations is in foods. The reasons may be to:

1. Provide foods with a *specific nutritive value:* This may concern products meant for specific groups of people, the most important example being infant formulas. Partially hydrolyzed proteins, i.e., peptide mixtures, are applied for people that are allergic to certain proteins. Another application is boosting protein content and nutritional quality of a product by adding milk protein preparations, e.g., to beverages or biscuits. See also Subsection 2.4.5.
2. *Replace more expensive proteins:* Most animal proteins are more expensive than vegetable proteins, but milk proteins can be comparatively cheap, as compared, e.g., to egg-white protein. Moreover, isolation of vegetable proteins such that these are sufficiently pure, flavorless and functional (e.g., soluble), often is expensive. Protein-rich whey products can (partially) replace skim milk in ices, desserts, beverages, calf milk replacers, etc.
3. Provide a product with *specific physical properties:* Examples are the preparation of stable emulsions (salad dressings, desserts) and of foam products (toppings, meringues), or the prevention of segregation of moisture and fat in meat products. This means that the proteins used must have specific functional properties.
4. *Make novel products:* An increasing trend in the food industry is the production of 'manufactured foods' from fairly pure and durable components. Examples are cheese-like sandwich spreads, coffee creamers, and meat analogues.

Of course, application of milk proteins may well be for more than one of the reasons mentioned.

Price and functional properties of milk protein preparations are of paramount importance for their possible use. In addition, keeping quality, flavor, or rather lack of flavor (pure proteins are virtually flavorless), and dispersibility are of importance.

Table 21.1 gives examples of milk protein products and their gross composition.

TABLE 21.1
Examples of Milk-Protein-Rich Preparations, Including Their Approximate Composition

Product	Method of Preparation	Isolated from	Approximate Gross Composition (%)				
			Protein	NPN[a]	Carbohydrate	'Ash'	Fat
Rennet casein	Renneting	Skim milk	83	~ 0	0.5	8	2
Acid casein	Acid	Skim milk	90	~ 0	0.5	2.5	2
Na-caseinate	Acid + NaOH	Skim milk	86	~ 0	0.5	5	2
Phosphocaseinate	MF/DF	Skim milk	83[b]	4[c]	1	8	1
Whey powder	Spray dry	Whey	10.5	1.5	71	9	1
WP concentrate[d]	UF	Whey	31	4	51	7	2
WP concentrate	UF	Whey	57	3	26	4	3
WP isolate	UF/DF	Whey	88	1	1	3	3
'Lactalbumin'	Heat + acid[e]	Whey	78	?	9	5	2
Coprecipitate	Heat + acid[e]	Skim milk	83	?	1	9	2

Note: Abbreviations: DF = diafiltration; MF = microfiltration; UF = ultrafiltration; WP = whey protein.

[a] 6.38 × nonprotein nitrogen.
[b] Casein.
[c] Noncasein protein.
[d] Also called skim milk replacer.
[e] And/or $CaCl_2$.

21.1 MANUFACTURE

Some milk protein products can be readily made by classical methods, e.g., acid casein from skim milk. Whey proteins can be obtained in a fully denatured form by heating acidified whey. The availability of a range of membrane processes has offered new possibilities, especially for obtaining undenatured proteins from whey (which has significantly increased the market value of cheese whey). Chapter 12 gives basic information on membrane processing.

The properties of the milk protein preparations may greatly depend on the pretreatment of the milk or the whey. Heat treatment is required to kill bacteria and to inactivate enzymes. It can cause denaturation, and thereby decreased solubility of serum proteins. Most of the denatured serum proteins are associated with the casein if the skim milk had been heated. The separation efficiency determines the fat content of the preparations. This content depends also on the extent to which fat globules have become covered with plasma proteins, caused by their disruption or loss of membrane (e.g., due to beating in of air). Such fat globules follow the protein during its separation and can hardly be removed from the protein preparation by the common purification methods. Bacterial spoilage and plasmin activity can cause proteolysis.

A wide variety of process schemes is applied for the accumulation of the proteins and for the further concentration and drying of the preparations. Some accumulation processes are described below. The proteins are obtained either as a coagulated mass or as a solution/colloidal dispersion. A coagulate is generally treated by: (1) 'washing' to (partly) remove undesirable solutes; (2) removal of moisture by pressing; (3) heat drying by any of a variety of processes, often leading to a granular product; and (4) grinding the latter to a fine powder. Composition of the product and processing conditions determine its solubility. Liquid preparations are generally concentrated by evaporation or reverse osmosis, and subsequently spray-dried. The resulting powder generally is well soluble.

When obtaining proteins from skim milk or whey, large quantities of liquid remain. These are often further processed to obtain other marketable products, especially lactose. Nevertheless, waste products remain, and a critical problem is whether these can be utilized or have to be discarded. In the latter case, expensive purification measures may be required.

21.1.1 Casein

Casein preparations are made from sharply skimmed milk. Preferably, the heat treatment of the (skim) milk is such that very little serum protein becomes denatured. Denatured protein ends up in the casein product. Casein is most often accumulated by rendering it insoluble. As discussed in Subsection 3.3.4, (skim) milk then tends to form a gel, still containing all of the liquid. To prevent this, the liquid should be stirred, causing the gel directly to be broken into pieces. The latter should lose most of their liquid, which can be achieved at raised temperatures, which greatly enhances syneresis rate (see Figure 24.12). Due to stirring and scalding (cooking), most of the liquid then is removed from the gel fragments.

Various types of casein preparations are being made.

1. *Rennet casein:* The casein is rendered insoluble by addition of calf rennet, followed by stirring at a temperature increasing to about 55°C. The fine syneresed curd particles so formed are separated by centrifugation or by using a vibrating sieve, washed with water, pressed to remove moisture, and then dried, for instance, in a drum or a belt drier. The resulting product is composed of calcium paracaseinate–calcium phosphate, with some impurities. It is insoluble in water and has a high 'ash' content. It is devoid of the caseinomacropeptide split off κ-casein by the rennet enzymes; this means a loss of about 4% by weight of the casein. On the other hand, it contains part of the proteose peptone.
2. *Acid casein:* Skim milk is acidified, while stirred, with hydrochloric acid (mostly), lactic acid, or sulfuric acid, until the isoelectric pH of casein (4.6) is reached. The casein then is insoluble. The temperature applied is quite critical. At a high temperature large lumps are formed, which are difficult to dry; a low temperature causes a fine voluminous

precipitate, which is hard to separate. The optimum temperature is about 50°C. The process is continued as described for rennet casein. Ideally, the preparation contains all of the casein and none of the colloidal calcium phosphate. The casein can be further purified by dissolving it in alkali, again precipitating it, etc. Acid casein is insoluble in water; the dried product is poorly dispersible in alkaline solutions because persistent lumps are formed.

3. *Caseinates:* Acid-precipitated casein can be dissolved in alkali (e.g., NaOH, KOH, NH_4OH, $Ca(OH)_2$, and $Mg(OH)_2$), and the resulting solution can be spray-dried. Na-caseinate is the most common product. K-caseinate sometimes is preferred for nutritional purposes. Ca-caseinate has somewhat different physicochemical characteristics as compared to Na- or K-caseinate. These products can be well soluble in water and be almost flavorless — if the pH during manufacture was never higher than 7. A combination of a high pH and a high temperature will result in formation of some lysinoalanine, which is considered undesirable (see Subsection 7.2.2).
4. *Micellar casein:* This can be obtained from skim milk by microfiltration; the product is generally called *phosphocaseinate*. When using a membrane of pore size 0.1 μm, by far most of the casein is retained (unless the separation temperature is quite low; Figure 3.20), whereas virtually all serum proteins can pass the membrane. Diafiltration with water is used to further remove dissolved substances. The micelles obtained appear to have properties close to those of natural casein micelles, also after drying and dispersion in, say, skim milk ultrafiltrate. Currently, phosphocaseinate is primarily used in research experiments.

21.1.2 Whey Protein

Large quantities of whey are used to obtain protein preparations and other products. It should be realized that the whey may vary substantially in composition. The main types are:

- Whey obtained in classical cheese making. Besides the soluble components of milk it contains: the caseinomacropeptide split off κ-casein; active rennet enzymes; starter bacteria, which have produced, and can produce additional, lactic acid from lactose, whereby the pH is decreased; and some globular fat, e.g. 0.3%, which generally is largely removed beforehand by centrifugal separation. The acidity of the whey greatly varies with the type of cheese made. Moreover, (part) of the whey may have been diluted with water and/or contain some added nitrate.
- Whey with a high NaCl content. This concerns, for instance, a small part of the whey resulting from Cheddar-type cheese making (see Subsection 27.4.1).

- Whey resulting from the manufacture of rennet casein; see Subsection 21.1.1. This whey is low in fat content and contains no starter bacteria, nor lactic acid. Otherwise, it is much the same as cheese whey.
- Whey resulting from the manufacture of acid casein (Subsection 21.1.1). The fat content is low and it contains no rennet or casein macropeptide. The pH is about 4.6, and it contains increased amounts of calcium and phosphate (see Figure 2.7).
- The permeate obtained by microfiltration of skim milk (mentioned above), although it is properly speaking not whey. It is very similar in composition to milk serum, although it contains a little casein. It is devoid of fat globules.

In all cases, about half of the protein in whey consists of β-lactoglobulin, and its properties tend to dominate those of the whey protein product. However, the difference in protein composition between the various whey types will be reflected in the whey protein preparations, and may significantly affect some properties.

Several types of protein preparations are obtained from whey. The main types will be briefly discussed.

Whey Protein Concentrate (WPC). The name is generally used for preparations containing 35 to 80% 'total protein.' As indicated in Table 21.1, the proportion of total N that consists of NPN depends on processing conditions; it also depends on the type of whey used. The preparations with a low protein content are often called *skim milk replacers.* The gross composition is similar to that of skim milk, although the protein composition is, of course, very different.

Traditionally, WPC is made of delactosed, desalted whey. The process starts by concentrating whey about tenfold by evaporation, to obtain lactose crystals. The mother liquid then is desalted in one way or another (see Section 12.4). Drying of the remaining liquid results in WPC. A considerable proportion of the nitrogen is NPN, generally more than 20%.

To obtain a higher protein concentration and a purer product, utltrafiltration is generally used. See Table 12.2 for the effect of some variables on the liquid obtained. To make still purer products, diafiltration can be applied.

The liquid is commonly spray-dried and the resulting WPC powders can be highly soluble. The part of the protein that does not dissolve because it is heat-denatured greatly depends on the pH and the Ca^{2+} activity during heating (Figure 21.1). Adding Ca salts after heating requires a higher concentration of Ca^{2+} to insolublilize the protein.

Whey Protein Isolate (WPI). The name is generally reserved for preparations in which 90% (or more) of the dry matter consists of whey protein. Preferably, the whey used for its manufacture is relatively pure; it is even better to use a microfiltration permeate of skim milk. WPI can be produced, like WPC, by ultrafiltration; at least one diafiltration step then is necessary.

Several other processes have been developed, and one example will be mentioned. This involves ion exchange in a stirred-bed reactor. The whey is acidified to pH 3.2, where the proteins present are predominantly positively charged. The ion exchange resin, which is in the form of porous granules, is negatively charged. Hence, the proteins become adsorbed and the solution can be removed by sieving. After washing the granules, alkali is added to a pH of about 8, leading to desorption of the protein. When, moreover, the raw material is an MF permeate, the protein contains no caseinomacropeptide. The protein is for almost 80% β-lactoglobulin and some 15% α-lactalbumin; it contains little of the other serum proteins. This composition markedly enhances gel forming properties. Besides, the protein is almost fully devoid of lipids, which is of advantage when using it for foam formation.

Lactalbumin. It has long been practice to heat sour cheese whey to precipitate its protein (see Figure 21.1) and to recover it. Obviously, the precipitate is impure. It is pressed, usually salted, and sometimes matured. This is 'whey cheese,' e.g., Ricotta or Ziger. A similar process can be applied to yield denatured whey protein. Often, $CaCl_2$ is added besides acid. The precipitate obtained is washed and dried, for example in a drum drier. This protein preparation is called 'lactalbumin' (not to be confused with the serum protein α-lactalbumin). It contains hardly any proteose peptone, caseinomacropeptide, or NPN. The high lactose content and the slowness of drying are responsible for extensive Maillard reactions. 'Lactalbumin' is insoluble in water.

21.1.3 Other Products

A few of the many other protein products will be mentioned:

Coprecipitate: This is made from skim milk or buttermilk in much the same manner as in the manufacture of 'lactalbumin'. Most of the skim milk proteins, except proteose peptone, are isolated in an insoluble state.

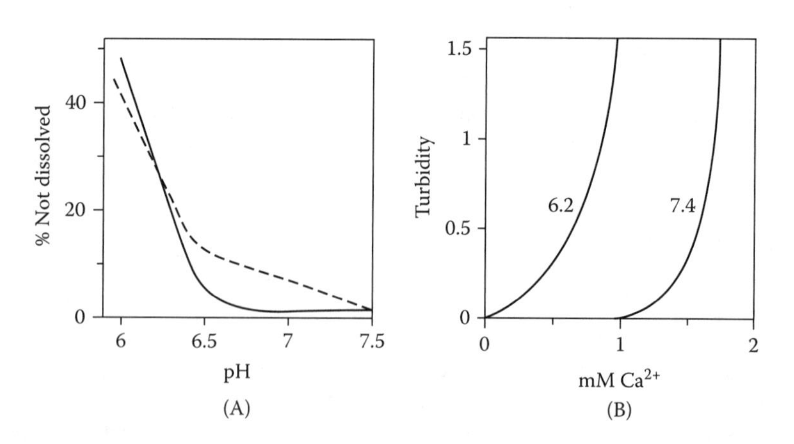

FIGURE 21.1 Solubility of β-lactoglobulin (—) and of purified whey protein (---) as a function of the pH and the concentration of dissolved calcium during heating for 10 min at 80°C. (A) Part of the protein that can be separated centrifugally; low Ca concentration. (B) Turbidity of 1% solutions as a measure for the protein aggregation; parameter is the pH. Approximate results. (Adapted from J.N. de Wit, *Neth. Milk Dairy J.*, **35**, 47, 1981.)

Nevertheless, the protein is easily digested, and has a high nutritive value. Most preparations contain much calcium. Coprecipitate may contain far less Maillard products than 'lactalbumin,' because its sugar content is lower.

Most of the protein of skim milk can also be precipitated by adding ethanol and heating to, say, 90°C.

Separate proteins: These can be obtained on relatively small scales by a range of methods, involving gel filtration, column electrophoresis, various kinds of chromatography, etc. A range of proteins is obtained from milk, also minor proteins like lactoferrin, lactoperoxidase, or specific immunoglobulins; these are used in some pharmaceuticals and cosmetics. Enrichment of certain proteins is also practiced. For instance, when applying ultrafiltration to a Na-caseinate solution of 4°C, most of the α-casein (i.e., $\alpha_{s1} + \alpha_{s2} + \kappa$) remains in a kind of micellar form and is retained. Much of the β-casein is dissociated at 4°C (see Figure 3.20) and passes the membrane. By heating the permeate to 45°C and ultrafiltering it again, β-casein is accumulated in the retentate.

Modification of proteins: This is generally restricted to their partial hydrolysis. Chemical hydrolysis, using acid or alkali, and enzymic hydrolysis are applied. Partially hydrolyzed casein can be used in foam products, especially if the foaming liquid is highly viscous. Site-directed enzymic hydrolysis can be applied, e.g., to β-lactoglubulin-rich preparations, to obtain peptides with reduced allergenic properties.

21.2 FUNCTIONAL PROPERTIES

A functional property of a material is its ability to produce a specified property in the product in which the material is applied. Proteins can produce nutritional value and a number of physico-chemical properties. The latter are considered here.

The functional performance of a protein primarily depends on its molecular structure and its concentration, but also on several other variables. The composition of the protein preparation is one aspect: what proteins are present, to what extent have these been modified during manufacture (denaturation, proteolysis, cross-linking, etc.), and what is the further composition. The performance naturally also depends on the environment of the protein during its application: temperature; pH; ionic strength and composition; solvent quality; presence of compounds, including enzymes that can chemically modify the protein, etc.

The literature gives various classifications of the physico-chemical functionalities, and different authors consider different aspects to be of importance. The confusion partly stems from the desire to have a few simple tests to evaluate the performance of proteins. Unfortunately, the results do often not correspond to practical experience. This may be because the situation during application of the protein is far from simple. For instance, a number of separate effects can often be identified, which depend in a different way on protein structure and environmental conditions. Or what is considered to be one and the same property, is observed to be a series of properties, according to the specific application that the user has in mind. Examples will be mentioned in the following text.

In this section, some groups of essential functional properties will be briefly discussed. The list is by no means exhaustive. It should be mentioned that the subject is huge and complicated, and that only a few aspects of each of the functionalities can be discussed.

21.2.1 Solution Properties

Rennet casein, 'lactalbumin,' and coprecipitate are preparations that are fully insoluble in water. They are applied in solid-like products such as biscuits. They can also be used to make 'texturized' products in an extruder, or for dry spinning.

In almost all other products the protein to be used must dissolve. Often, solubility is not primarily a functional property but a prerequisite for the other functional properties. See also Subsection 20.4.5.

21.2.1.1 Solubility

The solubility of proteins is discussed in Subsection 2.4.1.4. Milk proteins (with the exception of those in the fat globule membrane) are well soluble at physiological conditions, but not at all other conditions. The latter is illustrated for casein in Figure 21.2, which is almost fully insoluble at its IEP and also at very low ionic strength. The serum proteins remain soluble near the IEP, but less so, as is illustrated for β-lactoglobulin in Figure 2.23b.

However, the 'solubility of protein preparations' often means something different. To estimate this solubility, a dispersion of the preparation is made in a buffer solution. The mixture then is stirred in a standard way and the fraction of the protein that does not sediment in a centrifuge test is determined. A result of 50% 'soluble' can imply that half of the protein dissolves well in a little bit of solvent, whereas the remainder is insoluble; in other words, dilution then will

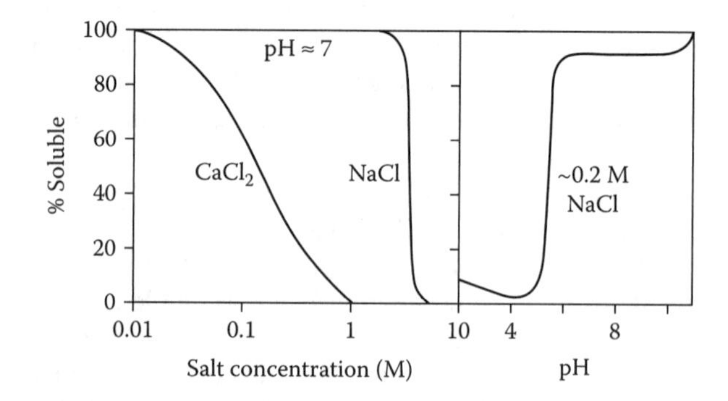

FIGURE 21.2 'Solubility' of Na-caseinate (3% in water) as a function of salt concentration and pH.

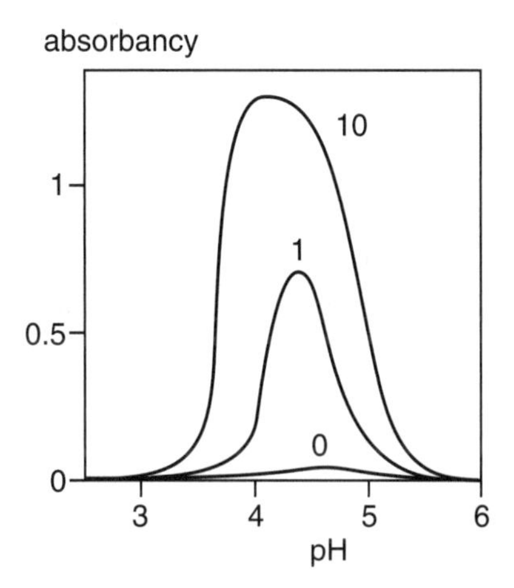

FIGURE 21.3 Turbidity (expressed as absorbancy) of solutions of a whey protein isolate, heated at 70°C for various times (indicated, in minutes), as a function of pH. (Adapted from results by H. Zhu and S. Damodaran, *J. Agr. Food Chem.*, **39**, 1555, 1988.)

not cause an increased dissolution of protein. If the result would represent a true solubility, however, all of the protein will dissolve if the double amount of solvent is used. In practice, the situation is mostly intermediate.

The rationale for the test seems to be that preparations often consist of a mixture of well soluble and poorly soluble protein, for instance, because part of it has been denatured. However, the result then may greatly depend on the test conditions, such as the time and intensity of centrifugation. Another manner to estimate protein aggregation caused by denaturation is to determine the turbidity of the dispersion. An example for whey proteins is given in Figure 21.3.

21.2.1.2 Viscosity

General aspects of the viscosity of solutions and dispersions are discussed in Subsection 4.7.1. From that discussion it follows that solutions of globular proteins

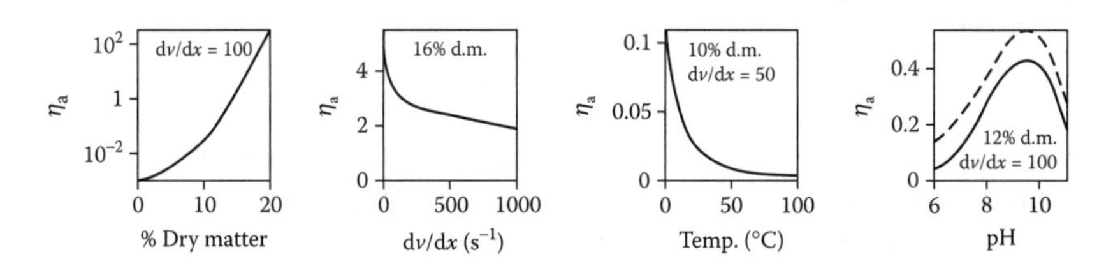

FIGURE 21.4 Apparent viscosity (η_a, Pa · s) of solutions of Na-caseinate in water (—) or 0.2 M NaCl (---), at pH ≈ 7 and room temperature, unless stated otherwise. d*v*/d*x* = shear rate (s^{-1}).

do not yield high viscosities, unless the protein concentration is very high. This is indeed observed in whey protein solutions. Heat denaturation tends to cause unfolding and aggregation of the proteins, both leading to an increase in viscosity; the results greatly depend on several conditions.

Casein molecules, on the other hand, tend to unfold to some extent in solution, whereby the viscosity is considerably increased (although not nearly as strongly as several polysaccharides can do). Examples are given in Figure 21.4. It is seen that the results significantly depend on the shear rate; therefore, apparent viscosities are determined. Ca-caseinate gives lower results than Na-caseinate at the same concentration and ionic strength. The viscosity tends to increase markedly at high salt contents, presumably due to 'salting out.' It appears that an essential factor in causing a high viscosity is association of the molecules, both by hydrophobic and electrostatic interactions.

21.2.2 Gels

A gel has a structure consisting of a space-filling network of strands of small particles or polymer molecules, holding a relatively large amount of liquid. In rheological terms, a gel is an elastic or viscoelastic soft solid.

Functionalities. An important functional property of milk proteins is that they can be used to make gels. The ability is generally characterized by the value of the elastic shear modulus G obtained (G equals the applied stress divided by the resulting shear deformation, as long as the latter is quite small). However, gels are mostly used in situations where large deformation and/or fracture prevails, especially during eating. The functionality then depends on the values of the stress and the strain at fracture, determined at the strain rate prevailing in practice (basic aspects are briefly discussed in Subsection 25.6.1). Unfortunately, such studies are rarely performed.

A gel can also be made in order to *immobilize the liquid* in the gel. (As mentioned in Subsection 2.4.1.4, only a very small fraction of the water can be considered to be bound.) The functionality then depends on the permeability B of the gel (see Equation 24.1). A smaller B value is obtained in a gel with smaller pores, therefore, with thinner strands and higher protein concentration. Another important gel property is its *stability*, especially against shrinking, which means the occurrence of syneresis (expulsion of liquid).

Caseinate Gels. In Subsection 3.3.4, skim milk gels are discussed in some detail. The results given for acid gels apply also to acid gels made of, say, Na-caseinate, although not precisely. Rennet gels can be made of Ca-caseinate, but that is not done in practice: the gels are too weak and are prone to extensive syneresis unless kept refrigerated. Acid casein gels may show syneresis at temperatures above 30°C.

Whey Protein Gels. If a solution of, for instance, β-lactoglobulin is heated, a gel forms that subsequently increases in modulus. The gelation is irreversible: upon cooling the gel does not disintegrate, it even increases in modulus. These events are illustrated in Figure 21.5a.

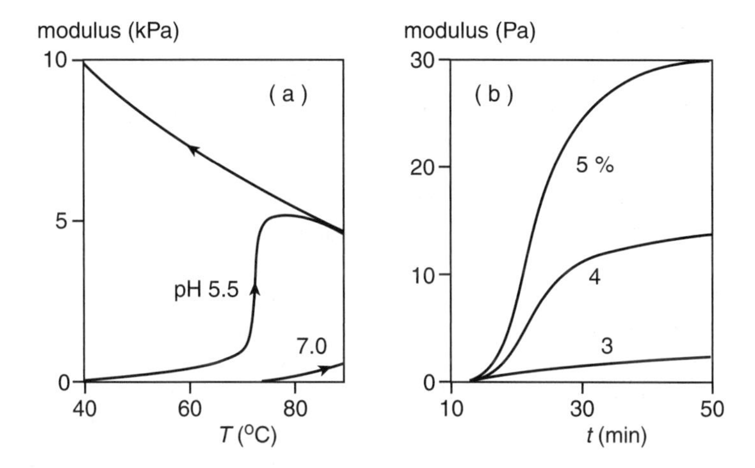

FIGURE 21.5 Gelation of β-lactoglobulin solutions. (a) Shear modulus as a function of temperature. Heating and cooling (the rates are not known) of a 10% solution at pH 5.5; arrows indicate the sequence. Heating at pH 7.0 is also shown. (b) Shear modulus as a function of time. Heating of solutions of various concentrations (indicated) to 90°C in 15 min, after which the temperature was kept constant. pH 7.0, 0.34 M NaCl.

This behavior is typical for many globular proteins. At high temperature the protein denatures and thereby becomes poorly soluble. It forms aggregates, and these eventually give a gel. Intermolecular cross-links are formed in the form of –S–S– bridges. This happens especially with β-lactoglobulin, which has an –SH group. However, the other globular proteins take part in the gelation, whereas caseinomacropeptide and proteose peptone do not.

According to physicochemical conditions, gels of different structures are formed. At a pH remote from the IEP (say, pH 7) and low ionic strength (say, 10 mM), the gels are *fine-stranded*: the strands are relatively long and thin, e.g., 30 nm. It may be noticed that this is nevertheless about 5 times the molecular diameter. Gelation temperature and rate are relatively high and slow, respectively. The gel is clear, its modulus is relatively low, and it is strongly deformable (large fracture strain). Near the isoelectric pH and at high ionic strength (e.g., 0.2 M), *coarse* gels are formed, resembling typical particle gels. The particles are about spherical and have a size of, for instance, 1 μm. These gels are turbid; the modulus is relatively high, and they are relatively brittle. At intermediate conditions, gels of intermediate types are formed. The rate of heating also affects gel properties.

Figure 21.5 gives examples of β-lactoglobulin gel properties. These are comparable, though not identical, to whey protein gels. Notice that fairly high protein concentrations are needed to obtain stiff gels.

Another phenomenon is *cold gelation*. If a whey protein solution is heated at a pH above 7 and very low ionic strength, little aggregation occurs and gelation does not occur. If the solution is subsequently acidified without stirring (this can be done by adding a slowly dissociating lactone), a gel is formed.

21.2.3 Emulsions

Milk proteins are often used to help making oil-in-water emulsions and to stabilize these against physical changes. This implies a number of functional properties. Often, simple tests are applied that yield, for instance, an 'emulsifying activity index'. Such tests make little sense unless the experimental conditions, especially homogenization, precisely mimic the practical situation.

Preferably, the following aspects are separately considered.

1. *Protein load:* The protein layer formed around the oil droplets can be characterized by the mass of protein per unit oil surface area. This surface load determines the amount of protein needed to produce an emulsion; it closely depends on the solubility of the protein. Globular proteins and well-dissolved Na-caseinate give a surface load of some 2.5 $mg \cdot m^{-2}$, while Ca-caseinate and micellar casein give far higher values. Poorly soluble proteins cannot be used for emulsification. Obviously, the load depends also on the amount of protein available per unit oil surface area created, as illustrated in Figure 21.6a and on factors affecting the conformation of the protein (pH, ionic strength, specific salts, temperature, and heat treatment).
2. *Droplet size:* The size of the droplets obtained depends on the emulsifying intensity and on the concentration and type of the protein. Figure 21.6b gives examples of the latter point. To obtain the same specific surface area, more protein is needed if its molar mass is higher or when it is in an aggregated state, but when the protein is relatively abundant, the differences between various proteins are small. However,

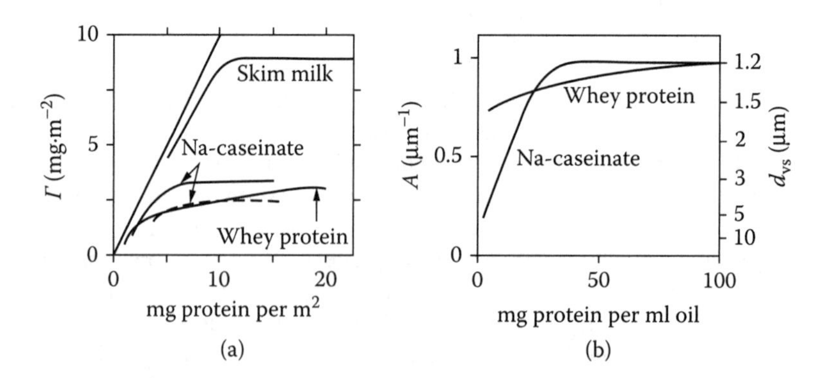

FIGURE 21.6 Emulsifying properties of milk proteins. (a) Surface load (Γ) as a function of protein initially available in the solution per m^2 oil surface area created by emulsification. Na-caseinate in 0.2 M NaCl (—) and in water (---). The straight line gives the Γ value that would be acquired if all protein available were adsorbed. (b) The specific surface area (A) obtained as a function of available protein per ml of oil; volume fraction of oil = 0.2. (Approximate examples at otherwise constant conditions.)

whey proteins are not very suitable for making emulsions near pH 5, where their solubility is small.

Proteins are not very good emulsifiers, because they do not give a very low interfacial tension between oil and water; with several small-molecule surfactants much smaller droplets can be obtained. This means that a relatively high homogenization pressure is needed to obtain small droplets (see Section 9.4). Small droplets are needed for emulsion stability, particularly against creaming and partial coalescence. Proteins cannot be used to make water-in-oil emulsions.

3. *Coalescence:* The stability of protein-covered oil droplets against coalescence tends to be very good, provided that the droplets do not aggregate, the protein load is at its plateau value, and the droplets are reasonably small (below, say, 10 μm). There are no great differences among proteins. Small peptides give a lower stability.

   Whey proteins are not suitable to stabilize emulsions that have to be heated; the denaturation of the protein presumably leads to bare patches on the droplets, and thereby to coalescence. Na-caseinate can be used to yield emulsions that remain stable during heat treatment, evaporation, spray drying, storage of the powder obtained, and when dissolving the powder.
4. *Partial coalescence:* This can occur in emulsions when part of the oil has crystallized. It is discussed in some detail in Subsection 3.2.2.2. Stable emulsions can be made both with caseinates and with whey proteins, provided that the globules are small, say less than 2 μm, and their protein load is at the plateau value.
5. *Aggregation:* Emulsion droplets tend to aggregate when conditions become such that the protein applied is insoluble. For instance, droplets coated with caseinate aggregate on adding rennet, bringing to a pH of about 4.6, or adding much $CaCl_2$.

21.2.4 Foams

Foams are generally made by beating a suitable liquid with air (or another gas). The important results of that process are the overrun, i.e., the relative increase in volume due to air inclusion, and the bubble size, often between 20 and 200 μm in food foams. Foams are relatively unstable, the lifetime being often less than an hour, as compared to at least a year for several emulsions. Three types of instability occur, although it may not be easy to distinguish between them. It concerns: (1) *drainage* of liquid from the foam layer; (2) *coalescence* of bubbles with each other and with the air above the foam; and (3) *Ostwald ripening* (see also Figure 3.1). Ostwald ripening and coalescence eventually lead to foam collapse. The instabilities mentioned are already occurring during beating, especially bubble coalescence. During storage of food foams, Ostwald ripening is generally the dominant process.

Proteins are the foaming agents of choice. If the protein is well soluble, its concentration is high, and the beating is vigorous, the overrun obtained can be considerable, up to 1000% or even more. Protein-stabilized foams are, however, quite sensitive to fast coalescence if the system contains even small quantities of lipids. The latter is especially true of casein. Na- and K-caseinate give copious and fairly stable foams in the absence of lipid. Ca-caseinate is less suitable.

Whey protein concentrates, if in an undenatured state, can also give copious and relatively stable foams. At a protein content of 3 or 4%, an overrun of 1000% can be reached. Whey proteins heat-denatured at high pH (e.g., 20 min at 85°C, pH 7.5) without much calcium being present, can also be used; they give a somewhat smaller overrun, but the foam is more stable. Furthermore, a suitable concentrated whey protein solution can be foamed and subsequently heated so that the whey protein gels, and a solid, stable foam is formed, e.g., a meringue.

Partial hydrolysis of proteins tends to enhance their foam-forming capacity, but it also tends to decrease foam stability. An obvious strategy, then, is to use a mixture of peptides and native proteins. The peptides largely prevent undue coalescence during beating. The proteins are more surface active and soon displace the peptides from the air–water interface; they provide greater stability to coalescence and Ostwald ripening of the bubbles during storage.

Suggested Literature

Physico-chemical properties of proteins and basic aspects of the various functional properties: P. Walstra, *Physical Chemistry of Foods,* Dekker, New York, 2003.

Several of the subjects mentioned discussed in this chapter: P.F. Fox and P.L.H. McSweeney, Eds., *Advanced Dairy Chemistry,* Vol. 1, *Proteins,* 3rd ed., Kluwer Academic, New York, 2003. (especially Chapter 26 to Chapter 29 in Part B).

22 Fermented Milks

22.1 GENERAL ASPECTS

Having a very old product history, fermented milks originated on the basis of the generally applicable rule that raw milk will spoil after storage because of microbial action. At moderate temperatures, lactic acid bacteria commonly are predominant, and the milk becomes spontaneously sour. When the sour milk has been used, and fresh milk is put in the same vessel without rigorous cleaning of that vessel, the fresh milk is inoculated with the remaining bacterial flora. The milk now sours more quickly, generally due to a smaller number of bacterial species and strains. If this process is repeated under fairly constant conditions (especially in regard to temperature), natural selection leads to an almost pure lactic acid fermentation, although some other bacteria may remain present. The process can be improved by rigorously cleaning the vessel, heat-treating the milk to kill undesirable microbes, and inoculating the milk with a little bit of the sour milk from the previous batch; this then acts as a starter for the fermentation. The fermented milk thus obtained has a longer keeping quality and, often, a pleasant flavor. It is also much safer to the consumer because pathogenic bacteria have been killed, and contamination with pathogens afterwards can almost never lead to growth of these organisms. The lactic acid bacteria alter the conditions of milk in such a way that most undesirable organisms cannot grow or will even die. These conditions include a low pH (4.6 to 4.0), a low redox potential, and growth inhibition by undissociated acids (e.g., lactic acid) and other metabolites such as H_2O_2 and compounds with an antibiotic activity.

As a result of variations in conditions, a great number of fermented milk types have developed. Variables include species of milch animal, heat treatment of the milk, percentage fat in the milk, concentration of the milk, fermentation temperature, and inoculation percentage. According to these conditions, various species of lactic acid bacteria become predominant, producing various flavor components. Most types of fermented milk contain two to four species of bacteria. In some products, yeasts or molds participate in the fermentation.

Nearly all types of fermented milks are the result of a very long evolution. Modern manufacture makes use of carefully selected and grown starters, and strictly hygienic processing is applied. Fermented milks are very popular products, and new varieties regularly enter the consumer market. In this chapter, several types of fermented milks will be discussed. They are classified according to the type of fermentation. The nutritive value of fermented milks is a special aspect and will also be addressed. Finally, the manufacture of two fermented milks, cultured buttermilk and yogurt, is treated in some detail.

22.2 TYPES OF FERMENTED MILKS

Fermented milks are classified into four different types: (1) products of lactic fermentation in which strains of mesophilic lactic acid bacteria are used, (2) products of lactic fermentation with thermophilic lactic acid bacteria, (3) products obtained through alcohol-lactic fermentation, involving yeasts and lactic acid bacteria, and (4) products where, in addition to fermentation type (1) or (2), growth of a mold occurs. Basic aspects of lactic fermentations and its starters have been described in Chapter 13.

22.2.1 Mesophilic Fermentation

22.2.1.1 Cultured Buttermilk

Cultured buttermilk is a pasteurized skim milk fermented by a mixture of mesophilic lactic acid bacteria. It has a mild acidic taste with an aromatic diacetyl flavor and a smooth viscous texture. *Lactococcus lactis* sspp. *cremoris* and *lactis* are responsible for the acid production, whereas *Lc. lactis* ssp. *lactis* biovar. *diacetylactis* and *Leuconostoc mesenteroides* ssp. *cremoris* are the primary sources of the characteristic aromatic flavor of the product because of their ability to produce diacetyl (see Subsection 13.1.2.3).

Most buttermilks are made with mixed cultures with strains of the species mentioned above, which actually are DL starters (see Subsection 13.5.1). After pasteurization, the milk is fermented at 20°C to 22°C to ensure a balanced growth of acid- and flavor-producing species. Incubation at higher temperature would favor the growth of *Lc. lactis* ssp. *lactis*, resulting in excess acid production and diminishing the flavor production by the aroma bacteria. Details of the manufacture of cultured buttermilk are described in Section 22.3.

22.2.1.2 Sour Cream

Cultured cream or sour cream is produced by the fermentation of high-pasteurized cream with a fat percentage of 18% to 20%, which is homogenized at a low temperature, to promote formation of homogenization clusters (see Section 9.7). The cream is inoculated with an aromatic starter (DL starter, see Table 13.4) and incubated at 20°C to 22°C until the pH has reached a value of 4.5. The functions of the starter culture are the same as in cultured buttermilk. During the acid production, the homogenization clusters aggregate, resulting in a highly viscous cream. To increase the firmness, a little rennet and/or a thickening agent are sometimes added before fermentation.

22.2.1.3 Fermented Milks

Several types of fermented milks produced by mesophilic lactic acid bacteria have been developed in countries that have cool climates. They constitute a group of products distinctly different from fermented milks made elsewhere, primarily owing to their unique physical properties, which are characterized by high viscosity and ropiness. If one applies a spoon to the milk surface, long strings appear when the spoon is lifted. In the cool climate of northern Europe, fermentation of raw milk is generally the result of spontaneous growth of mesophilic lactic acid bacteria, mainly *Lc. lactis* ssp. *cremoris*, that produce exopolysaccharides. Also, at the prevailing low temperatures, the viscous product stays homogeneous due to the limited syneresis. Many types of commercially manufactured products have been developed from this originally homemade Nordic fermented milk, almost all having the same attributes. Some examples are presented in the following text.

Långfil ('long milk,' 'ropy milk') is a representative of these highly viscous Nordic fermented milks. A DL or D starter with an exopolysaccharide-producing strain of *Lc. lactis* ssp. *cremoris* is used. A relatively low incubation temperature (about 18°C) enhances the growth of this organism. The high viscosity of the product prohibits bottle filling and, accordingly, the fermentation occurs in the package. It has a mild, sour taste. Långfil is produced in Sweden, and a similar fermented milk traditionally made in Norway is called *tettemelk*. The German *Dickmilch* has similar characteristics.

Filmjölk is a popular Swedish fermented milk characterized by a typical flavor derived primarily from diacetyl, a fairly high viscosity, and a fat content of 3%. High-pasteurized and homogenized milk is inoculated with an aromatic starter culture (DL) and incubated at around 20°C for 17 to 24 h. This sour milk (pH ≈ 4.6) is used as a drink, often consumed with meals. Variants of filmjölk with lower fat content are also on the market.

Ymer (Denmark) and *lactofil* (Sweden) are concentrated after fermentation of the milk by removal of a fixed percentage of whey or, alternatively, the milk is first concentrated by ultrafiltration and the retentate is subsequently fermented. For acidification, an aromatic starter is used. Ymer and lactofil contain at least 11% nonfat milk solids (including around 6% protein), and 3.5% and 5% fat, respectively. They are high in protein and relatively low in calories and have a fairly thick but pourable consistency. There is some resemblance with fresh cheese types like quarg (see Section 27.2).

22.2.2 Thermophilic Fermentation

22.2.2.1 Yogurt

Yogurt is probably the most popular fermented milk. It is made in a variety of compositions (fat and dry-matter content), either plain or with added substances such as fruits, sugar, and gelling agents. The essential flora of yogurt consists of the thermophiles *Streptococcus thermophilus* and *Lactobacillus delbrueckii* ssp. *bulgaricus*. For a satisfactory flavor to develop, approximately equal numbers of both species should be present. They have a stimulating effect on each other's growth. Volatile compounds produced by the yogurt bacteria include small amounts of acetic acid, diacetyl, and most importantly, acetaldehyde. Details on the manufacture of yogurt will be discussed in Subsection 22.4.2.

TABLE 22.1
Yogurt-Related Fermented Milks in Various Countries

Synonym	Country
Dahi	India
Dadih	Indonesia
Katyk	Kazakhstan
Laban, leben	Iraq, Lebanon, Egypt
Laben rayeb	Saudi Arabia
Mast	Iran, Iraq, Afghanistan
Matzoon, madzoon	Armenia
Roba, rob	Egypt, Sudan, Iraq
Tarho	Hungary
Tiaourti	Greece
Yaourt	Russia, Bulgaria
Zabady, zabade	Egypt, Sudan

Yogurt and yogurt-like products are made widely in the Mediterranean area, Asia, Africa, and central Europe. Synonyms for yogurt or related fermented milks throughout various countries are shown in Table 22.1. Zabady is traditionally made from ewes' milk. For the production of dahi and dadih, buffaloes' milk is often used, sometimes in combination with bovine milk.

22.2.2.2 Bulgarian Buttermilk

Bulgarian buttermilk is a high-acid fermented milk, made from pasterurized whole milk, inoculated with *Lb. delbrueckii* ssp. *bulgaricus* alone (at 2% inoculum), and incubated at 38°C to 42°C for 10 to 12 h, until a curd forms with about 150°N titratable acidity. The product has a sharp flavor and is popular only in Bulgaria.

22.2.2.3 Acidophilus Milk

Acidophilus milk is cultured with *Lb. acidophilus*, whose primary function is to produce lactic acid. Moreover, *Lb. acidophilus* is considered to be a probiotic bacterium, and has been claimed to confer various health benefits. It is not a natural representative of the milk flora and grows slowly in milk. Hence, contamination during the manufacture of acidophilus milk should be avoided. Sterilized milk is inoculated with a large amount of starter (2% to 5%) and incubated at about 38°C for 18 to 24h. Because *Lb. acidophilus* is fairly acid-tolerant, the lactic acid content of the milk can become high, i.e., 1% to 2%, if the product is stored at an insufficiently low temperature. The flavor of the milk then becomes sharp, and the number of living bacterial cells decreases quickly. This problem can be overcome by blending plain milk with a deep-frozen concentrated culture of

Lb. acidophilus and by keeping the mixture at low temperature (say, 4°C), which prevents the milk from souring.

22.2.2.4 Probiotic Fermented Milk

Probiotic fermented milks are made with various lactic acid bacteria, including bifidobacteria. *Lactobacillus acidophilus*, specific strains of *Lb. casei,* and *Bifidobacterium* spp. are the most commonly used probiotic bacteria in the manufacture of fermented milks. These and some other microorganisms are thought to confer health and nutritional benefits to the consumer, through their activity in the intestinal tract. The traditional yogurt starter cultures, *S. thermophilus* and *Lb. delbrueckii* ssp. *bulgaricus*, on the contrary, do not grow in the intestinal tract.

The number of types of fermented milks made with probiotic microorganisms has increased markedly over the past few decades. These products may contain a probiotic microorganism in addition to *S. thermophilus* and *Lb. delbrueckii* ssp. *bulgaricus*. Alternatively, *S. thermophilus* can be combined with one or two probiotics. The concentration of probiotics does not generally reach the level of that of the yogurt bacteria. The resulting products are commercialized under trade names like Bioghurt, Bifighurt, Biogarde, and Cultura, to name a few.

22.2.3 Yeast–Lactic Fermentation

22.2.3.1 Kefir

Kefir is made of ewes', goats', or cows' milk. During the fermentation, lactic acid and alcohol are produced. Originally, the milk drink was made in Russia and southwestern Asia. It is now being made in various countries on an industrial scale by using cows' milk.

The microflora of kefir is variable. Lactococci (*L. lactis* sspp. *lactis* and *cremoris*, and *L. lactis* ssp. *lactis* biovar. *diacetylactis*), leuconostocs (*Leuc. lactis* and *Leuc. cremoris*), and lactobacilli (*Lb. brevis*, *Lb. kefir*, sometimes also *Lb. delbrueckii* ssp. *bulgaricus* and *Lb. acidophilus*) can form lactic acid, whereas yeasts, including *Candida*, *Kluyveromyces*, and *Saccharomyces* species, produce alcohol. Kefir of a satisfactory quality is believed to contain acetic acid bacteria also. Typically, the organisms involved in the cultured product are present in structures (grains). During fermentation of the milk, the grains grow due to coagulation of protein, while they become connected by means of a formed polysaccharide (kefiran).

Kefir is a creamy, sparkling, acid milk drink. Its lactic acid content is 0.7% to 1%, and its alcohol content ranges from 0.05% to 1%, but is rarely over 0.5%. These levels depend on the incubation and storage conditions. Metabolites should be formed in certain proportions to obtain a good flavor. Some conversions are detrimental to the quality; an example is the formation of acetic acid from alcohol by the acetic acid bacteria after uptake of oxygen from the air.

In the traditional manufacture of kefir, milk with added active grains is first kept for some time at a temperature of 20°C to 25°C to enhance the lactic fermentation. Subsequently, the grains are sieved out of the milk, and the milk is further ripened at a temperature of 8°C to 10°C, which stimulates the alcoholic fermentation. Modern ways of processing use homogenized, pasteurized whole or standardized milk. The milk is not inoculated with the grains as such but with soured milk obtained by sieving a previously fermented culture of grains. A certain amount of L starter may also be added. The inoculated milk is put in well-closed packages and incubated. In this way, 'firm kefir' is obtained. A considerable amount of gas forms during the fermentation. Incubation time and temperature determine the properties of the final product, i.e., amounts of lactic acid, alcohol, and CO_2, and aroma. In the manufacture of 'stirred kefir,' the milk is fermented at a fairly high temperature, slowly cooled while stirred, further ripened at low temperature, and packaged. Modern packaging materials, e.g., aluminum foil-capped plastic cups, cannot resist a high CO_2 pressure and ballooning can readily occur. Accordingly, a hole is made in the foil, or the fermentation is stopped at an earlier stage at the expense of the traditional characteristics. Continuous production of kefir is also possible.

A surrogate for kefir can be obtained by adding sucrose to cultured buttermilk (e.g., 20 g/l) together with the yeast *Saccharomyces cerevisiae* and incubating it for 3 to 4 d at 18°C to 21°C in a closed firm package.

22.2.3.2 Kumiss

Kumiss is a well-known milk drink in Russia and western Asia. Formerly, the cultured milk was valued because of its supposed control of tuberculosis and typhus. The product is traditionally made of mares' milk. The fermenting flora is variable, as in kefir.

Kumiss is a sparkling drink. It contains 0.7% to 1% lactic acid, 0.7% to 2.5% alcohol, 1.8% fat, and 2% protein; it has a grayish color. During its manufacture, protein is substantially degraded. Together with the fermentation compounds formed, the proteolysis is responsible for a specific flavor. The fermentative processes must proceed in such a way that the metabolites are formed in certain proportions.

Traditional kumiss is not manufactured on an industrial scale. To raw mares' milk, at temperatures of 26°C to 28°C, 40% starter is added, which increases the acidity to 50 mM. (The starter is propagated as a kind of continuous culture in mares' milk.) The mixture is intensely stirred and subsequently left undisturbed, which raises the acidity to 60 mM. The milk is stirred for an additional hour to aerate it and to obtain dispersed protein particles, and then it is bottled. The bottles are kept for a few hours at 18°C to 20°C and then for a certain time at 4°C to 6°C, a temperature sequence that enhances the lactic and alcoholic fermentations.

An imitation product of kumiss is now being made on an industrial scale, starting from cows' milk. Compared to mares' milk, cows' milk has a high ratio of casein to serum proteins and a low lactose content (see Table 2.20). The composition of mares' milk is therefore simulated by mixing cows' milk and a heat-treated retentate of whey ultrafiltration; heat treatment of the whey is necessary to inactivate rennet. The starter contains *Lactobacillus delbrueckii* ssp. *bulgaricus* and *Candida kefir.*

22.2.4 Molds in Lactic Fermentation

Viili, a Finnish product, is made of pasteurized nonhomogenized milk. A polysaccharide-producing starter, comparable to that used for långfil, is employed. The milk is incubated at 18°C to 19°C for 18 to 20 h. A mold, *Geotrichum candidum*, is also added to the milk. A cream layer is formed, on which the mold creates a velvety layer and causes some hydrolysis of fat. The latter factor may increase the contribution of the lactic acid bacteria to lipolysis (see Section 25.4). The product is hermetically packaged, and the mold stops growing after the oxygen has been completely consumed. Most of the CO_2 formed dissolves in the product, leading to a slight underpressure in the package.

22.3 CULTURED BUTTERMILK

Conventional buttermilk is actually the aqueous liquid released during the manufacture of butter by churning of soured cream. This buttermilk has distinctive characteristics due to the presence of butter aroma (mainly diacetyl) and part of the natural fat globule membrane material that is released during churning. The latter implies a relatively high concentration of phospholipids — the more so, the higher the fat content in cream. These phospholipids are very sensitive to autoxidation because of their relatively high content of polyunsaturated fatty acid residues (see Subsection 2.3.4). Consequently, the buttermilk readily develops an off-flavor, often called metallic, which can become quite pungent. This implies that sour buttermilk from high-fat cream has a very short shelf life of, say, 2 d, even if ascorbic acid has been added as an antioxidant.

Conventional buttermilk has been replaced almost entirely by cultured or fermented buttermilk, which is produced by lactic acid fermentation of skim milk or low-fat milk. Mesophilic lactic acid bacteria are used as a starter, which contains *Lactococcus lactis* ssp. *lactis*, *L. lactis* ssp. *cremoris,* and *Leuconostoc mesenteroides* ssp. *cremoris*. The first two species produce mainly lactic acid and are often referred to as *acid producers*, in contrast, to the leuconostoc bacteria, which also ferment citric acid to important metabolites, such as CO_2, acetaldehyde and especially diacetyl. They are referred to as *aroma producers*. It is also possible to introduce *L. lactis* ssp. *lactis* biovar. *diacetylactis* as aroma producer. The balance between aroma and acid producers is very important, and not more than 20% of the total bacterial population should consist of aroma producers. Diacetyl, at a concentration of 2 to 5 $mg{\cdot}kg^{-1}$, is responsible for the characteristic aromatic flavor of cultured buttermilk. Acetaldehyde should not be present in excess, because it may then be responsible for the defect described as yogurt-like. Only relatively small amounts of acetaldehyde, less than 1 $mg \cdot kg^{-1}$, are required for a balanced flavor of the product. To achieve this, it is advantageous to use *Leuc. mesenteroides* ssp. *cremoris* rather than *L. lactis* ssp. *lactis* biovar. *diacetylactis* as a diacetyl producer.

Cultured buttermilk is made from pasteurized skimmed milk or homogenized, pasteurized low-fat milk, usually containing less than 1% fat. After pasteurization

TABLE 22.2
Composition of Conventional and Cultured Buttermilk in g/100 g

Component	Conventional Buttermilk[a]	Cultured Buttermilk[b]
Total solids	9.5–10.6	9.0–10.6
Fat	0.3–0.7	0.1–1.0
Phospholipids	0.07–0.18	~0.02
Protein	3.3–3.9	3.1–3.5
Lactose	3.6–4.3	3.6–4.3
Lactic acid	0.55–0.9	0.55–0.9

[a] Concentrations can be lower due to addition of water (<10%)
[b] Concentrations can be higher due to addition of skim-milk powder or whey solids

the milk is cooled to 22°C and inoculated with about 1% to 3% mesophilic starter. The milk is fermented at 19°C to 22°C for 15 to 20 h until a pH of 4.6 to 4.7 is reached. The coagulum formed is broken by gentle agitation and the product obtained is cooled and packaged. The important sensory characteristics of cultured buttermilk resulting from the lactic fermentation are the smooth and fairly thick body due to the coagulation of milk proteins and the aroma produced by the fermentation of citric acid and lactose. The texture depends on the concentration of total solids. There should be very little or no separation of whey. At refrigerator temperature, the keeping quality of cultured buttermilk is 2 to 3 weeks.

The composition of cultured buttermilk differs slightly from that of the conventional buttermilk (Table 22.2). It contains no, or hardly any, proteins and phospholipids derived from the milk fat globule membrane, which are characteristic for the latter. The total solids content depends to a large extent on the optional addition of milk powder/whey powder.

Cultured buttermilk may have defects in flavor and texture. Overacidification during storage is an occasional problem, but lack of flavor caused by reduction of diacetyl to acetoin is a more frequently occurring problem. Texture defects may occur due to the production of CO_2 by citric acid-fermenting bacteria, giving rise to disruption of the gel and the phenomenon of curd floating. Another defect is the separation of whey on storage. This can be avoided by adding a thickening agent such as pectin.

22.4 YOGURT

The manufacture of yogurt will be discussed here more extensively than that of cultured buttermilk to exemplify the problems met in the making of fermented milks. This discussion will start with a description of the starters for yogurt and their metabolic activity.

22.4.1 The Yogurt Bacteria

22.4.1.1 Growth

The yogurt bacteria, *Streptococcus thermophilus* and *Lactobacillus delbrueckii* ssp. *bulgaricus*, grow in milk better when present together than each alone (protocooperation). The proteolytic rods enhance growth of the streptococci by forming small peptides and amino acids, the main amino acid being valine. Milk contains too little of these amino acids and the cocci, which are only weakly proteolytic and form the acids too slowly. The cocci enhance the growth of the rods by forming formic acid out of pyruvic acid under anaerobic conditions and by a rapid production of CO_2 (see Section 13.1). The stimulatory effect of formic acid remains unnoticed in intensely heated milk because in this milk formic acid has been formed by decomposition of lactose. The production of formic acid by the cocci is, however, essential in industrial practice, where more moderate heat treatments of yogurt milk are applied, e.g., 5 to 10 min at 85°C. Due to mutual stimulation during combined growth of the yogurt bacteria in milk, lactic acid is produced much faster than would be expected on the basis of the acid production by the individual pure cultures. Some antibiosis also occurs in yogurt in that the cocci cannot grow after a certain acidity has been reached. The rods are less susceptible to acid and continue to grow. Protocooperation and antibiosis are of great importance in the growth of the yogurt bacteria as well as for the quality of yogurt (see also Figure 22.1).

The cocci as well as the rods contribute significantly to the properties of yogurt. The properties of the bacterial strains used should be matched to each other because not every combination of strains is suitable. Furthermore, both species should be present in large numbers in the product, and hence in the

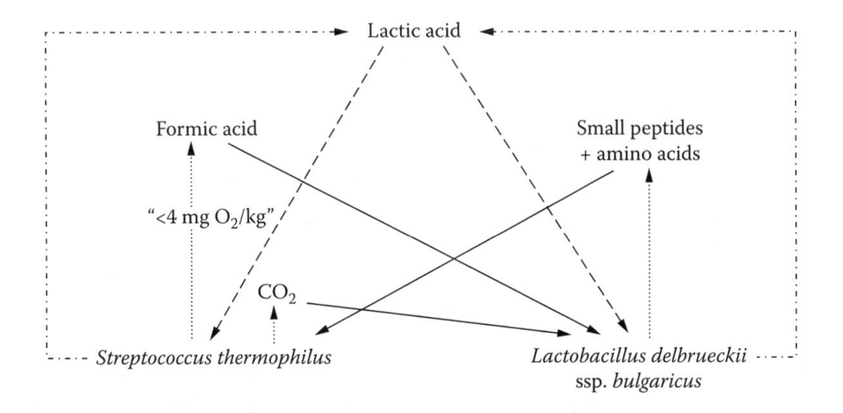

FIGURE 22.1 Outline of the stimulation and the inhibition of the growth of yogurt bacteria in milk. -·-·-·-, formation of lactic acid; ··············, formation of growth factors; ———— stimulation; – – – – – –, inhibition. (Adapted from F.M. Driessen, International Dairy Federation, Bulletin No. 179, 107–115, 1984.)

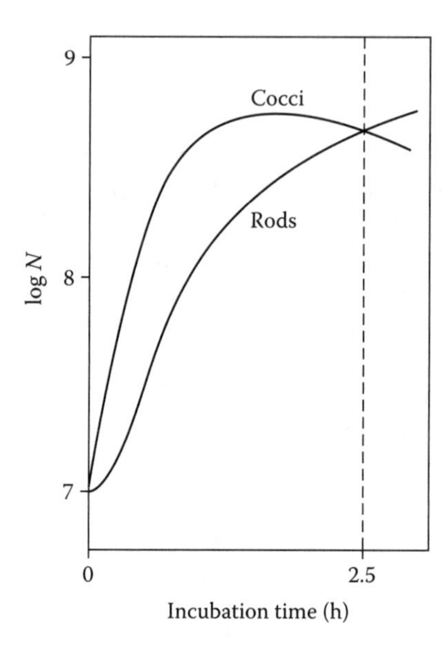

FIGURE 22.2 Growth of cocci and rods in yogurt (starter) cultured at 45°C in intensely heated milk. Inoculum percentage equals 2.5. N = count in ml^{-1}. Approximate results.

starter. The mass ratio of the two species depends on the properties of the strains and is often approximately 1:1. This ratio between the yogurt bacteria is best maintained if the inoculum percentage is, say, 2.5, the incubation time is 2.5 h at 45°C, and the final acidity is approximately 90 to 100 mM (pH ≈ 4.2). The growth of cocci and rods in yogurt incubated under these conditions is depicted in Figure 22.2. The ratio between the species keeps changing. Initially, the streptococci grow faster due to the formation of growth factors by the rods and probably also due to the latter compounds being added via the inoculum (especially in the manufacture of set yogurt). Afterwards, the cocci are slowed down by the acid produced. Meanwhile, the rods have started to grow faster because of the growth factors (CO_2 and formic acid) formed by the cocci. As a result, the original ratio is regained. The yogurt should then have attained the desired acidity. Continuing incubation or inadequate cooling causes the rods to become preponderant.

Varying the aforementioned conditions during incubation changes the ratio between the rods and the cocci as follows:

1. *Incubation time:* A shorter incubation time, which means a lower acidity, will cause too high a proportion of cocci. Transferring a yogurt starter repeatedly after short incubation times during the production of the starter may cause also the rods to disappear from the culture. Conversely, long incubation times will cause an increasing preponderance of the rods.

2. *Inoculum percentage:* Increasing the inoculum percentage will enhance the rate of acid production. The acidity at which the cocci are slowed down will thereby be reached earlier, resulting in an increased number of rods (incubation time being the same). At a smaller inoculum percentage, the ratio between the bacteria will shift in favor of the cocci.
3. *Incubation temperature:* The rods have a higher optimum temperature than the cocci. Incubation at a slightly higher temperature than 45°C will shift the ratio in favor of the rods; incubation at a lower temperature will enhance the cocci.

Obviously, a correct ratio between the species in the starter can be maintained, or be recovered if need be, by proper selection of the propagation conditions. Currently, concentrated starters are increasingly used, ensuring a correct bacterial composition of the starter.

22.4.1.2 Metabolites

S. thermophilus and *L. delbrueckii* ssp. *bulgaricus* form products that contribute to the flavor of yogurt as well as to its structure and consistency. The following are the main compounds involved:

1. *Lactic acid:* Both bacteria form lactic acid from glucose. Galactose, formed during the decomposition of lactose, is not converted. Hence, the molar concentration of galactose increases just as much as the lactose content decreases (see item 4). Most of the glucose is decomposed in a homofermentative way. *S. thermophilus* forms L(+) and *L. delbrueckii* ssp. *bulgaricus,* D(−) lactic acid. The isomers are produced in almost equal quantities. (Subsection 22.5.2 mentions physiological aspects with respect to the consumption of lactic acid.) CO_2, acetic acid, and ethanol are also produced, though in small amounts. The acetic acid content of yogurt is 30 to 50 $mg{\cdot}kg^{-1}$ (0.5 to 0.8 mM) and the ethanol content, 10 to 40 $mg{\cdot}kg^{-1}$ (0.2 to 0.7 mM). Ethanol has a relatively high flavor threshold and it probably does not contribute to the flavor of yogurt. The lactic acid content of yogurt is 0.7% to 0.9% w/w (80 to 100 mM).
2. *Acetaldehyde* (ethanal): This component is essential for the characteristic yogurt aroma. Most of it is formed by the rods. An important precursor is threonine (see Subsection 13.1.2), which is a natural component of milk, even though at low concentration. In addition, proteolysis by the lactobacilli yields threonine. The content of acetaldehyde of yogurt is about 10 $mg{\cdot}kg^{-1}$ (0.2 mM).
3. *Diacetyl* (CH_3–CO–CO–CH_3): *S. thermophilus* and, to a lesser degree, *L. delbrueckii* ssp. *bulgaricus* form diacetyl in a way that probably corresponds to the mechanism followed by leuconostocs and by *Lactococcus lactis* ssp. *lactis* biovar. *diacetylactis* (Subsection 13.1.2). The yogurt

bacteria do not decompose citric acid. Hence, pyruvic acid, formed during sugar fermentation, is the only precursor of diacetyl. The diacetyl content of yogurt ranges from 0.8 to 1.5 mg·kg^{-1} (0.01 to 0.02 mM).

4. *Polysaccharides:* The yogurt bacteria can form a 'hairy' layer or glycocalix, which predominantly consists of polysaccharide chains, made up of galactose and other glucides. They can be partially secreted into the liquid and are then called *exopolysaccharides* (see Subsection 13.1.2.8). The polysaccharides play an important role in yogurt consistency, especially of stirred yogurt (see following text). Although various strains show quite a variation in the amount of polysaccharide produced, this variation does not correlate with the consistency obtained. Presumably, the type of polysaccharide produced is of greater importance.

22.4.2 Manufacture

22.4.2.1 Natural Yogurts

The traditional product is *set yogurt,* made of concentrated milk. The milk was heated on an open fire until, say, one third of the water had evaporated. Then the milk was allowed to cool, and when a temperature of about 50°C was reached, the milk was inoculated with a little yogurt. After fermentation, a fairly firm gel was obtained. A similar process is still being used, but either the milk is evaporated under vacuum or milk powder is added. One may use the same process for making set yogurt from nonconcentrated milk (see Figure 22.3). The yogurt so obtained is less rich in flavor, far less firm, and prone to syneresis (wheying off). Generally, some gelling agent is added to prevent syneresis and to enhance firmness, especially if pieces of fruit are added. Another difference between both these products is their titratable acidity. Because a satisfactory flavor and texture are only obtained at a pH below, say, 4.5 and concentrated milk has a greater buffering capacity, the latter is fermented to an acidity of about 130 mM, as against 90 to 100 mM for nonconcentrated milk (see Section 13.2).

Another type is *stirred yogurt*, almost always made from nonconcentrated milk. After a gel is formed, it is gently stirred to obtain a smooth and fairly thick, but still pourable, product (see Figure 22.3). There are other differences in the manufacturing process. Set yogurt is fermented after being packaged, implying that final cooling has to be achieved in the package. Stirred yogurt is almost fully fermented before it is packaged. Another difference is that only certain strains of yogurt bacteria produce the correct consistency or thickness after stirring, and only so when incubating at a fairly low temperature. However, the bacteria make less of the desired flavor compounds at lower temperatures. In order to ensure that stirred yogurt has a distinct yogurt flavor, it is necessary that the starter be propagated under the same conditions as for set yogurt, i.e., at about 45°C and with such an inoculum size and incubation time as to reach about equal numbers of cocci and lactobacilli.

The rate of acidification greatly differs in set and stirred yogurt due to the differences in inoculum size and incubation temperature. Examples are given in

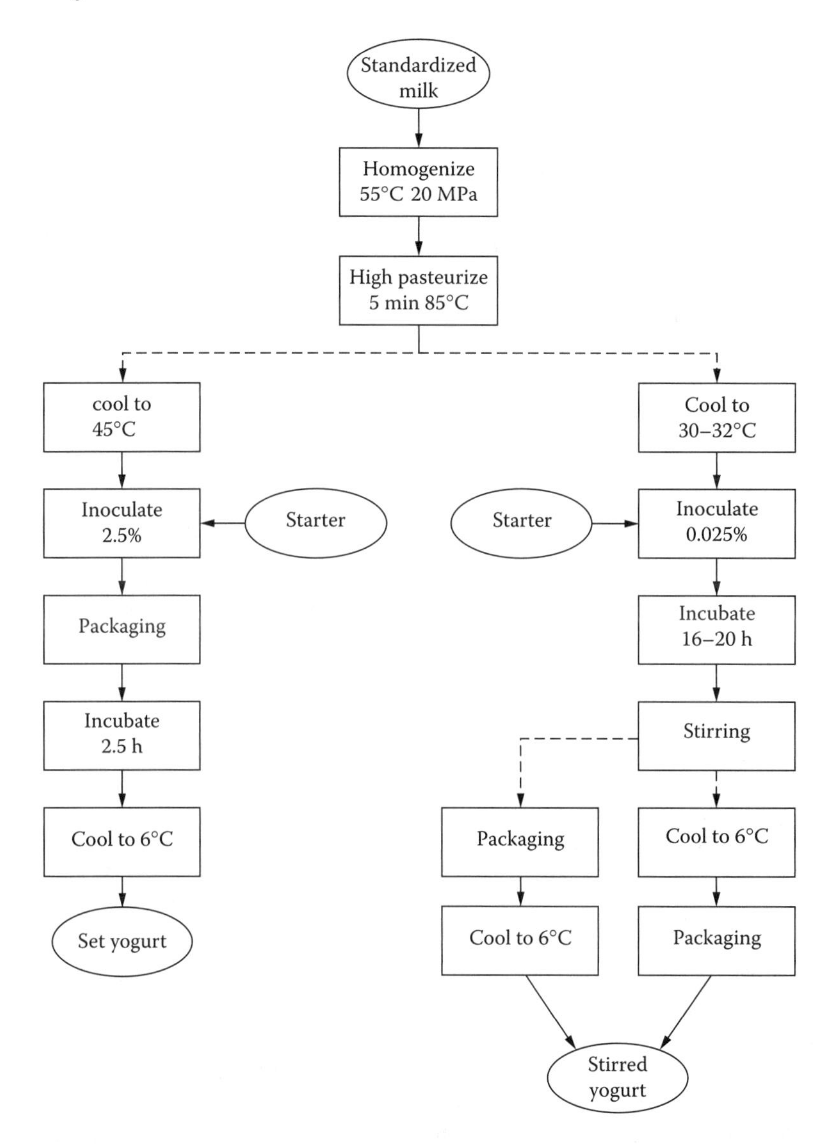

FIGURE 22.3 Examples of the manufacture of set yogurt and of stirred yogurt from whole milk. Set yogurt is often made from concentrated milk ($Q \approx 1.4$).

Figure 22.4, curves 1 and 3. Curve 2 gives the gelation as a function of time and, hence, of pH, for set yogurt. Gelation is seen to start when the pH reaches about 4.7 and that the stiffness then rapidly increases, to reach a fairly high value in about 20 min. When making stirred yogurt, gelation begins at about the same pH, but it takes a longer time before the gel has become sufficiently firm for the stirring to be started.

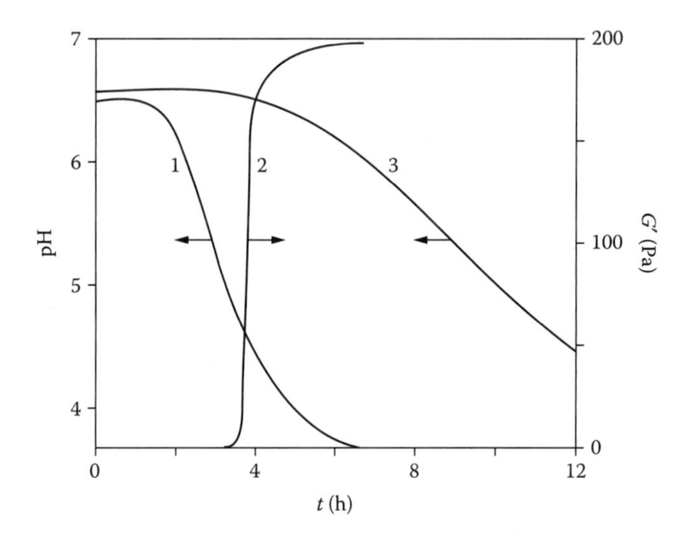

FIGURE 22.4 Relation between pH or elastic modulus (G') and incubation time (t) during yogurt manufacture. Curves 1 and 2, set yogurt; curve 3, stirred yogurt. Approximate examples.

Acidification will go on, albeit slowly, after the product has been cooled (Section 13.2). To minimize or even prevent ongoing acidification, stirred yogurts or yogurt-like products are sometimes pasteurized; this also prevents growth of any yeasts and molds present. To allow pasteurization without the product becoming inhomogeneous, it is necessary to add specific thickening agents (pectins, modified starch, gelatin).

22.4.2.2 Yogurt Drinks

From milk fermented with yogurt bacteria, several products can be derived, such as yogurt drinks, yogurt ice cream, and fruit yogurt. After fermentation, specific additives may be included, together with the appropriate processing, to obtain a product with the desired flavor, color, or consistency. The manufacture of yogurt drinks will be discussed here in some more detail.

The starting milk for the manufacture of most yogurt drinks is standardized skimmed milk, which is pasteurized for 15 min at 85°C to 95°C. The milk is fermented with the yogurt bacteria at 43°C till a pH of around 4.0 is reached. After cooling to approximately 20°C, fruit juice, sugar, and a dispersion of pectin in water are added. Also, flavoring and coloring agents may be added if required. The mixture is slowly agitated and adjusted to pH 3.8 to 4.2 with lactic acid. Subsequently, the mixture is homogenized at 15 to 20 MPa to disperse the pectin. The addition of a high methoxyl pectin is essential for the stabilization of the yogurt drink; it is primarily needed to allow heat treatment of the sour product, as it prevents separation of serum. Other stabilizing agents may be applied, such as carboxy-methyl cellulose or guar gum. Finally, the yogurt drink undergoes a heat treatment in order to extend its shelf life. This heat treatment may be pasteurization at 75°C for 20 s, after which the product is cooled and filled aseptically. Alternatively, the product may be UHT-treated (110°C, 5 s) and then cooled and filled aseptically. The latter product is essentially sterile and has a long shelf life. The packaging material should be impermeable to oxygen to avoid the development of oxidized flavor. Because many variations in the additions to yogurt and in the subsequent processing are possible, numerous varieties of yogurt drinks are on the market.

22.4.3 Physical Properties

As mentioned, the physical structure of yogurt is a network of aggregated casein particles (see also Subsection 3.3.3) onto which part of the serum proteins have been deposited due to their heat denaturation. The network encloses fat globules and serum. The largest pores of the network are on the order of 10 μm. The existence of a continuous network implies that yogurt is a gel, a viscoelastic material characterized by a fairly small yield stress (around 100 Pa). If the gel is broken up, as in the making of stirred yogurt, a fairly viscous non-Newtonian liquid can be formed; it is strongly shear rate thinning and thus has an apparent viscosity. Set and stirred yogurt have markedly different textures. See also Section 25.6, for a discussion of rheological properties.

22.4.3.1 Firmness of Set Yogurt

Firmness of set yogurt is often estimated by lowering a probe of a given weight and dimensions into the product for a certain time. The reciprocal of the penetration depth then is a measure of firmness. Firmness is not closely related to an elastic modulus but rather to a yield stress. Its value depends on the method of measurement, especially the timescale, and on several product and process variables (see also Figure 22.5):

1. *Casein content of the milk:* Firmness is approximately proportional to the cube of the casein content. Natural variation in casein content can

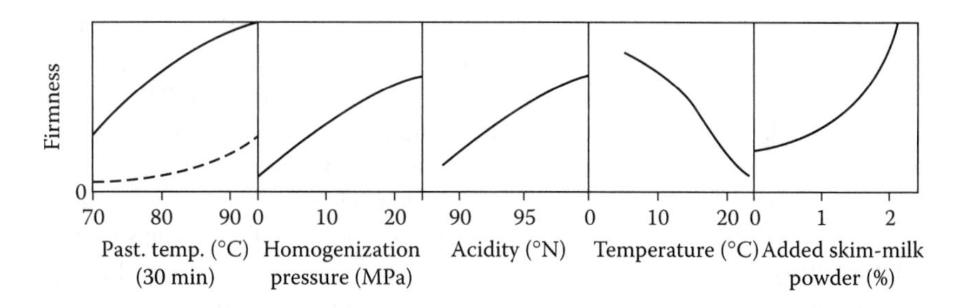

FIGURE 22.5 The influence of some product and process variables on the firmness (reciprocal of the penetration depth of a ball) of set yogurt. The broken line refers to yogurt made of nonhomogenized milk. Approximate examples.

thus have a marked effect. Evaporating the milk, adding skim-milk powder, or partial ultrafiltration increase firmness.

2. *Fat content:* The higher the fat content, the weaker the gel because the fat globules interrupt the network.
3. *Homogenizing:* Homogenization of the milk leads to a much enhanced firmness because the fat globules then contain fragments of casein micelles in their surface coat by which they can participate in the network upon acidification (see also Section 9.6). The volume fraction of casein is thus effectively increased. (Homogenization of skim milk makes no difference.)
4. *Heat treatment:* Heat treatment of the milk considerably enhances firmness. The deposition of denatured serum proteins increases the volume fraction of aggregating protein; it also may alter the number and the nature of the bonds between protein particles. Milk is generally heated for 5 to 10 min at 85°C to 90°C.
5. *Yogurt cultures:* These vary in the firmness they produce (at a given acidity), but as a rule, the differences are small.
6. *Acidity:* Generally, the yogurt is firmer at a lower pH. The preferred pH is between 4.1 and 4.6.
7. *Incubation temperature:* The lower it is, the longer it takes before a certain pH, and thereby a certain firmness, is reached, but the finished product is much firmer.
8. *Temperature of the yogurt:* For the same incubation temperature, a lower measuring temperature gives a greater firmness. The effect is quite strong (see Figure 22.5). The explanation is, presumably, that the casein micelles swell when the temperature is lowered (and vice versa); because the particles are essentially fixed in the network and the network cannot swell, this would imply that the contact or junction area between any two micelles is enlarged, by which a greater number of bonds are formed per junction.

22.4.3.2 Syneresis

Syneresis of casein gels is discussed for rennet-induced gels in Subsection 24.4.4. Briefly, syneresis is for the most part due to a rearrangement of the network, leading to an increase in the number of particle–particle junctions. The network then tends to shrink, thereby expelling interstitial liquid. Acid casein gels are not very prone to syneresis. In yogurt, syneresis is, of course, undesirable.

The tendency to exhibit syneresis greatly depends on the incubation temperature. If milk is incubated at 20°C (with a mesophilic starter because yogurt bacteria hardly grow at that temperature) so that the gel is formed at that temperature, absolutely no syneresis occurs, whereas when incubating at 32°C, syneresis is possible. When incubating at 45°C, syneresis can only be prevented if the milk has been intensively heated, if its casein content has been increased, and the storage temperature is low. However, if the package containing the product is even slightly shaken at a time when gel formation has just started and the gel is still weak, it may fracture locally with copious syneresis occurring subsequently. If the top surface of the set yogurt is wetted, possibly because water is condensed on the inside of the lid of the package and a few drops fall off, whey separation may be induced. If the pH of the yogurt has fallen below 4, some syneresis may also occur, especially if the temperature is fairly high and the package is shaken. Containers made of a material to which the formed gel does not stick will readily induce whey separation between the wall and the product.

In the manufacture of stirred yogurt, significant syneresis will lead to a poor product. The stirring breaks the gel into lumps, which then would immediately exhibit syneresis. An inhomogeneous mixture of lumps in whey is formed; further stirring would break down the lumps and make a smoother product, but it would then become insufficiently viscous. To prevent this, it is necessary to incubate the milk at a low temperature, e.g., 32°C or even lower, if the casein content of the milk is small.

22.4.3.3 Viscosity of Stirred Yogurt

Stirred yogurt should be smooth and fairly viscous. A good product also gives the impression of being 'long' or 'stringy': when slowly pouring it, a fairly thin thread readily forms that behaves somewhat elastically when it breaks. Viscosity is most easily determined by means of a Ford cup; a given amount of yogurt is allowed to flow from an opening at the conical lower end of a cup, and the time needed for that is a measure for the viscosity.

The product is strongly shear rate thinning, as illustrated in Figure 22.6. The figure also shows considerable hysteresis. After a high shear rate is applied, the apparent viscosity at lower shear rates is permanently decreased, and the viscous behavior becomes closer to Newtonian. This implies a lasting breakdown of structure. (Incidentally, the viscosity increases slightly on prolonged standing.) This is

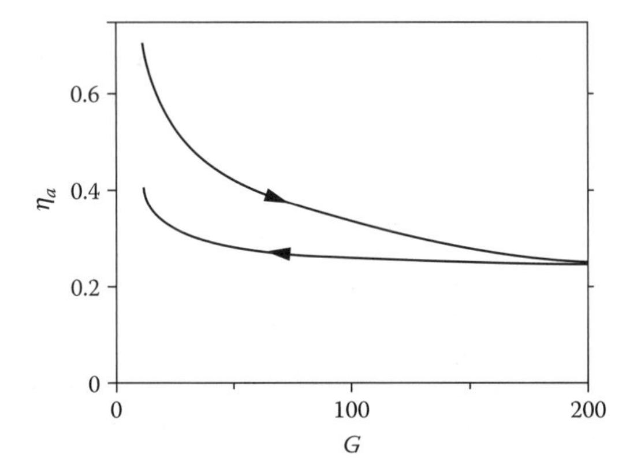

FIGURE 22.6 Example of the apparent viscosity (η_a in Pa·s) of stirred yogurt as a function of the shear rate (G in s^{-1}), before and after shearing at 200 s^{-1}.

all in agreement with the behavior of a liquid containing gel fragments. The viscosity increases with the viscosity of the continuous liquid ('solvent' or 'whey') and with the volume fraction φ of gel fragments. The latter is larger than the volume fraction of casein particles because the fragments contain a lot of interstitial solvent. More intensive stirring (a higher shear rate) further breaks down the gel fragments and also gives them a more rounded shape, thereby decreasing the effective φ.

The apparent viscosity at a given shear rate of stirred yogurt depends on:

1. *Firmness of the gel before stirring:* The higher it is, the larger φ is after stirring. The factors determining firmness have been listed and discussed earlier.
2. *Intensity of stirring:* The more vigorous the stirring, the lower the apparent viscosity but also the smoother the product. Consequently, a high gel firmness is needed to allow fairly vigorous stirring without the product becoming too thin.
3. *Syneresis:* The more syneresis occurs after stirring, the less viscous and more lumpy the product becomes. The tendency to show syneresis is less for a firmer gel and especially for a lower incubation temperature.
4. *Bacterial strains applied:* It is tempting to assume that a greater production of (exo)polysaccharides results in a higher viscosity of the solvent, and hence of the yogurt. The increase in solvent viscosity is, however, very small, and the increase in product viscosity does not correlate with polysaccharide production. Nevertheless, considerable variation occurs among strains. It appears that this is mainly due to a variation in inhomogeneity of the gel formed. An inhomogeneous gel readily gives large lumps on stirring, and the more homogeneous it is, the more viscous and smooth the stirred yogurt. How the bacterial polysaccharides affect the gel inhomogeneity is currently not quite clear.

Vigorous agitation of stirred yogurt during further processing must be avoided to prevent the product from becoming too thin. Packaging machines can be especially damaging.

22.4.4 Flavor Defects and Shelf Life

A main quality problem with yogurt is that souring tends to go on after delivery to the retailer, and the product may be too acidic when consumed; the acid flavor tends to be more pronounced in low-fat yogurt. Moreover, the yogurt may become bitter due to excessive proteolysis; this would also depend on the starter strains used. The development of these defects generally determines the shelf life. Of course, the product is cooled to slow down acidification, but it is difficult to cool it fast enough. Set yogurt is acidified in a package and cannot be stirred; stirred yogurt should not be stirred too vigorously because it would then become too thin. And even at refrigerator temperatures, acidification and other changes caused by the enzyme systems go on, albeit slowly.

Other defects may be caused by contaminating organisms, mainly yeasts and molds. The off-flavors may be characterized as yeasty, fruity, musty, cheesy, or bitter, and occasionally, soapy-rancid. A flavor threshold is generally reached at a count of about 10^4 yeasts and molds per ml. The growth of these microbes is largely determined by the amount of oxygen available, and hence by the headspace volume and the air permeability of the container.

Another defect is insufficient characteristic flavor due to reduced acetaldehyde formation (which is of less importance in yogurts with added fruits). It may be due to a low incubation temperature, an excessive growth of the streptococci, or the lactobacilli being weak aroma producers. Insufficient acidification, e.g., because the milk is contaminated with penicillin, also leads to a bland product. Finally, off-flavors in the milk used for manufacture may naturally cause flavor defects in the product.

22.5 NUTRITIONAL ASPECTS

Milk is a food of almost complete nutrition. Many changes occur to the components of milk during fermentation, although there is no significant difference between the gross composition of unfermented and fermented milk. Considerable progress has been made in demonstrating certain beneficial effects of fermented milk in animals, probably due to the changes occurring in milk during fermentation. However, unequivocal experimental or epidemiological evidence still needs to be gathered to substantiate claims of similar effects in humans. Some important health aspects comparing fermented milk with plain milk are discussed in the following text.

Several other health-improving and health-threatening effects of yogurt consumption have been suggested, but these have been shown to be insignificant or, at best, questionable. These will therefore not be included in the discussion.

22.5.1 COMPOSITION

1. *Lactose content:* Fermentation decreases the lactose content, but should not be allowed to continue to such a low pH that further sugar breakdown is impossible because the resulting product would become too acidic. At a lactic acid content of, say, 0.9% the fermentation is often slowed down by cooling. About 20% of the lactose in the milk has then been split if both glucose and galactose are fermented. In yogurt, twice as much lactose is split because most of the yogurt bacteria do not decompose galactose.
2. *Vitamin content:* Lactic acid bacteria often require certain B vitamins for growth and can produce other vitamins. Accordingly, the properties of the cultures involved largely determine the extent to which the

concentrations of vitamins in the fermented milk differ from those in the original milk. In yogurt, the level of most of the vitamins is somewhat reduced; the folic acid content may be increased. Also, some lactic acid bacteria can produce vitamin K_2. The vitamin content in fermented products is also affected by the storage conditions and especially by the pretreatment of the milk. For instance, heat treatment of milk results in a decrease of vitamins B_1, B_{12}, C, and folic acid (see Section 16.4).

3. Other changes due to bacterial action are nutritionally insignificant.
4. Composition can be changed by such process steps as standardization and ultrafiltration and by addition of skim milk powder, caseinates, stabilizers, flavorings, or fruit pulp.

22.5.2 Nutritional Value

1. *Edible energy:* The fermentation process *per se* does not cause a substantial change of the energy content of milk. The conversion of lactose to lactic acid reduces the energy value by only a small percentage.
2. *Lactose intolerance:* Lactose-intolerant users can digest a sour milk product like yogurt much better than plain milk. The lowered lactose content of sour milk plays a part. In addition, other factors must exist that cause easier digestion of lactose. The lactase activity of the yogurt bacteria as well as the stimulation of the lactase activity of the intestinal mucosa by yogurt have been held responsible. Alternatively, the depletion of the stomach contents into the duodenum may be retarded when fermented milks are consumed; thereby, the contact time of lactose hydrolyzing enzymes with the substrate would be extended, resulting in a better digestion of lactose.
3. *pH adjustment:* The consumption of fermented milks causes a smaller increase of the pH of the stomach contents and thereby diminishes the risk of passage of pathogens. This is of particular importance for people suffering from a weakened secretion of gastric juice, e.g., many elderly people and babies.
4. *Antimicrobial action:* Lactic acid bacteria can form antibiotic compounds that injure pathogens *in vitro*. The *in vivo* significance of these compounds in suppressing gastroenteritis is not clear (see Subsection 22.5.3).
5. *Lactic acid type:* The type of lactic acid formed has some physiological significance. Two stereoisomers of lactic acid exist: dextrorotatory L(+) lactic acid and levorotatory D(–) lactic acid. L(+) lactic acid can readily be metabolized in the body but D(–) at a slower rate. The latter acid is partly removed from the body through the urine. In traditional yogurt around 40% to 60% of the lactic acid is levorotatory, formed by *Lactobacillus delbrueckii* ssp. *bulgaricus*. Ingesting excessive quantities of D(–) lactic acid may cause acidosis, resulting in some tissue injury.

Young infants are more susceptible to acidosis than adults. Before 1974, the World Health Organization recommended a daily intake of D(−) lactate of less than 100 mg per kilogram of body weight. This limiting value is practically irrelevant to adults, because a 75-kg body weight would allow digestion of 1.5 l yogurt per day. The recommendation has been withdrawn; nevertheless, not too much D(−) lactic acid should be fed to infants younger than 3 months.

22.5.3 Probiotics

The term *probiotics* has evolved to describe cultures of live microorganisms that would favorably influence the health of the human or animal host by improving its indigenous microflora, especially in the gastrointestinal tract. An essential determinant for a probiotic microorganism is its ability to reach, survive, and persist in the environment in which it is intended to act. Probiotics mainly consist of species of lactic acid bacteria and bifidobacteria that are able to resist the low pH of the stomach and are resistant to bile acids present in the intestine. They should be of human origin and able to adhere to the intestinal epithelium. A probiotic microorganism cannot affect its environment unless its population reaches a certain minimum level, which is probably between 10^6 and 10^8 CFU per gram of intestinal content. To reach this level, probiotics should be able to colonize and grow in the intestinal tract. The terminal ileum and colon appear to be the preferential sites of colonization of intestinal lactobacilli and bifidobacteria, respectively.

Fermented milks are often used as vehicle for the intake of probiotics, and they should thus contain high numbers of these microorganisms. However, many probiotic strains do not grow well in milk and require the use of a supporter strain, often *S. thermophilus* and/or *Lb. delbrueckii* ssp. *bulgaricus*. This often imparts on probiotic fermented milks the character of yogurt.

The efficacy of a probiotic microorganism in producing a given health effect after adherence and colonization requires at least some additional conditions: production of antimicrobial substances, antagonism against pathogens, and competition for adhesion sites. With the aid of these mechanisms, probiotics can be successful in balancing the intestinal flora after disturbance by, for instance, antibiotic-associated diarrhea or viral diarrhea. Also, some probiotics have been shown to prevent rotavirus diarrhea. These effects have been documented in humans in well-designed clinical studies for specific strains of *Lb. acidophilus*, *Lb. rhamnosus*, *Lb. johnsonii*, *Lb. casei,* and *Bifidobacterium lactis*.

Another proposed effect of ingested lactic acid bacteria and bifidobacteria in human subjects is stimulation of the immune system. This host system appears to be enhanced mainly by nonspecific activation of phagocytes and increased activity of immune defense cells. However, in healthy subjects, an immune effect is usually not observed; hence, probiotics appear not to modify a well-balanced immune response. Also, clinical evidence for this immune effect is limited; more definitive proof of the value of this probiotic activity is required in human trials.

Some probiotics are believed to have anticarcinogenic properties. Most studies investigating antitumor and anticarcinogenic effects have been conducted in animal models, and there is little significant scientific evidence to support any of these responses in humans. Studies in animal models have shown that dietary intake of lyophilized bifidobacteria can reduce the carcinogenesis by a certain mutagen. Several enzymes of the indigenous microflora have the ability to generate mutagens from dietary compounds, and specific strains of probiotics can downregulate these enzyme activities.

Some studies have suggested a cholesterol lowering effect for milk-fermenting bacteria, including probiotics. Other studies, however, did not reproduce these results. Whether these bacteria can in general reduce serum cholesterol remains controversial.

In summary, health effects have been attributed to fermented milks, mainly on an anecdotal basis. However, some health effects of selected probiotic microorganisms have now been unequivocally established. Many claims of health effects still need further investigation.

22.5.4 Prebiotics

A *prebiotic* is defined as a nondigestible food ingredient that beneficially affects the consumer by selectively stimulating the growth and/or activity of one or a limited number of bacteria in the colon. It is a substance that modifies the composition of the intestinal microflora in such a way that a few of the potentially health-promoting bacteria (especially lactobacilli and bifidobacteria) become predominant in numbers. The prebiotics developed so far are short-chain carbohydrates that are poorly digested by human enzymes and that are often called nondigestible oligosaccharides. There are several examples of the oligosaccharides, fructo- and galacto-oligosaccharides being the most important prebiotics. The former are polymers of plant origin and the latter are produced from lactose by enzymatic transgalactosylation. Both stimulate the development of bifidobacteria in the colon. Also, human milk contains lactose-derived oligosaccharides, which stimulate the generation of a bifidus flora in breast-fed infants. In addition, these oligosaccharides are likely to have a role in preventing infection during breast feeding. Lactose and its directly derived disaccharides, lactulose and lactitol, do in fact also have prebiotic effects, particularly on the stool habit. They are hydrolyzed relatively slowly or not hydrolyzed by the human digestive enzymes and may reach the colon, where they affect the composition of the microflora.

The physiological effects of prebiotics have been shown in several human studies. They improve the frequency of the stool, which is probably due to the stabilization of the bifidobacteria-predominated colon microflora. Ingestion of certain oligosaccharides reduces the fecal concentration of ammonia and other degradation products. In an animal model, the suppressive effect of prebiotics on the activation of precarcinogens to products that are potential promoters of colon carcinogenesis has been shown. The modulation of the immune system by prebiotics has also been suggested. It is not likely that the prebiotics directly influence the immune system, but a change of the intestinal environment may enhance the immune system. Numerous studies conducted in animal models and with humans have demonstrated this effect of prebiotics. However, because not all of the suggested health effects have been substantiated by rigorous human clinical trials, and because the mechanisms underlying these effects are not always understood, it is obvious that more studies are required to confirm the claimed prebiotic effects.

Suggested Literature

Several aspects of manufacture and properties of fermented milks: *Fermented Milks: Science and Technology,* Bulletin of the International Dairy Federation No. 227, 1988.

Several fermented milk products: F.V. Kosikowski and V.V. Mistry, *Cheese and Fermented Milk Foods,* 3rd ed., two volumes, published by the authors, 1997.

A general treatment of yogurt and its manufacture: A.Y. Tamime and R.K. Robinson, *Yoghurt: Science and Technology,* 2nd ed., Wordhead, Cambridge, U.K., 1999.

Industrial yogurt manufacture: R.K. Robinson, Ed., *Modern Dairy Technology,* Vol. 2, *Advances in Milk Products,* 2nd edition, Elsevier, London, 1993.

The microbiology of fermented milks: A.Y. Tamime and V.M.E. Marshall, Microbiology and technology of fermented milks. In: B.A. Law, Ed., *Microbiology and Bio-Chemistry of Cheese and Fermented Milk,* Blackie Academic and Professional, London, pp. 57–152, 1997.

A critical review on probiotics: G.W. Tannock, Ed., *Probiotics: A Critical Review,* Horizon Scientific Press, Norfolk, 1999.

Part IV

Cheese

23 Principles of Cheese Making

Cheese making is approached differently in this part of the book from the organization of parts II and III. This is because processes and product properties are strongly interwoven and highly specific in the production of cheese (except for some of the processes applied to milk prior to clotting). Basic information is especially found in Subsections 2.4.4 and 2.4.5, and Section 3.3, Chapter 7 and Chapter 13.

In this chapter some of the rudiments of cheese manufacture and ripening are given to provide a framework for the detailed treatment in the ensuing chapters.

23.1 INTRODUCTION

When fresh milk is left to become sour, the casein aggregates. If souring occurs at not too low a temperature and without any stirring or shaking of the milk, a gel is formed. Some whey separation generally occurs when the gelled or clotted milk is kept for some time. This can be enhanced by heating and stirring; the mass then separates into curd grains and whey. By allowing more of the whey to drain out from the curd — for instance, by hanging the curd in a cloth — a primitive fresh cheese is obtained (baker's cheese, quarg, or simply 'curds').

This may have been the origin of cheese making. However, for centuries, milk has also been clotted by the addition of specific agents, especially rennet, an extract of calf stomach. It was possibly discovered when the stomach of a slaughtered animal was used to store milk. Various vegetable clotting agents were also used, for instance, from cardoon flowers, or fig tree latex. Altogether, the art of transforming milk into curd and whey is very old. It has always been accompanied by acidification, caused by the omnipresent lactic acid bacteria.

To make a real cheese, other process steps are needed: shaping (often by pressing), salting, and curing. These steps could readily evolve, and further evolution has led to a great variety of cheese types. However, all cheeses have a few things in common:

- The greater part of the casein and the fat of the milk are concentrated in the cheese, which is thus a very nutritious product.
- Cheese keeps much longer than milk, and also longer than fermented milks. During keeping there are changes in its properties: this is called *ripening* or *maturation*.

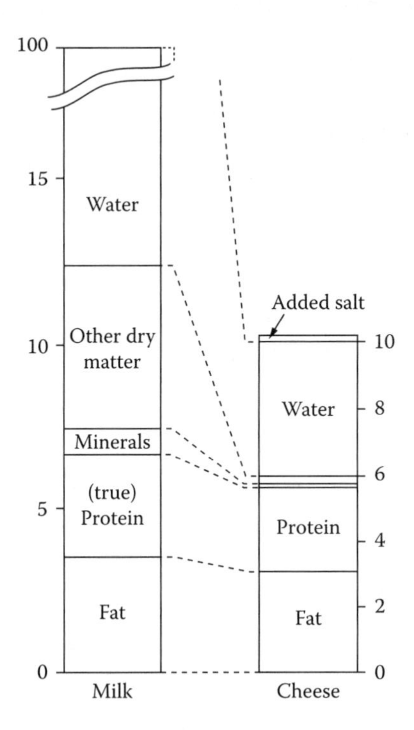

FIGURE 23.1 Example of the gross composition of milk and cheese and of the transfer of components from milk to cheese. (Scales are in kg.)

- Cheese generally has a distinct and characteristic flavor due to a great number of flavor compounds formed during ripening. The process of ripening, in particular, shows great variation.

When milk is made into cheese, casein and fat are concentrated, whereas the other milk components, especially water, are mainly removed along with the whey. None of the milk components is fully retained and new substances may be added, notably salt. This is illustrated in Figure 23.1. The yield and the composition of the cheese are determined by the properties of the milk, especially its composition, and by the manufacturing practice. Appendix A.12 gives the approximate composition of some cheese varieties.

Cheese making is a complicated process, involving many process steps and several biochemical transformations. All of these variables affect yield, composition, and quality of the cheese and its by-products (predominantly whey) and often in different ways. Consequently, optimization of cheese making is an intricate affair. Even the control of the composition of the end product, a rather straightforward activity for most dairy products, is not easy to achieve in cheese making. Moreover, the way of processing may strongly affect production costs: additives needed, labor, equipment, product loss, etc.

Overall, a full understanding of the various physical and chemical transformations, their interdependence, and the ways in which they can be affected is needed if one wants to base cheese making on scientific principles rather than empirical rules of thumb. Because several widely differing disciplines are involved — process engineering, physical chemistry, biochemistry, and microbiology, to name a few — our understanding of cheese making is still far from complete. Nevertheless, we will attempt to provide the reader besides with facts, with such understanding.

23.2 ESSENTIAL PROCESS STEPS

The manufacture of cheese involves several different processes, some of which are essential for (nearly) all cheese varieties:

1. *Clotting of the milk:* This is accomplished by means of enzymes or acid (or both). As discussed earlier, the enzymes involved remove the caseinomacropeptide 'hairs' from κ-casein; the resulting paracasein micelles will then aggregate. Acid, generally formed from lactose by lactic acid bacteria, dissolves the colloidal calcium phosphate of the micelles and neutralizes the electric charge on the resulting particles, which will then aggregate. The aggregation causes formation of a space-filling network, which encloses the milk serum and the fat globules.
2. *Removal of the whey:* The gel formed is prone to spontaneous syneresis, i.e., expulsion of whey. Whey expulsion is generally enhanced by cutting the gel into pieces and by stirring the curd–whey mixture that is thus formed. The curd obtained makes up 10 to 30% of the original volume of milk. The drier the curd, the firmer and the more durable the cheese will become.
3. *Acid production* in the cheese during its manufacture: This is due to the conversion of lactose into lactic acid by lactic acid bacteria. The resulting pH of curd and cheese affects such parameters as syneresis, consistency, and ripening of the cheese.
4. *Salting:* Cheese contains added NaCl, generally 1 to 4%. This does not apply to some fresh-type cheeses such as quarg. The salt affects durability, flavor, and consistency of the cheese, both directly and by its effect on ripening.
5. *Fusion of curd grains into a coherent loaf* that is easy to handle. Moreover, the cheese can acquire a rind, which protects the interior. Pressing enhances curd fusion and the formation of a closed rind.
6. *Curing:* That is, ensuring that conditions during storage and handling of the cheeses are such that ripening proceeds as desired. Ripening is the main factor determining the typical flavor and texture of a given cheese variety. To achieve this, the cheese is kept for a variable time under suitable conditions. The storage conditions vary widely with the type of cheese involved.

The last two processes are typical of ripened cheese; when these are not carried out, the product is referred to as fresh cheese.

Nowadays, some additional process steps are commonly applied, their main objective being to diminish variation in the conditions occurring during the manufacturing process and in the properties of the cheese. These concern:

Pasteurization of the cheese milk: This destroys microorganisms and enzymes that can be detrimental to ripening. It may also serve to kill any pathogens because some of these can survive for some time, especially in soft-type cheeses. To avoid recontamination after pasteurization, strict hygienic measures have to be taken.

Addition of cultures of microorganisms to the cheese milk, especially starters of lactic acid bacteria: The addition is essential if the cheese milk has been pasteurized but is also desirable for cheese made of raw milk. The composition of the starter depends on the type of cheese to be made. For some varieties, cultures of other specific microorganisms are also added.

Regulation of composition, i.e., water, fat, salt content, and pH of the cheese: The advantages are obvious but it calls for a detailed understanding of the various processes occurring during cheese making.

The main aim of modern cheese making is to process the milk rapidly, often maintaining a rigid time schedule, and to precisely adjust the composition of the cheese in order to control yield and quality. Moreover, mechanization and automation of most process steps have greatly altered the cheese-making operation. Nevertheless, the basic physical, chemical, and microbiological processes occurring remain largely the same.

23.3 CHANGES OCCURRING

Figure 23.2 illustrates in a highly schematic way the main changes occurring during making and ripening of cheese. The scheme is greatly simplified and applies fully only for some types of cheese. For instance, for Cheddar types, shaping and salting would occur in reverse order, and the timescale would be somewhat different. Nevertheless, the essential changes are given. Much of the figure speaks for itself.

Lactic acid bacteria (predominantly *Lactococcus* species), generally added in the form of a starter, play a key role in the development of flavor and texture. Their prime action is the conversion of lactose into lactic acid, which considerably lowers pH to, say, 5.1. An important consequence is the dissolution of the colloidal calcium phosphate, as discussed in Subsection 2.2.5. The pH decrease greatly affects cheese composition, the rate of syneresis, and the fusion of curd grains into a continuous mass. Moreover, specific flavor components and CO_2 can be formed, and the redox potential (E_h) of the cheese will be greatly reduced. In most types of cheese, no sugar remains. Some starter organisms also produce antibiotics.

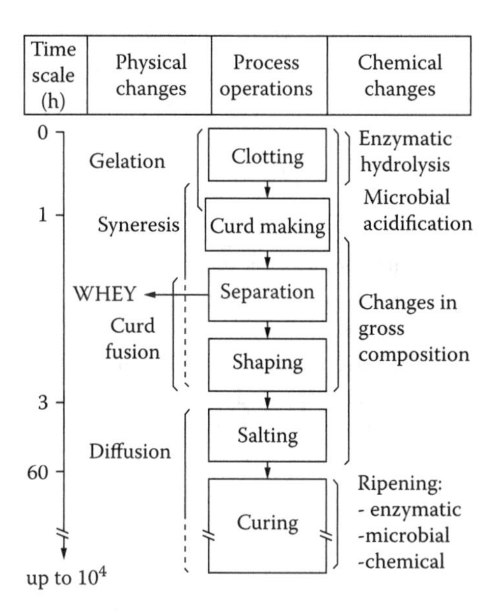

FIGURE 23.2 Schematic of the most essential physical and (bio)chemical changes occurring during the transformation of milk into cheese. Simplified example; the timescale is not linear.

A number of these factors are of great importance for limiting the growth of undesirable microorganisms and, therefore, for the *preservation* of the cheese. This concerns: (1) the high concentration of lactic acid combined with a low pH; (2) the low E_h, i.e., strictly anaerobic conditions; (3) the absence of a suitable carbon source (sugar) for most bacteria; and (4) the presence of other inhibiting substances. The formation of a closed rind around the cheese loaf, caused by fusion of curd particles and enhanced by local moisture loss, prevents further contamination of the interior of the cheese with microbes. Moreover, most cheese is solid-like, implying that bacteria are immobilized and that molecules (enzymes, reactants, etc.) diffuse sluggishly. Oxygen can diffuse into the loaf, but it is rapidly consumed by the enzyme system of the starter bacteria. Salting causes another important change in composition, which further limits microbial growth. The salt also affects flavor, texture, and the ripening processes.

What happens during ripening may be the most complicated and thereby the least understood part of cheese manufacture, although considerable progress has been made in recent years. Several enzymes, originating from milk (e.g., plasmin and lipase) or added to milk (especially chymosin), or from microorganisms (be it starter bacteria or a more specific flora), cause a wide range of biochemical reactions, generally followed or accompanied by purely chemical transformations. The main changes are that a considerable part of the protein is broken down, resulting in peptides of various size, free amino acids, and smaller breakdown products. A small part of the fatty acids is split off the triglycerides. All of this is discussed in Chapter 25. Moreover, diffusion of salt, water, and of the products resulting from the mentioned reactions occurs, especially from the rind into the interior and vice versa. In several types of cheese, a flora of yeasts and bacteria, and often molds as well, grow on the cheese rind, considerably altering cheese composition, e.g., by consuming lactic acid and by producing flavor compounds.

The various process steps are mutually dependent, and any variation in process conditions (temperature, amount of starter added, time allowed for process steps, and so on), influences several changes, not just one. The reactions themselves affect other reactions. Also, composition and pretreatment of the milk influence the results of the process steps and ripening. Altogether, cheese manufacture and curing is highly intricate, and changing one factor always has several consequences. On the other hand, this allows the production of cheese in a great number of varieties.

Suggested Literature

The best overview of fundamental aspects of cheese manufacture and properties: P.F. Fox, P.L.H. McSweeney, T.M. Cogan, and T.P. Guinee, Eds., *Cheese: Chemistry, Physics and Microbiology* Vol. 1: *General Aspects*, 3rd ed., Elsevier Academic Press, London, 2004.

Another general book: P.F. Fox, T.P. Guinee, T.M. Cogan, and P.L.H. McSweeney, *Fundamentals of Cheese Science,* Aspen, Gaithersburg, 2000.

24 Cheese Manufacture

In this chapter, nearly all process steps applied in cheese manufacture are discussed in some detail. It does not include cheese ripening.

24.1 MILK PROPERTIES AND PRETREATMENT

The treatments to be applied to the milk before curd making depend on the composition of the milk and on the properties that the milk should have. The latter vary according to the type of cheese to be produced.

24.1.1 THE RAW MILK

24.1.1.1 Chemical Composition

Milk composition has considerable effect on yield and composition of the cheese. The milk may vary substantially in composition, especially if it originates from only a few cows, as in farmhouse cheese making, or with the season if most cows calve at about the same time. However, cheese of satisfactory quality can be made of almost all milk, providing that the manufacturing process is adjusted. In most cases, the milk is standardized so as to yield the desired fat content in the dry matter of the cheese. The following are important aspects of the composition of the milk:

1. Casein and fat content largely determine the yield of cheese. Often, the crude protein content is considered. The ratio of casein N to total N is, however, somewhat variable and the serum proteins and NPN compounds are hardly or not retained in the cheese.
2. The ratio of fat to casein mainly determines the fat content in the dry matter of the cheese. It also slightly affects syneresis and thereby the ultimate water content of the cheese.
3. The lactose content, actually the lactose content of the fat- and casein-free milk, determines the potential lactic acid production and thereby markedly affects pH, water content, and ensuing properties of the cheese.
4. The pH of the cheese also depends on the buffering capacity of the dry matter. The only important variable is the ratio of colloidal calcium phosphate to casein. This ratio does not vary greatly; generally, it increases somewhat with the stage of lactation. (For Emmentaler-type cheeses, it can sometimes be too low to arrive at a satisfactory cheese quality.)
5. The rennetability of the milk and its ability to show syneresis may vary widely, mainly because of varying Ca^{2+} activity, but other components may have an effect as well.

6. Milk of cows suffering from severe mastitis has a low lactose content and a low ratio of casein N to total N (see Figure 2.34); often, it clots slowly, with the curd exhibiting poor syneresis.
7. Factors inhibiting bacterial growth may slow down the lactic acid production. Most natural factors (especially the lactoperoxidase system) do not vary widely in bulk milk. The presence of antibiotics can, however, be detrimental to acid production and maturation.
8. The milk should not be spoiled, e.g., be rancid or show other flavor defects. However, slightly sour milk may readily be made into cheese, at least certain types of (soft) cheese made from raw milk.

24.1.1.2 Microbial Composition

The requirements involved may vary widely. For cheese made of raw milk, coliforms and propionic acid bacteria are often harmful (Chapter 26). Certain lactic acid bacteria can also cause flavor defects, e.g., yeast-like or cabbage-type flavors, whereas fecal streptococci may cause H_2S flavor. Especially high counts of lactobacilli and/or pediococci are harmful to the flavor development in hard and semihard cheeses. Although most cheese milk is pasteurized, spores of *Clostridium tyrobutyricum* may be catastrophic for many cheese varieties, e.g., Gouda and Emmentaler (see Section 26.2). Heat-resistant lipases, originating from psychrotrophs, may cause undesirable soapy-rancid flavors in many cheese types; heat-resistant proteinases may cause a decrease in cheese yield.

24.1.2 Milk Treatment

Following are some important aspects of pretreatment:

1. *Thermalizing*, e.g., heating for 20 s at 65°C, if the milk is to be kept cool for some time: This is aimed at preventing the formation of considerable amounts of heat-resistant lipases and proteinases, and it may also reduce the count of some of the detrimental bacteria just mentioned.
2. *Removal of dirt particles*: This is done by filtering or centrifuging.
3. *Standardization of the fat content* of the milk: This will be discussed in Subsection 24.8.3.

   For some cheeses made from partly skimmed raw milk, e.g., Parmesan cheese, the cold milk is traditionally allowed to cream, and the cream layer removed. As discussed in Subsection 3.2.4.2, creaming, then, is fast due to cold agglutination. The cream so obtained contains most of the somatic cells and many of the bacteria originally present in the milk. This process may considerably improve the bacterial quality of the cheese milk. However, standardization of the fat content is not very accurate. In some cases, the milk is cooled in a heat exchanger to about 5°C to obtain a more efficient separation between cream and skim milk and the cream obtained is pasteurized; part of it is then added to the raw skim milk to obtain the desired fat content. This procedure allows accurate standardization and improves the bacterial quality even more.

4. *Adjusting the protein content* of the milk: This is done by some manufacturers, and it virtually always means increasing the protein concentration. It is generally achieved by addition of an ultrafiltration retentate of (skim) milk; occasionally, low-heat skim milk powder or plain concentrated milk is added. The main objective of boosting protein content is a more efficient use of the machinery, as a greater amount of cheese can be made with the same equipment in the same time (the concentration factor can be at most 1.4 or 1.5 when using traditional curd-making machinery). Moreover, standardization of the protein content allows a better process control. See Subsection 24.4.3 for ultrafiltration of cheese milk.
5. *Pasteurization*, sufficient to inactivate alkaline phosphatase: It serves to kill pathogenic and harmful organisms (see also Figure 24.1). A more intense pasteurization causes part of the serum proteins to become insoluble, leading to an increase in cheese yield; it decreases the rennetability and the syneresis (see Figure 24.15); it inactivates xanthine oxidase, thereby increasing the risk of bacterial spoilage. If a heat exchanger is operating for a long time without interruption (to thermalize or pasteurize the milk), growth of *Streptococcus thermophilus* may occur in it; eventually, large numbers develop, causing flavor defects (e.g., putrid and yeasty flavors).

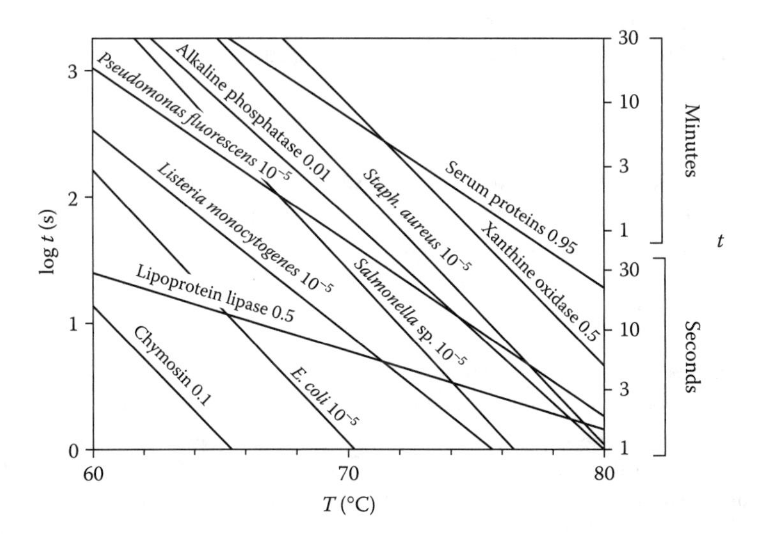

FIGURE 24.1 Time (t) needed as a function of temperature (T) to inactivate some enzymes, to kill some bacteria, and to denature serum proteins in milk or whey (at pH 6.7). The figures indicate the approximate fraction left unchanged after the treatment.

6. *Bactofugation*: This is sometimes applied to reduce the number of spores of *Clostridium tyrobutyricum* (to about 3%). The removal of the sediment obtained, containing the spores, causes about 6% reduction in cheese yield. Therefore, the sediment is UHT-heated and added again to the cheese milk.
7. Prevention of *recontamination*: See, for instance, Chapter 26.
8. Prevention of *damage to fat globules:* This must be considered in connection with potential lipase action and with 'churning,' i.e., the formation of visible lumps of fat; this involves partial coalescence (see Subsection 3.2.2.2). Such damage may occur when air is beaten in, especially if the milk is splashed from a height into the cheese vat.
9. *Bringing the milk to renneting temperature*: In many cases, the milk is pasteurized just before cheese making and directly cooled to renneting temperature, say, 30°C. In other cases, the (pasteurized) milk has been kept cool (e.g., 5°C) before cheese making. It then should be heated to 30°C. However, mere heating may be insufficient. It may cause the churning just mentioned. More importantly, it will cause renneting and especially syneresis to proceed slower than normal. The remedy is to heat the milk to, say, 50°C and then cool to 30°C.
10. *Homogenization* of the milk: This causes a different, in some cheeses an undesirable and sticky, texture and additional lipolysis. The latter may be desirable for blue-veined cheese.
11. *Addition of substances*, including:
    a. Calcium chloride: This speeds up the clotting or reduces the amount of rennet needed, and leads to a firmer gel; it especially diminishes the natural variation in rennetability.
    b. Potassium (or sodium) nitrate: If desired, to suppress fermentation of butyric acid and coliform bacteria (Chapter 26).
    c. Coloring matter (sometimes): Either annatto or carotene.

Starter addition will be discussed in Section 24.2. For some cheeses, some acid (e.g., HCl or citric acid) is added before clotting. The clotting agents to be added are discussed in Subsection 24.3.1.

24.2 STARTERS

Lactic acid bacteria appear to be indispensable in the manufacture of cheese. They are in most but not in all cases added to the milk as a starter. They cause several reactions to occur in the milk and in the curd, which are definitely important for the composition and the properties of cheese. General aspects of these reactions have been discussed in Sections 13.1 and 13.2. The choice of the type and the quantity of starter depend on the type of cheese to be made. Several variations are possible:

- The rate of acid formation in the milk can be varied and is dependent on the activity of the starter. The duration of the curd-forming process is influenced by the acidification (rennet is more active at a lower pH; see Figure 24.2) and also the composition of the cheese, such as the water and the calcium phosphate content and its pH, are affected by the acidification process. The milk is sometimes preacidified by adding the starter prior to rennet addition, which also affects the cheese composition.

- The incubation temperature of the starter before addition to the cheese milk determines the equilibrium between the different strains in a starter culture. A change in this equilibrium alters the activity and properties of the starter, and the latter affect the cheese composition.
- The starter is characterized by its proteolytic system, which is of paramount importance for the ripening of the cheese. It determines how the protein is hydrolyzed and how the peptides formed are subsequently converted into amino acids and flavor compounds. The proteolysis has also consequences for the texture of the cheese.
- The production of CO_2 by starter bacteria is essential for the texture of some cheeses in which the formation of openings is desirable, such as Gouda, Tilsiter, Camembert and Roquefort. This requires the choice of a specific DL-starter (see Section 13.5). Such a specific starter is also needed in fresh cheeses where the formation of diacetyl from citric acid is desired. In several other cheese types, notably Cheddar, formation of openings is undesirable and the starter used should produce little CO_2.
- The phage sensitivity of a starter may limit its use in certain circumstances and, therefore, knowing this property is a desired guide for choosing the right starter. The bacterium-phage interaction is extensively discussed in Section 13.3.

Details on the composition of starters containing lactic acid bacteria and their manufacture are given in Section 13.5.

For the manufacture of some cheese types, *secondary cultures* are used. The growth and development of these secondary cheese cultures are normally preceded by the fermentation of lactose to lactate by the primary starter of lactic acid bacteria. The various cheese cultures may contain species of yeasts, molds, or bacteria. They develop during the later stages of ripening on the surface of the surface-ripened cheeses or internally in the cheese matrix in Swiss-type cheeses and in blue-veined cheeses. These cheese varieties will be discussed in more detail in Chapter 27.

Starter Handling. Although there are still cheeses made without the use of starters, the manufacture of the vast majority of cheeses relies on the use of starter cultures. The way the starters are applied and handled, however, shows several variations from artisanal tradition to modern science-based technology. The traditional approach involves using some milk of a successful product batch or some whey derived from it after further incubation, as a starter for the next batch. This approach results in selective enrichment of microorganisms that survive and multiply under cheese-making conditions and that have the desirable properties. Such good artisanal cultures are the archive stocks for the production of undefined starters used in industrial cheese manufacture. These cultures are sequentially propagated in milk at the cheese factory before adding them to the cheese vat.

Alternatively, these cultures are preserved and propagated under controlled laboratory conditions and supplied to the cheese factory in concentrated and frozen form, to be used as inoculum for the bulk starter. The culture suppliers often use whey-based media, enriched with yeast extract or another vitamin source, instead of milk for the propagation of their cultures. This renders cultures with a higher concentration of cells in a shorter time, which ultimately lowers their price. Defined-strain starters consisting of purified strains, free of contaminants, are also propagated under controlled laboratory conditions and supplied in frozen form.

Undefined and defined-strain starters are nowadays provided in highly concentrated freeze-dried form as well. In this form they are applicable as a direct inoculum for the cheese vat (DVI), thus avoiding the on-site cultivation of a bulk starter. The usage of DVI is limited due to the high price of the freeze-dried starter. In particular, any plant producing more than 10,000 tons of cheese per annum would consider the use of DVI a major cost item. On the other hand smaller plants would consider DVI more convenient and more economical than using a bulk starter. Secondary cultures and adjunct starters are ideally suited to be supplied to the cheese vat in DVI form.

24.3 ENZYME-INDUCED CLOTTING

In this section we will especially deal with rennet coagulation, i.e., the clotting of milk by enzymes isolated from the abomasum of calves. The most important enzyme in calf rennet is chymosin.

Renneting enzymes cause a splitting of κ-casein in such a way that the 'hairs' protruding from the casein micelles (see Figure 3.17) disappear or, more precisely, become very much shorter. The released caseino-macropeptide dissolves, while para-κ-casein remains in the micelles. The altered casein is referred to as paracasein; it cannot be dissolved, nor dispersed, in milk serum. Because of this, the paracasein micelles in the milk aggregate, provided that the Ca^{2+} activity is high enough.

24.3.1 Enzymes Used

24.3.1.1 Chymosin

The prime enzyme used in cheese making is chymosin (EC 3.4.23.4, MW ≈ 35 600, isoelectric pH ≈ 4.65). It is an aspartate-proteinase, hence an endopeptidase, which means that it can split proteins into relatively large fragments. Chymosin is related to the common stomach enzyme pepsin (EC 3.4.23.1). Unlike pepsin, chymosin cannot hydrolyze the immunoglobulins of colostrum (which explains why the newborn calf produces chymosin and no pepsin). In the calf stomach the inactive pro-chymosin is excreted, which is converted into the active form by means of directed auto-proteolysis.

At pH = 6.7 especially the Phe–Met bond between residues 105 and 106 of κ-casein is split. This is a fast reaction. Presumably, the positive charge of that region of the peptide chain (see Figure 2.25) and the easy accessibility of that region account for the strong affinity toward the negatively charged active site of the enzyme. At lower pH values, chymosin can also split other bonds in the various caseins.

Synthetic peptides comprising some amino acid residues around the Phe–Met bond of κ-casein are also readily hydrolyzed by chymosin. The latter hydrolysis can be applied in tests for estimating the strength of rennet preparations.

Some chymosin becomes adsorbed onto paracasein, whereby part of it is included in the cheese, the amount increasing with decreasing pH. At pH 6.7 the adsorption is very weak.

Chymosin is readily inactivated under certain conditions (see Figure 24.2). At low pH this must be attributed to auto-proteolysis (the enzyme decomposes itself), at high pH to (heat) denaturation. In fresh milk, significant inactivation occurs already at a temperature as low as 45°C. Salt inhibits the inactivation; hence the commercial rennet is made with a high salt content.

24.3.1.2 Other Enzymes

Conventional rennet is produced by extraction of the stomach (abomasum) of young calves. The extract is treated in such a manner that the pro-chymosin is converted into chymosin. The extract is also purified, but it always contains *pepsin* as well. The pepsin activity is comparable to that of chymosin, and if the pepsin content of a rennet is not more than about 25% of the chymosin content, renneting, curd making, and cheese ripening are not significantly different as compared to the case when pure chymosin is used.

The latter is now produced by gene technology: the bovine gene is inserted into the DNA of microorganisms (bacteria, yeasts, or molds) that then produce the enzyme and excrete it into the environment. In some countries, this recombinant chymosin is the main clotting agent used in cheese manufacture; in other countries, conventional calf rennet is still applied.

Several other enzymes can be used, e.g., porcine pepsin. It acts much like bovine chymosin, but the pH of the milk has to be slightly decreased to achieve rapid clotting. Several vegetable rennets exist, either from various plants or from molds. The latter include acid proteinases obtained, e.g., from *Rhizomucor miehei* or *Cryphonectria* (formerly called *Endothia*) *parasitica*. These were developed as an alternative for (the expensive) calf rennet, but tend to be displaced by recombinant chymosin. Clotting agents from plants are traditionally used for local cheese varieties.

Nearly all of these alternative rennets have in common that they specifically split the Phe–Met bond of κ-casein at physiological pH, but their other properties can vary greatly. This may involve the factors affecting their stability; dependence

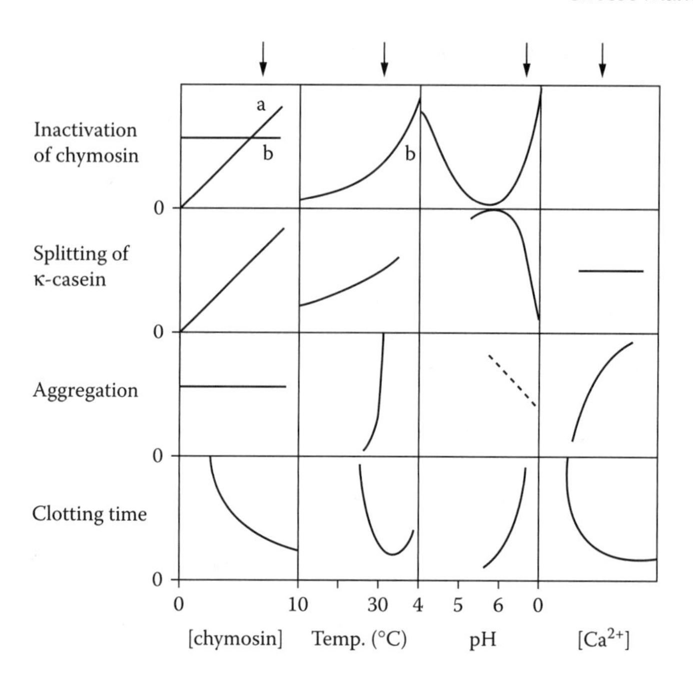

FIGURE 24.2 Effect of chymosin concentration, temperature, pH, and Ca^{2+} concentration on the rates of chymosin inactivation, splitting of κ-casein (in milk), aggregation of paracasein micelles (this implies that κ-casein has been completely split), and the clotting time of milk. Meant to illustrate trends. (a) at pH 3.5; (b) at pH 7; a blank space implies no or little effect, a broken line a rough estimate; arrows indicate the conditions as often used in making cheese from fresh milk.

of clotting activity on temperature, pH, and ionic strength; association of the enzyme with paracasein; and proteolytic activity during cheese ripening. In other words, some of the relations given in Figure 24.2 will not hold.

24.3.2 The Enzyme-Catalyzed Reaction

The rate of the enzyme reaction cannot be described on the basis of the Michaelis– Menten equation: the reaction is first order, both with respect to concentration and to time, at least at physiological pH. This must be ascribed to the great difference in size — therefore, in diffusion coefficient — between enzyme molecules and casein micelles. The latter do not move, as it were, and the enzyme must approach them by means of diffusion. During renneting, there is about one enzyme molecule per 30 casein micelles. Moreover, part of the diffusion path is in the hairy layer on the micelle, as the Phe-Met bond to be split is quite close to the micelle surface; this increases the diffusion time. The reaction velocity thus is diffusion-limited. The hairs are removed at

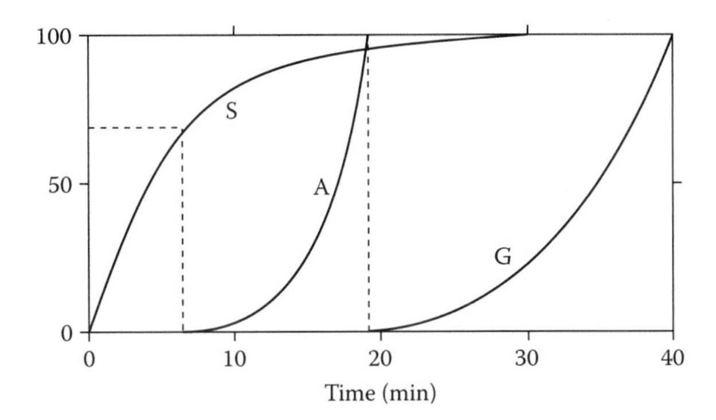

FIGURE 24.3 Renneting of milk. Percentage of κ-casein split (S), degree of aggregation of paracasein micelles (A, e.g., deduced from the viscosity increase), and shear modulus of the formed gel (G), as a function of the time after adding rennet. A and G, arbitrary scale. Only meant to illustrate trends. At physiological pH and about 30°C.

random, at least at physiological pH. The course of the reaction is depicted in Figure 24.3, curve S.

Some factors affecting the reaction velocity are illustrated in Figure 24.2. The effect of the temperature is relatively slight and corresponds to the dependence of the diffusion coefficient (through the viscosity) on the temperature. Serum proteins somewhat diminish chymosin activity. The calcium ion activity has little effect. A certain ionic strength is needed, and that of milk is appropriate. The pH has a considerable effect. At decreasing pH, the affinity of the enzyme for the micelles increases and leads to an increased reaction velocity. At a still lower pH the velocity is, however, smaller, presumably because the enzyme is now adsorbed so strongly onto the (partly denuded) casein micelle that it takes considerable time before an adsorbed molecule is released again and can diffuse further. This implies that the reaction is no longer diffusion-limited. An additional effect of the increased adsorption at low pH is that the hairs are not removed at random, but the enzyme tends to form 'bare patches' on a micelle, before desorbing and diffusing away.

24.3.3 Aggregation

Basic aspects are discussed in Subsection 3.1.3 and Subsection 3.3.3; see especially Figure 3.22. The micelles only aggregate (flocculate) when by far the greater part of the hairs has been removed, so that the steric repulsion of the micelles has been sufficiently diminished. Also, electrostatic repulsion is diminished, as illustrated in Figure 3.7, which gives the zeta potential of casein and paracasein micelles. As shown in Figure 24.3, aggregation starts when about 70% of the κ-casein has been split. Presumably, micelles must approach each other in such an

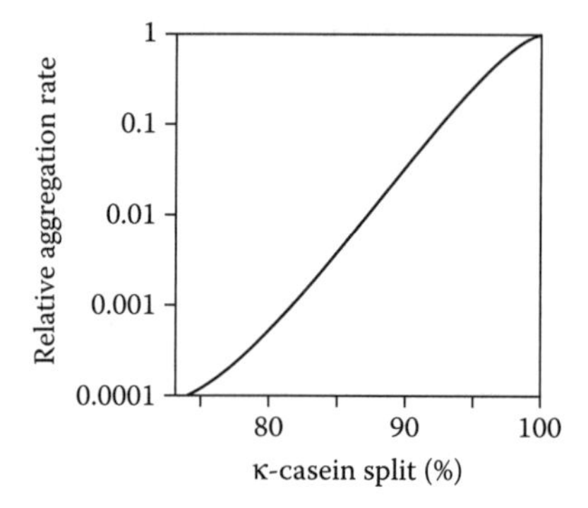

FIGURE 24.4 Approximate aggregation rate of micelles during clotting at 30°C as a function of the percentage of the κ-casein that has been split. 'Relative' implies as compared to true paracasein micelles at the given conditions (pH, temperature, Ca^{2+} activity, etc.).

orientation that each presents a bare, i.e., nonhairy, patch to the other. The more of the hairs have been removed, the greater will be number and size of bare patches on a micelle, and the faster the aggregation proceeds. This is illustrated in Figure 24.4, in which the estimated relative aggregation rate (which essentially is a measure for the reciprocal of the stability factor W; see Subsection 3.1.3.1) is given as a function of the degree of splitting.

The aggregation, i.e., the sticking together of paracasein micelles, is partly due to van der Waals attraction, but this attraction is in itself insufficient. The required Ca^{2+} activity already points to this. Presumably, the effect of the Ca ions is twofold. First, they diminish the electrostatic repulsion by neutralizing negative charges on the micelles. In the pH range concerned, Ca ions act more effectively than H^+ ions. Second, the Ca ions can make bridges (salt linkages) between negative sites on the paracasein micelles. It may further be noted that lowering the pH of milk considerably increases its Ca ion activity (see Figure 2.9).

Figure 24.2 illustrates some factors affecting the aggregation rate of paracasein micelles (κ-casein completely split). Besides the Ca ion activity, the temperature has a considerable effect. At 20°C, aggregation does not occur at all. At 60°C the rate is almost as fast as determined by the encounter frequency, as given in Equation 3.8 for $W = 1$ (at least if the Ca ion activity is sufficiently high). The mentioned stabilization at lower temperature presumably is to be ascribed to repulsion caused by protruding hairs of β-casein; such a protrusion will closely depend on temperature (see Subsection 3.3.3.3).

The pH has a twofold effect on the aggregation rate. First, a decrease of pH increases the Ca ion activity, as mentioned. Second, even at constant Ca^{2+} activity, the aggregation can be faster. This is because at low pH the enzyme does not remove the hairs at random, but tends to split off nearby hairs. Consequently bare patches on the micelle surface are formed at a lower degree of splitting κ-casein (see the previous section). It has been observed, for example, that aggregation started at a degree of splitting of 70%, 60%, and 40%, at pH values of 6.6, 6.2, and 5.6, respectively (at 30°C).

24.3.4 Gel Formation

Gel formation due to fractal aggregation of (para)casein micelles is discussed in Subsection 3.3.4. The gel formed is a particle gel; some properties are also given. Events occurring during renneting are further illustrated in Figure 24.5.

It is seen in Figure 24.3 that an elastic modulus can be detected, so a gel is formed when the viscosity approaches infinity. At that time, the aggregation has advanced so far that the aggregates occupy the whole volume. The modulus of the gel increases, at first because more micelles (or small aggregates) are incorporated in the gel network, and somewhat later because the junctions between any two micelles become stronger due to fusion of the paracasein micelles (Subsection 3.3.4). The gel consists of strands of casein particles, often being 3 to 4 particles in thickness and 10 to 20 particles long, alternated by some thicker nodes of particles. Since the micelles are only 0.1 to 0.3 μm in diameter, this means that considerable short-term rearrangement of the particles has occurred. The larger pores in the gel are some micrometers (up to 10) in width.

As is discussed in Subsection 3.3.2.4, the properties of casein micelles greatly depend on the pH. The properties of the gel more or less reflect this effect. This is further illustrated in Figure 24.6. At pH 5.25, where the colloidal phosphate in the micelles reaches a zero value, the bonds in the micelles are weakest, and the modulus (stiffness) of the gel then is lowest. The gel is clearly viscoelastic, and at pH 5.25, the loss tangent is at maximum. Its value of 0.6 implies that the gel then has a rather viscous character, meaning that it will flow when a stress is applied to it. This is an important property during curd shaping and fusion.

At a lower temperature the junctions between the particles of the gel are stronger. Presumably, this is because the particles are more swollen and are thereby connected to each other over a larger area, which means more bonds per junction.

Besides short-term rearrangement, long-term rearrangement occurs, leading to syneresis. This is discussed in Subsection 24.4.4.

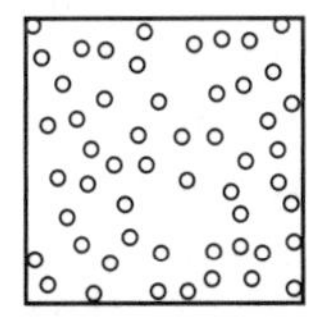
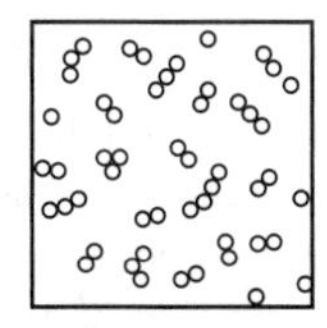
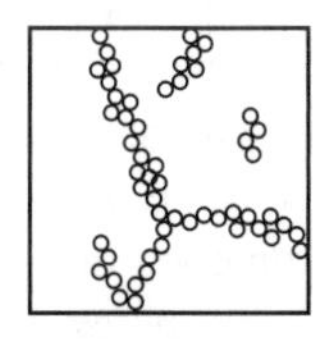
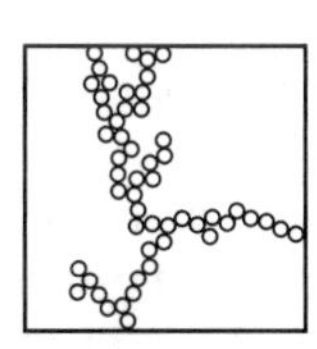
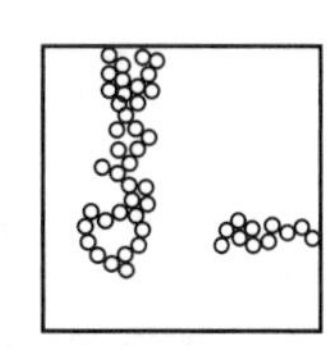

FIGURE 24.5 Representation of the aggregation of paracasein micelles, the formation of a gel, and the start of (micro)syneresis. Highly schematic.

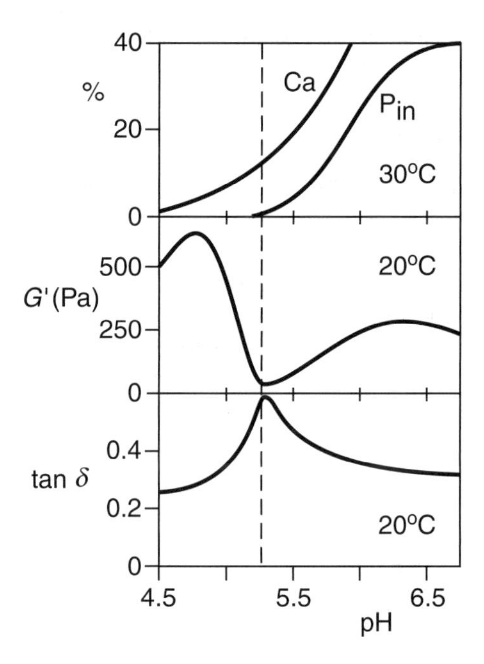

FIGURE 24.6 Some rheological properties of rennet gels as a function of pH. G' is the elastic part of the complex shear modulus and tan δ is the loss tangent (G''/G') at a frequency of 0.01 Hz (G'' is the viscous part of the modulus). For comparison, the proportions of calcium and of inorganic phosphate (P_{in}) in the casein micelles are also given.

24.3.5 The Renneting Time

When rennet is added to milk it takes a while before the micelles start to aggregate (Figure 24.3), but from that time on the aggregation rate increases rapidly (Figure 24.4). At a certain moment, small aggregates can be detected by eye. The time required for this may be defined as the renneting time. It is an important parameter in cheese making. Its dependence on some process variables is comparable to that of the clotting time given in Figure 24.2.

The renneting time is inversely proportional to the enzyme concentration. This relationship, known as the rule of Storch and Segelcke, does not fit precisely, nor can it be explained in a simple manner. The complicated combined action of the enzyme reaction and the aggregation (which increases in rate with time) can only be described by intricate formulas which, by chance, result in an almost linear relationship. By and large, the slower one of the two reactions mainly determines the renneting time, and it depends on conditions which reaction is the slower one.

The temperature especially affects the aggregation rate. Consequently, when milk of low temperature (say, 10°C) is provided with rennet, κ-casein is split, but the micelles fail to aggregate. When the milk is subsequently heated it clots very fast. The increase of the renneting time at temperatures above 35°C (Figure 24.2) is due to heat inactivation of the chymosin.

Milk shows considerable variation in renneting time, especially milk of individual cows. The variation is partly caused by variation in casein content: if it is high, clotting is faster. The main parameter, however, is the Ca^{2+} activity. If it is low, the aggregation is slow. The reaction can be accelerated by adding $CaCl_2$. Increasing the amount of $CaCl_2$ to more than a given level does not result in much change because the enzyme reaction will now determine the reaction velocity.

The influence of the pH is rather complicated. It appreciably affects the enzyme reaction, but little affects the aggregation rate of the paracasein micelles, provided that the Ca^{2+} activity is not too low. As mentioned (in Subsection 24.3.2), the hairs are not removed at random at low pH, which causes the aggregation of the micelles to start at an earlier stage of the enzyme reaction. This effect explains much of the dependence of the clotting time on the pH. It also causes curves as in Figure 24.3 to be different at low pH.

Factors that cause an increase in renneting time, generally also decrease the rate of firming of the gel, i.e., the rate of increase of the gel modulus. However, the changes are not proportional to each other; it also depends on what is taken as the characteristic time. This is illustrated in Figure 24.7; the aggregation time as defined in the figure is very close to the renneting time according to Storch and Segelcke. The cheese maker is more interested in the 'clotting time,' i.e., the time when the curd is (presumed to be) firm enough to start cutting the gel (see Subsection 24.4.4).

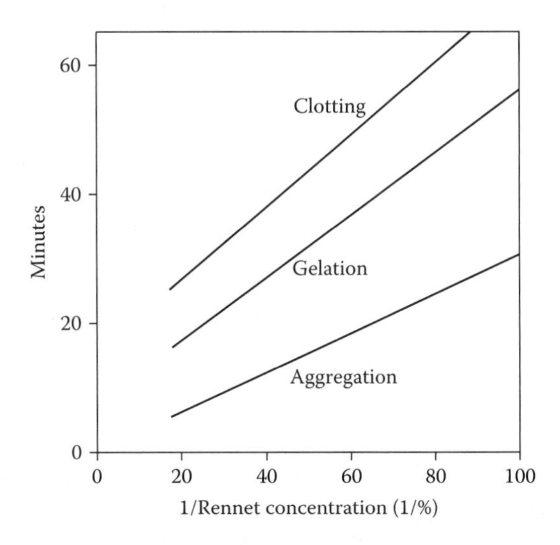

FIGURE 24.7 Time needed for visible aggregation, to start gelation, and to obtain sufficient firmness of the gel (clotting time), as a function of the rennet concentration. Approximate examples (30°C) for low-pasteurized milk; $CaCl_2$ is not added. (Data from A.C.M. van Hooydonk and G. van den Berg, *Bulletin of the IDF* 225, 2, 1988.)

An important point is determination of the *rennet strength* of a clotting agent. It is generally expressed in Soxhlet units, i.e., the number of grams of normal fresh milk that can be renneted by 1 g of rennet in 40 min at 35°C. The renneting time is defined as the time needed for the emergence of small aggregates in a tube with milk. The tube is held in a slanted position in a water bath while rotating; the aggregates are then readily spotted in the thin film of milk on part of the glass wall. A problem is that fresh milk varies in rennetability. Consequently, a synthetic peptide is used as a standard to set the renneting time of the milk used for testing at a specified value. Most rennet preparations are standardized at a strength of 10,000 Soxhlet units.

24.3.6 Clotting of Heat-Treated Milk

Heating milk more intensely than low pasteurization causes an increased clotting time, a weaker curd, and impaired syneresis. The renneting time increases steeply with the heating intensity. The rate of the enzyme reaction is little affected (at most by 20%), but the aggregation becomes much slower. Addition of a little $CaCl_2$, or lowering the pH restores the clotting time if the heating was not too intense. All the same, the effect of heating cannot primarily be explained by a decrease of the Ca^{2+} activity. This is because the heating does not change the clotting time if no β-lactoglobulin is present. The reduced rennetability and the reaction of this protein with the casein micelles (see Figure 7.9E) run closely parallel. Apparently, the stability of the paracasein micelles is considerably increased by the associated layer of serum protein. The detrimental effect of heat treatment can be partly undone by lowering the pH to below 6 and subsequently increasing it again, e.g., to 6.4.

See also Figure 24.8.

24.4 CURD MAKING

Traditionally, curd making is a batch process, involving clotting of the milk; cutting the gel into pieces; whey expulsion due to syneresis and to expression of the curd pieces formed; and separation of the curd from the whey; in the mean time, lactose is converted to lactic acid. This process essentially determines what the composition of the cheese will become: paracasein and most of the fat are collected, including a small amount of whey. Also the ultimate water content and acidity of the cheese are largely determined at this stage.

An additional and important objective is that the process be as short as possible, effective and economical; it should be well controllable and cause little loss of curd and fat. For the sake of efficiency, continuous processing would be preferable. Some processes will be mentioned, but for most semihard and hard cheeses, batch processing appears to be more economical, partly because the process conditions can often still be changed if something goes wrong with a batch. Batch size has greatly increased over the last decades.

To concentrate the protein and the fat, two courses are open: the classical one, in which the milk is clotted first, or concentrating the milk before it is clotted. The latter method will also be discussed.

(*Note on terminology*: We will call clotted milk a 'gel' and reserve the word 'curd' for pieces of gel that have at least lost part of the enclosed whey. Often, the word 'moisture' is used as a synonym for water, as in moisture content. We will not do so and reserve the word moisture for a low-viscosity aqueous solution; for instance, the moisture that can be pressed from a newly formed cheese loaf.)

24.4.1 Clotting

Clotting can be achieved with rennet or another enzyme preparation, with acid, by heating, or by combinations of these processes.

24.4.1.1 Rennet Clotting

The clotting process has been extensively discussed. To achieve it in practice, the milk should have the desired temperature (often 30°C) and be provided with starter and any other additives (e.g., $CaCl_2$) and, of course, with rennet. The milk is stirred to achieve homogeneous distribution of all the additives. The milk should then be left quiescent, to allow the paracasein network to form undisturbed.

As discussed, the clotting time can vary widely. Some variables such as rennet concentration, pH, and temperature can be adjusted. Nevertheless, lots of milk may vary in rennetability due to variation in calcium ion activity, in casein content, and in cold storage and/or heat treatment. Some examples are given in Figure 24.8. In practice, the milk often is a mixture of several milkings of a very high number of cows; this greatly decreases variability. The addition of $CaCl_2$ (about 1 mmolar) further diminishes variation in clotting time. The temperature history of the milk

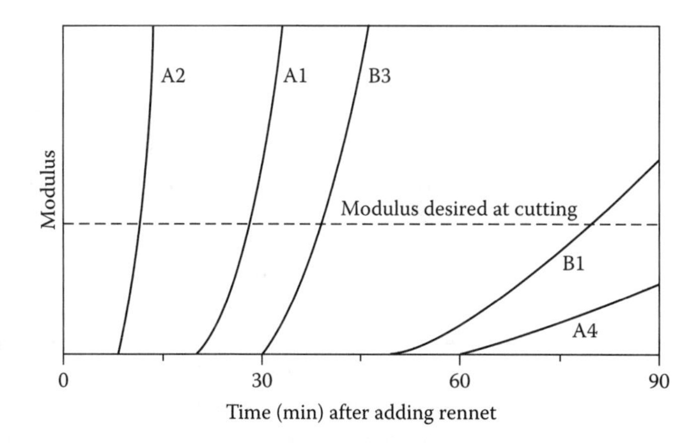

FIGURE 24.8 Firmness (expressed as the elastic shear modulus) of renneted milk as a function of time. Two milks: A is a rather average mixed milk, B a very slowly renneting milk of one cow. Renneting at 30°C, 0.03% rennet. Further conditions: 1 = unchanged milk, 2 = lowered pH (6.4), 3 = $CaCl_2$ added (1 mmolar), 4 = heated (15 s at 95°C). Approximate examples.

can be fixed. Currently, some dairies standardize the casein content of the cheese milk by addition of an ultrafiltration retentate of milk.

On the one hand, clotting should be fast, simply because it saves time. It will be faster when more rennet is added, but this is costly. Addition of $CaCl_2$ somewhat reduces clotting time, but this should not be overdone (the higher Ca content of the cheese then may result in an undesirable texture). After clotting, the gel should be cut. Figure 24.3 shows that the modulus of the gel keeps increasing for a long time. If the gel is quite weak at the moment of cutting, the knives tend to shatter the curd, leading to a lot of 'curd fines,' which are partly lost with the whey. If the gel is quite firm, it means that more energy is needed to cut the curd, hence the knives should move faster, which will also lead to a greater formation of fines. The cutting time should thus be carefully chosen.

In the manufacture of some cheese varieties, notably Cheddar and related types, the curd is allowed to acidify to a low pH before it is further processed. In the traditional process, this means that the curd grains sediment and form a fused layer. These aspects are discussed in Subsection 27.4.1. In the traditional method of making some soft cheese varieties (e.g., Camembert), the clotted milk is not cut into small pieces, but some lumps of gel are put in a perforated cheese vat at intervals to allow slow whey drainage and fusion of the resulting curd (see Subsection 27.6.1.)

A part of the rennet (say, 15%) is recovered in the cheese (see Subsection 24.4.2). This greatly affects the ripening, i.e., the protein degradation (see Section 25.3). In the manufacture of some cheese varieties (Emmentaler, most Italian cheeses) heating during the curd treatment (cooking or scalding) is sufficient to inactivate a considerable proportion of the rennet (see also Figures 24.1 and 24.2). The inactivation is much stronger at a relatively high pH (e.g., 6.4) than at a lower one (e.g., pH 5.3).

24.4.1.2 Acid Coagulation

Milk can be clotted by lowering the pH. This is mostly effected by lactic acid bacteria. Possibly, proteolytic enzymes of the bacteria also play a part because some of those can split κ-casein. However, the main mechanism is casein becoming insoluble near its isoelectric pH.

If the temperature of acid production is not too low (e.g., 30°C) and the milk is at rest, a gel forms, as during rennet coagulation. However, the gel is rather different (see Table 8.3). For instance, it is shorter and may be firmer than a rennet-induced gel. Its composition also differs. Rennet coagulation yields a coagulum of paracasein micelles including colloidal calcium phosphate, whereas an acid milk gel consists of casein particles of a similar size as in rennet coagulation, but without calcium phosphate. The acid gel can be cut and stirred like a rennet gel, but it shows little syneresis, especially in the pH range of 4.2 to 5. For removal of a sufficient amount of whey the temperature should be fairly high, but low-moisture cheese cannot be made via acid coagulation.

Alternatively, the milk may be acidified at some lower temperature while stirring, so that a voluminous precipitate forms, not a gel. Centrifugation separates the soured milk into whey and a pumpable curd slurry. The dry-matter content of such curd cannot be made higher than about 23%, or about 17% if skim milk is used. Therefore, the latter method is mainly applied in the manufacture of quarg; a fully continuous process can be applied.

Acid coagulation by bacterial growth requires a long time, even at optimum temperature. Of course, acid can also be added directly, for instance, lactic acid or hydrochloric acid. This causes clotting of the milk to start already during addition of the acid, which results in curd particles of widely varying shape and size. Rather than an acid, a lactone can be added, which is slowly hydrolyzed (e.g., over an hour) to form an acid. In this way, a homogeneous gel can be obtained. Alternatively, acid can be added to cold milk, i.e., at about 5°C. As in rennet clotting, acid coagulation does not occur at low temperature. After acidification, the milk is uniformly heated to ensure undisturbed gel formation. This process can be used in the manufacture of cottage cheese (Subsection 27.2.2).

Often, a combined acid and rennet coagulation is applied, especially in the manufacture of fresh and ripened soft cheeses. In fact, rennet coagulation is more or less enhanced by acid coagulation when a high percentage of starter is added. Further, the manufacture of some cheeses includes preculturing of the milk before rennet addition. The lower the pH, the faster the clotting (quantity of rennet being equal) (see Figure 24.2). This is not so much due to the pH itself as to the increased calcium ion activity. Syneresis also is faster.

24.4.1.3 Coagulation by Heating and Acidification

Heating of milk *per se* does not cause clotting or coagulation. To be sure, a considerable proportion of the serum proteins becomes insoluble, depending on both time and temperature of heating (Subsection 2.4.3 and Subsection 7.2.2). These proteins largely associate with the casein micelles and, consequently, are recovered in the cheese after coagulation with rennet or acid and syneresis. This may cause problems. The milk clots less satisfactorily with rennet (it takes a long time before a gel forms which, moreover, is not very firm) or even fails to clot (see Subsection 24.3.6). The water content of the curd remains high, even after extended curd treatment. Usually, bitter compounds form during maturation. Often, an increase of cheese yield by heat treatment of the milk is undesirable.

Heating can be applied to recover 'lactalbumin' from whey (Section 21.1.2). The protein is then added to cheese milk; this is sometimes applied in soft cheese manufacture. Alternatively, a protein-rich coagulum, obtained from acid whey by means of heat denaturation, can be processed as such (including molding, salting, and curing) to arrive at a protein-rich product, sometimes designated as cheese.

Queso Blanco is a Latin American cheese. It is made by heating milk to between 80 and 85°C. Then acid is added while gently stirring the milk, which leads to the formation of a coagulum. Various acids can be added: acetic, citric, phosphoric, etc., or simply lime juice. The coagulum is allowed to settle; the curd is gathered and pressed to remove whey. Salt is added, and the cheese is either consumed fresh or after a short ripening time. In the U.S. this cheese is called 'Hispanic' or 'Mexican.' In several tropical countries comparable types of cheese are made.

Mixtures of acid whey and skim milk, to which some lactic acid may be added to further lower the pH, can also be clotted by heat treatment. The curd then contains a considerable proportion of denatured serum protein. This is applied, for example, in the manufacture of Ricotta and Schabziger.

24.4.2 Accumulation of Various Components

Due to clotting a continuous network forms consisting of protein particles, usually paracasein micelles. The pores of the network are of the order of a few micrometers in width. Initially, the network encloses all of the milk, but it soon starts to contract, i.e., to show syneresis. This is illustrated in Figure 24.5. Thereby the moisture (i.e., water plus dissolved substances) is squeezed out. With respect to dissolved substances this seems to imply that at the moment of drainage the composition of the moisture in the curd is similar to that in the whey. On closer inspection, the ratio of water to dissolved substances in the curd will be higher. A certain quantity of nonsolvent water can be defined, as is discussed in Subsection 10.1.1, which is larger for solute molecules of larger size. For the serum proteins in milk, the quantity of nonsolvent water is about 2 to 3 g per gram of paracasein. Most cheese contains approximately this amount of water or less. It is indeed observed that semihard and hard cheeses are virtually devoid of serum proteins and also of the caseinomacropeptide split off κ-casein by rennet. This implies that enzymes that occur dissolved in the serum of milk will not reach the cheese. Cheese varieties of a high water content may contain some serum protein and serum enzymes.

On the other hand, some proteins, notably enzymes, tend to adsorb onto paracasein micelles, and these will accumulate in the cheese. This concerns the indigenous proteolytic enzymes plasmin and cathepsin D. It also occurs with chymosin and pepsin, and this is of paramount importance for cheese ripening (Section 25.3). The adsorption of chymosin is virtually nil at pH 6.7, but it greatly increases with decreasing pH. The amount of chymosin (and pepsin) retained in the curd, and thereby reaching the cheese, therefore depends on conditions during curd making, especially the pH at the drainage state. The amount retained in the cheese varies from 1% to 20% of the quantity added to the milk. This is one of the most important variables determining the rate of proteolysis in cheese. Enzymes that are added to cheese milk may also adsorb onto the paracasein micelles. An example is lysozyme, which adsorbs strongly, and which is sometimes added to inhibit growth of butyric acid bacteria (see Section 26.2).

Enzymes that are in the membrane of fat globules are also incorporated into the cheese; this primarily applies to xanthine oxidase and various aminotransferases and phosphatases. This is because all of the particles in the milk, i.e., fat globules, microorganisms, somatic cells, and dirt particles, are, for the greater part, mechanically entrapped in the network of aggregated paracasein micelles. The number of starter bacteria per gram of material, for instance, is nearly 10 times higher in cheese than in the cheese milk, not considering any growth.

Most important for the cheese composition is that not only protein is accumulated but so is fat. Treatment of the curd, i.e., cutting and stirring, causes loss of particles, especially at the cut surfaces. For instance, about 6% of the fat is lost with the whey. Most of this fat is recovered by centrifugation of the whey. A part of the curd, defined as *curd fines*, is also lost. Losses of fat and fines greatly depend on the firmness of the renneted milk at the moment of cutting and on the intensity of cutting and stirring. If the curd is too weak, the losses are great; if it is too firm, cutting needs excessive force, which can also enhance losses.

If the particles are not only entrapped in the network, but attach to it, or even form part of it, their loss during the mechanical curd treatment is lower. Fat globules become part of the casein network if they have casein in their surface layer. This can occur because of damage during processing of the milk (beating in of air, evaporation, or atomization) and especially by homogenization (see Section 9.5). In addition to a smaller loss of fat into the whey, a somewhat different consistency of the cheese results, often designated as 'sticky.'

24.4.3 Concentrating before Clotting

An alternative manner of curd making involves concentrating the protein before clotting by means of *ultrafiltration*. When diafiltration is also applied (see Section 12.2), it is even possible to obtain a liquid that nearly has the composition of unsalted cheese, provided that it is a cheese of high water content. The diafiltration is needed to remove part of the lactose and of the calcium phosphate. Unsalted soft cheese can so be made in a continuous process. The concentrated liquid is provided with starter and rennet and is slowly pumped through a long tube from which a cylinder of curd emerges, which can directly be cut into slices of the shape of a small soft cheese. Keeping these for a while at a temperature of, say, 20°C allows lactose conversion into lactic acid without undue whey expulsion. The cheese can then be salted and cured.

A variant of this process involves *cold renneting*, which greatly shortens the clotting process. The concentrate is provided with rennet at a low temperature, say 5°C, while stirring. The enzyme is allowed to split all κ-casein, which takes about twice the time needed at 30°C. The liquid is then slowly pumped through a tubular heat exchanger that heats the liquid to about 30°C, which results in rapid clotting. Due to the high casein content and the high viscosity, the gel obtained is sufficiently firm and homogeneous, despite the flow occurring during clotting.

The process is not feasible for semihard and hard cheeses: the milk cannot be concentrated sufficiently. The milk can be concentrated four- or fivefold, then clotted, and the emerging strand of curd is cut into small cubes that are subject to further syneresis, etc. Hence, a partial-UF semicontinuous curd-making process results.

The claimed advantages of UF-cheese making include increase of cheese yield due to incorporation of whey proteins; savings on rennet and other additives;

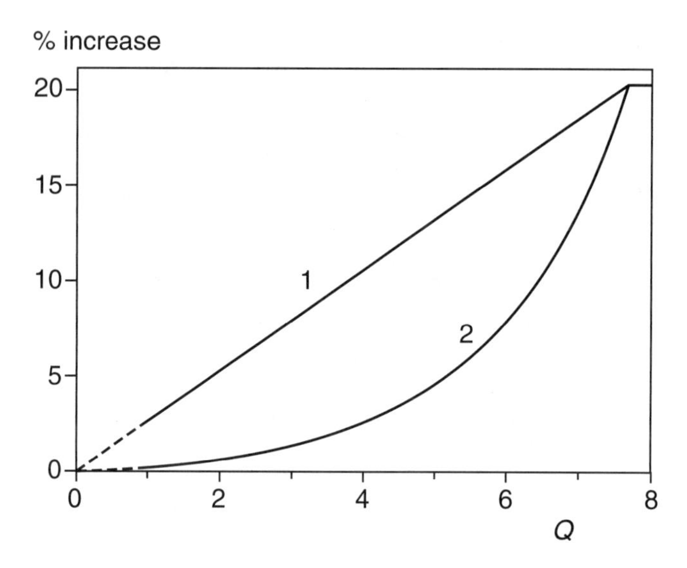

FIGURE 24.9 The relative increase in the yield of the yet unsalted cheese due to concentrating the cheese milk by ultrafiltration prior to clotting, as a function of the degree of concentration *Q*. Curve 1, assuming zero nonsolvent water. Curve 2, expected in practice. Calculated for a constant water-to-paracasein ratio of 2.5, and a constant fat content in the dry matter of 48%. Approximate results to illustrate the trend.

and a continuous process would be more efficient. Nevertheless, the UF process still has not met with general success for the following reasons:

- A profitable outlet has to be found for the ultrafiltration permeate produced, and this is often not possible.
- A substantial increase in yield is only obtained for full ultrafiltration, not for partial UF. This is due to the relatively large amount of nonsolvent water for most proteins mentioned in the previous section (about 2.5 g of water per gram of paracasein). Approximate results are given in Figure 24.9, and it is seen that for fivefold concentration of the milk — which is about the maximum possible — the yield increases by only 4%, as compared to a hypothetical 14% if nonsolvent water would be zero. (*Note*: The observation that some whey protein is included, the more so at a higher *Q*-value, must be due to a sieving effect: the higher *Q*, the smaller are the pores formed in the gel, and the greater the difference in flow rate through the gel between water and protein molecules.)
- The rate of proteolysis during cheese ripening is substantially reduced (see Subsection 25.3.4), resulting in lack of flavor.
- The formation of holes in the cheese due to CO_2 production by starter bacteria is not well possible (cf. Subsection 25.6.1). This is a disadvantage in some, especially soft, cheeses.

24.4.4 Syneresis

Many gels tend to shrink spontaneously, thereby expelling liquid. A rennet milk gel can do so, expelling whey, which will be further discussed in this section. Acid milk gels are also subject to syneresis by the same mechanism as rennet gels, though much slower (see Subsection 3.3.4).

24.4.4.1 The Mechanism of Syneresis

The aggregation of paracasein micelles causes formation of a particle gel with relatively large pores containing whey and fat globules. A micelle makes junctions with, on average, about three other micelles, but its total surface area is reactive, i.e., can form junctions with additional micelles. This would cause a gain in bond energy, which provides a driving force. However, making new junctions is for the most part sterically hindered, as the micelles are immobilized in the network. On the other hand, the immobilization is not complete, as strands of micelles can exhibit some Brownian motion. This leads to the occasional formation of new junctions, which may impart a tensile stress in the strands involved. This may, in turn, lead to breaking of such a strand. These events are illustrated in Figure 24.10. The breaking of a strand allows shrinkage to occur and, moreover, it may allow the formation of additional new junctions, etc.

A rennet gel often does not exhibit spontaneous expulsion of whey. This is because the gel may stick to the walls of the vessel, whereby it is confined. Nevertheless, the process illustrated in Figure 24.10 will go on, but it now merely will cause the local formation of larger pores and of denser regions elsewhere. This process of coarsening of the gel structure is called *microsyneresis*. Incidentally, the top layer of renneted milk in a vessel does neither show spontaneous whey expulsion, presumably because the milk surface is hydrophobic; putting a drop of water or whey on the surface will immediately cause syneresis to occur.

The local transport of whey through a gel can be described by the *Darcy equation*:

$$v \equiv \frac{Q}{A} = \frac{B}{\eta} \cdot \frac{\Delta p}{L} \qquad (24.1)$$

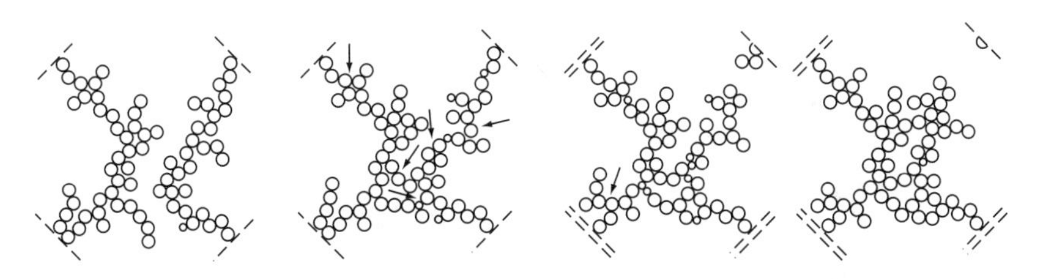

FIGURE 24.10 Schematic representation of strands of paracasein micelles forming new junctions, leading to the breaking of a strand and to (local) shrinkage of the gel network.

where v is the superficial velocity (m/s) of the whey, defined as its volume flow rate Q (m^3/s) over the cross-sectional area A (m^2); B is the *permeability* of the network (m^2), whose value increases with pore size and with number of pores per unit cross section; and η is the viscosity of the whey (Pa·s). These parameters characterize the gel.

The pressure acting on the whey and causing its flow, is given by

$$\Delta p = p_S + p_E \tag{24.1a}$$

where p_S is the *endogenous syneresis pressure*. It has quite a small value, on the order of 1 Pa (this corresponds to the gravitational pressure exerted by a water 'column' of 0.1 mm height). Hence, syneresis would be very slow, but its rate is always boosted by an external pressure p_E. It generally consists of two terms: a gravitational pressure exerted by the gel network (the density of the micelles is higher than that of the whey) and a mechanical pressure applied to the gel or the curd layer. The overriding effect of p_E is illustrated in Figure 24.11.

L is the distance over which the pressure acts and the whey has to flow. Cutting a gel into small pieces will thus greatly enhance syneresis rate (in terms of Q): the flow distance is smaller and, moreover, the surface area A is larger (about proportional to L^{-1}).

The values of the parameters B and p_S and, thereby, the rate of syneresis, depend on a number of product and process variables, as illustrated in Figure 24.12. The effect of the variable 'time' needs some elaboration. In the figure, it means the waiting time after which the macrosyneresis process is started, and all values of p_S are thus initial values ($p_{S,0}$). It is seen that B increases with

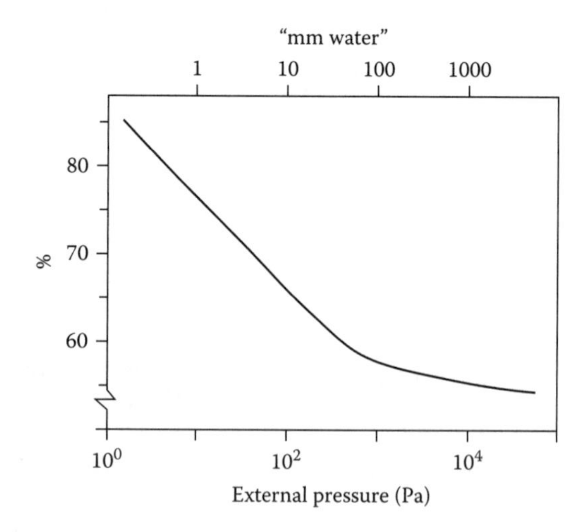

FIGURE 24.11 Water content (%) of curd made of whole milk after 1 h stirring and 1 h pressing under the whey at various pressures. pH and temperature were kept constant.

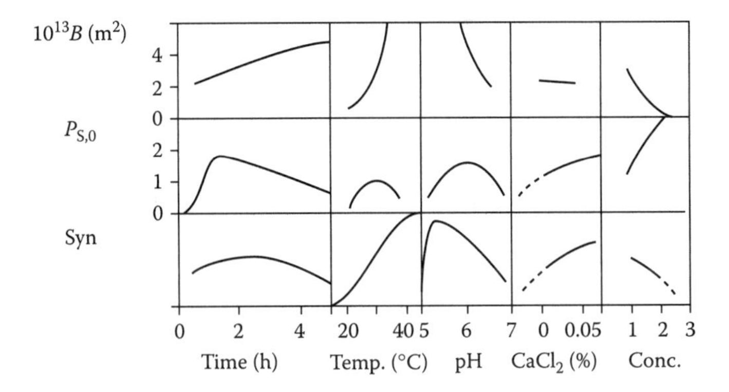

FIGURE 24.12 Effect of time after rennet addition, temperature, pH, added $CaCl_2$, and concentrating of the milk (by ultrafiltration) on syneresis: permeability (B), endogenous syneresis pressure ($p_{S,0}$), and approximate syneresis rate (Syn). The influence of the time refers to situations without (macro)syneresis, i.e., the syneresis is allowed to begin at the time indicated.

waiting time, undoubtedly due to ongoing microsyneresis. The value of $p_{S,0}$ at first increases due to the time needed for the gel structure to form. In the meantime, the junctions between paracasein micelles become stronger, which lowers the possibility for strands to break and for syneresis to occur. These changes also happen after syneresis has started, but the situation is more complicated. Due to the syneresis itself, the gel network becomes concentrated, i.e., more compact, and this lowers the permeability. Moreover, the value of B will now greatly vary with distance: close to the surface through which the whey flows out, B will be much smaller than further inside a curd particle.

Consequently, the kinetics of syneresis is intricate. Besides the points mentioned, the pH will generally decrease during the process, further altering the parameters, and the temperature may also be changed. In most cases, however, the syneresis rate will become ever slower during the process. When a fairly low water content has been reached, say 55%, the curd has become so firm that the gel network cannot readily comply with the deformation that further syneresis would cause. In other words, not the permeability, but the consistency of the gel (to be precise, the elongational viscosity at a low stress) becomes the limiting factor.

Finally, a distinction should be made between the *rate of syneresis*, which determines the time needed for the process, and the extent of syneresis, i.e., the *eventual water content*, which determines cheese properties. Fortunately, in the beginning of syneresis, when it is fast, much whey is removed without the water content of the curd greatly decreasing, while it is the other way round near the end, when syneresis is slow (see Figure 24.13). This allows fine tuning of the end point of the process. Syneresis can virtually be stopped by substantially lowering the temperature.

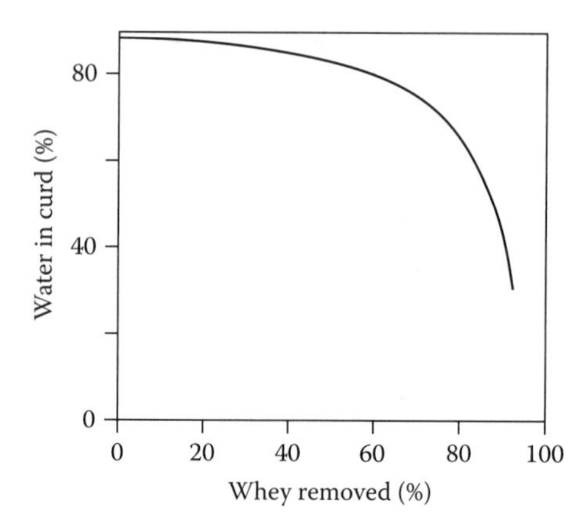

FIGURE 24.13 Calculated relation between the water content of curd (of whole milk of 87.7% water) and the quantity of whey (6.8% total solids) expelled as a percentage (w/w) of the original milk.

24.4.4.2 Factors Affecting Syneresis

The effect that some factors have on syneresis may vary widely according to conditions. The main cause is that syneresis pressure as well as permeability can be affected (see Figure 24.12). With high-moisture cheese, the syneresis should be slowed down or stopped after a certain time. The rate of syneresis then considerably affects the final result. However, the syneresis rate is less important if cheese is made with a low water content. The factors affecting syneresis are:

1. *Firmness of the gel at cutting:* If the gel is still weak at cutting, it tends to synerese slightly, but syneresis will increase rapidly. Of greater importance is that a large amount of curd fines is released, causing a reduced yield of cheese. As a rule of thumb, the shear modulus of the gel should be about 30 Pa when cutting starts.
2. *Surface area of the curd:* Initially, the syneresis is proportional to the area of the interface between curd and whey. Accordingly, the gel expels very little whey if it is not cut and keeps sticking to the wall of the vessel. (Clotted milk sticks to most hydrophilic materials, though not to copper; it sticks more or less to stainless steel, especially if it has been rinsed with an alkaline solution.) Therefore, the gel is generally cut into cubes. The smaller the cubes, the faster the syneresis. At the same time, more fat and curd fines are transferred to the whey. If the curd granules differ widely in initial size, the resulting cheese may become inhomogeneous. Small grains shrink fastest and become dryest. Moreover, most of these grains settle at the bottom of the layer of curd grains. Small differences in the average size of the curd grains have little effect on the final water content.

In the traditional manufacture of soft-type cheeses, large lumps are cut from the gel and ladled into molds, where the syneresis occurs. This leads to a high water content.

Fine curd grains are obtained and a dry cheese results if the milk is stirred during renneting and after. Of course, the loss of curd fines and fat is also markedly increased.

3. *Pressure:* Figure 24.11 has already been discussed briefly. The result holds for curd under the whey. Stirring exerts pressure, partly because of the pressure difference due to velocity gradients in the liquid, which, according to Bernoulli's law, equals $^1/_2\, \rho d^2 G^2$; ρ is liquid density, d the diameter of the curd grain, and G the velocity gradient. This causes pressures of the order of 10 Pa. Moreover, the curd grains collide and thereby compress one another for a short while. Furthermore, stirring prevents sedimentation of the curd; in a layer of curd grains, the surface of the grains available to release whey rapidly diminishes, thereby slowing down syneresis. Figure 24.14 gives an example of the influence of stirring. The effect markedly increases with the increasing ratio of curd to whey (by removal of whey) and stirring rate. Stirring or working the drained curd also causes a lower water content.
4. *Acidity:* The effect of the pH is shown in Figure 24.12. The explanation is not fully clear. Syneresis rate markedly increases with decreasing pH, but it decreases sharply if the pH is decreased to below 5.1. The permeability is scarcely affected, but the endogenous syneresis pressure falls to about zero. Incidentally, it makes some difference as to whether the pH is lowered before or after the gel

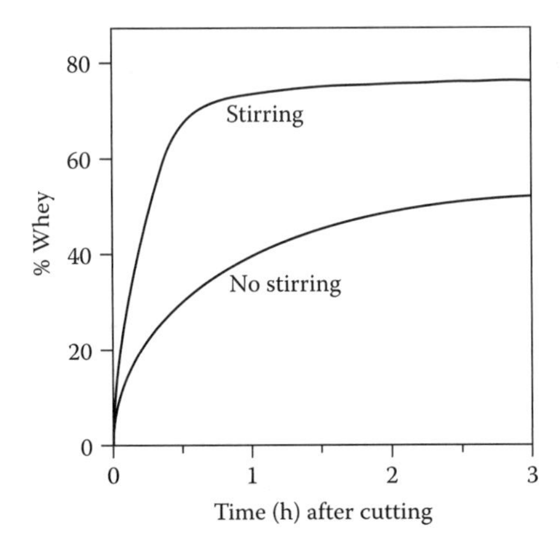

FIGURE 24.14 The volume of whey released (as percentage of the original volume of milk) as a function of time after the start of cutting, with or without stirring. Curd kept in the whey, temperature 38°C. (Data from A.J. Lawrence, *Austr. J. Dairy Technol.*, **14**, 169, 1959.)

has formed. In pracical cheese making with rennet coagulation, more acid production always implies a faster syneresis and a lower final water content.

5. *Temperature:* This has such a strong effect that syneresis can be stopped by cooling (see Figure 24.12). Increase of temperature markedly accelerates syneresis. However, a very quick rise of temperature (e.g., by adding hot water) causes the outer layer of the curd grains to shrink so fast that a 'skin' forms, i.e., a layer of such low permeability that it substantially slows down further syneresis; an additional reason may be that the skin has become so firm as to mechanically hinder further shrinkage of the curd grain.

   The temperature also affects swelling and shrinkage of the paracasein matrix. It appears that as the temperature is lowered, the casein becomes more swollen (see also Figure 3.20). This is of importance when the syneresis has almost been finished, i.e., when the paracasein micelles have reached a rather close-packed arrangement. Increasing the temperature of such a concentrated paracasein gel may cause further expulsion of moisture, whereas lowering the temperature may lead to absorption of moisture.
6. *Composition of the milk:* The higher the fat content, the less the curd can shrink; hence, a higher final water content in the fat-free cheese. (The water content in the cheese is lower because there is more fat, substituting fat-free cheese.) The fat also impedes the flow of whey out of the curd, i.e., the permeability is smaller, so fat reduces the syneresis rate.

   Furthermore, Ca^{2+} activity, pH, protein content, and concentration of colloidal calcium phosphate affect syneresis. Most milk of cows in early lactation clots faster and shows more syneresis. There is considerable variation among the milk of individual cows.
7. *Numerous other variables* also have an effect. Several factors that can be varied in practical cheese making are summarized in Figure 24.15.

24.4.5 Acid Production and Washing

The pH in the curd decreases because of the action of the starter bacteria. This decrease greatly enhances the syneresis. The rate of decrease in pH is determined by factors including starter (amount added, type, strain), composition and pretreatment of the milk, and temperature during curd treatment.

Almost all starter bacteria are entrapped in the curd, implying that the acid is mainly produced in the curd grains. As long as the grains are in the whey, lactic acid can diffuse from them into the whey and lactose in the opposite direction. In this way, acid is produced in the whole mixture. After drainage of the whey the accumulation of acid in the curd proceeds more quickly, and the lactose content in the curd decreases more rapidly.

In most cases, the rate of acid production, and so the pH at molding, has little effect on the final pH of the cheese. In most cheese types all lactose is converted,

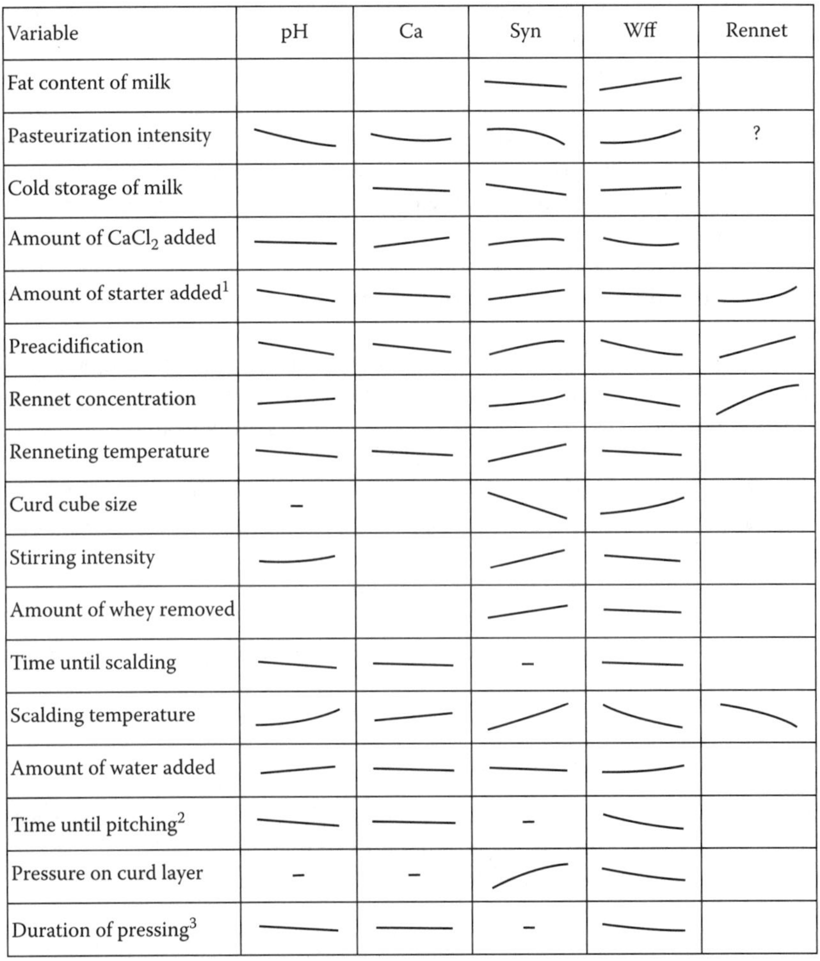

Variable	pH	Ca	Syn	Wff	Rennet
Fat content of milk					
Pasteurization intensity					?
Cold storage of milk					
Amount of $CaCl_2$ added					
Amount of starter added[1]					
Preacidification					
Rennet concentration					
Renneting temperature					
Curd cube size	-				
Stirring intensity					
Amount of whey removed					
Time until scalding			-		
Scalding temperature					
Amount of water added					
Time until pitching[2]			-		
Pressure on curd layer	-	-			
Duration of pressing[3]			-		

FIGURE 24.15 Effects of pretreatment of the milk and of conditions during curd making on the pH of the curd at the end of curd making (not the final pH of the cheese), the amount of Ca retained in the cheese, the rate of syneresis (Syn), the final water content of the fat-free cheese (Wff), and the quantity of rennet retained in the cheese. The relations are only meant to illustrate trends for cheese of average water content. (Blank): the relation is unknown but probably is weak. (Dash): no relation. Notes: (1) Mesophilic starter. (2) Total time between cutting and pitching (i.e., allowing the curd grains to sediment). (3) On the curd layer, e.g., by means of metal plates on top (not in the molds).

mainly to lactic acid. It implies that the ratio between lactic acid and buffering compounds determines the pH. The moisture content of the curd is of paramount importance. The higher it is, the more lactose, or its product lactic acid, is retained in the curd, and the more acidic the resulting cheese will be. The main buffering substances are paracasein and calcium phosphate.

Consequently, the rate of acid production can have a secondary effect on the pH of the cheese. It can affect syneresis and thereby the water content of the cheese. If the latter is held constant by means of additional measures, a small effect persists, e.g., minus 0.1 unit in pH. If the curd is more acidic at the moment of molding, more calcium phosphate has dissolved (Figure 24.15; see also Figure 24.24), and so less buffering substance is left in the cheese. Preacidifcation and inoculation with a large amount of starter have similar effects. Furthermore, a lower pH at molding causes a slightly lower yield of cheese dry matter. If cheese is salted at the curd stage, the bulk of the lactose must have been converted before molding because the lactic acid fermentation is markedly slowed down by the added salt.

To adjust the pH of the cheese independently of its water content, other steps should be taken. A smaller drop in pH is achieved by *washing*, i.e., adding water to the mixture of whey and curd. The lactose diffuses away from the curd grains until identical concentrations are attained in the water inside and outside the curd. The effect of the washing closely depends on the size of the grains and on the contact time. Equilibrium is rarely attained. In practice, the efficiency of reducing the lactose concentration in the curd approximates 90%.

The wash water is commonly also employed to raise the temperature of the curd and whey mixture, i.e., *scalding* or cooking. The higher temperature causes the syneresis to be stronger. There are also other methods for scalding the curd, such as direct heating.

24.4.6 Separation of Curd and Whey

Figure 24.16 illustrates that the water content of the curd sharply decreases when the curd is taken out of the whey. This effect should largely be ascribed to a greater gravitational pressure on the curd; the average increase will be by about 500 Pa or more. It is seen that extended stirring causes a smaller final water content. If the temperature of the curd mass could be kept constant after drainage of the whey, then the time of separation would affect the final water content only slightly. This illustrates that in practical cheese making the lowering of the temperature after the whey drainage rapidly restrains syneresis. The larger the block of curd, i.e., the larger the loaf of cheese, the slower it cools and the lower the final water content will be, although the transport distance of the moisture is longer in a larger loaf (see also Figure 24.18).

The curd grains can be allowed to settle and form a layer. Alternatively, the mixture of curd and whey is transferred to a vertical drainage cylinder in which the curd settles. As long as the curd is under the whey, any mechanical pressure has an effect on the water content (Figure 24.11) and so has the height of the column. The drainage of whey may be crucial in a column of curd in a drainage cylinder. Whey flows primarily between the curd grains, but these deform because of the pressure involved and also fuse, lowering the curd surface area, narrowing the pores between the grains and eventually almost closing them. A great spread in the size of the curd grains (e.g., presence of many curd

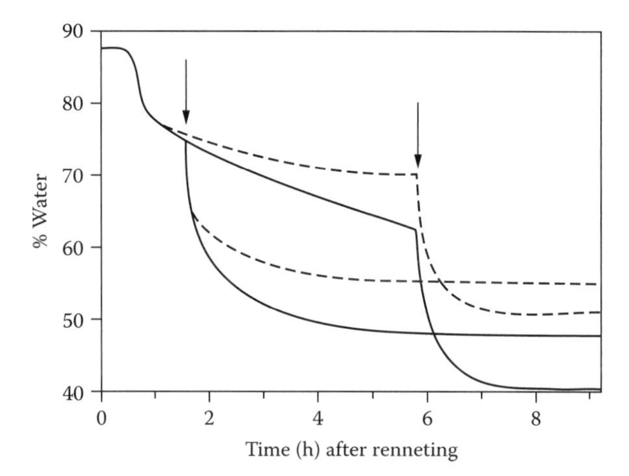

FIGURE 24.16 Examples of the course of the water content of curd during curd making. Cutting after 0.5 h. At two moments (indicated by arrows) curd was taken out of the whey and put into a cheese mold. The dotted lines refer to experiments without added starter. The curd and whey mixture was continually stirred. Temperature in the whey was 32°C throughout; temperature in the mold gradually fell to 20°C.

fines) enhances blockage of the pores, which decreases drainage. According to conditions, one or the other process will prevail, i.e., enhancement of syneresis by pressure or slowing it down by fusion, respectively. Figure 24.17 illustrates all of these effects.

In the traditional cheddaring process (Subsection 27.4.1), the drained curd is left for a long time, preferably without cooling it too much. Meanwhile, considerable acid production occurs. The long waiting time and the low pH cause the water content to become low. However, during cheddaring, the curd is allowed to spread, and this enhances closing of the pores between curd grains and fusing of grains. The latter two factors lead to a higher water content than would occur if the curd cannot spread. The difference amounts to some 1 to 2% water in the cheese.

If the curd has been made very dry before drainage of the whey, e.g., by means of prolonged stirring or a high scalding temperature, a higher temperature during drainage leads to a higher water content. Probably, this is caused by a more rapid closing of pores between curd grains which, in turn, is due to a more rapid deformation at the higher temperature. Correspondingly, in *this* case, a smaller loaf of cheese attains a lower water content.

To make curd very dry, the drained curd can be stirred again. This causes a considerable loss of fat and curd fines in the whey and marked mechanical openness in the cheese.

It will be clear that the various process steps should be matched to one another. Figure 24.16 shows this to be difficult if the curd, after it has been made batchwise in a tank, is subsequently drained, molded, and pressed in a continuous process,

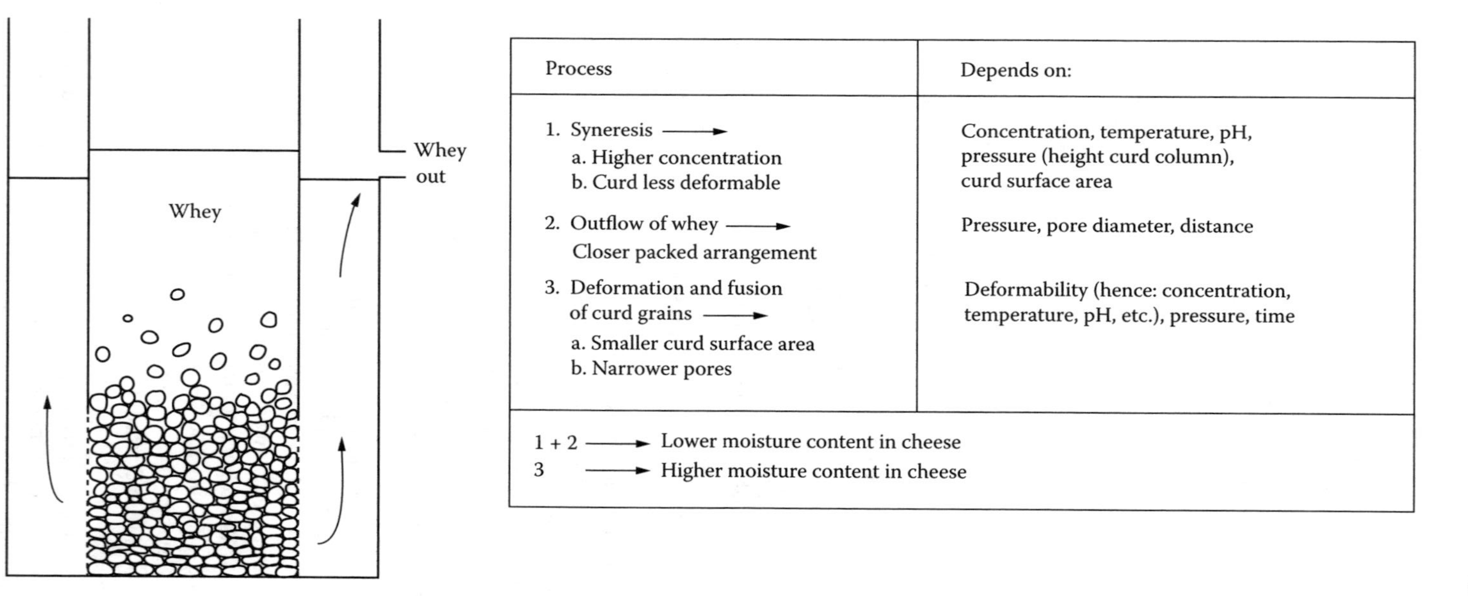

FIGURE 24.17 Processes in a draining column of curd and whey. Highly schematic. See Figure 24.6 for the effect of pH on curd deformability.

as syneresis goes on in the tank. It will lead to significant variation in water content of the cheeses. To overcome this problem the curd and whey mixture can be stirred gently in a buffer tank while being gradually decreased in temperature (see Figure 27.9). In the meantime it can be transferred from the buffer tank to the filling/molding device.

The mechanization and scaling up of the curd making that now have become normal practice has led to the use of a fixed time schedule for the various process steps. Adaptations during the manufacturing process, necessitated for instance by changing development of acidity or syneresis rate, are hard to achieve. One should therefore start from large quantities of milk (because this implies little variation in composition from one batch to the next) and precisely standardize the process conditions. Problems can still arise from varying rates of acid production, generally caused by contamination by bacteriophages.

24.5 SHAPING AND PRESSING

In most cases, it is desirable to make the curd into a coherent mass that is easy to handle, is of a suitable size, and has a certain firmness and a fairly smooth closed surface. To achieve this, the curd is shaped by putting it in molds or hoops; in hard and most semihard cheeses, the operation includes pressing of the curd.

Shaping of the curd can only be done if the grains deform and fuse (see also Figure 24.17). *Deformation* is needed because the whole mass of curd must adopt the shape of the mold and because the grains should ideally touch one another over their total surface area. Viscous deformation is needed, i.e., the mass of curd must approximately retain its obtained shape when the external force is released. The greater the force, the faster the deformation; pressing thus can help. The deformation is considerably affected by the composition of the curd. With decreasing pH, the deformability increases until pH 5.2 to 5.3 is attained; at still lower pH, the curd becomes far less deformable (cf. Figure 24.6). Furthermore, the deformability increases with the water content and especially with the temperature. At very high temperatures (e.g., 60°C) the curd can be kneaded into almost any shape and at a suitable pH it can even be stretched. This property is used in the manufacture of cheeses with Pasta Filata (Subsection 27.5.2). The high temperature and the kneading also affect the consistency, i.e., the cheese becomes tough and smooth. For a poorly deformable curd (low pH, low water content, and low temperature) holes can remain in the cheese, even if it is heavily pressed. This may be the case for traditional Cheddar cheese, but here it concerns fairly large pieces of curd, formed by cutting an already fused curd mass, and these pieces must undergo considerable deformation; moreover, the outside of these pieces of curd is firm due to the added salt. The applied pressure should therefore be high and the temperature not too low.

The *fusion* of the curd grains into a continuous mass is enhanced by increasing the area over which they touch one another. Obviously, conditions allowing easy deformation thereby enhance fusion. If, however, the curd still shows significant syneresis during molding, fusion is counteracted by the layer of whey forming between the grains. The fusion proceeds most easily if the pH is fairly low, e.g., 5.5. This may be caused by new bonds forming readily between the paracasein micelles. Soured curd fuses poorly. If curd is stirred until its pH has dropped to 5.0 and is also cooled, it cannot be pressed into a coherent mass: the loaf immediately disintegrates upon demolding.

Within a day after curd making, the fusion usually is complete, which means that no visible pores between curd grains are left in the mass of curd. The permeability coefficient has reduced to, say, 10^{-15} m^2. A few days may be needed to complete the fusion to the point where the mechanical properties of the cheese have become more or less homogeneous. Thus, if a piece of one-day-old cheese is strongly deformed, it fractures between the original curd grains, whereas a cheese that is 4 days old generally fractures through the grains. These observations do not apply to cheese that is salted at the curd stage.

As mentioned above, the *pressing* furthers the shaping; it is needed (except for soft cheeses) to achieve a closed surface, i.e., to form a rind. It is not so much meant to decrease the water content. Moisture can be released from a mass of curd that is already more or less coherent because free whey or whey moving away from the curd grains can flow through the pores between the grains. If, however, a rind has formed, outflow of whey is markedly hindered and, accordingly, one of the effects of pressing is that any further decrease of the water content is small. The earlier the pressing starts, and the higher the applied pressure, the higher the remaining water content of the cheese. All of this applies to a not-too-dry and not-too-acidic curd. In such cases, the pressure usually ranges between 5 and 50 kPa. In Cheddar cheese making, where the curd is much dryer and more acidic, pressures up to 200 kPa are used, and vacuum pressing may be applied to prevent holes from remaining in the cheese.

Before and during pressing, a rapid temperature decrease must be avoided because it hinders the deformation of the curd grains and the rind formation. Larger loaves cool more slowly. (They may even rise in temperature due to the heat produced by the starter bacteria.) If the curd is not yet very dry, it can still show syneresis, and this proceeds more vigorously at a higher temperature; therefore, the water content of the pressed cheese will be lower for a larger loaf size. Within an unsalted cheese mass, the water content is lower in the center than in the rind, with the difference amounting to, say, 6% water, as is illustrated in Figure 24.18. The above does not apply, however, to cheese that is made of curd stirred very dry. Applying a higher pressure to such curd results in a lower water content, as does a lower temperature. Presumably, the rind of this cheese only forms after a considerable proportion of the moisture has been squeezed out. In turn, the unevenness in the distribution of the water is much less.

Currently, Cheddar curd is often made by a stirred-curd process, after which the separated curd is provided with salt and put into large barrels or forms. After pressing, large cheese loaves result, weighing, e.g., 300 kg. Such a cheese then

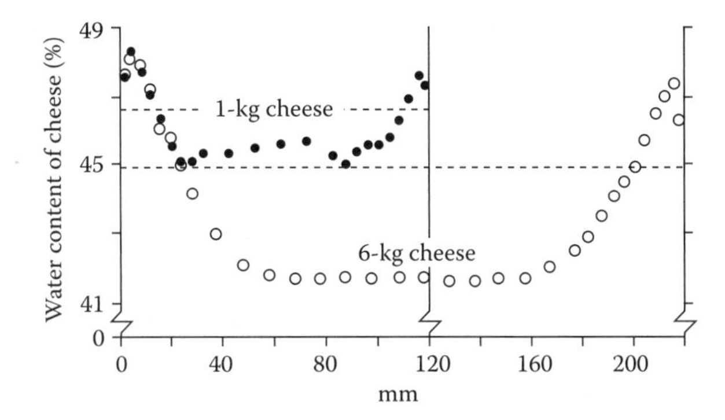

FIGURE 24.18 Distribution of the water throughout unsalted spherical loaves of cheese of 1 and 6 kg, respectively. The cheeses were shaped from one mass of curd, lightly pressed, and kept for a few days. The broken lines indicate the average water content in a similar cheese of the same batch. (From T.J. Geurts, *Neth. Milk Dairy J.*, **32**, 112–124, 1978.)

has a water distribution comparable to that shown in Figure 24.18; also in this case, the center of a loaf will cool slower than the outside.

The *formation of a rind* is also affected by the water content of the curd, the temperature and the pressure applied during pressing, and the duration of pressing. A main factor is the local drainage of whey. If no drainage can occur, as in a mold of smooth nonperforated steel, the surface will not become closed. A closed surface does form if moisture can be removed through perforations in the mold. Putting a cloth or a gauze between cheese and mold causes still better drainage and results in a true rind, i.e., a thin layer of cheese about 1 mm thick, with a reduced water content, forming a kind of skin. The subsequent evaporation of water markedly enhances the thickness and toughness of the rind. Very fast evaporation may cause cracks.

24.6 SALTING

Salting is an essential step in the manufacture of cheese. The primary functions of salt include preservation and its effect on cheese flavor, consistency, and ripening. Moreover, growth of lactic acid bacteria is inhibited at high salt content. Most cheese varieties contain about 2% salt, or 4 to 5% salt-in-water.

The methods as applied in the salting of cheese can be classified as follows:

1. *Dry salting:* Salt crystals are mixed with the curd grains, or with the milled curd pieces resulting from traditional cheddaring.
2. *Rubbing:* Salt or brine is rubbed onto the surface of the cheese. Currently, this is restricted to the salting of cheeses that develop a

microbial smear on the surface. The rubbing is repeated several times.

3. *Brining:* The cheese is kept immersed in a concentrated solution of NaCl (brine) until the desired amount of salt has been absorbed.

A combination of these methods may also be applied. For example, Gruyère cheese is brined, followed by rubbing dry salt onto the surface.

Salting affects the cheese yield: salt is taken up in the cheese, but at the same time a greater amount of water moves out, resulting in a (substantial) loss of weight. The loss of weight during brining amounts to about 3%.

Most cheeses are molded, and often pressed, before salting. Then, after most of the lactose has been converted into acid, brining takes place. This serves three purposes: cooling the cheese (brine temperature about 12°C); preventing the still-soft loaf from sagging under gravity; and, of course, salt uptake. It may take a long time before the salt reaches the interior of the cheese. Brining is a lengthy process; it often lasts several days. It requires much space, but is easy to perform and to control. A surplus of brine forms, however, leading to environmental problems.

The brine is not just an NaCl solution. Cheese contains several solutes, notably lactic acid and salts, which are partly leached out. This is of considerable importance. If cheese is put in a pure NaCl solution, it develops a soft and velvety rind that is readily damaged during handling and does not easily dry up. This is caused by a tendency of the cheese protein to dissolve in the brine (especially if the salt content is not very high), which is prevented by the presence of sufficient calcium ions, combined with a low pH. A low pH also assures that microorganisms cannot grow in the brine. In practice, brine is never renewed; of course, salt has to be added to keep the brine strength constant.

Direct mixing of salt with the curd grains has the advantage that the salt rapidly reaches the interior of the cheese. It also allows a fast evening-out of the salt content. The latter takes more time if the milled pieces of cheddared curd are dry salted. The salting is less easy to perform and to control, especially after cheddaring. Directly after salting the curd is molded and (strongly) pressed. This leads to considerable loss of salt with the 'press whey.' Hence, dry salting also causes an environmental problem.

24.6.1 Mass Transport during Salting

Transport of salt in cheese occurs primarily by diffusion. Diffusion of molecules in a liquid is due to their thermal or Brownian motion. Each molecule shows erratic and random movements, but the result is that different molecules become more evenly dispersed through the available space. This implies that salt ions or molecules will on average move from a region of high to one of low concentration; in other words, net transport of salt will occur. To be sure, it concerns mutual diffusion of salt and water: when salt is transported in a given direction, net water transport occurs in the opposite direction.

The transport processes will be discussed for the case of brining, but the relations given are also valid for other diffusion-driven transport processes, as, for instance, in dry salting.

24.6.1.1 Transport of Salt

Figure 24.19 illustrates how the salt has penetrated into a large loaf of cheese that has been brined. If the diffusion takes place through a plane surface (e.g., a cheese surface), application of Fick's second law results in:

$$\frac{C_b - C_x}{C_b - C_0} = \mathrm{erf}(y) = \frac{2}{\sqrt{\pi}} \int_0^y \exp(-\omega^2)\mathrm{d}\omega \tag{24.2}$$

where

$$y = x/(4\ D^* t)^{0.5}$$

and where C is salt content relative to water in the brine (C_b), in the unsalted cheese (C_0) and in the cheese at a distance x from the brine-cheese interface (C_x); t is brining time and ω is an integration variable. By the use of Equation 24.2,

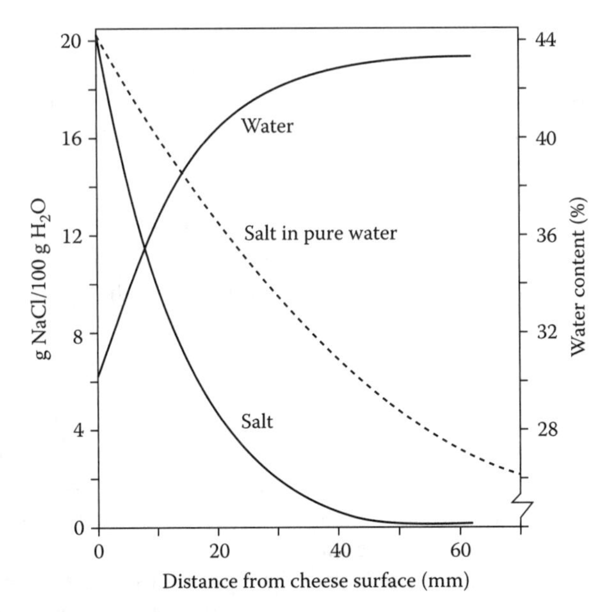

FIGURE 24.19 Distribution of salt and water in full-cream cheese after 8 d of salting in brine with 20.5 g NaCl/100 g H_2O. Also shown is what the salt distribution would be if the salt diffuses unhindered in pure water.

the effective diffusion coefficient D^* ($m^2 \cdot s^{-1}$) of the salt moving in the water in the cheese can be derived from results as depicted in Figure 24.19. It may be added that the value of D^* does not significantly depend on the salt concentration.

For example: $D^* \approx 2.3 \cdot 10^{-10}$ $m^2 \cdot s^{-1}$ for a full-cream Gouda cheese with an initial water content of 45%. D^* is smaller than D, the diffusion coefficient of NaCl in water, which is approximately $12 \cdot 10^{-10}$ $m^2 \cdot s^{-1}$. In both cases, however, it concerns transport of salt in water. The difference between D and D^* arises from the fact that the water in cheese is enclosed in a matrix; therefore, the salt moves in moisture held in that matrix. Neither salt nor water can diffuse unhindered through the cheese. Consequently, $D^* < D$. The main factors responsible for impeding NaCl diffusion in cheese are:

1. *Viscosity of the moisture:* The diffusion coefficient is inversely proportional to the viscosity of the medium available for transport. The viscosity of cheese moisture is higher than that of water, causing D^* to be as much as 10% smaller than D.
2. *Tortuosity:* The molecules (ions) diffusing in the water must travel by a circuitous route to bypass obstructing particles, so the effective distance covered by the molecules will be longer. The molecules must bypass the fat globules, therefore, D^* decreases with increasing fat content, and they must bypass the particles of the protein matrix in the fat-free cheese, so D^* decreases with decreasing water content in the fat-free cheese.
3. *Friction:* The pore width of the protein matrix in cheese is about 1 to 3 nm, depending on the water content. Because the diameter of the hydrated salt ions is at least 0.5 nm, a marked frictional effect is exerted on the diffusing ions.
4. *Counterflux:* During the salting process, salt penetrates into the cheese and at the same time much water moves out. The water flux considerably exceeds the salt flux, so the cheese shrinks (in the region into which salt has penetrated). The counterflux of water reduces the apparent rate of the salt diffusion.

The effective diffusion coefficient D^* greatly depends on the initial water content (see Figure 24.20). For example, in a full-cream cheese with 50% water, D^* is more than twice the value in a cheese with 39% water. Moreover, for the same water content, D^* increases with the fat content. This is because a higher fat content implies a higher water content in the fat-free cheese, and the latter variable has a stronger effect on D^* than the fat content has.

If the cheese is put into brine before the curd grains in the freshly pressed loaf have completely fused, the matrix will contain some rather large pores. In such a case, the diffusion coefficient of salt (and of water, for that matter) may be significantly larger during the first hours of the brining process than later on.

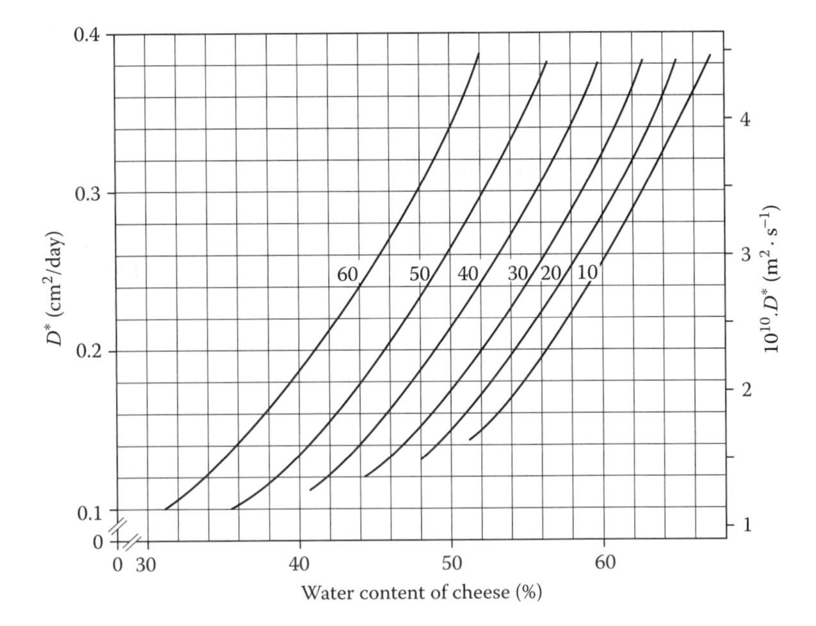

FIGURE 24.20 Diffusion coefficient of NaCl in the water in cheese (D^*) as a function of the initial water content of the cheese. Parameter is g fat/100 g dry matter in unsalted cheese.

24.6.1.2 Displacement of Water

Where salt penetrates, water diffuses out of the cheese (see Figure 24.19; the initial water content of the cheese was approximately 43%). The salt and water transports are quantitatively related. The proportionality factor or flux ratio p is defined as the ratio between the decrease in water content and the increase in salt content at distance x from the cheese–brine interface, and at time t; p may be fairly constant within a cheese, i.e., independent of x or t. During the process, the water content thus is lowest in the part of the cheese closest to the brine, whereas it remains unchanged where salt has not yet penetrated.

The flux ratio p approximates 2.5. The migration rate of the water surpasses the rate of the salt at every moment and at every location in the cheese. This is largely due to partial osmosis, because the frictional hindrance of H_2O diffusion in cheese by the narrowness of the pores in 'fat-free' cheese is less than that of NaCl diffusion. The outward migration of the water thus is a direct consequence of the penetration of the salt. The water is not displaced as a result of an independent shrinking of the matrix, but the reduction of cheese volume follows from the partial osmosis. The cheese matrix can fully or partly comply with the forces causing this reduction; if the cheese is fairly rigid (low pH, low water content, low temperature), the matrix resists shrinkage, and less water is lost than for a less rigid cheese matrix, i.e., p is smaller.

TABLE 24.1
Effect of Some Variables on the Difference (Δ) between the Initial Water Content and the Weighted Average Water Content of Cheese, and the Mass Flux Ratio (*p*) of Water to Salt

Fat in dry matter (g/100 g)	Initial water (g/100 g)	pH of cheese	Temperature (°C)	Salt in brine (g NaCl/ 100 ml)	Δ (g/100 g)	*p* (g/g)
10	59	5.0	12.5	20	10	2.7
40	49	5.0	12.5	20	8.5	2.5
60	39	5.0	12.5	20	6	2.3
50	36	5.0	12.5	20	5	2.5
50	45	5.0	12.5	20	7.5	2.4
50	50	5.0	12.5	20	8.5	2.2
50	45	4.7	12.5	20	6	1.7
50	45	5.7	12.5	20	8.5	2.8
50	45	5.0	20	20	10	2.8
50	45	5.0	12.5	14	4	1.8
50	45	5.0	12.5	31	11	3.2

The extent to which the average water content of the salted part of the cheese during salting is lower than the initial water content depends on several factors, but the length of the brining time does not affect it. Data are given in Tables 24.1 and 24.2.

24.6.1.3 Quantity of Salt Taken Up

The salt concentration of the cheese moisture as a function of x and t can be calculated by using Equation 24.2. The effective diffusion coefficient D^* is virtually independent of time t.

Provided that the salt has not yet penetrated into the center of the cheese, the quantity of salt absorbed from a flat cheese surface follows from:

$$M_t = 2\,(C_b - C_0)\,(D^* t/\pi)^{0.5}\,\overline{w} \qquad (24.3)$$

where M_t = quantity of salt absorbed over time, in kg NaCl/m^2; C_b = kg NaCl/m^3 brine; C_0 = kg NaCl/m^3 water in unsalted cheese; and $\overline{w}$ = average water content, expressed as a fraction of the cheese (kg/kg). For $\overline{w}$ the weighted average of the water content in that part of the cheese in which the salt penetrates should be taken; the weighting factor is the local salt uptake, and Table 24.1 gives the approximate difference Δ with the initial water content. Inserting the original water content of the cheese in Equation 24.3 would result in the calculated salt uptake being too high: $\overline{w}$ is much lower than the original water content because water moves out during the salting process.

TABLE 24.2
Influence of Several Important Factors on the Salting of Cheese

Factor	Effect on: Weighted Average Water Content	Quantity of Water Lost	D^*	Quantity of Salt Taken Up	p
Fat content	–	–	–	–	–
Water content	+	+	+ +	+ +	±/–
pH of cheese	–	+	0	–	+ +
Ratio of surface to weight	0	+ +	0	+ +	+
Temperature	–	+	+	+	+
Duration of brining	0	+ +	0	+ +	(a)
Salt content of brine	–	+ +	0/–[b]	+ +	+ +
pH of brine	+	–	?	+	–

Note: Weighted average water content refers to that part of the cheese in which the salt has penetrated; D^* is effective diffusion coefficient of the salt in the water in the cheese; p is mass flux ratio water/salt; quantities mentioned are per kg of cheese; + = positive correlation; + + = strong positive correlation; ± = correlation questionable, but at most slight; 0 = no correlation; – = negative correlation; and ? = not investigated

[a] Depends on time elapsed between pressing and brining, but generally ~0.
[b] D^* may be slightly smaller when the salt concentration of the brine is very high.

Obviously, the quantity of salt absorbed is not proportional to t, but to $t^{0.5}$. Soon, M_t increases even more slowly than would fit in with Equation 24.3 because of the limited dimensions of the cheese. The greatest limitation of salt absorption occurs where the surface is markedly curved, i.e., near edges. The smaller the cheese loaf (the more the surface is curved) and the higher the salt content of the cheese, the stronger is this effect.

The mass fraction of salt (Z) in the brine-salted cheese can be derived from:

$$Z = M_t A / G_s \tag{24.4}$$

where A and G_s are surface area (m^2) and weight (kg) of the cheese, respectively. Textbooks on process engineering deal with diffusion in bodies of various geometry, needed for adjustment of Equation 24.3. On the basis of the appropriate equations, the process variables to arrive at a desired salt content in the cheese can be calculated.

24.6.1.4 Water and Weight Loss

When the cheese takes up Z kg of salt per kg of (salted) cheese, it will lose pZ kg of water. Hence, its weight loss will amount to $Z(p - 1)$ kg per kg cheese. The value of $(p - 1)$ is on average about 1.5. The water content of the salted cheese can also be calculated from these parameters.

24.6.1.5 Dry Salting

The above applies to brine salting of cheese. By and large, it is also true for dry salting. If the curd is salted before pressing, the same processes occur but the quantitative relationships differ considerably and are hard to predict and control. In traditional Cheddar cheese manufacture, the curd first sediments and fuses into a compact mass, which is subsequently cut into strips ('milling'). Then dry salt is mixed with the curd. Salt diffuses inward, causing a counterflow of whey from the curd to the surface, which creates a brine solution around the curd strips. After approximately 10 min the absorption of the appropriate amount of salt has been achieved, but in addition to the curd, a fair amount of salt solution is present, which is discarded during subsequent pressing. In this way, up to 50% of the added salt may be lost, as well as an appreciable amount of cheese moisture, especially if the salt is inadequately mixed with the curd. An increase in salting rate increases the relative salt losses with the press whey, and it is very difficult to obtain high salt concentrations in the cheese. A more precise dosage is possible if the curd grains are not allowed to fuse before the addition of salt, as occurs after 'stirred curd' making.

24.6.2 Important Variables

Factors affecting D^*, p, salt content, and water content of the cheese after brining are given in Table 24.2. Figure 24.21 gives results as a function of brining time.

D^* is primarily affected by water and fat content (Figure 24.20). The temperature has a limited effect. Within the pH range of 4.7 to 5.7, D^* does not significantly depend on the pH. Flux ratio p is scarcely affected by the water content, but all other factors have an effect.

Salt and water content after the brining process greatly depend on initial water content, fat content, the ratio of surface to mass, salt content of the brine, and, of course, the duration of brining. Temperature and pH of cheese and brine have a small but significant effect.

It is not easy to ensure that all cheese loaves obtain the same amount of salt and take it up from all sides. This depends on the *brining operation*. Traditionally, the loaves are put in a shallow basin filled with brine. The loaves float and the upper sides are regularly sprinkled with salt, to ensure salt uptake (and water loss) from all sides; alternatively, the cheeses are regularly turned over. When enough salt has been acquired, the loaves are removed and left to dry before further handling. Currently, basins with a long and winding channel are often used; the loaves are slowly transported through the channel, while the brine is being stirred. Cheese can also be brined in deep basins. The cheeses are stacked in large wire cages that are fully submerged. The brine is stirred, and this allows a fairly even brining of all loaves. The method saves on floor space, but needs additional machinery to load and unload the cages, and to put the cages in and out of the basin.

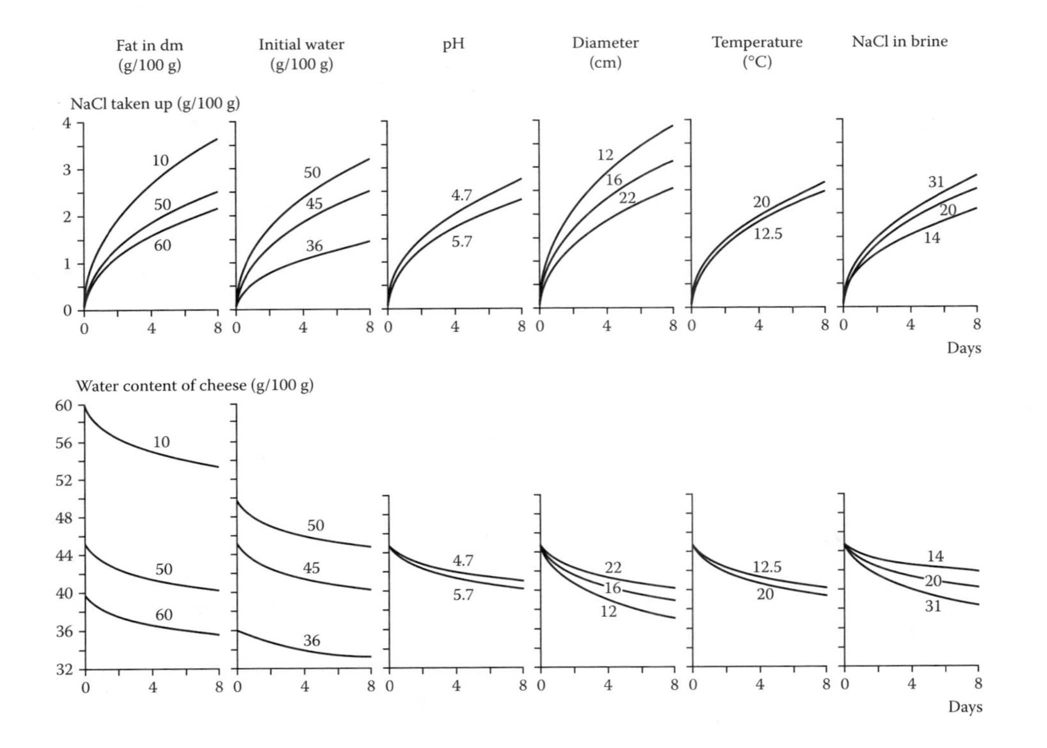

FIGURE 24.21 Predicted effects of some factors on the salt uptake and on the water content during salting of a spherical cheese with 50% fat in the dry matter, 62% water in the unsalted fat-free cheese (hence, a full-cream cheese contains 45% water), pH 5.0, diameter 22 cm (~6 kg), in brine of pH 5.0 containing 20 g NaCl/100 ml, and a temperature of 12.5°C (unless stated otherwise).

Salt has to be added to keep the brine at constant strength and also the pH should be regularly adjusted with HCl to about 4.8. Because the brine is diluted with water released by the cheese, the surplus has to be removed and discarded. The brine contains soluble components from the cheese, notably lactic acid and calcium salts. Moreover, it inevitably contains debris, which is regularly removed, preferably by microfiltration.

To save on operation time, cheese is often put into brine soon after pressing. This has some consequences. First, as mentioned, it leads to a faster and more variable salt uptake because the rind of the cheese often is not fully closed. Second, the cheese will initially still show some syneresis by which more soluble components can reach the brine. Third, the lactose in the cheese is not yet fully converted into lactic acid, which means that the brine will take up lactose.

The presence of lactose has some consequences. It implies less lactic acid in the brine, hence a higher pH, unless HCl is added. It may allow growth of salt-tolerant microorganisms in the brine; they can be removed by regular microfiltration. Moreover, it will take a long time before the lactose in the outer layer of the cheese is converted into lactic acid, as the starter bacteria are not very salt tolerant. In extreme cases, this may cause growth of undesirable microorganisms on the surface or in the outer layer of the cheese.

24.6.3 Distribution of Salt and Water after Salting

After the salting, water and salt become more or less evenly distributed throughout the cheese mass. For Edam cheese, this takes some 4 to 6 weeks; for 12-kg Gouda cheese, 8 weeks; for Camembert and Brie, 7 to 10 d; for Emmentaler, over 4 months. An example of the evolution of the salt-in-water distribution in a Gouda-type cheese is shown in Figure 24.22.

Water and salt distribution will never become completely even. This is because some different processes occur. Both salt and water diffuse, but the effective diffusion coefficient of water is larger than that of salt. With the long timescales involved, the cheese matrix can generally comply with local volume changes. Moreover, water is lost by evaporation through the rind, and this can be a substantial quantity (see Table 24.4). Finally, ripening processes can lead to local changes in water activity, hence, induce water diffusion.

If the cheese is salted at the curd stage, the salt initially is fairly unevenly distributed, especially when the curd is cut to relatively large strips, as in traditional Cheddar cheese manufacture. It then takes a fairly long time to arrive at an even distribution of the salt. As an example, after a week, the relative standard deviation between samples taken from one cheese was found to be 25%, whereas after 9 weeks it was still 10%. This must be ascribed to inadequate mixing of salt and curd because after such a long time, the differences between the salt contents at various spots in one strip of curd (diameter about 2 cm) should have disappeared.

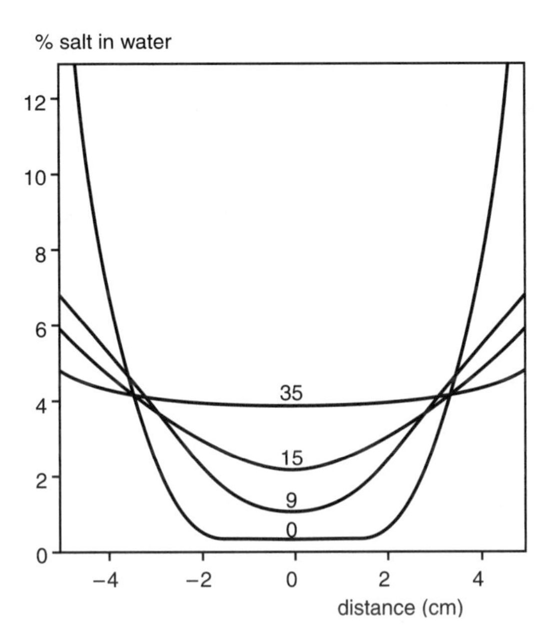

FIGURE 24.22 Distribution of salt in water as a function of distance from the center plane in a Gouda-type cheese of 10 cm thickness at various numbers of days after brining (indicated near the curves). Brining time was approximately 5 d. Approximate results to illustrate trends.

24.7 CURING, STORAGE, AND HANDLING

In this section, the basic principles of the storage of cheese are described. Types of cheese that are cured are considered, i.e., the types produced most. We can distinguish:

1. Types of cheese with a specific surface flora or internal molds, in addition to the normal flora of lactic acid bacteria
2. Types without a specific flora

The storage of cheese starts after its manufacture. Often, this is after the salting. In Cheddar cheese making salt is mixed with the curd before pressing. Feta-type cheeses are first salted, after which they are packaged and cured in brine or in acid whey with salt added. This holds also for the Domiati type of cheese; in its manufacture, the milk is provided with a high salt content (8% to 15% NaCl), and acid is produced in the cheese by salt-tolerant lactic acid bacteria.

The storage of ripening cheese is aimed at making and keeping it suitable for consumption. The product should develop its characteristic properties: flavor, consistency, body (cross-sectional appearance), and rind. Any loss, especially that caused by excessive evaporation of water, as well as deterioration of the rind

TABLE 24.3
Approximate Examples of Storage Conditions during Cheese Ripening and Storage, and of Ripening Time of a Number of Cheese Types

Type of Cheese	Stage	Temperature (°C)	Relative Air Humidity (%)	Ripening Time (Days)
Soft cheese without surface flora, e.g., Butterkäse		12–14	95	15–20
Soft cheese with surface smear, e.g., Munster		12–16	≥ 95	35
White-molded soft cheese,	1. (10 d)	11–14	85–90	
e.g., Camembert	2. (packaged)	4	85–90	35
Blue-veined cheese, e.g., Roquefort		7–10	95	100
Semihard cheese, clean surface, e.g., Gouda		12–16	85–90	50–300
Semihard cheese, surface flora, e.g., Tilsiter		12–16	90–95	150
Hard cheese, surface flora,	1. 2 weeks	10–14	≥ 95	
e.g., Gruyère	2. 5–10 weeks	16	85–90	
	3. Remainder	10–14	85	300
Hard cheese with large holes,	1. 2 weeks	10–14	80–85	
e.g., Emmentaler	2. 5–10 weeks	20–24	80–85	
	3. Remainder	10–14	85	90–200
Hard cheese, dry salted,	1. 2 weeks	12–16	75–80	
e.g., Cheddar	2. Remainder	5–7	75–80	60–300
Very hard cheese	1. 1 year	16–18	80–85	
e.g., Parmesan	2. 1 year	10–12	85–90	700

and/or of the texture due to undesirable microbes and cheese mites, should be prevented. For some cheeses, the handling during curing may require more labor than the manufacture proper.

The actual treatment significantly depends on the type of cheese involved and varies with the progress of maturation. Various types have a short ripening time and shelf life, whereas other cheeses are adapted for extended storage (Table 24.3). The following sections cover the main variables.

24.7.1 Temperature

Temperature affects the growth rate of microbes of a desirable specific flora and the activity of their enzymes as well as that of enzymes of foreign origin, especially rennet and starter; it thereby affects the rate of ripening. Generally, a higher temperature causes a quicker ripening, but at the same time, it enhances the risk of spoilage by undesirable microbial growth. Examples are undesired mold growth on the surface and butyric acid fermentation. Types of cheese that are susceptible to the latter fermentation may be cured at low temperature during the first ripening stage to allow the salt to become evenly distributed throughout the cheese (see Section 26.2). This method of working is especially applied for cheese to be used for making processed cheese.

If the temperature is too low, the rate of ripening is unsatisfactory. At very low temperatures the flavor remains flat and uncharacteristic. Storage at low temperature, after a preceding ripening time at a higher temperature, normally serves to slow down continuing ripening processes and to retard approaching defects. It thus extends the storage time. In particular, soft and prepackaged cheeses are treated in this way. Various types can also be frozen, especially if packaged in small containers. For example, Gouda and Cheddar cheeses can be stored for more than 6 months at –3°C. If stored at –20°C these cheeses turn mealy or even crumbly. A temperature of –20°C is suited to storing high-fat Gouda cheese (60% fat in the dry matter).

24.7.2 Air Conditions

Humidity, temperature, and velocity of the air affect the vaporization of water. The air humidity has a considerable effect on growth of microorganisms on the cheese rind. To allow the cheese to retain a satisfactory shape, the ripening cheese loaves should be turned, initially frequently. Such turning should also enhance the growth of any aerobic flora on the whole cheese surface and prevent, in cheese without a specific surface flora, the growth of microaerophilic microorganisms between loaf and shelf. The air humidity in the vicinity of the cheese surface (the 'microclimate') can appreciably differ from that elsewhere in the storage room.

Rate and extent of vaporization can be partly responsible for developing microbial defects in that the cheese does not become dry enough. On the other hand, the cheese should not dry too quickly, especially not just after brining, as this may cause cracks in the rind. (Sometimes the cheese loaf is rinsed with water after brining, causing the rind to become more supple.) Initially, the relative humidity may be taken somewhat lower and the air velocity higher, if the cheese has not been pressed or was pressed in such a way as to form a weak rind. Vaporization causes the rind to become firmer. If much water vaporizes the cheese rind turns into a closed horny layer that slightly retards the transport of water and gases. Surface-ripened soft cheese often develops a thin crust containing much calcium phosphate, especially when the pH of the rind becomes high.

Of course, vaporization implies loss of weight. This loss approximates 0.2% per day for the first 2 weeks in Gouda cheese (10-kg loaf). In Table 24.4 examples are given of the weight loss of cheese stored under various conditions.

24.7.3 Rind Treatment

24.7.3.1 Cheese with a Specific Flora

The reader is referred to Section 27.6. In this category we distinguish:

1. *Cheese with a surface smear:* The smear contains several microorganisms that do not grow if the pH in the cheese surface is too low. Lactic acid should first be decomposed, which is mainly effected by yeasts.

TABLE 24.4
Loss of Weight of Gouda Cheese Kept for 9 d under Various Conditions

Air Velocity (m/s)	Temperature (°C)	Relative Humidity (%)	Original Weight of Cheese (kg)	Loss of Weight (%)
0.1	14	85	10	1.7
0.2	14	85	10	2.2
0.4	14	85	10	2.6
1	14	85	10	3.2
0.2	12	85	10	2.1
0.2	16	85	10	2.3
0.2	14	82	10	2.6
0.2	14	86	10	2.1
0.2	14	90	10	1.4
0.2	14	85	4	2.7
0.2	14	85	15	2.1

Source: Data from S. Bouman, *Zuivelzicht* **69**, 1130–1133, 1977.

A supply of oxygen (fresh air) stimulates the growth of several bacteria. Regular smearing of the surface or washing the cheese with water or with weak brine aids in developing a uniform slimy layer. The necessary bacteria disappear if the cheese is washed too frequently or too intensely. The slimy layer inhibits mold growth. There are numerous types of soft cheese with a surface smear, e.g., Munster, Limburger, and Pont l'Évêque, which are all of small size. Examples of semihard cheeses with a smear on the rind are Tilsiter and Port Salut. An example of a hard cheese is Gruyère. As time passes the slimy layer is generally left to dry. Afterward, certain cheese types are coated with latex.

2. *White-molded cheese:* The cheese may be sprinkled with a mold culture after it has been salted and partly dried, or mold spores can be added to the cheese milk and/or the brine. Growth conditions can be enhanced by adjusting the temperature in the ripening room, by allowing contact with the air (including frequent turning of the loaves), and by a high relative humidity (which should, however, be lower than that for the preceding group of cheeses). During ripening, contamination of the cheese surface by undesirable molds should be prevented.
3. *Blue cheese:* Before the ripening starts, the cheese is perforated with needles. Cylindrical loaves may be put down on their round sides to stimulate the air supply into the pores formed, which enhances the growth of the blue mold. The cheese is cured at relatively low temperature and at high relative humidity. Most of the blue-veined cheeses should not develop a significant surface flora and are thus to be kept clean. Other types like Gorgonzola do have such a flora.

24.7.3.2 Cheese without a Specific Flora

Here we distinguish:

1. *Hard and semihard brine-salted cheese:* Microbial growth on the rind of the cheese may adversely affect cheese quality, especially flavor and appearance. Of particular importance is the growth of molds (some of which may produce mycotoxins), coryneforms, and yeasts. To avoid such growth, the cheese rind is supplied with a surface coating. Currently, a latex — often called plastic emulsion — is generally applied, i.e., a polymer latex of vinyl acetate, vinyl propionate, or dibutyl maleinate. On drying, a coherent plastic film forms that slows down vaporization of water and offers a better protection against mechanical damage than did earlier used expedients like linseed oil and paraffin oil. The latex coating allows the cheese rind to be much weaker. The mechanization and speeding up of the manufacture of many types of cheese would not have been possible without the introduction of these latex emulsions. The film mechanically hinders mold growth, be it incompletely. It may also contain fungicides, e.g., natamycin (= pimaricin), an antibiotic produced by *Streptomyces natalensis*, or calcium or sodium sorbate. In almost all European countries only natamycin is permitted. When compared to sorbates, it offers the advantages that its protective action is about 200 times as strong, that its migration into the cheese is limited to the outer few millimeters, and that it does not negatively affect the appearance, taste, and flavor of the cheese. Moreover, it is harmless. An acceptable daily intake of 0.3 mg natamycin per kg body weight has been proposed (note that the outermost cheese rind is rarely eaten). The amount applied to cheese rind is generally below 2 mg/dm^2.

   In practice, successive treatments with latex are applied to all sides of the cheese shortly after brining. During long curing, the treatment may be repeated. The surface should be sufficiently dry before each treatment. The conditions in the ripening room must permit the latex to dry quickly (not too quickly because cracks could form in the film, leading to mold growth in the cheese). Too slow a drying may cause growth of microorganisms, especially coryneforms and yeasts. Such growth is greatly enhanced by the high humidity between loaf and shelf, caused by moisture that is inevitably expelled by the cheese at the beginning of the ripening. To prevent this, as well as to allow the cheese to retain a satisfactory shape, the loaves are frequently turned during this stage; the frequency is reduced upon prolonged curing. Obviously, regular cleaning and drying of shelves should form part of a general program on hygiene in curing rooms.
2. *Cheese salted at the curd stage:* Traditionally, Cheddar and related cheeses were often stored in cheese cloth and only provisionally kept

clean. Currently, the cheese is usually formed into rectangular loaves, for example, weighing 20 kg. Shortly after pressing, the loaves are packaged under vacuum in plastic foil (e.g., Saran), requiring little if any further attention. At first, the loaves should be piled not too close together in order to allow cooling. If the cheese still shows some syneresis (high water content or high temperature), an aqueous layer forms between the cheese rind and the foil, in which deteriorative microorganisms can proliferate.

24.7.4 Packaging

Packaging is an important aspect of the curing of cheese. Several factors are involved in selecting a package: (1) type of cheese and its consequent resistance to mechanical damage, (2) presence of a specific flora, (3) wholesale or retail packaging, (4) permeability to water vapor, oxygen, CO_2, NH_3, and light, (5) labeling facilities, (6) migration of flavors from package to product, and (7) the system for storage, distribution, and sale (supermarket, specialist shop, and rate of turnover in the market). These aspects cannot be discussed in detail here but a few remarks will be made:

1. Formerly, semihard or hard cheese was often treated with paraffin wax, whereas currently many are coated with a latex which, of course, is also a kind of packaging. When the cheese is going to be waxed its surface should be very clean and dry; otherwise growth of bacteria between cheese rind and paraffin wax or latex coating will cause problems, especially because of gas production and off-flavors. Waxing thus can be applied for low-moisture cheese shortly after manufacture, whereas cheese with a higher water content may be waxed only after a suitable rind has developed.
2. Some cheese is cured while being packaged in an air- and water vapor-tight shrinking film, e.g., Saran foil. The cheese may be made in rectangular blocks of up to 300 kg, which are usually intended for sale in prepackaged portions or slices, or for the processed cheese industry. Compared to normally ripened cheese, important differences are as follows:
    a. The cheese has no firm rind.
    b. Its composition is more homogeneous due to moisture losses being quite small.
    c. The cheese has a lower water content immediately after manufacture because this content must meet the requirements for a ‘normal’ cheese after ripening (which loses more water during storage).
    d. The starter may not produce too much CO_2, as otherwise loosening of the wrapping would readily occur (‘ballooning’).
    e. The larger the blocks, the longer it takes to cool them to curing temperature, increasing the chance of microbial defects to occur.

f. Consequently, the curing temperature is often taken lower. Together with differences in composition, this causes the flavor development to be less than in normally ripened cheese of the same age.
g. After cooling, the blocks can be piled up closely and need not be turned.

24.8 CHEESE COMPOSITION AND YIELD

Cheese composition depends on milk composition and on numerous process variables. The composition of a cheese, together with its size and shape, largely determines *cheese properties*, hence type and quality. The gross composition, i.e., water, fat and salt content, can be varied deliberately over a wide range. For several cheese varieties, legal standards may apply, such as a maximum water content, and a minimum and maximum fat content in the dry matter. The fat content is generally calculated on the dry matter because the water content of the cheese can change considerably during storage. Similar standards may apply to salt (NaCl) content. Apart from additives such as spices, the further composition is more or less the outcome of cheese making and curing, and deliberate variation is often not easily realized. The fat content is generally controlled by standardizing the milk fat content.

Accurate *determination* of cheese composition may pose problems because of considerable random variation. Composition tends to vary (1) between batches, even if the same milk is used; (2) between cheese loaves within one batch (the extent of the variation strongly depends on the manner of curd making and loaf formation); (3) within a loaf, especially for brine-salted cheeses; and (4) with aging time. Ideally, some different cheese loaves should be sampled and from each of these a sector should be cut, which is ground and mixed before analysis.

The composition of the cheese determines, by and large, the *yield*, i.e., kg of cheese obtained per (100) kg of milk. Yield varies widely among cheese varieties (see Appendix A.12).

The manufacturer tends to be greatly interested in yield, as it may determine *profit*. However, a higher yield does not necessarily cause a corresponding increase in profit. For instance, a higher fat content of the cheese implies a higher cheese yield, but less milk fat remains to produce, say, butter; an increase in cheese yield by 1 kg would correspond to a decrease in butter yield by about 0.7 kg. Part of the fat in the milk is lost in the whey, but that is not truly lost: the whey can be separated by centrifuging and the whey cream added to the next batch of cheese milk. If the cheese is to have a legal water content and a legal fat content in the dry matter, an increase in the protein yield by a factor x implies that the proportion of included whey and the fat content also have to be increased by a factor of about x. The cost of whey is negligible, but that of milk fat is not, and for every kg increase in yield, about 0.3 kg of fat will be needed. Consequently, the profits obtained, if any, strongly depend on the price ratio of milk protein to milk fat. Moreover, measures taken to increase yield — for example, by increasing water content — may lead to quality loss and to increased processing or curing expenses.

24.8.1 Variables Involved

Here we discuss most of the factors that influence the extent to which various components will be incorporated in the cheese, which affects both composition and yield. (*Note:* the changes in yield given are relative percentages, not percentage units.):

Milk composition: This naturally affects cheese composition. The dry-matter constituents that reach the cheese can be classified as fat globules, casein micelles, and solutes. Fat globules contain, besides milk fat, nonlipid components of the fat globule membrane — almost 2 g per 100 g fat (for the most part, protein). This implies that about 2% of the protein in a full-cream cheese is not casein.

Although most cheeses contain more fat than protein, casein generally is considered the most important component: cheese can be made with very little fat, but not with less than about 20% casein (except for some fresh cheeses). Moreover, if the composition of the cheese has to be kept constant, a change in the amount of casein incorporated in the cheese will have to be accompanied by an about equal relative change in the amounts of fat and water incorporated.

To be precise, it is paracasein micelles that are incorporated. This means that the caseinomacropeptide (CMP) has been split off, which amounts to about 37% of the κ-casein. The proportion of the latter in whole casein varies, implying that 4 to 6% of the casein will not reach the cheese. On the other hand, the micelles contain colloidal calcium phosphate (CCP), making up 6 to 8% of the dry matter, as well as some other proteins, viz., part of the proteose peptones and some enzymes.

It has been argued that selection of cows producing milk with certain genetic variants of specific proteins may be useful to maximize cheese yield. Table 2.22, shows that milk with κ-casein variant B has a higher casein-to-protein ratio than variant A, which would mean a higher casein inclusion per unit protein; however, variant B also has a higher ratio of κ-casein to total casein, implying that more CMP is lost, which undoes the presumed increase in yield. The table also shows that milk with β-lactoglobulin variant B tends to give more casein (and relatively less κ-casein) than variant A, and an increased yield has indeed been observed. However, the increase is small and the correlation is not observed for all cows.

Solutes in milk serum can reach the cheese, insofar as serum is included, but not in proportion. This is because part of the water is not available as a solvent (see Subsection 10.1.1). As has been discussed in Subsection 24.4.3, the amount of serum proteins included is almost zero, except for high-moisture cheese varieties. For lactose, nonsolvent water in an average cheese amounts to about 35% of the water present.

The factors affecting milk composition have been treated in Subsection 2.7.1. Table 2.20 shows that milk of sheep and buffaloes has substantially higher fat and casein contents than milk of cows or goats; this is reflected in cheese yield and composition. For the cheese manufacturer, seasonal variation is of special importance. It may further be noted that starter is added to the cheese milk, and that the milk used for the starter may have a different composition; it often concerns skim milk, and whey-based starters are also used.

Proteolysis of casein: Several proteolytic enzymes can split peptides off casein, which go into solution; in particular, β- and α_{s1}-casein are attacked. If the proteolysis occurs before curd separation and pressing, nearly all of these peptides will be lost with the whey. Some enzyme sources can be distinguished.

Of the indigenous milk enzymes, it especially concerns plasmin (see Subsection 2.5.2.5), an enzyme that is also active at low temperature. Its activity in milk increases with the stage of lactation and, on average, with increasing somatic cell count. Part of the proteose peptones formed stay in the paracasein micelles. Loss of yield generally is 1 to 4%, depending on milk storage time and plasmin activity.

Several psychrotrophic bacteria that can occur in milk make heat-resistant proteinases (Subsection 5.2.2). If the contamination is high and the milk is stored for several days before heat treatment, the protein yield loss can be as high as 5%. Formation of these enzymes should thus be counteracted by hygienic measures at the farm, and by thermalizing the milk upon reception at the dairy if it is to be stored for more than some hours until cheese making.

Most starter organisms, especially the proteinase-positive (Prt +) lactococci, can hydrolyze casein (Subsection 13.1.2.6). This has occurred to a considerable extent in a milk-based starter and can occur to some extent during curd making; considerable proteolysis will occur during preacidification of the cheese milk. Depending on starter composition, amount of starter added, and the time elapsed between starter addition and pressing, the yield loss generally is 1 to 5%.

All milk-clotting enzymes can hydrolyze more bonds in casein than the Phe-Met bond in κ-casein — the more so as the pH is lower. This activity is almost negligible for chymosin as long as the pH stays above 6.4, but it is higher for pepsin and most microbial rennets. It thus depends on rennet type, rate of acid formation (or preacidification), and the time and temperature of curd making, to what extent yield is decreased; published results range between 0 and 1.5% loss.

Whey removal: The amount of whey removed is by far the most important variable determining composition and yield of cheese. It depends on the conditions during curd making, curd separation, and pressing. This is extensively discussed in Sections 24.4 and 24.5.

Curd washing: The whey can be diluted with water to regulate the pH of the cheese (see Subsection 24.4.5). This also affects the incorporation of whey dry matter in the cheese, especially lactose and its degradation products, and milk salts. Figure 24.23 gives an example of the decrease in yield as a function of the amount of water added. In practice, a typical yield loss would be 2 to 3%.

Acidification: Growth of starter bacteria goes along with proteolysis, as discussed above. It also causes acidification, which, in turn, leads to dissolution of colloidal calcium phosphate. The pH at the moment of pressing determines how

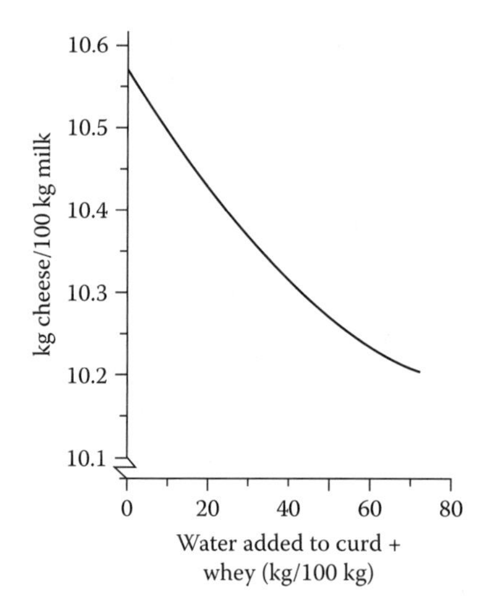

FIGURE 24.23 Example of the yield of Gouda cheese (12 d old, 41% water) as a function of the quantity of curd-washing water used. Water content and pH of the cheese are assumed constant.

much CCP will not be incorporated in the cheese. The amounts involved are illustrated in Figure 24.24. It follows that the difference between pressing at pH 6.5 and at 5.5, which is about the range for various cheese varieties, amounts to about 5% of the paracasein mass, or about 4% of the nonfat dry matter. This is a substantial effect. Moreover, the loss of solids due to curd washing will be

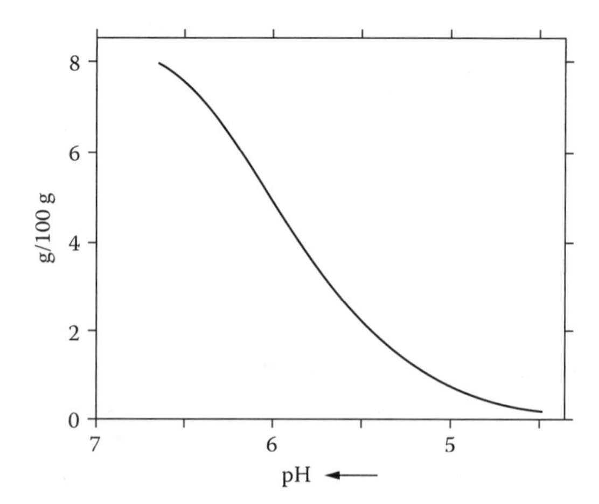

FIGURE 24.24 Quantity of mineral components associated with the casein micelles (the colloidal calcium phosphate) in grams per 100 g of paracasein as a function of pH. Approximate average values.

slightly greater when more CCP is dissolved. Finally, the amount of proteose peptone held in the paracasein micelles will decrease with decreasing pH.

Addition of calcium chloride: For several types of cheese about 1 mmolar $CaCl_2$ is added to the cheese milk to enhance clotting rate. This would lead to an increase in yield by about 0.3%, which has been experimentally confirmed.

Loss of curd fines and fat globules: When the milk has changed into a gel of some firmness, it is often cut into cubes of, say, 15 mm edge. Fat globules close to the cut surface tend to be transferred to the whey. After cutting, the curd–whey mixture is stirred, which may lead to some breaking and abrasion of curd grains, which causes additional transfer of fat globules. Moreover, small fragments of curd are broken from the grains, and also these curd fines turn up in the whey; the fines tend to have a low fat content. The amounts transferred to the whey vary widely: 5 to 12% of the fat and 0.2 to 1% of the paracasein originally in the curd.

Several factors affect the losses. If the gel is too soft at cutting, it becomes shattered to some extent, and the losses are high. If the gel is too firm, the pieces may be torn apart rather than cut, leading to relatively large and rough surfaces from which fat globules and curd fines can be detached; again, the losses are high. The latter may also happen if the gel firming rate is quite high; this can occur, for instance, when using UF-concentrated milk. The curd may have the desired firmness at the beginning of cutting, but before this is finished, the firmness has become too high. The remedy is to slow down the firming process, preferably by applying a lower clotting temperature. The loss of fat globules and fines strongly depends on the design of the curd making machinery and on the stirring intensity. For soft, i.e., high-moisture, cheese types, the losses tend to be smaller than for semihard and hard cheeses.

As mentioned, the fat in the whey is not really lost. The whey can be centrifuged, and the whey cream added to cheese milk or used for making whey butter. Occasionally, curd fines are recovered with the aid of sieves or hydrocyclones.

Ultrafiltration of the milk: This has been discussed in Subsection 24.4.3. Currently, low-concentration UF is often applied to standardize the milk as to protein content. It leads to a negligible increase of protein yield. It may, but need not, cause an increased loss of fines and fat globules, as discussed above. The higher caseinate concentration also causes a greater buffer capacity. This implies that acidification will be slower, unless corrective measures are taken such as adding more starter.

Heat denaturation of serum protein: As discussed in Subsection 7.2.2.2, heat-denatured β-lactoglobulin can react with κ-casein and become attached to casein micelles. Other serum proteins can also become attached due to heat treatment. Hence, denatured serum proteins can become incorporated in the curd upon clotting and curd making, increasing the cheese yield. However, the reaction mentioned impairs rennet clotting (see Subsection 24.3.6) and syneresis, the more so for a higher proportion denatured. It can be quite difficult to reach the desired water content of the cheese. Low pasteurization, e.g., 20 s at 72°C, as is often applied for cheese milk, leads to at most 2% denaturation. A more intense heat treatment is sometimes applied (if allowed) to increase yield. Up to about 20% denaturation clotting and syneresis can be satisfactory, provided that the curd making conditions are adjusted. An increase in protein yield by about 4% is then realized. However, the ripening of the cheese tends to be impaired, at least for hard and semihard aged varieties.

Addition of heat-denatured serum proteins recovered from intensely heat-treated whey is possible: it does not greatly affect curd making and results in a higher yield. However, the quality of this cheese is also, to some extent, impaired. Addition of a few per cent of intensely heated milk to the cheese milk, such as a milk-based starter or UHT-sterilized centrifugate from a bactofuge, does not cause problems with cheese making and quality.

Homogenization: Homogenization of milk results in smaller fat globules that are for a considerable part covered with micellar casein (Section 9.5). Upon clotting, by far most of these globules then participate in paracasein aggregation and gel formation. Upon cutting and stirring of the curd, far less fat is transferred to the whey than occurs for unhomogenized milk. The resulting decrease in fat loss does not provide an increase in profit: the cheese must have a normal fat content, and the fat in the whey from unhomogenized milk can be recovered. For most cheese varieties, homogenization causes an undesirable texture and flavor (it enhances lipolysis).

Salting: Salt is incorporated in the cheese, while more water is lost. The ratio between the amounts of water lost and salt taken up varies widely. The factors affecting these processes are discussed in Section 24.6.

Dry salting leads to a loss of whey upon mixing of curd and salt and upon pressing. This so-called white whey contains a lot of salt, whey solutes, and has a much higher fat content than the first whey, e.g., 2.6% vs. 0.4%, corresponding to almost 10% of the fat lost in the whey.

Curing: Cheese loses weight during curing, first, by vaporization of water to an extent depending on loaf size, coating, and storage conditions (see Subsection 24.7.2). Varieties that are subject to intense protein breakdown during ripening also lose NH_3 and CO_2; the weight loss involved is generally quite small. Moreover, the quantity of dry matter increases, as every peptide bond hydrolyzed results in the uptake of a molecule of water.

24.8.2 Yield

Yield can be expressed in various ways. The most common definition is: the mass of cheese in kg obtained per 100 kg of milk. The milk should include the added starter. A difficulty with this quantity is that the water content of cheese is quite variable. Hence, yield may be defined as the mass of dry matter in the cheese obtained from 100 kg of milk, or one may recalculate the yield to that of cheese with a standard water content. If it is desired to test the efficiency of the cheese making process, it may be useful to estimate the mass of cheese protein obtained per unit mass of milk (para)casein.

It is difficult to establish the mass of cheese obtained. First, it may be a problem to collect and weigh precisely all of the cheeses originating from a given batch of cheese milk. Second, the cheese tends to decrease in weight after pressing and salting; the rate at which this occurs varies greatly among cheese varieties and with storage conditions. Hence, it is desirable to weigh the cheeses at a fixed time, shortly after pressing or brining. Third, physical losses may occur (curd lost in the machinery, pieces broken from cheese loaves), and nondairy material may have been added (spices during cheese making, coating material afterwards).

It is often considered useful to have a predictive formula for cheese yield. Several equations have been derived, for instance

$$Y = \frac{aF + bC}{1 - W} + R \qquad (24.5)$$

Here Y = yield (kg per 100 kg of milk); F = fat content, and C = casein content of the milk (% w/w); a = fraction of the milk fat that is incorporated into the cheese, and b the same for casein. W = water content of the cheese (mass fraction), and R is the amount of other dry matter in the cheese (kg per kg cheese). F and C can be determined in the milk beforehand; Y and W must be had from the finished cheese. To calculate a standard yield Y^* for a standard water content W^* we use the formula

$$Y^* = Y(1 - W)/(1 - W^*) \qquad (24.6)$$

A problem is finding the values of a, b, and R. In principle, a is composed of two factors:

- To convert fat content to dry-matter content of fat globules; about 1.02
- To account for the fat loss to the whey; widely variable, for instance 0.91

Jointly, this would give $a = 0.93$.

The magnitude of the factor b depends on several variables, especially:

- The splitting of the CMP from κ-casein; factor of about 0.95
- Proteolysis before curd separation; widely variable factor, for instance 0.97
- The CCP and minor proteins associated with the micelles; widely variable and especially dependent on the pH of the curd at the moment of curd separation, for instance, a factor 1.055

Jointly, this would give $b = 0.97$.

The residual term R is the sum of two main (groups of) components:

- The dry matter of dissolved serum components reaching the cheese: This is, for the most part, lactose and its breakdown products, and also milk salts, organic acids, and some esters, etc. It may amount

to 0.02 kg/kg, but it depends on the value of W and on the extent of curd washing
- The salt taken up in the cheese: It is, for instance, 0.02 kg per kg, but varying among cheese varieties

The sum would then be $R = 0.04$ kg per kg of cheese.

Altogether, the 'constants' in the equation depend, in fact, strongly on cheese variety, a little on raw milk composition, and on variation in the conditions during processing. To check whether undue variation occurs, one may compare actual yield with predicted yield, where the values of the constants have been obtained from results of previous batches by regression analysis. It will be useful, also, to determine salt content of the cheese, as this allows better estimation of R. Such an analysis additionally yields an estimate of the standard error involved. When the difference between observed and predicted yield is larger than, say, twice the standard error, it may be useful to try to find out what has gone wrong. Subsection 24.8.1 gives basic information from which it would be possible to establish the cause.

24.8.3 Standardizing the Milk

As mentioned in Subsection 24.1.2, cheese milk may be standardized to protein (or, rather, casein) content by addition of a skim milk UF retentate; the aim is better process control.

Also, its fat content should be standardized to obtain the desired FDM, i.e., fat content in the dry matter of the cheese. According to Equation 24.5, FDM will be given by $aF/(1 - W)$, but W is not known beforehand. When regular yield control is applied, and if for a given cheese variety the manufacturing process is kept constant, the ratio of F/C in the milk can be correlated to FDM in the cheese, and the correlation can be quite good. (The value of C can be determined from infrared analysis of the raw milk and the whey obtained from it.) This allows standardization of fat content. If a legal minimum is required, a small safety margin should be observed, say 1% FDM. The margin is also needed because the nonfat dry matter in the cheese increases due to proteolysis.

Currently, computerized systems for standardization of cheese milk that take several variables into account are increasingly used.

Suggested Literature

Important aspects of various manufacturing steps: P.F. Fox, P.L.H. McSweeney, T.M. Cogan and T.P Guinee, Eds., *Cheese: Chemistry, Physics and Microbiology,* Vol. 1, *General Aspects.* 3rd ed., Elsevier Academic Press, London, 2004, Chapters on 'Rennet-induced coagulation of milk', 'The syneresis of rennet-coagulated curd', 'Starter cultures', 'Salt in cheese', and 'Application of membrane technology to cheese production'.

Some trends in manufacturing processes: R.K. Robinson, Ed., *Modern Dairy Technology,* Vol. 2, *Advances in Milk Products,* 2nd ed., Elsevier, London, 1993.

Aspects of standardization and yield, extensively discussed in an IDF report: Cheese yield and factors affecting its control, *Proceedings of the IDF Seminar,* Cork, April 1993, International Dairy Federation, Brussels, 1994; D.B. Emmons, Ed., *Practical Guide for Control of Cheese Yield,* International Dairy Federation, Brussels, 2000.

25 Cheese Ripening and Properties

Ripening of cheese includes all the chemical changes occurring in the cheese, some of which begin before the curd making is finished. The structure and composition of the cheese change, and hence its organoleptic properties. Biochemical and microbiological, as well as purely chemical and physical aspects are involved. Development of cheese properties, including consistency and flavor, is especially attributable to the conversion of lactose, protein, fat, and, in some cheeses, citrate.

25.1 LACTIC FERMENTATION

General aspects of lactic fermentation are treated in Section 13.1 and Section 13.2. Manufacture and maturation of cheese seem to be impossible without lactic acid bacteria; in most cases they are added to the milk as a starter. Mechanical inclusion in the curd (concentration factor of 5 to 12) and growth result in 10^9 to 10^{10} bacteria per gram of curd. Cheese milk is inoculated at a level of, say, 10^7 starter bacteria per milliliter, which implies that in the fresh cheese starter bacteria replicate only a few times. Furthermore, their growth in cheese stops at a fairly high pH, for example, 5.7; the actual pH depends on the species and the strains of bacteria involved. A full explanation of these phenomena is lacking. Although the growth stops, fermentation continues, further decreasing the pH. The lower the pH and the temperature, and the higher the salt content, the slower the decrease in pH. Growth of most lactic acid bacteria is also slowed, if not stopped, by the salt. Accordingly, the number of these bacteria in the cheese should be high before the salting starts. (This is not true of cheese varieties made of cheese milk to which a high salt content is added, which contain fairly salt-tolerant lactic acid bacteria.)

Starter bacteria may vary widely as to their growth rate, the number to which they grow in cheese, and the rate at which they lose viability during ripening. This is illustrated in Figure 25.1. In addition to starter bacteria, other lactic acid bacteria may grow in the cheese, and these often cause defects (Sections 26.3, 26.4, and 26.7).

Rapid acid production is generally desirable in cheese that is brined, in order to keep the time from molding to salting of the cheese as brief as possible. When the cheese is salted at the curd stage, the curd making takes far longer (because most of the acid should have been produced in the curd before salt is added) and rapid acid production is even more essential. The rate of acid production by the

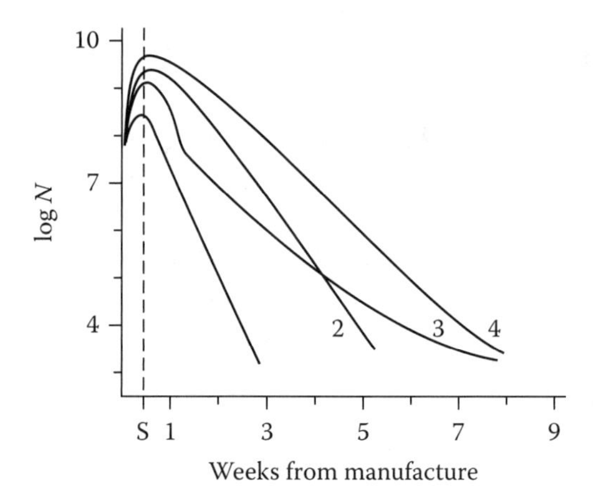

FIGURE 25.1 Number N of viable starter bacteria (cfu per gram of cheese) from various starters (1 to 4) during the manufacture and maturation of Gouda cheese. S is the moment at which brining starts. (Adapted from F.M.W. Visser, *Neth. Milk Dairy J.*, **31**, 120–133, 1977.)

starter, therefore, should be high. Satisfactory process control demands that the cheese always acidify at the same rate because a variable rate leads to variation in syneresis and, hence, in the water content of the cheese (see also Subsection 24.4.1, and Subsection 27.1.1).

The formation of lactic acid from lactose is vital for the preservation of the cheese, as it prevents growth of undesirable microorganisms. This conversion causes a low pH, thereby resulting in a high concentration of undissociated lactic acid and a considerably reduced lactose content. In many varieties of cheese, especially hard cheeses and those in which the curd is washed, lactose will be completely fermented. All these factors inhibit numerous microorganisms. An additional factor may be the anaerobic environment because the slow diffusion rate of oxygen into the cheese allows the enzyme system of the lactic acid bacteria to keep the redox potential of the cheese at the low level reached during the lactic acid fermentation. This effect is enhanced by the presence of a closed rind.

Starter bacteria strongly affect, either directly or indirectly, other properties of cheese such as its body (formation of holes, etc.), consistency, and flavor (Section 25.5 and Section 25.6). Hole (eye) formation can be caused by lactococci that produce CO_2 in the fermentation of citrate (Subsection 13.1.2).

25.2 ENZYME SOURCES

Proteolytic and lipolytic processes are all-important to obtain the characteristic properties of ripened cheese; the relative importance of any of these processes may vary widely with the type of cheese considered. The responsible enzymes can be classified into the following groups, based on the type of substrate and method of attack:

1. Proteolytic enzymes (EC 3.4), subdivided into:
   a. Endopeptidases or proteinases, which hydrolyze proteins to peptides.
   b. Exopeptidases, which split amino acids off peptides; this group includes aminopeptidases, carboxypeptidases, and di- and tripeptidases.
2. Enzymes that decompose amino acids produced by the exopeptidases, that is, decarboxylases, deaminases, transaminases, and demethiolases.
3. Lipases (EC 3.1.1), which break down triglycerides into free fatty acids, and di- and monoglycerides.
4. Enzymes that break down fatty acids or their derivatives, that is, dehydrogenases, decarboxylases, and esterases.
5. Acid phosphatases, which cause dephosphorylation of casein.

Potential enzyme sources in cheese are:

1. Rennet enzymes, insofar as they are transferred to the cheese during manufacture and remain active.
2. Endogenous milk enzymes, especially plasmin and lipoprotein lipase. The latter enzyme is largely inactivated by pasteurization.
3. Lactic acid (starter) bacteria.
4. An additional flora on the surface (e.g., white molds, salt-tolerant bacteria, etc.) or in the interior (such as blue molds).
5. Remaining nonstarter bacteria, originating in the raw milk and surviving pasteurization.
6. Extracellular proteinases and lipases originating from psychrotrophic bacteria growing in the raw milk. The bacteria are generally killed by pasteurization, but many of their enzymes are highly heat-resistant.
7. Recontaminating organisms in the pasteurized milk and undesirable organisms growing on the cheese rind (Section 24.7).
8. In some cases, specific enzyme preparations are added.

Rennet and milk proteinases cause proteolysis; other enzyme sources may also contribute to conversion of amino acids and to lipolysis (Section 25.4).

The enzyme system of a microorganism comprises several enzymes which, depending on the location in the cell, are classified as:

- Extracellular enzymes secreted into the substrate by the intact cell
- Cell-membrane-anchored enzymes, often protruding through the cell wall
- Intracellular enzymes exposed to the substrate after its uptake in, or after lysis of, the cell

The activity of an enzyme in cheese naturally depends on the enzyme concentration. Obviously, for microbial enzymes the final number of organisms in the cheese is an essential parameter, as is lysis of cells. Moreover, the activity of all enzymes depends on the conditions in the cheese, which may alter significantly during ripening. The following are important variables:

1. *Acidity*: Every enzyme has its optimum pH at which its activity is highest.
2. *NaCl content* of the moisture in the cheese: NaCl at fairly low concentration activates certain enzymes but inhibits others.
3. *Ripening temperature*: Under normal conditions, activity increases with temperature. The effect of temperature is stronger for lipolysis than for proteolysis.
4. *Water content* of the cheese: This affects the composition of the cheese moisture (for example, the calcium ion activity) and the conformation of proteins. This means that the enzyme activity may depend on the conformation of the enzyme and, for proteolytic enzymes, on that of the substrate. The lower the water content, the smaller are the diffusion coefficients, which may affect reaction velocities. The diffusion coefficient of enzymes may be smaller than 10^{-12} $m^2 \cdot s^{-1}$. By and large, protein degradation is faster for a higher water content, but specific knowledge is scarce. Furthermore, some enzymes are slowly or rapidly inactivated during ripening, whereas others (such as rennet) remain stable.

25.3 PROTEOLYSIS

During the ripening of cheese the protein is broken down by proteolytic enzymes into several products, ranging from large peptides to free amino acids and even ammonia. The concentration ratio of these products may vary widely according to the type of cheese considered.

25.3.1 METHODS OF CHARACTERIZATION

Several methods are available to assess proteolysis, and these may lead to widely varying results. The cheese may be extracted with water, with a solution of NaCl or $CaCl_2$, or at pH 4.6, that is, roughly the isoelectric pH of paracasein. The extracts so obtained may be centrifuged or filtered and analyzed for nitrogen. These 'soluble-N fractions' are heterogeneous in composition and may even contain intact protein, depending on pH, ionic strength, and Ca^{2+} activity of the solution. The estimated N content provides no information about the character of the proteolysis, because identical amounts of soluble N of various cheeses may be composed of widely varying substances. To obtain a better understanding, the extracts are fractionated. Less heterogeneous 'NPN (nonprotein nitrogen) extracts,' mainly containing low-molar-mass peptides and amino acids, are obtained by treating the extracts with trichloroacetic acid (TCA) solutions (for example, 12%) or with

70% aqueous ethanol; amino acids remain dissolved after treatment with phosphotungstic acid. Amino acid nitrogen can be determined using the Kjeldahl method, but the TCA or ethanolic extracts also contain di- and tripeptides. It is possible to estimate the individual free amino acids by ion exchange chromatography, which leads to lower values of amino acid N. Free amino groups can be determined with trinitro-benzenesulfonic acid or with *o*-phthalic dialdehyde. Ammonia nitrogen can also be determined spectrometrically.

By applying several forms of gel electrophoresis, the fractionated extracts can be used to further characterize the proteolysis semiquantitatively. Capillary electrophoresis is a very useful method for obtaining quantitative results for the analysis of protein degradation. For profiling of peptides and amino acids in a ripening cheese, reversed high-performance liquid chromatography is used. The sequence of amino acids in the peptides obtained from casein can subsequently be determined, thereby identifying the cleavage sites in the casein molecules of the enzymes involved. The use of specific online detection techniques (such as UV spectrometry, fluorimetry, and refractometry) enables direct identification of certain amino acids and various other analytes after chromatographic separation. A very powerful detection and identification tool is mass spectrometry, which determines the molar mass of the analytes. The molecules may be analyzed intact or after fragmentation, and for their identification the mass spectra obtained can be compared to those in a database. Mass-spectrometric detectors are extremely useful in the analysis of volatile flavor compounds in cheese after their gas chromatographic separation (see Section 25.5).

25.3.2 Milk Proteinases

Introductory information about enzymes is given in Subsection 2.5.2. In cheese, milk proteinases decompose α_{s1}-, α_{s2}-, and β-casein. The plasmin in milk is predominantly found as the inactive plasminogen, with only a small percentage being active. At the pH of milk the enzyme has a marked affinity for casein, whereas it dissociates from casein at low pH. It is essentially active at high pH and decomposes β-casein much faster than α_{s1}-casein. The hydrolysis of β-casein yields the γ-caseins: γ^1-casein (fragment 29–209), γ^2-casein (fragment 106–206) and γ^3-casein (fragment 108–209). The N-terminal parts of β-casein and γ-caseins split off by plasmin are proteose peptones. The detailed sequence of the 209 amino acids in β-casein is shown in the Appendix, Table A.6.

Acid milk proteinases also occur, especially cathepsin D, which are less important. Their optimum pH in cheese is 5.1 to 5.6. β-Casein is decomposed more slowly than α_{s1}-casein.

Milk proteinases are not inactivated by pasteurization of milk. In addition to pH, the following are important factors affecting enzyme activity in cheese:

1. *The proteinase content of the milk, the concentration of plasminogen activators, and the concentration of inhibitors of plasminogen activation*: All of these can vary among milkings of one cow and also among individual cows.

2. *The heat treatment of the cheese milk*: In raw milk the plasmin activity is less than in low-pasteurized milk, which may be ascribed to partial inactivation of compounds that inhibit plasminogen activators, resulting in increased plasmin activity.
3. *The scalding temperature during curd making*: High temperatures, as applied in the manufacture of several Swiss and Italian cheese varieties (Section 27.5), considerably enhance plasmin activity in the cheese.
4. *The salt content*: A low salt content in the cheese moisture (say, 2%) has a stimulating effect, but plasmin has very little activity at high NaCI concentration.
5. *The ripening temperature*: Many enzymes have an optimum temperature at about 35°C, but plasmin activity depends little on temperature (see Subsection 2.5.2.5).

The conditions in cheese, especially the pH, often are unfavorable for significant proteinase activity. A marked action may, however, occur in types of cheese with a high pH, such as Camembert (pH 6 to 7), and in types with a relatively high pH and a long ripening time, such as Emmentaler.

Proteolysis by milk proteinases increases the amount of soluble N compounds, mainly consisting of peptides; the production of amino acids is small.

25.3.3 Clotting Enzymes

These enzymes, which have a specific function in milk clotting (Section 24.3), also have a considerable effect on proteolysis in cheese and the ensuing properties of the product. The action of calf rennet (consisting of chymosin and 15 to 20% pepsin, as calculated on clotting activity) greatly depends on the amount of rennet retained in the cheese. It should be realized that the chymosin included in the cheese originates almost completely from adsorption onto the paracaseinate. It is not well known to what extent this is also true for other milk clotting enzymes.

The following factors determine the amount retained (see also Figure 24.15):

1. The *amount of rennet* added to the milk.
2. *The pH during curd making*: The lower it is, the higher the quantity of calf rennet that is adsorbed onto the paracasein. Hence, factors having an effect are the initial pH of the milk, rate of acid production by the starter, percentage of starter added, addition of $CaCl_2$, and preacidification of the milk as, for example, in the manufacture of Camembert.
3. *The scalding temperature of the curd*: At a high scalding temperature such as 55°C as applied for Emmentaler cheese, rennet (chymosin) is for the greater part inactivated. At low pH, however, chymosin is more heat-resistant (see Figure 24.2).

FIGURE 25.2 Proteolysis by calf rennet in aseptically made, starter-free Gouda cheese. Production of soluble N, expressed as a percentage of the nitrogen in the cheese. The rennet content of the cheese is 'normal' (1), corresponding to about 300 μl per kilogram cheese), half (0.5), and three times (3) the normal amount, respectively. Approximate examples. (Adapted from F.M.W. Visser, *Neth. Milk Dairy J.*, **31**, 210–239, 1977.)

The effect of the rennet content of cheese on the rate of proteolysis is illustrated in Figure 25.2. The results also indicate that the rennet retains most of its activity during the ripening. A relatively large part of the formed soluble nitrogen compounds consists of products with a relatively high molar mass; only a little, if any, amino acid forms, irrespective of the rennet content of the cheese.

The optimum pH for calf rennet action in cheese is about 5. α_{s1}-Casein degrades rapidly; β-casein far more slowly. The principle cleavage site in α_{s1}-casein is Phe23-Phe24, which is completely hydrolyzed in semihard cheeses within 4 months. The second most susceptible bond is Leu101-Lys102. Moderate salt content, up to about 4% NaCl in water, favors the degradation of α_{s1}-casein, whereas higher levels slow it down; the degradation of β-casein is slowed down already at low salt content.

The rate of proteolysis by calf rennet is little affected by the ripening temperature of cheese; at, say, 4°C it differs not much from that at 14°C. Because the enzyme is more active (on synthetic substrates) at higher temperatures, it must be that the substrate in cheese is more prone to attack at lower temperatures.

Shortage of calf stomachs, as well as increasing production of cheese, has resulted in the use of calf rennet substitutes. Bovine pepsin, blends of porcine pepsin and calf rennet, and preparations of the molds *Rhizomucor pusillus, Rhizomucor miehei,* and *Cryphonectria parasitica* are among those being used industrially. The properties of the cheese made by using a substitute should not differ significantly from the reference cheese. To achieve this, the means of manufacture must often be adapted because the rennets differ from calf rennet in several respects, that is:

1. *Susceptibility to temperature,* Ca^{2+}, *and pH during the clotting:* Porcine pepsin, for instance, inactivates rapidly above a pH of 6.5. A high rennetability is of paramount importance, as is formation of a firm gel that gives sufficient syneresis.
2. *Resistance to heat:* Too high a resistance can cause problems in the processing of whey.
3. *Distribution of enzyme activity between curd and whey:* For instance, this distribution is not affected by the pH for some mold rennets.
4. *Proteolytic activity:* Too high a level of activity during clotting of milk leads to a loss of yield and to excessive and abnormal degradation of the caseins during ripening, causing defects in consistency and flavor, such as a bitter flavor. Generally, microbial rennets are more proteolytic than calf rennet. This is especially true for the *C. parasitica* coagulant.

Bovine chymosin can also be produced by genetically modified microorganisms and used for cheese making. Its action is virtually identical to that of calf rennet.

25.3.4 Enzymes of Lactic Acid Bacteria

Proteolytic enzymes of the starter bacteria (such as mesophilic *Lactococcus* spp. for most cheese varieties) play a key role in the maturation of all ripened cheeses without a surface flora or internal blue molds. These enzymes attack the large peptides formed by rennet or plasmin action to produce the small peptides and free amino acids that contribute most to flavor, be it directly or as precursors of specific flavor compounds. Any variation in proteolysis within a type of cheese is for the most part due to variation in activity of enzymes of the starter bacteria. Although we have a reasonable understanding of the factors determining the activity, the factors in cheese are rather complex and, therefore, to predict the outcome of proteolysis quantitatively is not easy.

Some of the *Lactoccocus* strains are proteinase positive (Prt+), which means that they contain a cell envelope proteinase that is needed for the bacteria to grow in milk. Other strains are proteinase negative (Prt–), and these depend on the Prt+ strains present for the formation of peptides from milk proteins, especially casein. In cheese much the same happens; both α_{S1}- and β-caseins are attacked (mostly the former), but the rate of formation of peptides is slow. However, the rennet enzymes produce peptides at a much faster rate, and these are hydrolyzed further by bacterial enzymes. At least part of the peptides produced by rennet is not readily metabolized by the starter cells unless they are broken down to smaller peptides by the bacterial proteinase. Fairly small peptides are transported into the bacterial cell (see also Figure 13.6) by some specific energy-dependent transport mechanisms. The cell contains several specific peptidases that immediately hydrolyze the peptides into amino acids, and these can presumably diffuse out of the cell. These peptidases of lactic acid bacteria can be divided into endopeptidases, aminopeptidases, di- and tripeptidases, and proline-specific peptidases. The balance between the formation of peptides and their subsequent degradation into free amino acids is very important, because accumulation of peptides might lead to a bitter off-flavor in cheese. Various bitter-tasting peptides have been identified, and these peptides should be degraded rapidly. Some specific cultures have a high bitter-tasting-peptide degrading ability, and such cultures are frequently used in the manufacture of various types of cheese. The ability of cultures to degrade bitter-tasting peptides has been found to be positively correlated with the sensitivity of the cells to lysis.

Because the bacteria involved in peptidolysis require energy to transport the peptides into the cell and to convert them into amino acids, they should be metabolically active. However, because the main energy source, lactose, is fully consumed within 24 h in most cheeses, this is hardly achievable.

It is generally accepted that lysis of the starter bacteria is needed for their intracellular peptidases to be released in the cheese. For lysis to occur, the cell wall must be broken down, which generally is achieved by bacterial autolysins, and subsequently an osmotic shock would be needed; the osmotic pressure in the liquid around the bacterium must become either much lower or much higher than that in the cell. Presumably, a steep salt gradient may induce lysis in cheese. It is generally believed that the lysis slowly goes on, because the peptidase activity in the cheese appears to increase during maturation.

The amount of enzymes produced by the starter bacteria can vary according to their growth conditions. If the bacteria are grown in a medium containing a considerable quantity of small peptides or even amino acids, they would not need the mentioned enzymes for growth, and it is indeed observed that they produce far less of them. In the present case it primarily concerns growth in the cheese milk and the fresh curd, which would involve, say, five divisions, sufficient for the bacteria to adapt to the new medium (as compared to the starter). It has been observed that growth in 'cheese milk' that has been concentrated to a rather high degree by ultrafiltration (such as four times), or that has received intense heat treatment (for about 10 min at 120°C), leads to significantly reduced amounts of the cell envelope proteinase and of at least some of the peptidases.

There is a wide variation among starter strains in terms of the extent and specificity of proteolysis. The following variables can be distinguished:

1. The *types and amounts of enzymes* that the cells can produce: Generally, growth rate and production of proteolytic enzymes are correlated. Modern fast-growing starters produce fairly large amounts of soluble N in the cheese.
2. The *growth conditions* for the bacteria (hence, pretreatment of the milk, and so on) and the manner in which enzyme production depends on these conditions.
3. The *number of cells* reached in the cheese: In this respect, competition between various strains, as well as protocooperation, may be involved.

4. The extent to which *lysis* of the cells occurs in the cheese, thereby making the peptidases accessible: This varies greatly among strains and may be one of the most important factors determining a strain's action in cheese. Some strains can be induced to lyse by a temporary rise in temperature.
5. *Stability of the enzymes* in the cheese: Some are very stable (such as chymosin and plasmin), whereas others lose their activity fairly rapidly. This has been insufficiently studied.
6. The dependence of the *specific activity* of the enzymes on conditions such as pH, salt concentration, and temperature.

An important group of cheese varieties, discussed in Section 27.5, is made with thermophilic starters, generally involving mixtures of *Streptococcus thermophilus* and various *Lactobacillus* species, such as *L. helveticus*, *L. delbrueckii* ssp. *bulgaricus*, and *L. casei*. Although less is known about their proteolytic enzymes than about those of the lactococcus species, the pattern seems to be about the same. The bacteria generally have a cell envelope proteinase and several intracellular peptidases. The proteolytic activity of such starter bacteria, whether they are growing in milk or are present in cheese, generally is markedly stronger than that of the mesophilic starters. This is especially true of the lactobacilli; the streptococci can hardly produce free amino acids.

25.3.5 Enzymes of Nonstarter Organisms

Nonstarter lactic acid bacteria may enter the cheese milk by contamination and then grow out in the cheese; this generally concerns *Lactobacillus* species. Some of these do not need sugar as a carbon source but utilize certain amino acids or even the carbohydrate moiety of glycoproteins in the fat globule membrane. They can reach numbers of 10^7 to 10^8 colony-forming units (CFU) per gram of cheese. They are markedly proteolytic and can bring about a great variety of flavor compounds in cheese. In most cheeses these flavors are generally considered to be undesirable. Naturally, milk pasteurization and good hygienic practices during cheese making largely prevent growth of these organisms. In Cheddar cheese made of not very intensely heated milk, some lactobacilli can enhance what is considered a typical Cheddar flavor, and these organisms are sometimes added on purpose.

In cheese varieties with a specific flora, along with lactic acid bacteria, several other organisms are important, including yeasts, molds, coryneforms, and micrococci (see Section 27.6). The contribution to the proteolysis and to the production of flavor compounds (Section 25.5) naturally depends on the properties and number of the organisms present.

25.3.6 Interaction between Enzyme Systems

The various enzymes sources causing proteolysis, and their relative significance for protein degradation, can vary widely among cheese types. Some variables are the method of cheese manufacture; cheese composition; ripening conditions, especially ripening time and temperature; and the presence of a specific flora. Examples are given in Table 25.1; see also Section 27.3 to Section 27.6. Every type of cheese contains its specific enzyme systems that interact in an intricate way, resulting in its characteristic balanced ripening.

The extent and pattern of proteolysis in cheese largely depend on the contribution of each of the different agents participating in this process. Their specific roles have been estimated by analyzing the various products formed in biochemical assays and verifying the results obtained in experiments with cheese, using techniques as outlined in Subsection 25.3.1.

Most of the water-insoluble peptides produced in cheese are derived from α_{S1}-casein by chymosin and from β-casein by plasmin. These peptides are derived from the C-terminal portions of the molecules and some dominant representatives are fragments 24–199, 102–199, 33–199 from α_{S1}-casein and fragments 29–209, 106–209, 108–209 from β-casein. The water-soluble peptides are largely produced by the lactococcal envelope proteinase from the larger N-terminal peptides produced by chymosin or plasmin from α_{S1}- and β-casein, respectively. Many of these peptides reflect the known cleavage sites of the lactococcal proteinase. Some of the peptides are partially dephosphorylated, indicating acid phosphatase activity from milk or from the starter.

In low-scalded cheeses, such as Cheddar and Gouda, and in the interior of surface mold- and surface smear-ripened cheeses, such as Camembert and Munster, the primary proteolysis by chymosin and plasmin is generally similar (see Table 25.1). In high-scalded cheeses, such as Emmentaler and Gruyère, the rennet enzymes are extensively denaturated, and the primary proteolysis is due mainly to plasmin.

The secondary proteolysis in cheese relies very much on the proteolytic system of the starters used, which differ from cheese type to cheese type. These systems have been discussed in Subsection 25.3.4.

Probably, milk proteinases, rennet, and starter proteinases do not (or hardly ever) attack undenatured serum proteins and para-κ-casein in cheese.

Ripening of cheese alters the conditions, especially the pH; the change affects the action of enzymes and their interactions. Proteolysis, especially deamination (formation of NH_3) and decarboxylation of amino acids, as well as decomposition of lactic acid, causes an increase in pH. In most cheese varieties, the pH increases by only a few tenths of a unit, whereas in those with a distinct degradation of lactic acid and protein it increases markedly (see Section 27.6 and Figure 27.3). Some microbial defects also cause a considerable pH increase (Section 26.2).

25.3.7 Ultrafiltration of Cheese Milk

When making cheese from milk concentrated by ultrafiltration to a considerable degree, say, by a factor of five, proteolysis and the development of flavor and consistency are markedly slowed. The causes are complex, manifold, and insufficiently understood. One cause is 'dilution' of the casein by serum proteins,

TABLE 25.1
Relative Significance of Enzyme Sources for Proteolysis in Cheese

	Contribution to Proteolysis by						
			Lactic Acid Bacteria		Surface Flora		Internal
Cheese Type	Rennet	Plasmin	Mesophilic	Thermophilic	Coryneforms	White Mold	Blue Mold
Butterkäse, Meshanger	+++	±	++	–	–	–	–
Camembert (traditional)	++	±	++	–	++	++	–
Camembert ('modern')	+++	±	++	–	–	+++	–
Munster	+++	±	++	–	+++	–	–
Gouda-type	+++	±	+++	–	–	–	–
Cheddar-type	+++	–	+++	–	–	–	–
Emmentaler	±	++	±	++	–	–	–
Gruyère	±	++	±	++	+	–	–
Provolone	±	+	–	++	–	–	–
Roquefort	++	±	++	–	–	–	+++
Gorgonzola	++	±		++	+	–	+++

Note: Approximate examples; – = no; ± = a little; + = some; ++ = considerable; +++ = very much.

which are not prone to proteolysis. Furthermore, rennet activity is markedly decreased. This means that for the same concentration of chymosin in the cheese, less proteolysis occurs. It appears that some serum proteins can lower chymosin activity. Plasmin activity is decreased as well. Another factor appears to be the decreased production of peptidases by starter organisms growing in ultrafiltered milk; this is discussed in Subsection 25.3.4. Finally, it may be that the starter organisms are less prone to lysis in the cheese, but this has not been clearly shown.

25.4 LIPOLYSIS

Among the enzymes that may contribute to lipolysis are the following:

1. *Milk lipase* or lipoprotein lipase (see also Subsection 3.2.5): The concentration of active enzyme in milk closely depends on the pasteurization process (see Figure 16.2). Under the usual pasteurization conditions some 10 to 15% of the enzyme is left. In raw milk cheese the enzyme is relatively active, and it may eventually increase the acidity of the fat to some 20 or 30 mmol per 100 g of fat. Also, the lipase in raw milk is relatively stable, but in acidifying milk, curd, and cheese, the bulk of the enzyme activity may have been lost after 1 d. Considering the high salt content and the low pH, it is surprising that the enzyme can be active at all in cheese, although slowly. The optimum pH for the enzyme is above 8; the minimum generally is given as 6. In cheese, however, lipolysis often is stronger if the pH is lower. That can be partly, but not fully, ascribed to an inadequacy in the analytical methods. At lower pH more of the water-soluble, short-chain fatty acids are dissolved in the fat and subsequently titrated. For identical concentrations of fatty acids, the flavor caused by fatty acids is also more pronounced as the pH of the cheese is lower.

   Lipolysis considerably increases with temperature, at least in raw milk cheese. Homogenizing the cheese milk, or a part of it, or changing the surface layer of the fat globules in another way, results in a considerable increase in lipolysis, although only for a short time, that is, a few days. Subsequently, the rate of lipolysis appears to return to its normal (low) level. Hence, the acidity of the fat in the cheese can be adjusted as desired. In cheese made of pasteurized unhomogenized milk, with no additional lipolytic enzymes, the fat acidity usually does not exceed 1 or 2 mmol per 100 g of fat.
2. *Added enzymes*: Sometimes lipases are added, for example, to certain ripened Italian cheeses. Also the rennet may include some lipase (calf rennet does not).
3. *Microbial lipases*: Potential sources are:
   a. The flora of the milk, especially psychrotrophic bacteria in cold-stored milk.

b. Organisms constituting a specific surface or internal flora in some cheese varieties. Lipolysis can be considerable, especially in blue cheeses (see Subsection 27.6.2).
c. Lactic acid bacteria; these are not very lipolytic. They hardly decompose triglycerides, but do decompose mono- and diglycerides somewhat. Because of this, the contribution of these bacteria is mainly to enhance fatty acid production if other lipases are active.

25.5 DEVELOPMENT OF FLAVOR

25.5.1 Description

A cheese of satisfactory flavor always contains many different flavor compounds, well balanced. In curd, weak flavor compounds prevail, originating from fat and some other milk components. The initial sweetness in the taste of curd, due to lactose, disappears quickly. Lactic fermentation is responsible for the acid taste characteristic of almost all cheese varieties. In fresh-type cheeses, aroma compounds formed by the starter bacteria (for example, diacetyl) can play an important role. In ripened cheeses, salt is also an essential flavor component; the concentration varies widely among cheese types. The organoleptic saltiness of the cheese is about proportional to its salt content rather than to its salt-in-water content.

Large changes in flavor develop during ripening. Protein has no flavor, but many degradation products have. Free amino acids and short-chain peptides contribute to the basic flavor that is perceived in most cheese varieties. These compounds have specific tastes: sweet, bitter, and broth-like, in particular. The stage of ripening largely determines the intensity of the basic cheese flavor. The cheese may develop a bitter flavor if the protein is degraded in such a way that many short-chain hydrophobic peptides are formed.

The protein degradation also greatly affects the cheese consistency and thereby the mouthfeel. Presumably, the consistency also affects the flavor perception. Carbon dioxide, although without flavor *per se*, appears to affect the cheese flavor. Loss of CO_2 may contribute to the rapid loss of the typical flavor of grated cheese.

Fat plays an essential part in the flavor of cheese, although largely an indirect part. Reducing the fat content of several well-known ripened cheese varieties results in a much less satisfactory flavor perception, even if flavor compounds associated with the fat as such contribute little to the flavor. Probably, the distribution of aroma compounds over the fat and aqueous phases enhances a balanced flavor. The most important flavor compounds originating from the fat are the free fatty acids formed by lipolysis. The acids impart a somewhat pungent flavor. A pronounced and distinct flavor may be obtained when free fatty acids develop together with flavor compounds from protein degradation. In cheese lacking sufficient basic flavor from proteolysis, free fatty acids are considered undesirable because they impart a soapy-rancid flavor.

The methods currently used to evaluate the quality of cheese flavor are based on sensory evaluation by a panel of experts. These panels are able to monitor the flavor by descriptive tests, to compare samples to a standard, and to detect defects (off-flavors). This evaluation, using the sensory flavor descriptors, allows the establishment of the flavor profile of a cheese sample, a technique widely applied in the dairy industry and research. Instrumental analysis of cheese flavor compounds has concentrated on the volatile compounds, by a combined gas chromatography–olfactometry technique. The headspace of a cheese sample in a closed vial is injected into a gas chromatograph, and its eluate is sniffed to identify fractions containing the volatile key flavor compounds. Also, procedures other than dynamic headspace sampling are available to extract these compounds. Their molecular structure can be further identified by using a mass spectrometric detector coupled to a gas chromatograph. Evaluation of the concentration of volatiles emitted from cheese without their prior separation is also theoretically possible by using gas sensors (for example, based on mass spectrometry); a commercial instrument, the so-called electronic nose, is now available.

Mature cheese contains small amounts of several essential volatile flavor compounds. The compounds are predominantly degradation products of amino acids, including NH_3, various amines (in cheeses with a surface flora), methional (for example, in Cheddar cheese), H_2S, phenylacetic acid, and other compounds. Furthermore, the following components have been indicated: aldehydes, primary and secondary alcohols and their esters, short-chain fatty acids, and δ-lactones. In strongly flavored cheeses, blue-veined cheese in particular, methylketones are predominant, formed largely by microbial fat degradation. Propionic acid bacteria form propionic acid, which induces a sweetish taste, especially at a higher pH.

Finally, several off-flavors may occur such as yeasty flavor (alcohols and esters), unclean flavor, H_2S-like flavor, burnt flavor (by some lactic acid bacteria), cabbage-like flavor (by other lactic acid bacteria), bitter flavor (certain peptides), soapy-rancid flavor (free fatty acids), and many others (see also Chapter 26).

25.5.2 Formation of Flavor Compounds

The formation of flavor compounds in a ripening cheese is a complex and rather slow process involving various chemical and biochemical conversions of breakdown products of lactose, caseins, and fat. The enzymes involved in these conversions are predominantly derived from starter cultures used in the cheese manufacture.

Lactose is mainly converted to lactate by lactic acid bacteria (see Section 25.1), but a fraction of the intermediate pyruvate can alternatively be converted to various flavor compounds such as diacetyl, acetoin, acetaldehyde, or acetic acid. These products may also be formed from citrate via pyruvate (see also Subsection 13.1.2).

For flavor development from caseins, the peptides formed by the proteolysis system in cheese (Section 25.3) should be first converted to amino acids and these, subsequently, to various flavor compounds. Although amino acids themselves contribute to the basic cheese flavor, their conversion is undoubtedly the most important biochemical pathway for characteristic flavor formation in various cheese types, especially the hard and semihard types. They can be converted in

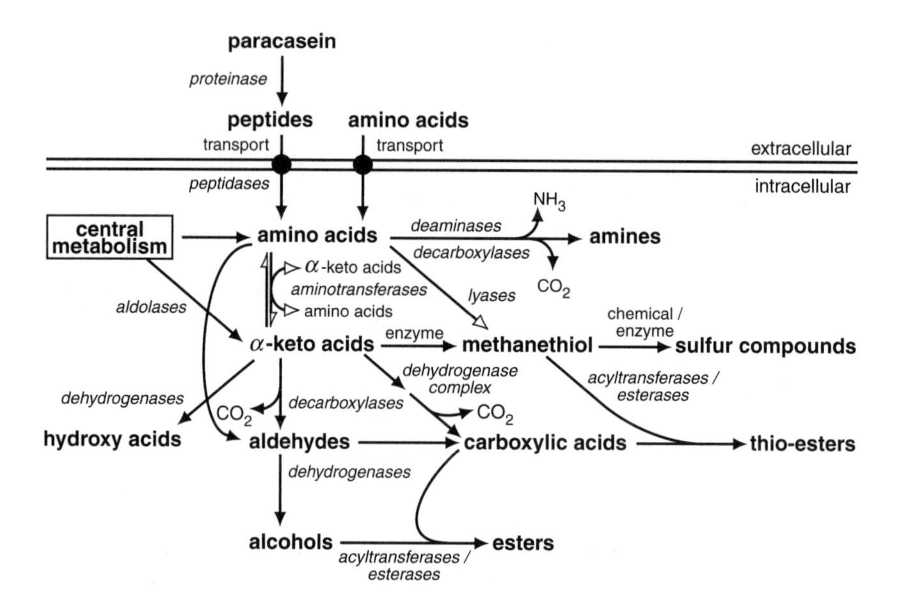

FIGURE 25.3 Summary of general pathways in lactic acid bacteria leading to intracellular amino acids and α-keto acids, and their degradation routes to potential flavor compounds in cheese. (Modified from Van Kranenburg et al., *Int. Dairy J.*, **12**, 111–121, 2002.)

many different ways by enzymes such as deaminases, decarboxylases, transaminases (aminotransferases), and lyases (Figure 25.3).

Deaminases and decarboxylases may yield amines and NH_3 from amino acids. Transamination of amino acids results in the formation of α-keto acids that can be converted into aldehydes by decarboxylation and, subsequently, into alcohols or carboxylic acids by dehydrogenation. Many of these compounds are odor active and contribute to the overall flavor of cheese. Moreover, other reactions may occur, for instance, by hydrogenase activity toward α-keto acids resulting in the formation of hydroxy acids, which do however hardly contribute to the flavor. The first step in the conversion of amino acids relies on the presence of enough acceptor α-keto acid. Transaminases are also used for the transformation of one amino acid into another, which is an essential step in the lactococcal protein biosynthesis.

Aromatic amino acids, branched-chain amino acids, and methionine are the most relevant substrates for cheese flavor development. Conversion of aromatic amino acids by lyases can result in formation of compounds such as *p*-cresol and indole, which contribute to off-flavors in cheese. Tryptophan and phenylalanine can also be converted into benzaldehyde, which contributes positively to the overall flavor of cheese.

Branched-chain amino acids are precursors of various aroma compounds, such as isobutyrate, isovalerate, 3-methylbutanal, 2-methylbutanal, and 2-methylpropanal. These compounds are found in many cheese types. Several enzymes that are able to convert these amino acids are found in *Lactococcus lactis* strains.

Volatile sulfur compounds derived from methionine, such as methanethiol, dimethylsulfide, and dimethyldisulfide, are regarded as essential components in many cheese varieties. In their formation, several lyases are involved, although some spontaneous chemical reactions are also thought to play a role. Other breakdown products of methionine are thiaalkanes, such as 2,4 dithiapentane (CH_3–S–CH_2–S–CH_3), which are responsible for a garlic flavor note in camembert cheese.

Free fatty acids formed during the lipolysis of fat are flavor compounds (Section 25.4) themselves. The short- and medium-chain (four to ten carbon atoms) fatty acids have relatively low perception thresholds, and each of them has a characteristic note. According to their concentration and perception threshold, these volatile fatty acids can contribute to the aroma of the cheese or to a rancid defect. A minor proportion of free fatty acids, having generally from two to six carbon atoms, comes from the degradation of lactose and amino acids. Long-chain fatty acids (having ten or more carbon atoms) play a minor role in flavor, owing to their high perception threshold. The nondissociated form of the acids is the only one that is aroma active. It is generally found in the fat phase of cheese; the water phase contains both forms, nondissociated and ionized. The equilibrium between these two forms is clearly pH-dependent and, therefore, the pH of the cheese is important for the role of the free fatty acids in flavor perception (Section 25.4).

The free fatty acids are the precursors of flavor compounds such as methylketones, alcohols, lactones, and esters (Figure 25.4). Lactic acid bacteria contribute relatively little to the conversion of fatty acids, but the fat-derived flavor compounds are very important in mold-ripened cheeses such as Camembert and Roquefort (Table 25.2). The molds in these cheeses possess an enzymatic system

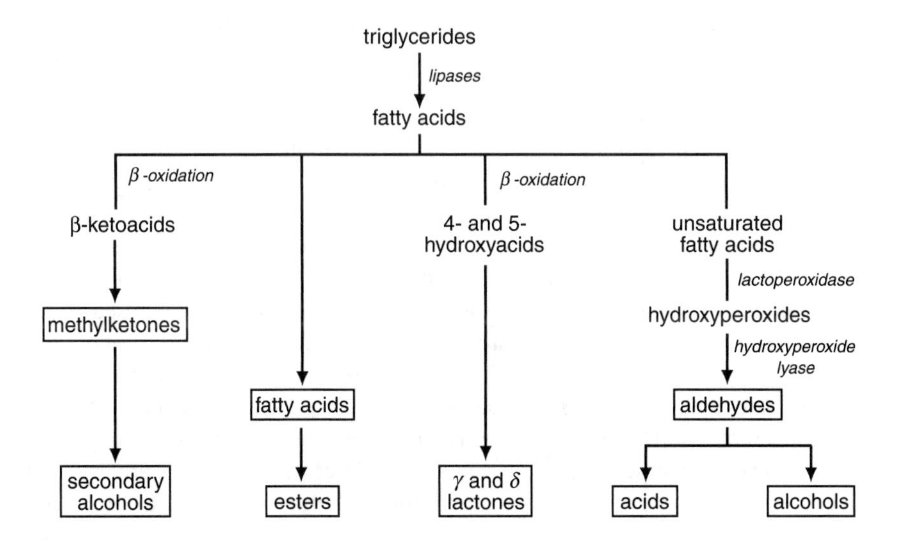

FIGURE 25.4 Formation of flavor compounds (framed) in cheese from fat. (Adapted from Molimard and Spinnler, *J. Dairy Sci.*, **79**, 169–184, 1996.)

TABLE 25.2
Some Groups of Flavor Compounds Formed by Specific Microorganisms Important in Cheese Ripening

Components Formed	*Penicillium*		Yeasts	Corneyforms (*B. linens*)	Propionic Acid Bacteria
	camemberti	*roqueforti*			
Amino acids	+	+	+	+	+
Amines				+	
NH_3	+	+		+	
Sulfur compounds	+			+	+
Fatty acids (short-chain)	+	+		+	+
Methylketones	+	+			
Secondary alcohols	+	+			
Oct-1-en-3-ol	+				
Fatty acid esters			+	+	+

Note: A plus sign indicates that the organism involved forms much more of the component concerned than the other organisms, including the organisms of the normal starter flora.

that permits the β-oxidation of fatty acids to a β-keto acid, which is rapidly decarboxylated to give a methylketone with one carbon fewer than the initial fatty acid. The notes of the methyl ketones are generally described as fruity, floral and musty.

The amount and type of compounds vary widely among the groups of organisms mentioned, as well as among strains of an individual group.

Methylketones can be reduced to form secondary alcohols, which along with ketones are considered to be the most important compounds in the aroma of soft and mold-ripened cheeses. The secondary alcohols found are heptan-2-ol and nonan-2-ol, which represent, together with the methylketones from which they are derived via a reductase activity, 20 to 30% of all aroma compounds in Camembert-type cheese. Primary alcohols in cheese are generally derived from the metabolism of lactose and amino acids. The alcohols generate mostly mild and fruity notes. Linoleic and linolenic acids are precursors of eight-carbon aroma compounds, particularly oct-1-en-3-ol. The principal enzymes supposed to be involved in this alcohol synthesis are a lipoxygenase and a hydroperoxide lyase found in molds. This compound is responsible for a mushroom note in cheese.

The lactones encountered in cheese are principally γ-decalactone, δ-decalactone, γ-dodecalactone, and δ-dodecalactone. Their precursors are 4- and 5-hydroxyacids, which can be present in the triglycerides of fat. After liberation they are cyclized to lactones. Hydroxyacids can also come from catabolism of fatty acids by certain eukaryotic microorganisms. Lactones are generally characterized by very pronounced fruity notes.

Cheeses contain a great diversity of esters, which are formed by esterification reactions between alcohols derived from lactose metabolism (ethanol) or from amino acid catabolism and short- to medium-chain fatty acids. The esterases involved are present in most microorganisms that contribute to cheese ripening. Most of the esters encountered in cheese are described as having fruity and floral notes.

The formation of specific flavor compounds in cheese clearly depends on the ripening organisms involved and on their enzymatic activity. They are largely responsible for the specific flavor notes, which add to the basic cheese flavor (Subsection 25.5.1). This is illustrated in Table 25.2. It can be inferred that cheeses with a specific flora can have high levels of amino acids and fatty acids. Types of cheese with a ripening flora of yeasts and molds show, in addition, higher concentrations of methylketones, secondary alcohols, NH_3, esters and, for Camembert cheese, oct-1-en-3-ol. Cheeses with a surface flora of coryneforms produce increased concentrations of amines, NH_3, and sulfur compounds. The main fatty acid formed by propionic acid bacteria (in Emmentaler cheese, for example; see Subsection 27.5.1) is propionic acid, which gives a sweetish taste. Ca and Mg salts of free amino acids and of low-molar-mass peptides are also among the compounds that produce the characteristic sweetish taste of Emmentaler cheese.

25.6 DEVELOPMENT OF TEXTURE

The word *texture* is meant to include physical inhomogeneity or structure, and rheological properties or consistency.

25.6.1 STRUCTURE

The *microstructure* of cheese during the first few hours after curd making, as observed by electron microscopy, reveals a matrix of paracasein micelles (diameter about 100 nm). The cavities in the matrix are largely filled with fat globules (~4 μm) and some whey. The moisture can still move fairly easily through the network. Within a day the matrix alters, that is, it becomes more homogeneous. From now on, fat globules and much smaller protein particles can be seen. In a cheese of $pH > 5.2$, particles of some 10 to 15 nm are observed; at $pH < 5.0$, particles of at most 4 nm; and at intermediate pH, particles of both sizes. Cavities filled with whey cannot be detected, and any displacement of the moisture has become increasingly difficult. The cause of this change is to be found in the dissolution of the calcium phosphate and, later on, also in proteolysis. When hard cheese is kept for a long time (for example longer than 4 months), the fat globules become partly fused, possibly due to enzymatic degradation of their membranes. Hence, in addition to a continuous aqueous phase, a continuous fat phase can develop. Often, crystals of free amino acids or their salts, and crystals of calcium lactate are formed. An overview of structural elements is given in Table 25.3.

When observing a freshly made cross section of a cheese, one perceives color and possible inhomogeneities, such as graininess or holes. In many cheese varieties, the cut surface looks yellowish, smooth (shiny), and slightly transparent;

TABLE 25.3
Size and Numbers of Some Structural Elements of Clotted Milk, and Semihard and Hard Cheese

Structural Element	Volume Fraction (–)	Number (m^{-3})	Size[a] (m)
Clotted Milk:			
Paracasein micelles	0.06	10^{20}	10^{-7}
Large cavities in network[b]	0.5	10^{16}	$5 \cdot 10^{-6}$
Fat globules	0.04	10^{15}	$4 \cdot 10^{-6}$
Cheese:			
Paracasein particles	0.5?	10^{24}	10^{-8}
Fat globules	0.3	10^{16}	$4 \cdot 10^{-6}$
Lactic acid bacteria	0.005	10^{15}	$2 \cdot 10^{-6}$
Curd particles	~1	10^{7}	$5 \cdot 10^{-3}$
Curd pieces[c] (Cheddar)	~1	10^{5}	$40 \cdot 10^{-3}$
Holes in Emmentaler	0.25	$5 \cdot 10^{4}$	$20 \cdot 10^{-3}$
Holes in Gouda	0.04	$2 \cdot 10^{5}$	$7 \cdot 10^{-3}$
Holes in Tilsiter	0.08	$2 \cdot 10^{7}$	$2 \cdot 10^{-3}$

Note: Approximate examples.

[a] Diameter of sphere of the same surface area.
[b] Network of aggregated paracasein micelles.
[c] 1.5 × 1.5 × 7 cm strips of curd, cut before salting.

in such a case, a piece of cheese is elastic and not very hard. Other varieties, especially those with a low pH or a high salt content, have a different appearance: The cheese looks white and dull or chalky, and it feels rather hard and brittle. In brine-salted varieties, a young cheese often shows the latter texture near the rind and the former texture in the center; the white rind portion then mostly disappears as the salt becomes more evenly distributed. Upon maturation, the mentioned differences disappear, especially in semihard and hard cheeses. In soft cheeses, a similar change in texture is even more conspicuous. At first the cheese has a low pH, and it looks white and dull, whereas upon maturation it changes into smooth, yellowish, and almost liquid-like (see Subsection 27.6.1.6). These differences in appearance may be linked to the differences in microstructure mentioned earlier.

Several varieties of cheese show *holes* in the cheese mass. These holes can form as a result of an imperfect fusion of curd particles combined with inclusion of air, and are referred to as 'mechanical holes.' Examples are Gouda cheese made from curd stirred after whey drainage, inadequately pressed Cheddar cheese, and several types of soft cheese that are not (or are only lightly) pressed. In many cases it concerns 'eyes,' which are spherical holes originating from gas production in the cheese. Usually, the gas is CO_2, produced by certain starter lactococci, predominantly from citrate (Gouda cheese), by propionic acid bacteria from lactic acid (Emmentaler cheese) or by lactobacilli from amino acids. In general, the CO_2 production *per se* is insufficient to form eyes if CO_2 exclusively forms from citrate. To that end N_2, or another gas such as H_2, is also needed. The milk often is almost saturated with air and, therefore, the fresh cheese is saturated with N_2 (because O_2 is consumed by the bacteria). If the milk is partly deaerated before the cheese making, 'blind' cheese is generally obtained. Currently, most cows' milk reaches a bulk milk tank through a closed circuit, and it may contain relatively little air.

Apart from supersaturation with $N_2 + CO_2$ by about 0.3 bar, nuclei would be needed to form gas bubbles, but homogeneous nucleation, that is, spontaneous formation of tiny gas cells, cannot occur. Small air cells remaining in the cheese after curd making can grow to form holes. However, air bubbles present in the milk disappear quickly, i.e., the larger ones rise to the surface, and the smaller ones tend to dissolve because of the high Laplace pressure in a small bubble (see Subsection 3.1.1.4). Presumably, some strongly shrunken air cells can remain, provided that the milk is about saturated with air. Alternatively, finely dispersed air can be blown into the mixture of whey and curd shortly before the molding of the cheese to obtain air cells.

Slits or cracks rather than eyes can be formed if much gas is produced in the cheese, especially if it concerns H_2. The type of hole formed depends partly on the consistency of the cheese (see Subsection 25.6.2.4).

25.6.2 Consistency

Rheologists define the consistency of a material as its resistance to permanent deformation. In other words, it is the relation between the force exerted onto a material and its resulting flow. In actual practice, the reversible (and therefore elastic) deformation may also play a part. Usually, when speaking of the consistency of cheese, one refers to all rheological and fracture properties.

Cheese consistency varies widely, from rigid, almost stony (such as old Edam) to nearly pourable (such as overripe Camembert), or from rubber-like (such as Emmentaler) to crumbly and spreadable (such as some goats' milk cheeses). Moreover, the consistency may vary within a cheese: Compare rind and center, presence of holes, acid spots, grains of crystals, and so forth. Such inhomogeneity makes precise determination of the consistency difficult.

25.6.2.1 Description

Rheological properties of cheese can be determined in various ways, leading to somewhat differing results. Often, a cylindrical piece of cheese is compressed between parallel plates. This can be done in two ways. In the first, the force is kept constant, for example, by putting a weight on top of the test piece, and the deformation is measured as a function of time; an example is given in Figure 25.5. The force applied can be varied. In the other method, the test piece is compressed at a constant rate, and the force is measured as a function of the compression.

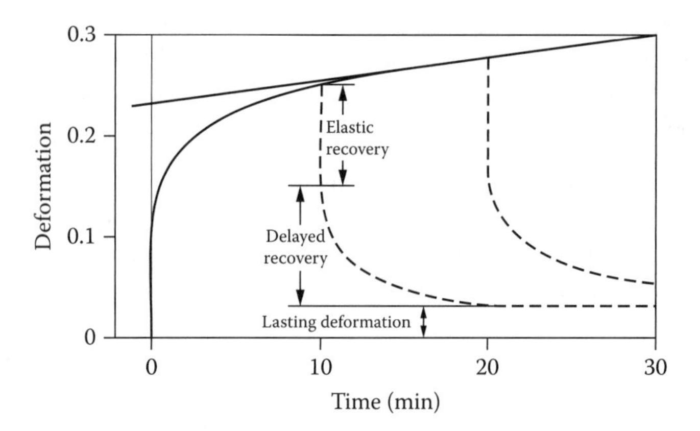

FIGURE 25.5 Example of the relative deformation of a piece of cheese after bringing it under a constant stress at time $t = 0$. The deformation after removal of the stress at $t = 10$ and 20 min, respectively, is also given (broken lines).

To obtain meaningful results, the force should be recalculated to a stress, that is, force over the cross-sectional area of the test piece, which alters during compression. Also, the deformation should be expressed as the true, natural, or Henky strain $\varepsilon = -\ln(h_t/h_0)$, where h is height of the test piece and t is the time. Examples are given in Figure 25.7. The strain rate $\dot{\varepsilon} = d\varepsilon/dt$ can be varied. If the test piece is a rather flat cylinder of the same diameter as the platens of the compression apparatus, and if there is, moreover, complete lubrication between the test piece and platens, the deformation of the test piece is true biaxial elongation. In that conformation, an (apparent) elongational viscosity (stress over $\dot{\varepsilon}$) can be determined; examples are given in Figure 25.6.

The behavior of cheese under stress is always *viscoelastic*, as is depicted in Figure 25.5. At first there is a purely elastic deformation, which is instantaneous and reversible. After some time there is a purely viscous deformation, which is permanent and proportional to the time for which the stress is applied. In between, the deformation behavior is more complicated. However, the magnitude of the rheological parameters (elastic modulus and apparent viscosity) and the timescale of the change from purely elastic to purely viscous vary widely. Over long timescales, viscous deformation is always predominant.

Based on other kinds of rheological measurements, both a true elasticity or storage modulus, and a true viscous or loss modulus, can be determined as a function of deformation rate. In practice, one may calculate an apparent elastic modulus E (stress divided by instantaneous deformation) and an apparent viscosity η_a (stress divided by rate of the lasting deformation) from data as illustrated in Figure 25.5

Most cheese has no yield stress (see Figure 4.5a): Even when subject to a very small stress it flows, although very slowly. Under small stresses, the elastic and the lasting deformations are linear, that is, proportional to the stress. Clearly, over longer timescales, cheese behaves as a highly viscous liquid. It sags under its own weight.

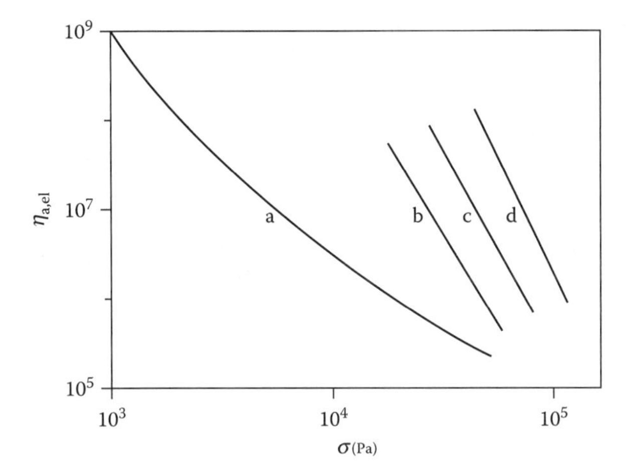

FIGURE 25.6 Influence of the stress (σ) applied on the apparent elongational viscosity ($\eta_{a,el}$) of some samples of Gouda cheese (full-cream) at 20°C. (a) Age 1 week, 42% water, pH 5.21. (b) Age 1 week, 50% water, pH 4.94. (c) Age 6 months, 38% water. (d) Age 6 months, 29% water.

The apparent viscosity greatly depends on the stress applied; see Figure 25.6, where elongational viscosity values are given. Such viscosity is relevant when the cheese is deformed in tension, which means that a velocity gradient occurs in the direction of the deformation. This type of deformation occurs during sagging of the cheese and during eye formation.

So far we have implicitly assumed that the cohesion of the cheese mass is preserved during the deformation. At a high deformation rate, however, most cheese fractures after a certain deformation has been reached. This is illustrated in Figure 25.7A and Figure 25.7B, where a sample of cheese is compressed until it fractures. In this way a modulus E is determined, as well as a fracture stress

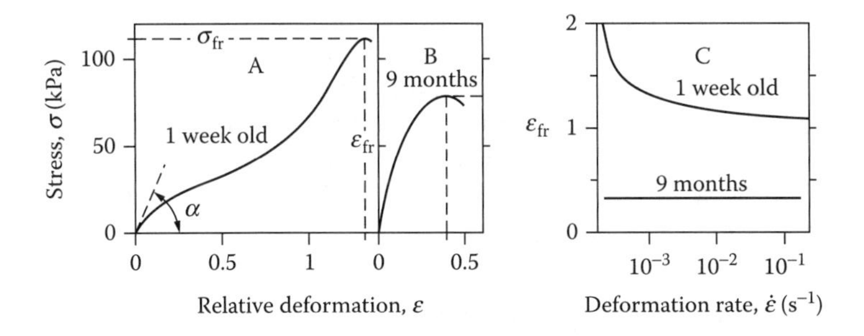

FIGURE 25.7 Fracture phenomena in Gouda cheese. (A, B) Examples of compression curves at $\dot{\varepsilon} \approx 10^{-2}$ s^{-1}; tan α is the modulus (E); σ_{fr} the fracture stress and ε_{fr} the relative deformation or strain at fracture. (C) Examples of the influence of the deformation rate on the strain at fracture.

σ_{fr} and a fracture strain ε_{fr}. The values of the respective quantities may vary widely among varieties of cheese: E from 10^4 to 10^6 Pa, σ_{fr} from 10^3 to $5 \cdot 10^5$ Pa, and ε_{fr} from 0.1 to 2. For a purely brittle material such as glass, σ is proportional to ε until fracture occurs and, hence, $\sigma_{fr}/\varepsilon_{fr} = E$. In cheese, however, this ratio always is smaller than E ($\sigma_{fr}/E\varepsilon_{fr} = 0.2$ to 0.7); the shape of the curve can vary considerably. Figure 25.7A and Figure 25.7B show two types of behavior. A somewhat S-shaped curve (A) is observed for semihard cheese that is hardly matured (up to a few weeks old) and has a pH above about 5.15; in such a cheese, the strain at fracture strongly increases with decreasing strain rate. All other semihard and hard cheeses (lower pH and/or far more matured) show a behavior as in Figure 25.7B, i.e., the slope of the stress–strain curve keeps decreasing with strain, and the strain at fracture is not or is hardly dependent on strain rate (C).

The mode of fracture can also vary. Often (as in case B), cracks occur in the interior of the test piece before the maximum in the compression curve has been reached; in other cases (as in case A), the test piece strongly bulges and vertical cracks appear at its outside. Furthermore, the modulus and the stress at fracture increase slightly with the strain rate. At very slow deformation, a cheese may not exhibit fracture but flow (see Figure 25.7C, upper curve, where ε_{fr} goes to infinity for very small $\dot{\varepsilon}$). This is often noted in a well-ripened soft cheese, which can be spread on bread. Prevention of fracture within a firmer cheese is only possible at extremely slow deformation rates.

The *firmness* of cheese can be defined either as the modulus or as the fracture stress. The latter seems more appropriate because one is usually interested in deformations causing the cheese to break into pieces. Methods applied in practice to characterize the firmness of cheese often determine a quantity related to a fracture stress, for example, the force needed to press a plunger of a given diameter at a given speed into the cheese.

In addition to the specifications 'firm' and 'soft,' an important characterization of the consistency of cheese is whether it is *short* or *long*. Wet sand and a rubber band are examples to illustrate the extremes on the scale of short to long. A long cheese is rubbery but not hard; a piece of it can be bent considerably before it fractures. A sample plug taken from such a cheese is thinner than the inner diameter of the cheese borer. A short cheese does fill the borer entirely because its elastic deformability is smaller. If, moreover, the latter cheese is soft, it is plastic and spreadable, whereas a firm and short cheese is stiff and almost crumbly. The properties, short and long, obviously are linked to the cohesion forces between cheese 'particles.' Shortness is best defined in terms of the deformation needed for fracture, and thus in terms of ε_{fr} (the smaller ε_{fr}, the shorter the cheese).

25.6.2.2 Factors Affecting Consistency

Several factors affect the consistency of cheese, but it is not easy to establish quantitative relationships: One cannot vary one factor without changing other factors. Moreover, the same variable can affect a young cheese and an aged one in different ways. Hence, the data in Figure 25.8 have to be considered with

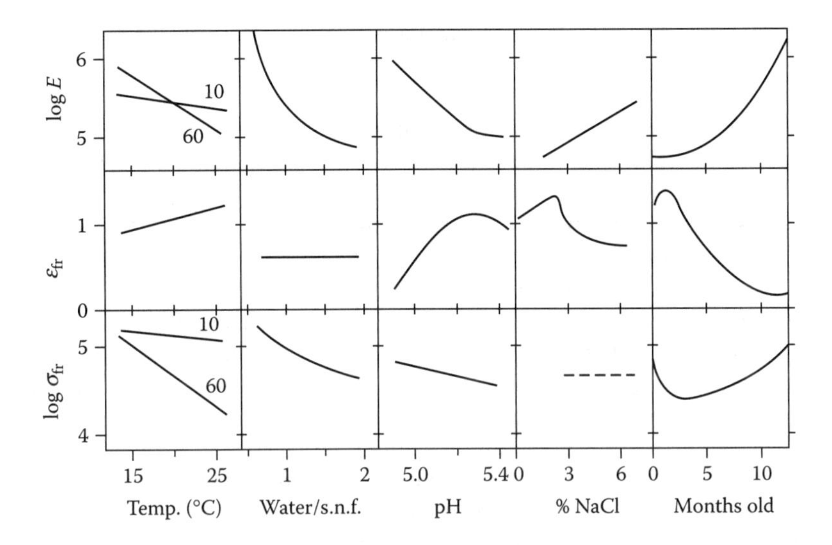

FIGURE 25.8 Rheological properties (modulus E, fracture strain ε_{fr}, fracture stress σ_{fr}) of cheese (excluding fresh and very soft cheeses). Effect of the temperature (water/solids-not-fat ≈ 1.3; 10% and 60% fat in the dry matter, respectively); the ratio of water to solids-not-fat; the pH (cheese that is 4 weeks old); the salt content (pH 5.2, water content 50%); and the age of the cheese. The latter applies specifically to Gouda cheese.

caution. We will now discuss the effects of several variables on modulus E, deformation (strain) at fracture ε_{fr}, and fracture stress σ_{fr}:

1. *Temperature*: If the cheese contains much fat, the temperature considerably affects E as well as σ_{fr}. The key factor thus is the melting of the fat. In a low-fat cheese E decreases by some 30% when the temperature is increased from 15 to 25°C, and by some 80% from 15 to 60°C.
2. *Fat content:* From what is mentioned under 1, it is clear that the effect of the fat content greatly depends on the temperature. Incidentally, a cheese of the same type but with a higher fat content in the dry matter generally has a higher water content in the fat-free cheese, and thus a lower E and σ_{fr}. The results in Figure 25.8 refer to identical water contents in the fat-free cheese for cheeses with 10% and 60% fat in the dry matter, respectively.
3. *Water content*: The water content in the fat-free cheese has a very great effect on E. Note the logarithmic scale in the figure. The network of paracasein is the main cause of the stiffness of the cheese and, hence, the paracasein concentration will have a considerable effect. A further decrease of the water content (by one percentage unit for example) has a stronger effect if the water content already is low. The water content hardly affects ε_{fr} (at least if the deformation rate is not too slow), but it can affect the fracture mode.

4. *Acidity*: The pH has a considerable effect on each of the consistency parameters and also on the fracture mode. The shortness is definitely affected. The cheese is 'longest' at a pH of about 5.3. The relationships between pH and consistency may, however, depend on other factors (see the following variables) and are therefore somewhat uncertain.
5. *Calcium phosphate*: The effect of the calcium phosphate content is insufficiently known because it cannot be varied independently. Presumably, calcium phosphate increases E and ε_{fr}, but its influence is not great unless the differences in concentration are large.
6. *Salt content*: Often, a higher salt content is associated with a far higher modulus, but the main cause may be that, in practice, a higher salt content mostly involves a lower water content. Even if the water content is kept constant, as in Figure 25.8, E distinctly increases with the ionic strength. There is an abrupt and strong decrease of ε_{fr} at about 2.5% salt, that is, 5% NaCl in the water. A cheese with a high salt content and a low pH is excessively short.
7. *Protein degradation*: Proteolysis has a considerable effect. However, the precise effect is poorly known because so many other factors also change when a cheese matures (see the following text). It is certain that the decomposition of casein causes a decrease of ε_{fr} and of σ_{fr}.

Figure 25.6 illustrates that the apparent viscosity of the cheese depends on the water content and perhaps on the age of the cheese. It is dominated, however, by the pH, especially in young cheese and at small stresses. The viscosity markedly decreases with increasing temperature, although reliable results are scarce. Around 70 or 80°C, the cheese appears to 'melt,' that is, it becomes a viscous liquid.

25.6.2.3 Changes during Ripening

Several changes in composition occur. The main ones can be outlined as follows:

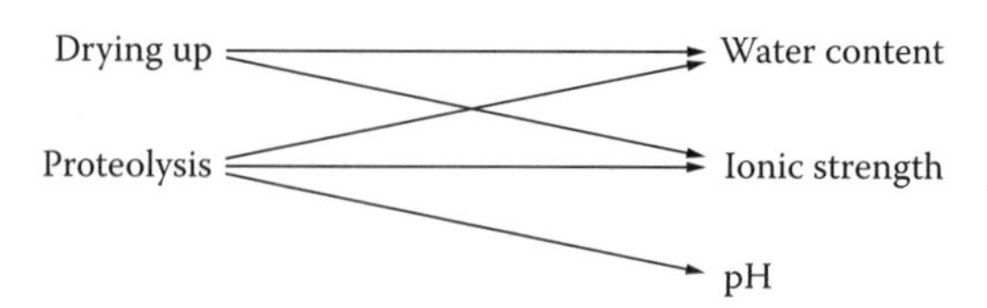

Proteolysis causes the uptake of water (it decreases the water content by, say, 1%) and the formation of ionic groups ($-COO^-$ and $-NH_3^+$). Generally, proteolysis increases the pH.

It will be clear that the changes involved and their rates closely depend on the prevailing conditions, including the size and shape of the cheese (which mainly affect the evaporation of water and the influence of a surface flora), amount of retained rennet, type and number of bacteria (or the enzymes released), pH, ionic strength, surface flora, and such ripening conditions as temperature and relative humidity. In Figure 25.8, examples are shown of the influence of the ripening time on the consistency of a cheese. The decrease of the water content is the main cause of the increase of the modulus. It may be noted, however, that several types of hard and semihard cheese are currently made as a 'rindless' loaf and packaged in a plastic foil that is almost impermeable to water. This implies that drying up of the cheese hardly occurs, especially when the loaves are large.

It is especially the proteolysis that causes the cheese to become shorte. As a result, the firmness (σ_{fr}) decreases slightly in the beginning and increases substantially afterward. Organoleptic observation reveals that the young cheese is somewhat rubbery, becoming plastic after aging.

The changes in consistency during ripening naturally depend on the type of cheese. A firm (that is, usually dry) cheese becomes crumbly. Most cheese with a high water content has a low initial pH, such as 4.8. It then is short and firm. At constant pH and water content, proteolysis causes little further change in consistency (the texture becomes somewhat shorter). If the pH of the same unripened cheese is increased (to, say, 5.4), the consistency becomes rubbery; proteolysis causes the cheese to become soft and plastic, even liquid-like, at this pH. This is essentially what happens when a surface flora grows on a cheese. The flora consumes lactic acid and thereby increases the pH of the cheese from the rind inward, causing the cheese mass to become soft, provided that sufficient proteolysis has occurred, predominantly due to rennet action. The typical soft and almost liquid consistency develops if:

- The ratio of water to solids-not-fat is greater than 2
- The pH is greater than 5.2
- At least 60% of the α_{s1}-casein has been degraded
- The calcium phosphate content is not very high

See also Subsection 27.6.1.6.

25.6.2.4 Important Consequences

During various stages of making, curing, handling, and eating cheese, different rheological properties are of importance:

1. *Curd fusion:* During molding, the curd grains must be sufficiently deformable to enable them to fuse into a coherent mass. Therefore, the apparent viscosity η_a must be low. This causes no problems if the water content and the temperature are not too low; η_a then is, for instance, 10^6 Pa·s. During classical 'cheddaring,' the curd has to deform at a fairly high rate, the velocity gradient being greater than 10^{-3} s^{-1}. The stress applied is at most 10^3 Pa. Hence, the elongational viscosity should be less than 10^6 Pa·s. This can only be the case at a fairly high temperature and is strongly pH dependent (cf. Figure 24.6).
2. *Shape retention:* The cheese has to retain its shape under its own weight. Because of this, soft cheese is made flat-shaped. For several

cheeses, the apparent viscosity as such is insufficient to resist considerable deformation. Consider, for example, cheese (a) in Figure 25.6. Assuming the height of the cheese to be 10 cm, the maximum stress due to its own weight (height × density × g) ≈ 10^3 Pa. We see that the apparent elongational viscosity then is approximately 10^9 Pa·s. Consequently, the elongation rate (stress/viscosity) would be about 10^{-6} s^{-1}, or roughly 0.1 d^{-1} (10% per day). Actually, the average stress is half the maximum value, and the relative change in height is about half the elongation rate; nevertheless, the height would decrease by a few percent per day. This does not normally happen, which is mostly due to the rind of the cheese being much firmer; this may, in turn, be due to its lower water content and its initially higher salt content. Still, the cheese may sag to a considerable extent, especially at high temperature. Sometimes, cheese loaves are supported by bands; very large blocks of Cheddar are kept in solid 'forms' or barrels.

3. *Hole formation:* If more gas is produced than can dissolve in the cheese at a given overpressure, holes will form (Subsection 25.6.1). If the cheese mass has a low viscosity, it may 'flow' due to the pressure in the hole, and the hole formed remains spherical, that is, an 'eye.' If the viscosity is high and gas production rate is high (rapid diffusion of gas to the hole), then the gas pressure in the hole will become high. If, moreover, the cheese mass has a low fracture stress (which often occurs in a short cheese, ε_{fr} being small), then the gas pressure in the hole becomes higher than the fracture stress and may cause fracture and formation of slits or cracks. Clearly, the risk involved is smallest if the pH is approximately 5.25 (see Figure 24.6), the salt content is less than 2.5%, and only a little protein has been degraded before hole formation (see Chapter 25, Figure 25.8).
4. *Handling:* It must be possible to handle the cheese, as it has to be cut, spread, or grated, according to usage. The cheese should mostly neither be too long nor too firm. Stickiness may be a problem in well-matured, soft, or semihard cheeses.
5. *Eating quality:* The cheese must be readily deformable in the mouth, and preferably be rather homogeneous. A long cheese that is also firm is considered unpleasant by many consumers. It is very difficult to chew a very hard cheese (very low water and low fat content); in such a cheese, the presence of weak spots, such as those provided by cumin seeds, eases chewing considerably. Usually, the consistency in the mouth is an important quality mark.
6. *Melting:* Several cheese varieties must melt down satisfactorily at high temperature, such as cheese used on pizzas. It should flow but retain its integrity.

The first three properties in the preceding list relate to slow deformation of the cheese mass, whereas items 4 and 5 relate to fast deformation and fracture.

Completely different rheological measurements must thus be performed to evaluate the various cheese properties.

25.7 ACCELERATED RIPENING

The quality of a cheese, as perceived by the consumer, will go through a maximum during ripening. This is illustrated in Figure 25.9. The characteristic properties, such as flavor and consistency, take a certain time to develop. Subsequently, ongoing decomposition leads often to undesirable flavors (such as a too-pronounced lipolysis). The product may also lose quality due to vaporization of water. As a rule, the process of rise and subsequent decline proceeds faster and is more pronounced in a cheese of a higher water content. The decline in quality can to some extent be checked by storing the cheese at a lower temperature after sufficient ripening has occurred.

Obviously, most semihard and hard cheeses need a long ripening time. Storage and maintenance of cheese are expensive because of investments in buildings and machinery, and costs of energy and labor. Reducing the ripening time is therefore attractive from an economical point of view, especially for slowly maturing cheese without a secondary flora. Conditions for an accelerated ripening process are as follows:

1. The properties of the cheese should not differ greatly from those of the reference product.
2. Overripening of the cheese should be prevented.
3. The additional processing costs should not exceed the economic gain of the shortened ripening time.
4. Legal and public health aspects should be upheld, for example, with respect to the permissibility of enzyme preparations and their possible toxicity.

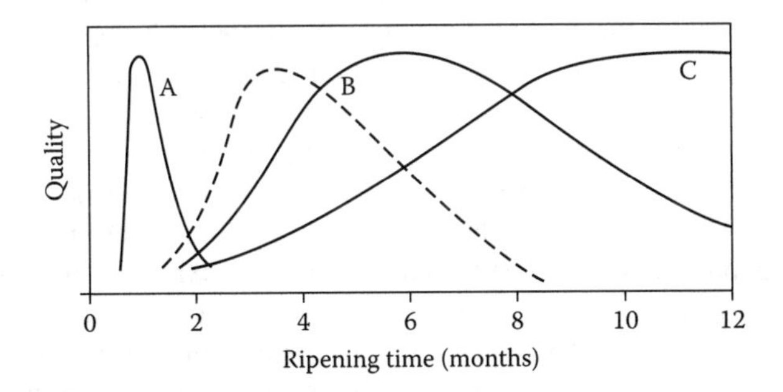

FIGURE 25.9 Approximate eating quality of cheese during maturation. A represents, for instance, a Camembert cheese; B a mild Gouda or Cheddar; and C a mature-type Gouda or Cheddar, made with low water content. The broken line is a hypothetical example of B made by applying accelerated ripening techniques.

Accelerated ripening processes are primarily aimed at accelerated flavor formation (especially a stronger protein break-down), while maintaining a satisfactory texture. Several methods have been tried, some of which are discussed briefly in the following text:

Increase of ripening temperature: The main effect of an increase in temperature is based on an increased proteolytic activity of the normal starter flora. The protein degradation by calf rennet is much less affected. Furthermore, any lipolysis is much more increased by an increase in temperature than is proteolysis. Increasing the ripening temperature is a simple and legal method that can be applied for cheese varieties with a traditionally low ripening temperature. Cheddar cheese, for example, traditionally matured at about 7°C, could be matured at 13°C, provided the bacteriological quality of the milk is satisfactory. In the case of cheese with a higher traditional ripening temperature (about 13°C), a further temperature increase will readily cause flavor and consistency defects, such as bitterness and butyric acid fermentation in Gouda-type cheese.

Use of enzyme preparations: Lipolytic and proteolytic enzymes can accelerate the production of flavor compounds. A successful use of preparations containing these enzymes is complicated by the need to attain a satisfactory balance among the various enzymes involved in the ripening process. An imbalance readily causes off-flavors, and too strong a proteolysis can lead to consistency defects. Because of this, the lipolytic enzyme preparations are exclusively applied to cheese that is characterized by a distinct lipolysis, such as Italian hard cheeses. Most proteolytic preparations originate from molds and bacteria. Usually, they display endopeptidase activity that should, however, be limited to avoid such defects as bitterness and a weak consistency. The number of suitable preparations is thus limited and they can only be used in fairly low concentrations. An extract of *Bacillus subtilis*, which contains a neutral proteinase, seems to be suitable. Moreover, it has the advantage of being barely active below 8°C, and it is even inactivated at lower temperatures. Hence, overripening can be prevented. Flavor defects can be kept to a minimum by combining the endopeptidase activity of enzyme preparations with the exopeptidase activity of extracts of lactic acid bacteria.

The enzyme preparations can be added in various forms. Addition to the cheese milk has the advantage of resulting in an even distribution throughout the cheese. The method is, however, expensive because most of the enzyme may be lost with the whey (although some proteolytic enzymes become associated with the paracasein micelles) and also because of a lowered cheese yield if strongly proteolytic preparations are used. In the manufacture of Cheddar cheese, the enzyme may be added to the curd with the salt, although even then as much as 40% may become lost. Enzymes may also be enclosed in or adsorbed onto various kinds of particles (liposomes and fat globules). Basically, it offers the opportunity to control the production of flavor compounds.

Increase of the number of starter bacteria: This method, essentially leading to an increase of the enzyme capacity of the starter flora, is an obvious one: The enzymes in the bacterial cells give a balanced flavor, and this balance is unlikely to be adversely affected by the number of cells. Therefore, the risk of defects is small. A higher number of bacteria cannot simply be attained by adding more starter; the resulting faster acid production would cause a more rapid syneresis, requiring adjustment of the curd-making process. The easiest way to solve the problem — also from an economical and technical point of view — is to heat-shock the starter bacteria, that is, heat-treat them to such an extent that their lactic-acid-producing capacity is largely lost, whereas their proteolytic capacity is retained. The heat-shocked starter then is added to the cheese milk, in addition to the normal starter.

Increasing the rate of lysis of starter cells: Because it has become known that lysis of starter organisms causes intracellular peptidases (and possibly other enzymes) to become accessible, which is very important for flavor development in many cheese varieties, an obvious method to enhance ripening rate is to enhance lysis. It is clear that starter strains vary widely in their rate of lysis in cheese, which means that proper selection of strains is required. Some strains carry a prophage that can become an active phage by a heat shock, leading to considerable lysis. As the heat shock has to be applied during curd making, enhanced syneresis may pose a serious problem. Some other methods to induce lysis have also been tried. Altogether, the practical results are still fairly uncertain. One difficulty is the uncertainty about the stage at which lysis should occur because of the largely unknown stability of the peptidases released.

Addition of other bacteria: Besides selecting highly proteolytic mesophilic starter strains, the addition of thermophilic lactic acid bacteria to cheese milk for Gouda- and Cheddar-type cheeses has been tried (see also Subsection 25.3.4). The acid-producing capacity of these cultures, often called adjunct starters, should be attenuated before adding them to the cheese milk. Different means of attenuation have been described, the most successful being treatment of the bacteria with a heat shock or a freeze shock. An increase in proteolysis and the level of amino acids in cheese is the result of using such cultures. Although accelerated ripening can certainly be achieved, the cheese so obtained has a flavor note that clearly differs from that of the reference cheese. The use of selected adjunct starters offers, however, the opportunity to develop new cheese varieties with special flavor notes.

In summary, the ripening of cheese can be accelerated in various ways. With respect to proteolysis much attention is given to the formation of low-molar-mass peptides and amino acids. The correlation with the development of flavor may, however, be poor. The mechanisms involved in aroma development in cheese are not yet fully known, as are the relationships between the genetic properties of lactic acid bacteria and the production of aroma compounds.

25.8 NUTRITIVE VALUE AND SAFETY

The nutritive value of cheese follows logically from its composition: much high-quality protein and often much fat. In most cheeses, nutritive value does not change significantly with protein degradation. The calcium content varies widely, according to the method of manufacture, from 1 $g \cdot kg^{-1}$ in quarg to 10 $g \cdot kg^{-1}$ in Gouda-type cheese (see also Figure 27.4). The concentration of the water-soluble B vitamins often is not high; that of vitamin B_{12} may be substantial.

A negative aspect of cheese composition can be the possible high salt content. Allergies caused by cheese are rarely encountered. Nitrate and nitrite contents are far below the acceptable levels, even when nitrate is added during the manufacture (see Section 26.2). The concentration of D(–) lactic acid may increase to half the lactic acid concentration, depending on the type of starter used (see Table 13.1) and on the presence of contaminating bacteria, for example, certain nonstarter lactic acid bacteria that can convert L(+) to D(–) lactic acid and vice versa. However, the concentration of D(–) lactic acid is always far below toxic levels. Most cheese varieties contain very little if any lactose, and cheese agrees well with people suffering from lactose malabsorption.

Very low concentrations of mutagenic *nitrosamines* [R_1–N(–N=O)–R_2] sometimes are found in cheese. The addition of nitrate was held responsible, but that proved incorrect. Nonvolatile nitrosamines can form in the stomach when cheese is eaten together with nitrate-containing food, e.g., some vegetables. These compounds are, however, hardly mutagenic, and that only if they are activated. Hence, their carcinogenicity remains very doubtful. Much more dangerous are the nitrosamides [R_1–CO–N(–N=O)–R_2], which have, however, not been found in cheese. On the contrary, cheese can inhibit the formation of nitrosamides. Cheese may act as an antimutagenic agent and, presumably, the paracaseinate is responsible.

In cheese, amines can be formed by decarboxylation of amino acids (see Table 25.4). Because the decarboxylation is performed by microorganisms, the amines are referred to as *biogenic amines*. Several of these amines are toxic. They can cause discomfort, such as a headache or dizziness, but the susceptibility varies with individuals and may depend on the use of particular drugs. Several species of bacteria, and probably other microorganisms also, can decarboxylate certain amino acids, but in all instances only some strains are involved. When these strains are not present (for example, because of pasteurization of the milk and adequate hygienic measures), no amines are formed. When, however, decarboxylating strains (especially some lactobacilli) are present, the production of amines is often determined by the production of free amino acids, and hence by proteolysis. Therefore, high concentrations may sometimes be found in mature cheese, often soft-type cheeses. The highest concentrations found that are given in Table 25.4

TABLE 25.4
Biogenic Amines and Their Potential Occurrence in Many Cheese Varieties

Amine	Toxic Dose[a] (mmol)	Formed from	Amount of Amino Acid in Casein ($mmol \cdot kg^{-1}$)	Responsible Bacteria	Highest Concentration in Cheese ($mmol \cdot kg^{-1}$)
Cadaverine $NH_2(-CH_2)_5-NH_2$	High[b]	Lysine	560	Lactobacilli, coliforms (e.g., *Hafnia alvei*)	35
Putrescine $NH_2(-CH_2)_4-NH_2$	High[b]	Arginine[c]	220	Lactobacilli, coliforms (e.g., *Hafnia alvei*)	11
Phenylethylamine $C_6H_5-CH_2-CH_2-NH_2$	?	Phenylalanine	330	*Enterococcus faecalis*[d]	12
Tyramine $HO-C_6H_4(-CH_2)_2-NH_2$	0.07[e]	Tyrosine	340	*Lactobacillus brevis* *Enterococcus faecalis*[d]	15
Tryptamine $C_6H_4\text{-}C_2NH_3(-CH_2)_2-NH_2$	Rather high	Tryptophan	60	Enterobacteriaceae ?	0.3
Histamine $C_3N_2H_3(-CH_2)_2-NH_2$	0.6	Histidine	190	*Lactobacillus büchneri; Escherichia* spp; *Micrococcus* spp.; Propionic acid bacteria	18

Note: The toxic level varies widely among individuals and eating habits. Of the bacteria mentioned, only some strains can decarboxylate amino acids. The highest concentrations found in cheese are very exceptional.

[a] For sensitive persons.
[b] But it can raise the toxicity of histamine.
[c] Through ornithine.
[d] If very high numbers are present.
[e] If particular drugs are being used.

TABLE 25.5
Pathogenic Bacteria That Can Occur in Raw Milk, and Their Potential Occurrence and Growth in Cheese (Excluding Fresh Cheese)

Designation	Growth in Raw Milk	Survive Pasteurization	Growth during Cheese Making	Growth in Cheese[a]	Die Off in Cheese[b]	Ever Isolated from Cheese	Remarks
Brucella abortus	No	No	No	No	Yes	?	
Pseudomonas aeruginosa	Yes	No	Yes	No	0.3	?	
Aeromonas hydrophila	?	No	No	No	0.1	?	
Enteropathogenic coliforms	Yes	No	Yes	Yes/No	1–5	Yes	Can grow in/on soft cheese, pH >5.3; no enterotoxin at $T < 15°C$
Salmonella spp.	Yes	No	Yes	No	1–3	Yes	
Yersinia enterocolitica	Yes	No	?	No	Yes	Yes	
Staphylococcus aureus	Yes	No	Yes	No	1–3	Yes	Grows on moist cheese surface; toxin resists heat and proteolysis
Mastitic streptococci	Yes	No	No	No	Probably	No	
Clostridium perfringens	No	Yes	No	No	?	No	
Clostridium botulinum	No	Yes	No	No	?	?	
Listeria monocytogenes	Yes	No	Slightly?	No	3–18	Yes	Grows on soft cheese; is psychrotrophic and fairly salt resistant
Mycobacterium tuberculosis	No	No	No	No	Yes	No	
Campylobacter jejuni	No	No	No	No	0.1	No	

[a] If pH < 5.4 and some salt is present.
[b] The figures represent the number of weeks needed for one decimal reduction in semihard cheese.

refer to blue-veined cheese. In general, fecal streptococci do not proliferate to such numbers that significant amounts of phenylethylamine form. Most attention is usually focused on histamine.

Mycotoxins can be toxic and carcinogenic. There are three potential sources in cheese:

1. *The milk:* It mostly concerns traces of aflatoxin M_1, which originate from growth of molds on the cattle feed.
2. *Functional molds:* Certain strains of *Penicillium roqueforti* on blue-veined cheese and of *P. camemberti* on white-molded cheese can produce toxins, which are, however, hardly toxic and are found in cheese in very small quantities at the most.
3. *Molds that grow on a cheese rind not coated with latex:* These mostly belong to *Penicillium* and *Fusarium* species. Toxins are, however, not observed in the cheese, although some strains are capable of forming the compounds. *Aspergillus versicolor* can still grow on cheese with a latex coat, and form the slightly toxic and mutagenic sterigmatocystine. Proper cleaning of the rind is the obvious remedy.

Occasionally, *pathogenic bacteria* are found in cheese, but most of the pathogens that can occur in milk are not (see also Subsection 5.2.1). A survey is given in Table 25.5. All potentially harmful bacteria are killed by pasteurization. Most of the types cannot grow in cheese due to its high acidity and salt content. Most problems are encountered in cheese made of raw milk if hygiene during the manufacturing process was poor and the curd did not acidify adequately. Pathogenic coliforms are sometimes found in soft cheese. *Salmonella* and *Yersinia* rarely occur in harmful numbers. In all cheese varieties, including hard cheese, toxin from *Staphylococcus aureus* can be present, provided conditions are such that these bacteria can grow in the milk or on a moist cheese surface. *Listeria monocytogenes* is of special concern because it can grow on a cheese surface, even at low temperature, and can survive for a relatively long time. The bacterium is killed by pasteurization. It is sometimes found on the rind of soft cheese made of raw milk.

Suggested Literature

Various aspects of the ripening of cheese, as well as sensory properties, texture, nutritional aspects, and the occurrence of pathogenic organisms and toxins: P.F. Fox, P.L.H. McSweeney, T.M. Cogan and T.P. Guinee, Eds., *Cheese: Chemistry, Physics and Microbiology*, Vol. 1, *General Aspects*. 3rd ed., Elsevier Academic Press, London, 2004.

A general review on accelerated ripening: P.F. Fox, J.M. Wallace, S. Morgan, C.M. Lynch, E.J. Niland, and J. Tobin, Acceleration of cheese ripening, *Antonie van Leeuwenhoek,* **70**, 271–297, 1996.

Aspects of flavor: P.L.H. McSweeney, H.E. Nursten, and G. Urbach, Flavours and off-flavours in milk and dairy products, Chapter 10 in: P.F. Fox, Ed., *Advanced Dairy Chemistry,* Vol. 3, *Lactose, Water, Salts and Vitamins,* 2nd ed., Chapman and Hall, London, 1997.

Aspects of texture: Rheological and fracture properties of cheese, *Bulletin of the International Dairy Federation No 268,* Brussels, 1991.

26 Microbial Defects

Several microorganisms present in a cheese or on its surface can cause flavor and or texture defects. Microorganisms in the milk are almost completely entrapped in the curd, and this raises their number per gram of cheese by a factor of 7 to 11, as compared to the count per milliliter of clotted milk. Nevertheless, growth of the microorganisms is nearly always needed to produce defects. The growth is determined by the following factors:

1. *The microbial composition of the milk*: Important aspects are:
   - Hygienic measures during milking and milk handling; this is especially important for cheese made of raw milk.
   - Pasteurization; low pasteurization of the milk kills most spoilage organisms, with the exception of *Clostridium* spores (Section 7.3.4).
   - Bactofugation; this can substantially reduce the microbial count, including the number of *Clostridium* spores.
   - Recontamination of the milk after pasteurization or bactofugation.
2. *Measures to prevent contamination* during manufacture and curing of the cheese: Regular cleaning of the equipment, treatment of the brine (e.g., by microfiltration) and overall hygiene are important.
3. *Measures to limit the growth of microorganisms* during manufacture and maturation of the cheese: This may involve addition of inhibitors to the milk or the curd, enhancing the rate and the completeness of the fermentation of lactose, and prevention of growth of organisms on the rind.
4. *Physicochemical properties of the cheese*: In particular, lactic acid content and pH, content of NaCl, presence of sugars, temperature, ripening time, and relative humidity of the air, determine whether or not microorganisms can grow in or on the cheese, as well as what undesirable metabolites they can produce.

These aspects will be discussed below. This applies to most types of cheese, but varieties with a specific internal or surface flora pose specific problems: see Section 27.6.

Table 26.1 gives an overview of defects that may occur. It is seen that bacterial growth can cause a wide range of off-flavors. Textural defects are generally due to excessive gas production, leading to large or numerous holes in the cheese. This can be 'early blowing,' which occurs within a few days after manufacture,

TABLE 26.1
Microorganisms That Can Cause Defects in Ripened Cheese

Microorganism	Important Source of Contamination	Carbon Source	Salt-Sensitive	Killed by Low Pasteurization	Off-Flavors	Gas in Cheese	Remarks
Yeasts	Sour whey	Lactose	Mostly	Yes	Yeasty, fruity	(CO_2)	Sometimes in raw milk cheese
Coliforms	Milk, curd, whey	Lactose, (citrate)	Yes	Yes	Yeasty, unclean	(H_2), CO_2	Early blowing[b]
Propionibacterium spp.	Dung → milk	Lactate	Yes	Yes	Sweet	CO_2	Desirable in some cheese varieties
Lactobacillus casei, paracasei	Milk, curd	Amino acids, (citrate, glucides)	Little	Yes	Unclean, sharp	CO_2	Widespread
Lactobacillus plantarum, brevis, buchneri	(Rennet), weak cheese brine	Amino acids (glucides)	No	Yes	Phenol, putrid	CO_2	Form cracks
Lactococcus lactis var. *maltigenes*	Milk, starter, sour whey	Lactose	Yes	Yes	Burnt, malty	—	
Enterococcus malodoratus	Dung → milk, curd	Lactose	Little	Partly	H_2S	—	
Streptococcus thermophilus	Heating equipment	Lactose	Little	No	Unclean, yeasty	CO_2	
Clostridium tyrobutyricum	Silage → dung → milk	Lactate	Yes	No	Sharp, putrid	H_2, CO_2	
Yeasts[a]	Brine, cheese shelf	Lactose, lactate	Variable	Yes	Various		Slimy rind
Coryneforms[a]	Cheese shelf	Lactose, lactate	Variable	Yes	(Cabbage)		Slimy, red rind
Aspergillus versicolor, etc.[a]	Air, cheese shelf	E.g., lactose	Little	Yes	Musty		Produces sterigmatocystine

Note: A compound between parentheses is not always used or formed.
[a] Grow on the surface (aerobically); see Section 27.6 for cheeses with a specific surface flora.
[b] Except coliforms, all bacteria producing gas in cheese can cause late blowing.

or 'late blowing,' which occurs after some weeks. The factors determining whether the holes formed will be eyes or slits are discussed in Subsections 25.6.1 and 25.6.4.

It should be stressed that microbial spoilage does *not* imply that the cheese presents a health hazard. Safety aspects are discussed in Section 25.8.

26.1 COLIFORM BACTERIA

Coliform bacteria rarely cause defects in cheese made from pasteurized milk, since they are killed by this heat treatment. In practice, however, a slight recontamination of the milk is almost inevitable; this generally concerns bacteria of nonfecal origin, e.g., *Enterobacter aerogenes.* To restrict contamination, hygienic measures during curd making have to be maintained.

Coliforms can only grow as long as sugar is available, since they cannot ferment lactic acid. Depending on the extent of contamination, they can rapidly grow to considerable numbers during cheese making, if temperature and pH are favorable. Important metabolites formed are CO_2 and H_2, and also lactic, acetic, succinic and formic acids, ethanol and 2,3-butanediol. Moreover, yeasty, putrid, and gassy off-flavors result, partly due to some strains that can attack protein degradation products.

Growth of coliforms can lead to early blowing, but not in all cases. If all sugar has been consumed, gas formation depends on the presence of strains that can ferment citric acid. Strains of *Escherichia coli* generally can not do this, whereas most strains of *Enterobacter aerogenes* can. However, growth of coliforms that do not cause blowing, nevertheless cause off-flavors to occur.

The growth of coliforms can be prevented by using a fast-souring starter that rapidly converts lactose, thereby decreasing the pH in a short time to a level that inhibits coliform growth. Moreover, after sufficient acidification, the temperature of the cheese should be lowered, and it should be salted as soon as possible.

The development of texture defects due to coliform bacteria can be counteracted by adding a sufficient amount of an oxidizing salt to the cheese milk, i.e., sodium or potassium nitrate (saltpeter). The nitrate suppresses the formation of the enzyme system that is normally involved in the production of H_2 (lactose $\rightarrow$ formic acid $\rightarrow H_2$, under sufficiently anaerobic conditions) and appears to induce the formation of nitrate- and nitrite-reducing enzyme systems. In effect, nitrate and nitrite act as hydrogen ion acceptors and thus, no H_2 is produced from formic acid. Nitrate and nitrite are converted into ammonia, which may be used for bacterial growth. The growth of coliform bacteria is not inhibited by nitrate, the production of CO_2 is not affected, and the development of off-flavors is not prevented.

Fermentation by coliforms has little effect on the pH of the cheese.

26.2 BUTYRIC ACID BACTERIA

Certain *Clostridium* spp. (i.e., anaerobic spore-forming bacteria) can grow in cheese and ferment lactic acid; these are called butyric acid bacteria. The pH of the cheese is increased by the fermentation. The main breakdown products are butyric acid, CO_2, and H_2:

$$2\ CH_3\text{–}CHOH\text{–}CO_2H \rightarrow CH_3\text{–}CH_2\text{–}CH_2\text{–}CO_2H + 2\ CO_2 + 2\ H_2$$

This butyric acid fermentation leads to texture and flavor defects. In serious cases, cracks or large spherical holes are formed in the cheese, as are very bad off-flavors. Because of this, growth of butyric acid bacteria in cheese is considered a serious defect. The butyric acid blowing manifests itself after several weeks, or even months, and is thus an example of late blowing. *Clostridium tyrobutyricum* is the main agent causing the defect; unlike other lactate-fermenting clostridia, it does not decompose lactose. *C. butyricum* can also cause defects.

Butyric acid fermentation in cheese depends on:

1. The *number of spores* of butyric acid bacteria present in the cheese milk and their virulence after germination: Silage of poor quality is the main source of contamination because it contains large numbers of spores, which survive the passage through the digestive tract of the cow and are accumulated in the manure. The number is further determined by the hygienic standards during milking. If silage is fed to cows, even modern milking methods cannot fully prevent contamination of milk by the spores via traces of manure on the udder surface. The spores survive pasteurization of cheese milk. Their numbers can be reduced to a small fraction by bactofugation of the milk. Rigorous hygienic standards during milk collection are, however, of paramount importance for those types of cheese in which a very low number of spores in the milk (some 5 to 10 per liter) can cause butyric acid blowing defects (see point 4).
2. *Lactic acid content of the cheese:* A high level of undissociated lactic acid has a growth-inhibiting effect; at a lower pH, a smaller fraction of the lactic acid is dissociated. It is uncertain as to what extent the pH, as such, is of importance. At a very low pH (e.g., 4.6), butyric acid fermentation has not been observed.
3. *The NaCl content of the cheese moisture:* Its growth-inhibiting effect closely depends on the lactic acid content. The higher the pH, the higher the salt content must be to avoid growth of *C. tyrobutyricum.* The rate at which growth-inhibiting salt levels are achieved in the cheese is also crucial. For example, butyric acid fermentation is almost never a problem in Cheddar cheese (initial pH usually 5 to 5.2) where salt is mixed with the curd. Much greater difficulties are encountered with cheese in which the salt becomes slowly distributed throughout the loaf, i.e., large cheeses that are brined.

4. *Addition of nitrate:* Since about 1830, nitrate has been added to cheese milk to control butyric acid fermentation in cheese, especially in those varieties that have a favorable initial pH for the fermentation and are subject to a slow salt absorption process as occurs in Gouda-type cheeses. Nitrate as such is ineffective; the inhibition mechanism requires the presence of the enzyme xanthine oxidase (EC 1.1.3.22), which reduces nitrate to nitrite. Nitrite, or one of its degradation products, prevents the germination of the bacterial spores, though only for a limited period. After that, the inhibitory action needs to be taken over by a sufficiently high salt-in-water content.

Traditionally, nitrate was added to the milk. Presently, it is often added to the mixture of curd and whey after most of the whey has been removed. Other conditions being equal, the required amount of nitrate depends on the number of spores of butyric acid bacteria in the milk. The critical concentration of spores is defined as the minimum number of spores per milliliter of milk (containing a certain amount of nitrate), capable of causing butyric acid blowing in the cheese made of that milk. For instance, the critical concentrations for the production of a 12-kg Gouda cheese, made from milk to which 15 or 2.5 g nitrate is added per 100 kg, are 20 and 0.25 to 1 spores per ml of milk, respectively. In Emmentaler and related cheeses, which have a fairly high initial pH and in which salt penetrates only very slowly, the critical spore concentration is virtually zero, even if some nitrate is to be added.

Addition of more nitrate causes a higher nitrate level in the ripening cheese. During ripening, the concentration keeps decreasing, falling more rapidly if the initial concentration is higher, but a certain residual amount is left. A higher level of nitrate only temporarily raises the nitrite concentration. The latter, however, remains low; in Gouda cheese the maximum amount of nitrite is already reached when only a small percentage of the amount of nitrate is lost (Figure 26.1). The causes of the losses of nitrate and nitrite are not precisely known.

Butyric acid blowing in cheese is thus determined by interaction of the various factors mentioned. Once the fermentation gets started, it continues at an ever-increasing rate because it causes the pH to rise, which, in turn, favors the conditions for bacterial growth.

A high number of coliforms may promote butyric acid fermentation because they rapidly consume the nitrate. Growth of mesophilic nitrate-reducing lactobacilli may have the same effect. Cheese made from milk heated to such an extent that xanthine oxidase is inactivated (see Figure 24.1) is very susceptible to butyric acid fermentation.

Occasionally, traces of nitrosamines are found in cheese (e.g., 0.2 μg per kg), but the levels are neither related to the amount of nitrate in the cheese nor to its degradation. Moreover, reducing the amount of nitrate added to the milk does not significantly diminish the human intake of nitrate. For example, a daily consumption of 30 g of nitrate-containing cheese contributes to less than 1% of the daily total intake of nitrate and nitrite. Nevertheless, in some countries the amounts of nitrate allowed in cheese are sharply restricted.

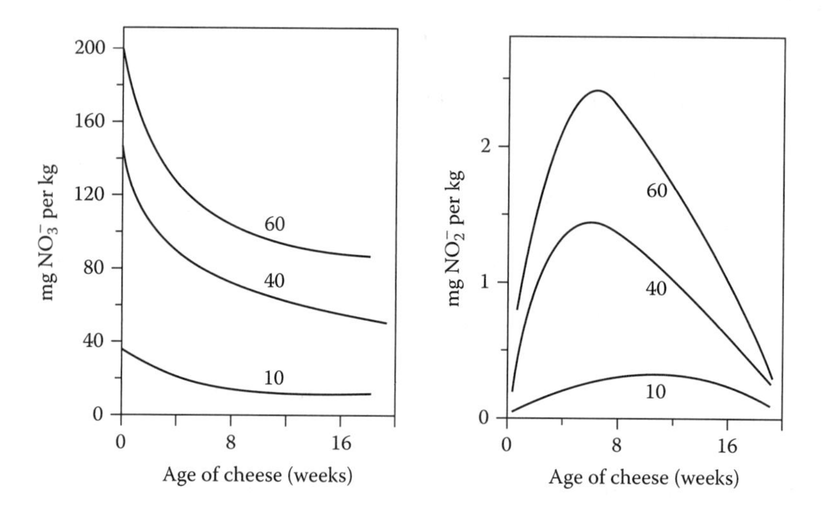

FIGURE 26.1 Contents of nitrate and nitrite of Gouda cheese during storage at 13°C. Parameter is the amount of $NaNO_3$ added to the cheese milk (g/100 l). Approximate results. (Adapted from Goodhead et al., *Neth. Milk Dairy J.*, **30**, 207, 1976.)

Consequently, alternative methods to prevent butyric acid fermentation have been developed, especially the following:

1. *Bactofugation* of the cheese milk: A certain amount of nitrate is still necessary, but it can be far smaller, e.g., 2.5 g rather than 15 g nitrate per 100 kg of milk.
2. Addition of *lysozyme*: Usually, 2.5 g of lysozyme is added per 100 l of cheese milk, which corresponds to 500 enzyme units per ml of milk (250 to 300 mg per kg of cheese); it is harmless to humans. Due to its association with casein micelles, lysozyme is almost entirely retained in the cheese. The action of the enzyme is based on the rupture of the peptidoglycan bonds in the bacterial cell wall, causing lysis. In the case of *C. tyrobutyricum*, lysis starts in the germinating spore. Lysozyme is not active when NaCl content is high (e.g., 5% NaCl in water).

   Gram-positive bacteria tend to be much more susceptible to lysozyme than Gram-negative ones. At moderate concentrations of the enzyme, lactic acid bacteria are, however, little affected or not affected at all. Propionic acid bacteria are also not very sensitive to lysozyme and so it can be used in Emmentaler cheese manufacture.

   The usual dosage of lysozyme to prevent butyric acid fermentation, does not suffice for every cheese because some strains of butyric acid bacteria (presumably *C. butyricum*) are not very sensitive to lysozyme. Simultaneous addition of some nitrate may therefore be necessary.

3. *Formaldehyde:* In some cases, formaldehyde is added, which is a powerful inhibitor of butyric acid fermentation. However, its use is illegal in most countries.
4. *Bacteriocins:* Some strains of the starter *Lactococcus lactis* produce a bacteriocin, e.g., nisin (see Subsection 13.1.4), which is active in controlling the growth of clostridia. However, many other starter bacteria are also sensitive to it. Therefore, a nisin-producing starter strain can only be used in combination with other starter bacteria that are nisin-insensitive.

Other factors affecting butyric acid fermentation are *ripening time* (if it is short enough, the defect will have no time to develop) and *ripening conditions*. A lower temperature causes a decrease in the growth rate of *C. tyrobutyricum* (minimum growth temperature is 7°C). However, in practice it is not possible to select a very low ripening temperature because of its influence on the maturation process. The high ripening temperature of Emmentaler cheese, together with a high pH and a low salt content, make this type of cheese very susceptible to butyric acid fermentation.

26.3 *LACTOBACILLI*

Growth of mesophilic lactobacilli may induce texture and flavor defects. The organisms involved are called nonstarter lactic acid bacteria and they occur in many cheese types as an adventitious flora, especially in raw milk cheeses. It concerns mainly facultatively heterofermentative lactobacilli, especially *Lactobacillus casei*, *Lb. paracasei* and *Lb. plantarum*, but obligately heterofermentative species, e.g., *Lb. brevis* and *Lb. fermentum*, are also found occasionally. Sometimes representatives of the closely related genus *Pediococcus* occur in cheese. These nonstarter lactic acid bacteria are able to grow under the hostile conditions of the cheese environment. These conditions include a lack of fermentable carbohydrates and of oxygen, a pH between 4.9 and 5.3, a low water content, low temperature, and 4.5 to 5.5% salt in water. Likely energy sources for these lactobacilli are the glucides in the glycoproteins of the milk fat globule membrane and the amino acid arginine. Their numbers may reach as high as 10^6 to 10^8 colony-forming units per gram of cheese after 4 to 6 weeks of ripening.

The formation of cracks in hard and semihard cheeses is often attributed to the presence of heterofermentative lactobacilli. These cracks are mostly caused by excessive formation of CO_2 from carbohydrate, citrate, or, most likely, from amino acids. During the latter decarboxylation reaction, amines may be formed (Section 25.8). Besides gas formation, the occurrence of flavor defects has also been related to amino acid metabolism products formed by lactobacilli. These flavor defects have been characterized as sulfurous, phenolic, putrid, and mealy. The obligately heterofermentative *Lb. brevis*, *Lb. buchneri*, and *Lb. fermentum* are recognized as the most common species causing defects in cheese. Carefully selected species from the facultatively heterofermentative lactobacilli, isolated from Cheddar cheese, are used as a secondary starter to make a positive contribution to flavor formation and to accelerate the ripening. The importance of these nonstarter lactic acid bacteria in the development of cheese flavor is not fully clear.

In cheeses, such as Cheddar, where the curd is not washed, the total lactate (usually the L(+) isomer) is at a concentration close to its saturation point. Thus, crystals of calcium lactate pentahydrate can be formed, especially at low temperature. This crystal formation is strongly stimulated in cheese when the L(+) isomer of lactate becomes a racemic mixture of L(+) and D(−) isomers. In cheese, this undesirable racemization of lactate is carried out by a number of lactobacilli, such as *Lb. fermentum* and *Lb. brevis*.

Raw milk and the factory environment are the major sources of lactobacilli in cheese. Although the bacteria are killed by low pasteurization, they may contaminate the continuously operating curd drainage equipment and after multiplication enter the cheese in sufficient numbers to proliferate during the ripening. Another important source of lactobacilli is the brine. Some of them, especially the obligately heterofermentative lactobacilli, are very salt resistant and can survive even in the presence of over 15% NaCl. Lactobacilli usually do not grow in brine, but they can grow in deposits on the walls of basins and cages just above the brine level. Growth conditions for the lactobacilli are more favorable in these deposits due to an increased pH as a result of the growth of salt-tolerant yeasts, a lower NaCl content due to absorption of moisture from the air, and a slightly higher temperature than that of the brine. Measures to keep the count of lactobacilli low include satisfactory hygienic measures in the brining room (e.g., removal of deposits), maintaining a NaCl content of the brine of at least 16%, and a pH < 4.5. Obviously, contamination of the cheese milk and the equipment by these bacteria should be avoided.

26.4 HEAT-RESISTANT *STREPTOCOCCI*

Although *S. thermophilus* is used as starter in the case of cheeses made with thermophilic cultures, some strains can cause defects in cheeses made with mesophilic starters. Unlike mesophilic lactococci, they grow at 45°C and survive thermalization and low pasteurization of milk. During these heat treatments they may become attached to the wall of the cooling section in the heat exchanger and multiply very rapidly (minimum generation time about 15 min). Continuous use of heat exchangers for a long time without intervening cleaning may cause heavy contamination of the cheese milk, with counts reaching as high as 10^6 bacteria per ml. Concentration in the curd and growth during the early stages of cheese making may increase the count to over 10^8 bacteria per gram of cheese. Unclean and yeasty flavors may develop.

26.5 PROPIONIC ACID BACTERIA

Growth of propionic acid bacteria is desirable in certain varieties of cheese (e.g., Emmentaler; see Section 27.5) to achieve a satisfactory quality; the bacteria are added to the cheese milk. In other cheeses, excessive growth of these bacteria causes defects (Table 26.1). The bacteria are killed by low pasteurization of the milk. Obviously, they are predominantly of interest in the manufacture of raw milk cheese.

The most important species is *Propionibacterium freudenreichii* ssp. *shermanii*. Most of the species ferment lactose, and all ferment lactic acid. In cheese the lactose fermentation goes unnoticed because the bacteria multiply slowly and cannot compete with the starter bacteria. Lactic acid is converted to propionic acid, acetic acid, CO_2, and water, according to the general formula:

$$3\ CH_3\text{–}CHOH\text{–}CO_2H \rightarrow 2\ CH_3\text{–}CH_2\text{–}CO_2H + CH_3\text{–}CO_2H + CO_2 + H_2O$$

The pH of the cheese slightly increases due to the fermentation.

A distinct propionic acid fermentation results in excessive gas formation and the development of a sweetish taste. In Gouda-type cheese such fermentation is considered a defect. Since the bacteria grow very slowly in cheese, any serious defects appear only after a prolonged ripening time and are a form of late blowing.

Conditions determining the growth of propionic acid bacteria in cheese are as follows:

1. *Acidity:* The organisms grow little or not at all at pH $\leq$ 5.0. The growth rate increases with increasing pH.
2. *NaCl content* in the cheese moisture: Increasing the concentration retards the growth of the bacteria; the NaCl content in most cheeses (<5% NaCl in the water) is too small to be effective, but inhibitory concentrations can occur in the rind of cheese shortly after brining.
3. *Storage temperature:* Increasing the temperature favors growth. (In the manufacture of Emmentaler cheese, advantage is taken of this effect to enhance the propionic acid fermentation; see Subsection 27.5.1).
4. *Presence of nitrate:* The fermentation is slowed down, probably due to the formation of nitrite.

When conditions allow growth of propionic acid bacteria in cheese, butyric acid bacteria, if present, may also be expected to flourish.

26.6 ORGANISMS ON THE RIND

Abundant growth of yeasts and coryneform bacteria on the cheese surface may cause a slimy rind and a parti-colored or pinkish appearance. Growth of these organisms is enhanced by inadequate drying of the rind after brining, and also by a significant lactose content in the rind due to insufficient souring of the cheese, by salting of cheese in weak brine of a high pH, and due to the use of poorly cleaned shelves in the curing room. Growth of molds causes discoloration and a musty flavor; under extreme conditions it may produce a health hazard because of mycotoxin formation. It particularly involves *Aspergillus versicolor*, which, under some conditions, can produce sterigmatocystine (see Section 25.8).

To prevent the growth of the organisms involved, special attention must be paid to the treatment of the cheese rind and to the hygiene and air conditioning in the curing room (see Section 24.7).

26.7 OTHER ASPECTS

Several microorganisms can cause *flavor defects*, especially in raw milk cheese. Among them are yeasts (yeasty, fruity flavor), *Lactococcus lactis* var. *maltigenes* (burnt flavor) and *Enterococcus malodoratus* (H_2S-like, gassy, unclean flavors); if present in increased numbers these enterococci can also cause texture defects due to the production of CO_2 from amino acids. Many organisms can cause bitterness in cheese.

Increased levels of psychrotrophs or of their thermostable lipases in the cheese milk may cause the cheese to turn rancid.

If gas formation is excessive, the smell and taste of the cheese may indicate the type of fermentation involved. To establish the type of fermentation in cheese with less serious defects, determination of the redox potential (E_h) is essential. In cheese of pH ~5.2, in which no gas is formed besides some CO_2, the E_h ranges from −140 to −150 mV, as measured with a normal hydrogen electrode. In cheese with H_2 formation and of the same pH, the E_h drops to −250 to −300 mV. If a more detailed classification is desired, microbiological assays can be performed (including growth in selective culture media, microscopic examination of the bacteria involved) as well as a determination of the content of butyric and propionic acid in the cheese.

Suggested Literature

An overview of health hazards caused by bacteria: P.F. Fox, T.P. Guinee, T.M. Cogan, and P.L.H. McSweeney, *Fundamentals of Cheese Science*, Aspen Publishers, Gaithersburg, 2000; P.F. Fox, P.L.H. McSweeney, T.M. Cogan and T.P. Guinee, Eds., *Cheese: Chemistry, Physics and Microbiology*, Vol. 1, *General Aspects*, 3rd ed., Elsevier Academic Press, Amsterdam, 2004, especially the chapters by C.W. Donnelly (Growth and survival of microbial pathogens in cheese) and by N.M. O'Brien, T.P. O'Connor, J.O'Callaghan and A.D.W. Dobson (Toxins in cheese).

Micobial defects, specific for particular cheeses, are discussed in the references given in Chapter 27.

27 Cheese Varieties

Cheese can vary greatly in composition and is manufactured and cured in widely varying manners. This results in a bewildering variety of types. In this chapter the variability as such is discussed first: What can be varied and what are the consequences? Subsequently, several cheese types are discussed in more detail. This illustrates the interaction between the various process steps and the need to integrate the steps into an efficient process, leading to a good-quality cheese of the type desired. Of course, it was necessary to select some types of cheese from among the numerous types produced. The main consideration in selecting a particular type was that it would allow the most important aspects of cheese making and cheese properties to be treated. Moreover, the economic importance of some cheese varieties has been taken into account.

27.1 OVERVIEW

The word 'cheese' is commonly used as a collective term for widely variable products such as fully or barely ripened cheese made with rennet, acid curd cheese, fresh cheese, and even processed cheese. Most of these fit the definition formulated by the FAO/WHO*:

> "Cheese is the fresh or matured solid or semisolid product obtained by coagulating milk, skimmed milk, partly skimmed milk, cream, whey cream, or buttermilk, or any combination of these materials, through the action of rennet or other suitable coagulating agents, and by partially draining the whey resulting from such coagulation."

The drainage of whey is an essential aspect of the definition. Concentrated products, obtained by removal of water only, are considered milk products. Also the Norwegian 'Mysost' (whey cheese) is not a cheese according to this definition. It is made by evaporating whey or a mixture of whey and milk or cream until a brown mass forms that coagulates on cooling. The main constituent is finely crystallized lactose.

Furthermore, cheese can be modified, e.g., to obtain processed cheese. In some countries, these and other preserves of cheese are not allowed to be referred to as cheese *per se*. They are briefly reviewed in Section 27.7.

Over the ages numerous varieties of cheese have evolved. The development of a type of cheese closely depended on the local conditions, including the amount and type of milk available, the climate, the presence of suitable caves for storage,

* FAO/WHO Standard No. A-6 (1978).

the need for a long shelf life (compare, for instance, cheese production in remote mountain meadows with that close to densely populated regions), and other economic and geographic factors. Furthermore, it was a matter of accumulating technological experience and of finding, by trial and error, the ideal manufacturing conditions that led to a product being tasty, easy to handle, and durable. Along with this, the general level of technological and hygienic know-how were of great importance. Various process steps that are now very common are relatively new, say 100 years old: the use of starter, use of industrially made rennet, washing of the curd, and pasteurization of the cheese milk, etc. Of more recent origin are the use of inocula for the surface flora, application of latex to the cheese rind, use of enzyme preparations to accelerate ripening, and so forth.

Over the ages, several types of cheese changed greatly in character. They usually retained their original names, e.g., because the name was that of a region or of its principal market town. Furthermore, when a particular variety of cheese was well reputed, other cheese manufacturers saw an advantage in borrowing its name. Often, technological know-how was 'exported.' At present the names of some varieties have been protected internationally, but most are not.

27.1.1 Variations in Manufacture

Following are the main variations applied in the manufacturing process:

1. *The kind of milk:* Milk of cows, goats, sheep (ewes), and buffaloes differ in composition. The fatty acid composition of the milk fat of the various animal species varies, which affects the flavor of the cheese. The short-chain fatty acids C_6–C_{10} cause a sharp flavor in goats' milk cheese. A similar flavor may form in ewes' milk cheese but only if growth of molds, etc., occurs (as in Roquefort cheese) or if some goats' milk has been added, because ewes' milk has very little lipase activity. Cheese made from sheep, goat, or buffalo milk has a fairly white color because these milks contain hardly any carotene (they do contain vitamin A). Other differences are given in Table 27.1. They affect cheese yield, fat content, and pH. Probably, the low level of α_{s1}-casein in goats' milk (varying from 0 to 20% of the total casein) is responsible for the short texture of goats' milk cheese. Curd of goats' milk exhibits stronger syneresis than that of cows' milk. Buffaloes' milk has a high ratio of calcium phosphate to casein, which gives rise to cheese of a fairly hard and elastic consistency.
2. *Standardization of the milk:* An important characteristic of cheese is its percentage of fat in the dry matter. This content is closely related to the percentages in milk. Changing the ratio between fat and fat-free dry matter in cheese milk affects clotting and syneresis, and hence water content and pH of the cheese. Flavor and consistency are linked to the fat content in the dry matter. Fresh cheese is often made of skim milk. In ripened cheese, fat content in the dry matter ranges from 20 to 60%.

TABLE 27.1
Compositional Data of Milk of Certain Animal Species of Importance for the Cheese Made of That Milk

Animal Species	Casein (%)	Fat/Casein	Lactose[a] (%)	Casein Composition[b]		
				α_s :	β :	κ
Cow	2.6	1.5	4.9	49	38	13
Goat	3.0	1.5	4.7	25	55	20
Sheep	4.5	1.7	5.2	54	34	12
Buffalo	3.3	2.2	5.4	47	40	13

Note: Approximate averages.

[a] In the fat-free and casein-free milk.
[b] Quite variable, especially in goats' milk.

Full-cream cheese, i.e., cheese made from whole milk, usually contains from 46 to 52% fat in the dry matter, primarily depending on the milk composition.

3. *Heat treatment of the cheese milk:* The heating and its intensity affect the type and extent of the bacterial flora of the milk, the growth rate, and the proteolytic activity of the starter bacteria, the activity of lipase, the rennetability of the milk, the tendency to show syneresis, and the retention of serum proteins in the cheese (see also Figure 24.1). In the manufacture of some varieties of fresh cheese the milk is high-heated during coagulation with acid.
4. *Preacidification of the cheese milk:* Preacidification or 'ripening' the milk over some 1 to 3 h (sometimes much longer) prior to rennet addition affects clotting and syneresis. In addition, preacidification has a direct effect on the pH of the cheese because a greater amount of colloidal phosphate dissolves during curd manufacture.
5. *Starter composition and inoculum percentage* (0 to 5%): Processing conditions, especially the desired rate of acid production and the scalding temperature, as well as the processes desired during maturation, determine what species and strains of lactic acid bacteria are selected. Considerable variation exists among species and strains, often resulting in cheeses of different type with respect to flavor, texture, gas production (and hence, hole formation), etc.
6. *Addition of a secondary microflora:* In the manufacture of many types of cheese, nonstarter microorganisms are added to the milk (adjunct starters) or to the cheese rind. All cheeses with large 'eyes' (Emmentaler, Gruyère, Jarlsberg) contain propionic acid bacteria in addition to lactic acid bacteria. For a desirable surface flora to develop, the cheeses or the milk are exposed to bacteria (coryneform bacteria) or molds (*Penicillium*).

7. *Manner of coagulation and type of coagulant:* Coagulation of the milk can occur by means of acid, rennet, or a combination of renneting enzymes and acid. The acid can originate from *in situ* production of lactic acid or may be added to the milk. Besides calf rennet, other renneting enzymes can be used. Amount, activity, and character of the renneting enzymes retained in the cheese greatly affect cheese composition and ripening.
8. *Curd making:* The rate and extent of syneresis depend on the size of the curd grains and on the intensity and duration (0 to 5 h) of the stirring of the curd–whey mixture. The treatment of the curd is among the important process variables that regulate the moisture content of the cheese.
9. *Scalding (cooking) temperature* (see Figure 27.1): This affects the rate of syneresis; the amount of rennet retained in the curd; inactivation of rennet; and inactivation of mesophilic starter bacteria.
10. *Washing of the curd:* Addition of water (up to 200%) decreases the concentration of lactose in the moisture of the cheese and hence, the ratio between lactic acid and buffering compounds. This ratio determines the pH of the cheese.

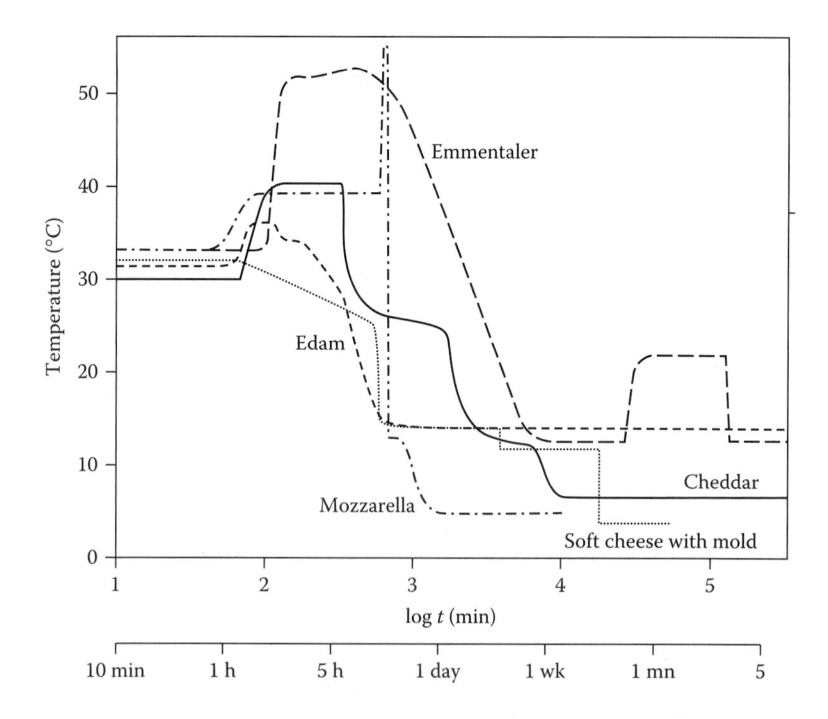

FIGURE 27.1 Course of the temperature during manufacture and ripening of some types of cheese; *t* = time after renneting. Approximate examples.

11. *Size and shape of the cheese:* Shape and weight are among the factors defining the variety of the cheese. Moreover, these factors are linked with the particular type of cheese because the specific surface area (which ranges from 1 to 20 dm^2/kg) greatly affects several changes during manufacture and curing. It affects the cooling rate, the rate of salt absorption, the effect of any surface flora on the ripening, and the vaporization of water during storage.
12. *Pressing:* Pressure is mainly applied to form a coherent loaf, which facilitates handling, and a closed rind, which extends shelf life.
13. *Resting:* The length of time between pressing and salting (ranging from 1 to 24 h), or that between pitching and salting for Cheddar types (see variable 14), affects the amounts of lactose left and of lactic acid formed at the moment that the starter bacteria are strongly slowed down by the salt, and thereby the growth of undesirable bacteria (e.g., coliforms).
14. *Method of salting:* Salting the cheese after pressing and resting fundamentally differs from salting by mixing salt with soured curd before pressing, as is done in the manufacture of Cheddar and Cantal cheese. Rubbing the cheese with salt or brine is the procedure used if a surface flora of yeasts and bacteria should develop.
15. *Salt content:* The salt content can vary widely, from 0.5 to 10% for ripened cheeses. White pickled cheese, a well-known example being Feta, is stored in brine. Keeping quality, flavor, consistency, and ripening of the cheese (especially proteolysis and lipolysis) are markedly affected by the salt content.
16. *Additives:* Spiced cheeses have been made of old. The spices added are chiefly cumin or caraway, usually mixed into the curd after the whey has been drawn off. The range of additives has increased in recent years, e.g., chives, green pepper, nettles, and parsley.
17. *Ripening temperature:* The ripening temperature is generally between 5 and 20°C. The higher it is, the faster all kinds of ripening processes occur. Temperature thereby affects flavor, texture, gas production and keeping quality (occurrence of microbial defects).
18. *Handling during storage:* Handling concerns the treatment of the cheese during curing, especially the treatment of the rind. Treatment variables include: (frequency of) turning the loaves and cleaning the rind; humidity and flow velocity of the air; and specific measures, such as piercing holes in the loaves to allow mold growth.
19. *Ripening time:* Ripening generally ranges from 1 d to 3 yr. Usually, the eating quality of a cheese (flavor and consistency) increases during a given ripening time and decreases thereafter. As a rule, the ripening time is not an independent variable because it depends on composition, size, and ripening conditions of the cheese. For many hard and semihard cheese varieties the ripening time can be varied widely, but not for most soft cheeses. See Table 24.3 for variables 17, 18, and 19.

20. *Covering the cheese rind and packaging:* Covering the surface with a latex will prevent, or at least slow down, growth of microorganisms on the rind. Covering of the rind (especially with paraffin) and packaging of the cheese (wrapping in Saran or other plastic foils) greatly reduce vaporization of water.
21. *Other modern techniques:* Processes can include continuous renneting, ultrafiltration, bactofugation, homogenization, and addition of enzyme preparations. They have not yet resulted in varieties that differ materially from the existing ones. However, they may greatly affect some properties of the cheese.

The sum of all different combinations of variables would lead to an impossibly high number of cheese types but that is not realistic. Many combinations make no sense, a great many achieve a poor result, whereas still others cause no appreciable differences. We will therefore try to introduce some order by considering the following *main variables* in the manufacture of (cows' milk) cheese:

1. *Adjustment of the water content*, i.e., essentially the ratio between water and protein (w/p): The moisture content is affected by many process steps. In effect the water content also roughly determines the size of the cheese (see Figure 27.2), because a large cheese inevitably is relatively dry (Section 24.5). In theory, a small cheese can be made dry, but it would soon become too hard because of water vaporization. The

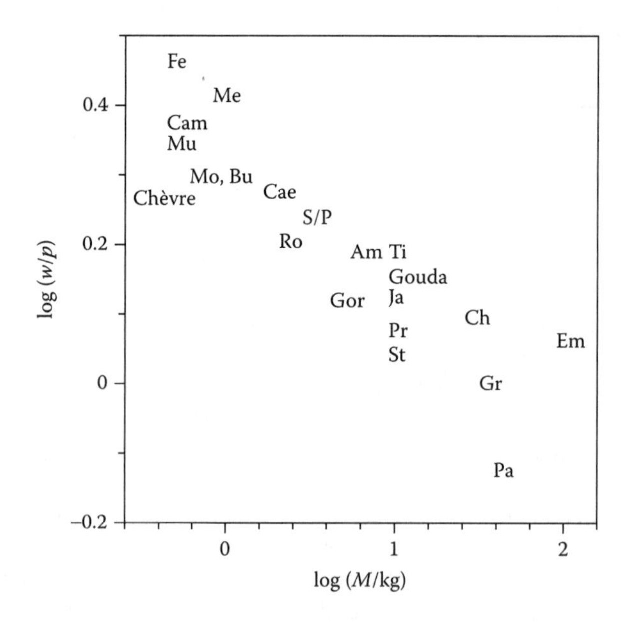

FIGURE 27.2 Approximate relation between the water/protein ratio (w/p) and the mass (M) of most of the cheeses in Figure 27.5. It concerns mature cheeses.

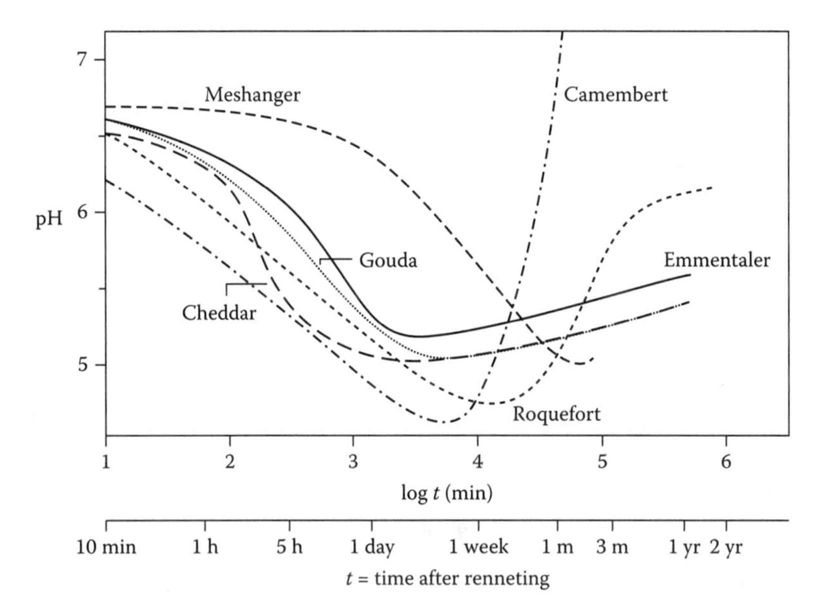

FIGURE 27.3 Course of the average pH during manufacture and ripening of some cheese varieties. Approximate examples.

most important point probably is that the water content determines the shelf life. Fresh cheeses ($w/p \approx 4$ or higher) keep for up to 1 week, soft cheeses ($w/p \approx 2.5$) for about 1 month, semihard cheeses ($w/p \approx 1.5$ to 2) up to several months, and hard cheeses ($w/p < 1.2$) for several years (e.g., Parmesan cheese).

2. *Adjustment of the pH* (see also Figure 27.3): Unless a quite dry cheese is made, the pH would generally become too low. This effect can be counteracted, however, by several methods. The curd can be washed, as is done for Gouda-type cheese. Furthermore, the acid production can be kept slow by applying little or no starter (and having milk of a low bacterial count) and it may be retarded further by a low temperature (Cottage cheese, Butterkäse, and Bel Paese) or by penetrating salt (Meshanger). Alternatively, lactic acid can be consumed by a surface flora (Camembert, Brie, and Munster), by internal molds (Roquefort, Bleu d'Auvergne, and Stilton), or by both (Gorgonzola). If the pH at the pressing stage is low, the cheese will have a low level of calcium (Figure 27.4) and phosphate.
3. The *dominating enzymes* largely originate from the cheese flora. This flora partly depends on factors 1 and 2 and on the manufacturing temperature. It mainly concerns mesophilic or thermophilic lactic acid bacteria. In addition, propionic acid bacteria, white molds, blue molds, or a smear flora of yeasts and bacteria may be present. The surface flora essentially is a complex and changing mixture of organisms. Besides microbial enzymes,

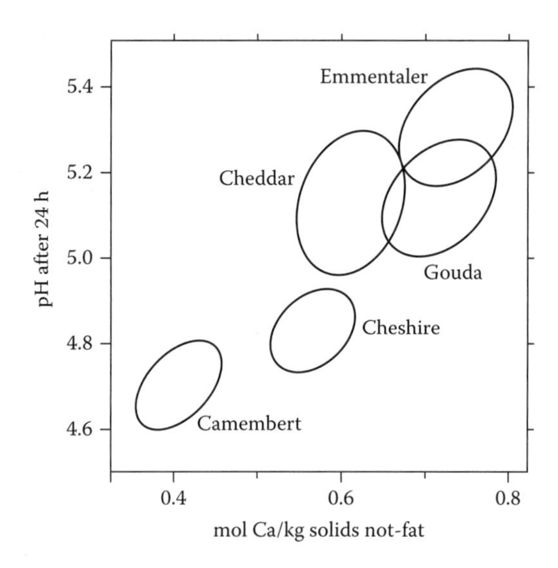

FIGURE 27.4 Approximate relation between the pH after 1 d and the amount of calcium in the dry matter of some cheese varieties.

> the amount and type of the renneting enzyme retained in the cheese is important. The contribution of the various enzymes to proteolysis can vary widely among cheese varieties, as is shown in Table 25.1.

Appendix A.12, gives the approximate gross composition of several cheese varieties.

27.1.2 Types of Cheese

How can the different varieties best be classified? The technologist will be inclined to take differences in the manufacturing process as the characteristics. Of course, perceptible properties of the cheese prevail for the consumer; these include size, shape, outer and inner appearance ('body'), and especially flavor and consistency. The latter two features depend on the composition in its broadest sense, including pH, free amino acids, and H_2S, etc. Some differences in texture, especially the number and distribution of eyes, depend on the means of manufacture and not so much on the composition; these differences are, however, of secondary importance. Furthermore, Gouda cheese made by a cheddar process may be hard to distinguish from a true Gouda, except for the eyes.

All in all, the main variables are the ratio between water and protein and the type of ripening, i.e., fresh or matured, type of starter, secondary organisms, and ripening phase. So we arrive at Figure 27.5, which shows some important varieties as well as some widely differing types. Some features not mentioned are fat content in the dry matter, pH, salt content, and ripening time. As shown in

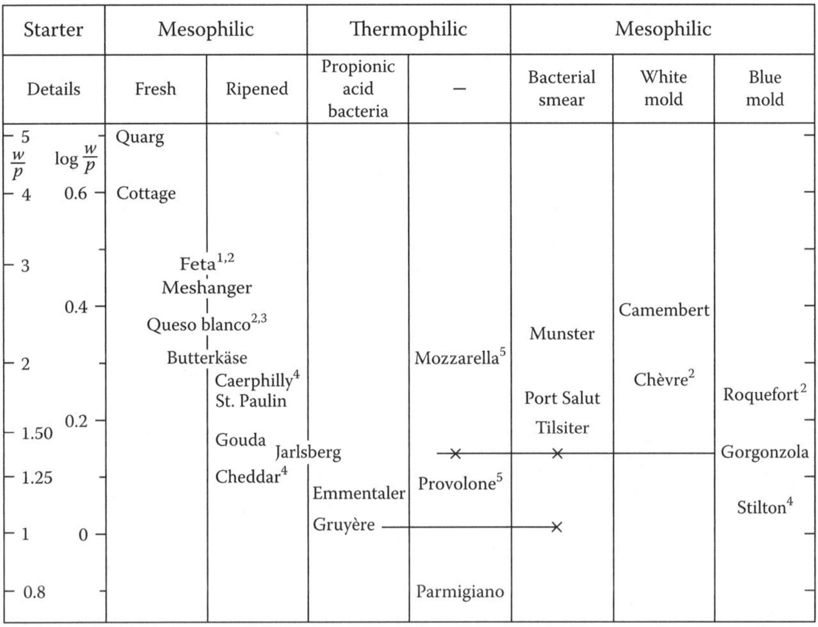

Starter	Mesophilic		Thermophilic		Mesophilic		
Details (w/p; $\log w/p$)	Fresh	Ripened	Propionic acid bacteria	–	Bacterial smear	White mold	Blue mold
5	Quarg						
4; 0.6	Cottage						
3	Feta[1,2]	Feta[1,2]					
0.4	Meshanger	Meshanger				Camembert	
	Queso blanco[2,3]	Queso blanco[2,3]			Munster		
2	Butterkäse	Butterkäse		Mozzarella[5]			
		Caerphilly[4]				Chèvre[2]	
		St. Paulin			Port Salut		Roquefort[2]
1.50; 0.2		Gouda			Tilsiter		
		Jarlsberg	Jarlsberg	×	×		Gorgonzola
1.25		Cheddar[4]	Emmentaler	Provolone[5]			
							Stilton[4]
1; 0			Gruyère	×	×		
0.8				Parmigiano			

FIGURE 27.5 Classification of cheese varieties based on type of maturation and on water content, i.e., ratio of water to protein, w/p. (1) Kept in brine. (2) Milk may differ from cows' milk. (3) Usually by acid coagulation. (4) Salting of the curd prior to pressing. (5) Stretched curd. × means that the parameter involved applies as well.

Figure 27.2, size and water content may be highly correlated, as they are both related to the duration of ripening.

How many cheese varieties exist? The answer is arbitrary because it remains undefined as to how large a difference must be for the cheese to be categorized as a different variety. Many different designations may be in use for what is essentially the same cheese variety, e.g., Gouda, Javor, Fynbo, Norvegia, Mazurski prästost, and Kostroma. On the other hand, the denomination of Gouda may include a 6-week-old, somewhat soft cheese with little flavor, as well as a more than 1-yr-old farmhouse cheese that is brittle and piquant; these two cheeses differ at least as much as do Saint Paulin and Gruyère. All in all, not more than a few dozen substantially different varieties can be distinguished. Most of these are shown in Figure 27.5.

Which cheeses constitute the (economically) most important varieties in the world? Most likely the following:

1. *Gouda-type varieties*, which are defined as those that:
   - Are made of cows' milk
   - Have 40 to 50% fat in the dry matter
   - Use mesophilic starters

- Are brine-salted after pressing
- Have a water content in the fat-free cheese below 63%
- Have no essential surface flora
- Are matured from 2 to 15 months

2. *Cheddar-type cheeses*, which are much like the cheese of the previous group, but they are made in a different way, in that they are salted in the curd stage. They are on average slightly drier and slightly more acidic than the Gouda-type varieties and have a different flavor note.
3. *The group of fresh cheeses*, which have a high water content and are either little matured or not matured at all.

Furthermore, fairly large quantities of the following cheeses are being produced:

Very hard cheeses (Parmesan, Sbrinz, and Pecorino Romano)
Cheese with propionic acid bacteria (Emmentaler and Jarlsberg)
Stretched-curd or pasta-filata cheeses (Provolone, Kashkaval, and Mozzarella)
A semisoft type of cheese with a very mild flavor (Saint Paulin, Monterey, and Amsterdammer)
White pickled cheeses (Feta and Domiati)
Soft cheeses with a white mold (Brie and Camembert, often poorly resembling the original cheese that had a more distinct flavor)
Blue-veined cheeses (Bleu d'Auvergne, Gorgonzola, Roquefort, and Stilton).

It may finally be noticed that cheese is often used as a food ingredient. This has led to the production of a wide variety of modified cheese products that have specific functional properties for use in the kitchen or in food manufacture.

27.2 FRESH CHEESE

Fresh or unripened cheeses are curd-like products that can be consumed immediately after manufacture. The products undergo little, if any, ripening, except that lactose is fermented. Generally, fresh cheese has a limited shelf life, say, 2 weeks under refrigeration.

Numerous varieties of fresh cheese exist and in many countries the quantities consumed are considerable. The cheeses may vary as to the kind of milk, fat content, and means of manufacture. Often, the milk is coagulated by acid or by a combination of heat and acid. Sometimes a little rennet is used. Generally, the fresh cheese has a spreadable (i.e., soft and short) or even granular texture. A kind of archetype is bag cheese. In its manufacture sour milk or buttermilk is boiled, which (further) coagulates the casein. The mixture is suspended in a cheesecloth bag to allow drainage of whey. In the manufacture of the Latin American Queso Blanco, the milk is provided with acid (e.g., vinegar, lemon juice, or HCl) and is heated with such intensity that most of the serum proteins are incorporated into the cheese. The

TABLE 27.2
Approximate Composition of Fresh Cheese and Some Comparable Products

Product	Water (g)	Protein (g)	Fat (g)	Ca (mg)	P (mg)	Energy (kJ)
Concentrated yogurt	82	5	5	175	140	380
Ymer	85	6.5	3	150	160	300
Quarg, low-fat	83	12	0.2	120	180	270
Quarg, high-fat	74	10	12	120	180	660
Cottage cheese, creamed	78	12	4	100	180	430
'White cheese'	60	16	19	450	300	1030

Note: Quantities are per 100 g of product.

Italian Ricotta is made by heating a mixture of sour milk and sour whey. A different type of fresh cheese is renneted 'white cheese.' Its manufacture is often similar to that of a small Gouda, except that the water content is high. It is lightly pressed and lightly brine-salted. The loaf is coherent.

It may be noted that there is a gradual transition from sour milk to ymer (see Subsection 22.2.1.3), to fresh acid cheese, to the white cheese just mentioned. Some typical examples of composition are given in Table 27.2.

In addition to traditional or domestic methods for manufacture, large-scale industrial processes have been developed that are often quite different. Preserves of fresh cheese are also made. Two often consumed products will be briefly considered.

27.2.1 Quarg

Quarg (also spelled quark) definitely has originated from bag cheese, and sometimes it is still made in that way. Speisequark, Neufchâtel, Tvorog, and Baker's cheese are closely related or identical products. Petit Suisse and Doppelrahmfrischkäse have an increased fat content. Due to its neutral flavor and its consistency (smooth, can be blended), quarg is a suitable ingredient in several dishes. At present, low-fat quarg is usually made. Figure 27.6 gives an outline of a manufacturing process.

An aromatic starter is commonly used for the sake of flavor. The acid production may be done at a higher temperature to speed it up. Preculturing of the milk is not always done. The breaking of the coagulum by stirring occurs when the pH has dropped to a value between 4.6 and 4.7. At that pH and the prevailing low temperature the curd exhibits little syneresis. Often a little rennet is added, causing the curd to become firmer and allowing the whey to be separated more readily. For the same reasons the milk is pasteurized at low temperature. Too high a level of rennet

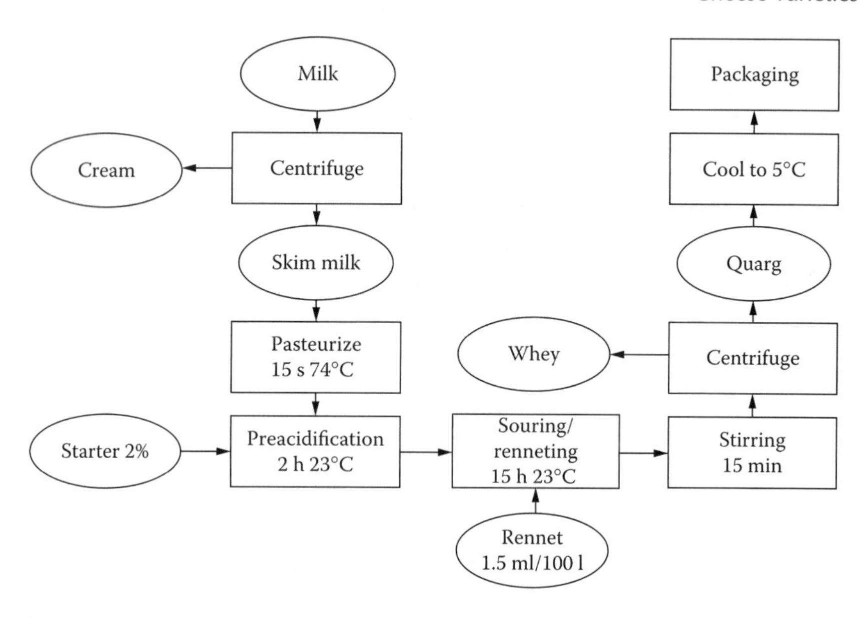

FIGURE 27.6 Example of the manufacture of low-fat quarg.

causes premature syneresis, and hence an inhomogeneous product; it can also cause the development of a bitter flavor during storage of the quarg.

Large-scale production units of quarg became feasible after introduction of the quarg separator. In this centrifuge, the clotted protein mass is separated from the whey by centrifugal force and discharged through small openings in the wall of the separator bowl. In this way a smooth product is obtained. The water content of the quarg can be satisfactorily adjusted by varying the flow rate. If the whey is separated in a different manner (usually by filtration), the quarg obtained should be smoothened (homogenized). Furthermore, centrifugal separation fails if quarg is made of whole milk because the difference in density is too small. Moreover, much of the fat is lost in the whey. Centrifugal force can be applied, however, by operating as follows: Diluted cream is homogenized at low temperature and high pressure. This causes the fat globules to be largely covered with micellar casein and to form homogenization clusters. After acidification, almost all of the fat globules participate in the protein network. Because the density of the protein and fat mixture involved is less than that of the whey, the product can be separated in a milk separator-type centrifuge. Then the desirable fat content of the final product can be adjusted by blending high-fat and low-fat quarg. However, it is much simpler to blend low-fat quarg with cream. This better masks the acid or astringent taste of low-fat quarg that many people consider unpleasant.

Alternatively, quarg is made by ultrafiltration of milk or skim milk (concentration factor 3 to 3.5). The retentate obtained is soured, cooled, and the coagulum broken by stirring. This results in a higher yield because serum proteins and additional calcium phosphate are enclosed. However, the product has a different, less firm consistency. Furthermore, some slight syneresis tends to occur during storage. Another method to increase the yield is to start from high-heated milk; sufficient whey can be separated by using the quarg separator. The product involved, again, is thin and somewhat sticky. Thickening agents may be added to arrive at a satisfactory consistency and to avoid wheying off during storage.

The shelf life of quarg is limited by proteolysis (which makes it bitter and 'cheesy') and by excessive acid production (causing a sharp, acidic taste). Contamination by yeasts and molds may reduce the keeping quality. Naturally, flavor deterioration and whey expulsion occur earlier when the product is kept at a higher temperature.

27.2.2 Cottage Cheese

Cottage cheese is an unripened, particulate, and slightly acidic cheese made of skim milk. A little salt and fresh or cultured cream are generally added to the cheese. Distinctive features of the product are its granular form and, in spite of the low fat content, its creamy flavor.

The traditional manufacture of cottage cheese (Figure 27.7) differs from that of quarg in some of the process steps:

1. The clotted milk is cut into cubes of a desirable size and is not broken by stirring.
2. A non-gas-forming starter is applied. Otherwise the curd would start to float and it would be hard to drain off the whey.
3. Cooking (scalding) enhances syneresis. This is necessary because the pH is quite low and because the stirring should not be too vigorous. Cooking also causes the curd to become firmer. To obtain a firm curd, the milk is low-pasteurized and a little rennet is added.
4. The curd is thoroughly washed to remove most of the lactose and lactic acid. At the same time, the mixture is cooled.
5. The low-fat curd is blended with cream.

Usually, the starter is added to the milk together with the rennet. Sometimes the rennet is added 1 to $1^1/_2$ h after inoculation with starter. As in quarg manufacture, the process conditions can vary widely, e.g., from acid production at 32°C for 4 to 5 h with 5 to 6% starter added (short-set method), to acid production at 22°C for 12 to 16 h with 0.25 to 1% starter (long-set method).

The moment of cutting considerably affects the properties of the product. The firmness of the coagulum closely depends on the pH and, to a lesser extent, on the rennet action. Usually, acid production is allowed until the pH reaches 4.6 to 4.8. Curd cut at a lower pH remains weaker, whereas curd of a pH higher than 4.9 becomes too firm and tough.

The milk may also be acidified directly by adding inorganic or organic acids, i.e., the direct-set method. The acid is added at low temperature, say, 7°C, while

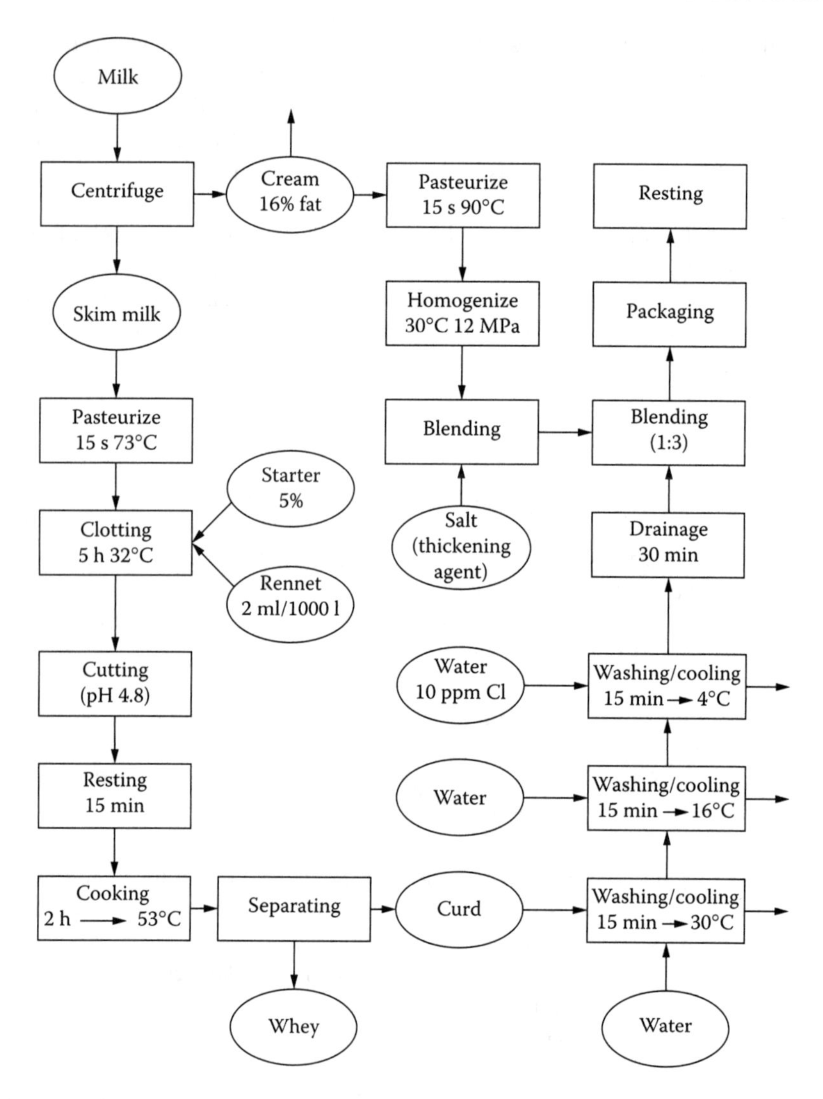

FIGURE 27.7 Example of the traditional short-set method for the manufacture of cottage cheese.

the milk is vigorously stirred until the pH is 4.6. Then, while the milk is at rest, its temperature is raised by ohmic heating (passing an electric current through the milk), which causes a uniform development of heat in the milk. Setting then takes about 12 min. Thereafter, the manufacturing process follows the usual procedures. The flavor of the product is often considered to be less satisfactory, presumably due to the lack of starter organisms that produce diacetyl.

Unlike the situation in pressed (mature) cheeses, the curd size and curd size distribution directly affect appearance and flavor perception of cottage cheese.

The freshly cut curd is soft and fragile. After cutting, it is left for some 10 to 15 min to expel a little whey. Thereby, the curd granules increase in firmness and suffer less damage during subsequent stirring. Consistency and firmness of cottage cheese greatly depend on the rate of heating during cooking. The slower it is, the more even the syneresis, whereas fast heating up results in curd grains with a dry and firm rind. To avoid fusing of the curd grains, the curd and whey mixture should be gently and continuously stirred. The cooking temperature depends on the desired water content. During the first stage of cooking (say, until 43°C) some additional acid is produced in curd and whey. If the curd is cooked to a relatively high temperature of 55 to 57°C and maintained at this temperature for some time, many bacteria are killed and most of the rennet is inactivated.

Usually, the curd is washed three times, the quantity of water added being equal to that of the whey drawn off. If washing is less intense, the acid flavor is insufficiently removed, which may be considered unacceptable, although preferences vary. Often, some 5 to 20 ppm of activated chlorine is added to the last wash water to prevent growth of undesirable microorganisms.

Blending the low-fat curd with cream considerably improves the flavor of cottage cheese. The fat content of the cream may range from 10 to 20%. The cream is often homogenized, which increases its viscosity (Section 9.7). To obtain a similar flavor, a much smaller amount of cream is needed if it is added to the curd rather than to the cheese milk. Moreover, the loss of fat into the whey is higher with the latter method. The plasma of the cream is believed to be 'absorbed' by the curd, so that, specifically, the fat globules remain on the outside of the curd granules to 'lubricate' them. The absorption appears to take approximately 30 to 40 min. Properties of the curd granules, including size and water content, as well as the properties of the cream (fat content, viscosity, pH, and homogenization), determine the quality of the product. The product should not whey off or separate 'free cream.' The low-fat curd can be blended with sweet or cultured cream. Cream with a high pH leads to more readily to separation of liquid than cream with a lower pH. Well-soured cream (in which the casein is thus coagulated) does not yield free cream. Sometimes a thickening agent is added to the cream to prevent this defect.

Attempts to preconcentrate the skim milk by ultrafiltration have met with little success; the curd acquires an undesirable consistency and fails to properly absorb the cream plasma. Continuous curd manufacture can be applied: The freshly cut curd is put onto a perforated conveyor belt, on which it first drains and is subsequently washed. Establishing the correct water content is accomplished by gentle pressing under rollers.

In spite of all precautions, cottage cheese does not have a long shelf life. Its composition permits growth of microorganisms. Very good hygiene during manufacture and packaging are essential. Sorbic acid may be added to the creaming mix to improve the shelf life. Alternatively, the product may be packaged under CO_2, whereby growth of Gram-negative psychrotrophs is especially inhibited.

27.3 GOUDA-TYPE CHEESES

Gouda-type cheeses are ripened, fairly firm, sliceable cheeses made from fresh cows' milk. Mesophilic starters are used that generally produce CO_2, mainly from citric acid, thus causing formation of holes in the cheese. The cheese is pressed, salted in brine, and has no essential surface flora. This type of cheese is made in large quantities. Gouda and Edam cheese are the archetypes, but there are numerous related or derived types in many countries. Accordingly, the properties vary widely as follows:

1. The water content in the (unripened) fat-free cheese ranges from 53 to 63%, and the consistency varies correspondingly from firm to rather soft.
2. The fat content in the dry matter usually ranges from 40 to 52%, but there are related low-fat and high-fat types as well.
3. The loaf size may be between 0.2 and 20 kg; the shape may be a sphere, a flat cylinder, a rectangular block, or like that of a loaf of bread.
4. The maturation may take from several weeks to years; the consistency and, above all, the flavor vary correspondingly.
5. Other ingredients are sometimes added in addition to salt, especially cumin seeds.

27.3.1 MANUFACTURE

Figure 27.8 is a flowchart for the manufacture of Edam cheese in a fairly traditional way, whereas Figure 27.9 is an example of Gouda cheese making by a modern large-scale method. Some process steps will be discussed more fully.

Thermalizing serves to prevent psychrotrophs from forming lipase; strong hydrolysis of fat can cause a distinct off-flavor, mainly in young, little matured, cheese. Adverse effects of the proteinases of psychrotrophs on flavor are not known.

Pasteurization is meant to kill pathogens and spoilage microorganisms (e.g., Figure 24.1); spores of butyric acid bacteria and some streptococci are not killed. Most enzymes need not be inactivated, except lipoprotein lipase in some instances (e.g., in young cheese). Xanthine oxidase should remain intact because it is essential for the action of nitrate against butyric acid bacteria (see Section 26.2). In some countries it is illegal to pasteurize the cheese milk to the extent of significantly decreasing the protein content of the resulting whey. Usually the milk is pasteurized shortly before renneting, which reestablishes its rennetability (which is considerably reduced by prolonged cold storage); moreover, pasteurization ensures that the fat is again liquid, thus preventing partial coalescence of fat globules.

Bactofugation and addition of nitrate serve to prevent butyric acid fermentation. At present, the nitrate often is not added to the milk but to the mixture of curd and whey after a good deal of the whey has first been removed. This is partly

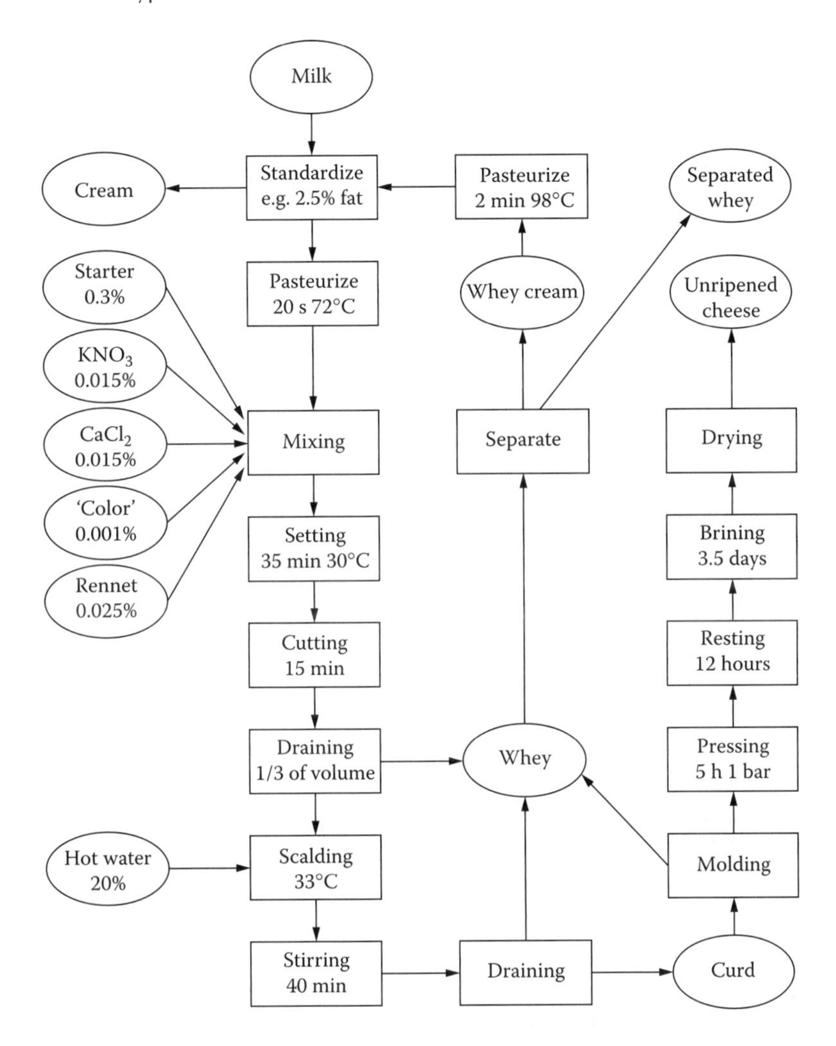

FIGURE 27.8 Example of a traditional method for the manufacture of Edam cheese.

to obtain the largest possible amount of whey without any added nitrate. A greater amount of nitrate should be added if bactofugation is omitted.

Starter (see also Section 13.5): Nowadays, starters are often made by transferring a deep-frozen concentrated culture into a closed vat containing high-pasteurized skim milk, in such a manner that contamination by phages is rigorously avoided. Subsequent culturing is for 16 to 18 h at 18 to 20°C. The quantity of starter added to the cheese milk is relatively large because a fast acid production in the curd is desirable. The added starter enriches the milk with, say, 10^7 bacteria (CFU) per ml; the latter are concentrated approximately 10 times by the cheese making, while they multiply again by a factor of 10, i.e., about 3 generation

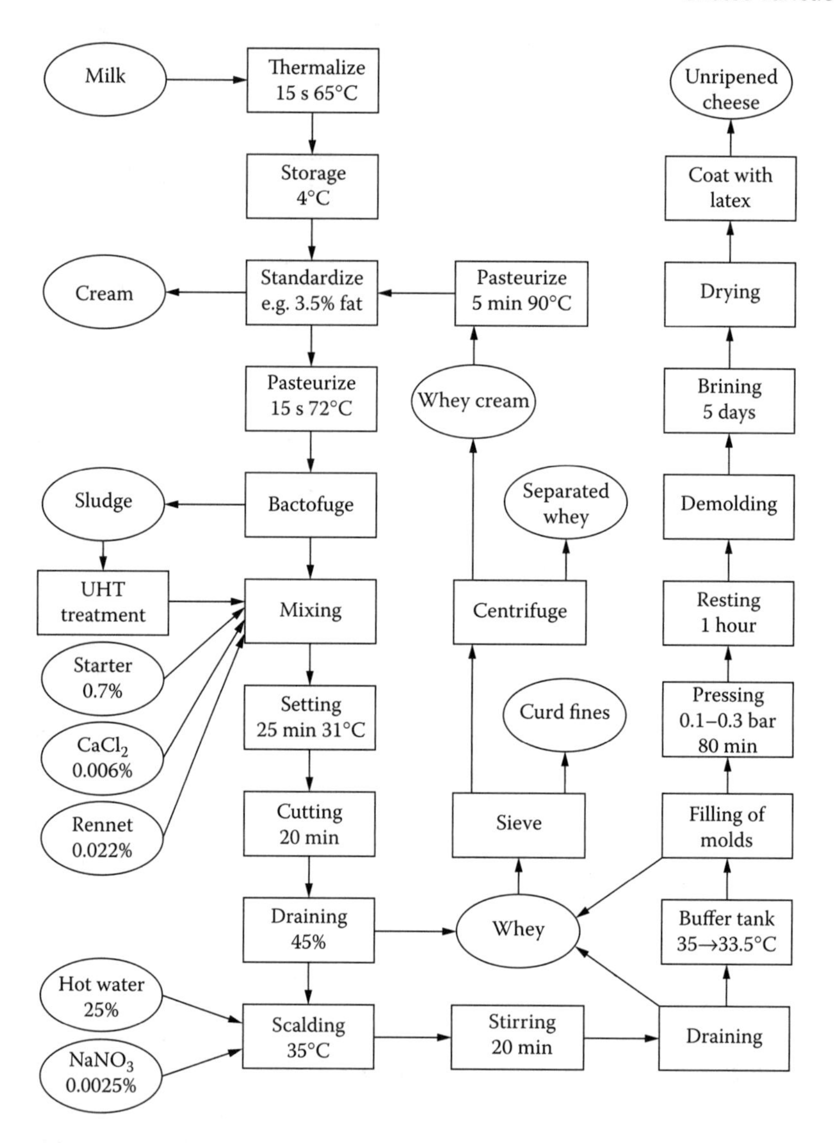

FIGURE 27.9 Example of the manufacture of 12-kg Gouda cheese by a modern method.

times. This leads to the cheese (at the pressing stage) containing about 10^9 starter bacteria per gram. Obviously, such a high count is needed because at brining the cheese often contains an appreciable amount of residual lactose. Because both the salt and the low temperature stop the growth of lactic acid bacteria, at least in the outermost part of the cheese (see Figure 27.10), the residual lactose has to be hydrolyzed by the enzyme system of bacterial cells already present.

Acidification of the curd should not only be fast, but reproducibly fast, to allow the making of cheese of a constant composition. In the modern manufacturing

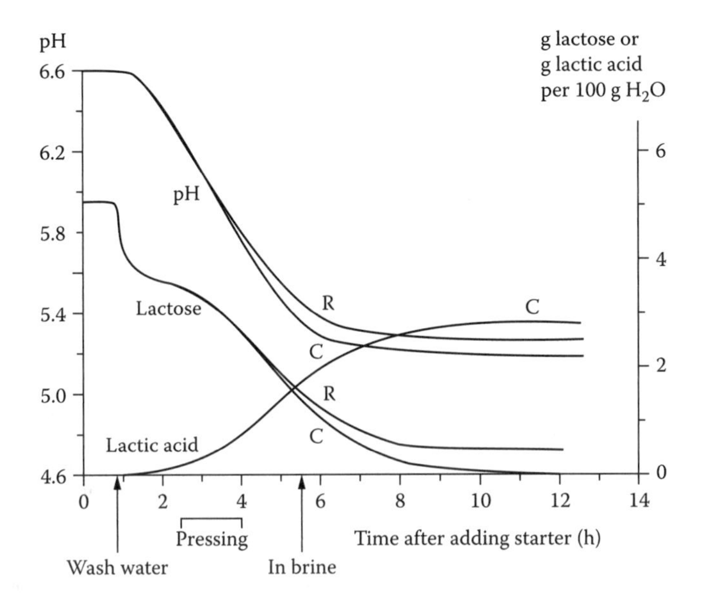

FIGURE 27.10 Acid production during Gouda cheese manufacture (conditions comparable to those in Figure 27.9). Quantities of lactose and lactic acid, and the pH of curd and cheese, as a function of the time after starter addition. C = center of the loaf; R = rind portion. (Data courtesy of M. D. Northolt and G. van den Berg, NIZO.)

process a fixed time schedule is maintained, making it virtually impossible to vary the process, e.g., to speed up or slow down syneresis. Because of this, the starter should be active and the cheese milk devoid of growth-inhibiting substances (such as antibiotics). No great harm will arise from a slight contamination of the cheese milk by bacteriophages, if starters are used that have a great number of different strains. In this situation it is highly improbable that more than a few of these strains will suffer from the phages, and most of the bacteria can always grow satisfactorily. Moreover, released phages can hardly spread throughout the bulk of the milk if it has been clotted (see also Subsection 13.5.5).

The activity of the starter is checked by inoculating the cheese milk, along with a standardized high-heated (and hence phage-free) skim milk, with the starter and by incubating these milks; the respective rates of acid production should be sufficiently high. Even if this is alright, problems can still arise due to accumulation of phages. Obviously, any contamination of cheese milk and of whey–curd with remnants of a previous batch should be rigorously avoided. Whey cream, used for adjusting the fat content of the cheese milk, should be intensely pasteurized; curd fines should not be put back into the vat; and the processing machinery should be thoroughly cleaned and disinfected after, say, 10 h of use. The starter should be cultured under strictly aseptic conditions.

Furthermore, the starters are selected on the basis of their capacity to produce appropriate amounts of CO_2 (presence of *Lactococcus lactis* ssp. *lactis* biovar.

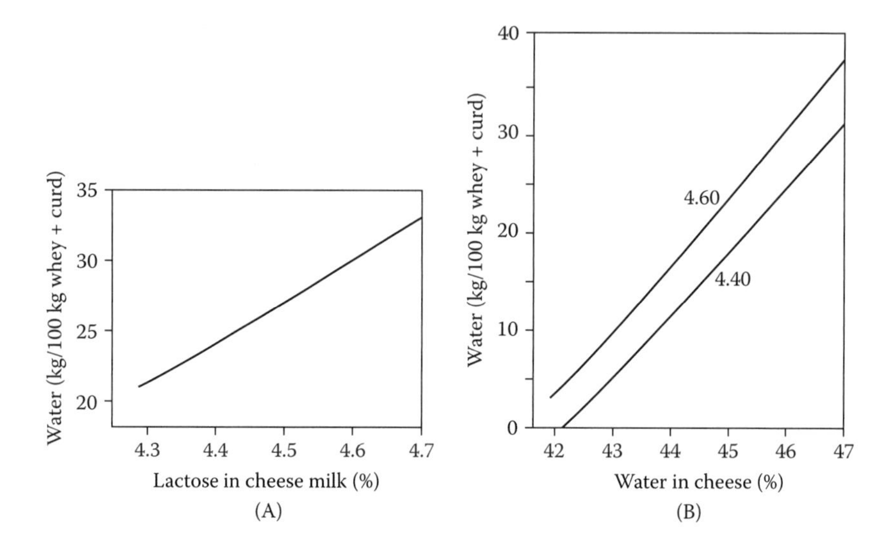

FIGURE 27.11 The amount of curd wash water needed to arrive at the pH desired for Gouda cheese (about 5.2). (A) Quantity of water to be added, as a function of the lactose content of the cheese milk. (B) Quantity of water to be added, as a function of the desired water content of the cheese before brining. Approximate averages for two different lactose contents of the milk (indicated near the curves); manufacturing conditions comparable to those in Figure 27.9. The curve in A refers to cheese with 46% water before brining. (Data from G. van den Berg and E. de Vries, *Zuivelzicht* **68**, 878, 924, 1976.)

diacetylactis and/or *Leuconostoc cremoris*) and on the basis of their proteolytic capacity (especially Prt+ variants of *L. lactis* ssp. *cremoris*). A proteolytic enzyme system that gives rise to bitter peptides, or too slow a breakdown of the bitter peptides formed, obviously is highly undesirable.

Washing of the curd serves to adjust the sugar content of the cheese and hence its ultimate pH (see Subsection 24.4.5). If washing is omitted (or if a fixed amount of water is used), every percentage unit increase in water content of the cheese — attained, for instance, by stirring for a shorter time or by using a lower scalding temperature — will lead to the pH of the cheese being approximately 0.1 unit lower.

Figure 27.11 gives the relation between the amount of curd wash water used and the lactose content of the cheese. This relation obviously depends on the ultimate water content of the cheese and on the lactose content of the milk. Note that here 'lactose' means the lactose present plus the lactose that has already been (partly) fermented to lactic acid. The 'lactose' content clearly should be adjusted so as to obtain the desired pH of the cheese after complete conversion of the sugar. This pH is lowered by about 0.03 unit if the lactose percentage in the fat-free dry cheese is increased by 0.2 (corresponding to a 5% lower amount of wash water). Such relations are approximative, because they also depend on other factors such as the rate of acid production and, of course, the efficiency of the above washing step. Usually, 90% of the lactose is supposed to be effectively washed out but that clearly depends on the duration of washing, the size of curd grains, and the intensity of stirring.

Control of the water content is most simply achieved by adjusting the duration of stirring or the scalding temperature. The stirring time is preferably kept brief for economic reasons and because of this, a fairly high scalding temperature is used. The activity of the starter bacteria significantly diminishes, however, if this temperature exceeds 35°C. Furthermore, the manufacturer will aim at keeping the water content of the cheese within the narrowest possible range. A complication arises when curd is made in large quantities because a considerable time elapses between the pressing of the first and the last loaf of the batch, and during this period the water content of the curd continues to decrease. To counterbalance this decrease, the temperature of the curd–whey mixture may gradually be lowered (see Figure 27.9). Alternatively, the molds filled latest can be the first ones put under the press (see Section 24.5).

Molding: Traditionally, the curd was allowed to settle to a layer of uniform thickness, and was then left for some time, resulting in a partial fusion of the curd grains. This process can be accelerated by lightly pressing with perforated metal plates laid on the layer of curd grains in the whey. When the mass of curd had become sufficiently coherent, it was cut into blocks. The blocks were wrapped in cloth, put into molds, and pressed for several hours at a pressure of 0.5 to 1 bar, causing a somewhat thick and firm rind to form. After pressing, the cheese was left upside down in the mold for several hours for 'shaping,' causing the loaf to attain a more symmetrical shape. The resting mainly functions, however, to allow the acid production in the cheese to continue, so that lactose is almost completely converted when brining starts, i.e., about 24 h after renneting. With time, this traditional method of manufacture (Figure 27.8) has increasingly been shortened (except in the manufacture of farmhouse Gouda cheese).

In the modern large-scale manufacture (Figure 27.9), the change is even more pronounced. The mixture of curd and whey is usually transferred to a vertical drainage and filling cylinder (Figure 27.12). The curd is allowed to settle for a while, which causes incipient fusion of the curd grains, especially in the bottom layer, where the pressure on the curd grains is highest. After a little while, the lower layer of the formed column of curd is allowed to fall into a mold and is cut off. The mold is provided with a lid and moved to the press. The plastic mold is lined with gauze to ensure formation of a closed rind. The pressure is applied for a much shorter time and is lower than in earlier days. The cheese is put into brine within 6 h of renneting of the milk.

In both manufacturing processes described, the curd is under the whey while being made into a coherent mass. In this way, inclusion of air is largely prevented. If the cheese is to be very dry, the drained curd can be stirred again, which causes considerable additional expulsion of whey (with a high fat content), but this also causes the inclusion of air pockets.

Brining is done to provide the cheese with salt and to cool the loaves and to give them a certain firmness. Newly made cheese that does not have a firm high-salt rind and would readily deform under its own weight, especially if its

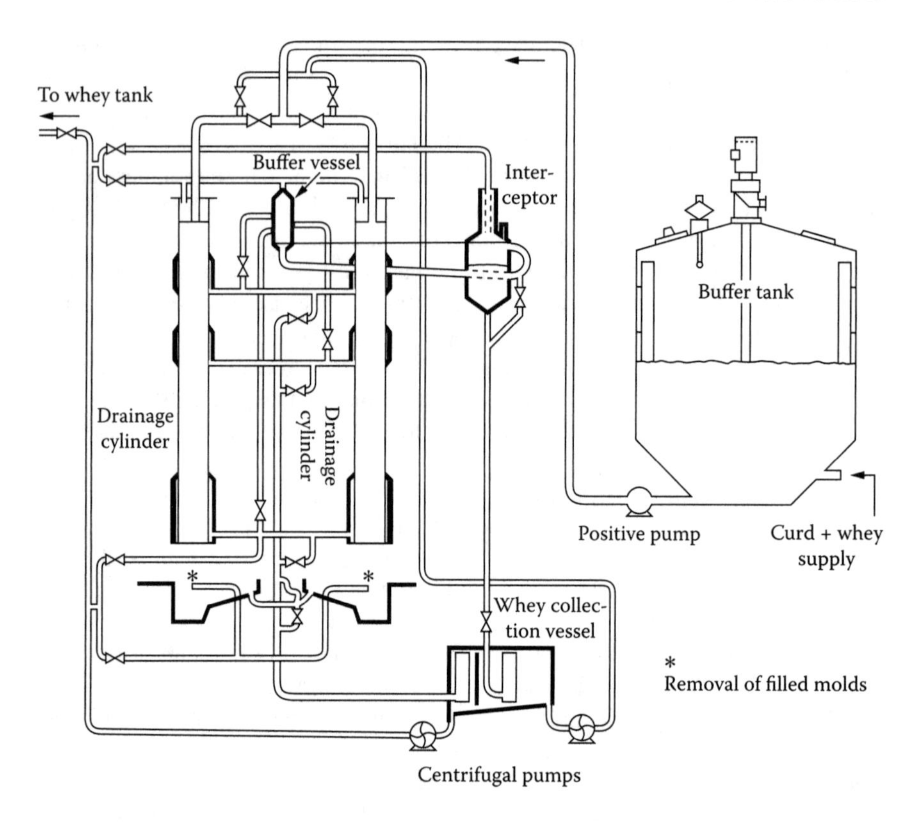

FIGURE 27.12 Diagram of machinery for the separation of whey and curd and for the formation of cylindrical loaves of coherent curd by means of drainage cylinders. Example according to the Casomatic system (Tetra Pak Tebel B. V.).

water content is high. Because the cheese is warmer by at least 15°C than the brine to which it is added, the brine must be regularly cooled. It has also to be subjected to microfiltration to remove debris and microbes.

An appreciable amount of lactose diffuses out of the cheese into the brine. This is because the cheese still contains lactose when salting starts; moreover, the conversion of lactose in the cheese rind is strongly impeded due to the high salt concentration. The lactose content of the brine amounts to about 0.5%.

Growth of microorganisms, mostly yeasts and lactobacilli (Section 26.3), may cause problems, especially if the temperature of the brine is too high (>14°C), the pH is too high (>4.6), or the concentration is too low (<16% NaCl). The pH is usually maintained below 4.6 by adding hydrochloric acid. Weak brine is sometimes used to allow the brining of Gouda cheese for exactly 1 week. This enables the use of a fixed daily time schedule in the plant. If the brine is too weak, especially if it contains too little Ca (as in freshly prepared brine), the rind of the cheese softens and swells; in more serious cases, it becomes slippery or slimy and the cheese starts to dissolve. Obviously, this must be avoided.

The cheese absorbs a certain amount of salt during brining, while at the same time losing a far greater quantity of moisture, say, by a factor of 2.5. The volume of the brine thus increases, leading to an excess of brine, which has to be discharged; this can cause environmental pollution. Discharging this excess, on the one hand, and adding salt to the remaining brine, on the other, essentially is an illogical operation because only the water released from the cheese into the brine is superfluous.

The rind treatment (see Subsection 24.7.3) generally involves providing the cheese with a layer of a latex. It primarily prevents mold growth on the rind, but it can also protect against mechanical damage, which implies that even a young cheese with a weak rind can be handled. Successive treatments with latex should be applied, and it should contain fungicides, e.g., natamycin.

Gouda cheese is sometimes made in rectangular loaves for curing while wrapped in shrink film (e.g., Saran). The water content of the cheese immediately after its manufacture should be slightly lower when such film is used (because less water vaporizes during maturation). The cheese is stored at a somewhat lower temperature (<10°C); otherwise an ill-balanced flavor tends to develop, the cause of which is not quite clear. Furthermore, a different starter is generally used.

27.3.2 Properties and Defects

Gouda-type cheese can vary widely with respect to loaf size, shape, water content, and maturation time. A smaller cheese is usually adjusted to have a higher water content and matures for a shorter time. At present, most Gouda-type cheese has a semihard consistency, which means that it is fairly firm and readily sliceable. It is not matured for very long (usually from 2 to 5 months), and hence the flavor is not very pronounced; especially near the rind, where the water content often is quite low, little proteolysis has occurred. Greater variation is especially found in raw milk cheese. Very mature (aged) cheese can have a very piquant flavor and has a short, often somewhat crumbly, consistency. If the cheese is to be matured for a long time it should be made differently (larger, somewhat drier) from a cheese that is to be consumed while young.

The pH of the young cheese often is between 5.0 and 5.5; that of some types of Edam is 4.9. Cheese manufacture usually aims at a pH, after brining, of 5.2. A higher pH implies an increased risk of defects due to growth of undesirable bacteria. The pH markedly affects maturation. At a low pH, hydrolysis of fat often is predominant, whereas the formation of flavor compounds due to protein degradation usually is more at a higher pH. Proteolysis is mainly due to the action of rennet and the enzymes of starter lactococci. Different starters cause different flavors. They may also cause differences in consistency.

Further see Sections 25.3, 25.5, and 25.6.

27.3.2.1 Eye Formation

The reader is also referred to Section 25.6. The cross section of most of the cheeses, except small loaves, should exhibit a few spherical and shiny holes of some 5 to 10 mm in diameter (eyes). 'Blind' cheese, cheese with many holes or with large holes that are irregular in shape, and cheese with slits or cracks are often considered defective. The formation of holes is mainly due to the production of CO_2 from citric acid by citric acid fermenting starter bacteria. Most CO_2 is formed within a week after manufacture, whereas the eyes develop from, say, 1 to 4 weeks, depending on the type of starter applied. L starters, in which leuconostocs are the only CO_2 producers, form less CO_2 than do DL starters (which also contain *Lactococcus lactis* ssp. *lactis* biovar. *diacetylactis*) and at a slower rate. Accordingly, the use of L starters more readily results in blind cheese.

The presence of cracks or slits in the cheese mass represents a problem, partly because cutting of the cheese into coherent slices is then hardly possible. Cracks form if the gas pressure in a hole exceeds the fracture stress of the cheese mass. This is enhanced by:

1. *A high gas pressure:* This can arise especially when gas is formed at a high rate (by a strongly gas-forming starter, by gas-forming contaminating bacteria, or due to a high temperature).
2. *A low fracture stress:* The fracture stress will usually be low if the cheese consistency is short, which means that the cheese fractures at a small deformation (Subsection 25.6.2). This can be expected when the cheese pH is low (say, <5.1), its salt content is high (because of this, cracks often form in the outermost part of the cheese), and the protein has been extensively degraded.

Excessive gas production (leading to cheese with many or very large holes) can be caused by abundant growth of contaminating bacteria (Chapter 26). As a rule, this is a serious defect, partly because it is usually associated with a marked off-flavor. Butyric acid fermentation often causes the cheese to become unsalable.

27.3.2.2 Defects

Some defects have already been mentioned (Section 25.8, and Chapter 26). Most defects are of microbial origin, but an altered cheese composition (due to errors in cheese manufacture) can considerably affect the growth potential of undesirable bacteria, e.g., salt, water, and sugar content, and pH. More bacteria can grow if the pH is higher, especially when unfermented sugar remains. Because of this, an inadequate acid production, which is usually caused by contamination by bacteriophages, favors defects. Strict hygiene is of paramount importance. Figure 27.13 identifies possible sources of contamination.

In cheese made of pasteurized milk the main defects include:

- Butyric acid fermentation (Section 26.2)
- Off-flavors, usually caused by excessive growth of lactobacilli (Section 26.3)

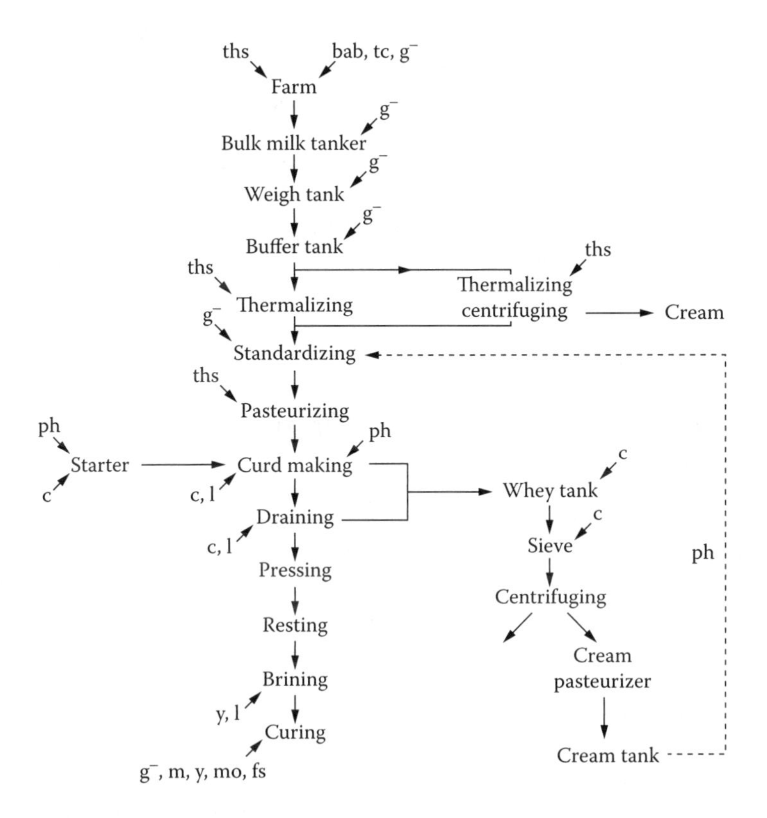

FIGURE 27.13 Points where contamination by microbes and disturbing phages can occur in the case of Gouda cheese manufacture, starting with modern milk collection (it thus involves cold storage of farmer-delivered milk) up to maturation. Abbreviations: bab = butyric acid bacteria; c = coliforms; fs = fecal streptococci; g^- = Gram-negative rods (including psychrotrophs); l = lactobacilli; m = micrococci; mo = molds; ph = bacteriophages; tc = total count; ths = thermophilic streptococci; y = yeasts. (Adapted from M.D. Northolt, *Voedingsmiddelentechnologie,* **20**, 17, 1987.)

- A bitter flavor, which can develop when the cheese retains much rennet and the starter bacteria form too many bitter peptides and decompose them insufficiently (see Section 25.5)
- Cracks in the cheese mass (see Subsection 27.3.2.1)
- Mold growth on or in the rind (Section 26.6)
- Shape and rind defects (e.g., due to rough handling)

In cheese made from raw milk several bacteria (including coliforms, propionic acid bacteria, and fecal enterococci) can grow and thereby cause defects. In addition to gas production (early and late blowing), off-flavors such as a sour, yeasty, putrid, fruity, and H_2S flavor can appear. The cheese can also turn soapy-rancid due to milk lipase activity, especially if the milk fat globule membranes have been badly damaged during milk handling. Young cheese made from pasteurized milk can have a rancid flavor if the milk was already rancid before pasteurization (which often happens due to bacterial lipases), but usually little lipolysis occurs.

Obviously, Gouda cheese will rarely have off-flavors if it is very hygienically made from good-quality pasteurized milk; but this also implies that its flavor may be rather flat (unless the cheese is ripened long enough) and, above all, with hardly any variation. Only the differences in the starters used then will bring about differences in flavor.

27.4 CHEDDAR-TYPE CHEESES

Cheddar-type cheeses are characterized by the mixing of salt with the curd before it is pressed into a coherent loaf. Salt considerably retards the growth of lactic acid bacteria. Because of this, most of the lactose in the curd should have been converted before the curd is salted, and curd making, therefore, requires a long time. Moreover, salted curd tends to fuse poorly during pressing if its pH is still too high (above, say, 5.6) because the curd flows insufficiently (Section 24.5).

Formerly, when cheese was made from skimmed milk, the milk was usually left for creaming for such a long time that it turned sour. Naturally, the curd was also acidic and could thus be salted before pressing; an example is Frisian cheese. However, currently most cheeses of this type are made of unsoured milk, such as Cantal and almost all British types. The cheese becomes relatively dry due to the long curd-making time and the low pH. Because the salt is relatively homogeneously dispersed through the fresh cheese, it can be made in large loaves, which is desirable to prevent water loss by vaporization and to minimize curing costs. On the other hand, it takes a long time for the interior of the loaf to cool.

These cheeses are typically hard with a long shelf life and without a surface flora. The best known is Cheddar: about 50% fat in the dry matter, not more than 38% water, originally of cylindrical shape, weighing about 30 kg. Nowadays, mostly rectangular blocks of variable (often large) size are made. Cheddar and derived varieties are now manufactured all over the world, though primarily in English-speaking countries. Cheshire is slightly more acidic and has a somewhat higher water content. This is also true of Caerphilly, but this cheese is eaten while young and is mainly used in cooking. Stilton is quite different. Its salted curd is not heavily pressed and is shaped into a cheese with an open texture; the cheese becomes veined with blue mold (Subsection 27.6.2).

Manufacture and properties of Cheddar will now be discussed in more detail.

27.4.1 MANUFACTURE

Figure 27.14 outlines the manufacturing process (see also Figure 27.1 and Figure 27.3). It represents a somewhat traditional method of manufacture, though the time from adding starter to the milling was originally often even longer. The endpoint of the curd treatment in the vat (including 'cheddaring') was assessed

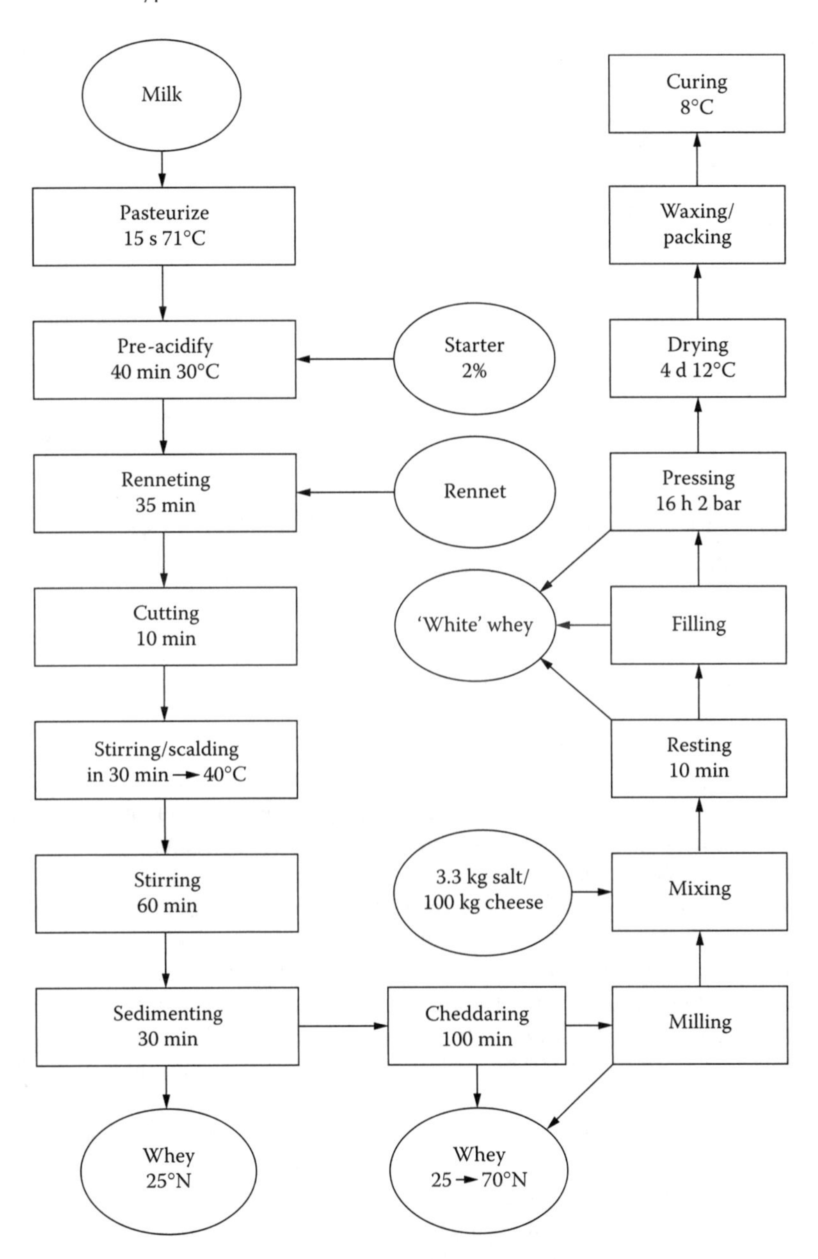

FIGURE 27.14 Example of a traditional method for manufacture of Cheddar. Simplified.

by determining the acidity of the whey. Nowadays, a fixed time schedule is usually maintained, and the processing time is much shorter, e.g., 3 h from renneting to milling. Figure 27.15 gives examples of the timescales involved in traditional and modern processes. Manufacture has also been mechanized to a high degree.

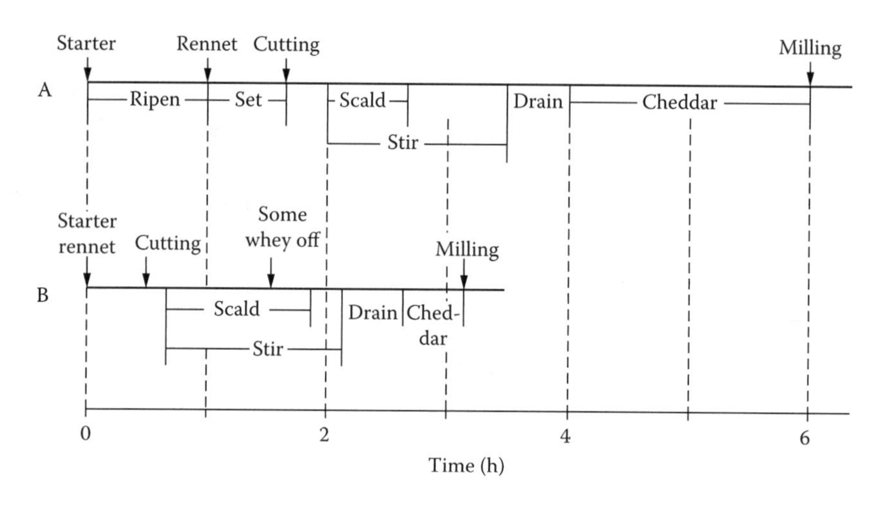

FIGURE 27.15 Time-scales involved in the manufacture of Cheddar, up to milling. (A) A fairly traditional process. (B) A modern and fast manufacturing scheme.

Presently, the cheese milk is generally pasteurized, though a slightly milder heat treatment can be applied. Preacidification is now little practiced; however, a high percentage of starter is added.

The *starter* must satisfy a number of requirements. The curd making should be as brief as possible for economic reasons, and this requires very fast acid production. Therefore, fast-growing, rapidly acid-producing, and phage-resistant bacteria are needed. This combination of requirements is hard to fulfill; for that reason, mixtures of single-strain, fast starters are generally employed and rigorous measures taken to prevent contamination of the starter by phages (see Chapter 13, Sections 13.3 and 13.5). Furthermore, the starter organisms should produce little or no CO_2 (i.e., be homofermentative), be fairly heat tolerant, and cause no bitterness in the cheese. Currently, a mixture of two strains is often applied. One of these, generally a strain of *Lactococcus lactis* ssp. *lactis*, is sufficiently heat resistant to keep growing during scalding. It is primarily responsible for a fast acid production, but it also tends to form bitter peptides. The other strain, generally of *L. lactis* ssp. *cremoris*, is less heat resistant; it does produce some acid during scalding but does not keep growing. However, it has considerable 'debittering' activity, while not producing many bitter peptides by itself. The proteolytic system of the cremoris strain is of great importance for a satisfactory ripening and, in particular, for the flavor.

To further enhance flavor development, *secondary starters* of selected lactobacilli can be added (see Section 26.3).

After cutting, stirring, and scalding, the curd settles and fuses into a rather compact mass. *Cheddaring* then starts, which is a process characteristic of Cheddar and of most of its related types. The whey is drained off and the curd mass cut into large strips that are piled up. The slabs fuse again and are allowed to spread slowly into thinner slabs that are turned, cut again into strips, and piled up, and so on. This used to require much labor because it had to be done in such a way as to prevent the curd from cooling unevenly and too fast, thereby ensuring a uniform and fast acid production. The current mechanized cheddaring processes require less time and far less labor than the traditional method.

The curd mass will only flow readily if its pH is lower than 5.8, and the temperature is not too low. The flow causes a 'fibrous' curd structure. It has long been assumed to be essential for obtaining a characteristic Cheddar and, accordingly, earlier manufacturing processes with mechanized curd manufacture included a spreading step. Later on, traditional cheddaring turned out to be unnecessary because the cheese does not show whether spreading has been applied or not. To be sure, the deformation and piling of the curd squeezes out entrapped air, which favors the cheese to obtain a close texture. However, this can also be achieved by pressing the curd under vacuum, which currently is common practice. Furthermore, the flow of curd slightly hinders syneresis. Obviously, this effect should be considered when adjusting the water content of the cheese.

The acid production during cheddaring is of paramount importance. When salting starts, the lactose content should not be over 0.6%, while the pH should preferably be 5.3 to 5.4. Water content and pH of the curd at that stage largely determine the composition of the cheese, i.e., water content, pH, and the amount of residual rennet and of calcium phosphate. (Cantal cheese is often made in a different way. The curd is pressed into a loaf when its acidity is low, and is subsequently kept for a few days at 12 to 15°C to allow conversion of lactose. The loaf then is cut into pieces and treated as milled and cheddared curd.)

Prior to *salting*, the curd is milled, i.e., cut into strips about the size of a finger. Milling too finely leads to excessive loss of fat and of curd fines in the press whey. Milling too coarsely causes the diffusion of the salt into the strips to take a long time, resulting in a nonhomogeneous cheese texture (presumably because lactic acid bacteria are killed locally). The salt is mixed with the curd, and 10 min is allowed for salt absorption ('mellowing'); otherwise, excess salt would be lost with the whey, which in normal cases already contains about 50% of the added salt (both milling and pressing cause a considerable expulsion of whey). The salt should be evenly distributed, but this is hard to achieve. Acid production in the curd is insufficient if the cheese contains over 5 to 5.5% salt in water, whereas at less than 4.5% salt concentration the lactic acid bacteria ferment too fast. In either case the flavor development is unsatisfactory and contaminating organisms have a greater chance of growing, which may cause strong off-flavors.

Formerly, the milled and salted curd was put into hoops (molds) and then put under the press. *Pressing* took a long time, and pressure was high (about 2 bar), which was needed to achieve a close texture. Currently, the curd is usually pressed under vacuum and lower pressures are applied. The lower the temperature and the pH during pressing, the more difficult it is to transform the salted curd into a coherent loaf. In Cheshire cheese making, the pH of the curd at salting and pressing is relatively low (about 5.1); accordingly, this cheese typically has a crumbly (very short) texture.

Soon after pressing, the cheese (often in the shape of large blocks) is generally wrapped in plastic foil, after which it needs little further care; the cheese loaves are often immediately packed in cases or boxes. Usually, the cheese is cured at low temperature (Table 24.3).

A modern variant of Cheddar cheese is *stirred curd* or *granular* cheese. The curd in the whey is continuously stirred until sufficient acid has been produced. Whey is separated from the curd under vacuum, the curd is salted, and the salt–curd mixture is pressed in molds. This procedure leads to a 'normal' Cheddar with a close texture. Types of cheese have also been developed that have a somewhat higher water content and pH, e.g., Colby (40% water) and Monterey (42%). Obviously, the curd should be washed; otherwise the pH becomes too low. Cold water should be used; otherwise the cheese becomes too dry. These cheeses are usually not pressed under vacuum and, accordingly, are open textured; the mechanical holes increase slightly in size as a result of CO_2 production. In fact, such cheese is much like a Gouda type. It has been claimed that even normal Gouda cheese can be made in a way similar to the Cheddar process, i.e., by means of salting the curd.

27.4.2 Properties

Traditionally, Cheddar was a fairly acidic cheese — pH about 4.9 — but presently a pH of 5.2 and even 5.3 is common (especially outside England). Its consistency is rather firm and short, at least if the pH is not too high and marked proteolysis has occurred. At the higher pH values, the consistency is more like that of Gouda cheese. The salt content also has a considerable effect: The cheese is too hard if the content is over 6% in the water, whereas at less than 4% it is too soft (almost spreadable).

Ripening. Cheddar contains little active milk proteinase, but it has active rennet and a large pool of proteolytic enzymes from lactic acid bacteria; most of the fast acid-producing strains are also strongly proteolytic. At the low curing temperature (usually below 10°C), large peptides are rapidly formed, but the degradation of these into smaller molecules is relatively slow (in Section 25.3, the effects of cheese composition and of temperature on proteolysis by rennet enzymes are discussed). For instance, in cheese that is several weeks old there is:

At pH 5.1: no α_{s1}-casein left, 55% β-casein left
At pH 5.3: 30% α_{s1}-casein left, 80% β-casein left

Cheddar is a cheese that may be cured for varying lengths of time, say, from 2 to 10 months. Naturally, curing time considerably affects the flavor. Presumably, short peptides and free amino acids play an important role in flavor development, but so do volatile compounds. Amino acids may be converted to short-chain fatty acids and to thiols (H_2S and CH_3–SH) (see Figure 25.3). Among the compounds formed via the pyruvate metabolism of the lactic acid bacteria are diacetyl, acetic acid, and ethanol, and probably also esters. CO_2 also is essential as a flavor enhancer; it is partly formed by decarboxylation of amino acids.

The fat is essential, not only for the consistency but also for the flavor. Low-fat cheese has been found to lack the typical Cheddar flavor. Probably the most important role of the fat is as a solvent for hydrophobic flavor compounds. In addition, lipolysis (if not too strong) and formation of ketones from free fatty acids have a role to play. It should be noted that there is no agreement at all among various investigators regarding the characteristic flavor of Cheddar.

Defects. Defects that may occur in Cheddar include the following:

- *Open texture*, which may lead to the formation of cracks upon gas production during maturation. In serious cases it causes mold growth or — if the cheese is kept in dry air — autoxidation of fat (tallowy discoloration).
- '*Seaminess*' refers to the appearance of whitish veins seen in a cross section of the cheese. Most Cheddar contains small crystals of $Ca(H_2PO_4)_2$, especially when its water content is low and the salt content high. These crystals are rapidly formed if the milled curd has a high pH at salting, is heavily salted, and not given enough time for adequate absorption of salt before pressing. The crystals are formed in the outermost layer of the curd pieces, causing this layer to turn white and hard. Seaminess is often accompanied by an open texture.
- *Incomplete acid production* is often responsible for insufficient flavor and abnormal consistency, and it also increases the risk of bacterial defects.
- *Contaminating bacteria* may cause defects, especially at high pH, low salt content, and high ripening temperature. It mostly concerns lactobacilli or pediococci, which cause several off-flavors (Section 26.3). Obviously, hygiene during cheese making is essential. Interestingly, the proteolysis caused by some *Lactobacillus* strains can lead to an improved cheese flavor (see Section 26.3).
- *White specks:* As discussed in Section 26.3, calcium lactate crystals can form in Cheddar cheese, especially when racemization of the originally formed L(+) isomer occurs, which is achieved by some lactobacilli. The lactate crystals can lead to the formation of so-called white specks, predominantly at the surface of the cheese.
- *Bland flavor:* Especially in the U.S., most Cheddar-type cheese is currently made in very large blocks. This makes it difficult to cool the interior of the cheese with sufficient speed; consequently, the bacterial defects just mentioned may readily occur. To prevent this, the cheeses are often kept in cold rooms, say, at 5°C. It goes without saying that this impedes normal flavor development.
- A *bitter flavor* defect often develops if the salt content is low and the curing temperature is high. Selection of strains of lactic acid bacteria that do not cause this bitter flavor is therefore important.

27.5 SWISS AND PASTA-FILATA TYPES

In the manufacture of these types of cheese high temperatures (50°C or higher) are applied during the curd treatment (see Figure 27.1). The use of such high temperatures has obvious effects on the properties of the cheese:

1. The cheese has a low water content and a relatively high pH, despite the absence of a washing step during curd making.
2. Many potentially harmful bacteria are killed at the high temperature that is maintained for a considerable time, often more than 1 h. For most of the bacteria, the decimal reduction time is 2 to 50 min at 50°C. This implies that thermalization or even pasteurization is essentially applied; hence, pasteurization of the cheese milk can often be omitted.
3. Mesophilic lactic acid bacteria are for the greater part inactivated. Accordingly, the starter bacteria involved are predominantly thermophilic.
4. A considerable proportion of the rennet is inactivated during curd treatment, which affects proteolysis in the cheese.

The starters are grown at high temperature, e.g., 40°C, and are mainly composed of strains of the thermophilic *Streptococcus thermophilus*, *Lactobacillus helveticus*, *L. lactis*, and sometimes also *L. delbrueckii* ssp. *bulgaricus*. The mesophilic *Lactococcus lactis* ssp. *lactis* may occur as a minority flora. The starter bacteria are homofermentative; they produce little, if any, CO_2 at the initial stages of fermentation; they cannot hydrolyze galactose or do so poorly (Subsection 13.1.2); and they develop a type of proteolysis (and thus a flavor in the mature cheese) that differs from that of mesophilic starters. The lactobacilli are fairly proteolytic; *S. thermophilus* is less so.

Originally, these types of cheese were made predominantly in the Alps and in Italy, but some types are now made in many other regions as well. There are several variants. A kind of archetype is Bergkäse (mountain cheese), also designated Alpkäse or Beaufort. In summer, the dairy cattle were transferred from the village to the elevated mountain meadows; it was here that the milk was made into cheese. Therefore, the manufacture of the cheese had to be simple; the cheese had to be easy to handle and very durable. These are dry, though still sliceable, quite large (20 to 40 kg), flat cylindrical cheeses that ripen for several months or longer. On the rind of some of these cheeses a layer of salt-tolerant yeasts and bacteria is grown (red smear, Rotschmiere, Ferments du Rouge; see Subsection 27.6.1), e.g., Gruyère (Switzerland) and Comté (France). The latter cheeses are often somewhat larger. The variant evolved in German-speaking Switzerland is Emmentaler (in the U.S. simply called Swiss), which is larger still (up to 130 kg), contains little salt, and does not have a surface smear; it has a distinct propionic acid fermentation. Still drier types, usually quite large high-cylindrical cheeses, are Sbrinz (Switzerland) and, in Italy, Parmesan (Parmigiano Reggiano and Grana Padana); a related type is Pecorino Romano, made of ewes' milk. These cheeses need to mature for years, becoming very hard over that time. They are grating cheeses to be used in cooking.

Parmesan cheese has a reduced fat content (say, 35% fat in the dry matter), with the reduction achieved by gravity creaming of the milk. Along with the cream layer, many bacteria and spores of butyric acid bacteria are removed.

Traditionally, in regions around the eastern part of the Mediterranean, pasta filata (i.e, stretched curd) cheeses are made. The curd is intensely heated and kneaded into loaves of small to medium size, and of several shapes. An almost unripened type is Mozzarella. Matured types are Provolone and Caciocavallo, known in the Balkans as Kashkaval. Some of these cheeses are smoked, which may prevent mold growth.

Two types of cheese that are nowadays made in large quantities will be considered in more detail.

27.5.1 Emmentaler

Emmentaler is a very large cheese, weighing 60 to 130 kg; the loaf shape is flat cylindrical, at least initially. The extensive gas production causes the cheese to bulge upward, whereas its great weight flattens it slightly leading to bulging sides. The cheese has many round eyes, 1 to 4 cm in diameter, and a smooth, long, somewhat rubbery texture. The flavor is mild and rather sweetish. The composition after 3 to 4 months of maturation is approximately as follows:

Protein	29%, including protein degradation products
Fat	31% (48% in the dry matter)
Water	35% (51% in the fat-free cheese)
NaCl	0.5% (1.4% in the water)
Propionic acid	0.7%
CO_2	0.2%

Figure 27.16 outlines the manufacturing process. The pH at 2 d usually is 5.1 to 5.3, slightly increasing afterward. For the course of the temperature, see Figure 27.1. The maximum temperature is often 52 to 54°C. The higher it is, the less trouble there is with defects caused by raw milk of poor microbial quality. The milk is hardly ever pasteurized. Traditionally, the curd was made in copper vats, causing considerable contamination of the milk by copper; hence, the cheese contained up to 15 mg of Cu per kg. The copper was believed to be necessary for a satisfactory flavor development, but this has never been shown conclusively. Presently, the curd is usually made in stainless steel vats (occasionally a copper salt is added). The copper contamination caused problems because the butter made of the whey rapidly developed autoxidation defects.

For Emmentaler cheese, the growth of propionic acid bacteria is of paramount importance. As a rule, a little of a culture of *Propionibacterium freudenreichii* ssp. *shermanii* is added to the cheese milk. These bacteria grow in 20 to 30 d to 10^8 to 10^9 cells per gram of cheese and ferment lactic acid approximately according to:

$$3\ CH_3\text{–}CHOH\text{–}COOH \rightarrow 2\ CH_3\text{–}CH_2\text{–}COOH + CH_3\text{–}COOH + CO_2 + H_2O$$

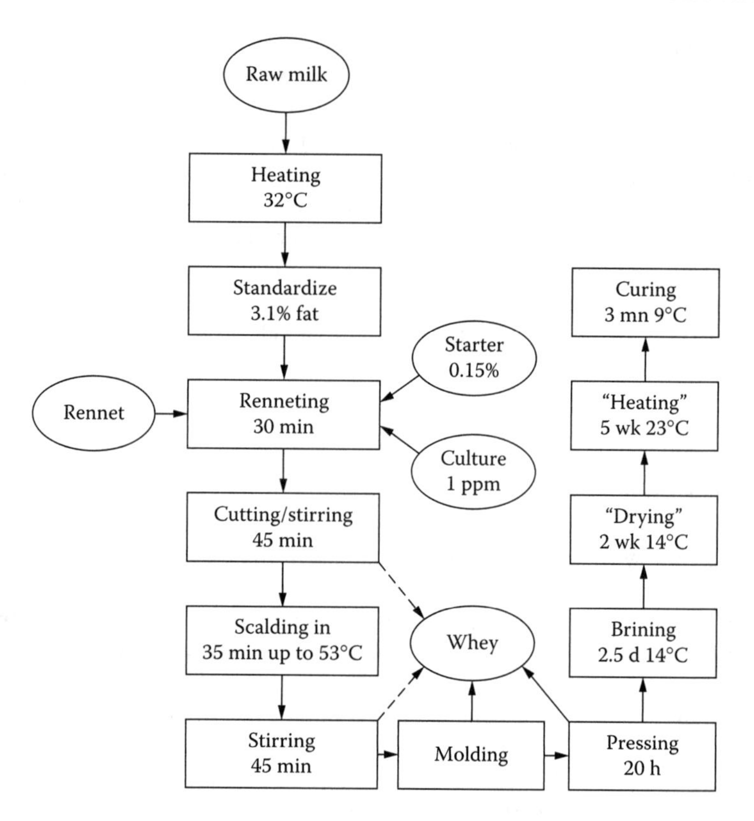

FIGURE 27.16 Example of the manufacture of Emmentaler cheese. The culture is of propionic acid bacteria. One thousand liters of milk yields one loaf of 81 kg. The ensuing whey yields about 10 kg whey butter.

As is shown in Figure 27.17, almost all of the lactic acid is consumed. This causes the flavor of the cheese to change (propionate gives a sweetish taste) and large eyes to form. To secure a satisfactory fermentation to propionic acid, the curing temperature should be raised to 22 to 25°C for about 5 weeks. The production of CO_2 is controlled by varying the temperature and the length of time in the curing room. The amount of CO_2 produced in an 80-kg cheese approximates 120 l, 60 l of which dissolves, 40 l of which diffuses out of the loaf, and 20 l of which finds its way into the eyes; this implies that the volume of the cheese increases by almost 25%. Before entering the curing room, the cheese is kept cool for some time, which allows the salt to partly diffuse throughout the cheese; furthermore, the keeping causes the cheese to attain a firm rind. During this time proteolysis should be limited because cracks could form rather than eyes during the ensuing CO_2 production.

Enzymes of the lactobacilli and plasmin from the milk are mainly responsible for proteolysis. Proteolysis is less than, and different from that, in Gouda- and

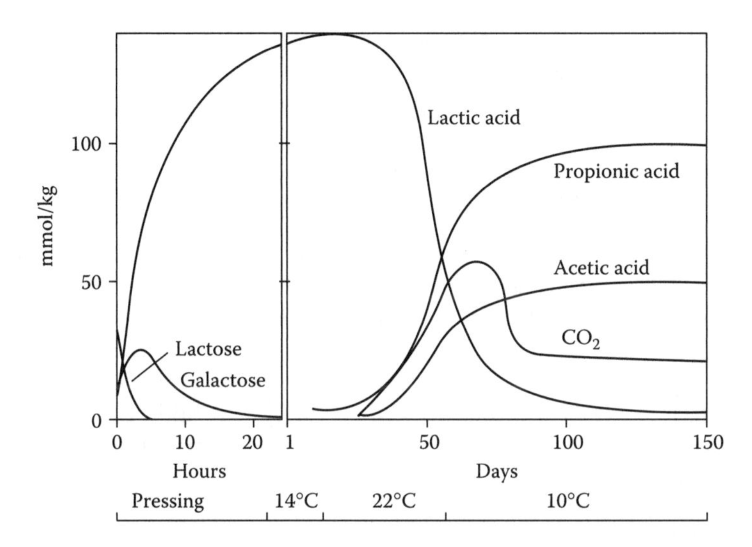

FIGURE 27.17 Fermentation of sugar and of lactic acid in Emmentaler cheese during and after pressing. All quantities are in mmol per kg cheese; quantities in the holes are not included. Storage temperature is indicated. (Data from Steffen et al., In P.F. Fox, Ed., *Cheese: Chemistry, Physics and Microbiology,* Vol. 2, *Major Cheese Groups,* 2nd ed., 1993.)

Cheddar-type cheeses. The cheese retains its long consistency during ripening; the true deformation (natural strain) at fracture in a mature cheese approximates 1.2 (Figure 25.8). Usually, the level of proteolysis is higher in the rind portion than in the center. This is ascribed to the long time needed for the center of the cheese to cool after pressing. Consequently, in the rind a greater proportion of rennet may remain active and mesophilic lactic acid bacteria may grow better. Accordingly, cracks, if present, are usually observed in the rind portion of the cheese. The latter phenomenon may also be affected by the higher salt content in the rind.

Possible defects are:

1. Growth of *Clostridium sporogenes* causes an offensive flavor. Hygiene during milking and milk collection is the remedy.
2. Lipolysis should be minimal. Too much lipolysis can cause flavor defects.
3. *C. tyrobutyricum* and, to a lesser extent, *C. butyricum* can grow very well in the cheese and spoil it completely. Control by means of addition of nitrate does not apply because the quantity of nitrate needed to suppress butyric acid fermentation would also suppress propionic acid fermentation. (Note that the cheese pH is fairly high and the salt content low.) Because of this, dairy farmers delivering milk for the manufacture of Emmentaler cheese are not allowed to have silage.
4. Cheese with numerous tiny holes can result from growth of heterofermentative lactic acid bacteria.

5. Enterobacteria can also cause cracks and an off-flavor. They are killed by adequate scalding.
6. Some lactic acid bacteria, especially mesophilic lactobacilli, can eventually produce a considerable amount of CO_2 by decarboxylation of amino acids. Cracks can form because in the meantime the cheese texture has become shorter. Moreover, biogenic amines can form.

Jarlsberg cheese is related to Emmentaler. The cheese resembles Gouda in many respects, i.e., size, water content, pasteurization of the cheese milk, use of mesophilic and partly heterofermentative starters, and a scalding temperature of about 35°C. However, propionic acid fermentation does occur (a culture is added). For this to happen, the pH should be relatively high, which is achieved by washing the curd. For a satisfactory flavor, the salt content should be fairly low. All in all, conditions are highly favorable for butyric acid fermentation. Three measures are taken to control it:

- Bactofugation of the cheese milk.
- Addition of $NaNO_3$.
- Addition of a culture of a special strain of *Lactobacillus delbrueckii* ssp. *bulgaricus*, which can reduce NO_3 to NO_2. NO_2 is reduced even further, which is necessary because too much NO_2 would inhibit the propionic acid fermentation.

27.5.2 Mozzarella

Mozzarella is a small cheese, weighing 50 to 400 g. The loaf shape is a somewhat flattened sphere. The flavor should not be pronounced. Originally, it was mainly made from buffaloes' milk. Nowadays it is usually made from cows' milk or from a mixture. The cheese contains 35 to 45% fat in the dry matter, 52 to 56% water, and about 1% salt. Being only a few days old, it is little matured and has a rather soft and long consistency.

An example of a manufacturing scheme is given in Figure 27.18. The starter is mainly composed of *S. thermophilus*, *L. delbrueckii* ssp. *bulgaricus*, and *L. helveticus*, although some mesophilic lactococci may be present. The lactobacilli can produce acid in the cheese even after the stretching process. A decrease of the pH of the curd to 5.2 to 5.3 before kneading starts is of prime importance because otherwise 'stretching' is not possible. If the pH is above 5.4, the cheese mass is too firm; if the pH is below 5.1, it is too crumbly. To arrive at the desired pH in the curd, two different manufacturing processes are applied, i.e., simple resting or a kind of cheddaring process. The long manufacturing time, the high temperature (needed to kill undesirable bacteria), and the significant acid production can readily cause the moisture content of the cheese to become too low. Because of this, the initial curd treatment is often very brief. After that, the curd settles and fuses, due to which the outflow of whey is considerably slowed down. The temperature is increased only after fusion is complete.

Then the curd is kneaded. Traditionally, the proper moment was fixed by testing if a string could be drawn from a heated curd piece, i.e., 'pasta filata'. The

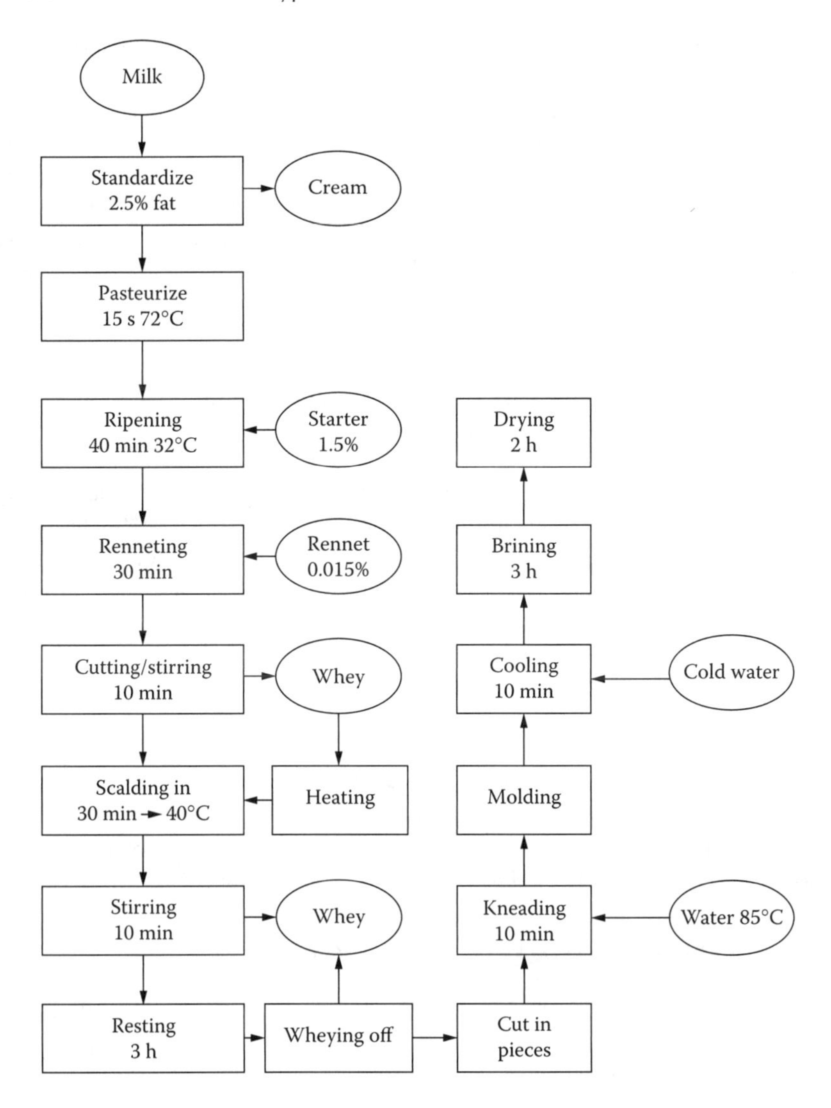

FIGURE 27.18 Example of the manufacture of traditional Mozzarella. 100-g loaves.

soured curd mass can be cut into strips, or it can be coarsely milled as in Cheddar cheese manufacture. The slabs or strips are put in hot water and kneaded, which may be done by hand. The kneading implies stretching and shearing of the curd mass; its main function is to rapidly increase the temperature of the whole mass to the desired level of about 55°C. The kneading also appears to slightly slow down syneresis (which may proceed rapidly at high temperature) and to affect the texture of the cheese: the cheese mass appears somewhat layered. Moreover, the milk fat globules may be disrupted into smaller ones, which can cause enhanced lipolysis. Finally, lumps of curd are kneaded into the desired shape and placed in cold water. Usually, the cheese is not pressed.

Low-moisture Mozzarella. The product just described is meant for direct consumption. Currently, large quantities of Mozzarella are made for cooking purposes, especially for use on pizzas. The cheese should have adequate melting properties, i.e., soften on heating, become smooth, and flow. However, the melted cheese should remain rather viscous and become solid again when the temperature decreases somewhat. This is only possible if the water content is lower, generally between 45 and 50%. The manufacturing procedure often is significantly modified; even coagulation with acid is applied. The product may differ greatly from the traditional Mozzarella; it is sometimes called 'pizza cheese'.

Some quality problems may occur. One may be due to proteolysis during storage, i.e., degradation of α_{s1}- and of β-casein. The relative contributions of chymosin and plasmin are uncertain because the extent to which rennet is inactivated is not quite clear. A low level of proteolysis enhances good melting characteristics. Often, attaining the desired melting quality is a problem, the cause of its variation being insufficiently known. Furthermore, the cheese mass may appear too transparent after melting, especially if its fat content is low. A further defect is brown discoloration. It is caused by Maillard reactions that occur during kneading, and later during melting, if galactose has not been sufficiently fermented by the starter bacteria.

Provolone is closely related to traditional Mozzarella as far as the method of manufacture is concerned. It is made slightly drier and into larger loaves that weigh, for example, 5 kg, and are of a different shape (e.g., an oblong pear). The salt content is somewhat higher, and the cheese is cured for several months during which time significant proteolysis and lipolysis occur. Often, a lipase preparation is added to the cheese milk, which may cause a rather sharp flavor.

27.6 CHEESES WITH A SPECIFIC FLORA

These types of cheese are characterized by their microbial ripening being controlled not only by a normal flora of lactic acid bacteria (sometimes also propionic acid bacteria), but also by either:

1. A surface flora in which *Penicillium camemberti* is the dominant organism. It mostly concerns cheeses with a soft consistency, e.g., Camembert and Brie.
2. A surface flora dominated by pigment-producing strains of coryneform bacteria such as *Arthrobacter* spp. and *Brevibacterium linens.* It concerns soft cheeses such as Pont l'Évêque, Munster, and Herve as well as some types of hard cheese, e.g., Gruyère, and semihard cheese, e.g., Tilsiter and Port du Salut.
3. An internal flora consisting of *Penicillium roqueforti.* Among the cheeses are the semihard Roquefort, many types designated 'Bleu', and also Stilton, which is a harder type of cheese.

These are the most important groups. Other types are Gamalost, with a blue surface mold (*Mucor* spp.), and intermediate types with *P. roqueforti* in the interior and *P. camemberti* (e.g., Lymeswold) or coryneforms (e.g., Gorgonzola) on the surface.

Below, some typical examples are discussed, i.e., soft cheeses with a surface flora and cheeses veined with blue mold. Characteristic aspects of cheese manufacture, factors determining the development of a specific flora, and the ensuing ripening characteristics will be considered.

27.6.1 Soft Cheese with Surface Flora

Traditional manufacturing methods are characterized by:

1. Acidification of the milk prior to renneting, which yields a gel with an acidic as well as a rennet character: The intensity of preculturing determines whether a rennet gel (e.g., in the manufacture of Pont l'Évêque) or an acid gel (e.g., old-fashioned Camembert) prevails.
2. Cutting of the curd into large lumps that are put into molds (hoops): The curd is not stirred and not washed; syneresis largely occurs after molding. The cheese attains a relatively high water content, usually 50 to 60%. The pH of the cheese shortly after its manufacture is low (say, 4.5 to 5) and the curd loses most of the calcium phosphate associated with the casein micelles and thereby much of its buffering capacity.
3. Consumption of much of the lactic acid by the surface flora, thereby raising the pH (see Figure 27.3): The latter factor is essential in obtaining a soft but not quite liquid consistency. Because the lactic acid has to diffuse to the surface, the cheese loaves are made quite flat, i.e., 3 to 4 cm thick; otherwise the inner part would remain firm and acidic. Often, the loaf diameter is also small, e.g., 10 cm.
4. Production by the surface flora of flavor compounds, which have to diffuse inward: If the flora is allowed to grow unchecked, the mature cheese will be fit to be eaten only for a short time (see also Figure 25.9). The shelf life can be lengthened by refrigeration.

27.6.1.1 Manufacture

Nowadays, curd making, in particular, differs markedly from the traditional method. Figure 27.19 gives an outline of an industrial manufacturing process for soft cheese having surface flora, made of preacidified milk.

Standardization and *pasteurization* are commonly applied. The fat content in the dry matter may range from 40 to 60%. If raw milk is used, it has to comply with the strictest bacteriological requirements because of the risk of growth of pathogenic microorganisms (e.g., *Listeria monocytogenes*), especially on the cheese surface, which is deacidified during maturation.

The milk is *preacidified* by means of a homofermentative mesophilic starter. A heterofermentative starter, which produces much more CO_2, can introduce

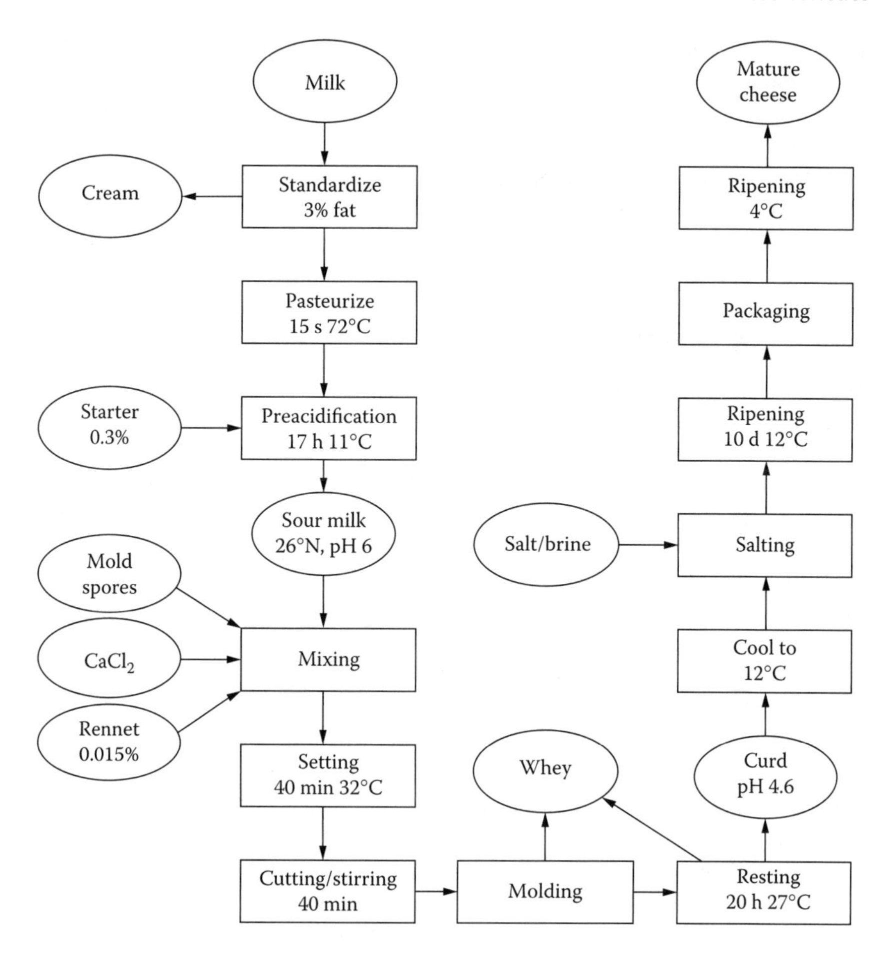

FIGURE 27.19 Example of the manufacture of soft cheese having a surface mold. Usually, the cheese is brine-salted. In the manufacture of cheese with a red smear, coryneforms are added instead of mold spores, and the cheeses are rubbed with weak brine in the first ripening stage.

defects due to growth of microaerophilic CO_2-tolerant blue molds that may have contaminated the milk. Nevertheless, most traditional Camembert has a fairly open texture (small irregular holes). Because of the preculturing the lactic acid bacteria hardly multiply any further during cheese manufacture. The great number of bacteria involved enhances a rapid drop of the curd pH, causing a considerable loss of buffering compounds during the whey drainage. Thereby the curd will be demineralized (cf. Figure 27.4) to such an extent that a satisfactory consistency of the cheese can be achieved.

Furthermore, the preculturing improves the rennetability of the milk, which has been kept cool during the preacidification stage, and it causes a firmer gel, thereby reducing the loss of fines during curd making.

Various methods are available to arrive at a desirable acid production during preculturing of the milk. The inoculum percentage of the starter as well as the temperature and time of incubation of the milk can be varied. A long incubation period at low temperature requires a very good bacterial quality of the milk (e.g., the number of psychrotrophs in the raw milk should be low); it also calls for a hygienic manufacturing process to prevent contamination by homologous bacteriophages and growth of spoilage organisms, especially enteropathogenic coliforms.

Additions. The amount of rennet added depends on the type of cheese and the acidity of the milk. Usually, it is much less than in the manufacture of harder cheeses, though more than in fresh cheeses. A relatively long clotting time at high temperature, i.e., 32 to 34°C, contributes to the formation of a firm gel and to a further decrease of the curd pH.

Usually, $CaCl_2$ is added to the milk, as also coloring matter. At the same time the milk is inoculated with organisms that should make up the surface flora, i.e., a suspension of spores of *P. camemberti* for mold-ripened cheese and coryneform bacteria, essentially *Brevibacterium linens*, for cheese with a red smear (Rotschmiere, ferment du rouge). Additional mold spores can be added to the cheese through the brine or by means of spraying in the ripening room. Coryneforms may additionally be added by means of the weak brine used to rub the cheese surface during the ripening.

Stirring and shaping Traditionally, the clotted milk was often not stirred, or only lightly agitated, before shaping the cheese. The curd was put in perforated molds, their capacity surpassing the volume of the cheese to be made. Repeated filling up with curd as long as whey continued to drain led to a satisfactory loaf size. In this way, the shaping took several hours and was not suitable for mechanization. Nowadays, this manufacturing protocol is still in use for farmhouse cheese (e.g., Camembert and Brie).

In mechanized manufacture, the curd is cut into pieces of, say, 2 to 2.5 cm. Furthermore, the curd is left in the whey for some 30 to 50 min to allow syneresis; thereby, a part of the whey (20 to 40%) is released, and mechanical damage during continued manufacture is restricted because the curd pieces already have become rather firm. After drainage of the whey, a mass of curd needed for one loaf of cheese is put into each mold. The water content of the curd at shaping is lower than in traditionally made curd. Sufficient demineralization is nevertheless possible by adjusting the rate of acid production. The curd is not washed (see, however, Subsection 27.6.1.2).

Syneresis and acid production during resting take a considerable time, about 15 to 20 h. These processes are enhanced by keeping the cheese at a fairly high temperature, i.e., 26 to 28°C. Cooling of the cheese should be prevented (note that the specific surface area of the loaves is large). In this period the loaf is turned a few times, and it shrinks in the mold to the desired height. The pH drops to 4.5 to 5. Not all of the lactose is converted because the curd is not washed, and even after a considerable time some sugar is left. The Ca content of the cheese is low, about 0.4% in the dry matter (Figure 27.4).

Salting. Nowadays, the cheese is usually salted in brine. Depending on water content, shape, and size of the cheese, the salting time varies from 30 min to several hours. Due to the high water content and the great specific surface area, the salt absorption is fast. After about a week the salt is evenly distributed throughout the cheese. The salt content of the cheese will be about 1 to 2%.

Ripening conditions depend on whether molds or coryneform bacteria are the dominant surface flora (see also Table 24.3). Many factors affect the growth of these microorganisms (see text following). The NaCl content at the cheese surface is critical. The humidity of the air is kept higher for coryneforms than for molds. Other variations in ripening conditions are:

1. To allow *mold* growth, the cheese is kept 'untouched' and dry. If need be, the salted cheese may be dried at a low relative humidity and a high temperature, e.g., RH 80% and 18°C, respectively. The cheese rests on a grid, allowing the mold to develop on the entire surface. Oxygen supply is essential. The air must constantly be refreshed and preferably be filtered to prevent contamination by undesirable microorganisms. The cheese is turned every 3 or 4 d. After about 10 d it is completely enveloped by the mold and is packaged. (Soft cheese with a surface flora is normally packaged in perforated foil, allowing any excess of NH_3 to escape.) The maturation is not complete but continues during subsequent storage at low temperature, e.g., 4°C.

   Extending the ripening period (at 11 to 13°C) by a few days causes a more intense maturation and results in more flavor. During this time coryneform bacteria can also grow, among which is the orange-red *Brevibacterium linens*. Their growth is enhanced by putting the cheese, which is enveloped by a full-grown surface mold, on a nonperforated support and turning it every day. Thereby the hyphae of the mold are damaged, giving the bacteria some competitive advantage. The presence of an additional bacterial flora was highly appreciated in earlier times because of the more intense flavor obtained, but currently most manufacturers ensure that the entire cheese surface is absolutely white.
2. To allow growth of *coryneform bacteria*, the cheese is put on closed supports and is initially turned and washed every day. Washing with a salt solution or water (inoculated with the bacteria, if desired) is done mechanically. The flora is well developed after, say, 10 d and from that time on the cheese is turned and washed every 3 or 4 d. The ripening time varies from 2 to 4 weeks, depending on shape, size, and desired maturity of the cheese. Also, these cheeses show continued maturation during cold storage and distribution. In some varieties, the mature cheese is allowed to dry somewhat and subsequently covered with a latex coating.

The industrial manufacture, including ripening conditions, of soft cheese types with a more definite rennet-gel character, e.g., Port du Salut, correspond largely with the description given above. Some differences in manufacture are as follows:

1. The milk is not, or is barely, precultured, and the pH at renneting is about 6.6.
2. The renneting temperature is higher (34 to 36°C), and more rennet is added, i.e., 30 to 40 ml per 100 l of milk. In this way a very firm coagulum is obtained.
3. The gel is cut finer and more intensely stirred. More whey is drained off before shaping. The water content is somewhat lower.
4. Less buffering material is lost during drainage, and the Ca content in the dry matter only decreases to about 1.1% (cf. Figure 27.4). These factors, together with the lower water content of the curd (corresponding to less potential lactic acid), are responsible for a higher pH after drainage, i.e., 4.9 to 5.1.
5. The cheese loaves often are larger.

27.6.1.2 Modern Variants

Currently, much of the surface-ripened soft cheese is made, cured, and handled in a rather different way than previously. Incentives to do so include the possibility of:

1. Decreasing production costs by mechanization, automation, increase of scale of production, and shortening of the time needed for cheese making. Technological innovations and adoption of some of the process steps applied in the making of semihard cheese have made this possible.
2. Making fewer, i.e., larger cheeses which generally means thicker loaves. This reduces cheese making and handling costs, and it allows portioning of a loaf in handy pieces.
3. Increasing the shelf life of the cheese, especially preventing it from becoming overripe.
4. Preventing gross inhomogeneity in flavor and consistency of a single loaf. Especially in a thick loaf, traditional cheese making leads to a quite soft, or even liquid, and well-ripened outer layer, whereas the inner part is still hard and sour. This is primarily because the lactic acid formed has to diffuse to the rind to be broken down, and the diffusion takes a considerable time (cf. Subsection 27.6.1.4).
5. Satisfying the preference of many consumers for a relatively mild-flavored cheese.
6. Controlling of the growth of pathogenic and other undesirable microorganisms. This is achieved by the use of equipment that allows better hygienic measures and by lowering the storage temperature of the cheese.

To achieve these aims, cheese making has been greatly altered. Apart from the use of other machinery for curd making, shaping, and brining, etc., the processing itself has been modified. Preacidification of the cheese milk is greatly shortened or even omitted, and the curd is washed to remove lactose and lactic acid. The result is that the minimum pH reached is generally over 5.0 (cf. Figure 27.3). This also means that syneresis is much slower, while the time available for syneresis needs to be shorter. Some measures are taken to achieve this: (1) more rennet is added, 30 to 35 ml per 100 l of milk; (2) the renneting temperature is higher, 36 to 39°C; (3) the curd is cut into smaller pieces, 0.7 to 1 cm; (4) stirring the curd–whey mixture is intensified and part of the whey is removed at an early stage.

Moreover, curing has been modified, especially the temperature at which it is done. Generally, the loaves are cooled fairly soon to about 5°C. In combination with the higher pH of the cheese, this causes the texture to be smooth and more or less even throughout a loaf shortly after manufacture. Biochemical processes are much slower, which means that flavor development is slow, but also that flavor deterioration is greatly postponed. However, a bitter flavor often develops because the formation of bitter peptides by the rennet is still relatively fast, while the debittering activity is quite small.

Another way to obtain a milder flavor, without decreasing maturation too much, is by making cheese with high fat content. This is especially applied in some Brie varieties (cream Brie), with up to 70% fat in the dry matter.

27.6.1.3 Development and Ecology of a Surface Flora

The importance of a surface flora is that it effectively determines the, often highly specific and pronounced, flavor and the characteristic texture of the cheese after ripening, at least for soft cheeses. Development of flavor is discussed in Section 25.5, that of texture in Subsection 25.6.2.3 and Subsection 27.6.1.6.

A surface flora represents a complicated ecological system. The dominant flora will depend on (1) which organisms are (initially) present and (2) how good the conditions are for growth, which — of course — varies greatly according to species and strain. The presence of microorganisms is highly dependent on any inoculation (type and number of organisms) and on contamination (and thus on hygiene). Contamination of the milk can be largely undone by pasteurization because virtually all microorganisms that can grow on the surface are heat labile.

Conditions affecting growth include temperature, pH, salt concentration, O_2 pressure, humidity of the air, availability of nutrients, and presence (and concentration) of substances that stimulate or inhibit growth. Some of these conditions may be very different at the surface as compared to those in the interior of the cheese or in the surrounding air. The conditions depend on cheese composition and on treatment (e.g., temperature, rubbing with weak brine etc.), and also on the action of the microorganisms themselves: Nutrients may become depleted, other nutrients may be formed (e.g., amino acids from protein) to be utilized by other organisms, and inhibiting substances (e.g., antibiotics) may be excreted. The most striking change often is an increase in pH.

All of these factors determine which organisms can grow fastest, and these will usually outnumber and even oust the others, unless the other bacteria are already present in very large numbers. Nevertheless, even in the latter case, the organisms will die off when conditions become very unfavorable for them. This implies that the dominant flora may change markedly during maturation.

The principles of microbial ecology of a surface flora are especially manifest in soft cheese with low initial pH (say, 4.7). These are discussed in the following subsection (see also Subsection 27.6.1.1).

27.6.1.4 Cheese with Surface Mold

The predominant organisms are yeasts, micrococci, coryneform bacteria (e.g., *Brevibacterium linens*), and the molds *Geotrichum candidum* and, in particular, *Penicillium camemberti*. *Penicillium* and *Geotrichum* are inoculated, the others are contaminating organisms. The following are essential parameters for their growth:

1. *Temperature:* The course of the temperature during maturation has already been mentioned (see also Figures 27.1 and 27.19). The optimum temperature for growth of yeasts and molds is 20 to 25°C; at the ripening temperature (about 12°C) the growth rate is far slower, with *G. candidum* being the most cold sensitive. If the other conditions are favorable, micrococci and coryneform bacteria grow fastest at the ripening temperature.
2. *pH and NaCl content:* The yeasts as well as the molds in the surface flora can grow over a rather broad pH range from 4.0 to 6.5. The molds, including *G. candidum*, tend to have a somewhat lower optimum pH than the yeasts. The yeasts are generally salt tolerant; some of them can withstand concentrations of up to 20% NaCl, *Debaryomyces hansenii* being the most osmophilic. Among the molds, *G. candidum* is highly salt sensitive, and *Penicillium camemberti* cannot grow at salt concentrations higher than 10%. The micrococci and coryneforms can grow at a pH above 5.5 and are relatively salt tolerant; most strains can withstand 10 to 15% NaCl.

A succession of surface organisms develops. The factors temperature, pH, and salt mainly determine the succession. Two stages can be distinguished:

1. *Before salting:* The temperature of the cheese is high (26 to 28°C), and the salt content is low; the pH drops quickly due to lactic acid production. These are favorable conditions for yeasts, which contaminate the cheese surface through air and machinery. Growth of molds is negligible at this stage because the spores have to germinate, which takes some time. Some of the yeasts ferment lactose, and all of them degrade lactic acid. A variety of yeasts can appear, depending on the actual flora in the dairy. *Saccharomyces lactis*, *S. fragilis*, *Kluyveromyces lactis*, *Torulopsis sphaerica*, and *Candida pseudotropicalis* are the most common species, but *Debaryomyces hansenii* or *Torulopsis candida* can also be

present. The yeast-like mold *Geotrichum candidum* can grow together with yeasts.

2. *After salting:* The brining causes the temperature of the cheese to decrease to about 12°C and the salt content in the rind to increase considerably. The quick salt penetration in the cheese and the fairly low equilibrium salt content in the moisture (4 to 5%) allow the yeast flora to grow further to a high final count, i.e., about 10^9 per cm^2 surface area. The pH of the cheese increases (see Figure 27.3) due to the degradation of lactic acid. This enhances the growth of molds. Soon, *P. camemberti* overgrows *G. candidum*, the former being much less sensitive to temperature and salt. A limited growth of *G. candidum* may improve the organoleptic properties of cheese ripened with *P. camemberti* because, for example, less of a bitter flavor develops. Abundant growth, however, interferes with the growth of *P. camemberti* and causes defects, e.g., a wrinkled surface or a poor flavor. After some 4 or 5 d, *Penicillium* becomes visible and after about 10 d it is full grown. The molds, including *G. candidum*, also consume lactose and especially lactic acid, which causes the pH to increase faster. This increase enables the lactic acid bacteria to ferment residual lactose. At a pH above about 5.5 non-acid-tolerant micrococci and coryneforms can grow. Usually, Gram-negative bacteria (e.g., *Pseudomonas* spp.) also develop and, if present in large numbers, cause flavor defects. As mentioned, growth of pigmented coryneforms is nowadays mostly considered undesirable. Strains of *P. camemberti* have therefore been selected that have a higher optimum pH for growth and thus can suppress the coryneforms. Occasionally, uncolored strains (mutants) of *Brevibacterium linens* are added.

Surface molds are endowed with a markedly greater enzymatic potential than bacteria. Consequently, some major processes during ripening, such as lipolysis and proteolysis, are generally more prominent in mold-ripened cheese than in other types. The lipolysis and proteolysis caused by *Penicillium* spp. and *G. candidum* is due to the synthesis of extracellular enzymes, whose activities release the precursors for typical flavor compounds, which play a major role in the uniqueness of these cheeses (see Subsection 25.5.2).

27.6.1.5 Cheese with a Flora of Coryneforms

As in the above-mentioned flora, surface organisms succeed one another. This applies to yeasts, micrococci, and especially coryneforms (added and contaminating organisms). Again, the cheese is deacidified. The growth of coryneform bacteria is greatly enhanced, whereas that of contaminating molds is suppressed. Domi-nance of mold growth can only occur if the mycelium can develop; any rubbing or washing destroys the hyphae. The cheese surface is therefore regularly washed with an NaCl solution or with water, especially at the beginning of curing. Furthermore, the cheese is matured at a high relative air humidity (Table 24.3).

A contiguous slimy layer forms around the cheese due to the production of microbial polysaccharides and because of the swelling of the protein matrix. Under these conditions, the slower-growing molds lose the competition.

The floras of corynebacteria and micrococci are influenced by the salt content. A relatively high NaCl content enhances the growth of the cream- or orange-colored *Staphylococcus equorum* and the orange-red strains of *Brevibacterium linens*, a lower content, the yellow-pigmented *Arthrobacter* strains. Several strains are scarcely or not at all pigmented in pure culture, but appear colored when growing on cheese. A high salt content also enhances the proteolytic activity of the bacteria. This is because they need a high concentration of proline to keep their internal osmotic pressure at a sufficiently high level compared to the environment. The proline can only be obtained by proteolysis of casein.

These principles are also valid when a flora settles on other types of cheese, including some hard varieties. The number of yeasts involved closely depends on the initial pH of the cheese. The higher the pH, the less important the yeasts are, and the earlier a dominating growth of molds or corynebacteria appears.

The flora of coryneforms is responsible for the development of volatile aromatic sulfur compounds originating from methionine and cysteine. These compounds are clearly key flavor compounds of cheeses with this type of surface flora (see Subsection 25.5.2).

27.6.1.6 Consistency

During ripening, cheeses with a high water content and an initially low pH must change in consistency from fairly hard and short to soft and spreadable. If the cheese is left to mature for a long time it may liquefy entirely. These changes start just below the cheese rind and move inward to the center. Formerly, proteolytic enzymes from the surface flora were held responsible; penetrating into the cheese they supposedly caused a moving boundary of protein degradation. The diffusion rate of the enzymes involved is, however, far too slow; the effective diffusion coefficient of proteins in cheese is at most 10^{-12} $m^2 \cdot s^{-1}$, which implies that it would take several months for the center of the cheese to acquire a significant enzyme concentration.

Therefore, the mechanism is different and the important factors are as follows:

1. *The composition of the cheese:* In particular the water content, or rather the water-to-paracasein ratio (which should be high), and the initial content of calcium phosphate (which should not be high).
2. *Sufficient protein breakdown:* This is necessary, and rennet action is especially important (see also Subsection 25.3.3). The maximum activity of calf rennet is at a pH of about 5.
3. *A sufficiently high pH:* At a lower pH (say, 4.7), proteolysis is still considerable but the consistency remains short and does not change until the pH is increased by the surface flora (see item 4). At a higher

pH, plasmin and the enzymes of starter bacteria also become increasingly active, but their importance in terms of consistency is not known.

4. *Deacidification of the cheese:* This is caused by the surface flora. Consumption of lactic acid by these floras causes migration of the acid to the surface and of basic protein degradation products of the flora, including NH_3, to the interior. This causes a pH gradient. More NH_3 is produced at a higher (initial) pH. The pH at the surface of mature cheese soon is about 7; in the center it may increase up to 6. (The effective diffusion coefficient of low-molar-mass compounds equals about 4×10^{-10} $m^2 \cdot s^{-1}$, allowing halving of a concentration difference over 1 cm in about 3 d).

The deacidification also leads to a 'demineralization' of the interior of the cheese, which enhances the solubility of the protein. A high pH at the surface leads to precipitation of insoluble di- and tricalcium phosphate. In turn, the precipitation causes diffusion of Ca and phosphate to the surface. In mature Camembert, about 80% of Ca and 50% of inorganic phosphate is therefore found in the outermost layer of a few millimeters.

Summarizing, a satisfactory consistency results from a complex of factors, i.e., cheese composition (water and salt content), sufficient protein breakdown, a fairly high pH (due to consumption of lactic acid and production of NH_3), and, presumably, a low content of calcium phosphate. Other conditions being equal and favorable, the softness increases with the NH_3 content; at a very high content, say, 0.3% NH_3, the cheese liquefies.

27.6.2 Blue-Veined Cheese

These cheeses are characterized by growth of *Penicillium roqueforti* in the interior. Many types of the cheese exist. Ewes' or cows' milk is generally used for the manufacture.

27.6.2.1 Manufacture

1. *Treatment of the milk:* Precultured raw or pasteurized milk is used. Usually, the milk is standardized. For some types of cheese, all or part of the milk is homogenized to obtain a whiter color (occasionally a whitener is added, e.g., chlorophyll), to enhance lipolysis, to improve the consistency, and to avoid creaming when the clotting time is long.
2. *Addition of starter:* To raw milk little or no mesophilic starter is added, e.g., in the manufacture of Roquefort. To pasteurized milk a mesophilic starter is added (e.g., for Stilton) or a thermophilic starter (e.g., for Gorgonzola). Gas formation by leuconostocs is desirable for the mold to become satisfactorily implanted, and most starters currently used contain these bacteria. If acidification is rapid, the leuconostocs cannot grow sufficiently and because of this, little starter is then used.

3. *Addition of spores of P. roqueforti* can be added as a suspension to the milk or to the curd. The selected mold can grow at low oxygen pressure (e.g., 5% O_2), can be added as a suspension to the milk, withstands a high CO_2 content, and grows readily at low temperature, i.e., 5 to 10°C.
4. *Renneting:* Up to 30 ml per 100 l of milk is used; the clotting temperature ranges from 30°C to 33°C and the clotting time from about 30 min to several hours. The latter is true for Roquefort, which is made from ewes' milk.
5. *Curd treatment in the vat:* This is done in such a manner that it enhances the implantation of the mold (see also item 6). Generally, the curd is coarsely cut and gently stirred for a short time. The mixture of curd and whey is scalded only a little or not at all and is not washed.
6. *Shaping and drainage:* Often, the curd particles are slightly cooled before shaping; the acid production is slow (allowing the leuconostocs to grow significantly), the drainage takes a long time, and the curd is not pressed. Most of these factors interfere with a quick and complete fusion of the curd particles. The formation of holes by the CO_2 produced by the leuconostocs, which is facilitated by mechanical air inclusion, readily causes an open texture, which enhances implantation of the mold. During drainage the cheese is turned from time to time, and the pH drops to 4.6 to 4.8.
7. *Salting:* Brining or rubbing is usually done. Roquefort is rubbed with dry salt; Stilton is salted at the curd stage, after cheddaring.
8. *Piercing:* To enhance mold growth, holes are pierced in the cheese, allowing ready penetration of air.
9. *Maturation:* The temperature should be 5 to 10°C and the RH about 90%. During the period of mold growth, the surface must be kept clean to prevent clogging of the holes; furthermore, the piercing of holes may be repeated. The ripening time ranges from, say, 3 weeks to several months. The pH of the mature cheese is about 6. After sufficient flavor has developed, the cheese is packaged in metal foil (tin or aluminum) to shut off the oxygen supply, which would cause too sharp a flavor to develop.

27.6.2.2 Microbial Interactions

The flora of blue cheese represents a complicated ecosystem. This has been extensively studied in Roquefort made from naturally contaminated raw ewes' milk (currently starter is added). The following are some of the conclusions of those investigations (see Figure 27.20).

At the end of drainage the flora consists of the following:

1. *Mesophilic lactococci: L. lactis* sspp. *lactis* and *cremoris*, about 10^9 per gram of cheese.

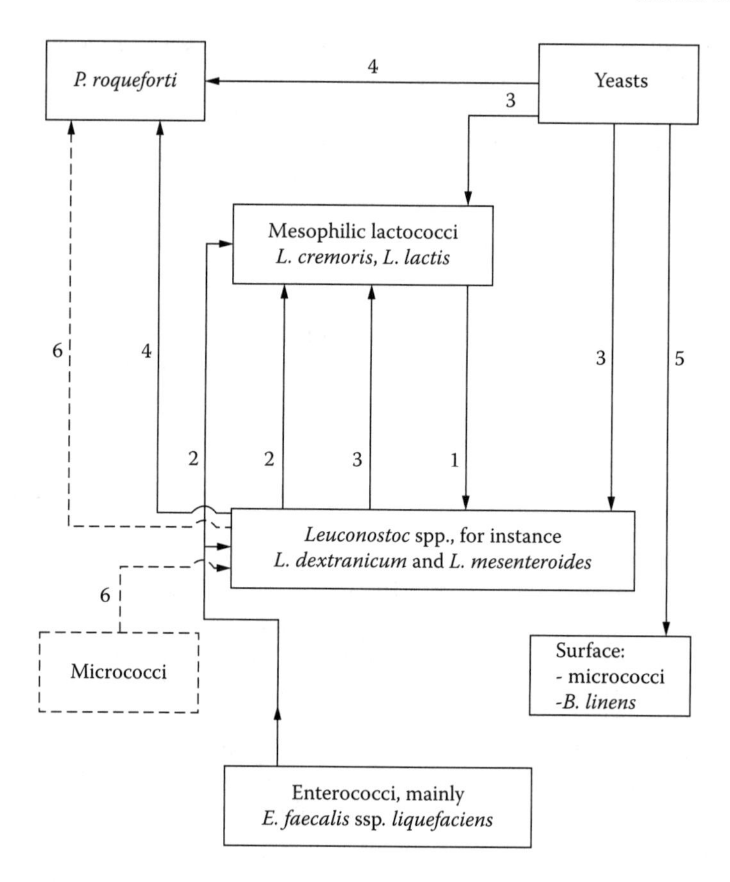

FIGURE 27.20 Microbial associations in Roquefort, made of naturally contaminated raw ewes' milk. ——→, enhances growth; ---→, impedes growth. (Adapted from Devoyed et al., *Le Lait,* **52**, 297, 1972.)

2. *Leuconostocs*, comprising over 10^6 per gram: Considerable growth occurs, enhanced by nitrogenous compounds formed by the more proteolytic lactococci (number 1 in Figure 27.20). It mainly concerns strains that do not ferment citrate and thus produce little, if any, CO_2 when growing in milk.
3. *Enterococci*, comprising about 10^7 per gram: Certain species, especially *E. faecalis* ssp. *liquefaciens*, enhance the growth of and acid production by the mesophilic lactococci, and enhance the production of CO_2 by the leuconostocs; metabolites produced by the enterococci may include low-molar-mass peptides and aromatic amino acids, e.g., tryptophan (number 2, Figure 27.20). CO_2 from leuconostocs can also stimulate the mesophilic lactococci.
4. *Yeasts*, comprising about 10^8 per gram: Before salting, it is mainly the lactose-fermenting salt-intolerant species, including *Saccharomyces,*

Candida, and *Torulopsis* spp. They also enhance the growth of the mesophilic lactococci and CO_2 production by leuconostocs (number 3, Figure 27.20). Due to the salting, the yeast flora shifts to non-lactose-fermenting, salt-tolerant species, including *Pichia*, *Hansenula*, and *Debaryomyces* spp. (The salt content of Roquefort is high, i.e., about 10% NaCl in water.)

5. *Other organisms:* These comprise especially micrococci, staphylococci, lactobacilli, and coliforms.

Some phenomena that occur during ripening are as follows:

1. CO_2 production by leuconostocs and yeasts causes holes in the cheese, creating an increased surface area for *P. roqueforti* (number 4, Figure 27.20). The mold can grow readily if air is supplied by piercing holes in the cheese.
2. Increase of the pH at the cheese surface, mainly caused by lactic acid consumption by yeasts, affords growth possibilities for salt-tolerant aerobic micrococci and corynebacteria, especially *Brevibacterium linens* (number 5, Figure 27.20).
3. Occasional contamination of the cheese milk by strongly proteolytic micrococci can result in cheese with insufficient openness and thereby unsatisfactory mold growth. This is caused by metabolites that do not affect the growth of the leuconostocs but impede their gas production (number 6, Figure 27.20).

The microbial interactions in the various blue-veined cheeses will roughly correspond to those found in Roquefort. These types of cheese will vary, however, according to composition and extent of contamination of the raw milk, heat treatment of the milk, recontamination, method of manufacture of the cheese, and ripening conditions. To ensure that a satisfactory open texture is achieved, citrate fermenting, CO_2 producing *Leuconostoc cremoris* bacteria are commonly added with the starter.

The specific flora of blue cheese largely determines the organoleptic quality of that cheese (see Section 25.5). *P. roqueforti* is highly lipolytic, and the free fatty acids released are essential for the typical flavor of blue-veined cheeses. Also, compounds derived from the fatty acids and protein degradation products contribute to this flavor. Moreover, variations in the manufacturing process and the kind of milk affect cheese quality. For instance, blue-veined cheeses made from cows' and ewes' milk differ considerably in flavor.

27.7 PROCESSED CHEESE

Some varieties of cheese have a long shelf life, but other varieties, mainly high-moisture cheeses, do not. Therefore, attempts have been made to preserve cheese. Flat pieces of cheese or curd can be left to dry in the sun. Alternatively, the cheese can be preserved by smoking, as is done for some Italian cheeses, e.g., Caciocavallo. Cheese can be stored in brine (Feta, Domiati) or even be boiled in brine. Small balls of (fresh) cheese can be candied. Pasteurization of cheese can also be tried, but that generally leads to melting of the cheese, followed by demixing. Oil separates and the proteinaceous mass may become lumpy, or a kind of serum may separate. Such demixing can be prevented by adding certain salts. In this way processed cheese evolved.

Processed cheese is made by grinding (shredding) and blending cheese, adding melting (emulsifying) salts, heating to, say, 80°C while stirring, stirring the melted mass for several minutes at that temperature, putting it into suitable containers, and cooling. The cooling causes setting of the mass. Continuous processing in scraped-surface heat exchangers or by steam injection heating is also applied. In addition to cheese, water, butterfat, and Na-caseinate is often added. The flavor of processed cheese distinctly differs from that of true cheese, partly due to the heat treatment, and partly due to the melting salts. Various additional ingredients, such as ham, celery, or nuts, are sometimes added.

The melting salts have two main functions. First, they increase the pH of the system, which is generally between 5.3 and 5.9. Second, the salts lower the calcium ion activity, presumably to below 0.2 mmolar. Generally, sodium salts of citric or polyphosphoric acids are used, because the calcium salts of these acids do not form much of a precipitate (a small amount of very small crystals is acceptable). Under these conditions, paracaseinate is 'soluble.' Electron micrographs of processed cheese show a homogeneous mass (with fat globules embedded), which at high magnification is seen to consist of very small particles, on the order of 10 nm in size. At high temperatures, the mass is liquid, presumably because the paracaseinate particles shrink, i.e., expel moisture; moreover, the viscosity of the moisture is quite low. The protein present is able to act as an emulsifier and electron micrographs show fat globules with a distinct surface layer, presumably of paracaseinate. As mentioned earlier, heating and stirring of cheese in the absence of melting salts causes considerable coalescence of fat globules. Upon cooling, the paracaseinate particles will swell (cf. Figure 3.20), presumably due to weakening of intraparticle hydrophobic bonds. A soft solid results that is firmer at lower temperatures.

During processing and storage, some other changes appear to occur but these are poorly understood. The protein mass tends to coagulate after stirring for a while at high temperature. Addition of a small amount of processed cheese — the remainder from a previous batch — as well as the use of well-ripened cheese enhances this coagulation. Processed cheese cannot be made of well-ripened cheese. An excess of barely ripened cheese is commonly used with some mature cheese added for flavor.

Two groups of processed cheese can be distinguished: spreads and sliceable blocks. The former contains, say, 58% water and 50% fat in the dry matter; it generally contains a higher proportion of mature cheese; citrate cannot be used in this type, because it causes a firm cheese mass. The slice-able cheese contains, e.g., 46% water. The quantity of melting salt required amounts to 10% (or even more) of the quantity of protein, which leads to levels of 2 to 3% in the cheese.

Growth of clostridia may spoil the product. Nisin can be added to prevent their growth. In modern continuous processing lines, a UHT treatment of the melted product can be applied to kill the spores.

Originally, processed cheese manufacture was often aimed at enhancing the market value of cheese of unsatisfactory quality. The blending of several lots, including cheese of satisfactory quality, resulted in an acceptable product. Processed cheese is now being made as a product with specific desirable properties, such as, having a long shelf life, being homogeneous, and easy to handle; it may also contain several added ingredients. In some countries it makes up an essential part of the cheese market. Moreover, several processed cheese analogues, e.g., with vegetable fats and other nondairy ingredients, are being produced.

A product related to processed cheese is Kochkäse (Germany) or Cancaillotte (France). Its manufacture begins with low-fat quarg of pH about 4.5 that is ripened for a few days at 23 to 30°C in a layer 5 to 10 cm thick. The layer is turned regularly to incorporate oxygen. Air humidity should be high to prevent water loss. Yeasts begin to grow and consume lactie acid, raising the pH to 5.6 to 6. The yeasts involved will be similar to those in a developing microbial smear (see Subsection 27.6.1.5); coryneform bacteria may also grow. The product so obtained allows subsequent melting without the addition of melting salts because the acidic quarg has a low content of Ca. The ripened curd, 1 to 2% salt, a fair amount of water or milk, and butter are blended and melted while being stirred for, say, 10 min at 90°C. Nowadays, a higher heating temperature is often applied, up to 115°C. The liquid mass is placed in cups and cooled. It is consumed as a smooth, light paste.

Suggested Literature

A comprehensive survey of cheese varieties: G. Burkhalter (Rapporteur), IDF Catalogue of Cheeses, IDF Bulletin, Document 141, 1981.

Several cheese varieties, including processed cheese, dealt with, generally in great depth: P.F. Fox, P.L.H. McSweeney, T.M. Cogan and T. Guinee, Eds. *Cheese: Chemistry, Physics and Microbiology,* Vol. 2, Major Cheese Groups, 3rd ed., Elsevier Academic Press, London, 2004.

Valuable information (in the encyclopedic part of the book): H. Mair-Waldburg, *Handbuch der Käse,* Volkswirtschaftlicher Verlag GmbH, Kempten, 1974.

Much information about fresh cheese: F.V. Kosikowski and V.V. Mistry, *Cheese and Fermented Milk Foods,* 3rd ed., published by the authors, 1997.

Part V

Appendix and Index

A.1 Often-Used Symbols*

LATIN

A	Specific surface area of particles (m^{-1})
a	Thermodynamic activity
a_w	Water activity (-)
B	Permeability (m^2)
C	Constant
c	Concentration
c_s	Relative standard deviation of a size distribution (-)
D	Diffusion coefficient ($m^2 \cdot s^{-1}$)
	Fractal dimensionality (-)
d	Diameter (m)
d_{vs}	Volume/surface average diameter (m)
E_h	Redox potential (V)
G	Elastic shear modulus (Pa)
g	Acceleration due to gravity (9.807 $m \cdot s^{-2}$)
	Generation time (s)
H	Sedimentation parameter (m^2)
I	Total ionic strength (molar)
M	Molar mass ($g \cdot mol^{-1}$ = Da)
N	Number concentration (m^{-3})
n	Refractive index (-)
pK	pH of 50% dissociation
p	Pressure (Pa)
Q	Degree of concentration (e.g., of milk) (–)
R	Gas constant (8.315 $J \cdot K^{-1} \cdot mol^{-1}$)
r	Radius (m)
S_n	n-th moment of a (size) distribution
T	(Absolute) temperature (K)
t	Time (s)
$t_{0.5}$	Time (s) needed to halve the value of a parameter
v	Linear velocity ($m \cdot s^{-1}$)
	Voluminosity ($ml \cdot g^{-1}$)
x, y, z	Linear coordinates (m)

* Units, for the most part according to the SI units system, are between parentheses; (-) means dimensionless.

GREEK

Γ	Surface load (surface excess) ($kg \cdot m^{-2}$; $mol \cdot m^{-2}$)
γ	Activity coefficient (-)
	Surface tension ($N \cdot m^{-1}$)
ε	Relative dielectric constant (-)
	Relative deformation (-)
η	Viscosity ($Pa \cdot s$)
η_a	Apparent viscosity ($Pa \cdot s$)
η_{rel}	Relative viscosity (-)
Π	Osmotic pressure (Pa)
ρ	Mass density ($kg \cdot m^{-3}$)
σ	Stress (Pa)
σ_{fr}	Fracture stress (Pa)
σ_y	Yield stress (Pa)
φ	Volume fraction (-)

OTHER

Re	Reynolds number
$\equiv$	Is by definition equal to
$\approx$	Is approximately equal to
$\propto$	Is proportional to
$\sim$	Approximately

A.2 Abbreviations*

BSA	Bovine serum albumin
CCP	Colloidal calcium phosphate
CFU	Colony-forming units
CMC	Critical micellization concentration
	Carboxymethyl cellulose
CMP	Caseinomacropeptide
DF	Diafiltration
D (−)	Levorotatory
DSC	Differential scanning calorimetry
DVI	Direct vat inoculation
FDM	Fat content in dry matter
FFA	Free fatty acids
HACCP	Hazard analysis/critical control points
HMF	Hydroxymethyl furfural
IEP	Isoelectric pH
L (+)	Dextrorotatory
MF	Microfiltration
NANA	N-acetylneuraminic acid (residue)
NPN	Nonprotein nitrogen (compounds)
PCBs	Polychlorinated biphenyls
PP3	Proteose peptone, fraction 3
R	Residue (in chemical formulas)
RH	Relative humidity (air)
r.p.m.	Number of revolutions per minute
SDS	Sodium dodecyl sulfate
sn	Stereospecific numbering
snf	Solids-not-fat
sp.	Species (singular)
spp.	Species (plural)
ssp.	Subspecies (singular)
sspp.	Subspecies (plural)
UF	Ultrafiltration
UHT	Ultra-high temperature short time (heating)
WP	Whey protein
WPC	Whey protein concentrate
WPI	Whey protein isolate

* For amino acid symbols, see Chapter 2, Table 2.12.

A.3 Conversion Factors

1 ångström (Å)	= 10^{-10} m = 0.1 nm
1 atmosphere (atm)	= 101325 Pa
1 bar	= 0.987 atm = 14.5 psi = 10^6 dyne·cm^{-2}
	= 10^5 Pa = 0.1 MPa
1 calorie	= 4.184 J = 0.003966 B.t.u
1 cP (centipoise)	= 1 mPa·s
1 dyne · cm^{-1}	= 1 mN·m^{-1}
1 erg	= 10^{-7} J
1 gallon (imperial)	= 4.546 l
1 gallon (U.S.)	= 3.785 l
1 molar (M)	= 1 mol·l^{-1}
1 pound (avoirdupois)	= 0.4536 kg
1 psi (pound/square inch)	= 6895 Pa
1%	= 1 kg/100 kg, unless stated otherwise

A.4

TABLE A.4
Physical Properties of Milk Fat

Temperature (°C)	Density ($kg \cdot m^{-3}$)	Refractive Index at $\lambda = 0.589$ μm	Viscosity (mPa·s)	Solubility of Water (% w/w)
10	922	(1.465)	—	(0.11)
20	915	1.462	70.8	(0.14)
30	909	1.458	45.7	0.17
40	902	1.454	30.9	0.20
50	895	1.451	22.1	0.24
60	889	(1.447)	16.6	0.27
70	882	(1.444)	12.5	0.32
80	876	(1.440)	9.8	0.36
90	(869)	(1.436)	7.6	0.41
100	(863)	(1.433)	6.2	0.46

Note: Some average results for liquid fat. Values between brackets are based on extrapolation.

Refractive index n_D (i.e., n at $\lambda = 0.589$ μm) is mostly 1.452 to 1.457 at 40°C.
Density varies by <1% with composition.
Heat conductivity: ~0.17 $J \cdot m^{-1} \cdot s^{-1} \cdot K^{-1}$ at room temperature.
Specific heat: ~2100 $J \cdot kg^{-1} \cdot K^{-1}$ at 40°C.
Total heat of fusion: ~85 to 100 $J \cdot g^{-1}$.
Total melting dilatation: ~0.055 to 0.06 $ml \cdot g^{-1}$.
Dielectric constant: ~3.1, but greatly depending on frequency.
Solubility of air in (liquid) fat: ~87 $ml \cdot kg^{-1}$, i.e., ~28 ml O_2 and ~59 ml N_2, all at room temperature in equilibrium with air at atmospheric pressure. Quantity of O_2 in fat ~0.004% w/w (liquid fat in equilibrium with air). Solubility of air in solid fat is negligible.

TABLE A.5
Amino Acid Composition of Milk Proteins (mol/kg Protein)[a]

Component	MW	Whole Casein	α_{s1}-Casein (B)	α_{s2}-Casein (A)	β-Casein (A^2)	κ-Casein (B)	α-Lactalbumin (B)	β-Lactoglobulin (B)	Serum Albumin (B)
N	14.007	11.30	11.22	11.37	11.17	11.67	11.42	11.27	11.72
P	30.974	0.26	0.34	0.44	0.21	0.05	0.00	0.00	0.00
S	32.06	0.23	0.21	0.24	0.25	0.22	0.63	0.49	0.59
Glycine (Gly)	75.1	0.25	0.38	0.08	0.21	0.11	0.42	0.22	0.24
Alanine (Ala)	89.1	0.36	0.38	0.32	0.21	0.79	0.21	0.82	0.69
Valine (Val)	117.2	0.62	0.47	0.55	0.79	0.58	0.42	0.49	0.54
Leucine (Leu)	131.2	0.74	0.72	0.52	0.92	0.42	0.92	1.20	0.92
Isoleucine (Ile)	131.2	0.47	0.47	0.44	0.42	0.68	0.56	0.55	0.21
Proline (Pro)	115.1	1.02	0.72	0.40	1.46	1.05	0.14	0.44	0.42
Phenylalanine (Phe)	165.2	0.33	0.34	0.24	0.38	0.21	0.28	0.22	0.41
Tyrosine (Tyr)	181.2	0.34	0.42	0.48	0.17	0.47	0.28	0.22	0.29
Tryptophan (Trp)	204.2	0.06	0.08	0.08	0.04	0.05	0.28	0.11	0.03
Serine (Ser)	105.1	0.68	0.68	0.67	0.67	0.68	0.49	0.38	0.42
Threonine (Thr)	119.1	0.38	0.21	0.59	0.38	0.74	0.49	0.44	0.51
Cysteine (Cys)	121.2	0.02	0.00	0.08	0.00	0.11	0.56	0.27	0.53
Methionine (Met)	149.2	0.21	0.21	0.16	0.25	0.11	0.07	0.22	0.06

Arginine (Arg)	174.2	0.22	0.25	0.24	0.17	0.26	0.07	0.16	0.35
Histidine (His)	155.2	0.19	0.21	0.12	0.21	0.16	0.21	0.11	0.26
Lysine (Lys)	146.2	0.56	0.59	0.95	0.46	0.47	0.85	0.82	0.89
Asparagine (Asn)	132.1	0.31	0.34	0.55	0.21	0.42	0.56	0.27	0.18
Aspartic acid (Asp)	133.1	0.22	0.30	0.16	0.17	0.16	0.92	0.55	0.63
Glutamine (Gln)	146.1	0.74	0.59	0.63	0.83	0.74	0.42	0.49	0.29
Glutamic acid (Glu)	147.1	0.87	1.06	0.95	0.79	0.68	0.49	0.88	0.91
g protein per gram of N[a]		6.32	6.36	6.28	6.39	6.12	6.25	6.34	6.09
MW protein[a]		23192	23618	25231	23986	19026	14181	18282	66277

[a] Calculated from the amino acid sequence of each protein, including (organic) phosphate but not carbohydrate groups.

A.6 Amino Acid Sequences of Caseins

(1) α_{S1}-CASEIN B

10 20

H-Arg-Pro-Lys-His-Pro-Ile-Lys-His-Gln-Gly-Leu-Pro-Gln-Glu-Val-Leu-Asn-Glu-Asn-Leu-

30 40

Leu-Arg-Phe-Phe-Val-Ala-Pro-Phe-Pro-Glu-Val-Phe-Gly-Lys-Glu-Lys-Val-Asn-Glu-Leu-

50 60

Ser-Lys-Asp-Ile-Gly-Ser-Glu-Ser-Thr-Glu-Asp-Gln-Ala-Met-Glu-Asp-Ile-Lys-Gln-Met-

P P

70 80

Glu-Ala-Glu-Ser-Ile-Ser-Ser-Ser-Glu-Glu-Ile-Val-Pro-Asn-Ser-Val-Glu-Gln-Lys-His-

P P P P P

90 100

Ile-Gln-Lys-Glu-Asp-Val-Pro-Ser-Glu-Arg-Tyr-Leu-Gly-Tyr-Leu-Glu-Gln-Leu-Leu-Arg-

110 120

Leu-Lys-Lys-Tyr-Lys-Val-Pro-Gln-Leu-Glu-Ile-Val-Pro-Asn-Ser-Ala-Glu-Glu-Arg-Leu-

P

130 140

His-Ser-Met-Lys-Glu-Gly-Ile-His-Ala-Gln-Gln-Lys-Glu-Pro-Met-Ile-Gly-Val-Asn-Gln-

150 160

Glu-Leu-Ala-Tyr-Phe-Tyr-Pro-Glu-Leu-Phe-Arg-Gln-Phe-Tyr-Gln-Leu-Asp-Ala-Tyr-Pro-

170 180

Ser-Gly-Ala-Trp-Tyr-Tyr-Val-Pro-Leu-Gly-Thr-Gln-Tyr-Thr-Asp-Ala-Pro-Ser-Phe-Ser-

190 199

Asp-Ile-Pro-Asn-Pro-Ile-Gly-Ser-Glu-Asn-Ser-Glu-Lys-Thr-Thr-Met-Pro-Leu-Trp-OH

(2) α_{S2}-CASEIN

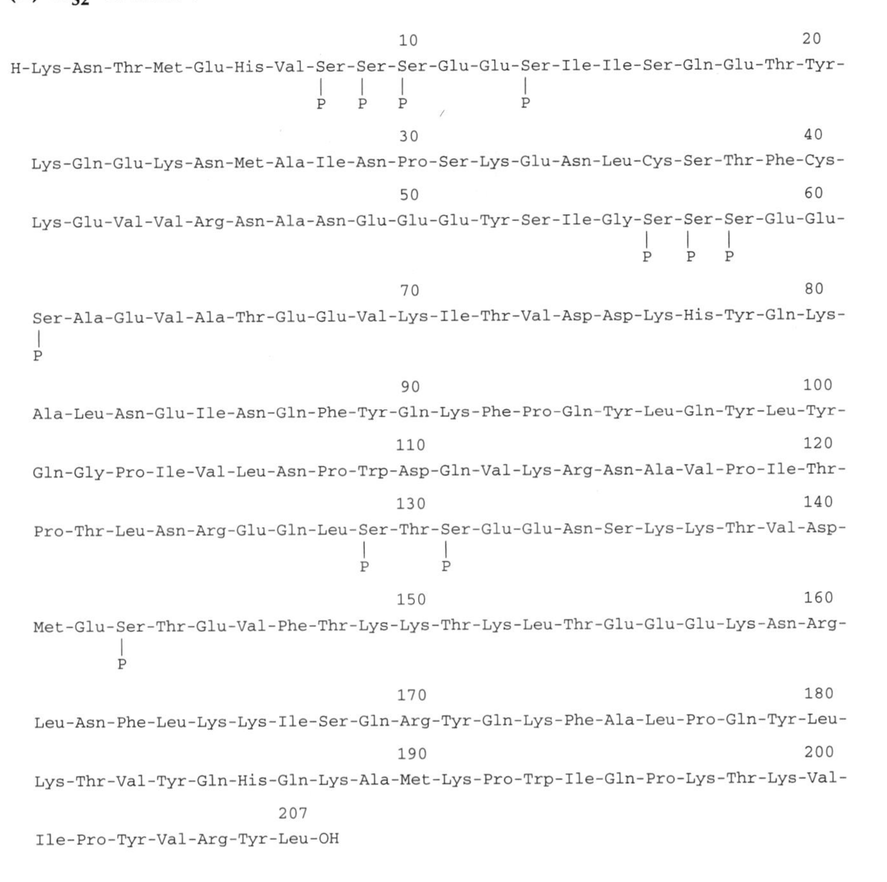

```
                                        10                                      20
H-Lys-Asn-Thr-Met-Glu-His-Val-Ser-Ser-Ser-Glu-Glu-Ser-Ile-Ile-Ser-Gln-Glu-Thr-Tyr-
                              |   |   |           |
                              P   P   P           P

                                        30                                      40
  Lys-Gln-Glu-Lys-Asn-Met-Ala-Ile-Asn-Pro-Ser-Lys-Glu-Asn-Leu-Cys-Ser-Thr-Phe-Cys-

                                        50                                      60
  Lys-Glu-Val-Val-Arg-Asn-Ala-Asn-Glu-Glu-Glu-Tyr-Ser-Ile-Gly-Ser-Ser-Ser-Glu-Glu-
                                                              |   |   |
                                                              P   P   P

                                        70                                      80
  Ser-Ala-Glu-Val-Ala-Thr-Glu-Glu-Val-Lys-Ile-Thr-Val-Asp-Asp-Lys-His-Tyr-Gln-Lys-
  |
  P

                                        90                                     100
  Ala-Leu-Asn-Glu-Ile-Asn-Gln-Phe-Tyr-Gln-Lys-Phe-Pro-Gln-Tyr-Leu-Gln-Tyr-Leu-Tyr-

                                       110                                     120
  Gln-Gly-Pro-Ile-Val-Leu-Asn-Pro-Trp-Asp-Gln-Val-Lys-Arg-Asn-Ala-Val-Pro-Ile-Thr-

                                       130                                     140
  Pro-Thr-Leu-Asn-Arg-Glu-Gln-Leu-Ser-Thr-Ser-Glu-Glu-Asn-Ser-Lys-Lys-Thr-Val-Asp-
                                  |       |
                                  P       P

                                       150                                     160
  Met-Glu-Ser-Thr-Glu-Val-Phe-Thr-Lys-Lys-Thr-Lys-Leu-Thr-Glu-Glu-Glu-Lys-Asn-Arg-
          |
          P

                                       170                                     180
  Leu-Asn-Phe-Leu-Lys-Lys-Ile-Ser-Gln-Arg-Tyr-Gln-Lys-Phe-Ala-Leu-Pro-Gln-Tyr-Leu-

                                       190                                     200
  Lys-Thr-Val-Tyr-Gln-His-Gln-Lys-Ala-Met-Lys-Pro-Trp-Ile-Gln-Pro-Lys-Thr-Lys-Val-

                          207
  Ile-Pro-Tyr-Val-Arg-Tyr-Leu-OH
```

(3) β-CASEIN A²

10 20
H-Arg-Glu-Leu-Glu-Glu-Leu-Asn-Val-Pro-Gly-Glu-Ile-Val-Glu-Ser-Leu-Ser-Ser-Ser-Glu-
P P P P
30 40
Glu-Ser-Ile-Thr-Arg-Ile-Asn-Lys-Lys-Ile-Glu-Lys-Phe-Gln-Ser-Glu-Glu-Gln-Gln-Gln-
P
50 60
Thr-Glu-Asp-Glu-Leu-Gln-Asp-Lys-Ile-His-Pro-Phe-Ala-Gln-Thr-Gln-Ser-Leu-Val-Tyr-
70 80
Pro-Phe-Pro-Gly-Pro-Ile-Pro-Asn-Ser-Leu-Pro-Gln-Asn-Ile-Pro-Pro-Leu-Thr-Gln-Thr-
90 100
Pro-Val-Val-Val-Pro-Pro-Phe-Leu-Gln-Pro-Glu-Val-Met-Gly-Val-Ser-Lys-Val-Lys-Glu-
110 120
Ala-Met-Ala-Pro-Lys-His-Lys-Glu-Met-Pro-Phe-Pro-Lys-Tyr-Pro-Val-Glu-Pro-Phe-Thr-
130 140
Glu-Ser-Gln-Ser-Leu-Thr-Leu-Thr-Asp-Val-Glu-Asn-Leu-His-Leu-Pro-Leu-Pro-Leu-Leu-
150 160
Gln-Ser-Trp-Met-His-Gln-Pro-His-Gln-Pro-Leu-Pro-Pro-Thr-Val-Met-Phe-Pro-Pro-Gln-
170 180
Ser-Val-Leu-Ser-Leu-Ser-Gln-Ser-Lys-Val-Leu-Pro-Val-Pro-Gln-Lys-Ala-Val-Pro-Tyr-
190 200
Pro-Gln-Arg-Asp-Met-Pro-Ile-Gln-Ala-Phe-Leu-Leu-Tyr-Gln-Glu-Pro-Val-Leu-Gly-Pro-
209
Val-Arg-Gly-Pro-Phe-Pro-Ile-Ile-Val-OH

(4) κ-CASEIN B*

```
                                          10                                       20
PyrGlu-Glu-Gln-Asn-Gln-Glu-Gln-Pro-Ile-Arg-Cys-Glu-Lys-Asp-Glu-Arg-Phe-Phe-Ser-Asp
                                          30                                       40
   Lys-Ile-Ala-Lys-Tyr-Ile-Pro-Ile-Gln-Tyr-Val-Leu-Ser-Arg-Tyr-Pro-Ser-Tyr-Gly-Leu-
                                          50                                       60
   Asn-Tyr-Tyr-Gln-Gln-Lys-Pro-Val-Ala-Leu-Ile-Asn-Asn-Gln-Phe-Leu-Pro-Tyr-Pro-Tyr-
                                          70                                       80
   Tyr-Ala-Lys-Pro-Ala-Ala-Val-Arg-Ser-Pro-Ala-Gln-Ile-Leu-Gln-Trp-Gln-Val-Leu-Ser-
                                          90                                      100
   Asn-Thr-Val-Pro-Ala-Lys-Ser-Cys-Gln-Ala-Gln-Pro-Thr-Thr-Met-Ala-Arg-His-Pro-His-
                                         110                                      120
   Pro-His-Leu-Ser-Phe-Met-Ala-Ile-Pro-Pro-Lys-Lys-Asn-Gln-Asp-Lys-Thr-Glu-Ile-Pro-
                                         130                                      140
   Thr-Ile-Asn-Thr-Ile-Ala-Ser-Gly-Glu-Pro-Thr-Ser-Thr-Pro-Thr-Ile-Glu-Ala-Val-Glu-
                                         150                                      160
   Ser-Thr-Val-Ala-Thr-Leu-Glu-Ala-Ser-Pro-Glu-Val-Ile-Glu-Ser-Pro-Pro-Glu-Ile-Asn-
                                   |
                                   P
                                   169
   Thr-Val-Gln-Val-Thr-Ser-Thr-Ala-Val-OH
```

* *Pyr*Glu means pyro-glutamic acid residue.

A.7

TABLE A.7
Some Properties of Lactose

A. **Solubility in water** (q, in g/100 g of water) as a function of temperature (T, °C):

α-Lactose: $\log q \approx 0.613 + 0.0128\,T$

β-Lactose: $\log q \approx 1.64 + 0.003\,T$

Equilibrium solution: $q = 12.48 + 0.2807T + 5.067 \times 10^{-3}T^2 + 4.168 \times 10^{-6}T^3 + 1.147 \times 10^{-6}T^4$.

Attention: $T \leq 93.5$°C

B. **Density, viscosity, and refractive index as a function of concentration**

Concentration (g Lactose/ 100 g Water)	Density of Solution at 20°C (kg·m^{-3})	Apparent Density of Lactose Dissolved in Water (kg·m^{-3})	Viscosity of Solution (mPa·s)		Refractive Index of Solution at 25°C, λ = 589 nm
			20°C	60°C	
0	998.2	—	1.00	0.47	1.3325
10	1043	1750	1.38	0.70	1.3484
20	1082	1629	2.04	0.90	1.3659
30	1124	1592	3.42	1.29	—
40	1173	1591	7.01	2.19	—

C. **Specific rotation α:**

α is expressed in degree of arc of rotation per cm pathlength in a hypothetical solution of 1 g anhydrous lactose per ml lactose solution. The rotation depends on temperature (T, °C), on the wavelength of the light (λ), and to some extent on concentration (c, g/100 ml solution), etc.

For equilibrium solutions:

$\alpha_D = 56.75 - 0.017c - 0.058\,T$ (D line, $\lambda = 589$ nm).

$\alpha_{Hg} = 66.25 - 0.007c - 0.054\,T$ (Hg line, $\lambda = 546$ nm).

For pure α-lactose: $\alpha_D = +91.1$ at 20°C.

For pure β-lactose: $\alpha_D = +33.2$ at 20°C.

A.8

TABLE A.8
Trace Elements in Cows' Milk ($\mu g \cdot l^{-1}$)

Element	Symbol	Contents Reported[a]	Average Content
Aluminum	Al	50–2100	500
Arsenic	As	10–400	50?
Barium	Ba	Traces–110	Traces
Boron	B	30–1000	200
Bromine	Br	20–25,000	600?
Cadmium	Cd	0.02–78	0.03
Cesium	Cs	3–46	?
Chromium	Cr	2–82	15?
Cobalt	Co	0–20	1
Copper	Cu	10–1200	25
Fluorine	F	20–700	180?
Iodine	I	5–800	50
Iron	Fe	100–2400	200?
Lead	Pb	1–65[b]	2.3
Lithium	Li	Traces–29	Traces
Manganese	Mn	3–370	20
Mercury	Hg	<0.1	Traces
Molybdenum	Mo	5–150	70
Nickel	Ni	0–180	30?
Rubidium	Rb	100–3400	2000?
Selenium	Se	4–1200	40?
Silicon	Si	1300–7000	1400?
Silver	Ag	Traces–54	Traces
Strontium	Sr	45–2000	170?
Tin	Sn	0–1000[c]	Traces
Titanium	Ti	2–500	?
Vanadium	V	Traces–310	0.1?
Zinc	Zn	220–19,000	3900

[a] Contents reported in literature vary widely, for the most part caused by contamination of the milk sample involved and inaccuracies of determinations.
[b] In milk packaged in cans up to 600 $\mu g \cdot l^{-1}$.
[c] In milk packaged in cans up to 200 $mg \cdot l^{-1}$.

A.9

TABLE A.9
Physical Properties of Milk and Milk Products (Average Values[a])

	Coefficient of Thermal Conductivity λ ($W \cdot m^{-1} \cdot K^{-1}$)	Density ρ^{20} ($kg \cdot m^{-3}$)	Specific Heat c_p ($J \cdot kg^{-1} \cdot K^{-1}$)
Water	0.60	998	4200
Whey	0.55	1025	—
Skim milk	0.54	1035	3800
Milk	0.52	1029	3900
Evaporated milk	0.45	1085	3200
Cream (20 %)	0.37	1009	3500
Cheese[b]	0.30–0.35	1060–1120[c]	2500
Butter	0.20	950	2300
Milk fat	0.17	916	2200

[a] Thermal diffusivity (D) can be calculated from $D = \lambda/\rho^{20}\, c_p$.

[b] Refers to hard and semihard cheeses.

[c] Cheese without holes. The density of ripened Emmentaler may amount to, say, 850 $kg \cdot m^{-3}$.

A.10

TABLE A.10
Mass Density (ρ) and Viscosity (η) of Some Milk Fractions as a Function of Temperature (T)

T (°C)	ρ_{plasma} (kg·m^{-3})	ρ_{serum}[a] (kg·m^{-3})	ρ_{waterr} (kg·m^{-3})	η_{plasma} (mPa·s)	η_{whey}[b] (mPa·s)	η_{water} (mPa·s)
0	—	—	999.9	3.45	—	1.787
5	1035.9	1027.3	1000.0	2.83	1.93	1.519
10	1035.2	1027.0	999.7	2.35	1.65	1.307
15	1034.4	1026.3	999.1	1.99	1.41	1.139
20	1033.3	1025.3	998.2	1.68	1.24	1.002
25	1031.9	1024.2	997.1	1.44	1.09	0.890
30	1030.0	1022.6	995.7	1.26	0.97	0.798
40	1026.1	1019.0	992.2	1.00	0.79	0.653
50	1019.8	1014.4	988.1	0.82	0.65	0.547
60	1016.6	1009.7	983.2	0.69	0.55	0.466
70	1011.2	1004.2	977.8	0.60	0.48	0.404
80	1005.0	998.0	971.8	0.56	—	0.355
90	—	—	965.3	—	—	0.315
100	—	—	958.4	—	—	0.282

[a] Calculated.

[b] *Source*: Adapted from R. Jenness and S. Patton, *Principles of Dairy Chemistry*, Chapman & Hall, London, 1959.

A.11
Heat Transfer

Some aspects on the subject of heat transfer are briefly summarized here. Numerical data are given in Tables A.11.1 to A.11.3.

The amount of heat that has to be transferred per unit time for heating a liquid from temperature T_1 to T_2 (without heat of fusion, heat of reaction, etc., occurring) is given by

$$q = (T_2 - T_1)\, Q c_p\, \rho \qquad \text{(A.11.1)}$$

where Q = liquid flow rate ($m^3 \cdot s^{-1}$), c_p = specific heat of the liquid, and ρ = liquid density. For example, heating milk from 10°C to 74°C at a flow rate of 7200 $l \cdot h^{-1}$ consumes about 5×10^5 W (according to data in Table A.11.1). [*Note*: If the temperature difference Δ varies throughout a heating or cooling section from Δ_1 to Δ_2, the average difference must be taken: $\Delta_{av} = (\Delta_1 + \Delta_2)/2$. If the change in Δ is relatively large, the so-called logarithmic average should be taken: $\Delta_{av} = (\Delta_1 - \Delta_2)/\ln(\Delta_1/\Delta_2)$.]

When two liquids are kept separate by a fixed wall, the heat transfer from one liquid to the other is given by

$$q = A \Delta T\, k_h \qquad \text{(A.11.2a)}$$

where A = surface area of the wall, ΔT = temperature difference between the liquids, and the total coefficient of heat transfer k_h is given by

$$(1/k_h) = (1/\alpha_1) + (\delta/\lambda) + (1/\alpha_2) \qquad \text{(A.11.2b)}$$

where α = individual coefficient of heat transfer from the wall to the liquid or vice versa, δ = thickness of the wall, and λ = coefficient of heat conductivity of the wall material. Assuming k_h to be 1 $kW \cdot m^{-2} \cdot K^{-1}$, then in our example a heating surface of 25 m^2 would be required if ΔT is 20 K, and 100 m^2 at 5 K. Obviously, to keep the heating surface as small as possible at a given ΔT, maximization of the total heat-transfer coefficient would be needed.

c_p scarcely depends on temperature, and λ and ρ only a little. But the α values greatly depend on conditions (see Table A.11.2). The media involved (milk product and heating or cooling agent) considerably affect α, but above all the flow does. In laminar flow, α is very small because all heat must be transferred through the fluids by thermal diffusion. Natural convection can increase α, but

TABLE A.11.1
Approximate Examples of the Effect of Composition and Temperature of a Milk Product on the Coefficient of Heat Conductivity λ ($W \cdot m^{-1} \cdot K^{-1}$), the Viscosity η (mPa·s), and the Specific Heat c_p ($kJ \cdot kg^{-1} \cdot K^{-1}$) (Any Heat of Fusion Not Included)[a]

	0°C		20°C			80°C	
Product/Material	λ	η	c_p	λ	η	λ	η
Water	0.57	1.79	4.2	0.60	1.00	0.66	0.36
Skim milk	—	3.45	3.8	0.54	1.68	0.63	0.56
Whole milk	0.45	—	3.9	0.52	1.93	0.61	—
Concentrated milk 1:1.9	—	—	3.5	0.48	3.1	0.56	—
Concentrated milk 1:2.5	—	—	3.2	0.45	6.3	0.53	—
25% Fat cream	0.32	—	3.5	0.37	4.2	—	—
45% Fat cream	0.28	—	3.2	0.32	13.5	—	—
Milk fat	0.13	—	2.2	0.17	71	—	—
Air	—	—	—	0.02	—	—	—
Stainless steel	—	—	—	17	—	—	—
Copper (red brass)	—	—	—	371	—	—	—

[a] η of concentrated milk depends strongly on preheating and η of cream on homogenization, if carried out. For comparison, λ of air and of two metals are also given.

TABLE A.11.2
Approximate Examples of the Individual Coefficient of Heat Transfer α of Some Media onto a Wall (under Various Conditions)

Medium	Condition	α in $W \cdot m^{-2} \cdot K^{-1}$
Air	Flowing	10–100
Water	Flowing	600–6000
Water	Boiling	2000–7000
Steam	Condensing	6000–17000
Whole milk	~38°C; Re = 10^4	800
	~38°C; Re = 10^5	2900
	~70°C; Re = 10^4	500
	~70°C; Re = 10^5	2000
25% Fat cream	~38°C; Re = 10^4	650
	~70°C; Re = 10^4	450

TABLE A.11.3
Coefficient of Total Heat Transfer k_h under Various Conditions[a]

Heating or Cooling Agent	Product	Condition	k_h (average) in $W{\cdot}m^{-2}{\cdot}K^{-1}$
Surrounding air	Whole milk	Tank with empty double jacket, no stirrer	3
Water	Water	Double-walled tank, stirrer and scraper	640
Water	35.5% Fat cream	Double-walled tank, stirrer and scraper	350
Water	Same, soured	Double-walled tank, stirrer and scraper	215
Water	35.5% Fat cream	Double-walled tank, stirrer but no scraper	230
Water	Yogurt	Double-walled tank, stirrer but no scraper	290
Water	Yogurt	Double-walled tank, stirrer rate halved and no scraper	140
Water	Water	Plate heat exchanger, regeneration section	3200
Water	Water	Plate heat exchanger, heating section	4900
Water	Water	Plate heat exchanger, cooling section	3500
Steam[b]	Whey	Evaporator[c] 40°C	1400
Steam[b]	Whey	Evaporator[c] 70°C	3300
Steam[b]	Concentrated whey 50% dry matter	Evaporator[c] 40°C	750
Steam[b]	Concentrated whey 50% dry matter	Evaporator[c] 70°C	2600
Steam[b]	Skim milk	Evaporator,[c] first effect	2300–2600
Steam[b]	Skim milk (concentrated)	Evaporator,[c] second effect	1900–2200
Steam[b]	Skim milk (concentrated)	Evaporator,[c] third effect	1000–1200
Steam[b]	Whole milk	Evaporator,[c] first effect	2000–2200
Steam[b]	Whole milk (concentrated)	Evaporator,[c] second effect	1700–1900
Steam[b]	Whole milk (concentrated)	Evaporator,[c] third effect	900–1100

[a] Approximate results.
[b] Condensing steam or water vapor.
[c] Falling film.

above all turbulent flow can lead to high α values. The higher the Reynolds number (Re), the higher α will be. This is because a higher Re causes the laminar surface layer to be thinner. For turbulent flow of fluids past a wall, α can be estimated from

$$\text{Nu} \approx 0.027\ \text{Re}^{0.8}\ \text{Pr}^{0.33} \tag{A.11.3}$$

where the Nusselt number is given by

$$\text{Nu} \equiv \alpha\, d/\lambda \tag{A.11.3a}$$

the Reynolds number by

$$\mathrm{Re} \equiv v\, d\, \rho/\eta \tag{A.11.3b}$$

and the Prandtl number by

$$\mathrm{Pr} \equiv c_p\, \eta/\lambda \tag{A.11.3c}$$

where v is average linear flow velocity, d is twice the distance between two plates or the diameter of a pipe, and η is viscosity of the liquid. Coefficients of heat transfer can be calculated by using these equations together with tabulated data. A difficulty involved is that α depends strongly on temperature (mainly because η is temperature dependent). Furthermore, for the variables in Equation A.11.3, the values at the wall, where the temperature is not precisely known, should be inserted. An additional problem may be that the viscosity can be dependent on flow rate for highly viscous products (see Subsection 4.7.1).

Under other conditions, α can have other values, e.g., for condensing steam. Usually, calculating the total heat transfer is not a simple task, especially when no forced convection (rapid forced circulation) is generated. Even such circulation as happens in heat exchangers of the common type causes problems in calculation, e.g., because the temperature difference involved may not be constant throughout the processing. In a falling film evaporator, the liquid velocity in the milk film v, and thereby Re, strongly depends on conditions, especially viscosity. Another difficulty is that the total coefficient of heat transfer may decrease during processing, due to fouling of the equipment, i.e., deposition of a layer of milk components.

TABLE A.12
Data on Some Cheese Varieties[a]

Cheese	Water	Protein	Fat	NaCl	*w/ffc*	*w/p*	*s/w*	*f/dm*	pH	kg[b]	Yield[c]
Appenzeller	42	25	29	2	0.59	1.7	0.05	0.5	6	7	10
Camembert	56	19	21	2	0.70	2.9	0.04	0.5	7	0.25	13
Cheddar	39	25	31	2	0.56	1.5	0.05	0.5	5.5	30	10
Cheshire	44	24	28	2	0.60	1.8	0.05	0.5	5.3	20	11
Cottage cheese	79	13	4	1	0.82	6.0	0.01	0.2	5	—	16
Cream cheese	45	6	44	1	0.80	7.5	0.02	0.8	4.6	—	—
Edam	46	26	24	2.5	0.60	1.8	0.06	0.45	5.3	2	10
Emmentaler	38	28	29	1	0.53	1.3	0.02	0.45	5.6	100	8
Feta[d]	56	16	23	3	0.73	3.5	0.05	0.5	4.5	1	15
Fontina	37	27	30	2	0.53	1.4	0.06	0.5	6	15	10
Gorgonzola	44	22	29	3	0.62	2.0	0.07	0.5	5.8	10	12
Gouda	43	24	29	2	0.60	1.8	0.04	0.5	5.4	10	11
Gruyère	38	29	30	1	0.54	1.3	0.03	0.5	5.7	40	8
Herve	52	20	24	2	0.68	2.6	0.04	0.5	7	0.3	13
Jarlsberg	44	25	28	1	0.60	1.8	0.03	0.5	5.6	10	10
Manchego[e]	40	27	29	2	0.56	1.5	0.05	0.5	5.1	3	16
Mozzarella[f]	55	23	19	1	0.68	2.4	0.02	0.4	5.2	0.3	13
Munster[g]	55	19	22	2	0.71	2.9	0.04	0.5		1	13
Parmesan[h]	33	36	26	2	0.45	0.9	0.06	0.4	5.5	35	7
Pecorino Romano[e]	33	33	26	6	0.45	1.0	0.18	0.4	5.4	25	10[i]
Pont-l'Évêque	53	20	22	3	0.68	2.6	0.06	0.45	7	0.3	12

Provolone[f]	40	27	29	2	0.56	1.5	0.05	0.5	5.3	5	9
Quarg	83	12	0.2	0.7	0.83	6.9	0.01	0.01	4.5	—	17[j]
Roquefort[e]	44	20	30	4	0.63	2.2	0.09	0.55	6.4	2	22
Sainte-Maure[k]	55	19	22	2	0.71	2.9	0.04	0.5		0.3	15
Saint Paulin[l]	55	20	22	2	0.70	2.8	0.04	0.5		2	13
Stilton	42	24	29	3	0.59	1.8	0.07	0.5	5.2	7	10
Tilsiter	45	24	27	2	0.61	1.8	0.04	0.5		5	10
Milk[m]	87.1	3.3	4.0		0.91	26.4		0.31			

[a] Average values, usually referring to minimum time of cheese ripening. It will be clear from Chapter 24 and Section 27.1.1, that cheeses of one variety vary considerably in composition. First, the manufacturing process as applied in one region may differ substantially from that in another region. Furthermore, the composition of a cheese usually will change during ripening. Finally, individual cheeses, even from one batch, differ. Hence, the data in the table are only meant to illustrate the main differences between varieties. Components are given in g per 100 g cheese; *f/dm* = fat/dry matter; *s/w* = NaCl/water; *w/ffc* = water/fat-free cheese; *w/p* = water/protein.

[b] Loaf weight.

[c] kg cheese per 100 kg of cheese milk.

[d] Made from cows' milk; it may also be made from ewes' or goats' milk, or mixtures of these milks.

[e] Made from ewes' milk.

[f] Stretched-curd cheese.

[g] The French cheese Munster should not be confused with American Muenster.

[h] Parmigiano-Reggiano.

[i] In addition to Pecorino Romano cheese, the production from 100 kg of milk may include 8 kg of Ricotta.

[j] kg quarg per 100 kg of skim milk.

[k] Made from goats' milk.

[l] Looks much like Port-Salut.

[m] Raw cows' milk data (see Table 1.1) given for comparison; components are in g per 100 g of milk.

Index

C

D

E

F

G

H

M

N

O

Q

R

S

X

Y

Z